Springer Proceedings in Physics

Volume 157

For further volumes:
http://www.springer.com/series/361

Jiří Bičák · Tomáš Ledvinka
Editors

Relativity and Gravitation

100 Years After Einstein in Prague

Springer

Editors
Jiří Bičák
Tomáš Ledvinka
Faculty of Mathematics and Physics
Charles University
Prague
Czech Republic

ISSN 0930-8989 ISSN 1867-4941 (electronic)
ISBN 978-3-319-06760-5 ISBN 978-3-319-06761-2 (eBook)
DOI 10.1007/978-3-319-06761-2
Springer Cham Heidelberg New York Dordrecht London

Library of Congress Control Number: 2014939531

Printed on acid-free paper

Springer is part of Springer Science+Business Media (www.springer.com)

Preface

A few days before leaving Prague, after 15 months spent at the German part of Charles-Ferdinand University, Albert Einstein submitted a paper titled *Relativity and Gravitation. Reply to a Comment by M. Abraham*. It was received by Annalen der Physik on July 4, 1912. Stimulated by Abraham's criticism, Einstein summarized the contemporary state of his relativistic theory of gravitation and, remarkably, anticipated what a future theory of gravity should look like.

The organizers of the conference "Relativity and Gravitation: 100 Years After Einstein in Prague," held in Prague on June 25–29, 2012, were inspired by the title, date, and significance of this last of Einstein's Prague papers. The aim of the conference was to review the present status of the general theory of relativity (both classical and quantum) and its applications in cosmology and astrophysics from a broad perspective. The articles based on the plenary lectures are published in a separate volume *General Relativity, Cosmology and Astrophysics: Perspectives 100 Years After Einstein's stay in Prague* (Springer 2014). The titles and the abstracts of these review talks are included in these *Proceedings* in the first chapter. The second aim of the conference was to present the newest results in each of the fields mentioned above. These were presented in afternoon sessions consisting of contributed talks and posters. This *Proceedings Volume* contains such contributions. For a better overview, they are divided into three parts:

- Classical General Relativity
- Quantum Fields and Quantum Gravity
- Cosmology and Relativistic Astrophysics

Within each of these parts, first articles based on the contributed talks are presented in alphabetical order, then follow articles based on the poster presentations.

Over 200 relativists and astrophysicists from 31 countries appeared in Prague. The abstracts of their talks and posters, and video recordings of all invited lectures are available on the conference website `http://www.ae100prg.mff.cuni.cz`.

The conference was organized under the auspices of the Rector of Charles University, the oldest university north of the Alps (founded in 1348). The lectures took place in the Blue Lecture Hall in the historical complex of the University in the heart of Prague's Old Town.

The Scientific Organizing Committee included Marek Abramowicz, Lars Andersson, Abhay Ashtekar, Julian Barbour, Jiří Bičák, Roger Blandford, Bernd Brügmann, Piotr Chruściel, Thibault Damour, Karsten Danzmann, Fernando de Felice, George Ellis, John Friedman, Helmut Friedrich, Valeri Frolov, Gary Gibbons, Gary Horowitz, Joseph Katz, Karel Kuchař, Jerzy Lewandowski, Gernot Neugebauer, Hermann Nicolai, Igor Novikov, Martin Rees, Oscar Reula, Luciano Rezzolla, Misao Sasaki, Gerhard Schäfer, Bernd Schmidt, Alexei Starobinsky, Paul Tod, Robert Wald, and Clifford Will.

The local organizing committee involved Jiří Bičák, Michal Bursa, Petr Hadrava, David Heyrovský, Vladimír Karas, David Kofroň, Pavel Krtouš, Jiří Langer, Tomáš Ledvinka, Jiří Podolský, Vojtěch Pravda, Oldřich Semerák, Zdeněk Stuchlík, Otakar Svítek, Václav Špička, and Martin Žofka, all of them either researchers from the Institute of Theoretical Physics of the Faculty of Mathematics and Physics of Charles University or former students of the Faculty.

The program of the conference included a ceremony in the Great Auditorium of the historical Carolinum building, during which the honorary degree of Doctor of Physical and Mathematical Sciences, *doctor honoris causa*, was awarded to Prof. D. Lynden-Bell, FRS from the University of Cambridge and to Prof. Dr. W. Domcke from Technical University Munich. A reception for all participants and accompanying persons was hosted by the Dean of the Faculty of Mathematics and Physics of Charles University in the foyer next to Aula Magna. The reception was followed by an impressive concert by Škampa Quartett performing pieces by Haydn, Dvořák, Suk, and Janáček in Aula Magna.

During the conference week, the participants could visit an adjacent exhibition "Einstein, Prague and Gravitation" prepared by Czech physicists and historians and significantly enriched by the "Einstein-Wellen Mobil," a mobile exhibition about gravitational waves brought from Tübingen by Dr. Hans-Peter Nollert. The exhibition was opened 10 days before the beginning of the conference and was visited by a great number of people, in particular by secondary school pupils.

On Wednesday, June 26, a public evening talk "Was Einstein right?—How cosmic time-keepers in space probe Einstein's strange world" by Prof. Michael Kramer from the Max-Planck Institute for Radioastronomy, Bonn, attracted more than 500 people into the Great Lecture Hall of the Faculty of Arts of Charles University despite a EURO 2012 semifinals match taking place at the same time. This presentation on the rich aspects of pulsar physics and astrophysics can be watched at: `http://www.edumeta.com/channels/MFFUK/38889120`.

On Thursday, a late-afternoon guided tour of the Old Town, taking participants to places related to great physicists and astronomers, was followed by a conference banquet in the foundry of Emperor Rudolph II at Prague Castle, decorated by various items related to Tycho Brahe, Johannes Kepler, goldsmiths, alchemists, artists, and others drawn to Rudolph's court. As a special "aperitif," an organ concert was served in St. Vitus Cathedral with the last evening rays penetrating through the cathedral's windows.

Music, in fact, framed the conference: as the participants were entering the Blue Lecture Hall for the first time on Monday morning, they were accompanied by

Madrigal stanzas by Bohuslav Martinů, a Czech composer who became a professor at Princeton and dedicated the stanzas to Albert Einstein in 1943. The same music played again on late Friday afternoon as the participants were leaving the Blue Hall at the end of the conference.

The conference was sponsored by the Faculty of Mathematics and Physics of Charles University, by the Wikow Company, and, above all, by the Karel Janeček's Foundation. Support was much appreciated by both the organizers and participants. We would also like to acknowledge the ongoing support from the Czech Science Foundation, at present realized by grant GAČR No. 14-37086G awarded to the Albert Einstein Center for Gravitation and Astrophysics.

All the contributions were refereed by the editors and by a number of colleagues at the Albert-Einstein Institute in Golm/Potsdam. In particular, we would like to thank Piotr Bizoń, Hermann Hamber, Daniele Oriti, and members of his group for their help with the refereeing process.

Finally, we would like to express our gratitude to Frank Schulz and Vera Osswald from Living Reviews in Relativity based at the Albert Einstein Institute for their great help with these volumes.

Prague, January 2014

Jiří Bičák
Tomáš Ledvinka

Contents

Abstracts of Plenary Talks xvii

Part I Classical General Relativity

Exact Hairy Black Holes 3
Andrés Anabalón

Black Hole Formation from a Complete Past for the Einstein–Vlasov System 11
Håkan Andréasson

How to Measure Deviation from the Kerr Initial Data: Recent Progress 19
Thomas Bäckdahl and Juan A. Valiente Kroon

Hidden Symmetries of the Dirac Equation in Curved Space-Time 25
Marco Cariglia

Geometrostatics: The Geometry of Static Space-Times 35
Carla Cederbaum

The Gravitational Equation in Higher Dimensions 43
Naresh Dadhich

Geometric Inequalities for Black Holes 51
Sergio Dain

Scalar Fields on Anti-de Sitter Background 53
Gyula Fodor, Péter Forgács and Philippe Grandclément

Canonical Superenergy Tensors in General Relativity: A Reappraisal 61
Janusz Garecki

Einstein's "Prague Field Equation" of 1912: Another Perspective 69
Domenico Giulini

Source Integrals of Asymptotic Multipole Moments 83
Norman Gürlebeck

Geodesic Equations and Algebro-Geometric Methods 91
Eva Hackmann

Illusory Horizons, Thermodynamics, and Holography Inside Black Holes 99
Andrew J. S. Hamilton

Shape Dynamics 111
Tim A. Koslowski

Superradiance or Total Reflection? 119
András László and István Rácz

Non-Linear Effects in Non-Kerr Spacetimes 129
Georgios Lukes-Gerakopoulos, George Contopoulos and Theocharis A. Apostolatos

The Conformal Einstein Field Equations for Trace-free Perfect Fluids 137
Christian Lübbe and Juan A. Valiente Kroon

Canonical Gravity, Non-Inertial Frames, Relativistic Metrology and Dark Matter 145
Luca Lusanna

Gravomagnetic Solenoids 155
Donald Lynden-Bell and Joseph Katz

Exact Dynamical AdS Black Holes and Wormholes with a Klein-Gordon Field 161
Hideki Maeda

2.5PN Kick from Black-Hole Binaries in Circular Orbit: Nonspinning Case 169
Chandra Kant Mishra, K. G. Arun and Bala R. Iyer

A Reference for the Covariant Hamiltonian Boundary Term 177
James M. Nester, Chiang-Mei Chen, Jian-Liang Liu and Gang Sun

On a Five-Dimensional Version of the Goldberg-Sachs Theorem 185
Marcello Ortaggio, Vojtěch Pravda, Alena Pravdová and Harvey S. Reall

Gravitomagnetism: From Einstein's 1912 Paper to the Satellites LAGEOS and Gravity Probe B 191
Herbert Pfister

Evolution of the Einstein Equations to Future Null Infinity 199
Oliver Rinne and Vincent Moncrief

Increase of Black Hole Entropy in Lanczos-Lovelock Gravity 207
Sudipta Sarkar

On the Stability Operator for MOTS and the 'Core' of Black Holes 215
José M. M. Senovilla

The Twin Paradox in Static Spacetimes and Jacobi Fields 223
Leszek M. Sokołowski

Geodesic Deviation in Kundt Spacetimes of any Dimension 229
Robert Švarc and Jiří Podolský

A Class of Conformal Curves on Spherically Symmetric Spacetimes 239
Juan A. Valiente Kroon and Christian Lübbe

Black Hole Collisions in Asymptotically de Sitter Spacetimes 247
Miguel Zilhão, Vitor Cardoso, Leonardo Gualtieri, Carlos Herdeiro, Ulrich Sperhake and Helvi Witek

On the Effects of Rotating Gravitational Waves 255
Jiří Bičák, Joseph Katz, Tomáš Ledvinka and Donald Lynden-Bell

Variations on Spacetimes with Boost-Rotation Symmetry 261
Jiří Bičák and David Kofroň

On the Existence and Properties of Helically Symmetric Systems 267
Jiří Bičák, Martin Scholtz and Paul Tod

Probing the Spacetime Structure Through Dynamics 275
Felipe T. Falciano

Analytical Conformal Compactification of Schwarzschild Spacetime 279
Jakub Haláček and Tomáš Ledvinka

Solutions in the 2 + 1 Null Surface Formulation 283
Tina A. Harriott and J. G. Williams

Electric and Magnetic Weyl Tensors in Higher Dimensions 287
S. Hervik, M. Ortaggio and L. Wylleman

Phase Structure of Five Dimensional Black Di-ring 291
Hideo Iguchi

The Null Geodesics in the Black Saturn Spacetime 295
Alicja Konieczny

Conformal Symmetries on the Horizon and Black Hole Entropy in Generic Dimensions 299
Jianwei Mei

Finsler Spacetimes and Gravity 305
Christian Pfeifer and Mattias Wohlfarth

Lagrangian Analysis of 'Trivial' Symmetries in Models of Gravity . . . 309
Debraj Roy

Quasi-normal Frequencies, Horizon Area Spectra and Multihorizon Spacetimes 315
Jozef Skákala

Asymptotically AdS Spacetimes and Isometric Embeddings 319
Steven Willison

Part II Cosmology and Relativistic Astrophysics

A Cosmological Concordance Model with Particle Creation 325
J. S. Alcaniz, H. A. Borges, S. Carneiro, J. C. Fabris, C. Pigozzo and W. Zimdahl

From 'Nothing' to Inflation and Back Again 331
Vladimír Balek

Quasinormal Modes from a Naked Singularity 339
Cecilia Chirenti, Jozef Skákala and Alberto Saa

Tracing a Relativistic Milky Way Within the RAMOD Measurement Protocol 347
Mariateresa Crosta

Is There a Flatness Problem in Classical Cosmology? 355
Phillip Helbig

Cosmology in *f*(*R*) Exponential Gravity 363
Luisa Jaime, Marcelo Salgado and Leonardo Patiño

Regular and Chaotic Motion in General Relativity: The Case of a Massive Magnetic Dipole 373
Ondřej Kopáček, Jiří Kovář, Vladimír Karas and Yasufumi Kojima

The Fitting Problem in a Lattice Universe 385
Julien Larena

Hair of Astrophysical Black Holes 393
Maxim Lyutikov

Backreaction Effects on the Luminosity-Redshift Relation in Inhomogeneous Cosmology 399
Giovanni Marozzi

Scalar Averaging in Szekeres Models 407
Roberto A. Sussman

On the Interplay Between Radial and Angular Reflection Emissivity from the Black Hole Accretion Disc 415
Jiří Svoboda, Michal Dovčiak, René W. Goosmann and Vladimír Karas

Critical-Curve Topologies of Triple Gravitational Lenses 423
Kamil Daněk and David Heyrovský

Modified Gravity Theories and Dark Matter Models Tested by Galactic Rotation Curves 427
Marek Dwornik, Zoltán Keresztes and László Árpád Gergely

Averaging Inside the LRS Family 431
Petr Kašpar, David Vrba and Otakar Svítek

Effect of Magnetic Fields on Equatorial Circular Orbits in Kerr Spacetimes 435
Ignacio F. Ranea-Sandoval and Héctor Vucetich

Exotic (Dark) Eigenspinors of the Charge Conjugation Operator and Cosmological Applications 439
Roldao da Rocha

On Motion of the Magellanic Clouds in the Milky Way Gravitational Field 443
Zdeněk Stuchlík and Jan Schee

Geodesic Chaos in Perturbed Black-Hole Fields 449
Petra Suková and Oldřich Semerák

Gravitational Waveforms for Black Hole Binaries with Unequal Masses 455
Márton Tápai, Zoltán Keresztes and László Árpád Gergely

Part III Quantum Fields and Quantum Gravity

Phenomenology of Quantum Gravity and its Possible Role in Neutrino Anomalies 461
Mario A. Acero and Yuri Bonder

Loop Quantum Cosmology: Anisotropy and Singularity Resolution . . . 469
Alejandro Corichi, Asieh Karami and Edison Montoya

Tensor Operators in Loop Quantum Gravity 479
Maïté Dupuis and Florian Girelli

Probability Distributions for Quantum Stress Tensors in Two and Four Dimensions 489
Christopher J. Fewster, L. H. Ford and Thomas A. Roman

Spontaneous Breaking of Lorentz Symmetry for Canonical Gravity 497
Steffen Gielen

The Transfer Matrix in Four-Dimensional Causal Dynamical Triangulations 505
Andrzej Görlich

Plane Gravitational Waves and Flat Space in Loop Quantum Gravity 515
Franz Hinterleitner and Seth Major

Unruh-DeWitt Detector on the BTZ Black Hole 523
Lee Hodgkinson and Jorma Louko

On the Observability of Quantum-Gravitational Effects in the Cosmic Microwave Background 531
Claus Kiefer and Manuel Krämer

Quantum Singularities in Conformally Static Spacetimes 539
Deborah A. Konkowski and Thomas M. Helliwell

Granularity in Angle: Observability in Scattering Experiments 547
Seth A. Major and Jake C. Zappala

Signature Change in Loop Quantum Cosmology 555
Jakub Mielczarek

Quantum Fields in Gravity 563
Giovanni Acquaviva

Classical and Quantum Scattering in Impulsive Backgrounds 567
Peter Aichelburg and Herbert Balasin

Effective Vacuum Bianchi IX in Loop Quantum Cosmology 573
Alejandro Corichi, Asieh Karami and Edison Montoya

Coupling Dimers to CDT to Obtain Higher Order Multicritical Behavior 579
Lisa Glaser

A Sheet of Graphene: Quantum Field in a Discrete Curved Space . . . 583
Nikodem Szpak

Abstracts of Plenary Talks

Introduction The texts based on 25 plenary talks from 32 given are included in the Volume "General, Relativity, Cosmology and Astrophysics: 100 Years After Einstein in Prague" (editors J. B. and T. L.), Springer Verlag 2013. In order to give a more complete picture of the whole conference, the abstracts of the 25 plenary talks are included in the first chapter of these Proceedings.

Part I Gravity and Prague

Kepler and Mach's Principle

Julian Barbour

The definitive ideas that led to the creation of general relativity crystallized in Einstein's thinking during 1912 while he was in Prague. At the centenary meeting held there to mark the breakthrough, I was asked to talk about earlier great work of relevance to dynamics done at Prague, above all by Kepler and Mach. The main topics covered in this paper are: some little known but basic facts about the planetary motions; the conceptual framework and most important discoveries of Ptolemy and Copernicus; the complete change of concepts that Kepler introduced and their role in his discoveries; the significance of them in Newton's work; Mach's realization that Kepler's conceptual revolution needed further development to free Newton's conceptual world of the last vestiges of the purely geometrical Ptolemaic world view; and the precise formulation of Mach's principle required to place GR correctly in the line of conceptual and technical evolution that began with the ancient Greek astronomers.

Einstein in Prague: Relativity Then and Now

Jiří Bičák

It was during his stay in Prague that Einstein started in earnest to develop his ideas about general relativity. I will recall those days in 1911 and 1912, discuss

Einstein's papers on gravitation from that period, and emphasize which new concepts and ideas he introduced. I also want to indicate how the main themes that preoccupied him then, the principle of equivalence, bending of light, gravitational redshift, and frame dragging effects, are alive in contemporary relativity.

Part II Classical General Relativity

Observers, Observables and Measurements in General Relativity

Donato Bini

To perform any physical measurement it is necessary to identify in a nonambiguous way both the observer and the observable. A given observable can be then the target of different observers: a suitable algorithm to compare among their measurements should necessarily be developed, either formally or operationally. This is the task of what we call "theory of measurement," which we discuss here in the framework of general relativity.

Some Links Between General Relativity and Other Parts of Physics

Gary W. Gibbons

Now that General Relativity has become such a central part of modern physics, its geometrical formalism being taught as part of almost all undergraduate physics courses, it is natural to ask: how can its basic concepts and techniques be used to illuminate areas of physics which have no connection with gravity? Another way of asking this question is: are the situations analogous to those occurring in General Relativity? The search for such analogs is of course an old one, but recently, because of advances in technology, these questions have become more topical. In this talk I will illustrate this theme by examples drawn from optics, acoustics, liquid crystals, graphene, and the currently popular topic of cloaking.

The General Relativistic Two Body Problem and the Effective One Body Formalism

Thibault Damour

A new analytical approach to the motion and radiation of (comparable mass) binary systems was introduced in 1999 under the name of Effective One-Body (EOB) formalism. We review the basic elements of this formalism, and discuss some of its recent developments. Several recent comparisons between EOB predictions and Numerical Relativity (NR) simulations have shown the aptitude of the EOB formalism to provide accurate descriptions of the dynamics and radiation of various binary systems (comprising black holes or neutron stars) in regimes that are inaccessible to other analytical approaches (such as the last orbits and the merger of comparable mass black holes). In synergy with NR simulations, post-Newtonian (PN) theory and Gravitational Self-Force (GSF) computations, the EOB formalism is likely to provide an efficient way of computing the very many accurate template waveforms that are needed for Gravitational Wave (GW) data analysis purposes.

Gravitational Self-Force: Orbital Mechanics Beyond Geodesic Motion

Leor Barack

The question of motion in a gravitationally bound two-body system is a long-standing open problem of General Relativity. When the mass ratio η is small, the problem lends itself to a perturbative treatment, wherein corrections to the geodesic motion of the smaller object (due to radiation reaction, internal structure, etc.) are accounted for order by order in η, using the language of an effective *gravitational self-force*. The prospect for observing gravitational waves from compact objects inspiralling into massive black holes in the foreseeable future has in the past 15 years motivated a program to obtain a rigorous formulation of the self-force and compute it for astrophysically interesting systems. I will give a brief survey of this activity and its achievements so far, and will identify the challenges that lie ahead. As concrete examples, I will discuss recent calculations of certain conservative post-geodesic effects of the self-force, including the $O(\eta)$ to the precession rate of the periastron. I will highlight the way in which such calculations allow us to make fruitful contact with other approaches to the two-body problem.

Hamiltonian Formalism for Spinning Black Holes in General Relativity

Gerhard Schäfer

A Hamiltonian treatment of gravitationally interacting spinning black holes is presented based on a tetrad generalization of the Arnowitt-Deser-Misner (ADM) canonical formalism of general relativity. The formalism is valid through linear order in single spins. For binary systems, higher order post-Newtonian Hamiltonians are given in explicit analytic forms. A next-to-leading order in spin generalization is presented, others are mentioned. Comparisons between the Hamiltonian formalisms by ADM, Dirac, and Schwinger are made.

Stability of Marginally Outer Trapped Surfaces and Geometric Inequalities

Marc Mars

Marginally outer trapped surfaces (MOTS) admit a notion of stability that in many respects generalizes a similar notion for minimal hypersurfaces. Stable MOTS play an interesting role in a number of geometric inequalities involving physical parameters such as area, mass, charge or, in the axially symmetric case, angular momentum. Some of those inequalities are global in nature while others are local, with interesting relationships between them. In this lecture the notion of stable MOTS will be reviewed and some of the geometric inequalities involving stable MOTS will be described.

Stationary Black-Hole Binaries: A Non-existence Proof

Gernot Neugebauer and Jörg Hennig

We resume former discussions of the question of whether the spin–spin repulsion and the gravitational attraction of two aligned black holes can balance each other. Based on the solution of a boundary problem for disconnected (Killing) horizons and the resulting violation of characteristic black hole properties, we present a nonexistence proof for the equilibrium configuration in question. From a mathematical point of view, this result is a further example of the efficiency of the inverse ("scattering") method in nonlinear theories.

Dynamic and Thermodynamic Stability of Black Holes and Black Branes

Robert M. Wald

I describe recent work with Stefan Hollands that establishes a new criterion for the dynamical stability of black holes in $D \geq 4$ spacetime dimensions in general relativity with respect to axisymmetric perturbations: Dynamical stability is equivalent to the positivity of the canonical energy, $\mathscr{E}$, on a subspace of linearized solutions that have vanishing linearized ADM mass, momentum, and angular momentum at infinity and satisfy certain gauge conditions at the horizon. We further show that $\mathscr{E}$ is related to the second order variations of mass, angular momentum, and horizon area by $\mathscr{E} = \delta^2 M - \sum_i \Omega_i \delta^2 J_i - (\kappa/8\pi)\delta^2 A$, thereby establishing a close connection between dynamical stability and thermodynamic stability. Thermodynamic instability of a family of black holes need not imply dynamical instability because the perturbations towards other members of the family will not, in general, have vanishing linearized ADM mass and/or angular momentum. However, we prove that all black branes corresponding to thermodynmically unstable black holes are dynamically unstable, as conjectured by

Gubser and Mitra. We also prove that positivity of $\mathscr{E}$ is equivalent to the satisfaction of a "local Penrose inequality," thus showing that satisfaction of this local Penrose inequality is necessary and sufficient for dynamical stability.

Instability of Anti-de Sitter Spacetime

Piotr Bizoń and Andrzej Rostworowski

In this talk we summarize our recent numerical and perturbative calculations which indicate that AdS spacetime is unstable. Namely, we study spherically symmetric Einstein-massless-scalar field equations with negative cosmological constant and show that this system is unstable against black hole formation for a large class of initial data arbitrarily close to the AdS solution. We conjecture that this instability is triggered by a resonant mode mixing which gives rise to diffusion of energy from low to high frequencies.

Higher-Dimensional Black Holes

Harvey S. Reall

This article reviews black hole solutions of higher dimensional General Relativity. The focus is on stationary vacuum solutions and recent work on instabilities of such solutions.

Black Holes, Hidden Symmetry and Complete Integrability: Brief Review

Valeri P. Frolov

This paper contains a brief review of the remarkable properties of higher dimensional rotating black holes with the spherical topology of the horizon. We demonstrate that these properties are connected with and generated by a special geometrical object, the Principal Conformal Killing-Yanc tensor (PCKYT). The most general solution, describing such black holes, Kerr-NUT-ADS metric, admits this structure. Moreover, a solution of the Einstein Equations with (or without) a cosmological constant which possesses PCKYT is the Kerr-NUT-ADS metric. This object (PCKYT) is responsible for such remarkable properties of higher dimensional rotating black holes as: (i) complete integrability of geodesic equations and (ii) complete separation of variables of the important field equations.

Part III Cosmology and Quantum Gravity

Cosmological Models and Stability

Lars Andersson

Principles in the form of heuristic guidelines or generally accepted dogma play an important role in the development of physical theories. In particular, philosophical considerations and principles figure prominently in the work of Albert Einstein. As mentioned in the talk by Jiří Bičák at this conference, Einstein formulated the equivalence principle, an essential step on the road to general relativity, during his time in Prague 1911–1912. In this talk, I would like to discuss some aspects of cosmological models. As cosmology is an area of physics where "principles" such as the "cosmological principle" or the "Copernican principle" play a prominent role in motivating the class of models which form part of the current standard model, I will start by comparing the role of the equivalence principle to that of the principles used in cosmology. I will then briefly describe the standard model of cosmology to give a perspective on some mathematical problems and conjectures on cosmological models, which are discussed in the later part of this paper.

Inflation and Birth of Cosmological Perturbations

Misao Sasaki

We review recent developments in the theory of inflation and cosmological perturbations produced from inflation. After a brief introduction of the standard, single-field slow-roll inflation, and the curvature and tensor perturbations produced from it, we discuss possible sources of nonlinear, non-Gaussian perturbations in other models of inflation. Then we describe the so-called δN formalism, which is a powerful tool for evaluating nonlinear curvature perturbations on super Hubble scales.

Loop Quantum Gravity and The Planck Regime of Cosmology

Abhay Ashtekar

The very early universe provides the best arena we currently have to test quantum gravity theories. The success of the inflationary paradigm in accounting for the observed inhomogeneities in the cosmic microwave background already illustrates this point to a certain extent because the paradigm is based on quantum field theory on the curved cosmological space-times. However, this analysis excludes the Planck era because the background space-time satisfies Einstein's equations all the way back to the big bang singularity. Using techniques from loop quantum gravity, the paradigm has now been extended to a self-consistent theory from the Planck regime to the onset of inflation, covering some 11 orders of magnitude in curvature. In addition, for a narrow window of initial conditions, there are departures from the standard paradigm, with novel effects, such as a modification of the consistency relation involving the scalar and tensor power spectra and a new source for non-Gaussianities. The genesis of the large-scale structure of the universe can be traced back to quantum gravity fluctuations in the Planck regime. This report provides a bird's eye view of these developments for the general relativity community.

The Inflationary Origin of the Seeds of Cosmic Structure: Quantum Theory and the Need for Novel Physics

Daniel Sudarsky

The Inflationary account for the emerging of the seeds of cosmic structure from quantum fluctuations is a central part of our current views of cosmology. It is, on the one hand, extremely successful at the phenomenological level, and yet it retains an aspect that is generally regarded as controversial: The exact mechanism by which quantum fluctuations transmute into actual inhomogeneities. We will review the considerations that lead us to conclude that the fully satisfactory resolution of the issue requires novel physics and we will discuss an option we have been considering in this regard.

Quantum Gravity: The View From Particle Physics

Hermann Nicolai

This lecture reviews aspects of and prospects for progress towards a theory of quantum gravity from a particle physics perspective, also paying attention to recent findings of the LHC experiments at CERN.

Part IV Numerical Relativity and Relativistic Astrophysics

Three Little Pieces for Computer and Relativity

Luciano Rezzolla

Numerical relativity has made big strides over the last decade. A number of problems that have plagued the field for years have now been mostly solved. This progress has transformed numerical relativity into a powerful tool to explore fundamental problems in physics and astrophysics, and I present here three representative examples. These "three little pieces" reflect a personal choice and describe work that I am particularly familiar with. However, many more examples could be made.

Instabilities of Relativistic Stars

John L. Friedman and Nikolaos Stergioulas

Stable relativistic stars in uniform rotation forma two-parameter family, parametrized by mass and angular velocity. Limits on each of these quantities are associated with relativistic instabilities. A radial instability to gravitational collapse or explosion sets upper and lower limits on their mass, and an instability driven by gravitational waves may set an upper limit on their spin. Our summary of relativistic stability theory given here is based on and includes excerpts from the book *Rotating Relativistic Stars*, by the present authors.

Gravity Talks: Observing the Universe with Gravitational Waves

Bernard F Schutz

When the current upgrade of the large ground-based gravitational wave detectors LIGO and VIRGO is completed, the new science of gravitational wave astronomy will begin. In this overview I review the current status of the detector projects on the ground and in space (LISA), the kinds of signals and sources they expect to observe, and the science returns that are anticipated.

LISA in 2012 and Beyond: 20 Years After the First Proposal

Gerhard Heinzel and Karsten Danzmann

After 20 years of study as a joint ESA-NASA mission, LISA had to be redesigned as an ESA-only mission in 2011/2012 to meet programmatic and budgetary constraints of the space agencies. The result is a mission concept called "eLISA" or "NGO" with two arms instead of three and 1 million km armlengths instead of 5, which results in smaller launch mass but still provides revolutionary science. Nevertheless, even the reduced science is expected to be revolutionary for the study of black holes and other astrophysical and cosmological questions. "eLISA"/"NGO" was not selected in ESA's call for the first ("L1") large mission in the Cosmic Vision program, but is a strong candidate for the L2 call, with possible international contributions from the US and/or China.

Einstein's Gravity as Seen by a Cosmic Lighthouse Keeper

Michael Kramer

The last years have seen continuing activities in the exploration of our understanding of gravity, motivated by results from precision cosmology and new precision astrophysical experiments. At the center of attention lies the question as to whether general relativity is the correct theory of gravity. In answering this question, we work not only towards correctly interpreting the phenomenon of "dark energy" but also towards the goal of achieving a quantum theory of gravity. In these efforts, the observations of pulsars, especially those in binary systems, play an important role. Pulsars do not only provide the only evidence for the existence of gravitational waves so far, but they also provide precision tests of general relativity and alternative theories of gravity. This talk summarizes the current state of art in these experiments and looks into the future.

The Astrophysical Signatures of Black Holes: The Horizon,
The ISCO, The Ergosphere and The Light Circle

Marek A. Abramowicz

Three advanced instruments planned for a near future (`LOFT, GRAVITY, THE EVENT HORIZON TELESCOPE`) provide unprecedented angular and time resolutions, which allow to probe regions in the immediate vicinity of black holes. We may soon be able to search for the signatures of the superstrong gravity that is characteristic of black holes: the event horizon, the ergosphere, the innermost stable circular orbit (ISCO), and the photon circle. This review discusses a few fundamental problems concerning these theoretical concepts.

Energy Extraction from Spinning Black Holes Via Relativistic Jets

Ramesh Narayan, Jeffrey E. McClintock and Alexander Tchekhovskoy

It has for long been an article of faith among astrophysicists that black hole spin energy is responsible for powering the relativistic jets seen in accreting black holes. Two recent advances have strengthened the case. First, numerical general relativistic magnetohydrodynamic simulations of accreting spinning black holes show that relativistic jets form spontaneously. In at least some cases, there is unambiguous evidence that much of the jet energy comes from the black hole, not the disk. Second, spin parameters of a number of accreting stellar-mass black holes have been measured. For ballistic jets from these systems, it is found that the radio luminosity of the jet correlates with the spin of the black hole. This suggests a causal relationship between black hole spin and jet power, presumably due to a generalized Penrose process.

Part I
Classical General Relativity

Exact Hairy Black Holes

Andrés Anabalón

Abstract This contribution reviews the recent discovery of a certain class of—regular on and outside the horizon—exact hairy black hole solutions in four dimensional general relativity. Their construction follows from the integrability of a cohomogeneity two Weyl rescaling of the Carter–Debever ansatz in the presence of an arbitrary number of scalar fields with an arbitrary self interaction and an arbitrary non-minimal coupling to the scalar curvature. Two field equations, independent of the specific form of the energy momentum tensor, are used to integrate the metric. The remaining ones fix the form of the scalar field self interaction. The cohomogeneity one black holes are described and are shown to encompass all the exact—regular in the domain of outer communications—uncharged, black holes with a minimally coupled scalar hair, available in the literature.

1 Introduction and Discussion

The field of exact solutions in gravity, as well as their interpretation, is as old as general relativity and the research group at Charles university, and their collaborators, are well known for their contributions to this subject. Many of them can be found in the review [1] or the book [2]. From the black hole uniqueness theorems it is already well known that at least in four dimensions, the asymptotically flat, stationary and regular black holes in the electrovacuum case are exhausted, for references see [3]. Therefore, it is natural to attempt to extend these studies when other matter

A. Anabalón (✉)
Departamento de Ciencias, Facultad de Artes Liberales, Facultad de Ingeniería y Ciencias, Universidad Adolfo Ibáñez, Viña Del Mar, Chile
e-mail: andres.anabalon@uai.cl

J. Bičák and T. Ledvinka (eds.), *Relativity and Gravitation*,
Springer Proceedings in Physics 157, DOI: 10.1007/978-3-319-06761-2_1,

fields are included. Indeed, the studies of the minimally coupled scalar field have a prominent role in the construction of black holes. In the static, asymptotically flat case, the minimally coupled no-hair conjecture was shown to be true for convex potentials [4], and, more generally, for potentials satisfying the strong [5] and weak energy condition [6] . These studies have their counterpart in Brans–Dicke [7] and more generally in scalar-tensor theories [8], showing that whenever the scalar field potential satisfies the weak energy condition in the Einstein frame and the black hole spacetime is stationary and asymptotically flat it must be Kerr. When the scalar field satisfies the null energy condition an exact family of spherically symmetric black hole solutions has been recently constructed [9].

When the cosmological constant is negative exact uncharged AdS_4 hairy black hole solutions have been extensively studied [10–14]. There is a precise conjecture on the non-existence of spherically symmetric black holes in AdS for scalar field potentials that comes from "the right" superpotential [15]. These solutions are interesting in the light of the AdS/CFT conjecture. In particular, in four dimensions, and when the scalar field is charged, they define the setting for the AdS/Condensed matter correspondence [16]. When the cosmological constant is positive the black holes have also attracted some attention of the community [17].

This paper intends to shortly summarize my recent contributions to the subject. I have followed the idea that stationary and axisymmetric spacetimes that have a hidden symmetry, in the form of a conformal Killing tensor, should allow for a complete integrability of some form of a non-trivial self interaction of the scalar field. Therefore, in [14] I explicitly showed that, starting with the ansatz that contains all the vacuum Petrov type D solutions, it is possible to integrate the system in the presence of a non-minimally coupled scalar field or a non-linear sigma model. It is very interesting to note that the self interaction of the scalar field is completely fixed by the form of the metric ansatz and, therefore, the scalar field potential is an output of the analysis. While these results are presented in the Einstein frame, their extension to a Scalar-Tensor theories in some Jordan frame or $F(R)$ theory is straightforward.

The scalar field potential turns out to be contained as special case of all the exact hairy (A)dS black holes available in the literature. The static solutions are black holes continuously connected with the Schwarzschild (A)dS solution, and can be generalized to include non-minimally coupled gauge fields [18]. For the asymptotically AdS black holes, with cosmological constant $\Lambda = -3/l^2$, the scalar field mass is $m^2 = -2/l^2$, which is above the Breitenlohner–Freedman bound, $m^2 = -9/4l^2$, ensuring the perturbative stability of AdS_4. This mass is the one of the scalar fields of the $U(1)^4$ truncation of gauged $N = 8$ supergravity [19] and a sub class of the solutions can be embedded in this supergravity theory.

The content of the paper is as follows. In Sect. 2 the general integrability of the ansatz with two Killing vectors is reviewed and in Sect. 3 the static case and its special limits are presented.

2 The Integrable System with Two Killing Vectors

The conventions are given by the action principle

$$S(g,\phi)=\int d^4x\sqrt{-g}\left[\frac{R}{2\kappa}-\frac{1}{2}g^{\mu\nu}\partial_\mu\phi\partial_\nu\phi-\frac{\xi}{12}\phi^2R-V(\phi)\right], \tag{1}$$

where $\kappa = 8\pi G$. We are interested in studying a cohomogeneity two Weyl rescaling of the Carter–Debever [20, 21], also studied by Plebański [22]:

$$\begin{aligned}ds^2 = S(q,p)\Big(&\frac{1+p^2q^2}{Y(q)}dq^2+\frac{1+p^2q^2}{X(p)}dp^2-\frac{Y(q)}{1+p^2q^2}\left(p^2d\tau+d\sigma\right)^2\\&+\frac{X(p)}{1+p^2q^2}\left(d\tau-q^2d\sigma\right)^2\Big).\end{aligned} \tag{2}$$

When $S(q,p)=p^{-2}$, this metric contains the Kerr–Newman hole with a cosmological constant. Letting $S(q,p)$ free, this metric can be integrated in vacuum, and with the same Maxwell field as in the Kerr–Newman case; the Plebański–Demiański spacetime arises [23].

The observation made in [14] is that for stationary and axisymmetric scalar fields, $\phi=\phi(q,p)$, the energy momentum tensor of a minimally coupled scalar field

$$T_{\mu\nu}=\partial_\mu\phi\partial_\nu\phi-\frac{1}{2}g_{\mu\nu}\left(\partial\phi\right)^2-g_{\mu\nu}V(\phi), \tag{3}$$

is such that components $T^\tau_\sigma = 0 = T^\sigma_\tau$ and, therefore, the Einstein equations, $R^\mu_\nu-\frac{1}{2}\delta^\mu_\nu R=\kappa T^\mu_\nu$, imply $R^\tau_\sigma=0=R^\sigma_\tau$. These two equations are enough to completely determine the metric functions; the solution is

$$X(p)=C_0+C_2p^2+C_4p^4+C_1p^{-\nu+2}+C_3B_3p^{\nu+2}, \tag{4}$$

$$Y(q)=C_4-C_2q^2+C_0q^4+C_3C_1q^{-\nu+2}+B_3q^{\nu+2}, \tag{5}$$

$$S(q,p)=C\frac{p^{\nu-1}q^{\nu-1}}{(C_3p^\nu+q^\nu)^2}. \tag{6}$$

This solution reduces to the Plebański–Demiański spacetime when $\nu=\pm1$. The remaining Einstein equations fix the scalar field and the scalar field potential to a very precise form. The same process can be done when the scalar field is non-minimally coupled to gravity and, more generally, when a non-linear sigma model is the source of the Einstein equations.

To extract more physical information let us study the cohomogeneity one black holes.

3 The Static Black Holes

The static limit of the previous configuration is

$$ds^2 = \Omega(r)\left(-F(r)dt^2 + \frac{dr^2}{F(r)} + d\Sigma_k^2\right), \tag{7}$$

$$\Omega(r) = \frac{\nu^2\eta^{\nu-1}r^{\nu-1}}{(r^\nu - \eta^\nu)^2}, \qquad \phi = l_\nu^{-1}\ln(r\eta^{-1}), \tag{8}$$

$$\begin{aligned} F(r) = &\frac{r^{2-\nu}\eta^{-\nu}(r^\nu - \eta^\nu)^2 k}{\nu^2} \\ &+ \left(\frac{1}{\nu^2 - 4} - \left(1 + \frac{\eta^\nu r^{-\nu}}{\nu - 2} - \frac{\eta^{-\nu}r^\nu}{\nu + 2}\right)\frac{r^2}{\eta^2\nu^2}\right)\alpha - \frac{\Lambda}{3}, \end{aligned} \tag{9}$$

where $l_\nu = \left(\frac{2\kappa}{\nu^2-1}\right)^{\frac{1}{2}}$ and $d\Sigma_k^2$ is the line element of a surface of constant curvature $k = \pm 1$ or 0. η is the only integration constant of the black hole. The solution and theory are invariant under the transformation $\nu \to -\nu$.

The scalar field potential is

$$\begin{aligned} V(\phi) = &\frac{\Lambda(\nu^2 - 4)}{6\kappa\nu^2}\left(\frac{\nu - 1}{\nu + 2}e^{-(\nu+1)\phi l_\nu} + \frac{\nu + 1}{\nu - 2}e^{(\nu-1)\phi l_\nu} + 4\frac{\nu^2 - 1}{\nu^2 - 4}e^{-\phi l_\nu}\right) \\ &+ \frac{\alpha}{\nu^2\kappa}\left(\frac{\nu - 1}{\nu + 2}\sinh((1 + \nu)\phi l_\nu) + \frac{\nu + 1}{\nu - 2}\sinh((1 - \nu)\phi l_\nu)\right. \\ &\left. + 4\frac{\nu^2 - 1}{\nu^2 - 4}\sinh(\phi l_\nu)\right). \end{aligned} \tag{10}$$

It has two coupling constants, namely α and Λ, and the parameter ν. Indeed, $V(\phi = 0) = \Lambda/\kappa$ which allows to identify it as a cosmological constant.

It is easy to see from the form of the metric, and without any reference to the details of the solution itself, that it is possible to introduce Eddington–Finkelstein coordinates $u_\mp = t \pm \int \frac{dr}{F(r)}$, which allow to cover either the black hole (u_-) or the white hole (u_+). The asymptotically flat solution has a single horizon from which it follows that the Penrose diagram is the same as for the Schwarzschild black hole.

The energy momentum of the scalar field, in a static tetrad, has the form $T^{ab} = \mathrm{diag}(\rho, p_1, p_2, p_2)$ and, in the static regions of the spacetime, defined by $F(r) > 0$, satisfies the null energy condition

$$\rho + p_2 = 0, \qquad \rho + p_1 = \frac{(\nu^2 - 1)(r^\nu - \eta^\nu)^2 F(r)}{2r\nu^2\eta^{\nu-1}r^\nu} > 0. \tag{11}$$

In the hairless limit, $\nu = 1$, the change of coordinates $r = \eta - 1/y$ brings the hairy solution (7–9) to the familiar Schwarzschild–de Sitter black hole (in this case $p_1 = \Lambda$)

$$ds^2 = -\left(k - \frac{2M}{y} - \frac{\Lambda}{3}y^2\right)dt^2 + \frac{dy^2}{k - \frac{2M}{y} - \frac{\Lambda}{3}y^2} + y^2 d\Sigma_k^2, \tag{12}$$

where $M = \frac{3\eta^2 k + \alpha}{6\eta^3}$.

The parameterization of the black holes has been chosen such that the leading order at $r = \eta$ is either Minkowski, anti-de Sitter or de Sitter in the following form

$$ds^2_{r=\eta} = \frac{1}{(r-\eta)^2}\left(-\left(k(r-\eta)^2 + \frac{\Lambda}{3}\right)dt^2 + \frac{dr^2}{(k(r-\eta)^2 + \frac{\Lambda}{3})} + d\Sigma^2\right). \tag{13}$$

The easiest way to see that there is always α such that $F(r)$ has a simple zero is to see that the equation $F(r_+) = 0$ is linear in α

$$0 = \frac{r_+^{2-\nu}\eta^{-\nu}\left(r_+^\nu - \eta^\nu\right)^2 k}{\nu^2} + \left(\frac{1}{\nu^2 - 4} - \left(1 + \frac{\eta^\nu r_+^{-\nu}}{\nu - 2} - \frac{\eta^{-\nu} r_+^\nu}{\nu + 2}\right)\frac{r_+^2}{\eta^2\nu^2}\right)\alpha - \frac{\Lambda}{3}; \tag{14}$$

therefore it is possible to solve this equation for α for any value of the other parameters.

As a final remark it is instructive to compare the behaviour of these solutions in AdS, with the asymptotic form given in [24] . When the backreaction is ignored, a scalar field with mass m minimally coupled to an AdS background has the well known fall-off $\phi \sim \frac{a}{\rho^{\Delta_-}} + \frac{b}{\rho^{\Delta_+}}$ where $\Delta_\pm$ are the roots of $\Delta(3 - \Delta) + m^2 l^2 = 0$. When $-\frac{9}{4l^2} \leq m^2 < -\frac{5}{4l^2}$, both branches are normalizable but the a-branch contributes to the surface charges of the system. The form of the potential makes it possible to see that the mass is $m^2 = -\frac{2}{l^2}$. When the mass is exactly $\frac{\Delta_+}{\Delta_-} = 2$ then the scalar field develops a logarithmic branch that, again, has a non-trivial contribution to the charges at infinity. However this logarithmic branch only appears if the expansion of the potential contains a cubic term. Indeed, it is possible to verify that with the change of coordinates, $r = \eta\exp(\frac{1}{\eta\rho} - \frac{1}{2\rho^2\eta^2} - \frac{\nu^2 - 9}{24\eta^3\rho^3})$, the scalar field takes the form

$$\phi = l_\nu^{-1}\left(\frac{1}{\eta\rho} - \frac{1}{2\rho^2\eta^2} - \frac{\nu^2 - 9}{24\eta^3\rho^3}\right), \tag{15}$$

and the departure from the AdS metric, defined by

$$ds^2 = -(k + \frac{\rho^2}{l^2})dt^2 + \left(k + \frac{\rho^2}{l^2}\right)^{-1} d\rho^2 + \rho^2 d\Sigma_k^2, \tag{16}$$

is

$$h_{mn} = \frac{\nu^2 - 4}{6\eta^3 \rho} g_{mn} + O(\rho^{-2}), \tag{17}$$

$$h_{tt} = \frac{\Lambda(\nu^2 - 4)}{18\eta^3 \rho} + \frac{k\,(\nu - 1) + 6M\nu^2\eta^\nu}{3\eta\rho} + O(\rho^{-2}), \tag{18}$$

$$h_{\rho\rho} = \frac{3\left(\nu^2 - 1\right)}{4\eta^2 \Lambda \rho^4} + O(\rho^{-5}), \tag{19}$$

where g_{mn} are the components along $d\Sigma_k^2$. This coincides exactly with (6.2) of [24] with $\Delta_- = \Delta = 1$, $a = \frac{1}{\eta l_\nu}$ and $b = -\frac{1}{2\eta^2 l_\nu}$. The case $\nu^2 = 4$ is peculiar in the sense that the deformation of the metric at infinity is subleading as for generic ν.

The cases with $\nu = 2$ and $\nu = \infty$ are special, and can be treated by a simple limiting procedure.

3.1 The Case $\nu = 2$

Indeed, the potential (10) has a smooth limit when $\nu = 2$, which is given by

$$\begin{aligned} V(\phi) =& \frac{\alpha}{16\kappa} \left(\sinh(3\phi l_2) + 9\sinh(\phi l_2) - 12\phi l_2 \cosh\left(\phi l_2\right)\right) \\ &+ \frac{\Lambda}{2\kappa\nu^2}\left(e^{\phi l_2} + e^{-\phi l_2}\right), \end{aligned} \tag{20}$$

where $l_2 = \sqrt{\frac{2\kappa}{3}}$. The metric functions also have a smooth limit

$$\Omega(r) = \frac{4\eta r}{\left(r^2 - \eta^2\right)^2}, \tag{21}$$

$$F(r) = \frac{\eta^{-2}\left(r^2 - \eta^2\right)^2}{4} k + \left(\frac{3}{16} + \left(\frac{r}{2\eta}\right)^4 - \left(\frac{r}{2\eta}\right)^2 + \frac{1}{4}\ln\frac{r}{\eta}\right)\alpha - \frac{\Lambda}{3}. \tag{22}$$

The potential (10) has been considered in the context of the existence of topological AdS black holes in [11]. When $\alpha = 0$ and $k = -1$ this is the MTZ black hole [10].

3.2 The Case $\nu = \infty$

The $\nu = \infty$ case is a bit more subtle. First, it is necessary to rescale the area of the unit sphere as $d\Sigma \longrightarrow \nu^{-2} d\Sigma$ which implies that the metric function F rescales accordingly,

$$F(r) = r^{2-\nu} \eta^{-\nu} \left(r^{\nu} - \eta^{\nu}\right)^2 k + \left(\frac{1}{\nu^2 - 4} - \left(1 + \frac{\eta^{\nu} r^{-\nu}}{\nu - 2} - \frac{\eta^{-\nu} r^{\nu}}{\nu + 2}\right) \frac{r^2}{\eta^2 \nu^2}\right) \alpha, \tag{23}$$

and the solution is now

$$ds^2 = \Omega(r) \left(-F(r) dt^2 + \frac{dr^2}{F(r)} + \nu^{-2} d\Sigma_k^2\right), \tag{24}$$

$$\Omega(r) = \frac{\nu^2 \eta^{\nu-1} r^{\nu-1}}{\left(r^{\nu} - \eta^{\nu}\right)^2}, \qquad \phi = l_{\nu}^{-1} \ln(r\eta^{-1}). \tag{25}$$

Let us introduce the changes of coordinates $r = \rho^{\frac{1}{\nu}}$, $t = \frac{\tau}{\nu}$, and the reparameterization $\eta \to \eta^{\frac{1}{\nu}}$, $\alpha \longrightarrow \nu^3 \alpha$. The $\nu = \infty$ limit is then easily seen to give

$$ds^2 = \Omega_{\infty}(\rho) \left(-F_{\infty}(\rho) d\tau^2 + \frac{d\rho^2}{F_{\infty}(\rho)} + d\Sigma^2\right), \tag{26}$$

$$\Omega_{\infty}(\rho) = \frac{\eta\rho}{(\rho - \eta)^2}, \qquad \phi = \frac{1}{\sqrt{2\kappa}} \ln(\rho\eta^{-1}), \tag{27}$$

$$F_{\infty}(\rho) = \rho^{-1} \eta^{-1} (\rho - \eta)^2 k + \left(2 \ln\left(\frac{\eta}{\rho}\right) + \frac{\rho}{\eta} - \frac{\eta}{\rho}\right) \alpha - \frac{\Lambda}{3}, \tag{28}$$

$$V_{\infty}(\phi) = \frac{2\alpha}{\kappa} \left(2\phi l_P + \phi l_P \cosh(\phi l_P) - 3 \sinh(\phi l_P)\right) + \frac{\Lambda}{3} \left(4 + 2\cosh(l_P \phi)\right), \tag{29}$$

where $l_P = \sqrt{2\kappa}$ is proportional to the Planck length. The potential (29) plus the corresponding limit of the part proportional to Λ of (10) was considered in the context of de Sitter hairy black holes compatible with inflation in [17].

Acknowledgments The author would like to thank the organizers of the conference "*Relativity and Gravitation—100 years after Einstein in Prague*" for its excellent environment and organization. Research of Andrés Anabalón is supported in part by the Conicyt grant Anillo ACT-91: "Southern Theoretical Physics Laboratory" (STPLab) and by the FONDECYT grant 11121187.

References

1. Bičák, J.: Selected solutions of Einstein's field equations: Their role in general relativity and astrophysics. In: Schmidt, B. (ed.) Einstein's Field Equations and Their Physical Implications: Selected Essays in Honour of Jürgen Ehlers. Lecture Notes in Physics, vol. 540, pp. 1–125. Springer, Berlin; New York (2000)
2. Griffiths, J., Podolský, J.: Exact Space-Times in Einstein's General Relativity. Cambridge Monographs on Mathematical Physics. Cambridge University Press, Cambridge (2012)
3. Chruściel, P., Costa, J., Heusler, M.: Stationary black holes: uniqueness and beyond. Living Rev. Relativ. **15**(7), lrr-2012-7 (2012). http://www.livingreviews.org/lrr-2012-7
4. Bekenstein, J.: Transcendence of the law of baryon-number conservation in black hole physics. Phys. Rev. Lett. **28**, 452 (1972). doi:10.1103/PhysRevLett.28.452
5. Heusler, M.: A no hair theorem for selfgravitating nonlinear sigma models. J. Math. Phys. **33**, 3497 (1992). doi:10.1063/1.529899
6. Sudarsky, D.: A simple proof of a no hair theorem in Einstein Higgs theory. Class. Quant. Grav. **12**, 579 (1995). doi:10.1088/0264-9381/12/2/023
7. Hawking, S.: Black holes in the Brans-Dicke theory of gravitation. Commun. Math. Phys. **25**, 167 (1972). doi:10.1007/BF01877518
8. Sotiriou, T., Faraoni, V.: Black holes in scalar-tensor gravity. Phys. Rev. Lett. **108**, 081103 (2012). doi:10.1103/PhysRevLett.108.081103
9. Anabalón, A., Oliva, J., Oliva, J.: Exact hairy black holes and their modification to the universal law of gravitation. Phys. Rev. D **86**, 107501 (2012). doi:10.1103/PhysRevD.86.107501
10. Martinez, C., Troncoso, R., Zanelli, J.: Exact black hole solution with a minimally coupled scalar field. Phys. Rev. D **70**, 084035 (2004). doi:10.1103/PhysRevD.70.084035
11. Kolyvaris, T., Koutsoumbas, G., Papantonopoulos, E., Siopsis, G.: A new class of exact hairy black hole solutions. Gen. Relativ. Gravit. **43**, 163 (2011). doi:10.1007/s10714-010-1079-0
12. Anabalón, A., Canfora, F., Giacomini, A., Oliva, J.: Black holes with primary hair in gauged $N = 8$ supergravity. J. High Energy Phys. **2012**(06), 010 (2012). doi:10.1007/JHEP06(2012)010
13. Toldo, C., Vandoren, S.: Static nonextremal AdS_4 black hole solutions. J. High Energy Phys. **2012**(09), 048 (2012). doi:10.1007/JHEP09(2012)048
14. Anabalón, A.: Exact black holes and universality in the backreaction of non-linear sigma models with a potential in $(A)dS_4$. J. High Energy Phys. **2012**(06), 127 (2012). doi:10.1007/JHEP06(2012)127
15. Hertog, T.: Towards a novel no-hair theorem for black holes. Phys. Rev. D **74**, 084008 (2006). doi:10.1103/PhysRevD.74.084008
16. Hartnoll, S., Herzog, C., Horowitz, G.: Building a holographic superconductor. Phys. Rev. Lett. **101**, 031601 (2008). doi:10.1103/PhysRevLett.101.031601
17. Zloshchastiev, K.: On co-existence of black holes and scalar field. Phys. Rev. Lett. **94**, 121101 (2005). doi:10.1103/PhysRevLett.94.121101
18. Salvio, A.: Holographic superfluids and superconductors in dilaton-gravity. J. High Energy Phys. **2012**(09), 134 (2012). doi:10.1007/JHEP09(2012)134
19. Duff, M., Liu, J.: Anti-de Sitter black holes in gauged $N = 8$ supergravity. Nucl. Phys. B **554**, 237 (1999). doi:10.1016/S0550-3213(99)00299-0
20. Carter, B.: Hamilton-Jacobi and Schrodinger separable solutions of Einstein's equations. Commun. Math. Phys. **10**, 280 (1968)
21. Debever, R.: Sur les espaces de Brandon Carter, Bulletin de la Classe des Sciences. 5e série. Académie Royale de Belgique **55**, 8 (1969)
22. Plebański, J.: A class of solutions of Einstein-Maxwell equations. Ann. Phys. (N.Y.) **90**, 196 (1975). doi:10.1016/0003-4916(75)90145-1
23. Plebański, J., Demianski, M.: Rotating, charged, and uniformly accelerating mass in general relativity. Ann. Phys. (N.Y.) **98**, 98 (1976). doi:10.1016/0003-4916(76)90240-2
24. Henneaux, M., Martinez, C., Troncoso, R., Zanelli, J.: Asymptotic behavior and Hamiltonian analysis of anti-de Sitter gravity coupled to scalar fields. Ann. Phys. (N.Y.) **322**, 824 (2007). doi:10.1016/j.aop.2006.05.002

Black Hole Formation from a Complete Past for the Einstein–Vlasov System

Håkan Andréasson

Abstract A natural question in general relativity is to find initial data for the Einstein equations whose past evolution is regular and whose future evolution contains a black hole. In [1] initial data of this kind is constructed for the spherically symmetric Einstein–Vlasov system. One consequence of the result is that there exists a class of initial data for which the ratio of the Hawking mass $\mathring{m} = \mathring{m}(r)$ and the area radius r is arbitrarily small everywhere, such that a black hole forms in the evolution. This result is analogous to the result [2] for a scalar field. Another consequence is that there exist black hole initial data such that the solutions exist for all Schwarzschild time $t \in (-\infty, \infty)$. In the present article we review the results in [1].

1 Introduction

In the study of gravitational collapse it is important to identify physically admissible initial data, and it is natural to require that the past evolution of the data is regular. However, in numerical relativity it is often the case that the given initial data, which form black holes to the future, also result in a singular past due to topological reasons. Moreover, most of the existing mathematical results which ensure a regular past also ensure a regular future which rules out the study of the formation of black holes. The exceptions being the classical result for dust [3], and the recent result [4] for a scalar field. In the latter work, which in part rests on the studies [2, 5], initial data whose past evolution is regular and whose future evolution forms a black hole is constructed. Neither dust nor a scalar field are realistic matter models in the sense that they are used by astrophysicists. Dust is a perfect fluid where the pressure is assumed to be zero, and a scalar field is merely a toy model. Thus, there is so far

H. Andréasson (✉)
Department of Mathematics, University of Gothenburg and Chalmers
University of Technology, 42196 Göteborg, Sweden
e-mail: hand@chalmers.se

J. Bičák and T. Ledvinka (eds.), *Relativity and Gravitation*,
Springer Proceedings in Physics 157, DOI: 10.1007/978-3-319-06761-2_2,

no example of a solution to the Einstein-matter system for a realistic matter model possessing a regular past and a singular future.

Here we consider Vlasov matter, or collisionless matter, governed by the Vlasov equation, cf. [6] for an introduction. Although this is a simple matter model, it has rich dynamics and many features that are desirable of a realistic matter model. For instance, there is a large number of stable and unstable spherically symmetric and axially symmetric stationary solutions, there is numerical support that time periodic solutions exist, it behaves as Type I matter in critical collapse, and it is used by astrophysicists, cf. [7]. The following theorem is the main result in [1].

Theorem 1 *There exists a class of initial data $\mathscr{I}$ for the spherically symmetric Einstein–Vlasov system with the property that black holes form in the future time direction and in the past time direction spacetime is causally geodesically complete.*

The following corollary is analogous to a result in [2] for a scalar field. Let $\mathring{m}$ be the initial Hawking mass. We then have

Corollary 1 *Given $\epsilon > 0$, there exists a class $\mathscr{I}_r$ of initial data for the spherically symmetric Einstein–Vlasov system which satisfy*

$$\sup_r \frac{\mathring{m}(r)}{r} \leq \epsilon,$$

for which black holes form in the evolution.

Another consequence of our result is that there exists a class of black hole initial data such that the corresponding solutions exist for all Schwarzschild time $t \in (-\infty, \infty)$, cf. Corollary 2 in [1].

Theorem 1 relies in part on the previous studies [8–10]. In [9] global existence in a maximal time gauge is shown for a particular class of initial data where the particles are moving rapidly outwards. One of the restrictions imposed on the initial data is that

$$\sup_r \frac{2\mathring{m}(r)}{r} < k_0, \tag{1}$$

where the constant k_0 is roughly $1/10$. The situation considered in [10] is in a sense the reverse since the initial data is such that the particles move rapidly inwards and the quantity $\sup_r 2\mathring{m}/r$ is required to be close to one. The main result of [10] is that data of this kind guarantee the formation of black holes in the evolution. Particles that move inward in the future time direction move outward in the past time direction. It is thus natural to try to combine these two results with the goal of constructing solutions with a regular past and a singular future. The conditions on the ratio $2\mathring{m}/r$ are clearly very different in [9] compared to [10], and moreover, the Cauchy hypersurfaces are different since a maximal time gauge and a polar time gauge are imposed in the respective cases. The reason a maximal time gauge is used in [9] is due to the difficulties related to the so called pointwise terms which appear in the characteristic equations in a polar time gauge. In [8] the problem of global existence for general initial data is investigated under conditional assumptions on the solutions. The analysis along characteristics is applied to a modified quantity for which the problem with the pointwise terms do not appear.

The proof of Theorem 1 is obtained by combining the strategies in [8] and [9], and a sketch of proof is given in Sect. 3. The spherically symmetric Einstein–Vlasov system is introduced in Sect. 2.

2 The Einstein–Vlasov System

For an introduction to the Einstein–Vlasov system and kinetic theory we refer to [6, 11]. In Schwarzschild coordinates the spherically symmetric metric takes the form

$$ds^2 = -e^{2\mu(t,r)}dt^2 + e^{2\lambda(t,r)}dr^2 + r^2(d\theta^2 + \sin^2\theta d\varphi^2). \tag{2}$$

The Einstein equations read

$$e^{-2\lambda}(2r\lambda_r - 1) + 1 = 8\pi r^2 \rho, \tag{3}$$

$$e^{-2\lambda}(2r\mu_r + 1) - 1 = 8\pi r^2 p, \tag{4}$$

$$\lambda_t = -4\pi r e^{\lambda+\mu} j, \tag{5}$$

$$e^{-2\lambda}(\mu_{rr} + (\mu_r - \lambda_r)(\mu_r + \frac{1}{r})) - e^{-2\mu}(\lambda_{tt} + \lambda_t(\lambda_t - \mu_t)) = 8\pi p_T. \tag{6}$$

The indices t and r denote partial derivatives. The Vlasov equation for the density function $f = f(t, r, w, L)$ is given by

$$\partial_t f + e^{\mu-\lambda}\frac{w}{E}\partial_r f - (\lambda_t w + e^{\mu-\lambda}\mu_r E - e^{\mu-\lambda}\frac{L}{r^3 E})\partial_w f = 0, \tag{7}$$

where

$$E = E(r, w, L) = \sqrt{1 + w^2 + L/r^2}. \tag{8}$$

Here $w \in (-\infty, \infty)$ can be thought of as the radial component of the momentum variables, and $L \in [0, \infty)$ is the square of the angular momentum. The matter quantities are defined by

$$\rho(t, r) = \frac{\pi}{r^2}\int_{-\infty}^{\infty}\int_0^{\infty} Ef(t, r, w, L)\, dwdL, \tag{9}$$

$$p(t,r) = \frac{\pi}{r^2}\int_{-\infty}^{\infty}\int_0^{\infty}\frac{w^2}{E}f(t,r,w,L)\,dwdL, \tag{10}$$

$$j(t,r) = \frac{\pi}{r^2}\int_{-\infty}^{\infty}\int_0^{\infty} wf(t,r,w,L)\,dwdL, \tag{11}$$

$$p_T(t,r) = \frac{\pi}{2r^4}\int_{-\infty}^{\infty}\int_0^{\infty}\frac{L}{E}f(t,r,w,L)\,dwdL. \tag{12}$$

Here ρ, p, j and p_T are the energy density, the radial pressure, the current and the tangential pressure respectively. The following boundary conditions are imposed to ensure asymptotic flatness

$$\lim_{r\to\infty}\lambda(t,r) = \lim_{r\to\infty}\mu(t,r) = 0, \tag{13}$$

and if a regular centre is required we set

$$\lambda(t,0) = 0. \tag{14}$$

As initial data it is sufficient to prescribe a density function $\mathring{f} = \mathring{f}(r,w,L) \geq 0$ such that

$$\int_0^r 4\pi\eta^2\mathring{\rho}(\eta)\,d\eta < \frac{r}{2}. \tag{15}$$

Here we denote by $\mathring{\rho}$ the energy density induced by the initial distribution function $\mathring{f}$. This condition ensures that no trapped surfaces are present initially. We now introduce a couple of notations. From (4) and (13) we have

$$\mu(t,r) = -\int_r^{\infty}\frac{m(t,\eta)}{\eta^2}e^{2\lambda} - \int_r^{\infty}4\pi\eta p e^{2\lambda}\,d\eta =: \hat{\mu} + \check{\mu}. \tag{16}$$

Moreover, the Hawking mass $m = m(t,r)$ is given by

$$m(t,r) = 4\pi\int_0^r \eta^2\rho(t,\eta)d\eta. \tag{17}$$

Finally, we note that in [6, 12] local existence theorems are proved for compact and non-compact initial data respectively and it will be used below that solutions exist on some time interval $[0,T[$, which is assumed to be maximal.

3 Global Existence for Outgoing Matter

The aim in this section is to consider initial data of the type constructed in [10], which guarantee the formation of black holes to the future, and show that global existence holds to the past for such data. We remark that the time direction is reversed in this section so that the particles move outwards initially and the global existence to the past refers to the time interval $[0, \infty[$. Furthermore, in [1] two different classes of initial data are given adapted to the two corollaries of Theorem 1 mentioned above. Here we only consider one of these classes of data.

Let $0 < r_0' < r_0 < r_1$ be given and put $M = r_1/2$. Let $\mathring{f}_s$ be data of a steady state supported in $[r_0', r_0]$ and let

$$M_{\text{in}} := \int_{r_0'}^{r_0} 4\pi r^2 \mathring{\rho}(r)dr. \tag{18}$$

The results in [13, 14] guarantee that there are such steady states and moreover that

$$\sup_{0\le r\le r_0} \frac{2\mathring{m}(r)}{r} < \frac{8}{9},$$

and in particular $2M_{\text{in}}/r_0 < 8/9$ so that $M > M_{\text{in}}$. Let $M_{\text{out}} := M - M_{\text{in}}$. Let $R_1 > r_1$ be such that

$$R_1 - r_1 < \frac{r_1 - r_0}{6}, \tag{19}$$

and define

$$R_0 := \frac{1}{2}(r_1 + R_1).$$

Let $L_+ > 0$ and let $W_* > 0$ be such that

$$|W_*| \ge 1 + \frac{\sqrt{L_+}}{R_0}. \tag{20}$$

Let $W_- > 0$ satisfy

$$|W_-|\, e^{\frac{-5M}{2R_0(1-\frac{2M}{R_0})}} (1 - \frac{2M}{R_0})^{3/2} \ge 3|W_*|. \tag{21}$$

We can now specify the initial data. Let $\mathring{f} = \mathring{f}_s + \mathring{f}_m$ be initial data of ADM mass M, and such that

$$\operatorname{supp} \mathring{f}_m \subset [R_0, R_1] \times [W_-, \infty[\times [0, L_+],$$

and

$$\int_{r_0}^{\infty} 4\pi r^2 \mathring{\rho}(r)dr = \int_{R_0}^{R_1} 4\pi r^2 \mathring{\rho}_m(r)dr = M_{\text{out}}. \tag{22}$$

In view of [10] the initial data $\mathring{f}$ guarantee the formation of black holes. Hence, Theorem 1 follows from the following global existence theorem.

Theorem 2 *Assume that r_0', r_0, M_{in}, M, L_+, R_0, R_1, W_*, W_- and $\mathring{f}$ are given as above, and consider a solution f of the system (3)-(6), launched by $\mathring{f}$, on its maximal existence interval $[0, T[$. Then $T = \infty$, and there is a $\kappa_* > 0$ such that*

$$\text{supp } f_m(t) \subset [R_0 + |t\,\kappa_*|, \infty[\times[W_*, \infty[\times[0, L_+], \tag{23}$$

and the resulting spacetime is future causally geodesically complete.

Sketch of proof: We focus here on the main idea of the proof and refer to [1] for the complete argument. It is shown in [1] that f, λ and μ_r remain time independent for $r \leq r_0$ and therefore the present arguments only concern the outer matter given by f_m. The steady state is needed to guarantee the formation of black holes.

Let $[0, t_1[$ be the maximal time interval such that for $t \in [0, t_1[$ and $(r, w, L) \in \text{supp } f_m(t)$, $w > W_*$. By continuity $t_1 > 0$. Suppose that $t_1 \in]0, T[$, then we must have $w = W_*$ for some $w \in \text{supp } f_m(t_1)$, but we will show that $w > W_*$ for all $w \in \text{supp } f_m(t_1)$. Thus $t_1 = T$ and since the matter stays strictly away from $r = 0$ it follows that $T = \infty$ in view of [8].

Consider a characteristic $(R(s), W(s), L)$ with $R(0) \in [R_0, R_1]$ and define

$$G(s) := E(R(s), W(s), L) + W(s).$$

Below we suppress the arguments but it should be clear that $R = R(s)$, $\mu_r = \mu_r(s, R(s))$ etc. The main idea of the proof is to consider the evolution of the quantity

$$G(t)e^{\hat{\mu}(t,R(t))}(1 - 2M/R(t))$$

along the characteristic $(R(s), W(s), L)$. The following inequality is then obtained in [1]:

$$\frac{d}{ds}\Big(Ge^{\hat{\mu}}(1 - \frac{2M}{R})\Big) \geq -\Big[\lambda_t \frac{W}{E} + \mu_r e^{\mu-\lambda} - \hat{\mu}_t\Big] Ge^{\hat{\mu}}\Big(1 - \frac{2M}{R}\Big). \tag{24}$$

This implies that

$$G(t_1)e^{\hat{\mu}(t_1,R(t_1))}\left(1 - \frac{2M}{R(t_1)}\right) \geq e^{-\int_0^{t_1}\left[\lambda_t(s,R(s))\frac{W}{E} + \check{\mu}_r(s,R(s))e^{(\mu-\lambda)(s,R(s))} - \hat{\mu}_t(s,R(s))\right]ds}$$
$$\times\, G(0)e^{\hat{\mu}(0,R(0))}(1 - \frac{2M}{R(0)}). \tag{25}$$

Let γ be the curve

$$\gamma := \{(t,r) : 0 \le t \le t_1,\ r = R(t)\}.$$

The time integral in (25) can be written as

$$\int_\gamma e^{(-\mu+\lambda)(t,r)}\lambda_t(t,r)\,dr + \left(e^{(\mu-\lambda)(t,r)}\breve{\mu}_r(t,r) - \hat{\mu}_t(t,r)\right)dt. \tag{26}$$

An application of Green's formula in the plane to this curve integral, making crucial use of the second order Einstein equation (6) and the Vlasov equation, leads in [1] to the inequality

$$\int_\gamma e^{-\mu+\lambda}\lambda_t\,dr + (e^{\mu-\lambda}\breve{\mu}_r - \hat{\mu}_t)\,ds \le \frac{5M}{2R_0\left(1-\frac{2M}{R_0}\right)}.$$

Inserting this into the main inequality (25) we get

$$G(t_1)e^{\hat{\mu}(t_1,R(t_1))}\left(1-\frac{2M}{R(t_1)}\right) \ge e^{\frac{-5M}{2R_0(1-\frac{2M}{R_0})}}G(0)e^{\hat{\mu}(0,R(0))}\left(1-\frac{2M}{R(0)}\right).$$

Noticing that $\hat{\mu}$ is monotone in r and nonpositive, and that $R(0) \ge R_0$, we find that

$$\begin{aligned} G(t_1) &\ge e^{\frac{-5M}{2R_0(1-\frac{2M}{R_0})}}G(0)e^{\hat{\mu}(0,R_0)}(1-\frac{2M}{R_0}) \\ &\ge e^{\frac{-5M}{2R_0(1-\frac{2M}{R_0})}}G(0)\sqrt{\frac{R_0-2M}{R_0}}(1-\frac{2M}{R_0}). \end{aligned} \tag{27}$$

Here we made use of the estimate

$$\hat{\mu}(t,R_0) \ge -\int_{R_0}^\infty \frac{M\,d\eta}{\eta^2(1-\frac{2M}{\eta})} = \frac{1}{2}\log\left(1-\frac{2M}{R_0}\right). \tag{28}$$

We have that $G(0) > W(0) \ge W_-$, and in view of (20) we also have $3W(t) \ge G(t)$ on $[0,t_1]$. We now use the condition (21) and obtain

$$3W(t_1) \ge G(t_1) > 3W_*.$$

Thus $W(t_1) > W_*$, and necessarily we have $t_1 = T$. As was pointed out in the beginning of the proof, since matter stays strictly away from the centre of symmetry, it follows that $T = \infty$, cf. [8]. For the remaining statements in the theorem we refer to [1].

□

References

1. Andréasson, H.: Black hole formation from a complete regular past for collisionless matter. Ann. Henri Poincaré **13**, 1511 (2012). doi:10.1007/s00023-012-0164-1
2. Christodoulou, D.: The formation of black holes and singularities in spherically symmetric gravitational collapse. Commun. Pure Appl. Math. **44**, 339 (1991). doi:10.1002/cpa.3160440305
3. Oppenheimer, J., Snyder, H.: On continued gravitational contraction. Phys. Rev. **56**, 455 (1939). doi:10.1103/PhysRev.56.455
4. Dafermos, M.: Black hole formation from a complete regular past. Commun. Math. Phys. **289**, 579 (2009). doi:10.1007/s00220-009-0775-7
5. Christodoulou, D.: Bounded variation solutions of the spherically symmetric Einstein-scalar field equations. Commun. Pure Appl. Math. **46**, 1131 (1993). doi:10.1002/cpa.3160460803
6. Andréasson, H.: The Einstein-Vlasov system/kinetic theory. Living Rev. Relativ. **14**(4), lrr-2011-4 (2011). http://www.livingreviews.org/lrr-2011-4
7. Binney, J., Tremaine, S.: Galactic Dynamics. Princeton Series in Astrophysics. Princeton University Press, Princeton (1987)
8. Andréasson, H.: Regularity results for the spherically symmetric Einstein-Vlasov system. Ann. Henri Poincaré **11**, 781 (2010). doi:10.1007/s00023-010-0039-2
9. Andréasson, H., Kunze, M., Rein, G.: Global existence for the spherically symmetric Einstein-Vlasov system with outgoing matter. Commun. Part. Diff. Equ. **33**, 656 (2008). doi:10.1080/03605300701454883
10. Andréasson, H., Kunze, M., Rein, G.: The formation of black holes in spherically symmetric gravitational collapse. Math. Ann. **350**, 683 (2011). doi:10.1007/s00208-010-0578-3
11. Rendall, A.: An introduction to the Einstein-Vlasov system. In: Chruściel, P. (ed.) Mathematics of Gravitation, Part I: Lorentzian Geometry and Einstein Equations, Banach Center Publications, vol. 41, pp. 35–68. Polish Academy of Sciences, Institute of Mathematics, Warsaw (1997)
12. Rein, G., Rendall, A.: Global existence of solutions of the spherically symmetric Vlasov-Einstein system with small initial data. Commun. Math. Phys. **150**(561), 1996 (1992). doi:10.1007/BF02096962.Erratum:ibid.176,475-478
13. Andréasson, H.: Sharp bounds on $2m/r$ of general spherically symmetric static objects. J. Differ. Equ. **245**, 2243 (2008). doi:10.1016/j.jde.2008.05.010
14. Andréasson, H.: On static shells and the Buchdahl inequality for the spherically symmetric Einstein-Vlasov system. Commun. Math. Phys. **274**, 409 (2007). doi:10.1007/s00220-007-0285-4

How to Measure Deviation from the Kerr Initial Data: Recent Progress

Thomas Bäckdahl and Juan A. Valiente Kroon

Abstract In these proceedings we will present recent progress concerning a construction of a geometric invariant for initial data sets for the Einstein vacuum field equations. This geometric invariant vanishes if and only if the initial data set corresponds to data for the Kerr spacetime, and thus, it characterizes this type of data. The construction was initially based on Killing spinors, but here we translate the results to tensor language. We can now handle both compact domains and domains reaching the asymptotically flat ends.

1 Introduction

Given a solution $(\mathscr{S}, h_{ab}, K_{ab})$ to the Einstein vacuum constraint equations, how do we know if it is a slice of the Kerr spacetime? If not, can we measure how much it differs? These are the questions we will consider here. We will introduce a geometric invariant on the slice, which will measure this deviation from Kerr data. The invariant is constructed as an L^2-norm constructed from global information on the slice, but it only depends on information from one slice, and is therefore local in time.

It is expected that a dynamical vacuum black hole will always settle down to a Kerr black hole. To make a proper mathematical formulation of this statement, one will need a good way to measure how close data on a slice is to Kerr data. This gives a clear motivation for our work. Our invariant is coordinate independent and straightforward to compute. It is therefore well suited for studying how the non-

T. Bäckdahl (✉)
Max Planck Institut für Gravitationsphysik, Albert Einstein Institut, Am Mühlenberg 1, 14476 Golm, Germany
e-mail: thomas.backdahl@aei.mpg.de

J. A. Valiente Kroon
School of Mathematical Sciences, Queen Mary University of London, Mile End Road, London E1 4NS, UK

J. Bičák and T. Ledvinka (eds.), *Relativity and Gravitation*,
Springer Proceedings in Physics 157, DOI: 10.1007/978-3-319-06761-2_3,

Kerrness evolves for numerically computed spacetimes. It can also be used as a tool for more theoretical work, for instance when one studies non-linear stability of the Kerr solution. For this purpose a coordinate independent integral over a slice as we have is well suited.

We have developed the theory in four papers [1–4]. Here we only make a summary and concentrate on some aspects of the problem. In the first paper [1] we present the general ideas and the invariant for general non-boosted slices with two asymptotic ends. The second paper [2] contains all technical details and generalizations so even boosted slices can be handled. The third paper [3] deals with the same problem, but only using data exterior to a surface. In the fourth paper [4] we also introduce an outer boundary so our domain becomes compact. This requires some extra conditions that we in the previous papers got from the asymptotic behaviour.

2 Characterization of the Kerr Spacetime

To construct the non-Kerrness invariant of initial data on a slice, we begin with a spacetime characterization of the Kerr solution. We then make a $3+1$ splitting of the equations involved. This is done in such a way that one can reconstruct the spacetime objects that we used for the spacetime characterization. Therefore, we get an initial data characterization of Kerr initial data. The initial data equations one obtains are then used to construct the non-Kerrness invariant that measures the deviation from Kerr data.

In our papers we used a spacetime characterization based on Killing spinors, but for ease of presentation we will here translate this to Killing-Yano tensors.

2.1 Killing-Yano Tensors

Throughout, we will assume that $(\mathscr{M}, g_{\mu\nu})$ is an orientable and time orientable globally hyperbolic vacuum spacetime, and we let ∇_μ denote the Levi-Civita connection of $g_{\mu\nu}$. Here we will use the $(-+++)$ sign convention to obtain a positive definite spatial metric. Observe that this differs from the one used in the references [1–4].

Definition 1 A *conformal Killing-Yano tensor* is an antisymmetric tensor $Y_{\mu\nu} = Y_{[\mu\nu]}$ that satisfies

$$\nabla_{(\mu} Y_{\nu)\lambda} = \tfrac{1}{3} g_{\lambda(\mu} \nabla^\sigma Y_{\nu)\sigma} - \tfrac{1}{3} g_{\mu\nu} \nabla^\sigma Y_{\lambda\sigma}. \tag{1}$$

It is called a *Killing-Yano tensor* if it satisfies

$$\nabla_{(\mu} Y_{\nu)\lambda} = 0. \tag{2}$$

Given a (conformal) Killing-Yano tensor, one automatically gets a Killing vector $\xi_\mu = \varepsilon_\mu{}^{\nu\gamma\lambda}\nabla_\nu Y_{\gamma\lambda}$ of the spacetime. We also get an integrability condition that will strongly restrict the Weyl tensor. In the tensorial picture, this condition is given by

$$0 = -C_{[\mu\nu]}{}^{\delta}{}_{[\lambda}Y_{\rho]\delta} - C_{[\lambda\rho]}{}^{\delta}{}_{[\mu}Y_{\nu]\delta} + C^{\delta}{}_{[\mu\nu][\lambda}Y_{\rho]\delta} + C^{\delta}{}_{[\lambda\rho][\mu}Y_{\nu]\delta}. \quad (3)$$

The spinorial version of the above condition is simpler. From it, it is easy to conclude that the spacetime has to be of Petrov type D, N or O.

2.2 Spacetime Characterization

We have the following theorem which is a translation of Theorem B.3 in [3]

Theorem 1 *A smooth spacetime $(\mathscr{M}, g_{\mu\nu})$ is locally isometric to the Kerr spacetime if and only if the following conditions are satisfied:*

(i) *there exists a Killing-Yano tensor $Y_{\mu\nu}$, with associated Killing vector ξ_μ;*
(ii) *the spacetime $(\mathscr{M}, g_{\mu\nu})$ has a stationary asymptotically flat 4-end with non-vanishing mass in which ξ_μ tends to a time translation.*

2.3 Initial Data Characterization

To get an initial data characterization, we make a $3 + 1$ splitting of the conformal Killing-Yano equation and the integrability condition to obtain the (4a)–(4b) below. After some work we see that one can propagate solutions to this system to obtain a conformal Killing-Yano tensor of the spacetime. See [2] and [4] for details.

Theorem 2 *Let $(\mathscr{S}, h_{ab}, K_{ab})$ be a vacuum initial data set, where $\mathscr{S}$ is a Cauchy hyper-surface. The development of the initial data set will have a conformal Killing-Yano tensor in the domain of dependence of $\mathscr{S}$ if and only if*

$$\zeta_{ab} \equiv D_{(a}\kappa_{b)} - \tfrac{1}{3}h_{ab}D_d\kappa^d - i\varepsilon_{(a}{}^{dl}K_{b)d}\kappa_l = 0. \quad (4a)$$

$$F_{ab} \equiv -C_{(a}{}^{c}\varepsilon_{b)c}{}^{d}\kappa_d = 0, \quad (4b)$$

are satisfied on $\mathscr{S}$. Here, $C_{ab} \equiv E_{ab} + iB_{ab}$, where E_{ab} and B_{ab} are the electric and magnetic parts of the Weyl tensor. Furthermore, these conditions give a complex spacetime Killing vector. Reality of this Killing vector gives a Killing-Yano tensor.

Remark The 1-form κ_a is the pull-back of $-\frac{i}{2}t^\nu Y_{\mu\nu} + \frac{1}{4}\varepsilon_{\mu\nu\lambda\delta}t^\nu Y^{\lambda\delta}$, where t^μ is the normal to $\mathscr{S}$ with normalization $t^\mu t_\mu = -1$. Observe that we use a different normalization in references [1–4]. The lapse and shift of the Killing vector initial data is constructed from κ_a via

$$\zeta = D^a \kappa_a, \quad \zeta_a = -\tfrac{3i}{2}\varepsilon_a{}^{bc} D_c \kappa_b + \tfrac{3}{2} K_{ab}\kappa^b - \tfrac{3}{2} K^b{}_b \kappa_a. \tag{5}$$

The vacuum constraint equations are

$$R = K_{ab}K^{ab} - (K^a{}_a)^2, \qquad D^a K_{ab} = D_b K^a{}_a. \tag{6}$$

The electric and magnetic parts of the Weyl tensor can be written entirely in terms of initial data via

$$E_{ab} = -K_a{}^c K_{bc} + K_{ab}K^c{}_c - R_{ab}, \quad B_{ab} = -\varepsilon_{(a}{}^{cd} D_{|c|} K_{b)d}. \tag{7}$$

3 Non-Kerrness Invariant

The idea behind our non-Kerrness invariant is to measure the L^2-norms of the left hand sides of (4a) and (4b) for a clever choice of κ_a. We have to choose κ_a in a unique coordinate independent way, and we need it to coincide with the solution to the system (4a) and (4b) if such a solution exists. It turns out that minimizing the L^2-norm of left hand side of (4a) while specifying the asymptotic behaviour, gives rise to a good choice.

3.1 Approximate Killing-Yano Tensors

Let J denote the L^2-norm of the left hand side of (4a), that is

$$J = \int_{\mathscr{S}} \zeta_{ab}\bar{\zeta}^{ab} \mathrm{d}\mu. \tag{8}$$

To minimize this we need to solve the corresponding Euler–Lagrange equation, which reads

$$L(\kappa_a) \equiv D_b \zeta_a{}^b + i\varepsilon_{acf} K^{bc} \zeta_b{}^f = 0. \tag{9}$$

The operator L is a linear second order self adjoint elliptic operator. A solution, κ_a, to the elliptic equation (9) is called an *approximate spatial Killing-Yano tensor*. Clearly, any solution to $\zeta_{ab} = 0$ is also a solution to (9).

Definition 2 An initial data set $(\mathscr{S}, h_{ab}, K_{ab})$ will be called *asymptotically Schwarzschildean* at an end if there is a Schwarzschild initial data set with the same asymptotic expansion up to and including the mass order term.

For non-boosted data and asymptotically Cartesian coordinates this condition reads

$$h_{ij} = \left(1 + 2mr^{-1}\right)\delta_{ij} + o_\infty(r^{-3/2}), \tag{10a}$$

$$K_{ij} = o_\infty(r^{-5/2}), \tag{10b}$$

where $o_\infty(r^\delta)$ denotes functions in a weighted Sobolev space, and m denotes the ADM-mass of the asymptotic end. See [2] for definitions of the Sobolev spaces and asymptotics for boosted data.

Theorem 3 *Given an initial data set* $(\mathscr{S}, h_{ab}, K_{ab})$ *with two asymptotically Schwarzschildean ends, there exists a smooth unique solution to* (9) *with the same asymptotic behaviour as the solution for Kerr from Theorem 1.*

This result was proven in [2]. In [3] we replaced one asymptotic end with an inner boundary. Appropriate data for the boundary was then constructed from the Weyl tensor. Finally in [4] we replaced all asymptotic ends with boundaries so the result extends to compact domains. In the latter case however, we needed to add extra conditions on one point to replace the asymptotic conditions in Theorem 1.

Now we can define the geometric invariant. Let κ_a be a solution to (9) as given by Theorem 3. With

$$J = \int_{\mathscr{S}} \zeta_{ab}\bar{\zeta}^{ab}\mathrm{d}\mu, \tag{11a}$$

$$I_1 \equiv \int_{\mathscr{S}} F_{ab}\bar{F}^{ab}\mathrm{d}\mu, \tag{11b}$$

the geometric invariant is defined by

$$I \equiv J + I_1. \tag{12}$$

By construction I is coordinate independent and non-negative. It can furthermore be verified that it is finite. More importantly we have

Theorem 4 *Let* $(\mathscr{S}, h_{ab}, K_{ab})$ *be a vacuum initial data set with two asymptotically Schwarzschildean ends. Let* I *be the invariant defined above, where* κ_a *is the only solution to* (9) *with the same asymptotic behaviour as the solution for Kerr from Theorem 1. The invariant* I *vanishes if and only if the development of* $(\mathscr{S}, h_{ab}, K_{ab})$ *is locally isometric to the Kerr spacetime.*

For the proof, see Theorem 28 in [2] and Theorem B.3 in [3] together with Theorem 4 in [4]. In [3] this was extended to the case with one asymptotic end and an inner boundary, and in [4] this was worked through for compact domains.

Acknowledgments T.B. is funded by the Max-Planck-Institute for Gravitational Physics, Albert Einstein Institut.

References

1. Bäckdahl, T., Valiente Kroon, J.: Geometric invariant measuring the deviation from Kerr data. Phys. Rev. Lett. **104**, 231102 (2010). doi:10.1103/PhysRevLett.104.231102
2. Bäckdahl, T., Valiente Kroon, J.: On the construction of a geometric invariant measuring the deviation from Kerr data. Ann. Henri Poincaré **11**, 1225 (2010). doi:10.1007/s00023-010-0063-2
3. Bäckdahl, T., Valiente Kroon, J.: The 'non-Kerrness' of domains of outer communication of black holes and exteriors of stars. Proc. R. Soc. Lond. Ser. A **467**, 1701 (2011). doi:10.1098/rspa.2010.0535
4. Bäckdahl, T., Valiente Kroon, J.A.: Constructing 'non-Kerrness' on compact domains. J. Math. Phys. **53**, 042503 (2012). doi:10.1063/1.3702569

Hidden Symmetries of the Dirac Equation in Curved Space-Time

Marco Cariglia

Abstract These are introductory notes on the study of the Dirac equation in curved spacetime and its relation to hidden symmetries of the dynamics. We present general results on the relation between special spacetime tensors and hidden symmetries, both for the full Dirac equation and for its semi-classical limit, the spinning particle. A concrete application of the general results is provided by the case of rotating higher dimensional black holes with cosmological constant, which we discuss. For these metrics the Dirac equation is separable and the relation between this and hidden symmetries is explained.

1 Introduction

The Dirac equation, since its derivation in 1928, has successfully described the relativistic hydrogen atom and phenomena such as the existence of anti-particles. A natural evolution of the theory, stimulated by the progress in General Relativity, is that of studying relativistic spin $\frac{1}{2}$ particles on a curved background, such as for example the Schwarzschild and Kerr spacetimes. Further development in unification theories such as String/M-Theory and in cosmological models lead to the additional ingredient of considering extensions of General Relativity and quantum field theories to higher dimensions than four.

Parallel to this, the natural interest of physicists in solutions of the Dirac equation that can be obtained by separation of variables leads to the mathematical problem of finding a theory of separation of variables for this system of first order equations. In the case of the classical Hamilton–Jacobi equations, the Schrödinger equation and the Klein Gordon equation, there exists a well understood theory of separation of vari-

M. Cariglia (✉)
Departamento de Física. Campus Morro do Cruzeiro, Universidade Federal de Ouro Preto, ICEB, Ouro Preto, Minas Gerais 35400-000, Brazil
e-mail: marco@iceb.ufop.br

J. Bičák and T. Ledvinka (eds.), *Relativity and Gravitation*,
Springer Proceedings in Physics 157, DOI: 10.1007/978-3-319-06761-2_4,

ables [1–5]. In this theory the main objects playing a role are special tensors, namely Killing vectors and rank 2 Killing–Stäckel tensors. With these one can build either conserved quantities in the classical theory, or symmetry operators in the quantum mechanical one. In the case of the Dirac equation instead a complete theory of separation of variables is still lacking. Several known cases of separation of variables for the Dirac equation involve only symmetry operators of first order in the derivatives, such as the Dirac equation in the Kerr metric or the higher dimensional Kerr-NUT-(A)dS metrics that we will discuss in this review. First order symmetry operators have been built in 4 dimensions first and successively in arbitrary dimensions and signature [6–10]. However, Fels and Kamran have shown that there exist cases where the Dirac equation is separable but the separability is underlain by the existence of symmetry operators of order higher than one. In some cases second order symmetry operators have been built, see [11] and references therein, but neither the general construction is known for an arbitrary dimension, nor there is control on necessary and sufficient conditions for separability.

In this review we present the current knowledge on the first order symmetry operators of the Dirac equation. We show how these are in exact correspondence with special Conformal Killing–Yano tensors, and the relation between spacetime differential forms and Clifford algebra valued operators. We discuss one important application, the separability of the Dirac equation in Kerr-NUT-(A)dS metrics, which is fully accounted for by a complete set of linear symmetry operators that are mutually commuting and admit common separable spinorial eigenfunctions. We also briefly discuss the semi-classical limit of the Dirac equation, the theory of the supersymmetric spinning particle. In this theory the analogue of linear symmetry operators is given by phase space functions that are linear in the momenta and correspond to generators of extra supersymmetries. For Kerr-NUT-(A)dS metrics it is possible to show that the bosonic sector of the theory is integrable, and its integrability is related to the presence of Killing vectors and of a set of new conserved quantities that are quadratic in the momenta and generalise the quadratic conserved quantities of a scalar particle to the case with spin.

2 Gamma Matrices and Differential Forms

In this section we discuss the one-to-one map between differential forms on a spin manifold and sections of its Clifford bundle. The application of interest for this review is the fact that the properties of Conformal Killing–Yano tensors, which are differential forms, automatically lift to those of appropriate differential operators defined on the Clifford bundle.

We model spacetime as a (pseudo-)Riemannian spin manifold M of dimension n with metric $g_{\mu\nu}$ and local coordinates $\{x^\mu\}$. We use lowercase Greek indices to denote 'curvy' components of spacetime tensors, associated with general diffeomorphism transformations, and lowercase Latin indices to denote 'flat' components, associated with $SO(n)$ or $SO(1, n-1)$ transformations. Each fiber of the Clifford bundle has the structure of the Clifford algebra generated by the gamma matrices γ^μ, which connect the Clifford bundle with the tangent space. The gamma matrices satisfy

$$\gamma^\mu \gamma^\nu + \gamma^\nu \gamma^\mu = 2g^{\mu\nu}, \tag{1}$$

which allows to reduce any element $\alpha\!\!\!/$ of the Clifford algebra to a sum of antisymmetric products $\gamma^{\mu_1 \ldots \mu_p} := \gamma^{[\mu_1} \ldots \gamma^{\mu_p]}$ with appropriate coefficients:

$$\alpha\!\!\!/ = \sum_p \frac{1}{p!} \alpha^{(p)}_{\mu_1 \ldots \mu_p} \gamma^{\mu_1 \ldots \mu_p}. \tag{2}$$

This representation is unique and the coefficients are given by anti-symmetric forms $\alpha^{(p)}_{\mu_1 \ldots \mu_p} \in \Omega^{(p)}(M)$, thus providing an isomorphism γ_* of the Clifford bundle with the exterior algebra $\Omega(M) = \bigoplus_{p=0}^n \Omega^p(M)$ of inhomogeneous antisymmetric forms: $\alpha\!\!\!/ = \gamma_* \alpha$, where $\alpha = \sum_p \alpha^{(p)}$ is an inhomogeneous form. In the rest of the review, whenever the context makes it clear we will write α instead of $\alpha\!\!\!/$ to describe an element of the Clifford algebra, with the action of the isomorphism implied.

The metric allows to raise and lower indices (musical isomorphism): if α is a 1-form and v a vector, we denote the corresponding vector and 1-form as $\alpha^\sharp$ and $v^\flat$, respectively. This generalises to higher rank tensors. We define two operations on $\Omega(M)$. The 'hook' operation (*inner derivative*) is an action of a vector v on any antisymmetric form α. In components:

$$(v \lrcorner \alpha)_{a_1 \ldots a_{p-1}} = v^b \alpha_{b a_1 \ldots a_{p-1}}. \tag{3}$$

For a scalar φ, we set $v \lrcorner \varphi = 0$. The second operation is the wedge product. When it acts on a p-form α and a q-form β it is defined so that in components $(\alpha \wedge \beta)_{a \ldots b \ldots} = \frac{(p+q)!}{p!q!} \alpha_{[a \ldots} \beta_{b \ldots]}$.

The Clifford algebra relation (1) means that a product of any two rank p and q gamma matrices $\gamma^{\mu_1 \ldots \mu_p}$ and matrices $\gamma^{\nu_1 \ldots \nu_q}$ can be decomposed in terms of other gamma matrices. In particular it can be shown that for $\alpha \in \Omega^p(M)$, $\beta \in \Omega^q(M)$ Clifford bundle forms, with $p \leq q$, the Clifford product expands as

$$\alpha\beta = \sum_{m=0}^{p} \frac{(-1)^{m(p-m)+[m/2]}}{m!} \alpha \underset{m}{\wedge} \beta, \tag{4}$$

where the $\underset{m}{\wedge}$ contraction operator is defined recursively as

$$\begin{aligned} &\alpha \underset{0}{\wedge} \beta = \alpha \wedge \beta, \\ &\alpha \underset{k}{\wedge} \beta = (X_a \lrcorner \alpha) \underset{k-1}{\wedge} (X^a \lrcorner \beta) \qquad (k \geq 1), \\ &\alpha \underset{k}{\wedge} \beta = 0 \qquad\qquad\qquad\qquad\quad (k < 0). \end{aligned} \tag{5}$$

Given a set of n-beins $\{e^a_\mu\}$, we can build n 1-forms $e^a = e^a_\mu dx^\mu$, with $X^a = (e^a)^\sharp$ a dual vector basis. The e^a are mapped under γ_* to 'flat' gamma matrices that satisfy (1) with the flat metric η^{ab} instead of g. Flat and curvy indices can be transformed one into the other using either e^a_μ or its inverse E^μ_a, such that $e^a_\mu E^\mu_b = \delta^a_b$ and $e^a_\mu E^\nu_a = \delta^\nu_\mu$. We can also group the e^a together and consider a single object $\gamma_*(e^a)$ which transforms as an $SO(n)$ tensor. We lift the covariant derivative given on $\Omega(M)$ to one on $\gamma_*(\Omega(M))$ by asking that for any α in the Clifford bundle

$$\nabla_a \alpha = \partial_a \alpha - \omega_a \underset{1}{\wedge} \alpha, \tag{6}$$

where $\partial_a \alpha = X_a[\alpha] = E^\mu_a \partial_\mu \alpha$, and ω_a is the connection 2-form $\omega_a = \frac{1}{2}\omega_{abc} e^b \wedge e^c$ and ω_{abc} are the components of the spin connection. For a form which is also an $SO(n)$ tensor such as e^a the covariant derivative becomes

$$\nabla_a e^b = \partial_a e^b + \omega_a{}^b{}_c\, e^c - \omega_a \underset{1}{\wedge} e^b. \tag{7}$$

In particular, for the n-bein tensor itself, $\nabla_a e^b = 0$.

Lastly we introduce the degree operator π that acts on an inhomogeneous form $\alpha = \sum_p \alpha^{(p)}$ as $\pi\alpha = \sum_{p=0} p\, \alpha^{(p)}$.

In this formalism the Dirac operator is written as $D \equiv e^a \nabla_a = \nabla_a e^a$, the exterior derivative acting on forms as $d = e^a \wedge \nabla_a = \nabla_a\, e^a \wedge$, and the co-differential $\delta = -X^a \lrcorner \nabla_a = -\nabla_a\, X^a \lrcorner$. All these expressions are to be understood as operators acting on the right.

3 Special Killing–Yano Tensors

Conformal Killing–Yano tensors (CKY) are forms $f_{\mu_1 \ldots \mu_p} = f_{[\mu_1 \ldots \mu_p]}$ such that

$$\nabla_\lambda f_{\mu_1 \ldots \mu_p} = \nabla_{[\lambda} f_{\mu_1 \ldots \mu_p]} + \frac{p}{D-p+1} g_{\lambda[\mu_1} \nabla^\rho f_{|\rho|\mu_2 \ldots \mu_p]}, \tag{8}$$

or equivalently, without using components,

$$\nabla_X f = \frac{1}{\pi+1} X \lrcorner df - \frac{1}{n-\pi+1} X^\flat \wedge \delta f, \tag{9}$$

for any vector X. When $p = 1$ this reduces to the Killing equation.

The formula above generalises automatically to the case of inhomogeneous forms. When f is co-closed, $\delta f = 0$, f is called a Killing–Yano form (KY), and when it is closed, $df = 0$, it is called a closed conformal Killing–Yano form (CCKY). Equation (9) is invariant under Hodge duality, interchanging KY and CCKY tensors. A symmetry operator for the Dirac equation is an operator S that R-commutes with

D, or in other words such that $DS = RD$ for some operator R. It maps solutions of the massless Dirac equation into solutions. Benn, Charlton, and Kress [9, 10] have shown the important result that, in all dimensions n and arbitrary signature, R-symmetry operators of the massless Dirac operator that are first-order in derivatives are in one to one correspondence with CKY forms. Any such operator S can be written as

$$S = S_f + \alpha D, \tag{10}$$

where α is an arbitrary inhomogeneous form, and S_f is given in terms of an inhomogeneous CKY form f obeying (9) according to

$$S_f = X^a \lrcorner f\, \nabla_a + \frac{\pi - 1}{2\pi} df - \frac{n - \pi - 1}{2(n - \pi)} \delta f. \tag{11}$$

The freedom of adding an arbitrary form α is unavoidable. In [12] it was shown that, as a special case, the most general first-order operator S that strictly commutes with D splits into the Clifford even and Clifford odd parts

$$S = S_{\rm e} + S_{\rm o}, \tag{12}$$

where

$$S_{\rm e} = K_{\omega_{\rm o}} \equiv X^a \lrcorner \omega_{\rm o} \nabla_a + \frac{\pi - 1}{2\pi} d\omega_{\rm o}, \tag{13}$$

$$S_{\rm o} = M_{\omega_{\rm e}} \equiv e^a \wedge \omega_{\rm e} \nabla_a - \frac{n - \pi - 1}{2(n - \pi)} \delta\omega_{\rm e}, \tag{14}$$

where $\omega_{\rm o}$ is an odd KY form and $\omega_{\rm e}$ is an even CCKY form.

Given a KY p-form f it is possible to see that the tensor

$$K^{\mu\nu} = f^{\mu}{}_{\lambda_1 \ldots \lambda_{p-1}} f^{\nu\lambda_1 \ldots \lambda_{p-1}}, \tag{15}$$

is Killing–Stäckel tensor. Such a tensor satisfies $\nabla^{(\lambda} K^{\mu_1 \ldots \mu_p)} = 0$ and is associated with conserved quantities of higher order in the momenta for the theory of the classical particle.

In [13] it was shown that CCKY tensors form an algebra under the wedge product. In particular, closed conformal Killing–Yano tensors of rank 2 that are non-degenerate are called *Principal Conformal Killing–Yano (PCKY) tensors*. They are crucial for the integrability of various systems in four and higher dimensional black hole spacetimes.

Generalisations of these concepts to the case of metrics with torsion and fluxes have been treated in [14, 15].

4 Kerr-NUT-(A)dS Black Holes

While a classification of Lorentzian metrics with a PCKY tensor is not available, the analogue problem in Riemannian signature has been solved [16, 17]. The most general *canonical metric* admitting a PCKY tensor in $n = 2N + \varepsilon$ dimensions, $\varepsilon = 0, 1$, is given by

$$ds^2 = \sum_{\mu=1}^{N}\left[\frac{dx_\mu^{\ 2}}{Q_\mu} + Q_\mu\Big(\sum_{j=0}^{N-1} A_\mu^{(j)} d\psi_j\Big)^2\right] + \varepsilon S\Big(\sum_{j=0}^{N} A^{(j)} d\psi_j\Big)^2. \tag{16}$$

Here, coordinates x_μ ($\mu = 1, \ldots, N$) stand for the (Wick rotated) radial coordinate and longitudinal angles, and Killing coordinates ψ_k ($k = 0, \ldots, N - 1 + \varepsilon$) denote time and azimuthal angles associated with Killing vectors $\xi^{(k)} = \partial_{\psi_k}$. We have further defined the functions

$$Q_\mu = \frac{X_\mu}{U_\mu}, \quad U_\mu = \prod_{\nu\neq\mu}(x_\nu^2 - x_\mu^2), \quad S = \frac{-c}{A^{(N)}}, \tag{17}$$

$$A_\mu^{(k)} = \sum_{\substack{\nu_1,\ldots,\nu_k \\ \nu_1<\cdots<\nu_k,\ \nu_i\neq\mu}} x_{\nu_1}^2 \cdots x_{\nu_k}^2, \quad A^{(k)} = \sum_{\substack{\nu_1,\ldots,\nu_k \\ \nu_1<\cdots<\nu_k}} x_{\nu_1}^2 \cdots x_{\nu_k}^2. \tag{18}$$

The quantities X_μ are functions of a single variable x_μ, and c is an arbitrary constant. The vacuum (with a cosmological constant) black hole geometry is recovered by setting

$$X_\mu = \sum_{k=\varepsilon}^{N} c_k\, x_\mu^{2k} - 2b_\mu\, x_\mu^{1-\varepsilon} + \frac{\varepsilon c}{x_\mu^2}. \tag{19}$$

This choice of X_μ describes the most general known Kerr-NUT-(A)dS spacetimes in all dimensions [18]. The constant c_N is proportional to the cosmological constant and the remaining constants are related to angular momenta, mass and NUT parameters.

The PCKY tensor reads [19]

$$h = db, \quad b = \frac{1}{2}\sum_{j=0}^{N-1} A^{(j+1)} d\psi_j. \tag{20}$$

The $2j$-forms $h^{(j)}$, which are the jth wedge power of the PCKY tensor h, $h^{(j)} = h \wedge \cdots \wedge h$, form the tower of associated closed conformal Killing–Yano tensors. Their Hodge duals are KY forms and can be 'squared' to rank 2 Killing–Stäckel tensors.

5 Separability of the Dirac Equation in the Kerr-NUT-(A)dS Metric

According to the results of Sect. 3, for the $N+\varepsilon$ Killing vectors $\xi^{(0)}, \ldots, \xi^{(N-1+\varepsilon)}$ there are associated the operators $K_{\xi^{(0)}}, \ldots, K_{\xi^{(N-1+\varepsilon)}}$, which commute with the Dirac operator D. And for the N CCKY forms $h^{(j)}$ there are associated operators $M_{h^{(1)}}, \ldots, M_{h^{(N-1)}}$, which also commute with D. In [12] it has been shown that all these operators are in fact mutually commuting. Thus it is possible to diagonalise them simultaneously and to look for common spinorial eigenfunctions.

The Dirac equation in this metric had been shown to be separable by a direct calculation in [20]. A geometrical understanding of the result has been given in [21], where it has been shown that the solution to the eigenvalue problem

$$K_{\xi^{(k)}}\chi = i\Psi_k\chi, \quad M_j\chi = m_j\chi, \tag{21}$$

can be found in the tensorial R-separated form

$$\chi = R\exp\big(i\textstyle\sum_k \Psi_k\psi_k\big)\displaystyle\bigotimes_\nu \chi_\nu, \tag{22}$$

where $\{\chi_\nu\}$ is an N-tuple of 2-dimensional spinors and R is an appropriate Clifford bundle-valued prefactor. χ_ν depends only on the variable x_ν, $\chi_\nu = \chi_\nu(x_\nu)$, and satisfies the equation

$$\begin{aligned}\Bigg[\left(\frac{d}{dx_\nu} + \frac{X'_\nu}{4X_\nu} + \frac{\tilde{\Psi}_\nu}{X_\nu}\iota_{\langle\nu\rangle} + \frac{\varepsilon}{2x_\nu}\right)\sigma_{\langle\nu\rangle}& \\ - \frac{\left(-\iota_{\langle\nu\rangle}\right)^{N-\nu}}{\sqrt{|X_\nu|}}\left(\varepsilon\frac{i\sqrt{-c}}{2x_\nu^2} + m_\nu\right)\Bigg]\chi_\nu = 0\,,&\end{aligned} \tag{23}$$

where

$$\tilde{\Psi}_\mu = \sum_k \Psi_k(-x_\mu^2)^{N-1-k}\,, \tag{24}$$

and

$$m_\nu = \sum_j (-i)^j m_j\left(-\iota_{\langle\nu\rangle}x_\nu\right)^{N-1-j}. \tag{25}$$

The operator $\iota_{\langle\nu\rangle}$ acts as a σ_3 Pauli matrix on the 2-spinor χ_ν while leaving the other spinors χ_μ, $\mu \neq \nu$, invariant, and similarly $\sigma_{\langle\nu\rangle}$ acts as σ_1.

The solution (22) is the same as that given in [20]. The arbitrary integration constants found there are related to the eigenvalues Ψ_k and m_j.

6 The Spinning Particle

The spinning particle theory can be thought of as a semi-classical description of a Dirac fermion. The degrees of freedom are coordinates x^μ and Grassmannian variables θ^a related to the spin. The Hamiltonian is given by

$$H = \frac{1}{2}\Pi_\mu \Pi_\nu g^{\mu\nu}, \ \Pi_\mu = p_\mu - \frac{i}{2}\theta^a\theta^b\omega_{\mu ab} = g_{\mu\nu}\dot{x}^\nu, \tag{26}$$

where p_μ is the momentum canonically conjugate to x^μ and Π_μ is the covariant momentum. Poisson brackets are defined as

$$\{F, G\} = \frac{\partial F}{\partial x^\mu}\frac{\partial G}{\partial p_\mu} - \frac{\partial F}{\partial p_\mu}\frac{\partial G}{\partial x^\mu} + i(-1)^{a_F}\frac{\partial F}{\partial \theta^a}\frac{\partial G}{\partial \theta_a}, \tag{27}$$

where a_F is the Grassmann parity of F.

The theory is worldsheet supersymmetric and the generator of supersymmetry is given by

$$Q = \theta^a e_a{}^\alpha \Pi_\alpha, \tag{28}$$

which obeys

$$\{H, Q\} = 0, \quad \{Q, Q\} = -2iH. \tag{29}$$

Equations of motion are accompanied by two physical (gauge) conditions

$$2H = -1, \quad Q = 0, \tag{30}$$

In the Kerr-NUT-(A)dS spacetimes there are $(N+\varepsilon)$ Killing vectors $\xi_{(k)}$. It is possible to show that with these one can construct bosonic super invariants linear in velocities, given by

$$Q_{\xi_{(k)}} = \xi^\alpha_{(k)}\Pi_\alpha - \frac{i}{4}\theta^a\theta^b(d\xi_{(k)})_{ab}. \tag{31}$$

These can be used to express some components of the velocities Π in terms of the conserved quantities and of the θ variables. In [22] it was shown that it is possible to find N further bosonic supersymmetric conserved quantities $\mathcal{K}_{(j)}$, this time quadratic in the velocities. These new quantities will be neither conserved nor supersymmetric in a general metric, but they are for Kerr-NUT-(A)dS. The $(N+\varepsilon)+N = n$ quantities are all independent and using them it is possible to express all the components of Π, thus showing that the bosonic sector of the theory is integrable.

The quantities $\mathcal{K}_{(j)}$ are written as

$$\begin{aligned}\mathcal{K}_{(j)} &= K^{\mu\nu}_{(j)}\Pi_\mu\Pi_\nu + \mathcal{L}^\mu_{(j)}\Pi_\mu + \mathcal{M}_{(j)},\\ \mathcal{L}^\mu_{(j)} &= \theta^a\theta^b L_{(j)ab}{}^\mu, \quad \mathcal{M}_{(j)} = \theta^a\theta^b\theta^c\theta^d M_{(j)abcd}.\end{aligned} \tag{32}$$

The tensors K, L and M are given by

$$\begin{aligned} K^{\mu\nu} &= f^{\mu\kappa_1\ldots\kappa_{p-1}} f^{\nu}{}_{\kappa_1\ldots\kappa_{p-1}}\,, \\ L_{\mu\nu}{}^{\rho} &= -\frac{2i}{p+1} f_{[\mu|\kappa_1\ldots\kappa_{p-1}|}(df)_{\nu]}{}^{\rho\kappa_1\ldots\kappa_{p-1}} \\ &\quad -\frac{2i}{p+1}(df)_{\mu\nu\kappa_1\ldots\kappa_{p-1}} f^{\rho\kappa_1\ldots\kappa_{p-1}}\,, \\ M_{\mu\nu\rho\sigma} &= -\frac{i}{4}\nabla_{[\mu} L_{\nu\rho\sigma]}\,. \end{aligned} \tag{33}$$

Here $f_{\mu_1\ldots\mu_p}$ is the appropriate rank-p Killing–Yano tensor present in the spacetime.

Acknowledgments The author is partially funded by Fapemig under the project CEX APQ 2324-11.

References

1. Benenti, S.: Intrinsic characterization of the variable separation in the Hamilton Jacobi equation. J. Math. Phys. **38**, 6578 (1997). doi:10.1063/1.532226
2. Benenti, S., Francaviglia, M.: Remarks on certain separability structures and their applications to general relativity. Gen. Relativ. Gravit. **10**, 79 (1979). doi:10.1007/BF00757025
3. Demiański, M., Francaviglia, M.: Separability structures and Killing-Yano tensors in vacuum type-D space-times without acceleration. Int. J. Theor. Phys. **19**, 675 (1980). doi:10.1007/BF00670402
4. Kalnins, E., Miller, W.: Killing tensors and nonorthogonal variable separation for Hamilton Jacobi equations. SIAM J. Math. Anal. **12**, 617 (1981). doi:10.1137/0512054
5. Fatibene, L., Ferraris, M., Francaviglia, M., McLenaghan, R.: Generalized symmetries in mechanics and field theories. J. Math. Phys. **43**, 3147 (2002). doi:10.1063/1.1469668
6. McLenaghan, R., Spindel, P.: Quantum numbers for Dirac spinor fields on a curved space-time. Phys. Rev. D **20**, 409 (1979). doi:10.1103/PhysRevD.20.409
7. Carter, B., McLenaghan, R.: Generalized total angular momentum operator for the Dirac equation in curved space-time. Phys. Rev. D **19**, 1093 (1979). doi:10.1103/PhysRevD.19.1093
8. Kamran, N., McLenaghan, R.: Symmetry operators for neutrino and Dirac fields on curved space-time. Phys. Rev. D **30**, 357 (1984). doi:10.1103/PhysRevD.30.357
9. Benn, I., Charlton, P.: Dirac symmetry operators from conformal Killing-Yano tensors. Class. Quant. Grav. **14**, 1037 (1997). doi:10.1088/0264-9381/14/5/011
10. Benn, I., Kress, J.: First-order Dirac symmetry operators. Class. Quantum Grav. **21**, 427 (2004). doi:10.1088/0264-9381/21/2/007
11. Carignano, A., Fatibene, L., McLenaghan, R., Rastelli, G.: Symmetry operators and separation of variables for Dirac's equation on two-dimensional spin manifolds. SIGMA **7**, 57 (2011). doi:10.3842/SIGMA.2011.057
12. Cariglia, M., Krtouš, P., Kubizňák, D.: Commuting symmetry operators of the Dirac equation, Killing-Yano and Schouten-Nijenhuis brackets. Phys. Rev. D **84**, 024004 (2011). doi:10.1103/PhysRevD.84.024004
13. Krtouš, P., Kubizňák, D., Page, D., Frolov, V.: Killing-Yano tensors, rank-2 Killing tensors, and conserved quantities in higher dimensions. J. High Energy Phys. **2007**(02), 004 (2007). doi:10.1088/1126-6708/2007/02/004

14. Houri, T., Kubizňák, D., Warnick, C., Yasui, Y.: Symmetries of the Dirac operator with skew-symmetric torsion. Class. Quantum Grav. **27**, 185019 (2010). doi:10.1088/0264-9381/27/18/185019
15. Kubizňák, D., Warnick, C., Krtouš, P.: Hidden symmetry in the presence of fluxes. Nucl. Phys. B **844**, 185 (2011). doi:10.1016/j.nuclphysb.2010.11.001
16. Houri, T., Oota, T., Yasui, Y.: Closed conformal Killing-Yano tensor and Kerr-NUT-de Sitter space time uniqueness. Phys. Lett. B **656**, 214 (2007). doi:10.1016/j.physletb.2007.09.034
17. Krtouš, P., Frolov, V., Kubizňák, D.: Hidden symmetries of higher-dimensional black holes and uniqueness of the Kerr-NUT-(A)dS spacetime. Phys. Rev. D. **78**, 064022 (2008). doi:10.1103/PhysRevD.78.064022
18. Chen, W., Lü, H., Pope, C.: General Kerr NUT AdS metrics in all dimensions. Class. Quantum Grav. **23**, 5323 (2006). doi:10.1088/0264-9381/23/17/013
19. Kubizňák, D., Frolov, V.: The hidden symmetry of higher dimensional Kerr NUT AdS spacetimes. Class. Quantum Grav. **24**, F1 (2007). doi:10.1088/0264-9381/24/3/F01
20. Oota, T., Yasui, Y.: Separability of Dirac equation in higher dimensional Kerr NUT de Sitter spacetime. Phys. Lett. B **659**, 688 (2008). doi:10.1016/j.physletb.2007.11.057
21. Cariglia, M., Krtouš, P., Kubizňák, D.: Dirac equation in Kerr-NUT-(A)dS spacetimes: intrinsic characterization of separability in all dimensions. Phys. Rev. D **84**(2), 024008 (2011). doi:10.1103/PhysRevD.84.024008
22. Kubizňák, D., Cariglia, M.: Integrability of spinning particle motion in higher-dimensional rotating black hole spacetimes. Phys. Rev. Lett. **108**(5), 051104 (2012). doi:10.1103/PhysRevLett.108.051104

Geometrostatics: The Geometry of Static Space-Times

Carla Cederbaum

Abstract We present a new geometric approach to the study of static isolated general relativistic systems for which we suggest the name *geometrostatics*. After describing the setup, we introduce localized formulas for the ADM-mass and ADM/CMC-center of mass of geometrostatic systems. We then explain the pseudo-Newtonian character of these formulas and show that they converge to Newtonian mass and center of mass in the Newtonian limit, respectively, using Ehlers' frame theory. Moreover, we present a novel physical interpretation of the level sets of the canonical lapse function and apply it to prove uniqueness results. Finally, we suggest a notion of force on test particles in geometrostatic space-times.

1 Introduction

Static isolated general relativistic systems have been studied from a number of perspectives including their regularity, compactification and asymptotic considerations, symmetry classifications, construction of explicit solutions etc. They serve as models of static stars and black holes. Also, they play an important role in R. Bartnik's definition of quasi-local mass and his associated conjecture on static metric extensions [1].

Here, we present a new geometric approach to the study of static isolated systems and their physical properties for which we suggest the name *geometrostatics*. We consider space-times that are *static* (possess a smooth global time-like Killing vector field that is hypersurface-orthogonal) and *isolated* (see below). Static space-times generically possess a $3 + 1$-decomposition with vanishing shift vector. In this *canonical* decomposition, the *canonical* lapse function is time-independent and

C. Cederbaum (✉)
Department of Mathematics, Eberhard Karls Universität Tübingen, Auf der Morgenstelle 10, 72076 Tübingen , Germany
e-mail: cederbaum@math.uni-tuebingen.de

J. Bičák and T. Ledvinka (eds.), *Relativity and Gravitation*,
Springer Proceedings in Physics 157, DOI: 10.1007/978-3-319-06761-2_5,

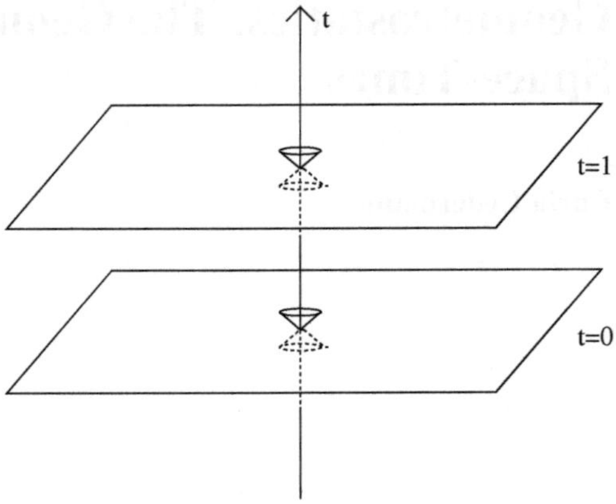

Fig. 1 The time-slices of a canonically decomposed static space-time

coincides with the Lorentzian length of the time-like Killing vector field. The space-like time-slices orthogonal to the time-like Killing vector field are all isometric and have vanishing extrinsic curvature, see Fig. 1. Their induced Riemannian metric is time-independent. We will subsequently identify all canonical time-slices.

For our purposes, static systems are called *isolated* if the Riemannian metric and the lapse function on the time-slice decay suitably fast to the flat metric and the constant 1, respectively, at spacelike infinity see [2] for a precise definition of our asymptotic flatness condition in the language of weighted Sobolev spaces. Moreover, we request that the space-time satisfies the *vacuum* Einstein equations outside some spatially compact tube in the space-time (or, in other words, outside some compact set in the time-slice). This can be interpreted as a (spatially) finite extension of the sources, whether they are matter sources and/or black holes.

This article is structured as follows: In Sect. 2, we will introduce the central equations of geometrostatics and summarize a few of their central analytic properties. In Sect. 3, we present a novel physical interpretation of the level sets of the lapse function of a geometrostatic system and discuss some applications of this insight. In Sect. 4, we will perform a conformal transformation into what we suggest to call *pseudo-Newtonian* variables. Moreover, we will define and analyze localized surface integral expressions for the mass and center of mass of geometrostatic systems. Finally, in Sect. 5, we will discuss the Newtonian limit of geometrostatics.

Further details can be found in my thesis [2].

2 Geometrostatics

The Lorentzian metric of a generic static space-time $\mathbb{R} \times M^3$ can globally be decomposed as

$$ds^2 = -N^2c^2dt^2 + g, \tag{1}$$

where $N := \sqrt{-ds^2(\partial_t, \partial_t)}$ is the (canonical) lapse function arising as the Lorentzian length of the time-like Killing vector field ∂_t, c is the speed of light, and g is the Riemannian metric induced on the time-slice M^3. Observe that N is non-negative and vanishes only along Killing horizons.

In the vacuum region outside the matter, the (vacuum) Einstein equations imply that these variables satisfy the so-called *vacuum static metric equations*

$$N \operatorname{Ric} = \nabla^2 N, \tag{2}$$
$$\Delta N = 0, \tag{3}$$

where Ric is the Ricci curvature tensor of g, $\nabla^2 N$ denotes the covariant Hessian, and ΔN denotes the covariant Laplacian of N with respect to g. It is well-known that solutions to the vacuum static metric equations are real analytic in suitable coordinates [3].

We define a *geometrostatic system* to be an asymptotically flat Riemannian 3-manifold (M^3, g) endowed with a smooth positive lapse function N so that the vacuum static metric equations (2) and (3) are satisfied. Hence, geometrostatic systems model the vacuum region outside the support of the matter and the horizons of all black holes within a slice of an asymptotically flat static space-time. The lapse function N describes the lapse of time in the space-time.

3 The Level Sets of the Lapse

In Newtonian gravity, the relevant gravitational variable is the Newtonian potential. The gradient of the potential defines the force on a unit mass test particle. This has a well-known consequence for the *equipotential surfaces* (or level sets of the Newtonian potential): if a test particle is constrained into one of these surfaces then the gravitational force does not have a tangential component and hence the test particle does not tangentially accelerate within the surface, see Fig. 2.

Surprisingly, the "same" is true for level sets of the lapse function in a geometrostatic system. In order to make this rigorous, we make the following definitions: Consider a closed smooth surface $\Sigma \subset M^3$ in a geometrostatic system (M^3, g, N) arising as the $n = n_0$ level set of a smooth function $n : M^3 \to \mathbb{R}$. A time-like[1] curve

[1] Here, time-like curves and the time functional are taken with respect to the static space-time metric $ds^2 = -N^2c^2dt^2 + g$ induced by (M^3, g, N).

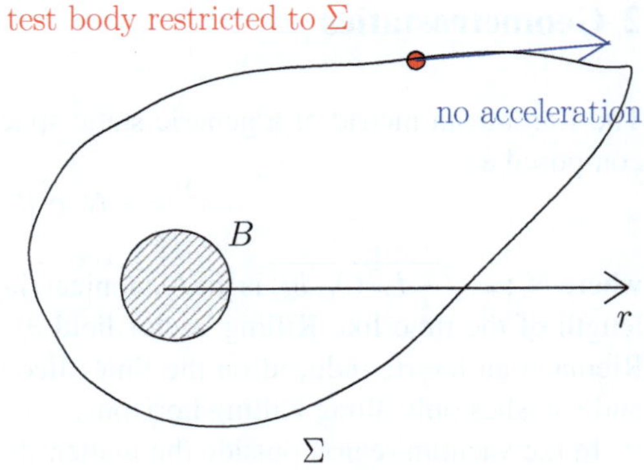

Fig. 2 A test particle constrained to a surface Σ

$$\mu(\tau) = \Big(t(\tau), x(\tau)\Big) \tag{4}$$

satisfying $x(\tau) \in \Sigma$ is called a *test particle constrained to* Σ if it is a critical point of the time functional

$$T(\mu) := \int_{\tau_0}^{\tau_1} \Big\{ |\dot{\mu}(\tau)| + \sigma (n \circ x(\tau) - n_0) \Big\}\, d\tau, \tag{5}$$

where $\sigma \in \mathbb{R}$ is a Lagrange multiplier ensuring that all comparison curves are also constrained to Σ. With this notion of constrained test particle, we say that a smooth closed surface Σ is an *equipotential surface* if every test particle constrained to Σ is a geodesic in Σ with respect to the induced 2-metric, see Fig. 2. Analyzing the geodesic equation, we find that a surface $\Sigma \subset M^3$ is an equipotential surface in (M^3, g, N) if and only if Σ is a level set of N. Thus, the level sets of the lapse function N in geometrostatics play precisely the same role as those of the Newtonian potential in Newtonian gravity.

In a static vacuum space-time, the Einstein constraint equations reduce to Scal $=$ 0. In particular, the lapse function does not appear in this constraint equation. As a consequence, Choquet-Bruhat's theorem (see e.g. [4]) on the local existence and uniqueness of solutions to the Einstein equations implies that the space-time induced by (M^3, g, N) is in fact *independent*[2] of the lapse function N. Combining this view of the lapse function with the physical interpretation of its level sets of the lapse as well as with the vacuum static metric equations (2) and (3) and the assumed asymptotic conditions for g and N, we obtain that the lapse function is indeed unique if it exists. This is to say that if (M^3, g, N) and $(M^3, g, \widetilde{N})$ are geometrostatic systems, then $N = \widetilde{N}$. We interpret this result as saying that "there is only one way of synchronizing time at different locations in a geometrostatic space-time such that one sees staticity" just as, for a geodesic, "there is only one way of walking along a

[2] This assumes that the lapse function *exists* in the first place.

geodesic such that one does not accelerate (up to affine transformations of the curve parameter)". The affine freedom of the parameter along the geodesic does not make an appearance in the geometrostatic space-time picture because we fixed the lapse function to asymptotically converge to 1 at spacelike infinity and therewith fixed the time unit.

4 Pseudo-Newtonian Gravity

The geometrostatic variables g and N are ideal for investigating geometric and relativistic effects influencing test particle behavior and the behavior of light rays. In order to better understand asymptotic and analytic properties of solutions, however, it is more convenient to perform a conformal change and consider the new variables

$$\gamma := N^2 g, \tag{6}$$

$$U := c^2 \ln N. \tag{7}$$

These variables have been used by many authors[3], see e.g. [5]. We suggest to call them *pseudo-Newtonian metric and potential*, respectively. The vacuum static metric (2), (3) translate into

$$\mathrm{Ric}_\gamma = 2c^{-4}\, dU \otimes dU, \tag{8}$$

$$\Delta_\gamma U = 0, \tag{9}$$

where Ric_γ denotes the Ricci curvature tensor of γ and Δ_γ denotes the γ-covariant Laplacian on M^3.

N was assumed to converge to 1 asymptotically at spacelike infinity, so U must asymptotically tend to 0. Indeed, Kennefick and O'Murchadha [5] showed that the asymptotic flatness assumptions incorporated into the above definition of a geometrostatic system induce the decay conditions

$$\gamma_{ij} = \delta_{ij} + \mathcal{O}(r^{-2}), \tag{10}$$

$$U = -\frac{mG}{r} + \mathcal{O}(r^{-2}), \tag{11}$$

as $r \to \infty$ in suitable coordinates. Here, m is the ADM-mass of the slice (M^3, g), see [6–8]. In [2], we prove asymptotic estimates in weighted Sobolev spaces that improve this fall-off result. In particular, we find that

[3] Albeit without explicit reference to the speed of light.

$$\gamma_{ij} = \left(1 - \frac{M^2}{r^2}\right)\delta_{ij} + \frac{2M^2 x_i x_j}{r^4} + \mathcal{O}(r^{-3}), \tag{12}$$

$$U = -\frac{mG}{r} - \frac{mG\mathbf{z}_A \cdot \mathbf{x}}{r^3} + \mathcal{O}(r^{-3}), \tag{13}$$

as $r \to \infty$, where $M = mG/c^2$ and $\mathbf{z}_A \in \mathbb{R}^3$ is a fixed vector. This decay occurs in asymptotically flat γ-harmonic coordinates. As a matter of fact, these coordinates coincide with the asymptotically flat (spatial) wave-harmonic coordinates on $(\mathbb{R} \times M^3, ds^2 = -N^2c^2dt^2 + g)$. The vector $\mathbf{z}_A$ can be interpreted as the coordinate vector of the *asymptotic center of mass* of the system, see below.

In terms of the pseudo-Newtonian variables γ and U and inspired by Newtonian gravity, we suggest the following quasi-local definitions of *pseudo-Newtonian mass and center of mass*[4] of a geometrostatic system (M^3, g, N) with associated pseudo-Newtonian variables (γ, U):

$$m_{PN}(\Sigma) := \frac{1}{4\pi G}\int_\Sigma \frac{\partial U}{\partial \nu}\, d\sigma, \tag{14}$$

$$\mathbf{z}_{PN}(\Sigma) := \frac{1}{4\pi G m}\int_\Sigma \left(\frac{\partial U}{\partial \nu}\mathbf{x} - U\frac{\partial \mathbf{x}}{\partial \nu}\right) d\sigma, \tag{15}$$

where Σ is any surface enclosing the support of the matter, ν and $d\sigma$ are the outer unit normal to and area measure of Σ with respect to γ, and $\mathbf{x}$ is the vector of γ-harmonic coordinates.

Surprisingly, both of these expressions are *independent* of the particular choice of surface Σ (as long as the surface encloses the support of the matter, imagine for example a large coordinate sphere). For the pseudo-Newtonian mass (14), this independence of the surface can be seen by combining (9) with the divergence theorem. For the center of mass (15), the independence of the surface follows from (9) combined with Green's formula and the fact that the coordinates are γ-harmonic such that $\Delta_\gamma \mathbf{x} = \mathbf{0}$. We will thus drop the explicit reference to the surface when referring to pseudo-Newtonian mass and center of mass.

Using the asymptotic decay (10), (11), we find that

$$m_{PN} = m_{ADM}. \tag{16}$$

The total mass of a geometrostatic system is thus *localized*. It can be read off *exactly* on any surface enclosing the matter. Applying formula (14) to any smooth surface $\Sigma \subset M^3$, we immediately obtain a notion of mass for an arbitrary part of the system (namely the part bounded by the surface Σ). By the divergence theorem and (9), the masses of all components in a multi-component system add up to the total mass of the system just as Newtonian masses do.

If we combine the asymptotic decay (12), (13) with formula (15) defining the pseudo-Newtonian center of mass, we find that $\mathbf{z}_{PN} = \mathbf{z}_A$. We claim that this vector

[4] We note that our discussion of center of mass only applies to systems with non-vanishing mass.

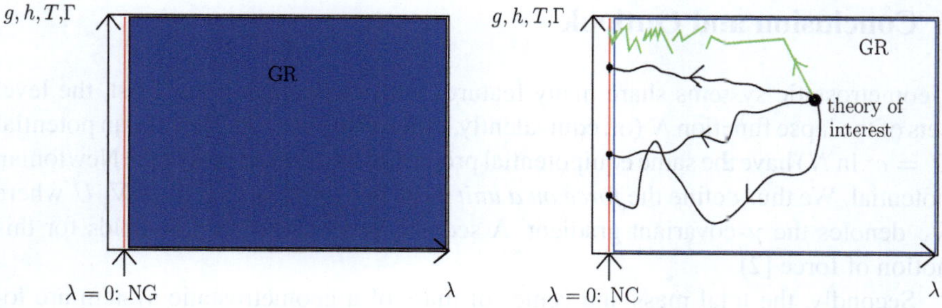

Fig. 3 The universe of Ehlers' frame theory and the Newtonian limit

can indeed be physically interpreted as the coordinate vector of the total center of mass of the system. For this, we exploit Huang's work [9] showing that the ADM-center of mass [6] coincides[5] with the CMC-center of mass constructed via a constant mean curvature (CMC) foliation near infinity by Huisken and Yau [10] and generalized by Metzger [11]. Using again (12) and (13), we obtain

$$\mathbf{z}_{PN} = \mathbf{z}_A = \mathbf{z}_{ADM} = \mathbf{z}_{CMC} \tag{17}$$

which justifies the name center of mass for the quantities $\mathbf{z}_{PN}$ and $\mathbf{z}_A$. Moreover, this shows that the center of mass of a geometrostatic system is also *localized.* As above, we obtain a notion of center of mass for an arbitrary part of the system (namely the part bounded by the surface Σ). By Green's formula, γ-harmonicity of the coordinates, and (9), the centers of mass of all components in a multi-component system add up to the total center of mass of the system just as Newtonian centers of mass do.

5 The Newtonian Limit of Geometrostatics

Intuitively, the mass and center of mass of a relativistic system should converge to the Newtonian mass and center of mass of its Newtonian limit $c \to \infty$. To make this precise, we use Ehlers' frame theory [12] which unifies general relativity (GR) and Newton-Cartan gravity (NC) into a common geometric framework with geometric variables g, h, Γ and matter tensor T.

In frame theory, taking the Newtonian limit corresponds to taking a parametric curve of solutions of GR with parameter $\lambda = c^{-2}$ to its limit $\lambda \to 0$, see Fig. 3. Modeling Killing vector fields, staticity, and asymptotic flatness in frame theory, we show that the pseudo-Newtonian potential converges to the Newtonian potential and the metric γ converges to the flat metric along any family of geometrostatic systems that possesses a static Newtonian limit. As the localized pseudo-Newtonian formulas (14) and (15) are nearly identical with the Newtonian formulas, this proves that indeed the relativistic mass and center of mass converge to their Newtonian counterparts.

[5] Under precise fall-off conditions at spacelike infinity that are satisfied here.

6 Conclusion and Outlook

Geometrostatic systems share many features with Newtonian ones. First, the level sets of the lapse function N (or, equivalently, those of the pseudo-Newtonian potential $U = c^2 \ln N$) have the same equipotential properties as the level sets of the Newtonian potential. We thus define the *force on a unit mass test particle* as $\mathbf{F} := -\nabla_\gamma U$ where ∇_γ denotes the γ-covariant gradient. A second Newtonian type law holds for this notion of force [2].

Secondly, the total mass and center of mass of a geometrostatic system are localized. We put forward explicit geometric formulas for them that also allow for the computation of the notions of mass and center of mass of individual regions. We applied these formulas to prove consistence of ADM-mass and ADM/CMC-center of mass with the Newtonian limit.

This fact and the notion of force might turn out useful for the study of the well-known static n-body problem and of Bartnik's conjecture [1].

References

1. Szabados, L.: Quasi-local energy-momentum and angular momentum in general relativity. Living Rev. Relativ. **12**(4), lrr-2009-4 (2009). http://www.livingreviews.org/lrr-2009-4
2. Cederbaum, C.: The Newtonian limit of geometrostatics. Ph.D. thesis, Free University Berlin, Berlin (2011)
3. Müller zum Hagen, H.: On the analyticity of static vacuum solutions of Einstein's equations. Proc. Camb. Phil. Soc. **67**, 415 (1970). doi:10.1017/S0305004100001237
4. Choquet-Bruhat, Y., Jork, J.: The Cauchy problem. In: Held, A. (ed.) General Relativity and Gravitation: One Hundred Years after the Birth of Albert Einstein, vol. 1, Chap. 4, pp. 99–172. Plenum Press, New York (1980)
5. Kennefick, D., O'Murchadha, N.: Weakly decaying asymptotically flat static and stationary solutions to the Einstein equations. Class. Quantum Grav. **12**(1), 149 (1995). doi:10.1088/0264-9381/12/1/013
6. Arnowitt, R., Deser, S., Misner, C.: Coordinate invariance and energy expressions in general relativity. Phys. Rev. **122**(3), 997 (1961). doi:10.1103/PhysRev.122.997
7. Bartnik, R.: The mass of an asymptotically flat manifold. Commun. Pure Appl. Math. **39**, 661 (1986). doi:10.1002/cpa.3160390505
8. Chruściel, P.: On the invariant mass conjecture in general relativity. Commun. Math. Phys. **120**, 233 (1988). doi:10.1007/BF01217963
9. Huang, L.H.: Foliations by stable spheres with constant mean curvature for isolated systems with general asymptotics. Comm. Math. Phys. **300**(2), 331 (2008)
10. Huisken, G., Yau, S.T.: Definition of center of mass for isolated physical systems and unique foliations by stable spheres with constant mean curvature. Invent. Math. **124**, 281 (1996). doi:10.1007/s002220050054
11. Metzger, J.: Foliations of asymptotically flat 3-manifolds by 2-surfaces of prescribed mean curvature. J. Differ. Geom. **77**(2), 201 (2007)
12. Ehlers, J.: Examples of Newtonian limits of relativistic spacetimes. Class. Quantum Grav. **14**, A119 (1997). doi:10.1088/0264-9381/14/1A/010

The Gravitational Equation in Higher Dimensions

Naresh Dadhich

Abstract Like the Lovelock Lagrangian which is a specific homogeneous polynomial in Riemann curvature, for an alternative derivation of the gravitational equation of motion, it is possible to define a specific homogeneous polynomial analogue of the Riemann curvature, and then the trace of its Bianchi derivative yields the corresponding polynomial analogue of the divergence free Einstein tensor defining the differential operator for the equation of motion. We propose that the general equation of motion is $G_{ab}^{(n)} = -\Lambda g_{ab} + \kappa_n T_{ab}$ for $d = 2n+1,\ 2n+2$ dimensions with the single coupling constant κ_n, and $n = 1$ is the usual Einstein equation. It turns out that gravitational behavior is essentially similar in the critical dimensions for all n. All static vacuum solutions asymptotically go over to the Einstein limit, Schwarzschild-dS/AdS. The thermodynamical parameters bear the same relation to horizon radius, for example entropy always goes as r_h^{d-2n} and so for the critical dimensions it always goes as $r_h,\ r_h^2$. In terms of the area, it would go as $A^{1/n}$. The generalized analogues of the Nariai and Bertotti–Robinson solutions arising from the product of two constant curvature spaces, also bear the same relations between the curvatures $k_1 = k_2$ and $k_1 = -k_2$ respectively.

1 Introduction

What stands gravity apart from rest of the physics is its universal character that it links to everything including massless particles and hence it can only be described by the spacetime curvature, and its dynamics has therefore to follow from the geometric properties of the Riemann curvature tensor [1]. The Einstein gravitational equation

N. Dadhich (✉)
Centre for Theoretical Physics, Jamia Millia Islamia, New Delhi 110025, India
e-mail: nkd@iucaa.ernet.in

N. Dadhich
Inter-University Centre for Astronomy and Astrophysics, Post Bag 4, Pune 411007, India

J. Bičák and T. Ledvinka (eds.), *Relativity and Gravitation*,
Springer Proceedings in Physics 157, DOI: 10.1007/978-3-319-06761-2_6,

could be deduced from the geometric property of Riemann curvature, known as the Bianchi identity, implying vanishing of its Bianchi derivative identically. Its trace yields the divergence-free second rank symmetric Einstein tensor. It defines the differential operator on the left hand side of the equation while the gravitational source – energy momentum distribution described by a second rank symmetric tensor with the condition of vanishing divergence—appears on the right hand side. This is the case for Einstein gravity which is linear in Riemann curvature, and its vacuum is trivially flat in 3 dimensions and it becomes dynamically non-trivial in 4 dimensions.

The question is, could this be generalized to a polynomial analogue of the Riemann tensor? Consider a tensor with the same symmetry properties as the Riemann which is a homogeneous polynomial of degree n in Riemann, and then demand that the trace of its Bianchi derivative vanishes. This will fix the coefficients in the polynomial and will give the divergence free second rank symmetric tensor $G^{(n)}_{ab}$, the nth order analogue of the Einstein tensor, which is the same as what one would get from the variation of the nth order Lovelock Lagrangian [2]. Thus we have the generalized polynomial Riemann curvature, $R^{(n)}_{abcd}$, which would describe gravitational dynamics in $d = 2n + 1, 2n + 2$ in the same manner as Riemann does for $d = 3, 4$. We can define corresponding vacuum as $R^{(n)}_{ab} = 0$, would it also be trivial in $d = 2n + 1$ dimension? The answer is indeed, yes [3]. It would be $R^{(n)}_{abcd}$ flat but not Riemann flat, and for that it would describe a global monopole [4].

What should be the gravitational equation in dimension > 4? Should it continue to be the Einstein equation which is linear in Riemann or should it include the one following from the higher order Riemann, $R^{(n)}_{abcd}$ yet giving the second order quasi-linear equation? A general abiding principle is that the equation be second order quasi-linear so that the initial value problem is well formulated giving unique evolution. This uniquely identifies the Lovelock polynomial Lagrangian or equivalently the above discussed polynomial Riemann curvature [2]. Should all orders that are non-trivial in the equation be included like the linear Einstein, quadratic Gauss-Bonnet, and so on, or the only highest one? Should it be $\sum G^{(n)}_{ab}$ or $G^{(n)}_{ab}$? In the former, each order will have its own coupling and so there would be n of them, and there is no obvious way to fix them. Since there is only one force which allows determination of only one coupling parameter by experimentally measuring its strength, gravity should therefore have only one dimensional coupling parameter and its dimension would however depend upon the spacetime dimension. Thus we propose the gravitational equation should in general be written as

$$G^{(n)}_{ab} = -\Lambda g_{ab} + \kappa_n T_{ab} \tag{1}$$

for $d = 2n + 1, 2n + 2$ dimensions. Note that Λ, which characterizes dynamics free spacetime, is part of the structure of spacetime on the same footing as the velocity of light [5]. In what follows we wish to demonstrate that this equation imbibes beautifully the general vacuum character [3] while the static vacuum solutions asymptotically go over to the right Einstein limit, even though the linear Einstein term is not included. This means higher order terms in curvature are only pertinent to the high energy end near the black hole horizon while their effect weans out asymptotically at the low energy end approximating to the linear order Einstein solution, Schwarzschild-dS/AdS in d dimension [6, 7]. It is remarkable that the thermodynamical parameters, temperature and entropy bear universal relation to the horizon radius for static black holes in $d = 2n + 1, 2n + 2$, and interestingly this property also marks the characterization of this class of black holes [7, 8].

2 The Lovelock Curvature Polynomial and the Equation of Motion

Following Ref. [2], we define the Lovelock curvature polynomial

$$R^{(n)}_{abcd} = F^{(n)}_{abcd} - \frac{n-1}{n(d-1)(d-2)} F^{(n)}(g_{ac}g_{bd} - g_{cd}g_{bc}),$$
$$F^{(n)}_{abcd} = Q_{ab}{}^{mn} R_{cdmn}, \tag{2}$$

where

$$Q^{ab}{}_{cd} = \delta^{aba_1b_1...a_nb_n}_{cdc_1d_1...c_nd_n} R_{a_1b_1}{}^{c_1d_1}, \ldots, R_{a_{n-1}b_{n-1}}{}^{c_{n-1}d_{n-1}},$$
$$Q^{abcd}{}_{;d} = 0. \tag{3}$$

It follows that the trace of the Bianchi derivative yields the divergence-free $G^{(n)}_{ab}$; i.e.

$$g^{ac} g^{bd} R^{(n)}_{abcd;e} = G^{(n)b}{}_{e;b} = 0, \tag{4}$$

where the analogue of nth order Einstein tensor is given by

$$G^{(n)}_{ab} = n(R^{(n)}_{ab} - \frac{1}{2} R^{(n)} g_{ab}). \tag{5}$$

Note that

$$R^{(n)} = \frac{d-2n}{n(d-2)} F^{(n)}, \tag{6}$$

which vanishes for $D = 2n$ while $F^{(n)}$, the Lovelock action polynomial, is non-zero but its variation, $G^{(n)}_{ab}$ vanishes identically. Since $R^{(n)} = g^{ab} R^{(n)}_{ab} = 0$ for $d = 2n$ for arbitrary g_{ab}, it implies $R^{(n)}_{ab} = 0$ identically as it involves apart from the metric its first and second derivatives which are arbitrary.

Since $G^{(n)}_{ab}$ is divergence free, we could write

$$G^{(n)}_{ab} = \kappa_n T_{ab} - \Lambda g_{ab}, \ T^{ab}_{;b} = 0. \tag{7}$$

This is the gravitational equation for $d = 2n + 1, 2n + 2$ dimensions with κ_n as the gravitational constant, and $n = 1$ is the Einstein equation for 3 and 4 dimensions. What degree of polynomial in Riemann should the equation have is thus determined by the spacetime dimension. It is linear for 3, 4, quadratic for 5, 6, and so on.

3 Universal Features

The first universal feature studied was that of gravitational field inside a uniform density sphere and it was shown that it was always given by the Schwarzschild interior solution in Einstein as well as in Einstein–Gauss–Bonnet/Lovelock theories [9]. Here we shall consider the cases of static black holes, and product spaces describing the Nariai and Bertotti–Robinson spacetimes.

3.1 Static Black Holes

The static spherically symmetric solution of the vacuum (1) is given by

$$g_{tt} = -1/g_{rr} = V = 1 - r^2(\Lambda + M/r^{d-1})^{1/n}, \tag{8}$$

which asymptotically takes the form of the Schwarzschild-dS/AdS solution in d dimensions showing the correct Einstein limit. The solution for the general case of the Einstein–Lovelock equation can also be written in terms of the nth order algebraic polynomial equation which cannot be solved in general for $n > 4$. It is therefore clear that we cannot carry on with arbitrarily large number of coupling parameters. For the case of dimensionally continued black holes [10], it was proposed that all the couplings are determined in terms of the unique ground state Λ, and the solution is then given by $V = 1 - r^2\Lambda - M/r^{d-1/2}$ which clearly does not go over to the Einstein solution for large r. This corresponded to the algebraic polynomial being degenerate. It turns out that the proper Einstein limit could be brought in simply by considering the polynomial to be derivative degenerate [7]. Then the solution agrees near the horizon with the dimensionally continued black hole and asymptotically with the proper Einstein limit, and it is the solution of equation (1).

Further, the thermodynamical parameters, temperature and entropy bear the universal relation to the horizon radius for the critical $d = 2n + 1,\ 2n + 2$ dimensions [8]. For instance, the entropy always goes as r_h^{d-2n} which for the critical dimensions would always go as $r_h,\ r_h^2$. In terms of the area, it would however go as $A^{1/n}$, and hence the entropy is proportional to area only for the $n = 1$ Einstein theory. Interestingly, this universality is also the characterizing property of this class of pure Lovelock black holes [7, 8].

We would like to conjecture that the above universality property would also be true for the rotating black hole solution as and when it is found.

3.2 Product Spaces: Nariai and Bertotti–Robinson Solutions

The Nariai and Bertotti–Robinson solutions arise as product of two constant curvature spaces. When the two curvatures are equal, $k_1 = k_2$, it is the Nariai solution of (1) with $T_{ab} = 0$ for $n = 1$, and when the curvatures are equal and opposite, $k_1 = -k_2$, it is the Bertotti–Robinson solution describing the uniform electric field. The former is the Λ vacuum spacetime but is not conformally flat while the latter is the Einstein–Maxwell solution for uniform electric field which is conformally flat. It turns out the generalized pure Lovelock solutions of (1) for any n bear out the same curvature relations for the Nariai vacuum ($k_1 = k_2$) and Bertotti–Robinson uniform electric field ($k_1 = -k_2$), and the condition for conformal flatness is also $k_1 k_2 = 0$ [11].

In $d = 2n+2$ dimensions, we have the following general relation connecting the two curvatures, Λ and the electric field E,

$$(k_1 + k_2)E^2 = -4(k_1 - k_2)\Lambda\,. \tag{9}$$

This clearly indicates $k_1 = k_2$ for $E = 0$, the Nariai vacuum spacetime and $k_1 = -k_2$ for $\Lambda = 0$, the Bertotti–Robinson uniform electric field spacetime. The metric is given by

$$ds^2 = (1 - k_1 r^2)dt^2 - \frac{dr^2}{1 - k_1 r^2} - \frac{1}{k_2} d\Sigma^2_{(d-2)}\,. \tag{10}$$

4 Discussion

We have proposed that equation (1) is the proper equation for gravity in higher dimensions. The correct equation should have the following properties: (a) it should be second order quasi-linear, (b) for a given dimension, it should be of degree $n = [(d-1)/2]$ in the Riemann curvature, (c) it should have only one coupling constant which could be determined by experimentally measuring the strength of the force, and (d) since higher order curvature contributions are the high energy corrections to the linear order in Riemann Einstein gravity which should wean out asymptotically, hence solutions should tend to the corresponding Einstein solution for large r. The proposed equation satisfies all these properties. The latter feature of the asymptotic Einstein limit is verified for the static black hole solutions which, however, is also true for the Einstein–Gauss–Bonnet black hole. What is remarkable here is that the equation is free of the Einstein term, yet asymptotically solutions go over to the proper Einstein limit. This means high energy effects which come through the higher order curvature terms are fully and properly taken care by the highest order

$n = [(d-1)/2]$ term, and they could be realized only in higher dimensions [12]. It is interesting that gravity asks for higher dimensions for realization of its high energy effects. This is because inclusion of higher orders in Riemann curvature and the demand that the equation continues to be second order quasi-linear naturally lead to higher dimensions. This does not happen for any other force that one has to consider higher dimension for realization of its high energy corrections. It happens for gravity because the spacetime curvature is the basic field variable, and hence high energy effects involve higher orders in it and their contribution in the equation, if it continues to retain its second order quasi-linear character, can be realized only in higher dimensions [12]. We would like to emphasize that higher dimensions and high energy effects seem to be intimately connected. Since high energy effects ask for higher dimensions, quantum gravity should also involve higher dimensions. This is because quantum gravity should approach the classical limit via the high energy intermediate limit.

One of the problems with the Einstein–Lovelock solutions is number of coupling constants and there is no way to fix them. For the dimensionally continued static black holes, all the couplings were prescribed in terms of the unique ground state Λ [10]. These solutions were, however, not asymptotically Einstein, Schwarzschild-dS/AdS. Instead, the corresponding solutions of (1) have the right limits at both ends, nearer to horizon agreeing with the dimensionally continued and asymptotically to Schwarzschild-dS/AdS. This is indicative of the inherent correctness of the equation. The universal character of gravity in the critical dimensions is another very attractive feature of the equation. That the vacuum, $G^{(n)}_{ab} = 0$, in the odd critical dimension is always trivial, $R^{(n)}_{abcd} = 0$ [3]. All this taken together points to the fact that (1) is right equation for gravitation in higher dimensions.

For a given order n in the Riemann curvature, the critical dimensions are $d = 2n+1,\ 2n+2$ and it is trivial/kinematic in the former and it becomes dynamic in the latter. This is a universal general feature. In the critical dimensions, gravity has the similar behavior as indicated by universality of the thermodynamic parameters in terms of the horizon radius and of the Nariai and Bertotti–Robinson solutions. It is interesting to note that in terms of black hole area, entropy is always proportional to $A^{1/n}$ and so it is proportional to area only for the $n = 1$ Einstein gravity. This is an interesting general result that entropy always goes as the nth root of area of the black hole. In an intuitive sense we can say that it is nth root of the Einstein gravity for the critical $d = 2n+1,\ 2n+2$ dimensions.

All this we have established for the simple case of static black hole but we believe that it is indeed a general feature and hence should be true for the stationary rotating black hole as well. So far there exists no rotating pure Lovelock black hole solution, and this conjecture would be verified as and when a solution is found.

References

1. Dadhich, N.: Subtle is the gravity. ArXiv e-prints arXiv:gr-qc/0102009 (2001)

2. Dadhich, N.: Characterization of the Lovelock gravity by Bianchi derivative. Pramana **74**, 875 (2010). doi:10.1007/s12043-010-0080-1
3. Dadhich, N., Ghosh, S., Jhingan, S.: The Lovelock gravity in the critical spacetime dimension. Phys. Lett. B **711**, 196 (2012). doi:10.1016/j.physletb.2012.03.084
4. Barriola, M., Vilenkin, A.: Gravitational field of a global monopole. Phys. Rev. Lett. **63**, 341 (1989). doi:10.1103/PhysRevLett.63.341
5. Dadhich, N.: On the measure of spacetime and gravity. Int. J. Mod. Phys. D **20**, 2739 (2011). doi:10.1142/S0218271811020573
6. Dadhich, N.: On Lovelock vacuum solution. Math. Today **26**, 37 (2011)
7. Dadhich, N., Pons, J., Prabhu, K.: On the static Lovelock black holes, ArXiv e-prints arXiv:1201.4994 [gr-qc] (2012)
8. Dadhich, N., Pons, J., Prabhu, K.: Thermodynamical universality of the Lovelock black holes. Gen. Relativ. Gravit. **44**, 2595 (2012). doi:10.1007/s10714-012-1416-6
9. Dadhich, N., Molina, A., Khugaev, A.: Uniform density static fluid sphere in Einstein-Gauss-Bonnet gravity and its universality. Phys. Rev. D **81**, 104026 (2010). doi:10.1103/PhysRevD.81.104026
10. Bañados, M., Teitelboim, C., Zanelli, J.: Black hole in three-dimensional spacetime. Phys. Rev. Lett. **69**, 1849 (1992). doi:10.1103/PhysRevLett.69.1849
11. Dadhich, N., Pons, J.: On universality of the pure Lovelock gravity for the generalized Nariai and Bertotti-Robinson solutions, ArXiv e-prints arXiv:1210.1109 [gr-qc] (2012)
12. Dadhich, N.: On the Gauss-Bonnet Gravity, ArXiv e-prints arXiv:hep-th/0509126 (2005)

Geometric Inequalities for Black Holes

Sergio Dain

Abstract It is well known that the parameters that characterize the Kerr black hole satisfy several important geometric inequalities. Remarkably enough, some of these inequalities also hold for dynamical black holes. This kind of inequalities play an important role in the characterization of the gravitational collapse, they are closely related with the cosmic censorship conjecture. We briefly review recent results in this subject.

The Kerr black hole is characterized by two parameters, the mass m and the angular momentum J. These parameters satisfy the following inequality

$$m \geq \sqrt{|J|}. \tag{1}$$

It is important to emphasize that the Kerr metric is a solution of Einstein vacuum equation for any choice of m and J, but it only describes a black hole if (1) holds. From Newtonian considerations, we can interpret this inequality as follows (see [1]): in a collapse the gravitational attraction ($\approx m^2/r^2$) at the horizon ($r \approx m$) dominates over the centrifugal repulsive forces ($\approx J^2/mr^3$).

Black holes are very simple macroscopic objects. The black hole uniqueness theorem ensures that Kerr is the only stationary black hole in vacuum, and hence stationary black holes are characterized by the two parameters, m and J. However, black holes are not stationary in general. Astrophysical phenomena like the formation of a black hole by gravitational collapse or a binary black hole collision are highly dynamical. For such systems, the black hole cannot be characterized by few parameters as in the stationary case.

S. Dain (✉)
Facultad de Matemática, Astronomía y Física, FaMAF, Universidad Nacional de Córdoba,
Instituto de Física Enrique Gaviola, IFEG, CONICET, Ciudad Universitaria,
(5000) Córdoba, Argentina
e-mail: dain@famaf.unc.edu.ar

J. Bičák and T. Ledvinka (eds.), *Relativity and Gravitation*,
Springer Proceedings in Physics 157, DOI: 10.1007/978-3-319-06761-2_7,

Remarkably, inequality (1) extends (under appropriate assumptions) to fully dynamical axially symmetric black holes. Moreover, this inequality is deeply connected with properties of the global evolution of Einstein equations, in particular with the cosmic censorship conjecture.

Inequality (1) is a global inequality for two reasons. First, it involves the total mass m of the spacetime. Second it assumes global restrictions on the initial data: axial symmetry and vacuum.

The area A and the angular momentum J in axial symmetry are quasi-local quantities. Namely they carry information on a bounded region of the spacetime. In contrast with a local quantity like a tensor field which depends on a point of the spacetime or a global quantities (like the total mass) which depends on the whole initial conditions. The area of the horizon of the Kerr black hole satisfies the following inequality

$$A \geq 8\pi |J|. \tag{2}$$

A natural question is whether dynamical black holes satisfy purely quasi-local inequalities. The relevance of this kind of inequalities is that they provide a much finer control on the dynamics of black holes than the global versions. It turns out that inequality (2) also holds for dynamical, axially symmetric black holes.

The equality in (1) and (2) is achieved only for the extreme Kerr black hole. In the dynamical regime, this rigidity also holds. Extreme black holes lie in the boundary between black holes and naked singularities. They play an important role as minimizer in the corresponding energy in the proof of these inequalities.

Both inequalities (1) and (2) can be extended to include an electric charge. The quasi-local inequality (2) is also valid when generic matter sources are present at the horizon of black holes. For more details see the recent review article [2] on the whole subject.

References

1. Wald, R.M.: Final states of gravitational collapse. Phys. Rev. Lett. **26**, 1653 (1971). doi:10.1103/PhysRevLett.26.1653
2. Dain, S.: Geometric inequalities for axially symmetric black holes. Class. Quantum Grav. **29**, 073001 (2012). doi:10.1088/0264-9381/29/7/073001

Scalar Fields on Anti-de Sitter Background

Gyula Fodor, Péter Forgács and Philippe Grandclément

Abstract The study of scalar fields coupled to gravity when there is a negative cosmological constant gives important insight on the possible instability of anti-de Sitter spacetime. In this short paper we consider the question how different the scalar field evolution is when the background is a fixed AdS metric. It is known that self-interacting massive real scalar fields on flat Minkowski background can form long living oscillating localized objects, named oscillons. In the flat background case these objects radiate energy extremely slowly, in a rate which is exponentially suppressed in terms of the central amplitude. However, on AdS background there are localized exactly time-periodic non-radiating solutions.

1 Introduction

In a recent influential paper Bizoń and Rostworowski [1] studied a real scalar field coupled to gravity when there is a negative cosmological constant. In this case the geometry approaches asymptotically the anti-de Sitter spacetime. They considered the time evolution of a spherically symmetric massless scalar field, and observed that the energy is continuously shifted to small wavelength high frequency modes. This phenomenon is generally called weak turbulence in the literature. The shift of energy to high frequency modes continues until a black hole forms at the symmetry center.

G. Fodor (✉) · P. Forgács
Institute for Particle and Nuclear Physics, Wigner Research Centre for Physics, Budapest 114, Budapest, P.O.Box 49, 1525 Hungary
e-mail: fodor.gyula@wigner.mta.hu

P. Forgács
LMPT, CNRS-UMR 6083, Université de Tours, Parc de Grandmont, 37200 Tours France

P. Grandclément
LUTH, CNRS-UMR 8102, Observatoire de Paris-Meudon place Jules Janssen, 92195 Meudon Cedex France

J. Bičák and T. Ledvinka (eds.), *Relativity and Gravitation*,
Springer Proceedings in Physics 157, DOI: 10.1007/978-3-319-06761-2_8,

In this short paper we consider the question what changes when the background is not dynamical but a fixed AdS spacetime. For a Klein-Gordon field on a fixed background the field equations are linear and consequently there is no weak turbulence. For this reason we consider self-interacting scalar fields.

We consider $3+1$ dimensional anti-de Sitter spacetime in the conformal coordinate system

$$ds^2 = \frac{1}{k^2 \cos^2 x} \left(-d\tau^2 + dx^2 + \sin^2 x \, d\Omega^2 \right), \tag{1}$$

where $d\Omega$ is the metric of a unit two-sphere. In these coordinates $x = 0$ corresponds to the center of symmetry, and $x = \pi/2$ to infinity. All timelike geodesics emanating from a point meet again at another point. A light ray can travel to infinity and back in finite coordinate time, if we assume that infinity acts as a mirror for null rays. The behavior of geodesics indicates that the AdS background corresponds to an effective attractive force.

2 Evolution of a Scalar Field on AdS Background

A spherically symmetric self-interacting scalar field on $3 + 1$ dimensional AdS background evolves according to the field equation

$$-\phi_{,\tau\tau} + \phi_{,xx} + \frac{4}{\sin(2x)} \phi_{,x} = \frac{U'(\phi)}{k^2 \cos^2 x}, \tag{2}$$

where k is related to the cosmological constant by $\Lambda = -3k^2$, and $U(\phi)$ is the potential describing the self-interaction of the scalar field. In order to show that weak turbulence is likely to occur even in this simpler system, let us consider a specific example. We choose $k = 1$ and the scalar potential as $U(\phi) = \frac{1}{2}\phi^2 - \frac{1}{4}\phi^4 + \frac{1}{6}\phi^6$. For initial data we take a finite width spherically symmetric shell, for which

$$\phi = c \exp \frac{b^2 d}{(x-a)^2 - b^2} \tag{3}$$

for $|x - a| < b$ and $\phi = 0$ otherwise. For the concrete example that we present here we have chosen the constants as $a = 0.4$, $b = 0.2$, $c = 100$ and $d = 4$. At the beginning the shell separates into ingoing and outgoing shells. The ingoing shell approaches the center and then becomes outgoing. Later both shells get reflected back from infinity in a finite time. It takes approximately $\pi/2$ time interval in the coordinate time τ for a shell to go from the center to infinity, or to come back, quite similarly to how null geodesics behave. On the top panel of Fig. 1 we show the time evolution of the scalar field ϕ at the center for the first few reflections. The bottom panel shows the evolution of the central value of the energy density,

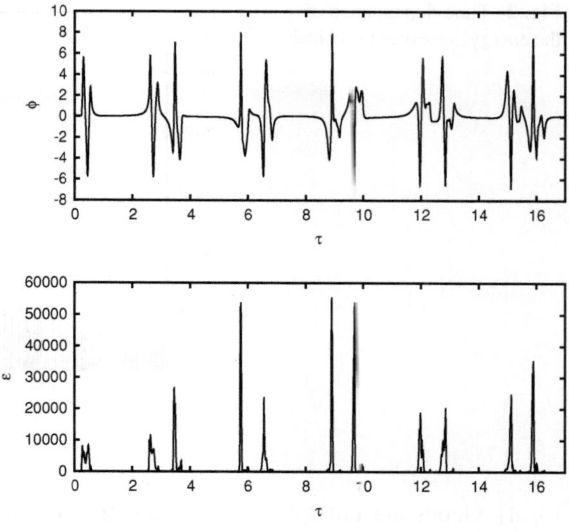

Fig. 1 Time dependence of the scalar field and its energy density at the center

$$\varepsilon = k^2 \cos x \left(\frac{1}{2} \left(\phi_{,\tau}\right)^2 + \frac{1}{2} \left(\phi_{,x}\right)^2 + \frac{U(\phi)}{k^2 \cos^2 x} \right). \tag{4}$$

On the energy density plot there are peaks when the shells come to the center. Looking at a much longer time interval, it can be seen that the amplitude of the peaks increases very quickly. On Fig. 2 we show the central energy density for a longer time interval, on a logarithmic plot. The energy density increases about five magnitudes by the time the shells are reflected about one hundred times. It seems very likely that there is week turbulence in this system. It would take further numerical work to study how the energy density increases for other types of initial data and for different choices of the scalar potential. For massless self-interacting fields, such as that with a potential $U(\phi) = \phi^4$, we could not observe a significant increase in the central energy density. For the massive case there is density increase, but we could not see any simple scale invariance property of the time evolution depending on the amplitude of the initial data, which has been, however, observed for the self-gravitating massless Klein-Gordon case in [1].

3 Periodic Solutions

In a recent paper of Dias et al. [2], vacuum spacetimes have been considered when there is a negative cosmological constant. Using perturbation theory the existence of resonant modes has been shown, which indicates weak turbulence. It has been also claimed that the nonlinear generalization of a single perturbative mode is a localized periodic vacuum solution, which is generally called geon in the literature

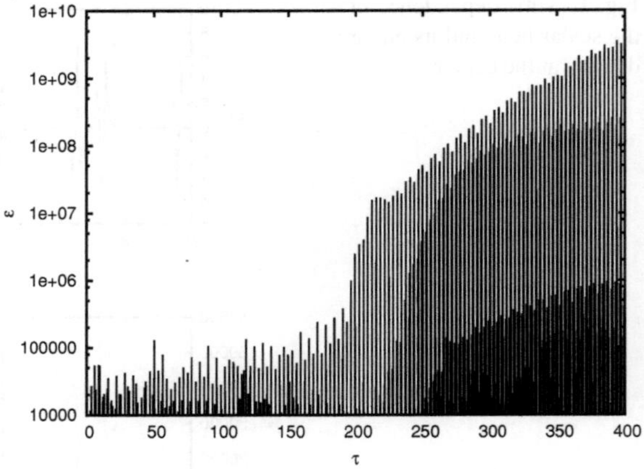

Fig. 2 Time dependence of the energy density at the center

[3, 4]. Geons are not spherically symmetric, but if one includes a scalar field in the system, then spherically symmetric localized periodic solutions are expected to exist, which are called oscillatons in the asymptotically flat case [5, 6]. Oscillatons are similar to boson stars [7], but for boson stars the scalar field is complex, and the metric is static. Similar periodic localized solutions already exist for scalar fields on a fixed AdS background. On flat Minkowski background those objects are known as breathers or oscillons [8].

For the case of a massive or massless Klein-Gordon field on AdS background the periodic solutions are explicitly known. In this case the scalar potential is $U(\phi) = \frac{1}{2}m^2\phi^2$, and there is a family of breather solutions [9] labeled by a non-negative integer n, which gives the number of the nodes of the solution

$$\phi^{(n)} = \cos[(\mu + 2n)\tau]\,(\cos x)^{\mu}\, P_n^{(1/2,\mu-3/2)}(\cos(2x))\,, \tag{5}$$

where

$$\mu = \frac{3}{2} + \sqrt{\frac{9}{4} + \frac{m^2}{k^2}}\,, \tag{6}$$

and $P_n^{(a,b)}(x)$ denotes the Jacobi polynomial. All finite energy solutions can be expressed as sums of $\phi^{(n)}$ with appropriate phases.

Self-interacting scalar fields with any potential $U(\phi)$ also admit periodic localized solutions on AdS background. We can look for solutions oscillating with frequency ω by Fourier decomposing the scalar field in the form

$$\phi = \sum_{n=0}^{N} \phi_n \cos(n\,\omega\,\tau)\,, \tag{7}$$

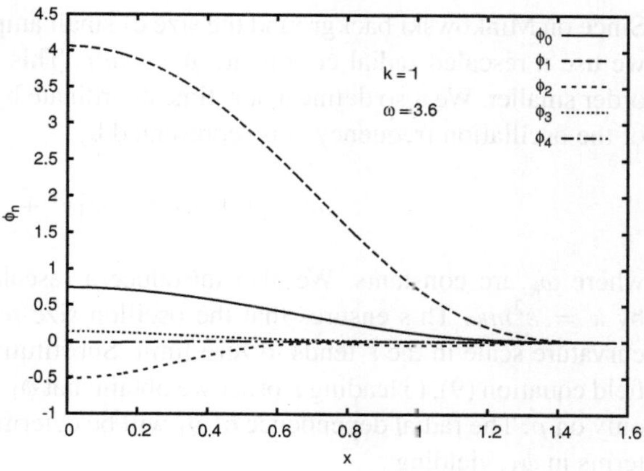

Fig. 3 The first few Fourier modes of a periodic localized solution in case of a ϕ^4 potential

where ϕ_n are functions of x and the system is truncated at order N. We have solved the resulting system of ordinary differential equations using the spectral code Kadath developed by Philippe Grandclément [10]. On Fig. 3 the Fourier modes of an example configuration are given for the standard ϕ^4 potential $U(\phi) = \phi^2(\phi - 2)^2/8$, when $k = 1$ and $\omega = 3.6$. Unlike in the linear Klein-Gordon case, the frequency ω changes when the oscillation amplitude grows. In the limit of small oscillations ω tends to the Klein-Gordon value $\omega_0 = \mu = (3 + \sqrt{13})/2 \approx 3.30278$ given by (6).

A small-amplitude expansion procedure has been successfully applied in the past for oscillons [8], asymptotically flat oscillatons [11], and also for oscillatons when there is a small positive cosmological constant [12]. In this paper we present the small-amplitude expansion of oscillons on a fixed AdS background when $|\Lambda|$ is small. We use Schwarzschild area coordinates,

$$ds^2 = -(1 + k^2r^2)dt^2 + \frac{dr^2}{1 + k^2r^2} + r^2 d\Omega^2 . \tag{8}$$

Then the field equation takes the form

$$-\frac{1}{1 + k^2r^2}\phi_{,tt} + (1 + k^2r^2)\phi_{,rr} + \left[\frac{D-1}{r} + (D+1)k^2r\right]\phi_{,r} = U'(\phi) . \tag{9}$$

We describe the scalar potential by its expansion coefficients g_k,

$$U(\phi) = m^2\left(\frac{1}{2}\phi^2 + \frac{g_2}{3}\phi^3 + \frac{g_3}{4}\phi^4 + \dots\right), \tag{10}$$

where m is the mass of the scalar field. We expand the scalar field in powers of a small parameter ε

$$\phi = \varepsilon\phi_1 + \varepsilon^2\phi_2 + \varepsilon^3\phi_3 + \dots . \tag{11}$$

Since on Minkowski background the size of small amplitude oscillons scales as $1/\varepsilon$, we use a rescaled radial coordinate $\rho = \varepsilon m r$. This makes spatial derivatives one order smaller. We also define a new time coordinate by $\tau = m\omega t$. The ε dependence of the oscillation frequency ω is represented by

$$\omega^2 = 1 + \omega_2 \varepsilon^2 + \omega_4 \varepsilon^4 + \ldots , \tag{12}$$

where ω_k are constants. We also introduce a rescaled cosmological parameter κ by $k = \varepsilon^2 m \kappa$. This ensures that the oscillon size remains small compared to the curvature scale in the ε tends to zero limit. Substituting the expansion (11) into the field equation (9), to leading ε order we obtain that $\phi_1 = p_1 \cos \tau$, where p_1 depends only on ρ. The radial dependence of p_1 will be determined by the absence of secular terms in ϕ_3, yielding

$$p_{1,\rho\rho} + \frac{D-1}{\rho} p_{1,\rho} + \left(\omega_2 - \rho^2 \kappa^2\right) p_1 + \lambda p_1^3 = 0 , \tag{13}$$

where $\lambda = \frac{5}{6} g_2^2 - \frac{3}{4} g_3$. This gives the spatial profile of the oscillon to leading order. For Minkowski background $\kappa = 0$, and localized solutions can exist only if λ is positive. If $\lambda > 0$ we may consider the potential as an "attractive potential". In this case the potential is more flat near its minimum than the same mass harmonic potential, and the oscillation period becomes longer. For anti de Sitter background we can rescale ρ, and consequently ε in $\rho = \varepsilon m r$, in order to set

$$\kappa = 1 , \tag{14}$$

which we assume from now on.

If $\lambda = 0$, then (13) is linear, and there are localized solutions only if $\omega_2 = 3 + 4n$, for $n \geq 0$ integer. The solutions can be written in terms of generalized Laguerre polynomials,

$$p_1^{(n)} = \exp\left(-\frac{\rho^2}{2}\right) L_n^{1/2}(\rho^2) . \tag{15}$$

The integer n gives the number of nodes. These solutions correspond to the small k limit of the Klein-Gordon breathers given earlier by (5).

If λ is nonzero, then defining $S = p_1/\sqrt{|\lambda|}$, (13) can be written as

$$S_{,\rho\rho} + \frac{D-1}{\rho} S_{,\rho} + (\omega_2 - \rho^2) S \pm S^3 = 0 , \tag{16}$$

where the positive sign is valid for $\lambda > 0$ and the negative for $\lambda < 0$. The solutions are labeled by the single parameter ω_2. If $\lambda > 0$, there are localized nodeless solutions for any $\omega_2 < 3$. In this case high amplitude solutions are more localized than small amplitude ones because of the attraction represented by the scalar potential. The radial profiles of a few such solutions are shown on the upper panel of Fig. 4. For

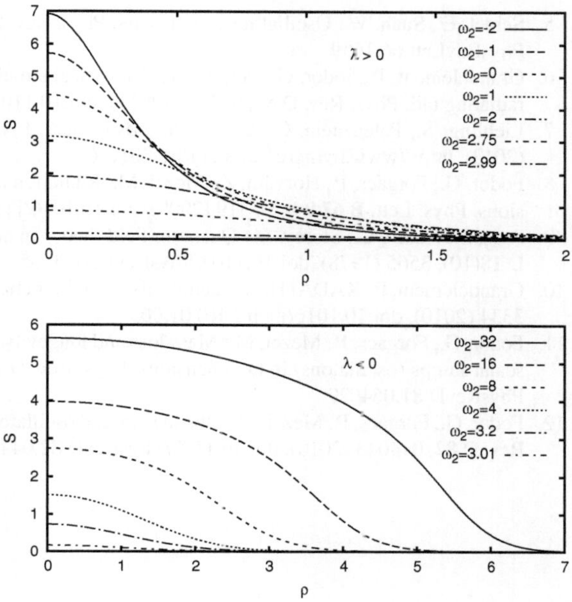

Fig. 4 Radial behavior of solutions of (16) for $\lambda > 0$ attractive, and $\lambda < 0$ repulsive potentials

$\lambda < 0$ localized nodeless solutions exist for any $\omega_2 > 3$. Then higher amplitude solutions have larger size, and it is natural to call potentials with $\lambda < 0$ "repulsive potentials". Corresponding solutions are shown on the lower panel of Fig. 4. Since by the choice $\kappa = 1$ we have $\varepsilon^2 = k/m$, using (11) and (12) the leading order behavior of the scalar field can be written as

$$\phi = \sqrt{\frac{k|\lambda|}{m}}\, S \cos\left(m\sqrt{1+\frac{k}{m}\omega_2}\, t\right), \tag{17}$$

where S depends on $\rho = r\sqrt{km}$ according to (16).

Acknowledgments This research has been supported by OTKA grant K 101709.

References

1. Bizoń, P., Rostworowski, A.: Weakly turbulent instability of anti-de Sitter spacetime. Phys. Rev. Lett. **107**, 031102 (2011). doi:10.1103/PhysRevLett.107.031102
2. Dias, Ó., Horowitz, G., Santos, J.: Gravitational turbulent instability of anti-de Sitter space. Class. Quantum Gravity **29**(19), 194002 (2012). doi:10.1088/0264-9381/29/19/194002
3. Wheeler, J.: Geons. Phys. Rev. **97**, 511 (1955). doi:10.1103/PhysRev.97.511
4. Brill, D., Hartle, J.: Method of the self-consistent field in general relativity and its application to the gravitational geon. Phys. Rev. **135**, B271 (1964). doi:10.1103/PhysRev.135.B271

5. Seidel, E., Suen, W.: Oscillating soliton stars. Phys. Rev. Lett. **66**, 1659 (1991). doi:10.1103/PhysRevLett.66.1659
6. Grandclément, P., Fodor, G., Forgács, P.: Numerical simulation of oscillatons: Extracting the radiating tail. Phys. Rev. D **84**, 065037 (2011). doi:10.1103/PhysRevD.84.065037
7. Liebling, S., Palenzuela, C.: Dynamical boson stars. Living Rev. Relativ. **15**(6), lrr-2012-6 (2012). http://www.livingreviews.org/lrr-2012-6
8. Fodor, G., Forgács, P., Horváth, Z., Mezei, M.: Radiation of scalar oscillons in 2 and 3 dimensions. Phys. Lett. B **674**(4–5), 319 (2009). doi:10.1016/j.physletb.2009.03.054
9. Avis, S., Isham, C., Storey, D.: Quantum field theory in anti-de Sitter space-time. Phys. Rev. D **18**(10), 3565 (1978). doi:10.1103/PhysRevD.18.3565
10. Grandclément, P.: KADATH: A spectral solver for theoretical physics. J. Comput. Phys. **229**(9), 3334 (2010). doi:10.1016/j.jcp.2010.01.005
11. Fodor, G., Forgács, P., Mezei, M.: Mass loss and longevity of gravitationally bound oscillating scalar lumps (oscillatons) in D dimensions. Phys. Rev. D **81**(6), 064029 (2010). doi:10.1103/PhysRevD.81.064029
12. Fodor, G., Forgács, P., Mezei, M.: Boson stars and oscillatons in an inflationary universe. Phys. Rev. D **82**, 044043 (2010). doi:10.1103/PhysRevD.82.044043

Canonical Superenergy Tensors in General Relativity: A Reappraisal

Janusz Garecki

Abstract We discuss the role of the canonical superenergy tensors.

1 Introduction

In the framework of general relativity (**GR**), as a consequence of the Einstein Equivalence Principle (**EEP**), the gravitational field *has non-tensorial strengths* $\Gamma^i_{kl} = \{^{\,i}_{kl}\}$ and *admits no energy-momentum tensor*. One can only attribute to this field *gravitational energy-momentum pseudotensors*. The leading object of such a kind is the *canonical gravitational energy-momentum pseudotensor* ${}_Et_i^{\ k}$ proposed already in past by Einstein. This pseudotensor is a part of the *canonical energy-momentum complex* ${}_EK_i^{\ k}$ in **GR**.

The canonical complex ${}_EK_i^{\ k}$ can easily be obtained by rewriting Einstein equations to the superpotential form

$$ {}_EK_i^{\ k} := \sqrt{|g|}\left(T_i^{\ k} + {}_Et_i^{\ k}\right) = {}_FU_i^{\ [kl]}{}_{,l}\,, \tag{1}$$

where $T^{ik} = T^{ki}$ is the symmetric energy-momentum tensor for matter, $g = det[g_{ik}]$, and

J. Garecki (✉)
Institute of Mathematics University of Szczecin and Cosmology Group University of Szczecin, Wielkopolska 15, 70-451 Szczecin, Poland
e-mail: garecki@wmf.univ.szczecin.pl

J. Bičák and T. Ledvinka (eds.), *Relativity and Gravitation*,
Springer Proceedings in Physics 157, DOI: 10.1007/978-3-319-06761-2_9,

$$
\begin{aligned}
{}_E t_i{}^k = \frac{c^4}{16\pi G}\{&\delta_i^k g^{ms}(\Gamma^l_{mr}\Gamma^r_{sl} - \Gamma^r_{ms}\Gamma^l_{rl}) \\
&+ g^{ms}{}_{,i}[\Gamma^k_{ms} - \frac{1}{2}(\Gamma^k_{tp}g^{tp} - \Gamma^l_{tl}g^{kt})g_{ms} \\
&- \frac{1}{2}(\delta^k_s\Gamma^l_{ml} + \delta^k_m\Gamma^l_{sl})]\};
\end{aligned}
$$

$$
{}_F U_i{}^{[kl]} = \frac{c^4}{16\pi G}\frac{g_{ia}}{\sqrt{|g|}}[(-g)(g^{ka}g^{lb} - g^{la}g^{kb})]_{,b}. \tag{2}
$$

${}_E t_i{}^k$ are components of the canonical energy-momentum pseudotensor for gravitational field, and ${}_F U_i{}^{[kl]}$ is von Freud superpotential.

$$
{}_E K_i{}^k = \sqrt{|g|}(T_i{}^k + {}_E t_i{}^k) \tag{3}
$$

are components of the *Einstein canonical energy-momentum complex for matter and gravity*.

In consequence of (1) the complex ${}_E K_i{}^k$ satisfies local conservation laws

$$
{}_E K_i{}^k{}_{,k} \equiv 0. \tag{4}
$$

In very special cases one can obtain reasonable integral conservation laws from these local conservation laws. Additionally, one can also introduce *the canonical superenergy tensors*. This was done in past in a series of our articles (see, e.g., [1–11] and references therein).

It appears that the idea of the superenergy tensors is universal: to any physical field having an energy-momentum tensor or pseudotensor one can attribute a corresponding superenergy tensor.

2 The Canonical Superenergy Tensors

Here we give a short description of the general, constructive definition of the superenergy tensor $S_a{}^b$ applicable to gravitational field and to any matter field. The definition uses *locally Minkowskian structure* of the spacetime and, therefore, it fails in a spacetime with torsion, e.g., in a Riemann-Cartan spacetime.

In the normal Riemann coordinates **NRC(P)** we define (pointwise)

$$
S_{(a)}^{(b)}(P) = S_a{}^b := (-)\lim_{\Omega\to P}\frac{\int\limits_{\Omega}\left[T^{(b)}_{(a)}(y) - T^{(b)}_{(a)}(P)\right]d\Omega}{1/2\int\limits_{\Omega}\sigma(P;y)d\Omega}, \tag{5}
$$

where

$$T_{(a)}^{(b)}(y) := T_i^{\ k}(y)e^i_{\ (a)}(y)e_k^{(b)}(y),$$
$$T_{(a)}^{(b)}(P) := T_i^{\ k}(P)e^i_{\ (a)}(P)e_k^{\ (b)}(P) = T_a^{\ b}(P)$$

are *physical or tetrad components* of the pseudotensor or tensor field which describes an energy-momentum distribution, and $\{y^i\}$ are normal coordinates. $e^i_{\ (a)}(y), e_k^{\ (b)}(y)$ denote an orthonormal tetrad $e^i_{\ (a)}(P) = \delta^i_a$ and its dual $e_k^{\ (a)}(P) = \delta^a_k$, parallelly propagated along geodesics through P (P is the origin of the **NRC(P)**).

We have

$$e^i_{\ (a)}(y)e_i^{\ (b)}(y) = \delta^b_a. \tag{6}$$

For a sufficiently small 4-dimensional domain Ω which surrounds **P** we require

$$\int_\Omega y^i d\Omega = 0, \quad \int_\Omega y^i y^k d\Omega = \delta^{ik} M, \tag{7}$$

where

$$M = \int_\Omega (y^0)^2 d\Omega = \int_\Omega (y^1)^2 d\Omega = \int_\Omega (y^2)^2 d\Omega = \int_\Omega (y^3)^2 d\Omega \tag{8}$$

is a common value of the moments of inertia of the domain Ω with respect to the subspaces $y^i = 0, \quad (i = 0, 1, 2, 3)$.

As Ω we can take, e.g., a sufficiently small ball centered at P:

$$(y^0)^2 + (y^1)^2 + (y^2)^2 + (y^3)^2 \le R^2, \tag{9}$$

which for an auxiliary positive-definite metric

$$h^{ik} := 2v^i v^k - g^{ik}, \tag{10}$$

can be written in the form

$$h_{ik} y^i y^k \le R^2. \tag{11}$$

A fiducial observer **O** is at rest at the origin **P** of the Riemann normal coordinates **NRC(P)** and its four- velocity is $v^i = * \ \delta^i_o$, where $*$ means that equation is valid only in special coordinates. In [3] $\sigma(P; y)$ denotes the two-point *world function* introduced by J.L. Synge [12]:

$$\sigma(P; y) = *\frac{1}{2}\left(y^{o^2} - y^{1^2} - y^{2^2} - y^{3^2}\right). \tag{12}$$

The world function $\sigma(P; y)$ can be defined covariantly by the *eikonal-like* expression [12]

$$g^{ik}\sigma_{,i}\sigma_{,k} = 2\sigma, \qquad \sigma_{,i} := \partial_i \sigma, \tag{13}$$

together with requirements

$$\sigma(P; P) = 0, \qquad \partial_i \sigma(P; P) = 0. \tag{14}$$

The ball Ω can also be given by the inequality

$$h^{ik}\sigma_{,i}\sigma_{,k} \le R^2. \tag{15}$$

Tetrad components and normal components are equal at **P**, so, we will write the components of any quantity attached to **P** without tetrad brackets, e.g., we will write $S_a{}^b(P)$ instead of $S_{(a)}^{(b)}(P)$ and so on.

If $T_i{}^k(y)$ are the components of an energy-momentum tensor of matter, then we get from (5)

$${}_mS_a{}^b(P; v^l) = \left(2\hat{v}^l\hat{v}^m - \hat{g}^{lm}\right)\nabla_l\nabla_m\hat{T}_a{}^b = \hat{h}^{lm}\nabla_l\nabla_m\hat{T}_a{}^b. \tag{16}$$

Hat over a quantity denotes its value at **P**, and ∇ means covariant derivative.

Tensor ${}_mS_a{}^b(P; v^l)$ is called *the canonical superenergy tensor for matter*.

For the gravitational field, substitution of the canonical Einstein energy-momentum pseudotensor as $T_i{}^k$ in (5) gives

$${}_gS_a{}^b(P; v^l) = \hat{h}^{lm}\hat{W}_a{}^b{}_{lm}, \tag{17}$$

where

$$\begin{aligned} W_a{}^b{}_{lm} = \frac{2\alpha}{9}\Big[&B^b{}_{alm} + P^b{}_{alm} \\ &- \frac{1}{2}\delta^b_a R^{ijk}{}_m\left(R_{ijkl} + R_{ikjl}\right) + 2\delta^b_a\beta^2 E_{(l|g}E^g{}_{|m)} \\ &- 3\beta^2 E_{a(l|}E^b{}_{|m)} + 2\beta R^b{}_{(a|g|l)}E^g{}_m\Big]. \end{aligned}$$

Here $\alpha = \frac{c^4}{16\pi G} = \frac{1}{2\beta}$, and

$$E_i{}^k := T_i{}^k - \frac{1}{2}\delta_i^k T \tag{18}$$

is the modified energy-momentum tensor of matter.[1]

[1] In terms of $E_i{}^k$ Einstein equations read $R_i{}^k = \beta E_i{}^k$.

On the other hand

$$B^b{}_{alm} := 2R^{bik}{}_{(l|}R_{aik|m)} - \frac{1}{2}\delta^b_a R^{ijk}{}_l R_{ijkm} \tag{19}$$

are the components of the *Bel-Robinson tensor* (**BRT**), while

$$P^b{}_{alm} := 2R^{bik}{}_{(l|}R_{aki|m)} - \frac{1}{2}\delta^b_a R^{jik}{}_l R_{jkim} \tag{20}$$

is the Bel-Robinson tensor with "transposed" indices (ik).

In vacuum ${}_gS_a{}^b(P; v^l)$ takes the simpler form

$${}_gS_a{}^b(P; v^l) = \frac{8\alpha}{9}\hat{h}^{lm}\big(\hat{C}^{bik}{}_{(l|}\hat{C}_{aik|m)} - \frac{1}{2}\delta^b_a \hat{C}^{i(kp)}{}_{(l|}\hat{C}_{ikp|m)}\big). \tag{21}$$

Here $C^a{}_{blm}$ denotes components of the *Weyl tensor*.

Some remarks:

1. In vacuum the quadratic form ${}_gS_a{}^b v^a v_b$, where $v^a v_a = 1$, *is positive-definite*. This form gives the gravitational *superenergy density* ε_g for a fiducial observer **O**.
2. In general, the canonical superenergy tensors are uniquely determined only along the world line of an observer **O**. But in special cases, e.g., in Schwarzschild spacetime or in Friedmann universes, when there exists a physically and geometrically distinguished four-velocity field $v^i(x)$, one can introduce, in a unique way, unambiguous fields ${}_gS_i{}^k(x; v^l)$ and ${}_mS_i{}^k(x; v^l)$.
3. It can be shown that the superenergy densities ε_g, ε_m, which have dimension $\frac{Joule}{(meter)^5}$, exactly correspond to the Appel's *energy of acceleration* $\frac{1}{2}\mathbf{aa}$. The Appel's energy of acceleration plays the fundamental role in Appel's approach to classical mechanics [13–15].
4. We have proposed, in our previous papers, to use the tensor ${}_gS_i{}^k(P; v^l)$ as gravitational energy-momentum tensor.
5. We have used the canonical superenergy tensors ${}_gS_i{}^k$ and ${}_mS_i{}^k$ to local (and also to global) analysis of some well-known solutions to the Einstein equations like Schwarzschild, Kerr, Friedmann, Gödel, Kasner, Bianchi I, de Sitter and anti-de Sitter solutions. The obtained results were interesting (see [1–5, 7, 8, 11]), e.g., in Gödel universes the sign of the superenergy density $\varepsilon_s := \varepsilon_g + \varepsilon_m$ depends on causality ($\varepsilon_s < 0$) and non-causality ($\varepsilon_s > 0$), and, in Schwarzschild spacetime the integral exterior superenergy S is connected with Hawking temperature T of the Schwarzschild black hole: $S = \frac{8\pi k c^3}{9\hbar G}T$. We have also studied the transformation rules for the canonical superenergy tensors under conformal rescaling of the metric $g_{ik}(x)$ [8, 16].
6. The idea of the superenergy tensors can be extended to angular momentum (see, [3, 10]). The angular supermomentum tensors *do not depend* on a radius vector and, in gravitational case, they depend only on "spinorial part" of the suitable gravitational angular momentum pseudotensor.

7. As a result of averaging, the tensors ${}_gS_a^{\ b}(P;v^l)$ and ${}_mS_a^{\ b}(P;v^l)$, in general, do not satisfy any local conservation laws. Only in a symmetric spacetime or in a spacetime which has constant curvature one can get

$$\left[{}_gS_a^{\ b}(P;v^l)\right]_{,b} = 0. \tag{22}$$

8. There exists an exchange of the canonical superenergy between gravity and matter in the following sense. Let us consider the consequence of (4)

$$\left(\Delta^{(4)}_E K_i^{\ k}\right)_{,k} = \left[\left(\Delta^{(4)}(\sqrt{|g|}_E t_i^{\ k}\right) + \Delta^{(4)}\left(\sqrt{|g|}T_i^{\ k}\right)\right]_{,k} = 0, \tag{23}$$

where $\Delta^{(4)} := (\partial_0)^2 + (\partial_1)^2 + (\partial_2)^2 + (\partial_3)^2$.
The quantities (with total balance equal to zero)

$$\Delta^{(4)}\left(\sqrt{|g|}_e t_i^{\ k}\right),\ \ \Delta^{(4)}\left(\sqrt{|g|}T_i^{\ k}\right) \tag{24}$$

have dimensions of the canonical superenergy and, when taken at the origin **P** of the **NRC(P)** and written covariantly, then they coincide with the canonical superenergy tensors ${}_gS_i^{\ k}(P;v^l),\ {}_mS_i^{\ k}(P;v^l)$ respectively.
9. Recently we have noticed that the total superenergy density is positive-definite or null for known stable solutions to the Einstein equations and negative-definite for unstable solutions. The physical meaning of this fact is under study.

Changing the constructive definition (5) to the form

$$< T_a^{\ b}(P) > := \lim_{\varepsilon \to 0} \frac{\int\limits_{\Omega} \left[T_{(a)}^{\ \ (b)}(y) - T_{(a)}^{\ \ (b)}(P)\right] d\Omega}{\varepsilon^2/2 \int\limits_{\Omega} d\Omega}, \tag{25}$$

where $\varepsilon := \frac{R}{L} > 0$ (equivalently $R = \varepsilon L$) is a real parameter and L is a dimensional constant:$[L] = m$, one obtains *the averaged relative energy-momentum tensors*. Namely, from (25) one obtains:

$$<_m T_a^{\ b}(P;v^l) >=_m S_a^{\ b}(P;v^l)\frac{L^2}{6} \tag{26}$$

for matter and

$$<_g t_a^{\ b}(P;v^l) >=_g S_a^{\ b}(P;v^l)\frac{L^2}{6} \tag{27}$$

for gravity.

The components of the averaged relative energy-momentum tensors have correct dimensions but they depend on a dimensional parameter L. How to choose the dimensional parameter L?

In [7] we have proposed a universal choice of the parameter L. Namely, we have proposed $L = 100L_P \approx 10^{-33}$ m. Here $L_P := \sqrt{\frac{\hbar G}{c^3}} \approx 10^{-35}$ m is the Planck length.

Such choice of L gives the averaged gravitational relative energy-momentum tensors components which are negligible in comparison with the components of the energy-momentum tensor for matter. Consequently, with such choice of the parameter L, these tensors play no role in evolution of the material objects and in evolution of the Universe. On the other hand, the choices can be made such that $L = \frac{2GM}{c^2}$ for a closed system with mass M, $L = \lambda$ for a gravitational wave of wave length λ, and $L = \frac{2GM_U}{c^2} = \frac{c}{H_0} = ct_0$ in cosmology, which lead us to the averaged relative energy densities of the same order as ordinary energy density of matter. Here M_U, H_0, t_0 mean mass of the observed part of the Universe, actual value of the Hubble constant and an approximate age of the Universe respectively.

Of course, there exist other possibilities of choosing the length parameter L.

3 Conclusion

On the *superenergy level*, or on the *averaged relative energy-momentum* level, there seems to be no problem with a suitable expression for energy. However, canonical superenergy tensors seem more fundamental than the corresponding averaged relative energy-momentum tensors because they do not depend on the choice of L.

Acknowledgments This paper was mainly supported by Polish Ministry of Science and Higher Education Grant No 505-4000-25-0976 (years 2011–2013). Author also would like to thank Professor Jiří Bičák for possibility to deliver talk during the Conference "Relativity and Gravitation – 100 years after Einstein in Prague".

References

1. Garecki, J.: Energy and superenergy of a closed universe. Rep. Math. Phys. **33**, 57 (1993). doi:10.1016/0034-4877(93)90040-L
2. Garecki, J.: Energy-momentum and angular momentum of isotropic homogeneous cosmological models in a gauge theory of gravity. Int. J. Theor. Phys. **34**, 2307 (1995). doi:10.1007/BF00673845
3. Garecki, J.: Canonical angular supermomentum tensors in general relativity. J. Math. Phys. **40**, 4035 (1999). doi:10.1063/1.532941
4. Garecki, J.: Do gravitational waves carry energy-momentum and angular momentum? Ann. Phys. (Berlin) **11**, 442 (2002). doi:10.1002/1521-3889(200206)11:6<442::AID-ANDP442>3.0.CO;2-A
5. Dąbrowski, M., Garecki, J.: Superenergy and supermomentum of Gödel universes. Class. Quantum Gravity **19**, 1 (2002). doi:10.1088/0264-9381/19/1/301
6. Garecki, J.: On the gravitational energy of the Bonnor spacetime. Class. Quantum Gravity **22**, 4051 (2005). doi:10.1088/0264-9381/22/19/015

7. Garecki, J.: The tensors of the averaged relative energy momentum and angular momentum in general relativity and some of their applications. Found. Phys. **37**, 341 (2007). doi:10.1007/s10701-007-9107-y
8. Dąbrowski, M., Garecki, J.: Statefinders and observational measurement of superenergy. Phys. Lett. B **686**, 6 (2010). doi:10.1016/j.physletb.2010.02.019
9. Garecki, J.: A superenergetic analysis of the plane and plane-fronted gravitational waves. Rep. Math. Phys. **40**, 485 (1997). doi:10.1016/S0034-4877(97)85897-1
10. Garecki, J.: Superenergy and angular supermomentum tensors in general relativity. Rep. Math. Phys. **44**, 95 (1999). doi:10.1016/S0034-4877(99)80149-9
11. Garecki, J.: Superenergy, conformal transformations, and Friedman universes. Ann. Phys. (Berlin) **19**, 263 (2010). doi:10.1002/andp.201010424
12. Synge, J.: Relativity: The General Theory. North-Holland, Amsterdam (1960)
13. Appel, P.: Sur une forme générale des équations de la dynamique. J. reine angew. Math. **1900**(121), 310 (1900). doi:10.1515/crll.1900.121.310
14. Appel, P.: Sur une forme générale des équations de la dynamique et sur le principe de gauss. J. reine angew. Math. **1900**(122), 205 (1900). doi:10.1515/crll.1900.122.205
15. Białkowski, G.: Classical Mechanics. PWN, Warsaw (1975). In Polish
16. Dąbrowski, M., Garecki, J., Blaschke, D.: Conformal transformations and conformal invariance in gravitation. Ann. Phys. (Berlin) **18**, 13 (2009). doi:10.1002/andp.200810331

Einstein's "Prague Field Equation" of 1912: Another Perspective

Domenico Giulini

Abstract I reconsider Einstein's 1912 "Prague-Theory" of static gravity based on a scalar field obeying a non-linear field equation. I point out that this equation follows from the self-consistent implementation of the principle that *all* forms of energy are the source of the gravitational field according to $E = mc^2$. This makes it an interesting toy-model for the "flat-space approach" to General Relativity (GR), as pioneered by Kraichnan and later Feynman. Solutions modelling stars show features familiar from GR, e.g., Buchdahl-like inequalities. The relation to full GR is also discussed. This lends this toy theory also some pedagogical significance.

1 Introduction

Ever since he wrote his large 1907 review of Special Relativity [1] for the *Jahrbuch der Radioaktivität und Elektronik*, Einstein reflected on how to extend the principle of relativity to non-inertial motions. His key insight was that such an extension is indeed possible, provided gravitational fields are included in the description. In fact, the last chapter (V) of [1], which comprises four (17-20) out of twenty sections, is devoted to this intimate relation between acceleration and gravitation. The heuristic principle Einstein used was his "Äquivalenzhypothese" (hypothesis of equivalence) or "Äquivalenzprinzip" (principle of equivalence),[1] which says this: Changing the description of a system from an inertial to a non-inertial reference frame is equivalent

[1] In his Prague papers Einstein gradually changed from the first to the second expression.

D. Giulini (✉)
Institute for Theoretical Physics, Riemann Center for Geometry and Physics,
Leibniz University of Hannover, Hannover, Germany
e-mail: giulini@itp-uni-hannover.de

D. Giulini
Center of Applied Space Technology and Microgravity, University of Bremen,
Bremen, Germany

J. Bičák and T. Ledvinka (eds.), *Relativity and Gravitation*,
Springer Proceedings in Physics 157, DOI: 10.1007/978-3-319-06761-2_10,

to not changing the frame at all but adding a *special* gravitational field. This principle is *heuristic* in the sense that it allows to deduce the extension of physical laws, the forms of which are assumed to be known in the absence of gravitational fields, to the presence of at least those special gravitational fields that can be "created" by mere changes of reference frames. The idea behind this was, of course, to postulate that the general features found in this fashion remain valid in *all* gravitational fields. In the 1907 review Einstein used this strategy to find out about the influence gravitational fields have on clocks and general electromagnetic processes. What he did not attempt back in 1907 was to find an appropriate law for the gravitational field that could replace the Poisson equation of Newtonian gravity. This he first attempted in his two "Prague papers" from 1912 [2, 3] for static fields. The purpose of my contribution here is to point out that the field equation Einstein arrived at in the second of these papers is not merely of historical interest.

After 1907 Einstein turned away from gravity research for a while, which he resumed in 1911 with a paper [4], also from Prague, in which he used the "Äquivalenzhypothese" to deduce the equality between gravitational and inertial mass, the gravitational redshift, and the deflection of light by the gravitational field of massive bodies. As is well known, the latter resulted in half the amount that was later correctly predicted by GR.

In the next gravity paper [2], the first in 1912, entitled "*Lichtgeschwindigkeit und Statik des Gravitationsfeldes*", Einstein pushed further the consequences of his heuristics and began his search for a sufficiently simple differential equation for static gravitational fields. The strategy was to, first, guess the equation from the form of the special fields "created" by non inertial reference frames and, second, generalise it to those gravitational fields generated by real matter. Note that the gravitational acceleration was to be assumed to be a gradient field (curl free) so that the sought-after field equation was for a scalar field, the gravitational potential.

The essential idea in the first 1912 paper is to identify the gravitational potential with c, the local velocity of light.[2] Einstein's heuristics indicated clearly that Special Relativity had to be abandoned, in contrast to the attempts by Max Abraham (1875–1922), who published a rival theory [5, 6] that was superficially based on Poincaré invariant equations (but violated Special Relativity in abandoning the condition that the four-velocities of particles had constant Minkowski square). In passing I remark that Einstein's reply [7] to Abraham, which is his last paper from Prague before his return to Zürich, contains in addition to his anticipation of the essential physical hypotheses on which a future theory of gravity could be based (here I refer to Jiří Bičák's contribution), also a concise and very illuminating account of the physical meaning and limitation of the special principle of relativity, the essence of which was totally missed by Abraham.

[2] Since here we will be more concerned with the mathematical form and not so much the actual derivation by Einstein, we will ignore the obvious objection that c has the wrong physical dimension, namely that of a velocity, whereas the proper gravitational potential should have the dimension of a velocity-squared.

Back to Einstein's first 1912 paper, the equation he came up with was

$$\Delta c = kc\rho \,, \tag{1}$$

where k is the "universal gravitational constant" and ρ is the mass density. The mathematical difference between (1) and the Poisson equation in Newtonian gravity is that (1) is homogeneous (even linear) in the potential. This means that the source strength of a mass density is weighted by the gravitational potential at its location. This implies a kind of "red-shift" for the active gravitational mass which in turn results in the existence of geometric upper bounds for the latter, as we will discuss in detail below. Homogeneity was Einstein's central requirement, which he justified from the interpretation of the gravitational potential as the local velocity of light, which is only determined up to constant rescalings induced from rescalings of the timescale.

Already in a footnote referring to (1) Einstein points out that it cannot be quite correct, as he is to explain in detail in a follow-up paper [3]. This second paper of 1912 is the one I actually wish to focus on in my contribution here. It appeared in the same issue of the *Annalen der Physik* as the previous one, under the title "*Zur Theorie des statischen Gravitationsfeldes*" (on the theory of the static gravitational field). In it Einstein once more investigates how the gravitational field influences electromagnetic and thermodynamic processes according to what he now continues to call the "*Aquivalenzprinzip*", and derives from it the equality of inertial and gravitational mass.[3]

After that he returns to the equation for the static gravitational field and considers the gravitational force-density **f**, acting on ponderable matter of mass density ρ, which is given by (Einstein writes σ instead of our ρ)

$$\mathbf{f} = -\rho \nabla c \,. \tag{2}$$

Einstein observes that the space integral of **f** does not necessarily vanish on account of (1), in violation of the principle that *actio* equals *reactio*. Terrible consequences,

[3] Einstein considers radiation enclosed in a container whose walls are "massless" (meaning vanishing rest-mass) but can support stresses, so as to be able to counteract radiation pressure. Einstein keeps repeating that equality of both mass types can only be proven if the gravitational field does not act on the stressed walls. That remark is hard to understand in view of the fact that unbalanced stresses add to inertia, as he well knew from his own earlier investigations [8]. However, as explained by Max Laue a year earlier [9], the gravitational action on the stressed walls is just cancelled by that on the stresses of the electromagnetic field, for both systems together form a "complete static system", as Laue calls it. A year later, in the 1913 "Entwurf" paper with Marcel Grossmann [10], Einstein again used a similar Gedankenexperiment with a massless box containing radiation immersed in a gravitational field, by means of which he allegedly shows that any Poincaré invariant scalar theory of gravity must violate energy conservation. A modern reader must ask how this can possibly be, in view of Noether's theorem applied to time-translation invariance. A detailed analysis [11] shows that this energy contains indeed the expected contribution from the tension of the walls, which may not be neglected.

like self-acceleration, have to be envisaged.[4] He then comes up with the following non-linear but still homogeneous modification of (1):

$$\Delta c = k\left\{c\rho + \frac{1}{2k}\frac{\nabla c\cdot\nabla c}{c}\right\}. \tag{3}$$

In the rest of this paper we will show how to arrive at this equation from a different direction and discuss some of its interesting properties as well as its relation to the description of static gravitational fields in GR.

2 A self-consistent modification of Newtonian Gravity

The following considerations are based on [12]. We start from ordinary Newtonian gravity, where the gravitational field is described by a scalar function ϕ whose physical dimension is that of a velocity-squared. It obeys

$$\Delta\varphi = 4\pi G\,\rho\,. \tag{4}$$

The force per unit volume that the gravitational field exerts upon a distribution of matter with density ρ is

$$\mathbf{f} = -\rho\nabla\varphi\,. \tag{5}$$

This we apply to the force that the gravitational field exerts upon its own source during a real-time process of redistribution. This we envisage as actively transporting each mass element along the flow line of a vector field $\boldsymbol{\xi}$. To first order, the change $\delta\rho$ that ρ suffers in time δt is given by

$$\delta\rho = \frac{-L_{\delta\boldsymbol{\xi}}\left(\rho d^3x\right)}{d^3x} = -\nabla\cdot(\delta\boldsymbol{\xi}\,\rho)\,, \tag{6}$$

where $\delta\boldsymbol{\xi} = \delta t\,\boldsymbol{\xi}$ and $L_{\delta\boldsymbol{\xi}}$ is the Lie derivative with respect to $\delta\boldsymbol{\xi}$. We assume the support $\mathrm{supp}(\rho) =: B \subset \mathbb{R}^3$ to be compact. In general, this redistribution costs energy. The work we have to invest for redistribution is, to first order, just given by

$$\delta A = -\int_{\mathbb{R}^3}\delta\boldsymbol{\xi}\cdot\mathbf{f} = -\int_B\varphi\nabla\cdot(\delta\boldsymbol{\xi}\,\rho) = \int_B\varphi\,\delta\rho\,, \tag{7}$$

where we used (6) in the last step and where we did not write out the Lebesgue measure d^3x to which all integrals refer. Note that in order to obtain (7) we did

[4] "*Anderenfalls würde sich die Gesamtheit der in dem betrachteten Raume befindlichen Massen, die wir auf einem starren, masselosen Gerüste uns befestigt denken wollen, sich in Bewegung zu setzen streben*" ([3], p. 452).

not make use of the field equation. Equation (7) is generally valid whenever the force-density relates to the potential and the mass density as in (5).

Now we make use of the field Eq. (4). We assume the redistribution-process to be adiabatic, that is, we assume the instantaneous validity of the field equation at each point in time throughout the process. This implies

$$\Delta\delta\varphi = 4\pi G\,\delta\rho\,. \tag{8}$$

Hence, using (7), the work invested in the process of redistribution is (to first order)

$$\delta A = \int_B \varphi\,\delta\rho = \delta\left\{-\frac{1}{8\pi G}\int_{\mathbb{R}^3}(\nabla\varphi)^2\right\}. \tag{9}$$

If the infinitely dispersed state of matter is assigned the energy-value zero, then the expression in curly brackets is the total work invested in bringing the infinitely dispersed state to that described by the distribution ρ. This work must be stored somewhere as energy. Like in electro-statics and -dynamics, we take a further logical step and assume this energy to be spatially distributed in the field according to the integrand. This leads to the following expression for the energy density of the static gravitational field

$$\varepsilon = -\frac{1}{8\pi G}(\nabla\varphi)^2\,. \tag{10}$$

All this is familiar from Newtonian gravity. But now we go beyond Newtonian gravity and require the validity of the following

Principle *All forms of energy, including that of the gravitational field itself, shall gravitate according to $E = mc^2$*. This principle implies that if we invest an amount of work δA in a system its (active) gravitational mass will increase by $\delta A/c^2$.

Now, the (active) gravitational mass M_g is defined by the flux of the gravitational field to spatial infinity (i.e. through spatial spheres as their radii tend to infinity):

$$M_g = \frac{1}{4\pi G}\int_{S^2_\infty}\mathbf{n}\cdot\nabla\varphi = \frac{1}{4\pi G}\int_{\mathbb{R}^3}\Delta\varphi\,. \tag{11}$$

Hence, making use of the generally valid Eq. (7), the principle that $\delta A = M_g c^2$ takes the form

$$\int_B \varphi\,\delta\rho = \frac{c^2}{4\pi G}\int_{\mathbb{R}^3}\Delta\delta\varphi\,. \tag{12}$$

This functional equation relates φ and ρ, over and above the restriction imposed on their relation by the field equation. However, the latter may—and generally will—be inconsistent with this additional equation. For example, the Newtonian field Eq. (4) is easily seen to manifestly violate (12), for the right-hand side then becomes just the integral over $c^2\delta\rho$, which always vanishes on account of (6) (or the obvious remark that the redistribution clearly does not change the total mass), whereas the left hand side will generally be non-zero. The task must therefore be to find field equation(s) consistent with (12). Our main result in that direction is that the unique generalisation of (4) which satisfies (12) is just (3), i.e. the field equation from Einstein's second 1912 paper.

Let us see how this comes about. A first guess for a consistent modification of (4) is to simply add ε/c^2 to the source ρ:

$$\Delta\varphi = 4\pi G\left(\rho - \frac{1}{8\pi G c^2}(\nabla\varphi)^2\right). \tag{13}$$

But this cannot be the final answer because this change of the field equation also brings about a change in the expression for the self-energy of the gravitational field. That is, the term in the bracket on the right-hand side is not the total energy according to *this* equation, but according to the original equation (4). In other words: equation (13) still lacks *self*-consistency. This can be corrected for by iterating this procedure, i.e., determining the field's energy density according to (13) and correcting the right-hand side of (13) accordingly. Again we have changed the equation, and this goes on ad infinitum. But the procedure converges to a unique field equation, similarly to the convergence of the "Noether-procedure" [5] that leads from the Poincaré invariant Pauli-Fierz theory of spin-2 mass-0 fields in flat Minkowski space to GR [13–15].

In our toy model the convergence of this procedure is not difficult to see. We start from the definition (11) and calculate its variation δM_g assuming the validity of (13). From what we said above we know already that this is not yet going to satisfy (12). But we will see that from this calculation we can read off the right redefinitions.

We start by varying (11):

$$\delta M_g = \frac{1}{4\pi G}\int \Delta\delta\varphi\,. \tag{14}$$

We replace $\Delta\delta\varphi$ with the variation of the right-hand side of (13). Partial integration of the non-linear part gives us a surface term whose integrand is $\propto \varphi\nabla\delta\varphi = O(r^{-3})$ and hence vanishes. The remaining equation is

$$\delta M_g = \int_B \delta\rho + \frac{1}{4\pi G}\int_{\mathbb{R}^3}\left(\frac{\varphi}{c^2}\right)\Delta\delta\varphi\,. \tag{15}$$

Playing the same trick (of replacing $\Delta\delta\varphi$ with the variation of the right-hand side of (13) and partial integration, so as to collect all derivatives on $\delta\varphi$) again and again, we arrive after N steps at

[5] Pioneered by Robert Kraichnan in his 1947 MIT Bachelor thesis "Quantum Theory of the Linear Gravitational Field".

$$\delta M_g = \int_B \sum_{n=0}^{N-1} \frac{1}{n!}\left(\frac{\varphi}{c^2}\right)^n \delta\rho + \frac{1}{N!c^{2N}} \frac{1}{4\pi G} \int_{\mathbb{R}^3} \varphi^N \delta(\Delta\varphi) . \tag{16}$$

As φ is bounded for a regular matter distribution, and the spatial integral over $\delta\Delta\varphi$ is just $4\pi G\delta M_g$, the last term tends to zero for $N \to \infty$. Hence

$$\delta M_g = \int_B \delta\rho \, \exp(\varphi/c^2) . \tag{17}$$

This *is* of the desired form (12) required by the principle, provided we redefine the gravitational potential to be Φ rather than φ, where

$$\Phi := c^2 \exp(\varphi/c^2) . \tag{18}$$

Saying that Φ rather than φ is the right gravitational potential means that the force density is not given by (5), but rather by

$$\mathbf{f} = -\rho\nabla\Phi . \tag{19}$$

As we have made use of Eq. (13) in order to derive (17), we must make sure to keep *that* equation, just re-expressed in terms of Φ. This leads to

$$\Delta\Phi = \frac{4\pi G}{c^2}\left[\rho\Phi + \frac{c^2}{8\pi G}\frac{(\nabla\Phi)^2}{\Phi}\right], \tag{20}$$

which is precisely Einstein's improved "Prague equation" (3) with $k = 4\pi G/c^2$. Note from (18) that the asymptotic condition $\varphi(r \to \infty) \to 0$ translates to $\Phi(r \to \infty) \to c^2$. Note also that for $r \to \infty$ the $1/r^2$-parts of $\nabla\varphi$ and $\nabla\Phi$ coincide, so that in the expressions (11) for M_g we may just replace φ with Φ:

$$M_g = \frac{1}{4\pi G}\int_{S^2_\infty} \mathbf{n}\cdot\nabla\Phi = \frac{1}{4\pi G}\int_{\mathbb{R}^3} \Delta\Phi . \tag{21}$$

The principle now takes the form (12) with φ replaced by Φ. It is straightforward to show by direct calculation that (12) is indeed a consequence of (20), as it must be. It also follows from (20) that the force density (19) is the divergence of a symmetric tensor:

$$f_a = -\nabla^b t_{ab} , \tag{22a}$$

where

$$t_{ab} = \frac{1}{4\pi G c^2}\left\{\frac{1}{\Phi}\left[\nabla_a\Phi\nabla_b\Phi - \tfrac{1}{2}\delta_{ab}(\nabla\Phi)^2\right]\right\} . \tag{22b}$$

This implies the validity of the principle that actio equals reactio that Einstein demanded. *This* was Einstein's rationale for letting (3) replace (1).

Finally we mention that (20) may be linearised if written in terms of the square-root of Φ:

$$\Psi := \sqrt{\frac{\Phi}{c^2}}\,. \tag{23}$$

One gets

$$\Delta\Psi = \frac{2\pi G}{c^2}\,\rho\,\Psi\,. \tag{24}$$

This helps in finding explicit solutions to (20). Note that Ψ is dimensionless.

3 Spherically symmetric solutions

In this section we discuss some properties of spherically symmetric solutions to (24) for spherically symmetric mass distributions ρ of compact support. In the following we will simply refer to the object described by such a mass distribution as "star".

In terms of $\chi(r) := r\Psi(r)$ Eq. (24) is equivalent to

$$\chi'' = \frac{2\pi G}{c^2}\,\rho\,\chi\,. \tag{25}$$

The support of ρ is a closed ball of radius R, called the star's radius. For $r < R$ we shall assume $\rho(r) \geq 0$ (weak energy condition). We seek solutions which correspond to everywhere positive and regular Ψ and hence everywhere positive and regular Φ. In particular $\Phi(r = 0)$ and $\Psi(r = 0)$ must be finite. For $r > R$ Eq. (25) implies $\chi'' = 0$, the solution to which is

$$\chi_+(r) = r\Psi_+(r) = r - R_g\,, \quad \text{for } r > R\,, \tag{26}$$

where R_g denotes the gravitational radius

$$R_g := \frac{GM_g}{2c^2}\,. \tag{27}$$

R_g comes in because of (21), which fixes one of the two integration constants, the other being fixed by $\Psi(\infty) = 1$.

Let χ_- denote the solution in the interior of the star. Continuity and differentiability at $r = R$ gives $\chi_-(R) = R - R_g$ and $\chi'_-(R) = 1$. We observe that $\chi_-(R) \geq 0$. For suppose $\chi_-(R) < 0$, then (25) and the weak energy condition imply $\chi''(R) \leq 0$. But this implies that for $r \in [0, R]$ the curve $r \mapsto \chi_-(r)$ lies below the straight line $r \mapsto r - R_g$ and assumes a value less than $-R_g$ at $r = 0$, in contradiction to the finiteness of $\Psi(r = 0)$ which implies $\chi_-(r = 0) = 0$. Hence we have

Theorem 1 *The gravitational radius of a spherically symmetric star is universally bound by its (geometric) radius, $R_g \leq R$. Equivalently expressed in terms of M_g we may say that the gravitational mass is universally bound above by*

$$M_g < \frac{2c^2 R}{G} . \tag{28}$$

This may be seen in analogy to Buchdahl's inequality in GR [15], which, using the isotropic (rather than Schwarzschild) radial coordinate, would differ from (28) only by an additional factor of 8/9 on the right-hand side. The Buchdahl bound is optimal, being saturated by the interior Schwarzschild solution for a homogeneous star.

So let us here, too, specialise to a homogeneous star,

$$\rho(r) = \begin{cases} \frac{3M_b}{4\pi R^3} & \text{for } r \leq R \\ 0 & \text{for } r > R , \end{cases} \tag{29}$$

where M_b is called the bare mass (integral over ρ). It is convenient to introduce the radii corresponding to bare and gravitational masses, as well as their ratio to the star's radius R:

$$R_b := \frac{GM_b}{2c^2} , \quad x := \frac{R_b}{R} , \tag{30a}$$

$$R_g := \frac{GM_g}{2c^2} , \quad y := \frac{R_g}{R} . \tag{30b}$$

We also introduce the inverse length

$$\omega := \frac{1}{R} \cdot \sqrt{\frac{3R_b}{R}} , \tag{31}$$

so that (25) just reads $\chi'' = \omega^2 \chi$. From this the interior solution is easily obtained. If written in terms of Ψ it reads

$$\Psi_-(r) = \frac{1}{\cosh(\omega R)} \frac{\sinh(\omega r)}{\omega r} , \quad \text{for } r < R . \tag{32}$$

As a result of the matching to the exterior solution given in (26), R_g is determined by R and ω, i.e. R and R_b. In terms of x and y this relation takes the simple form

$$y = 1 - \frac{\tanh\left(\sqrt{3x}\right)}{\sqrt{3x}} , \tag{33}$$

which convex-monotonically maps $[0, \infty)$ onto $[0, 1)$. The fact that $y < 1$ for all x is just the statement of the Theorem applied to the homogeneous case.

If $x = R_b/R \ll 1$ we have $y = x - \frac{6}{5}x^2 + \cdots$, which for $E_{\text{total}} := M_g c^2$ reads

$$E_{\text{total}} = M_b c^2 \left(1 - \tfrac{3}{5}x + O(x^2)\right) . \tag{34}$$

We note that $-3M_b c^2 x/5 = -\frac{3}{5}GM_b^2/R$ is just the Newtonian binding energy of a homogeneous star. In view of our Principle it makes good sense that to first order just this amount is subtracted from the bare mass in order to obtain the active gravitational mass. In Newtonian gravity this negative amount is just identified with the field's self-energy, but here the interpretation is different: The two terms that act as source for the gravitational field in (20) are the matter part, which is proportional to ρ but diminished by Φ, and the field's own part, which is proportional to $(\nabla\Phi)^2/\Phi$ and positive definite! Their contributions are, respectively,

$$E_{\text{matter}} = \int_B \rho\Phi = M_b c^2 \left(1 - \tfrac{6}{5}x + O(x^2)\right) , \tag{35}$$

$$E_{\text{field}} = \frac{c^2}{8\pi G}\int_{\mathbb{R}^3} \frac{(\nabla\Phi)^2}{\Phi} = M_b c^2 \left(\tfrac{3}{5}x + O(x^2)\right) . \tag{36}$$

Hence even though the total energy is decreased due to binding, the gravitational field's self energy *increases* by the same amount. Twice that amount is gained from the fact that the matter-energy is "red-shifted" by being multiplied with Φ, so energy is conserved (of course).

Two more consequences, which are related, are noteworthy:

- Unlike in Newtonian theory, objects with non-zero gravitational mass cannot be modelled by point sources. In the spherically symmetric case this is an immediate consequence of (28), which implies $M_g \to 0$ for $R \to 0$. Hence there are no δ-like masses.
- Unlike in Newtonian gravity, unlimited compression of matter does not lead to unlimited energy release. Consider a sequence of homogeneous (just for simplicity) stars of fixed bare mass M_b and variable radius R, then the gravitational mass M_g as function of $x = R_b/R$ is given by

$$M_g(x) = M_b \cdot \left\{ \frac{1}{x} \cdot \left(1 - \frac{\tanh\left(\sqrt{3x}\right)}{\sqrt{3x}} \right) \right\} . \tag{37}$$

  The function in curly brackets[6] is a strictly monotonically decreasing function $[0,\infty] \mapsto [1,0]$. This shows that for infinitely dispersed matter, where $R \to \infty$ and hence $x \to 0$, we have $M_g(x=0) = M_b$, as expected, and that for infinite compression $M_g(x \to \infty) = 0$. As the gained energy at stage x is $(M_b - M_g(x))c^2$, we can at most gain $M_b c^2$.

[6] Its Taylor expansion at $x = 0$ is $1 - 6x/5 + 51x^2/35 + \cdots$.

4 Relation to General Relativity

Finally I wish to briefly comment on the relation of Eq. (3) or (20) to GR. Since Einstein's 1912 theory was only meant to be valid for static situations, I will restrict attention to static spacetimes (M, g). Hence I assume the existence of a timelike and hypersurface orthogonal Killing field K. My signature convention shall be "mostly plus", i.e. $(-, +, +, +)$.

We choose adapted coordinates (t, x^a), $a = 1, 2, 3$, where the level sets of t are the integral manifolds of the foliation defined by K and $K = \partial/\partial(ct)$. We can then write the metric in a form in which the coefficients do not depend on t (called "time"),

$$g = -\Psi^2(x)\, c^2\, dt \otimes dt + \hat{g}_{ab}(x)\, dx^a \otimes dx^b\,. \tag{38}$$

Clearly $c^2\Psi^2 = -g(K, K)$. From now on, all symbols with hats refer to the spatial geometry, like the spatial metric $\hat{g}$.

The t-component of the geodesic equation is equivalent to $\Psi^2\dot{t} = \text{const}$, where an overdot refers to the derivative with respect to an affine parameter. This equation allows us to eliminate the affine parameter in favour of t in the spatial components of the geodesic equation. If we set[7]

$$\Psi = \sqrt{\frac{2\Phi}{c^2}}, \tag{39}$$

they read

$$\frac{d^2x^a}{dt^2} + \hat{\Gamma}^a_{bc}\frac{dx^b}{dt}\frac{dx^c}{dt} = -\Phi_{,b}\hat{g}^{ab} + \Phi_{,b}\left[\frac{1}{\Phi}\frac{dx^a}{dt}\frac{dx^b}{dt}\right], \tag{40}$$

where the $\hat{\Gamma}^a_{bc}$ are the Christoffel coefficients for $\hat{g}$, and $\Phi_{,a} = \partial_a\Phi$. This should be compared with (19) together with Newton's second law, which give $d^2\mathbf{x}/dt^2 = -\nabla\Phi$. As we did not attempt to include special relativistic effects in connection with high velocities, we should consistently neglect terms v^2/c^2 in (40). This results in dropping the rightmost term. The rest has the pseudo-Newtonian form in arbitrary (not just inertial) spatial coordinates. A non-zero spatial curvature would, of course, be a new feature not yet considered.

The curvature and Ricci tensors for the metric (38) are readily computed, most easily by using Cartan's structure equations:

$$\mathrm{Ric}(n, n) = \Psi^{-1}\,\hat{\Delta}\Psi\,, \quad R_{ab} = \hat{R}_{ab} - \Psi^{-1}\,\hat{\nabla}_a\hat{\nabla}_b\Psi\,. \tag{41}$$

[7] This differs by a factor of 2 from (23) which we need and to which we return below.

Here $n = \Psi^{-1}\partial/c\partial t$ is the unit timelike vector characterising the static reference frame, $\hat{\nabla}$ is the Levi-Civita covariant derivative with respect to $\hat{g}$, and $\hat{\Delta}$ is the corresponding Laplacian.

Using this in Einstein's equations

$$R_{\mu\nu} = \frac{8\pi G}{c^4}\Big(T_{\mu\nu} - \tfrac{1}{2}g_{\mu\nu}T^{\lambda}_{\lambda}\Big) \tag{42}$$

for pressureless (we neglect the pressure since it enters multiplied with c^{-2}) dust at rest and of mass-density ρ in the static frame, i.e.

$$T_{\mu\nu} = \rho c^2 n_\mu n_\nu\,, \tag{43}$$

we get

$$\hat{\Delta}\Psi = \frac{4\pi G}{c^2}\rho\Psi \qquad \text{time component}\,, \tag{44a}$$

$$\hat{\nabla}_a\hat{\nabla}_b\Psi = \hat{R}_{ab}\Psi \qquad \text{space components}\,. \tag{44b}$$

We note that, apart from the space curvature, (44a) is almost—but not quite—identical to (24). They differ by a factor of 2! Rewriting (44a) in terms of Φ according to (39), we get

$$\hat{\Delta}\Phi = \frac{8\pi G}{c^2}\left[\rho\Phi + \frac{c^2}{16\pi G}\frac{\hat{g}^{ab}\hat{\nabla}_a\Phi\hat{\nabla}_b\Phi}{\Phi}\right]. \tag{45}$$

This differs from (20) by the same factor of 2 (i.e., $G \to 2G$). Note that we cannot simply remove this factor by rescaling Ψ and Φ, as the equations are homogeneous in these fields. Note also that the overall scale of Φ is fixed by (40): It is the gradient of Φ, and not a multiple thereof, which gives the acceleration. But then there is another factor of 2 in difference to our earlier discussion: If the metric (38) is to approach the Minkowski metric far away from the source, then Ψ should tend to one and hence Φ should asymptotically approach $c^2/2$ according to (39). In (20), however, Φ should asymptotically approach c^2, i.e. twice that value. This additional factor of 2 ensures that both theories have the same Newtonian limit. Indeed, if we expand the gravitational potential Φ of an isolated object in a power series in G, this implies that the linear terms of both theories coincide. However, the quadratic terms in GR are twice as large as in our previous theory based on (19) and (20). This is not quite unexpected if we take into account that in GR we also have the space curvature that will modify the fields and geodesics in post Newtonian approximations. We note that the spatial Einstein equations (44b) prevent space from being flat. For example, taking their trace and using (44a) shows that the scalar curvature of space is, in fact, proportional to the mass density.

Finally we show that the total gravitational mass in GR is just given by the same formula (21), where Φ is now that used here in the GR context. To see this we recall that for spatially asymptotically flat spacetimes the overall mass (measured at spatial infinity) is given by the ADM-mass. Moreover, for spatially asymptotically flat spacetimes which are stationary and satisfy Einstein's equations with sources of spatially compact support, the ADM mass is given by the Komar integral (this is, e.g., proven in Theorem 4.13 of [17]). Hence we have

$$M_{\mathrm{ADM}} = \frac{c^2}{8\pi G}\int_{S^2_\infty} \star dK^\flat \,. \tag{46}$$

Here $K = \partial/\partial(ct)$, and $K^\flat := g(K,\cdot) = -\Psi^2 c dt$ is the corresponding 1-form. The star, $\star$, denotes the Hodge-duality map. Using (39) and asymptotic flatness it is now straightforward to show that the right hand side of (46) can indeed be written in the form of the middle term in (21). This term only depends on Φ at infinity, i.e. on the Newtonian limit, and hence it gives a value independent of the factor-2 discrepancy discussed above. In that sense the active gravitational mass M_g defined earlier corresponds to M_{ADM} in the GR context.

This ends our discussion of Einstein's 1912 scalar field equation, which is thus seen to contain many interesting features we know from GR, albeit in a pseudo-Newtonian setting.

Acknowledgments I sincerely thank the organisers and in particular Jiří Bičák for inviting me to the most stimulating and beautiful conference *Relativity and Gravitation—100 years after Einstein in Prague.*

References

1. Einstein, A.: Über das Relativitätsprinzip und die aus demselben gezogenen Folgerungen. Jahrbuch der Radioaktivität und Elektronik **4**, 411 (1907). [Erratum, ibid. **5**, 98–99 (1908)]
2. Einstein, A.: Lichtgeschwindigkeit und Statik des Gravitationsfeldes. Ann. Phys. **343**, 355 (1912). doi:10.1002/andp.19123430704
3. Einstein, A.: Zur Theorie des statischen Gravitationsfeldes. Ann. Phys. **343**, 443 (1912). doi:10.1002/andp.19123430709
4. Einstein, A.: Über den Einfluß der Schwerkraft auf die Ausbreitung des Lichtes. Ann. Phys. **340**, 898 (1911). doi:10.1002/andp.19113401005
5. Abraham, M.: Zur Theorie der Gravitation. Physikalische Zeitschrift **13**, 1 (1912)
6. Abraham, M.: Das Elementargesetz der Gravitation. Physikalische Zeitschrift **13**, 4 (1912)
7. Einstein, A.: Relativität und Gravitation. Erwiderung auf eine Bemerkung von M. Abraham. Ann. Phys. **343**, 1059 (1912). doi:10.1002/andp.19123431014
8. Einstein, A.: Über die vom Relativitätsprinzip geforderte Trägheit der Energie. Ann. Phys. **328**, 371 (1907). doi:10.1002/andp.19073280713
9. Laue, M.: Zur Dynamik der Relativitätstheorie. Ann. Phys. **340**, 524 (1911). doi:10.1002/andp.19113400808
10. Einstein, A., Grossmann, M.: Entwurf einer verallgemeinerten Relativitätstheorie und einer Theorie der Gravitation. Zeitschrift für Mathematik und Physik **62**, 225 (1914)
11. Giulini, D.: What is (not) wrong with scalar gravity? Stud. History Philis. Mod. Phys. **39**, 154 (2008). doi:10.1016/j.shpsb.2007.09.001

12. Giulini, D.: Consistently implementing the field self-energy in Newtonian gravity. Phys. Lett. A **232**, 165 (1997). doi:10.1016/S0375-9601(97)00369-1
13. Kraichnan, R.H.: Special-relativistic derivation of generally covariant gravitation. Phys. Rev. **98**(4), 1118 (1955). doi:10.1103/PhysRev.98.1118
14. Feynman, R.P., Morínigo, F.B., Wagner, W.G., Hatfield, B.: Feynman Lectures on Gravitation. Westview Press, Boulder (2002)
15. Deser, S.: Self-interaction and gauge invariance. Gen. Rel. Grav. **1**, 9 (1970). doi:10.1007/BF00759198
16. Buchdahl, H.A.: General relativistic fluid spheres. Phys. Rev. **116**, 1027 (1959). doi:10.1103/PhysRev.116.1027
17. Choquet-Bruhat, Y.: General Relativity and the Einstein Equations. Oxford Mathematical Monographs. Oxford University Press, Oxford (2009)

Source Integrals of Asymptotic Multipole Moments

Norman Gürlebeck

Abstract We derive source integrals for multipole moments that describe the behaviour of static and axially symmetric spacetimes close to spatial infinity. We assume isolated non-singular sources but will not restrict the matter content otherwise. Some future applications of these source integrals of the asymptotic multipole moments are outlined as well.

1 Introduction

In experiments that measure general relativistic effects, some parameters characterizing the spacetime have to be determined. The multipole moments are one set of such parameters. They are measured in the exterior region of astrophysical objects like neutron stars or galaxies but also planets and describe the gravitational field near spatial infinity. They will be called here *asymptotic multipole moments* (AMM). The bending of light and the gravitational lensing proved particularly useful for their measurement, see, e.g., [1–3] and references therein. But what information can be gathered about the matter distribution and the metric in its interior by their measurement? What does it mean to measure a certain value of the quadrupole moment? In Newtonian theory, the answers are provided by the source integrals of the AMM. These integrals determine the asymptotic multipole moments by an integration over the mass density. In general relativity, similar expressions for the AMM are only known in approximations to general relativity, see e.g. [4]. Here we will present

N. Gürlebeck (✉)
ZARM, University of Bremen, Am Fallturm, 28359 Bremen, Germany
e-mail: norman.guerlebeck@zarm.uni-bremen.de

N. Gürlebeck
EU and Institute of Theoretical Physics, Charles University, V Holešovičkách 2,
18000 Praha, Czech Republic

J. Bičák and T. Ledvinka (eds.), *Relativity and Gravitation*,
Springer Proceedings in Physics 157, DOI: 10.1007/978-3-319-06761-2_11,

source integrals of the AMM for static and axially symmetric spacetimes in full general relativity. At the same time these provide quasi-local definitions of *all* asymptotic multipole moments.

2 Preliminaries

In this section, we shortly review several concepts necessary in the derivation of the source integrals. We use throughout this article geometric units $G = c = 1$ and the signature of the metric is $(-, +, +, +)$.

2.1 The Line Element and the Field Equations

We concentrate on axially symmetric and static spacetimes of the Weyl form, i.e.,

$$ds^2 = \mathrm{e}^{2k-2U}\left(d\rho^2 + d\zeta^2\right) + W^2\mathrm{e}^{-2U}d\varphi^2 - \mathrm{e}^{2U}dt^2. \tag{1}$$

We do not restrict the type of matter except in that the line element (1) can be introduced, see [5]. The metric functions e^{2U} and W can be expressed by the timelike Killing vector $\xi^\alpha = (\partial_t)^\alpha$ and the Killing vector of the axial symmetry $\eta^\alpha = (\partial_\varphi)^\alpha$

$$\mathrm{e}^{2U} = -\xi_\alpha\xi^\alpha, \quad W^2 = -\eta_\alpha\eta^\alpha\xi_\beta\xi^\beta. \tag{2}$$

Let us choose a sphere $\mathscr{S}_0$ of finite radius $r = R_0$ ($\rho = r\sin\theta$, $\zeta = r\cos\theta$) that covers the entire matter distribution, cf. Fig. 1. Outside of $\mathscr{S}_0$, canonical Weyl coordinates ($W = \rho$) are introduced by virtue of one of the vacuum field equations. This allows still a shift in the $\zeta-$coordinate, which enables us later to move the origin with respect to which the AMM are measured. The vacuum field equations in canonical Weyl coordinates read

$$\Delta U = 0, \quad k_{,\zeta} = 2\rho U_{,\rho}U_{,\zeta}, \quad k_{,\rho} = \rho\left((U_{,\rho})^2 - (U_{,\zeta})^2\right), \tag{3}$$

where $\Delta = \left(\frac{\partial^2}{\partial\rho^2} + \frac{1}{\rho}\frac{\partial}{\partial\rho} + \frac{\partial^2}{\partial\zeta^2}\right)$. The function k is determined via a line integration, cf. the last two equations of (3), once U is known. Hence, only a Laplace equation for U remains to be solved in practice.

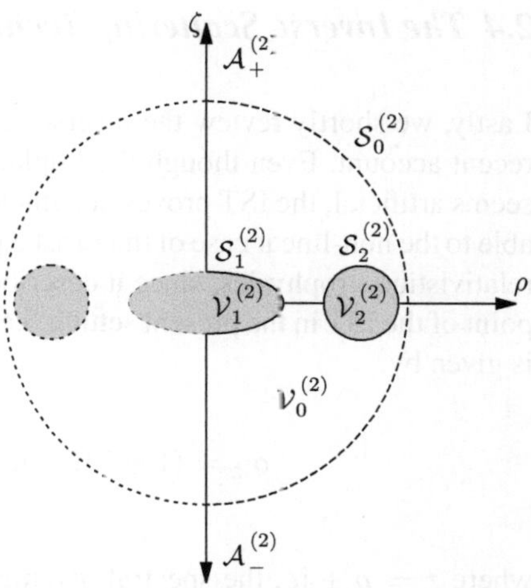

Fig. 1 An example of the physical situations discussed here: The surfaces of the individual matter components are denoted by $\mathscr{S}_i^{(2)} = \partial\mathscr{V}_i^{(2)}$ with $i \geq 1$ and their respective volumes by $\mathscr{V}_i^{(2)}$. The surface $\mathscr{S}_0^{(2)}$ describes a *circle* with sufficiently large but finite radius enclosing all matter components, cf. Sect. 2.1

2.2 The Physical Setting

We depict in Fig. 1 an example of a physical situation that will be covered by the subsequent considerations. The relevant surfaces and volumes are defined there as well. For simplicity, we allow only non-singular sources. However, such can be incorporated into the formalism as we showed in [6]. The 3-dimensional projection of the matter region into an hypersurface of constant Killing-time t is obtained by a rotation around the ζ−axis in Fig. 1. In this way, the quantities $\mathscr{A}_\pm^{(3)}$, $\mathscr{S}_i^{(3)}$ and $\mathscr{V}_i^{(3)}$ are defined starting from $A_\pm^{(2)}$, $\mathscr{S}_i^{(2)}$ and $\mathscr{V}_i^{(2)}$, respectively.

2.3 The Multipole Moments

For asymptotically flat and static spacetimes a geometric definition of AMM was given by Geroch in [7]. This definition was generalized and applied by many authors, see the reviews [4, 8] and references therein. In the axially symmetric case with the line element (1), Geroch's multipole moments M_r can be obtained by an expansion of U along the symmetry axis in $|\zeta|^{-1}$:

$$U(\rho = 0, \zeta) = \sum_{r=0}^{\infty} U^{(r)} |\zeta|^{-r-1}. \tag{4}$$

The M_r follow uniquely from Weyl's multipole moments $U^{(r)}$ and vice versa as was shown in [9]. Therefore, we consider only the $U^{(r)}$ here.

2.4 The Inverse Scattering Technique

Lastly, we shortly review the inverse scattering technique (IST), see e.g. [10] for a recent account. Even though the Laplace equation is linear and the use of the IST seems artificial, the IST proves nonetheless beneficial, because it is easily generalizable to the non-linear case of the Ernst equation. This equation is of special interest in relativistic astrophysics, since it describes the exterior of rotating stars. The starting point of the IST in the present setting, i.e. the linear problem of the Laplace equation, is given by

$$\sigma_{,z} = (1+\lambda)U_{,z}\sigma, \quad \sigma_{,\bar{z}} = \left(1+\frac{1}{\lambda}\right)U_{,\bar{z}}\sigma, \tag{5}$$

where $z = \rho + \mathrm{i}\zeta$, the spectral parameter $\lambda = \sqrt{\frac{K-\mathrm{i}\bar{z}}{K+\mathrm{i}z}}$, $K \in \mathbb{C}$ and a bar denotes complex conjugation. The complex valued function σ depends on z, $\bar{z}$ and λ. The integrability condition of (5) is the Laplace equation for U. Therefore, having a solution σ of (5) yields also a solution U of the Laplace equation and vice versa. The main technical steps of the IST as described in [10] are to integrate (5) along $\mathscr{A}_{\pm}^{(2)}$, along a circle with sufficiently large radius and along a compact curve connecting $\mathscr{A}_{+}^{(2)}$ with $\mathscr{A}_{-}^{(2)}$. This scheme can be carried out partially and we quote only the results (simplified to the static case), which are relevant for us, from [10]:

$$\begin{aligned}
(0,\zeta) \in \mathscr{A}^+ : \ & \sigma\left(\lambda = +1, \rho = 0, \zeta\right) = F(K)\mathrm{e}^{2U(\rho=0,\zeta)}, \\
& \sigma\left(\lambda = -1, \rho = 0, \zeta\right) = 1, \\
(0,\zeta) \in \mathscr{A}^- : \ & \sigma\left(\lambda = +1, \rho = 0, \zeta\right) = \mathrm{e}^{2U(\rho=0,\zeta)}, \\
& \sigma\left(\lambda = -1, \rho = 0, \zeta\right) = F(K).
\end{aligned} \tag{6a}$$

The function $F : \mathbb{C} \to \mathbb{C}$ is given for $K \in \mathbb{R}$ with $(\rho = 0, \zeta = K) \in \mathscr{A}^{\pm}$ by

$$F(K) = \begin{cases} \mathrm{e}^{-2U(\rho=0,\zeta=K)} & (0,K) \in \mathscr{A}^+ \\ \mathrm{e}^{2U(\rho=0,\zeta=K)} & (0,K) \in \mathscr{A}^- \end{cases} . \tag{6b}$$

The integration along $\mathscr{S}_0^{(2)}$ does not enter the derivation of these formulas and it forms the crucial part of our considerations in the next section.

3 Source Integrals of Weyl's Multipole Moments

The derivation of the source integrals consists of several steps. First, the AMM are expressed as line integrals along $\mathscr{S}_0^{(2)}$. This is the most important step, because it makes it possible to determine the AMM quasi-locally. Then these integrals will be rewritten in a coordinate independent form as surface integrals over $\mathscr{S}_0^{(3)}$ by virtue of the axial symmetry. Subsequently, Stokes' theorem is used to rewrite these as volume integrals over $\mathscr{V}^{(0)}$. In the final step, it is shown that the contributions in the vacuum regions vanish. Thus, the steps from before can be retraced to obtain the contributions in source integral form of each individual matter component. We suppress the details of these derivations and show only the crucial steps.

The linear problem (5) is well-defined along $\mathscr{S}_0^{(2)}$ and reads:

$$\sigma_{,s} = \left[U_{,s} + \frac{1}{2}\left(\left(\frac{1}{\lambda} + \lambda\right) U_{,s} + \mathrm{i}\left(\frac{1}{\lambda} - \lambda\right) U_{,n}\right)\right]\sigma, \tag{7}$$

where $U_{,s}$ and $U_{,n}$ are the tangential and the (outward pointing) normal derivative of U along $\mathscr{S}_0^{(2)}$ with respect to a parametrisation $[s_N, s_S] \to \mathscr{S}_0^{(2)}$. The indices N and S refer to the values of a parameter or function at the "north" and "south" pole of $\mathscr{S}_0^{(2)}$, i.e., to the intersection points ($\rho = 0$, $\zeta = \zeta_{N/S}$) of $\mathscr{S}_0^{(2)}$ and the symmetry axis. Equation (7) is easily integrated using the boundary values from (6):

$$U(0, K) = \frac{U_N - U_S}{2} + \frac{1}{4}\int_{s_N}^{s_S} \left(\left(\lambda^{-1} + \lambda\right) U_{,s} + \mathrm{i}\left(\lambda^{-1} - \lambda\right) U_{,n}\right) \mathrm{d}s. \tag{8}$$

If we expand this equation in $|K|^{-1}$, we obtain expression for Weyl's multipole moments in terms of a line integration. Let us introduce the abbreviations $N_+^{(r)}$ and $N_-^{(r)}$ for the expansion coefficients $(\lambda^{-1} + \lambda)^{(r)}$ and $\mathrm{i}(\lambda^{-1} - \lambda)^{(r)}$ to order $r + 1$, respectively. After a lengthy but straightforward calculation they evaluate to

$$\begin{aligned} N_-^{(r)} &= \sum_{k=0}^{\lfloor \frac{r}{2} \rfloor} \frac{2(-1)^{k+1} r! \rho^{2k+1} \zeta^{r-2k}}{4^k (k!)^2 (r-2k)!}, \\ N_+^{(r)} &= \sum_{k=0}^{\lfloor \frac{r-1}{2} \rfloor} \frac{2(-1)^{k+1} r! \rho^{2k+2} \zeta^{r-2k-1}}{4^k (k!)^2 (r-2k-1)!(2k+2)}. \end{aligned} \tag{9}$$

The $r = -1$ order in $|K|^{-1}$ of (8) is satisfied trivially and will not be considered subsequently. The orders $r \geq 0$ of (8) yield the desired quasi-local definitions of Weyl's multipole moments:

$$U^{(r)} = \frac{1}{4}\int_{\mathscr{S}_0^{(2)}} \left(N_+^{(r)} U_{,\hat{s}} + N_-^{(r)} U_{,\hat{n}}\right) \mathrm{d}\mathscr{S}_0^{(2)}, \tag{10}$$

where $U_{,\hat{s}}$ and $U_{,\hat{n}}$ are the tangential and normal derivatives along $\mathscr{S}_0^{(2)}$ with respect to the unit tangent vector and the unit normal vector, which are defined with the induced metric on $\mathscr{S}_0^{(2)}$; $\mathrm{d}\mathscr{S}_0^{(2)}$ denotes the proper distance along $\mathscr{S}_0^{(2)}$. The functions $N_\pm^{(r)}$ and U are to be read as functions along $\mathscr{S}_0^{(2)}$, i.e. as functions of $(\rho(s), \zeta(s))$.

To make the coordinate independence apparent, we express ρ and ζ by scalars built from the Killing vectors. Firstly, observe that (2) holds everywhere and that in vacuum we have $W = \rho$. Additionally, the 1-form

$$Z_\alpha = \varepsilon_{\alpha\beta\gamma\delta} W^{,\beta} W^{-1} \eta^\gamma \xi^\delta \tag{11}$$

is well-defined and hypersurface orthogonal everywhere as well as exact in the vacuum region. Hence, there exist a potential Z and an integrating factor X such that $Z_{,\alpha} = X Z_\alpha$, where $X = 1$ in the exterior of $\mathscr{S}_0^{(3)}$. In the vacuum region and in canonical Weyl coordinates, we find $Z = \zeta + \text{const}$. Since we can shift the ζ-coordinate freely, we can drop the constant of integration, which specifies the origin with respect to which the AMM are measured. Thus, W and Z coincide with ρ and ζ in the vacuum region and can be used as their continuation into the interior of the matter. This choice is not unique and other continuations are possible, although they do not alter the values of the source integrals, which we present below.

Using W and Z along $\mathscr{S}_0^{(2)}$ instead of ρ and ζ, respectively, we can rewrite (10) as surface integrals:

$$U^{(r)} = \frac{1}{8\pi} \int\limits_{\mathscr{S}_0^{(3)}} \frac{\mathrm{e}^U}{W} \left(N_-^{(r)} U_{,\hat{n}} - N_{+,W}^{(r)} Z_{,\hat{n}} U + N_{+,Z}^{(r)} W_{,\hat{n}} U \right) \mathrm{d}\mathscr{S}_0^{(3)}. \tag{12}$$

An integration by parts, the axial symmetry and the vacuum field equations are necessary for this step.

Using Stokes' theorem and the field equations we obtain

$$\begin{aligned} U^{(r)} =& \frac{1}{8\pi} \int\limits_{\mathscr{V}_0^{(3)}} \mathrm{e}^U \left[-\frac{N_-^{(r)}}{W} R_{\alpha\beta} \frac{\xi^\alpha \xi^\beta}{\xi^\gamma \xi_\gamma} + N_{+,Z}^{(r)} U \left(\frac{W^{,\alpha}}{W} \right)_{;\alpha} \right. \\ & \left. - N_{+,W}^{(r)} U \left(\frac{Z^{,\alpha}}{W} \right)_{;\alpha} + N_{+,WZ}^{(r)} \frac{U}{W} \left(W^{,\alpha} W_{,\alpha} - Z^{,\alpha} Z_{,\alpha} \right) \right] \mathrm{d}\mathscr{V}_0^{(3)} \\ =& \frac{1}{8\pi} \sum_i \int\limits_{\mathscr{V}_i^{(3)}} \mathrm{e}^U \left[8\pi \frac{N_-^{(r)}}{W} (T g_{\alpha\beta} - T_{\alpha\beta}) \frac{\xi^\alpha \xi^\beta}{\xi^\gamma \xi_\gamma} + N_{+,Z}^{(r)} U \left(\frac{W^{,\alpha}}{W} \right)_{;\alpha} \right. \\ & \left. - N_{+,W}^{(r)} U \left(\frac{Z^{,\alpha}}{W} \right)_{;\alpha} + N_{+,WZ}^{(r)} \frac{U}{W} \left(W^{,\alpha} W_{,\alpha} - Z^{,\alpha} Z_{,\alpha} \right) \right] \mathrm{d}\mathscr{V}_i^{(3)}. \end{aligned} \tag{13}$$

The $\mathrm{d}\mathscr{V}_i^{(3)}$ are the proper volume elements of $\mathscr{V}_i^{(3)}$ and a semicolon denotes the covariant derivative with respect to the line element (1). The last equality is due to Einstein's equations, which imply that the integrand vanishes in vacuum. The integrals (13) are the desired source integrals. They determine the AMM from the geometry inside the matter regions alone. Of course, Stokes' theorem can again be used to rewrite the source integrals as surface integrals over $\mathscr{S}_i^{(3)}$ of the respective matter component. In turn, these can be reformulated as line integrals, cf. Sect. 4. The fact that the contributions of the individual matter components, $\mathscr{V}_i^{(3)}$, to the asymptotic multipole moments superpose linearly is due to the choice of Weyl's multipole moments. If we employ the method from [9] to calculate Geroch's multipole moments M_r from Weyl's multipole moments $U^{(r)}$, we obtain a mixing of the contributions $U_i^{(k)}$ of the individual matter components with $k < r$ in the M_r. This is already apparent for the quadrupole moment M_2, which depends non-linearly on $U^{(0)}$:

$$M_2 = U^{(2)} - \frac{1}{3} U^{(0)3}. \tag{14}$$

The Geroch mass M_0 equals $U^{(0)}$ and is given by the (negative) Komar integral. This follows also from (12) with $r = 0$.

4 Applications

We conclude the paper by discussing one possible application of the source integrals (12). Assume a matter distribution is given, where the metric is known in the interior or the Dirichlet and the Neumann data for U are known at the surface. Even then it is far from trivial (at least in the stationary case, see [11]) to obtain a global asymptotically flat solution, if it exists. The source integrals for the AMM provide a tool to solve this task. As a simple example serves here the case of static dust without any surface distributions. In Weyl coordinates (not necessarily canonical) the energy momentum tensor is given by

$$T_{\alpha\beta} = \mu e^{2U} \delta^t_\alpha \delta^t_\beta. \tag{15}$$

The contracted Bianchi identities imply $U =$ const. in the interior and, thus, the gradient of U vanishes at $\mathscr{S}_i^{(3)}$ in all coordinates. Using the line integrals for Weyl's AMM, which follow from (13), we get:

$$U^{(r)} = \frac{1}{4} \sum_i \int_{\mathscr{S}_i^{(2)}} \left(N_-^{(r)} U_{,\hat{n}} + N_+^{(r)} U_{,\hat{s}} \right) \mathrm{d}\mathscr{S}_i^{(2)} = 0. \tag{16}$$

Thus, all AMM vanish and the spacetime is flat in the exterior. This contradicts the presence of a dust distribution with positive mass density. Of course, this result is already known and more general non-existence results for dust including the rotating case can be found in [12, 13] and references therein. Although the non-existence is proved here, this example shows in a concise way how the source integrals can be applied in more difficult physical situations like rotating stars. This and other applications, e.g. to tidal distortions of black holes, will be investigated in future work.

Acknowledgments N.G. gratefully acknowledges support from the DFG within the Research Training Group 1620 "Models of Gravity" and from the Grant GACR-202/09/0772. The author thanks C. Lämmerzahl, V. Perlick and O. Svítek for helpful discussions.

References

1. Wambsganss, J.: Gravitational lensing in astronomy. Living Rev. Relativ. **1**(12), lrr-1998-12 (1998). http://www.livingreviews.org/lrr-1998-12
2. Perlick, V.: Gravitational lensing from a spacetime perspective. Living Rev. Relativ. **7**(9), lrr-2004-9 (2004). http://www.livingreviews.org/lrr-2004-9
3. Kopeikin, S., Makarov, V.: Gravitational bending of light by planetary multipoles and its measurement with microarcsecond astronomical interferometers. Phys. Rev. D **75**, 062002 (2007). doi:10.1103/PhysRevD.75.062002
4. Thorne, K.: Multipole expansions of gravitational radiation. Rev. Mod. Phys. **52**, 299 (1980). doi:10.1103/RevModPhys.52.299
5. Stephani, H., Kramer, D., MacCallum, M., Hoenselaers, C., Herlt, E.: Exact Solutions of Einstein's Field Equations, 2nd edn. Cambridge Monographs on Mathematical Physics. Cambridge University Press, Cambridge (2003)
6. Gürlebeck, N.: Source integrals for multipole moments in static and axially symmetric spacetimes, ArXiv e-prints [http://arXiv:1207.4500 [gr-qc]] (2012)
7. Geroch, R.: Multipole moments. II. Curved space. J. Math. Phys. **11**, 2580 (1970). doi:10.1063/1.1665427
8. Quevedo, H.: Multipole moments in general relativity – Static and stationary vacuum solutions. Fortschr. Phys. **38**, 733 (1990). doi:10.1002/prop.2190381002
9. Fodor, G., Hoenselaers, C., Perjés, Z.: Multipole moments of axisymmetric systems in relativity. J. Math. Phys. **30**, 2252 (1989). doi:10.1063/1.528551
10. Neugebauer, G., Meinel, R.: Progress in relativistic gravitational theory using the inverse scattering method. J. Math. Phys. **44**, 3407 (2003). doi:10.1063/1.1590419
11. Ansorg, M., Kleinwächter, A., Meinel, R., Neugebauer, G.: Dirichlet boundary value problems of the Ernst equation. Phys. Rev. D **65**, 044006 (2002). doi:10.1103/PhysRevD.65.044006
12. Gürlebeck, N.: The interior solution of axially symmetric, stationary and rigidly rotating dust configurations. Gen. Relativ. Gravity **41**, 2687 (2009). doi:10.1007/s10714-009-0796-8
13. Pfister, H.: Do rotating dust stars exist in general relativity? Class. Quantum Gravity **27**, 105016 (2010). doi:10.1088/0264-9381/27/10/105016

Geodesic Equations and Algebro-Geometric Methods

Eva Hackmann

Abstract For an investigation of the physical properties of gravitational fields the observation of massive test particles and light is very useful. The characteristic features of a given space-time may be decoded by studying the complete set of all possible geodesic motions. Such a thorough analysis can be accomplished most effectively by using analytical methods to solve the geodesic equation. In this contribution, the use of elliptic functions and their generalizations for solving the geodesic equation in a wide range of well known space-times, which are part of the general Plebański-Demiański family of solutions, will be presented. In addition, the definition and calculation of observable effects like the perihelion shift will be presented and further applications of the presented methods will be outlined.

1 Introduction

The observation of massive particles and light is a very important tool for exploring the features of gravitational fields and also for tests of general relativity. The motion of massive and massless test particles is described by the geodesic equation, which is a coupled system of ordinary differential equations dependent on the metric of the considered gravitational field. A wide range of exactly known solutions of Einstein's field equations possesses certain symmetries, which allow to decouple the geodesic equations. Here we discuss metrics within the Plebański-Demiański family of solutions (see [1]), which is a seven parameter solution with mass, rotation, cosmological constant, electric and magnetic charge, NUT charge, and acceleration and which comprises the Schwarzschild and the Kerr metric as special cases. In this family of solutions the equations of motion considerably simplify due to the separability of the

E. Hackmann (✉)
ZARM, Universität Bremen, 28359 Bremen, Germany
e-mail: eva.hackmann@zarm.uni-bremen.de

J. Bičák and T. Ledvinka (eds.), *Relativity and Gravitation*,
Springer Proceedings in Physics 157, DOI: 10.1007/978-3-319-06761-2_12,

Hamilton-Jacobi equation (for lightlike geodesics and, if the acceleration vanishes, also for timelike geodesics).

Due to this simplification, the analytic solutions for the complete set of geodesics in Schwarzschild space-time was already found in 1931 by Hagihara [2] in terms of elliptic functions. With essentially the same methods also the geodesic equations in Kerr-Newman-Taub-NUT space-times and subcases can be handled (and also an additional acceleration for massless particles). For the case of the Kerr metric, this was first done in the equatorial plane (see [3] for a review) and later, after the introduction of Mino time [4] which allows to fully decouple the equations of motion, by Fujita and Hikida for bound orbital motion [5]. For a nonvanishing cosmological constant, the structure of the equations of motions is more complex but can still be solved analytically by using hyperelliptic functions as demonstrated for the Schwarzschild-de Sitter metric [6] and for general axially symmetric space-times [7]. Here, we will explain these general methods to analytically solve the geodesic equations in Kerr-Newman-Taub-NUT-de Sitter space-times (and the C-metric for lightlike geodesics).

For observational purposes also explicit expressions for the deviations of relativistic orbits from the Kepler orbits are of interest. Here we concentrate on the observables for bound orbital motion, namely the periastron shift and the Lense-Thirring effect. However, in a strong gravitational field concepts like the orbital plane and the orbital ellipse are no longer valid. A fully relativistic treatment of these effects in the Kerr gravitational field was given by Schmidt [8] and combined with the Mino time by Drasco and Hughes [9] as well as Fujita and Hikida [5]. In this article, we will show how these concepts can be generalized to the above mentioned space-times.

2 Equations of Motion

The motion of test particles is described by the geodesic equation

$$\frac{d^2x^\mu}{ds^2} + \Gamma^\mu_{\rho\sigma}\frac{dx^\rho}{ds}\frac{dx^\sigma}{ds} = 0 \tag{1}$$

where $\Gamma^\mu_{\rho\sigma} = \frac{1}{2}g^{\mu\alpha}(\partial_\rho g_{\sigma\alpha} + \partial_\sigma g_{\rho\alpha} - \partial_\alpha g_{\rho\sigma})$ is the Christoffel symbol and $\mu = 0, 1, 2, 3$. This system of coupled ODE's can be simplified if the underlying space-time has certain symmetries. In particular, for the Plebański-Demiański solutions (with vanishing acceleration for massive test-particles), there exist four constants of motion which can be used for decoupling: the normalization constant $\varepsilon = g_{\mu\nu}\frac{dx^\mu}{ds}\frac{dx^\nu}{ds}$ with $\varepsilon = 0$ for light and $\varepsilon = 1$ for massive test-particles, the energy E, the angular momentum L in direction of the symmetry axis, and the Carter constant K.

For the considered family of solutions in standard Boyer-Lindquist coordinates the radial and latitudinal equations of motion can be reduced to the form [7]

$$\left(x^i \frac{dx}{d\lambda}\right)^2 = P(x; p), \tag{2}$$

where x is the radius or the (squared) cosine of the latitude, λ is an affine parameter, the Mino time [4], P is a polynomial in x of degree $2g + 1$ or $2g + 2$, $0 \leq i < g$ is an integer, and $p = \{p_1, \ldots, p_n\}$ is a set of parameters of the space-time and the test-particle. For example, in Kerr space-time we get in geometrized units ($G = 1$, $c = 1$) [3, 4]

$$\left(\frac{dr}{d\lambda}\right)^2 = \left((r^2 + a^2)E - aL\right)^2 - (r^2 + a^2 - 2r)(\varepsilon r^2 + K) =: R(r), \tag{3}$$

$$\left(\frac{d\xi}{d\lambda}\right)^2 = 4\xi\left[(1 - \xi)(K - \varepsilon a^2 \xi) + (aE(1 - \xi) - L)^2\right], \tag{4}$$

where $\xi = \cos^2\theta$ and all quantities are normalized such that they are dimensionless.

3 Algebro-Geometric Methods

The equation of motion (2) should be solved for $x(\lambda)$, i.e. we consider the inversion problem

$$\int_{x_0}^{x} \frac{x^i dx}{\sqrt{P(x; p)}} = \lambda - \lambda_0 , \tag{5}$$

where $x(\lambda_0) = x_0$ are initial values. The solution of this problem should be independent from the chosen integration path. This implies that for a closed path γ with $\omega := \oint_\gamma \frac{x^i dx}{\sqrt{P(x;p)}} \neq 0$ the solution $x(\lambda)$ has to have the period ω, $\int_{x_0}^{x} \frac{x^i dx}{\sqrt{P(x;p)}} = \lambda - \lambda_0 - \omega$. This can be taken into account automatically if (2) is considered as an algebraic curve

$$w^2 = P(x; p), \tag{6}$$

where $w = x^i \frac{dx}{d\lambda}$. For the considered space-times one of the following two situations occurs

- P is of order 3 or 4: then (6) is an elliptic curve of genus 1,
- P is of order 5 or 6: then (6) is an hyperelliptic curve of genus 2.

Topologically, (hyper-) elliptic curves can be considered as Riemann surfaces. The genus g corresponds to the number of 'holes' in the Riemann surface, see Fig. 1. If there are g holes this implies that there are $2g$ independent closed integration paths

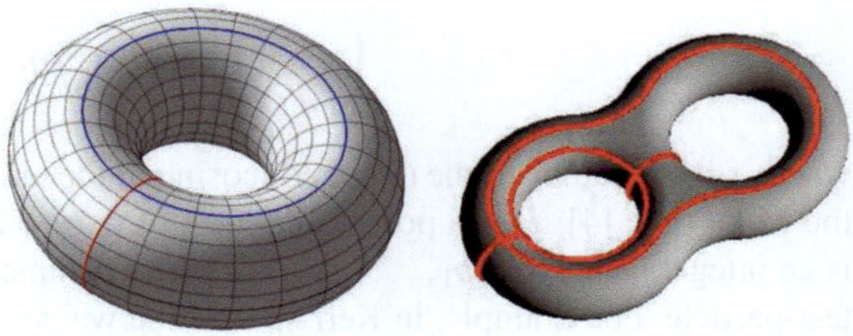

Fig. 1 Riemann surface of a genus one curve (*left*) and a genus two curve (*right*) with two or four independent paths

whose integrals do not vanish and, therefore, the solution $x(\lambda)$ needs to have $2g$ independent periods [10].

In the case of an elliptic curve, (6) can be reduced to the Weierstrass form by a rational substitution,

$$\tilde{w}^2 = 4\tilde{x}^3 - g_2\tilde{x} - g_3 \,. \tag{7}$$

In this standard form, $\tilde{w}$ and $\tilde{x}$ are parametrized by the Weierstrass elliptic function and its derivative, $\tilde{w} = \wp'(z)$ and $\tilde{x} = \wp(z)$. In the above example of Kerr space-time, the substitution for the radial equation of motion for a timelike geodesic is $r = \frac{a_3}{4x - \frac{a_2}{3}} + r_K$, where $R(r_K) = 0$ and $a_j = \frac{1}{(4-j)!}\frac{d^{(4-j)}R}{dr^{(4-j)}}(r_K)$. The resulting equation is given by (7) with

$$g_2 = \frac{1}{4}\left(\frac{1}{3}a_2^2 - a_1a_2\right), \quad g_3 = \frac{1}{16}\left(\frac{1}{3}a_1a_2a_3 - \frac{2}{27}a_2^3 - a_0a_3^2\right). \tag{8}$$

The analytical solution for the timelike radial equation in Kerr space-time is then given by

$$r(\lambda) = \frac{a_3}{4\wp(\lambda - c) - \frac{a_2}{3}} + r_K \,, \tag{9}$$

where $c = c(r_0, \lambda_0)$ is a constant which depends only on the initial conditions.

With a completely analogous procedure all equations of motion which reduce to elliptic curves may be solved. This includes geodesics in Schwarzschild, Reissner-Nordström, Kerr-Newman, and Taub-NUT space-times as well as the C-metric (see e.g. [11]) for massless particles.

The situation gets more complicated if the equation of motion is described by a hyperelliptic curve of genus two. This is due to the fact that the solution has to have four independent periods, which is impossible for functions in a single complex variable. Therefore, it is necessary to consider a function in $g = 2$ complex variables. However, as we have only one degree of freedom, we need to restrict the space of variables again to a one-dimensional submanifold. This is done by considering the equation of motion as part of the Jacobi inversion problem for $g = 2$, which is to solve the system of equations

$$w_1 = \int_\infty^{x_1} \frac{dx}{\sqrt{P(x)}} + \int_\infty^{x_2} \frac{dx}{\sqrt{P(x)}}, \qquad (10)$$
$$w_2 = \int_\infty^{x_1} \frac{xdx}{\sqrt{P(x)}} + \int_\infty^{x_2} \frac{xdx}{\sqrt{P(x)}},$$

for x_1, x_2 as functions of w_1, w_2. If P is transformed to a standard form $P(x) = 4x^5 + \sum_{i=0}^4 a_i x^i$, the solution to this problem is known in terms of generalized $\wp$-functions, $x_1 x_2 = \wp_{12}(w_1, w_2)$, $x_1 + x_2 = \wp_{22}(w_1, w_2)$ [10]. Here $\wp_{ij}(w_1, w_2) = -\frac{\partial}{\partial w_i}\frac{\partial}{\partial w_j} \log \sigma(w_1, w_2)$ with the generalized σ function. By letting x_2 go to infinity, we can get rid of the second integral on the right hand side and simultaneously restrict (w_1, w_2) to the one-dimensional sigma divisor, i.e. the set of zeros of the two-dimensional σ-function. The solution for $x = x_1$ is then given by

$$x = -\frac{\sigma_1}{\sigma_2}(w_1, w_2), \qquad \text{with} \quad \sigma(w_1, w_2) = 0. \qquad (11)$$

where σ_i denotes derivative of σ w.r.t. the i-th variable. For example, in the case of Schwarzschild-de Sitter space-time, the solution for the radius r in terms of the angle φ is given by [6]

$$r(\varphi) = -M\frac{\sigma_2}{\sigma_1}(f(\varphi), \varphi), \qquad \text{with} \quad \sigma(f(\varphi), \varphi) = 0. \qquad (12)$$

This solution method can be applied to all geodesic equations which reduce to hyperelliptic curves. This comprises the radial and latitudinal equations of motion in Kerr-de Sitter space-time as well as all the de Sitter-versions of Kerr-Newman-Taub-NUT space-times and subcases [7]. For lightlike geodesics also the C-metric with an additional cosmological constant can be treated this way.

4 Observables

For the bound orbital motion of massive test-particles there are two main gravitational effects: the periastron shift Δ_{P} and the Lense-Thirring effect Δ_{LT}. The first is a precession of the orbital ellipse within its orbital plane and the latter a precession of the orbital plane itself.[1] In this picture, the periastron shift is defined as the angle between two consecutive periastrons and the Lense-Thirring effect as the angle between two consecutive minimal latitudes. Mathematically this means that they are given by the difference between the period of $r(\varphi)$ or $\theta(\varphi)$, respectively, and 2π. If $2\omega_r$ denotes the period of $r(\varphi)$ and $2\omega_\theta$ the period of $\theta(\varphi)$ then

[1] In the original paper [12] a combined effect was computed: the timely variation of the ascending node and the argument of periapsis. With 'Lense-Thirring effect' we refer only to the first correction.

$$\Delta_{\mathrm{P}} = 2\omega_r - 2\,\mathrm{sign}(L)\,\pi, \qquad \Delta_{\mathrm{LT}} = 2\omega_\theta - 2\,\mathrm{sign}(L)\,\pi, \tag{13}$$

where the sign of L is included to distinguish between prograde and retrograde motion.

In Schwarzschild space-time, where $r(\varphi)$ is directly known, the exact analytical expression for the periastron shift is given by

$$\Delta_{\mathrm{P}} = \frac{4LK(k)}{\sqrt{(E^2-1)r_{\mathrm{p}}(r_{\mathrm{a}}-r_1)}} - 2\pi, \tag{14}$$

where $K(k)$ is the complete elliptic integral of the first kind,

$$K(k) = \int_0^1 \frac{dx}{\sqrt{(1-x^2)(1-k^2x^2)}}, \tag{15}$$

which is implemented in standard mathematical software. Here $k^2 = \frac{r_1(r_{\mathrm{a}}-r_{\mathrm{p}})}{r_{\mathrm{p}}(r_{\mathrm{a}}-r_1)}$ with the zeros $0 < r_1 < r_{\mathrm{p}} < r_{\mathrm{a}}$ of $R(r) = \left(\frac{dr}{d\varphi}\right)^2$.

The expression for the perihelion shift can be generalized to axially symmetric space-times, where only $r(\lambda)$ and $\varphi(\lambda)$ is known. Writing $\varphi(\lambda)$ as a part linear in λ, given in the form of an infinite Mino time average Υ_φ, plus oscillatory deviations [5, 8], we may use $\lambda(\varphi) = \Upsilon_\varphi^{-1}\varphi$. Then the period of $r(\varphi)$ is given by

$$r(\lambda(\varphi + 2\Lambda_r\Upsilon_\varphi)) = r(\Upsilon_\varphi^{-1}\varphi + 2\Lambda_r) = r(\Upsilon_\varphi^{-1}\varphi) = r(\lambda(\varphi)), \tag{16}$$

where $2\Lambda_r$ is the period of $r(\lambda)$. Accordingly, $2\omega_r = 2\Lambda_r\Upsilon_\varphi$ and the perihelion shift can be written as

$$\Delta_{\mathrm{P}} = 2\Lambda_r\Upsilon_\varphi - 2\,\mathrm{sign}(L)\,\pi. \tag{17}$$

In a totally analogous way the Lense-Thirring effect can be found. If $2\Lambda_\theta$ denotes the period of $\theta(\lambda)$, we may write again $\theta(\lambda(\varphi + 2\Lambda_\theta\Upsilon_\varphi)) = \theta(\lambda(\varphi))$ and, therefore,

$$\Delta_{\mathrm{LT}} = 2\Lambda_\theta\Upsilon_\varphi - 2\,\mathrm{sign}(L)\,\pi. \tag{18}$$

In the case of Kerr-Newman-Taub-NUT-de Sitter space-times, $\Lambda_{r,\theta}$ and Υ_φ are given in terms of (hyper-)elliptic integrals. For the elliptic case, they can be rewritten in terms of the three standard Jacobian elliptic integrals $K(k)$, $E(k)$, and $\Pi(n,k)$. For the case of genus two or higher, to our knowledge such a standard form does not exist. A possible generalization of the first Jacobian elliptic integral would be

$$K_{\mathbf{A}}(\mathbf{k}) = \int_0^1 \frac{\sum_{i=1}^{g} A_i t^{i-1} dt}{\sqrt{t(1-t)\prod_{i=1}^{2g-1}(1-k_i^2 t)}}, \tag{19}$$

where $\mathbf{A}$ is a vector of length g, which reflects the fact that there are g independent differentials of the first kind, and $\mathbf{k}$ is of length $2g-1$.

For example, in terms of these integrals the perihelion shift for Schwarzschild-de Sitter space-time is given by

$$\Delta_{\mathrm{P}} = \frac{r_4 K_{\left(\frac{1}{r_4}, \frac{(r_4-r_3)}{r_3 r_4}\right)}(k_1, k_2, k_3)}{\sqrt{L^2 \Lambda r_3 (r_4 - r_0)(r_4 - r_2)(r_5 - r_4)}} - 2\pi, \tag{20}$$

where $r_0 < 0 < r_1 < r_2 < r_3 < r_4 < r_5$ are the zeros of the defining polynomial with $r_3 = r_{\min}$ and $r_4 = r_{\max}$ for small positive Λ and

$$k_1^2 = \frac{r_0(r_4 - r_3)}{r_3(r_4 - r_0)}, \quad k_2^2 = \frac{r_2(r_4 - r_3)}{r_3(r_4 - r_2)}, \quad k_3^2 = -\frac{r_5(r_4 - r_3)}{r_3(r_5 - r_4)}. \tag{21}$$

5 Outlook

The methods presented here are powerful tools for the analytic integration of the geodesic equation in a wide range of space-times. Beside the space-times we focused on here, geodesics in higher-dimensional spherically symmetric space-times [13, 14] and the Myers-Perry space-time [15] can be treated. It may also be possible to extend these methods to the equations of motion in space-times with given multipole moments like the Erez-Rosen space-time.

The hyperelliptic curves which we used here to represent the equations of motions are a special case of Abelian curves, which allow for higher orders of w in (6) and mixed terms. These more general curves appear for example in Hořava-Lifshitz and Gauss-Bonnet gravity, which may be represented by quartic curves of the form $(w - P(x))^2 = Q(x)$. A generalization of the presented methods to these cases is in preparation.

Analogously to the analytic expressions presented here for observables of bound orbital motion also the bending of light and the gravitational time delay may be considered. Linked to that, we plan the development of an analytical timing formula for pulsars orbiting a black hole.

Concerning the numerical calculation of the analytical expressions, the complete elliptic integrals can be computed very efficiently by using the arithmetic geometric mean. This can be generalized to genus two hyperelliptic integrals, see [16, 17].

References

1. Griffiths, J., Podolský, J.: A new look at the Plebański-Demiański family of solutions. Int. J. Mod. Phys. **15**, 335 (2006). doi:10.1142/S0218271806007742
2. Hagihara, Y.: Theory of relativistic trajectories in a gravitational field of Schwarzschild. Jpn. J. Astron. Geophys. **8**, 67 (1931)
3. Chandrasekhar, S.: The Mathematical Theory of Black Holes. International Series of Monographs on Physics. Oxford University Press, Oxford (1983)
4. Mino, Y.: Perturbative approach to an orbital evolution around a supermassive black hole. Phys. Rev. D **67**, 084027 (2003). doi:10.1103/PhysRevD.67.084027
5. Fujita, R., Hikida, W.: Analytical solutions of bound timelike geodesic orbits in Kerr spacetime. Class. Quantum Grav. **26**, 135002 (2009). doi:10.1088/0264-9381/26/13/135002
6. Hackmann, E., Lämmerzahl, C.: Complete analytic solution of the geodesic equation in Schwarzschild-(anti) de Sitter space-times. Phys. Rev. Lett. **100**, 171101 (2008). doi:10.1103/PhysRevLett.100.171101
7. Hackmann, E., Kagramanova, V., Kunz, J., Lämmerzahl, C.: Analytic solutions of the geodesic equation in axially symmetric space-times. Europhys. Lett. **88**, 30008 (2009). doi:10.1209/0295-5075/88/30008
8. Schmidt, W.: Celestial mechanics in Kerr spacetime. Class. Quantum Grav. **19**, 2743 (2002). doi:10.1088/0264-9381/19/10/314
9. Drasco, S., Hughes, S.: Rotating black hole orbit functionals in the frequency domain. Phys. Rev. D **69**, 044015 (2004). doi:10.1103/PhysRevD.69.044015
10. Farkas, H., Kra, I.: Riemann Surfaces, Graduate Texts in Mathematics, vol. 71. Springer, Berlin (1992)
11. Bičák, J., Schmidt, B.: Asymptotically flat radiative space-times with boost-rotation symmetry: the general structure. Phys. Rev. D **40**, 1827 (1989). doi:10.1103/PhysRevD.40.1827
12. Lense, J., Thirring, H.: Über den Einfluß der Eigenrotation der Zentralkörper auf die Bewegung der Planeten und Monde nach der Einsteinschen Gravitationstheorie. Phys. Z. **19**(8), 156 (1918)
13. Hackmann, E., Kagramanova, V., Kunz, J., Lämmerzahl, C.: Analytic solutions of the geodesic equation in higher dimensional static spherically symmetric spacetimes. Phys. Rev. D **78**, 124018 (2008). doi:10.1103/PhysRevD.78.124018
14. Enolski, V., Hackmann, E., Kagramanova, V., Kunz, J., Lämmerzahl, C.: Inversion of hyperelliptic integrals of arbitrary genus with application to particle motion in general relativity. J. Geom. Phys. **61**(5), 899 (2011). doi:10.1016/j.geomphys.2011.01.001
15. Kagramanova, V., Reimers, S.: Analytic treatment of geodesics in five-dimensional Myers-Perry space-times. Phys. Rev. D **86**, 084029 (2012). doi:10.1103/PhysRevD.86.084029
16. Bost, J.B., Mestre, J.F.: Moyenne Arithmético-géométrique et Périodes des Courbes de genre 1 et 2. Gaz. Math. Soc. France **38**, 36 (1988)
17. Braden, H., D'Avanzo, A., Enolski, V.: On charge-3 cyclic monopoles. Nonlinearity **24**, 643 (2011). doi:10.1088/0951-7715/24/3/001

Illusory Horizons, Thermodynamics, and Holography Inside Black Holes

Andrew J. S. Hamilton

Abstract There is persistent and endemic confusion between the true (future) horizon and the illusory (past) horizon of a black hole. The illusory horizon is the redshifting surface of matter that fell into the black hole long ago. A person who free-falls through the horizon of a black hole falls through the true horizon, not the illusory horizon. The infaller continues to see the illusory horizon ahead of them, all the way down to the classical singularity. The illusory horizon is the source of Hawking radiation, for both outsiders and infallers.The entropy of a black hole is 1/4 of the area of the illusory horizon, for both outsiders and infallers. The illusory horizon holographically encodes states hidden behind it, for both outsiders and infallers. The endpoint of an infaller approaching the classical singularity is to merge their states with the illusory horizon. The holographic boundary of the black hole is then the union of the illusory horizon and the classical spacelike singularity. When an infaller reaches the classical singularity, any entanglement of the infaller with outsiders or other infallers is transferred to entanglement with the states of the black hole, encoded on the illusory horizon. Locality holds between an infaller and a spacelike-separated outsider or other infaller as long as their future lightcones intersect before the singularity, but breaks down when the future lightcones no longer intersect.

1 Introduction

There is persistent and endemic confusion in the literature between the true (future) horizon and the illusory (past) horizon of a black hole. The confusion has led to the misconception that Hawking radiation is emitted from the true horizon, and that the states of a black hole are encoded on the true horizon.

A. J. S. Hamilton (✉)
JILA, Box 440, U. Colorado, Boulder, CO 80309, USA
e-mail: Andrew.Hamilton@colorado.edu

J. Bičák and T. Ledvinka (eds.), *Relativity and Gravitation*,
Springer Proceedings in Physics 157, DOI: 10.1007/978-3-319-06761-2_13,

The presence of a black hole introduces a bifurcation boundary to spacetime, separating the spacetime into a region that an observer can see, and a region that is invisible to the observer. This bifurcation horizon is the illusory horizon, and it is observer-dependent. The illusory horizon is the boundary of the past lightcone of an observer watching the black hole.

When an observer measures thermodynamic variables such as temperature or entropy, they must measure degrees of freedom that are actually available to them, which is to say, degrees of freedom along their past lightcone. Thus a consistent description of generalized thermodynamics by an actual observer must involve the observer's illusory horizon, not the true horizon.

The purpose of this paper is to set forward a number of proposals regarding generalized thermodynamics from the perspective of observers who fall through the true horizon.[1] The proposals are motivated by the classical appearance of the illusory horizon seen by an infaller. The classical appearance suggests that the principles of generalized thermodynamics and holography extend to infallers in the simplest and most obvious way.

For simplicity, this paper considers only a spherically symmetric, uncharged (Schwarzschild) black hole.

2 The Illusory Horizon

Figure 1 shows the familiar Penrose diagram of a Schwarzschild black hole, with the illusory (past) and true (future) horizons labelled. In the analytically extended Schwarzschild geometry, the illusory horizon is a true horizon, the horizon of a white hole and parallel universe. In a real black hole however, the Schwarzschild past horizon is replaced by the exponentially dimming and redshifting image of the star that collapsed to the black hole long ago.

As the Penrose diagram of the Schwarzschild black hole shows, when an observer outside the black hole looks at the black hole, they are looking at the illusory horizon. When an observer free-falls through the horizon of the black hole, they fall through the true horizon, not the illusory horizon. The true horizon becomes visible to the observer only after the observer has passed through it. The illusory horizon continues to appear ahead of the observer even after they have passed through the true horizon.

Figure 2 illustrates three frames from a visualization of the scene seen by an observer who free-falls into a Schwarzschild black hole [1, 5]. These scenes are general relativistically ray-traced, not artist's impressions. The illusory and true horizons of the black hole are painted with grids of latitude and longitude, so that they can be seen. The illusory horizon is of course infinitely redshifted in the Schwarzschild

[1] Editors' footnote: The author did not follow the requests of referees to distinguish clearly between the established results and vague conjectures/proposals. The editors decided to include the paper in its original form because of its possible inspiring role.

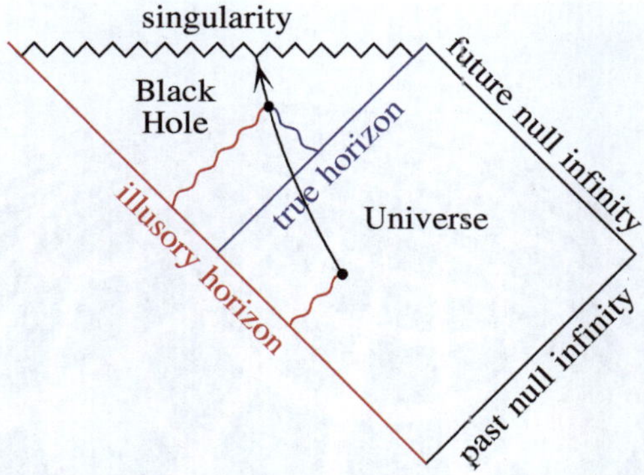

Fig. 1 Penrose diagram of a Schwarzschild black hole. The *arrowed line* represents the trajectory of an observer, while the *wiggly lines* represent light rays perceived by the observer from the illusory (*red*) and true (*blue*) horizons

geometry, but it is nevertheless possible to ray-trace light rays from an infinitesimal distance off the illusory horizon.

The visualization confirms the expectation from the Penrose diagram. When the observer falls through the horizon, they do not fall through the illusory horizon, which continues to appear a finite distance ahead of the observer. Instead, the observer falls through a new entity, the true horizon, which was invisible until the observer passed through it. At the moment the observer passes through the true horizon, it forms a line extending down to the illusory horizon. As the observer falls inward, the true horizon expands into a bubble over the observer's head. The circle where the illusory and true horizons intersect expands.

Are visualizations of the Schwarzschild geometry a reliable guide to visualizations of real spherical black holes? Yes. Figure 3 shows three frames from the collapse of a spherical, uniform density, pressureless star that starts from zero velocity at infinity, a problem first solved by Oppenheimer and Snyder [3]. The frames are as seen by an observer at radius 20 geometric units. Again, these frames are general relativistically ray-traced, not artist's impressions. The frames take into account the differential light travel time from different parts of the star's surface to the observer. As the star approaches its horizon, the star freezes, and takes on the appearance of a Schwarzschild black hole.

3 The Illusory Horizon is the Source of Hawking Radiation, for Outsiders and Insiders

At its most fundamental level, Hawking [6] or Unruh [7, 8] radiation arises when an observer watches an emitter that is accelerating relative to the observer. When waves that are pure negative frequency (positive energy) in the emitter's frame are propagated to the observer, the acceleration causes the waves to appear to be a mix of

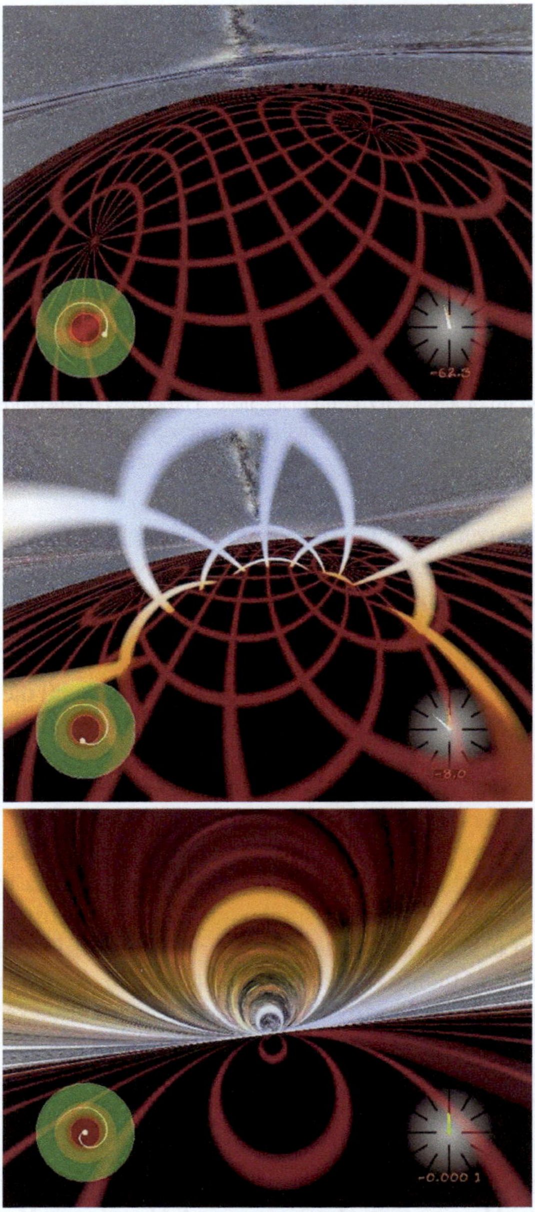

Fig. 2 Visualization of the scene seen by an observer falling into a Schwarzschild black hole on a geodesic with specific energy and angular momentum $E = 1$ and $L = 3.92$ geometric units, from [1]. In the *upper panel*, the observer is at a radius of 3.000, outside the true horizon; in the *middle panel* the observer is at a radius of 1.613, inside the true horizon; in the *bottom panel* the observer is at a radius of 0.045, near the central singularity. The illusory horizon is painted with a *dark red grid*, as befits its infinitely redshifted appearance, while the true horizon is painted with an appropriately *red-* or *blue-*shifted blackbody color. Further frames and details of this visualization are at [1]. The background is Axel Mellinger's Milky Way [2] (with permission)

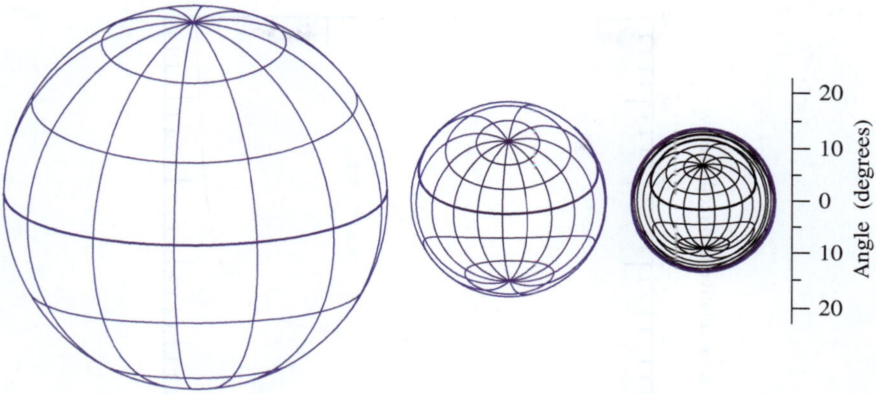

Fig. 3 Three frames in the Oppenheimer-Snyder collapse of [3]. a star, as seen by an outside observer at rest at a radius of 20 geometric units. As time goes by, from *left* to *right*, the collapsing star appears to freeze at its horizon, and take on the appearance of a Schwarzschild black hole. An animated version of this visualization is at [4]

negative and positive frequencies in the observer's frame. In particular, the emitter's vacuum ("in" vacuum) is not the same as the observer's vacuum ("out" vacuum). A classic calculation (e.g. [9, 10]) shows that if the acceleration is approximately constant over several acceleration timescales, then the observer will see the emitter's vacuum as a thermal state with temperature proportional to the acceleration.

An observer watching a black hole sees Hawking radiation because matter that collapsed to the black hole long ago appears classically frozen at the illusory horizon, apparently accelerating away from the observer, redshifting and dimming into the indefinite future. When an infaller free-falls through the true horizon, they do not encounter the redshifting surface at the true horizon. Rather, the infaller sees the redshifting surface of the collapsed matter continue to remain on the illusory horizon ahead of them, as illustrated by Fig. 2.

An exact calculation of the Hawking emission seen by an infaller is difficult, as illustrated by the efforts of [11] reported at this conference. The reason for the difficulty is that, whereas for a distant observer only the monopole mode of emission is important, for an infaller all angular modes contribute. However, it is possible to predict the qualitative character of the Hawking radiation from a classical calculation of the acceleration at the illusory horizon, as witnessed by an infaller.

The acceleration, hence the Hawking or Unruh radiation, that an infaller sees depends on the state of motion of the infaller. The simplest case is that of an observer who free-falls radially from zero velocity at infinity, and who fixes their gaze in a particular direction (that is, the infaller's detector is non-rotating). Figure 4 shows the acceleration on the illusory horizon seen by such an infaller well inside the true horizon, at a radial position $r = 0.01$ geometric units. Note that the observer here is staring at a fixed angular direction relative to their own locally inertial frame, not at a fixed angular position on the black hole. Figure 4 shows that the acceleration is

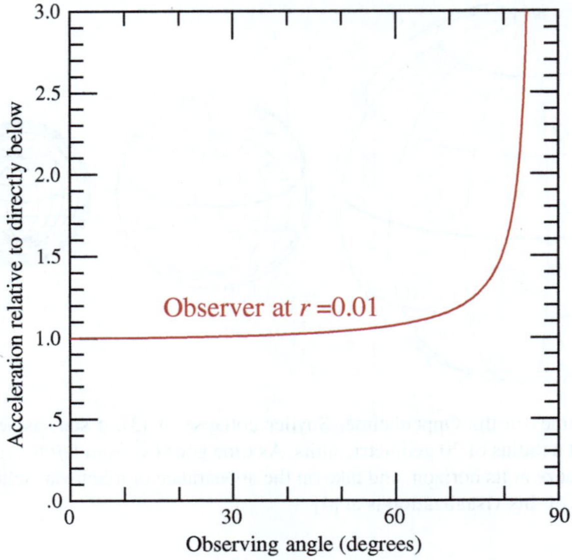

Fig. 4 Acceleration κ on the illusory horizon seen by a radially free-falling non-rotating infaller, relative to the acceleration κ_0 directly below (towards the black hole), as a function of the viewing angle relative to directly below. The example *curve* shown is as seen by an infaller well inside the horizon, at radius 0.01 geometric units. The acceleration is constant out to near the perceived edge of the black hole, where the acceleration diverges. *Curves* at other radii are similar

approximately constant out to near the perceived edge of the black hole, indicating that the acceleration directly below is representative of the black hole as a whole.

Figure 5 shows the acceleration on the illusory horizon directly below, as seen by the radially free-falling infaller as a function of their radial position r. The acceleration is approximately constant (1/4 geometric units) far from the black hole, but increases inward, diverging as the infaller approaches the classical singularity, $r \to 0$. The figure shows that the acceleration changes on a timescale comparable to the proper time left for the infaller to hit the singularity. Thus the usual connection between acceleration and temperature (which requires the acceleration to remain approximately constant over several acceleration times) fails. Nevertheless, the calculation does suggest that the Hawking radiation witnessed by an infaller might diverge as the infaller approaches the singularity. The calculation suggests of order one Hawking quantum per time remaining, or a logarithmically diverging total number of quanta. Rigorous calculation will be required to test this proposal.

Figure 5 also shows the acceleration on the distant sky directly above, as seen by the radially free-falling infaller. The acceleration is negligible when the infaller is far from the black hole, but increases inward. Interestingly, the acceleration on the sky above approaches the same diverging value as that on the illusory horizon below as the infaller approaches the singularity. This suggests that the infaller approaching the singularity might see logarithmically diverging Hawking radiation from all directions.

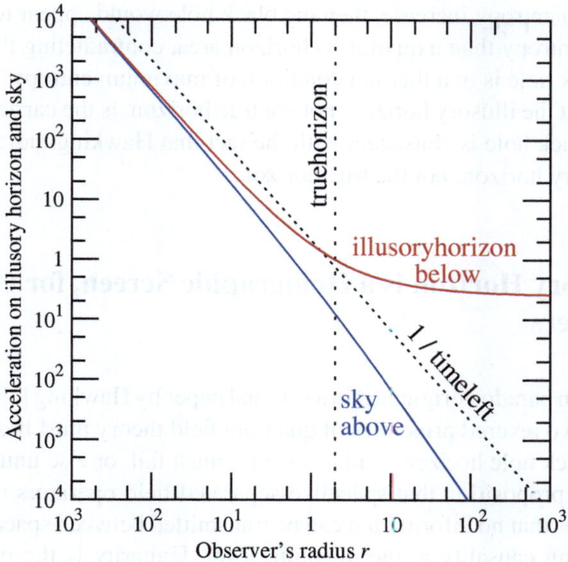

Fig. 5 Acceleration at the illusory horizon directly below, and at infinity directly above, seen by a radially free-falling infaller at radius r. The *dashed line* shows the reciprocal of the proper time left until the infaller hits the singularity. The acceleration diverges towards the singularity $r \to 0$, suggesting a logarithmic divergence in the total number of Hawking quanta observed by an infaller reaching the singularity

4 The Entropy of a Black Hole is 1/4 the Area of the Illusory Horizon, for Outsiders and Insiders

Generalized thermodynamics (e.g. [12]) postulates that from the perspective of an observer outside the true horizon, a black hole that has reached near stationarity should be treated as an object in near thermodynamic equilibrium, with an entropy equal to 1/4 of its horizon area in Planck units, and a temperature equal to $1/(2\pi)$ times the acceleration at the illusory horizon.

Generalized thermodynamics may reasonably be expected to hold also for infallers. For example, it would be quite extraordinary if an infaller witnessed a violation of the second law of thermodyamics. As remarked in the Introduction, an observer must count entropy that is visible to them, that is, entropy along their past lightcone. The boundary of the observer's past lightcone towards the black hole is the illusory horizon. Generalized thermodynamics teaches that entropy must be associated with the boundary, the illusory horizon.

Figure 2 shows that the appearance of the illusory horizon is seamless for infallers who free-fall through the true horizon. It is natural therefore to propose that the entropy of the black hole is 1/4 the area of the illusory horizon not only for outsiders, but also for infallers. Indeed, if an infaller saw the horizon entropy decrease when they fell inside, then that would violate the second law. Conversely if the infaller saw the horizon entropy increase, then the black hole would appear to the infaller to contain more entropy than a quarter its horizon area, contradicting the notion that a stationary black hole is in a thermal condition of maximum entropy.

The idea that the illusory horizon, not the true horizon, is the carrier of the hidden states of the black hole is consistent with the fact that Hawking radiation originates from the illusory horizon, not the true horizon.

5 The Illusory Horizon is a Holographic Screen, for Outsiders and Insiders

The information paradox originated in a seminal paper by Hawking [13]. The paradox is that one of two revered principles of quantum field theory must break down in the presence of black hole horizons: either locality must fail, or else unitarity must fail. Locality is the proposition that spacelike-separated field operators must commute. Locality ensures that no information can be transmitted between spacelike-separated points, enforcing causality at the quantum level. Unitarity is the proposition that dynamics is reversible at the quantum level. Hawking tacitly assumed that locality holds, and showed that the Hilbert space of states inside a black hole is then disjoint from those of an observer to the future of when the black hole has evaporated. Consequently information is destroyed, violating unitarity.

The most widely accepted resolution of the information paradox is holography, an idea originally proposed by t'Hooft [16] and Susskind [17]. Holography asserts that the quantum states seen by an insider are seen by an outsider as residing on the horizon of the black hole. Holography violates locality because the Hilbert spaces of spacelike-separated regions, far from being disjoint, are identified with each other. Information about what happens inside the black hole is encoded on its horizon, and eventually radiated to the outside as Hawking radiation, preserving unitarity. Holography has received impetus from gauge/gravity dualities that arise in string theory, whereby a strongly gravitating system is dual to a conformal gauge theory residing on the boundary of the system.

Arguments favouring a breakdown of locality become stark when one considers not just one insider, but a succession of infallers. As shown by [14], if a black hole accretes gas, increasing its Bekenstein-Hawking entropy by some amount, then processes of dissipation inside the black hole can potentially increase the entropy of the gas not merely by the increase in the Bekenstein-Hawking entropy, but rather by some fraction of the total Bekenstein-Hawking entropy of the entire black hole. If locality held, then it would be legitimate to accumulate the entropy from multiple parcels of infalling gas, leading to a total entropy inside the black hole many orders of magnitude greater than its Bekenstein-Hawking entropy. This would imply a gross violation of the second law when the black hole subsequently evaporated, as illustrated by Fig. 6. To save the second law of thermodynamics from the [14] argument, locality must be abandoned not only across the horizon, but between a multiple succession of infallers.

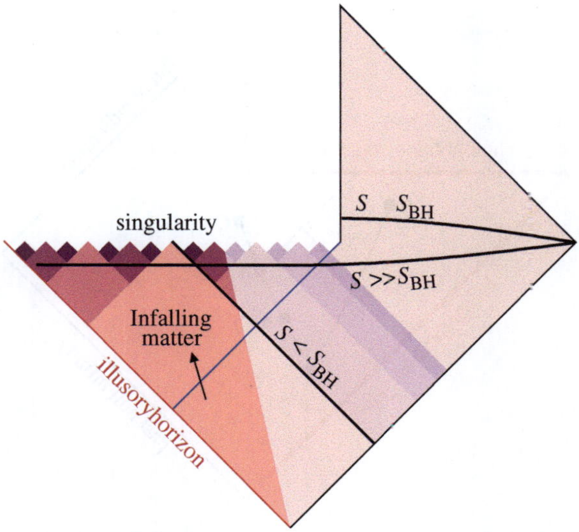

Fig. 6 Near its singularity, a black hole contains numerous regions whose future lightcones do not intersect. If locality held inside a black hole, then it would be legimitate to accumulate entropy along a spacelike surface slicing through these causally disconnected regions. Dissipative processes inside a black hole can potentially cause the entropy accumulated along the spacelike surface to exceed greatly the Bekenstein-Hawking entropy of the black hole [14], leading to a violation of the second law when the black hole evaporates. This argument strongly supports the idea that locality must break down inside black holes. Whereas entropy passing through a spacelike surface inside the black hole may exceed the Bekenstein-Hawking entropy, the entropy passing through any null surface inside the black hole is always less than the Bekenstein-Hawking entropy, consistent with Bousso's [15] covariant entropy bound

Holography produces just the kind of breakdown of locality that is needed to save the second law of thermodynamics inside black holes. Just as an outsider must count states hidden behind their illusory horizon as being holographically encoded on their illusory horizon, so also an infaller must count states hidden behind their illusory horizon as being holographically encoded on their illusory horizon. In this view, an infaller should not count the entropy production witnessed by earlier infallers if that entropy production occurred behind the later infaller's illusory horizon.

6 An Infaller Merges States with the Illusory Horizon at the Classical Singularity

The bottom panel of Fig. 2 shows that, as an infaller approaches the classical singularity, they have the impression of reaching the illusory horizon, which gives the appearance of a flat plane. Any quantitative measure of distance to the illusory horizon, such as the affine distance (the affine parameter normalized to measure proper distance in the observer's frame), or the angular diameter distance (the distance

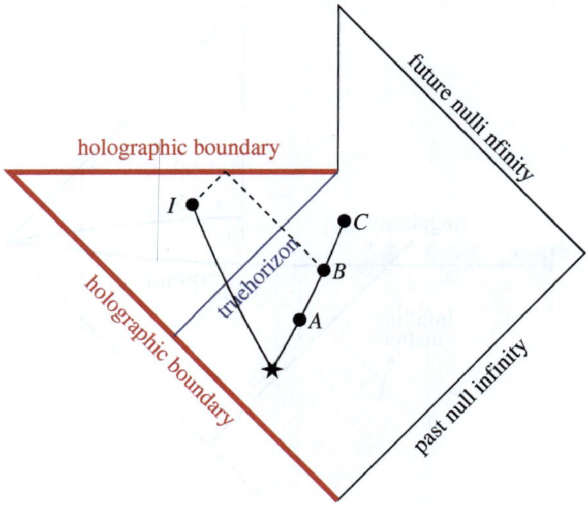

Fig. 7 The illusory horizon and the singularity constitute the holographic boundary of an evaporating black hole. The diagram illustrates the delocalization of an entangled pair created at the *star point*. Locality holds between an inside observer I and an outside observer as long as their future lightcones intersect, so that they can communicate before I hits the singularity. Thus locality holds between I and A, is at the brink of failure between I and B, and fails between I and C

inferred from the apparent angular separation of objects a known distance apart, such as lines of constant latitude and longitude), indeed goes to zero as the observer approaches the singularity.

In the light of the classical appearance, it is natural to propose that an infaller who reaches the singularity merges their states with the illusory horizon. It has been argued in this paper that prior to the singularity, the experience of an infaller can be described by general relativity coupled with a natural extension of generalized thermodynamics. Such a description must fail at the singularity, where the tidal force diverges, and, as argued in Sect. 3, the Hawking radiation may also diverge. The proposal is that the description of physics at the singularity should be replaced by a holographic dual description. In this picture, as illustrated in the Penrose diagram in Fig. 7, the complete holographic boundary of the black hole consists of the union of the illusory horizon and the singularity.

7 Where Locality Breaks Down Inside Black Holes

The simplest possibility is that the transition from a classical to a dual holographic description at the singularity is so rapid as to be effectively instantaneous. If so, then any quantum entanglement between an infaller and an outsider or other infallers will be replaced "instantly" by entanglement with the holographic image of the black hole when the infaller hits the singularity.

Figure 7 illustrates how locality between a pair of particles created in an entangled state (e.g. a spin-zero singlet of spin-up and spin-down particles) breaks down as one of the pair falls inside the black hole towards the singularity. Locality holds between an insider who observes the inside particle at I, and an outsider who observes the outside particle at A, because their future lightcones intersect, so they can compare their measurements of spin. But locality fails between I and an outsider who observes the outside particle at C, because their future lightcones do not intersect, so it is too late to compare measurements. The transition between locality and non-locality takes place at B, where the future lightcones just intersect at the singularity.

8 Summary

In this paper I have presented several arguments and proposals about generalized thermodynamics and holography from the point of view of observers who fall through the true horizon of a black hole. The proposals are motivated by the classical appearance of a black hole as seen by an infaller. The proposals are consistent with, and extend, prevailing popular ideas about generalized thermodynamics and holography from the point of view of observers who remain outside the horizon.

An important point is that observers see Hawking radiation not from the true (future) horizon, but from the illusory (past) horizon, which is the redshifting surface of matter that fell into the black hole long ago. The illusory horizon is the boundary of the past lightcone of an observer, and is observer-dependent. The illusory horizon is the holographic screen of the black hole for both outsiders and insiders, encoding for each observer the states hidden behind their illusory horizon.

An infaller who nears the singularity has the impression that they actually reach the illusory horizon. This motivates the most speculative proposal in this paper, that an infaller who hits the singularity merges their states with the illusory horizon, the holographic image of the black hole. In this picture, the holographic boundary of the black hole is the union of the illusory horizon with the spacelike singularity.

Acknowledgments I thank Gavin Polhemus for numerous helpful conversations.

References

1. Hamilton, A., Polhemus, G.: Stereoscopic visualization in curved spacetime: seeing deep inside a black hole. New J. Phys. **12**, 123027 (2010). doi:10.1088/1367-2630/12/12/123027
2. Mellinger, A.: A color all-sky panorama image of the Milky Way. Publ. Astron. Soc. Pac. **121**, 1180 (2009). doi:10.1086/648480
3. Oppenheimer, J., Snyder, H.: On continued gravitational contraction. Phys. Rev. **56**, 455 (1939). doi:10.1103/PhysRev.56.455
4. Hamilton, A.: Collapse to a Black Hole (2011). http://casa.colorado.edu/ajsh/collapse.html
5. Hamilton, A.: Journey into a Schwarzschild Black Hole (2010). http://jila.colorado.edu/ajsh/insidebh/schw.html

6. Hawking, S.: Particle creation by black holes. Commun. Math. Phys. **43**, 199 (1975). doi:10.1007/BF02345020
7. Unruh, W.: Notes on black-hole evaporation. Phys. Rev. D **14**, 870 (1976). doi:10.1103/PhysRevD.14.870
8. Crispino, L., Higuchi, A., Matsas, G.: The Unruh effect and its applications. Rev. Mod. Phys. **80**, 787 (2008). doi:10.1103/RevModPhys.80.787
9. Visser, M.: Essential and inessential features of Hawking radiation. Int. J. Mod. Phys. D **12**, 649 (2003). doi:10.1142/S0218271803003190
10. Padmanabhan, T.: Gravitation: Foundations and Frontiers. Cambridge University Press, Cambridge (2010)
11. Hodgkinson, L., Louko, J.: Static, stationary and inertial Unruh-DeWitt detectors on the BTZ black hole. Phys. Rev. D **86**, 064031 (2012). doi:10.1103/PhysRevD.86.064031
12. Wald, R.: The thermodynamics of black holes. Living Rev. Relativ. **4**(6), lrr-2001-6 (2001). http://www.livingreviews.org/lrr-2001-6
13. Hawking, S.: Breakdown of predictability in gravitational collapse. Phys. Rev. D **14**, 2460 (1976). doi:10.1103/PhysRevD.14.2460
14. Wallace, C., Hamilton, A., Polhemus, G.: Huge entropy production inside black holes. ArXiv e-prints arXiv:0801.4415 [gr-qc] (2008)
15. Bousso, R.: The holographic principle. Rev. Mod. Phys. **74**, 825 (2002). doi:10.1103/RevModPhys.74.825
16. 't Hooft, G.: Dimensional reduction in quantum gravity. In: Ali, A., Ellis, J., Randjbar-Daemi S. (eds.) Salamfestschrift: A Collection of Talks from the Conference on Highlights of Particle and Condensed Matter Physics, World Scientific Series in 20th Century Physics, vol. 4. World Scientific, Singapore (1994)
17. Susskind, L.: The world as a hologram. J. Math. Phys. **36**, 6377 (1995). doi:10.1063/1.531249

Shape Dynamics

Tim A. Koslowski

Abstract Barbour's formulation of Mach's principle requires a theory of gravity to implement local relativity of clocks, local relativity of rods and spatial covariance. It turns out that relativity of clocks and rods are mutually exclusive. General Relativity implements local relativity of clocks and spatial covariance, but not local relativity of rods. It is the purpose of this contribution to show how Shape Dynamics, a theory that is locally equivalent to General Relativity, implements local relativity of rods and spatial covariance and how a BRST formulation, which I call Doubly General Relativity, implements all of Barbour's principles.

1 Introduction

A reflection on Mach's principle lead Barbour to postulate that rods and spatial frames of reference should be locally determined by a procedure that he calls "best matching," while clocks should be locally determined by what he calls "objective change". (For more, see [1].) More concretely, Barbour's principles postulate local time reparametrization invariance, local spatial conformal invariance and spatial covariance. The best matching algorithm for spatial covariance and local spatial conformal invariance turns out to be equivalent to the imposition of linear diffeomorphism and conformal constraints

$$H(\xi) = \int_\Sigma d^3x\, \pi^{ab}(\mathcal{L}_\xi g)_{ab}, \quad C(\rho) = \int_\Sigma d^3x\, \rho\, \pi, \tag{1}$$

T. A. Koslowski (✉)
Perimeter Institute for Theoretical Physics, 31 Caroline St. N, Waterloo, ON, Canada
e-mail: t.a.koslowski@gmail.com

T. A. Koslowski
Department of Mathematics and Statistics, University of New Brunswick, Fredericton, NB E3B 5A3, Canada

J. Bičák and T. Ledvinka (eds.), *Relativity and Gravitation*,
Springer Proceedings in Physics 157, DOI: 10.1007/978-3-319-06761-2_14,

where we use a compact Cauchy surface Σ without boundary with Riemannian metric g_{ab} and metric momentum density π^{ab} with trace π. The vector field ξ and the scalar field ρ are Lagrange multipliers. A more involved procedure, which I will not explain here, leads to the implementation of local time reparametrization invariance through quadratic Hamilton constraints

$$\hat{S}(N) = \int_\Sigma d^3x\, N \left(\pi^{ab} F_{abcd} \pi^{cd} - V\right), \tag{2}$$

where $F_{abcd}(x)$ and $V(x)$ are constructed from $g_{ab}(x)$ and its derivatives at x and N denotes a Lagrange multiplier. There is no reason for $\hat{S}$ to have homogeneous conformal weight, so a system containing the constraints $\hat{S}(N)$ and $C(\rho)$ will not be first class except for very special choices of F_{abcd}, V. This means that time reparametrization symmetry and spatial conformal symmetry generically exclude one another. An interesting situation occurs when we choose the F_{abcd}, V to reproduce the Hamilton constraints of General Relativity

$$S(N) = \int_\Sigma d^3x\, N \left(\frac{\pi^{ab}(g_{ac}g_{bd} - \frac{1}{2} g_{ab}g_{cd})\pi^{cd}}{\sqrt{|g|}} - (R - 2\Lambda)\sqrt{|g|}\right), \tag{3}$$

where the constraint system $S(N)$, $Q(\rho)$ is completely second class, while the constraint system $S(N)$, $H(\xi)$ is first class as is the constraint system $Q(\rho)$, $H(\xi)$. We will shortly see that this is the reason, why a Shape Dynamics and Doubly General Relativity can be constructed [2].

2 Symmetry Trading

Gauge theories describe a physical system using redundant degrees of freedom. The physical degrees of freedom are identified with orbits of the action of the gauge group. This redundant description is very useful in field theory, because it is often the only local description of a given system. The canonical description of gauge theories (see e.g. [3]) is provided by a regular irreducible set of first class constraints $\{\chi_\alpha\}_{\alpha \in \mathscr{A}}$, whose elements χ_α are smooth functions on a phase space Γ with Poisson bracket $\{.,.\}$. First class means that the constraint surface $\mathscr{C} = \{x \in \Gamma : \chi_\alpha(x) = 0, \forall \alpha \in \mathscr{A}\}$ is foliated into gauge orbits, whose infinitesimal generators are the Hamilton vector fields $v_\alpha : f \mapsto \{\chi_\alpha, f\}$. For simplicity we will assume that the system is generally covariant, so it has vanishing on-shell Hamiltonian. Observables $[O]$ of the system are equivalence classes of smooth gauge-invariant functions O on Γ, where two functions are equivalent if they coincide on $\mathscr{C}$ and where gauge invariance means that O is constant along gauge orbits. This means that an observable is completely determined by determining its dependence on a reduced phase space Γ_{red}, which contains one and only one point out of each gauge orbit. There is

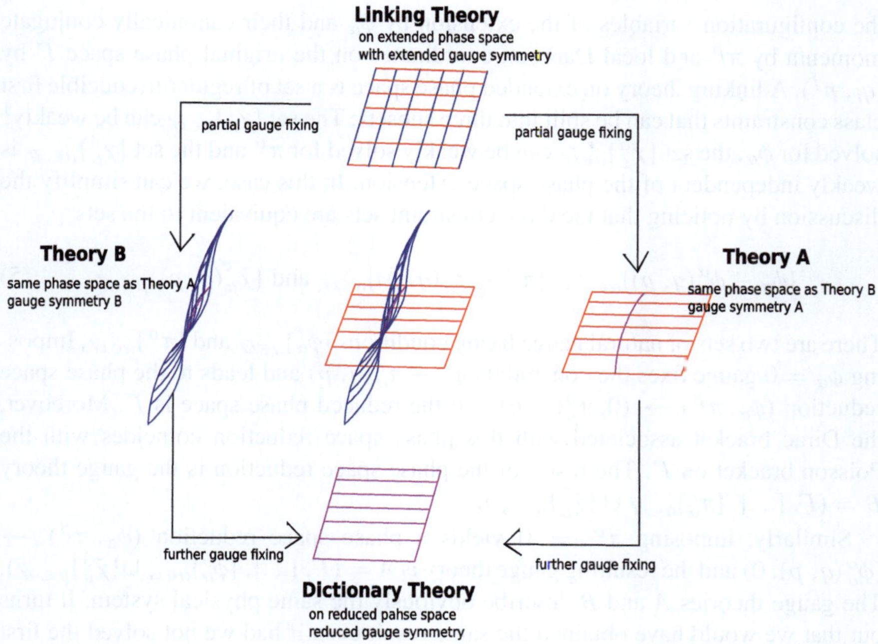

Fig. 1 Construction of equivalent gauge theories from a linking theory

no unique choice of Γ_{red} and the simplest description is through a regular irreducible set of gauge fixing conditions $\{\sigma^\alpha\}_{\alpha\in\mathscr{A}}$, such that a proper reduced phase is defined through $\Gamma_{red} = \{x \in \Gamma : \chi_\alpha(x) = 0 = \sigma^\alpha(x),\ \forall\alpha \in \mathscr{A}\}$. The observable algebra can then be identified with the Dirac algebra on reduced phase space, where the Dirac bracket takes the form

$$\{f, g\}_D := \{f, g\} - \Big(\{f, \chi_\alpha\}M^\alpha_\beta\{\sigma^\beta, g\} - \{f, \sigma^\beta\}M^\alpha_\beta\{\chi_\alpha, g\}\Big), \tag{4}$$

where M denotes the inverse of the linear operator $\{\chi, \sigma\}$.

The condition that Γ_{red} contains one and only one point out of each gauge orbit poses important restrictions on the gauge fixing conditions, but the set of gauge fixing conditions is *not* required to be first class. A very interesting situation arises when the set of gauge fixing condition is itself first class: In this case one can switch the role of gauge fixing conditions and constraints and describe the same observable algebra, and thus the same physical system, with the gauge theory $A = (\Gamma, \{.,.\}, \{\chi_\alpha\}_{\alpha\in\mathscr{A}})$ or with the gauge theory $B = (\Gamma, \{.,.\}, \{\sigma^\alpha\}_{\alpha\in\mathscr{A}})$. The manifest equivalence of the two theories is established by gauge fixing theory A with the gauge fixing set $\{\sigma^\alpha\}_{\alpha\in\mathscr{A}}$ and gauge theory B with the gauge fixing set $\{\chi_\alpha\}_{\alpha\in\mathscr{A}}$. One thus trades one gauge symmetry for another.

A very useful tool for the construction of equivalent gauge theories is a *linking theory*, see [4] and Fig. 1. Let us start with a phase space extension and denote the configuration variables of the extension by ϕ_α and their canonically conjugate momenta by π^β and local Darboux coordinates on the original phase space Γ by (q_i, p^j). A linking theory on extended phase space is a set of regular irreducible first class constraints that can be split into three subsets: The set $\{\chi^1_\alpha\}_{\alpha\in\mathscr{A}}$ can be weakly[1] solved for ϕ_α, the set $\{\chi^\alpha_2\}_{\alpha\in\mathscr{A}}$ can be weakly solved for π^α and the set $\{\chi^3_\mu\}_{\mu\in\mathscr{M}}$ is weakly independent of the phase space extension. In this case, we can simplify the discussion by noticing that the three constraint sets are equivalent to the sets

$$\{\phi_\alpha - \phi^\alpha_o(q,p)\}_{\alpha\in\mathscr{A}}, \quad \{\pi^\alpha - \pi^o_\alpha(q,p)\}_{\alpha\in\mathscr{A}} \text{ and } \{\tilde{\chi}^3_\mu(q,p)\}_{\mu\in\mathscr{M}}. \tag{5}$$

There are two sets of natural gauge fixing conditions $\{\phi_\alpha\}_{\alpha\in\mathscr{A}}$ and $\{\pi^\alpha\}_{\alpha\in\mathscr{A}}$. Imposing $\phi_\alpha = 0$ gauge fixes the constraints $\pi^\alpha - \pi^o_\alpha(q,p)$ and leads to the phase space reduction $(\phi_\alpha, \pi^\beta) \to (0, \pi^o_\beta(q,p))$, so the reduced phase space is Γ. Moreover, the Dirac bracket associated with this phase space reduction coincides with the Poisson bracket on Γ. The result of the phase space reduction is the gauge theory $B = (\Gamma, \{.,.\}, \{\pi^o_\alpha\}_{\alpha\in\mathscr{A}} \cup \{\tilde{\chi}^3_\mu\}_{\mu\in\mathscr{M}})$.

Similarly, imposing $\pi^\alpha = 0$ yields a phase space reduction $(\phi_\alpha, \pi^\beta) \to (\phi^\alpha_o(q,p), 0)$ and the resulting gauge theory is $A = (\Gamma, \{.,.\}, \{\phi^o_\alpha\}_{\alpha\in\mathscr{A}} \cup \{\tilde{\chi}^3_\mu\}_{\mu\in\mathscr{M}})$. The gauge theories A and B describe obviously the same physical system. It turns out that we would have obtained the same result even if had we not solved the first two subsets of constraints for the phase space extension.

3 Shape Dynamics

Let us now extend the phase space of General Relativity by a conformal factor ϕ and its conjugate momentum density π_ϕ. The linking theory between General Relativity on a compact manifold Σ without boundary and Shape Dynamics can be obtained by canonical best matching General Relativity in the ADM formulation with respect to conformal transformations that do not change the total spatial volume. This yields the following set of constraints

$$TS(N) = \int_\Sigma d^3x N\left(\frac{\sigma^a_b\sigma^b_a}{\sqrt{|g|}}e^{-6\hat{\phi}} + (2\Lambda - \frac{1}{6}\langle\pi\rangle^2)\sqrt{|g|}e^{6\hat{\phi}} - R(e^{6\hat{\phi}}g)\sqrt{|g|}e^{2\hat{\phi}} + a\right), \tag{6a}$$

$$Q(\rho) = \int_\Sigma d^3x\, \rho\left(\pi_\phi - 4\pi + 4\langle\pi\rangle\sqrt{|g|}\right), \tag{6b}$$

$$H(\xi) = \int_\Sigma d^3x\left(\pi^{ab}(\mathscr{L}_\xi g)_{ab} + \pi_\phi \mathscr{L}_\xi \phi\right), \tag{6c}$$

where $\hat{\phi} := \phi - \frac{1}{6}\ln\langle e^{6\phi}\rangle$, $\sigma^a_b = \pi^{ac}g_{cb} - \frac{1}{3}\pi\delta^a_b$ and where triangle brackets denote the mean w.r.t. $\sqrt{|g|}$ and where the term a vanishes when $\pi = \langle\pi\rangle\sqrt{|g|}$. Imposing the gauge fixing condition $\phi = 0$ results in a phase space reduction $(\phi, \pi_\phi) \to$

[1] "Weakly" means on the constraint surface.

$(0, 4(\pi - \langle\pi\rangle))$ and reduces the system back to the ADM formulation of General Relativity. Imposing the gauge fixing condition $\pi_\phi = 0$ gauge fixes all $TS(N)$, except for one. The simplest way to see this is to observe that $TS = 0$ is equivalent to imposing that ϕ solves the Lichnerowicz-York equation

$$8\Delta_g \Omega = \left(\frac{1}{6}\langle\pi\rangle^2 - 2\Lambda\right)\Omega^5 + R\,\Omega - \frac{\sigma^a_b \sigma^b_a}{|g|}\Omega^{-7} \tag{7}$$

for $\Omega = e^\phi$, but with the reducibility condition that the conformal factor is volume preserving $\int_\Sigma d^3x\sqrt{|g|}\,(1 - e^{6\phi}) = 0$. The left-over constraint is thus equivalent to the constraint

$$\int_\Sigma d^3x\sqrt{|g|}\left(1 - e^{6\phi_o[g,\pi]}\right) = 0, \tag{8}$$

where $\phi_o[g, p; x)$ denotes the positive solution to the Lichnerowicz-York equation, which is known to uniquely exist on physical phase space [5]. The phase space reduction thus yields the constraint system

$$H_{SD} = \int_\Sigma d^3x\sqrt{|g|}\left(1 - e^{6\phi_o[g,\pi]}\right), \tag{9a}$$

$$\hat{C}(\rho) = \int_\Sigma d^3x\,\rho\left(\pi - \langle\pi\rangle\sqrt{|g|}\right), \tag{9b}$$

$$H(\xi) = \int_\Sigma d^3x\pi^{ab}(\mathcal{L}_\xi g)_{ab}. \tag{9c}$$

This is *not exactly* Shape Dynamics, because the total conformal transformations generated by $\hat{C}(\rho)$ preserve the total spatial volume. One can however obtain a *true* theory of Shape Dynamics by observing that the only nonlinear constraint H_{SD} has the form of a reparametrization constraint $p_t - H(t) \approx 0$ of parametrized dynamics. Thus, after identifying the total volume V with the momentum conjugate to York time $\tau = \frac{3}{2}\langle\pi\rangle$ and *deparametrizing* the theory one obtains the physical Hamiltonian

$$H_{phys} = \int_\Sigma \sqrt{|g|}e^{6\phi[g,\pi]}. \tag{10}$$

The $\pi(x)$ is constrained to $\langle\pi\rangle\sqrt{|g|(x)}$ and the conformal factor of the metric is pure gauge, except for the total volume. The physical phase space is thus coordinatized by the conformal metric $\rho_{ab} = |g|^{-1/3}g_{ab}$ and σ^a_b and $V, \langle\pi\rangle$. The only physical phase space coordinate that is affected differently by volume preserving conformal transformations as opposed to unrestricted conformal transformations is V. But after reinterpreting $\frac{3}{2}\langle\pi\rangle$, V as time and its momentum there is *no* difference on the physical phase space volume preserving and unrestricted conformal transformations. The Shape Dynamics Hamiltonian H_{phys} comes with the constraints

$$C(\rho) = \int_\Sigma d^3x\,\pi, \quad H(\xi) = \int_\Sigma d^3x\,\pi^{ab}(\mathscr{L}_\xi g)_{ab}. \tag{11}$$

The dictionary between Shape Dynamics and General Relativity is established as follows: Given a solution $\rho_{ab}(\tau), \sigma^a_b(\tau)$ to the Shape Dynamics equations of motion, one finds that this is also a solution to the equations of motion of General Relativity in constant mean curvature (CMC) gauge.

4 Doubly General Relativity

An important difficulty of Shape Dynamics is that the physical Hamiltonian is non-local. One can improve this by working with the BRST formalism. One obtains the BRST formalism by adjoining to each of the regular irreducible first class constraints χ_α a ghost η^α and canonically conjugate ghost momentum P_α with opposite statistics. A result from cohomological perturbation theory then shows that the first class property of the constraints allows one to construct a nilpotent ghost number one BRST generator $\Omega = \eta^\alpha \chi_\alpha + \mathscr{O}(\eta^2)$, which provides a resolution of the observable algebra. This means that $\{\Omega, .\}$ is a differential, whose cohomology at ghost number zero is precisely the classical observable algebra.

The prerequisite for symmetry doubling to work was that there were two first class surfaces that gauge fixed one another. One can thus construct two nilpotent BRST generators: A ghost number $+1$ generator Ω form the first set of first class constraints and a ghost number -1 generator Ψ from the second set of first class constraints. For generally covariant theories, i.e. theories with vanishing on-shell Hamiltonian, one finds that the BRST-gauge fixed Hamiltonian can be written as $H_{BRST} = \{\Omega, \Psi\}$. The Jacobi identity and nilpotency of the generators then implies that H_{BRST} is annihilated by two BRST transformations: the ones generated by Ω and the ones generated by Ψ. One can thus not only relate the observables of the two theories with one another (Ω provides a resolution for the first and Ψ for the second), but one sees that the BRST gauge-fixed actions of the two theories can be chosen to coincide.

It goes beyond the scope of this contribution to discuss a detailed application of this construction for the duality between General Relativity and Shape Dynamics (for details see [6]), so I will only illustrate what the two BRST charges can be chosen to be:

$$\Omega = \int_\Sigma d^3x\Big(S(\eta) + H^a(\eta_a) + \mathscr{O}(\eta^2)\Big), \tag{12a}$$

$$\Psi = \int_\Sigma d^3x\left(P\frac{\pi}{\sqrt{|g|}} + \frac{H^a(g_{ab}P^b)}{\sqrt{|g|}} + \mathscr{O}(P^2)\right), \tag{12b}$$

where the higher orders in ghosts are chosen such that the two generators are nilpotent. It is evident that the construction will lead to a canonical action that is left invariant invariant by ADM- and Shape Dynamics BRST transformations, hence the name Doubly General Relativity. I conclude this short description with the warning that the ghost-number zero term of this Hamiltonian is *not* the CMC gauge-fixing of ADM.

5 Interpretation

Although local relativity of clocks and local relativity of rods seem to be incompatible as gauge symmetries, they are reconcilable. This reconciliation can be seen in the canonical formalism, where it appears that gravitational dynamics is equivalently described either by the ADM system or by the Shape Dynamics system. This means that both theories have the same solutions and make the same predictions for all observables. It can also be seen in the BRST formalism, where not only the predictions for observables coincide, but where it turns out that the gauge fixed gravity action has two BRST invariances, one corresponding to the on-shell spacetime symmetries of the ADM description of gravity, the other corresponding to the conformal symmetries of the Shape Dynamics description of gravity.

I want to conclude with considering the ideas behind the construction of Shape Dynamics. The construction of Shape Dynamics requires two steps: symmetry trading and the identification of a parametrized dynamical system. There is a lot of literature on the second step (e.g. Kuchař's perennials), but the idea behind symmetry trading seems to have avoided extensive discussion, although the idea itself is obvious.

Gauge theories are formulated with redundant degrees of freedom and gauge invariance, to have a local[2] (and thus comprehensible) field theory. This is a special instance of the more general fact that a comprehensible description of the real world often requires auxiliary concepts, which are not really part of reality. Which auxiliary concept is chosen is not unique; in general there is an infinite number of internally consistent descriptions. But although all descriptions are required to accurately describe the real world and be internally consistent, it can happen that two descriptions are mutually exclusive because the auxiliary concepts are incompatible. We see this in the duality between General Relativity and Shape Dynamics: General Relativity teaches that gravity is spacetime geometry and not a conformal theory, while Shape Dynamics teaches that gravity is a conformal theory without spacetime.

This can be very disturbing and leads to the question: "How can we discriminate better from worse descriptions?" I can only think of one criterion, provided two descriptions are accurate and internally consistent. The criterion is: Which description has more explanatory power? However, this may be the wrong question. I think one should rather embrace the fact there are many possibly equally good consistent but mutually exclusive descriptions of the real world and one should use whichever

---

[2] Note that the Shape Dynamics Hamiltonian, although nonlocal, can be described locally through the linking theory.

is most adapted to answer a particular problem. For example, a question regarding spacetime has most likely a simple answer in a covariant description, while a question regarding the observable algebra has most likely a simple answer in the Shape Dynamics description.

Acknowledgments Research at the Perimeter Institute is supported in part by the Government of Canada through NSERC and by the Province of Ontario through MEDT.

References

1. Barbour, J.: Shape dynamics: an introduction. In: Finster, F., Müller, O., Nardmann, M., Tolksdorf, J., Zeidler, E. (eds.) Quantum Field Theory and Gravity: Conceptual and Mathematical Advances in the Search for a Unified Framework, pp. 257–298. Basel, Birkhäuser (2012)
2. Gomes, H., Gryb, S., Koslowski, T.: Einstein gravity as a 3D conformally invariant theory. Class. Quantum Grav. **28**, 045005 (2011). doi:10.1088/0264-9381/28/4/045005
3. Henneaux, M., Teitelboim, C.: Quantization of Gauge Systems. Princeton University Press, Princeton, NJ (1992)
4. Gomes, H., Koslowski, T.: The link between general relativity and shape dynamics. Class. Quantum Grav. **29**, 075009 (2012). doi:10.1088/0264-9381/29/7/075009
5. Ó Murchadha, N., York, J.: Existence and uniqueness of solutions of the Hamiltonian constraint of general relativity on compact manifolds. J. Math. Phys. **14**, 1551 (1973). doi:10.1063/1.1666225
6. Gomes, H., Koslowski, T.: Symmetry Doubling: Doubly General Relativity. ArXiv e-prints arXiv:1206.4823 [gr-qc] (2012)

Superradiance or Total Reflection?

András László and István Rácz

Abstract Numerical evolution of massless scalar fields on Kerr background is studied. The initial data is chosen to have compact support separated from the ergoregion and to yield nearly monochromatic incident wave packets. The initial data is also tuned to maximize the effect of superradiance. We give evidence indicating that instead of the anticipated energy extraction from the black hole the incident radiation fails to reach the ergoregion and instead it suffers a nearly perfect reflection.

1 Introduction

To motivate our investigations let us mention first that the stability of the Kerr family of black hole solutions within the space of the vacuum solutions to the Einstein equations is one of the most important unresolved issues in general relativity. The ultimate goal is to provide boundedness and decay statements for solutions of the vacuum Einstein equations around the members of the Kerr family.

It may be a surprise that—even nowadays when numerical simulations of binary black hole systems become a daily routine—essentially all work concerning the aforementioned black hole stability problem has been confined to the linearized setting. Indeed, considerations are restricted to study of the solutions to the Klein-Gordon equation

$$\Box_K \Phi = 0 \tag{1}$$

A. László
CERN, 1211 Genéve 23, Switzerland
e-mail: laszlo.andras@wigner.mta.hu

A. László · I. Rácz (✉)
Wigner RCP, Budapest 1121, Hungary
e-mail: racz.istvan@wigner.mta.hu

J. Bičák and T. Ledvinka (eds.), *Relativity and Gravitation*,
Springer Proceedings in Physics 157, DOI: 10.1007/978-3-319-06761-2_15,

on Kerr background. This is done with the hope that the understanding of these simplified scalar perturbations is a good preparation to the study of the more complicated problem of complete, but yet linear, gravitational perturbations.

It should also be mentioned that all the available analytic proofs justifying the linear stability, with respect to scalar perturbations subject to (1), are known to be restricted to the case of slowly rotating subextremal Kerr black holes [1, 2].

2 Superradiance

Recall first that superradiance, as a new phenomenon, was discovered in the early 1970s and it may be associated with the names of Misner, Zel'dovich and Starobinskii [3–5]. It is also considered to be the wave analog of the Penrose process and it is supposed to allow energy to be extracted from black holes.

The common belief related to the interaction of black holes with incident radiation is summarized as "...if scalar, electromagnetic or gravitational wave is incident upon a black hole, part of the wave (the "transmitted wave") will be absorbed by the black hole and part of the wave (the "reflected wave") will escape to infinity" [6]. Recall that by using Teukolsky's equation [7] the evolution of scalar, electromagnetic and gravitational perturbations can be investigated within the same setting. It is also important to be mentioned that all the conventional arguments ending up with superradiance, including the ones based on Teukolsky's equation, refer to properties of individual modes [8].

Interestingly, as first pointed out by Bekenstein [9], whenever superradiance occurs it can be seen to be completely consistent with the laws of black hole thermodynamics. It is also worth to mention some of the expectations concerning scalar perturbations. As claimed in [10]: "Starobinskii made an asymptotic expansion for the reflection coefficient and found a relative gain of energy of about 5 % for $m = 1$ and less than 1 % for $m \geq 2$".

3 Superradiance in Mode Analysis

It was realized first by Carter [11] that the temporal Fourier transform, $\mathscr{F}\Phi = \frac{1}{\sqrt{2\pi}} \int_{-\infty}^{+\infty} \Phi\, e^{i\omega t} dt$, of a solution to (1) may be decomposed as

$$\mathscr{F}\Phi(\omega, r_*, \vartheta, \varphi) = \frac{1}{\sqrt{r^2 + a^2}} \sum_{\ell=0}^{\infty} \sum_{m=-\ell}^{\ell} R^m_{\ell,\omega}(r_*) S^m_{\ell,a\omega}(\vartheta, \varphi), \tag{2}$$

where $t, r_*, \vartheta, \varphi$ are local coordinates, while ω is the frequency in the time translation direction. In (2) $S^m_{\ell,a\omega}$ denotes the *oblate spheroidal harmonic functions*, with oblateness parameter $a\omega$, and with angular momentum quantum numbers ℓ, m. The functions $S^m_{\ell,a\omega}$ are eigenfunctions of a self-adjoint operator.

For the radial functions $R^m_{\ell,\omega}$ in (2) a *one-dimensional Schrödinger equation* of the form

$$\frac{d^2 R^m_{\ell,\omega}}{dr_*^2} + \left[\left(\omega - \frac{ma}{r^2 + a^2} \right)^2 + (r - r_H) \cdot V^m_{\ell,\omega}(r_*) \right] R^m_{\ell,\omega} = 0\,, \tag{3}$$

can be derived from (1), with suitable real potentials $V^m_{\ell,\omega}(r_*)$.

The "physical solutions" to (3) are supposed to possess the asymptotic behavior

$$R^m_{\ell,\omega} \sim \begin{cases} e^{-i\omega r_*} + \mathscr{R}\, e^{+i\omega r_*} & \text{as } r \to \infty \\ \mathscr{T}\, e^{-i(\omega - m\Omega_H) r_*} & \text{as } r \to r_H\,, \end{cases} \tag{4}$$

where Ω_H stands for the angular velocity of the black hole with respect to the asymptotically stationary observers, while $\mathscr{R}$ and $\mathscr{T}$ denote the reflection and transmission coefficients. Notice that these boundary conditions presume the existence of a transmitted wave submerging into the ergoregion.

By evaluating the Wronskian of the associated fundamental solutions, "close" to infinity and "close" to the horizon, the relation

$$(\omega - m\Omega_H)\, |\mathscr{T}|^2 = (1 - |\mathscr{R}|^2)\, \omega \tag{5}$$

can be seen to hold. In virtue of this relation it follows then that whenever $|\mathscr{R}| > 1$—or equivalently whenever $|\mathscr{T}|$ does not vanish and the inequality

$$0 < \omega < m\Omega_H \tag{6}$$

is satisfied—energy is supposed to be acquired by the backscattered scalar mode due to its interaction with the Kerr black hole.

4 Numerical Studies of Superradiance

So far our considerations have been restricted to the study of individual modes. However the investigation of the linear stability problem [1, 2, 12, 13] taught us the lesson that statements which are valid at the level of individual modes typically do not imply statements for finite energy solutions composed of infinitely many modes.

This section is to reveal some of our pertinent numerical results. Before proceeding let us mention that the first time domain studies of superradiance were carried out long time ago in [14, 15]. The scale of energy extraction was found to be smaller than the estimates recalled above. The numerical results reported below were derived by making use of our code called GridRipper which is fully spectral in the angular directions while the dynamics in the complementary 1 + 1 Lorentzian spacetime is followed by making use of a fourth order finite differencing scheme [16–18].

4.1 The Initial Data

To have an incident scalar wave—to study the way in which a possible superradiant solution acquires extra energy by submerging into the ergoregion—in a sufficiently small neighborhood of the initial data surface in the asymptotic region, the solution was assumed to possess the form

$$\Phi(t, r_*, \vartheta, \tilde{\varphi}) \approx e^{-\mathrm{i}\,\omega_0\,(r_* - r_{*0} + t)}\, f(r_* - r_{*0} + t)\, Y_\ell^m(\vartheta, \tilde{\varphi}), \tag{7}$$

where $f : \mathbb{R} \to \mathbb{C}$ is a smooth function of compact support and ω_0, r_{*0} are real parameters. This suggests the use of initial data

$$\begin{aligned} \phi(r_*, \vartheta, \tilde{\varphi}) &= e^{-\mathrm{i}\,\omega_0\,(r_* - r_{*0})}\, f(r_* - r_{*0})\, Y_\ell^m(\vartheta, \tilde{\varphi})\,, \\ \phi_t(r_*, \vartheta, \tilde{\varphi}) &= -\mathrm{i}\,\omega_0\, \phi(r_*, \vartheta, \tilde{\varphi}) + e^{-\mathrm{i}\,\omega_0\,(r_* - r_{*0})}\, f'(r_* - r_{*0})\, Y_\ell^m(\vartheta, \tilde{\varphi}), \end{aligned}$$

where f' denotes the first derivative of $f : \mathbb{R} \to \mathbb{C}$. The Fourier transform, $\mathscr{F}\Phi$, of the approximate solution (7) reads

$$\mathscr{F}\Phi(\omega, r_*, \vartheta, \tilde{\varphi}) \approx e^{-\mathrm{i}\,\omega\,(r_* - r_{*0})}\, \mathscr{F}f(\omega - \omega_0)\, Y_\ell^m(\vartheta, \tilde{\varphi}), \tag{8}$$

where ω is the temporal frequency and $\mathscr{F}f$ stands for the Fourier-transform of f. Notice that $\mathscr{F}f$ plays the role of a frequency profile, which guarantees that whenever $\mathscr{F}f$ is chosen to be sufficiently narrow the approximate solution (7) has to be close to a monochromatic wave packet, which for suitable value of ω_0 becomes superradiant (for more details see [17]).

4.2 Numerical Results

The plots shown below refer to the evolution of pure quadrupole type initial data with a radial profile function $f : \mathbb{R} \to \mathbb{C}$

$$f_w(x) = \begin{cases} e^{\left[-\left|\frac{w}{x+\frac{w}{2}}\right| - \left|\frac{w}{x-\frac{w}{2}}\right| + 4\right]}, & \text{if } x \in [-\frac{w}{2}, \frac{w}{2}] \\ 0, & \text{otherwise}\,, \end{cases} \tag{9}$$

which is a smooth function of the real variable x with compact support $[-\frac{w}{2}, \frac{w}{2}]$, and with initial parameters $M = 1$, $a = 0.99$, $\ell = m = 2$, $\omega_0 = \frac{1}{2} m \Omega_H$, $r_{*0} = 31.823$.

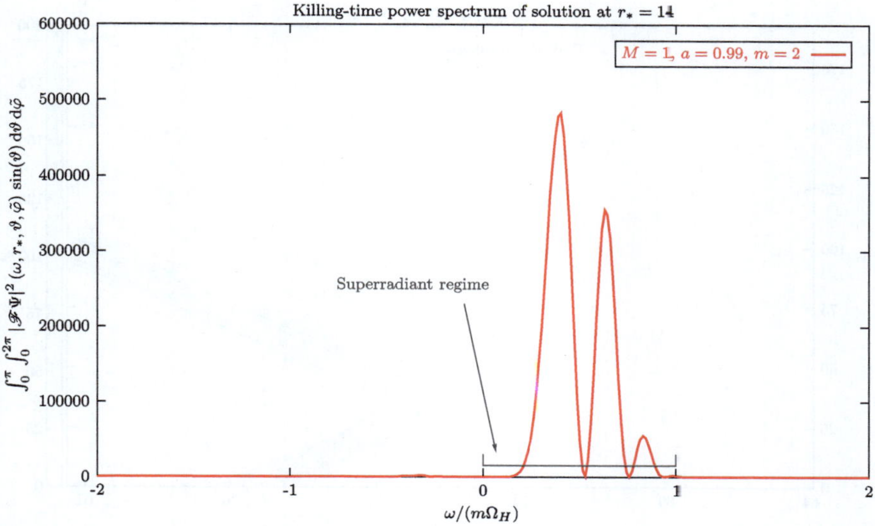

Fig. 1 The power spectrum of the to be superradiant solution at $r_* = 14$, located between the compact support and the black hole

It is important to be sure (see Fig. 1) that the above choice yields a solution with the expected frequency profile.

The time dependence of the radial energy and angular momentum distributions, along with the complete power spectrum, are shown[1] in Figs. 2 and 3 below. These figures indicate that the reported nearly perfect reflection really does happen for the considered solution.

It is also informative to have a look at the corresponding figures (see Figs. 4 and 5) for an "almost superradiant" solution yielded by shifting the compact support towards the black hole—decreasing thereby slightly the ratio of the angular momentum of the radiation to its energy—while all the other parameters were kept intact.

Notice that, in virtue of Fig. 5, the frequency content of the part of the incident wave packet, submerging into the ergoregion, gets completely evacuated from the superradiant domain.

It is also important to be mentioned that to the accuracy of our code, no energy extraction—or, at least, not more than 10^{-3} times the initial energy—happened in either of these (or analogous) simulations.

[1] Note that on all the included 2-dimensional plots the indicated quantities are integrated with respect to the radial degrees of freedom.

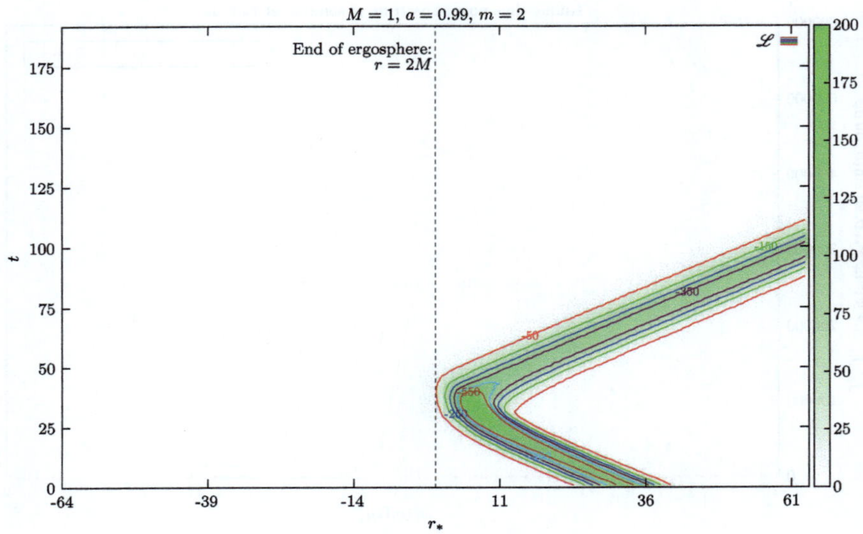

Fig. 2 The radial energy and angular momentum distributions

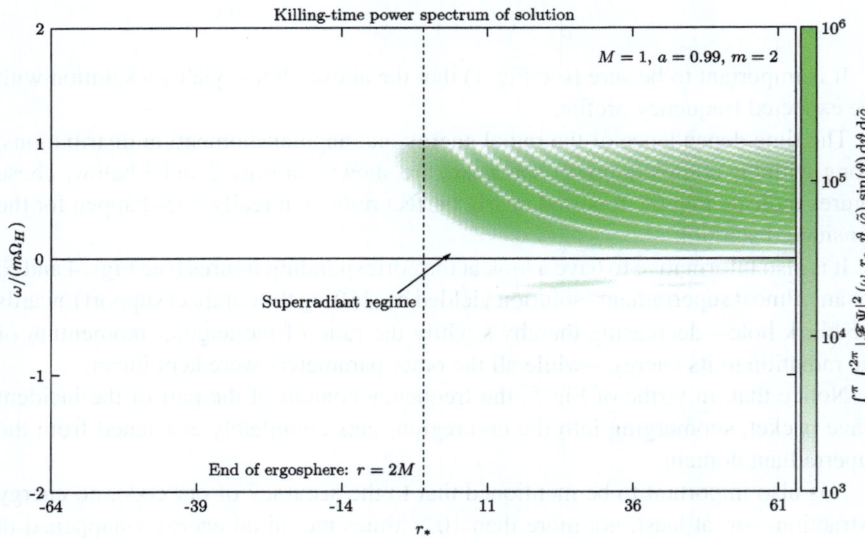

Fig. 3 The radial distribution of the power spectrum

5 Summary

The numerical evolution of massless Klein-Gordon field on Kerr background, arising from initial data with compact support in the asymptotic region, was considered. The incident wave packet was tuned to maximize the effect of superradiance.

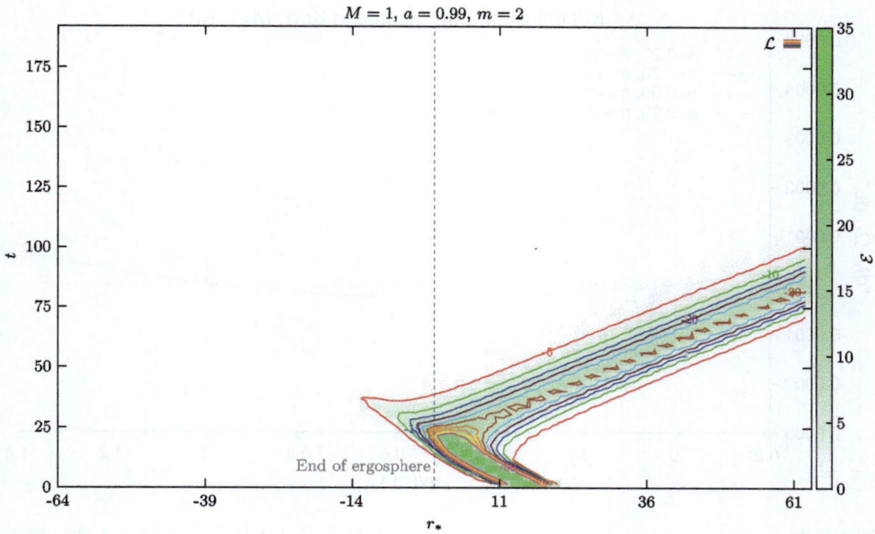

Fig. 4 The frequency ω_0 of this nearly superradiant solution is the same as before. Only the support is shifted to get a submerging part

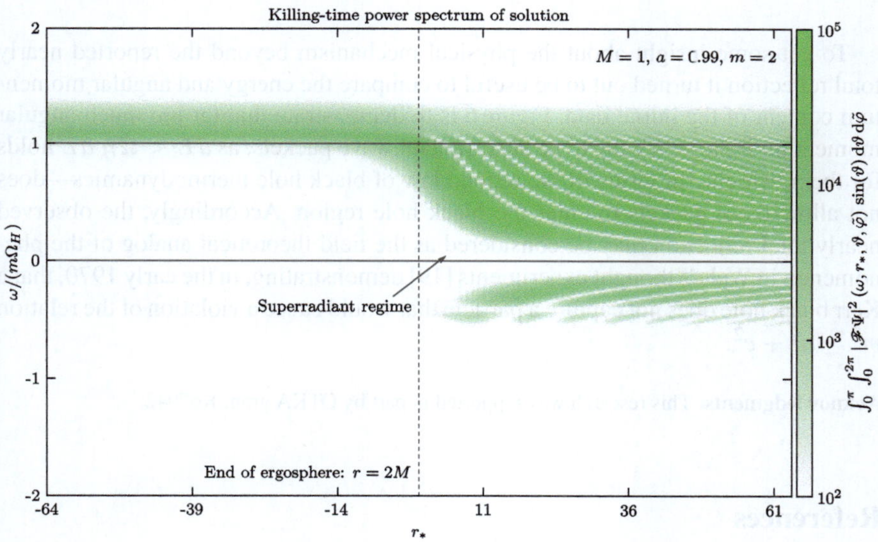

Fig. 5 The untuned solution has a submerging part but its power spectrum jumps out of the superradiant domain on reaching the ergoregion

For perfectly tuned initial data no energy extraction could be observed. Significant part of the incident radiation fails to reach the ergoregion and the time evolution mimics the phenomenon of a total reflection.

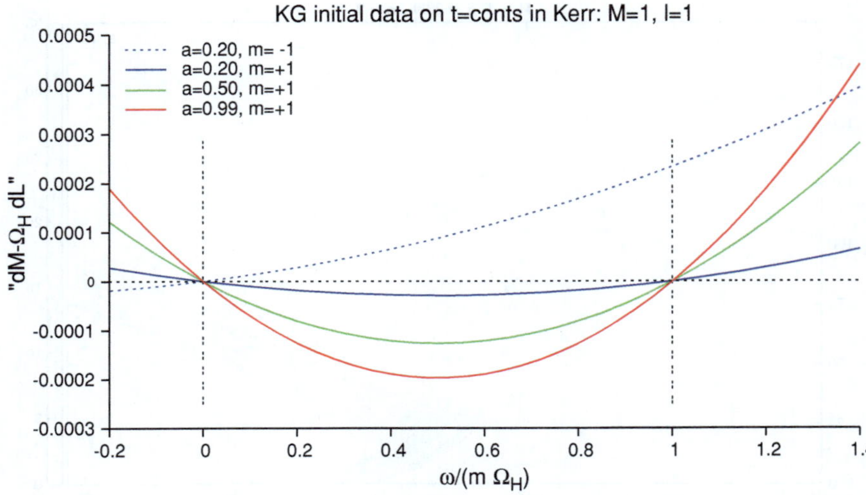

Fig. 6 The energy and angular momentum content of the initial data are compared. The to be superradiant configurations cannot deliver their full energy and angular momentum to the black hole without violating the second law of black hole thermodynamics

To get some insight about the physical mechanism beyond the reported nearly total reflection it turned out to be useful to compare the energy and angular momentum content of the initial data. Figure 6 is to demonstrate that far too much angular momentum is stored by the to be superradiant wave packets, as $dE < \Omega_H\, dL$ holds for them, which—in virtue of the second law of black hole thermodynamics—does not allow these packets to enter the black hole region. Accordingly, the observed nearly total reflection may be considered as the field theoretical analog of the phenomenon in Wald's thought experiments [19] demonstrating, in the early 1970, that a Kerr black hole does not capture a particle that would cause a violation of the relation $m^2 \geq a^2 + e^2$.

Acknowledgments This research was supported in part by OTKA grant K67942.

References

1. Andersson, L., Blue, P.: Hidden symmetries and decay for the wave equation on the Kerr spacetime. ArXiv e-prints arXiv:0908.2265 (2009)
2. Dafermos, M., Rodnianski, I.: Decay for solutions of the wave equation on Kerr exterior spacetimes I–II: the cases $|a| \ll M$ or axisymmetry. ArXiv e-prints arXiv:1010.5132 [gr-qc] (2010)
3. Misner, C.W.: Interpretation of gravitational-wave observations. Phys. Rev. Lett. **28**, 994 (1972). doi:10.1103/PhysRevLett.28.994
4. Zel'dovich, Y.B.: Amplification of cylindrical electromagnetic waves reflected from a rotating body. Zh. Eksp. Teor. Fiz. **62**, 2076 (1971)

5. Starobinskiĭ, A.A.: Amplification of waves during reflection from a rotating "black hole". Zh. Eksp. Teor. Fiz. **64**, 48 (1973)
6. Wald, R.M.: General Relativity. The University of Chicago Press, Chicago (1984)
7. Teukolsky, S.: Rotating black holes—separable wave equations for gravitational and electromagnetic perturbations. Phys. Rev. Lett. **29**, 1114 (1972). doi:10.1103/PhysRevLett.29.1114
8. Press, W.H., Teukolsky, S.A.: Perturbations of a rotating black hole. II. Dynamical stability of the kerr metric. Astrophys. J. **185**, 649 (1973). doi:10.1086/152445
9. Bekenstein, J.D.: Extraction of energy and charge from a black hole. Phys Rev. D **7**, 949 (1973). doi:10.1103/PhysRevD.7.949
10. Finster, F., Kamran, N., Smoller, J., Yau, S.T.: A rigorous treatment of energy extraction from a rotating black hole. Commun. Math. Phys. **287**, 829 (2009). doi:10.1007/s00220-009-0730-7
11. Carter, B.: Hamilton-Jacobi and Schrödinger separable solutions of Einstein's equations. Commun. Math. Phys. **10**, 280 (1968)
12. Wald, R.M.: Note on the stability of the Schwarzschild metric. J. Math. Phys. **20**, 1056 (1979). doi:10.1063/1.524181
13. Kay, B.S., Wald, R.M.: Linear stability of Schwarzschild under perturbations which are non-vanishing on the bifurcation 2-sphere. Class. Quantum Gravity **4**, 893 (1987). doi:10.1088/0264-9381/4/4/022
14. Krivan, W., Laguna, P., Papadopoulos, P., Andersson, N.: Dynamics of perturbations of rotating black holes. Phys Rev. D **56**, 3395 (1997). doi:10.1103/PhysRevD.56.3395
15. Andersson, N., Laguna, P., Papadopoulos, P.: Dynamics of scalar fields in the background of rotating black holes. II. A note on superradiance. Phys. Rev. D **58**(8), 087503 (1998). doi:10.1103/PhysRevD.58.087503
16. The GridRipper 3+1d PDE solver. http://www.rmki.kfki.hu/gridripper
17. Csizmadia, P., László, A., Rácz, I.: On the use of multipole expansion in time evolution of nonlinear dynamical systems and some surprises related to superradiance. Class. Quantum Gravity **30**(1), 015010 (2013). doi:10.1088/0264-9381/30/1/015010
18. Csizmadia, P., László, A., Rácz, I.: Linear waves on fixed Kerr background and their relevance in jet formation. J. Phys. Conf. Ser. **218**(1), 012007 (2010). doi:10.1088/1742-6596/218/1/012007
19. Wald, R.: Gedanken experiments to destroy a black hole. Ann. Phys. **82**, 548 (1974). doi:10.1016/0003-4916(74)90125-0

Non-Linear Effects in Non-Kerr Spacetimes

Georgios Lukes-Gerakopoulos, George Contopoulos and Theocharis A. Apostolatos

Abstract There is a chance that the spacetime around massive compact objects which are expected to be black holes is not described by the Kerr metric, but by a metric which can be considered as a perturbation of the Kerr metric. These non-Kerr spacetimes are also known as bumpy black hole spacetimes. We expect that, if some kind of a bumpy black hole exists, the spacetime around it should possess some features which will make the divergence from a Kerr spacetime detectable. One of the differences is that these non-Kerr spacetimes do not posses all the symmetries needed to make them integrable. We discuss how we can take advantage of this fact by examining EMRIs into the Manko–Novikov spacetime.

1 Introduction

We expect that a star which at the end of its life becomes a compact object with mass greater than three solar masses is a Kerr black hole. However, this anticipation should be somehow tested by observations.

One way to test the Kerr hypothesis is to study the gravitational wave signal produced by an inspiraling relatively light compact object (e.g., stellar) into the spacetime background of a supermassive compact object. This kind of motion is called Extreme Mass Ratio Inspiral (EMRI). Such binary systems should exist in the center of galaxies which we believe are occupied by supermassive black holes

G. Lukes-Gerakopoulos (✉)
Theoretical Physics Institute, University of Jena, 07743 Jena, Germany
e-mail: gglukes@gmail.com

G. Contopoulos
Research Center for Astronomy, Academy of Athens, Soranou Efesiou 4, 11527 Athens, Greece

T. A. Apostolatos
Section of Astrophysics, Astronomy, and Mechanics, Department of Physics, University of Athens, Panepistimiopolis Zografos, 15783 Athens, Greece

J. Bičák and T. Ledvinka (eds.), *Relativity and Gravitation*,
Springer Proceedings in Physics 157, DOI: 10.1007/978-3-319-06761-2_16,

(10^5–10^9 solar masses). In EMRIs the lighter object basically traces the background spacetime by following approximately geodesic orbits. Ryan in [1, 2] showed that we could extract the multipole moments of the background from the gravitational wave signal, and Collins and Hughes [3] produced a perturbed Schwarzschild black hole spacetime, which they called "bumpy" black hole spacetime, in order to perform the first tests of the Kerr hypothesis. Since then several such tests have been proposed, see e.g., [4–6] and references therein.

The bumpy black hole spacetimes are axisymmetric and stationary, but in general lack a Carter-like constant [7]. It has been shown that even in an axisymmetric stationary Newtonian potential, a higher order Killing tensor connected to a Carter-like constant cannot be found [8], contrary to conjectures that were initially postulated [9]. The lack of a Carter-like constant implies that the bumpy black hole systems are non-integrable, which in turn suggests that non-linear effects like chaos should be present. In a series of publications [10–12] we have studied what implications these non-linear effects will bring to a gravitational wave signal coming from an EMRI into a non-Kerr spacetime background. In the present article we present briefly these findings.

The article is organized as follows. Section 2 introduces some basic theoretical elements about a bumpy black hole spacetime and the geodesic motion in such spacetime. Section 3 discusses the non-integrability imprints of the non-Kerr background in gravitational wave signals. Our conclusions are given in Sect. 4.

2 Theoretical Elements

2.1 The Manko–Novikov Spacetime

The bumpy black hole spacetime we used in [10–12] is a spacetime which belongs to the so-called Manko–Novikov (MN) metric family [13]. Manko and Novikov found an exact vacuum solution of Einstein's equations which describes a stationary, axisymmetric, and asymptotically flat spacetime with arbitrary mass-multipole moments [13]. The MN metric subclass we used was introduced in [14] and deviates from the Kerr at all moments higher or equal to the quadrupole one. The new spacetime is characterized by one more parameter q than the ones describing a Kerr metric. Namely, the quantity q measures how much the MN quadrupole moment Q departs from the Kerr quadrupole moment $Q_{Kerr} = -S^2/M$ (that is $q = (Q_{Kerr} - Q)/M^3$), where M and S are the mass and the spin of a Kerr black hole respectively. If $q = 0$ the MN solution becomes exactly a Kerr solution. The line element of the MN metric in the Weyl-Papapetrou cylindrical coordinates (t, ρ, φ, z) is

$$ds^2 = -f(dt - \omega d\varphi)^2 + f^{-1}\left[e^{2\gamma}(d\rho^2 + dz^2) + \rho^2 d\varphi^2\right], \qquad (1)$$

where $f,\ \omega,\ \gamma$ are considered functions of the prolate spheroidal coordinates v, w, while the coordinates ρ, z can be expressed as functions of v, w as well. Namely,

$\rho = k\sqrt{(v^2-1)(1-w^2)}$, $z = kvw$, where $k = M\frac{1-\alpha^2}{1+\alpha^2}$, $\alpha = \frac{-1+\sqrt{1-\chi^2}}{\chi}$, while χ is the dimensionless spin parameter $\chi = S/M^2$. The exact formulae of f, ω, γ are lengthy, and can be found in [11, 14].

2.2 Geodesic Motion in the Manko–Novikov Spacetime

The geodesic orbits of a test particle of mass μ are described as equations of motion of the Lagrangian $L = \frac{1}{2}\mu\, g_{\mu\nu}\, \dot{x}^\mu \dot{x}^\nu$, where the dots denote derivatives with respect to the proper time. The MN metric has two integrals of motion, namely the energy (per unit mass)

$$E = -\frac{\partial L}{\partial \dot{t}}/\mu = f(\dot{t} - \omega\,\dot{\varphi}), \tag{2}$$

and the z-component of the angular momentum (per unit mass)

$$L_z = \frac{\partial L}{\partial \dot{\varphi}}/\mu = f\omega(\dot{t} - \omega\,\dot{\varphi}) + f^{-1}\rho^2\dot{\varphi}, \tag{3}$$

The Kerr metric has one more integral of motion, the so-called Carter constant [7], thus it is an integrable system. However, the MN model lacks in general (as long as $q \neq 0$) such constant which means that MN is a non-integrable system, and therefore chaos should appear.

We can reduce the four degrees of freedom of the MN system to two, by using the two integrals of motion E, and L_z, and thus, restrict the motion to the meridian plane $(\rho,\ z)$. By rewriting the metric (1) we see that the motion in the meridian plane satisfies the relation

$$\frac{1}{2}(\dot{\rho}^2 + \dot{z}^2) + V_{eff}(\rho, z) = 0, \tag{4}$$

where the effective potential $V_{eff}(\rho, z)$ depends on the parameters q, χ, E, and L_z. $V_{eff} \leq 0$ for all possible orbits. On the boundary $V_{eff} = 0$ the velocity vanishes: $\dot{\rho} = \dot{z} = 0$; this is the so called curve of zero velocity (CZV). Inside the CZV lie the non-escaping orbits.

In the Kerr spacetime case ($q = 0$) every non-plunging geodesic orbit confined by the CZV lies on a two dimensional torus in the phase space. On such a torus each orbit is described by two characteristic frequencies ω_1, ω_2. If the ratio of these frequencies ν_θ is an irrational number, the motion is quasiperiodic, and the corresponding torus is covered densely by the orbit. If the ratio is a rational number, the motion is periodic, and the corresponding torus is called resonant. A resonant torus is covered by an infinite number of periodic orbits, all having the same frequency ratio ν_θ.

By setting $q \neq 0$ we perturb the integrable system (Kerr), and the transition to the non-integrable system (MN) is described basically by two theorems: the KAM theorem, and the Poincaré-Birkhoff theorem. The first theorem states that after the perturbation most of the non-resonant tori will survive deformed. These surviving tori are called KAM tori. The second theorem implies that from a resonant torus only a finite even number of periodic orbits will survive; half of them will be stable and the other half unstable. Around the stable orbits small islands of stability are formed, while the asymptotic manifolds emanating from the unstable periodic orbits fill a region of chaotic orbits. The above formation is known as a Birkhoff chain.

One way to study the aforementioned different structures in a non-integrable system of two degrees of freedom is to take a section through the foliation of the tori. Such section is known as a surface of section, or as a Poincaré section. Another way is provided by the frequency ratio ν_θ, by which we can detect the different types of orbits and it is known as the rotation number (e.g., [15]). If ν_θ corresponds to an irrational number, we have a KAM curve; if it corresponds to a rational number, we are on a Birkhoff chain of islands of stability; if the value of ν_θ is indefinite, and does not correspond to a particular number, then the orbit is chaotic.

3 Non-Integrability Imprints on the Gravitational Wave Signal

3.1 The Plateau Effect of the Resonances

One possible imprint of the non-integrability of a bumpy black hole spacetime on the corresponding gravitational wave signal is the effect of the resonances. The left panel of Fig. 1 shows a part of the surface of section $z = 0$ ($z > 0$) of a MN spacetime for the parameter set $E = 0.95$, $Lz = 3\ M$, $\chi = 0.9$, $q = 0.95$. The first impression one might get is that the surface of section indicates an integrable system, because no straightforward signs of chaos are prominent. However, the islands of stability (left panel of Fig. 1) imply the existence of Birkhoff chains, which in turn indicate that chaos is also present.

If we take initial conditions along a straight line in the phase space, like the $\dot{\rho} = 0$ line starting from the center $\mathbf{u}_0$ of the main island of stability shown in the left panel of Fig. 1, and evaluate the rotation number for each of these initial conditions, then we get a rotation curve (right panel of Fig. 1). This curve seems to be smooth and strictly monotonic (in a Kerr spacetime this is the case), however a more detailed look reveals that this is not exactly true. At the resonances, plateaus appear. For instance in the embedded plot of the right panel of Fig. 1, we can see a plateau at the $2/3$ resonance. This happens because all the orbits belonging to the same chain of islands of stability share the same rotation number, i.e., the same frequency ratio. Such plateaus do not appear in the case of a Kerr metric.

However, geodesic orbits are simply an approximation of real EMRI orbits. A more realistic approximation demands the inclusion of the radiation reaction. Since there are no reliable computations describing the radiation reaction in a bumpy black hole spacetime, we used the same trick as the authors of [14]. Namely, we used the hybrid approximative method [16] (Eqs. (44) and (45) in [16]), where we added by

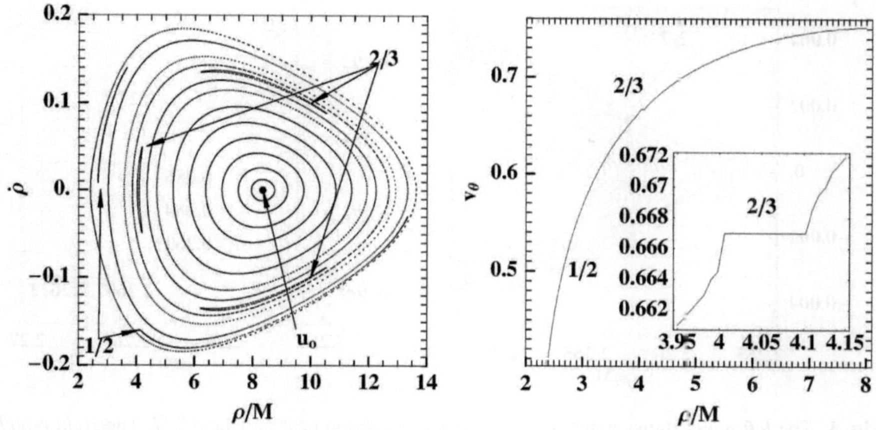

Fig. 1 The *left panel* shows a part of the surface of section in the plane $(\rho, \dot{\rho})$ focusing on the main island of stability, where $\mathbf{u}_0$ indicates the center of the main island. The *right panel* shows the rotation number along the line $\dot{\rho} = 0$ (starting from $\mathbf{u}_0$ and moving leftwards) on the surface of section shown in the *left panel*. Embedded in the *right panel* is a detail of the *rotation curve* around the 2/3-resonance. The parameters used are $E = 0.95$, $Lz = 3\ M$, $\chi = 0.9$, $q = 0.95$

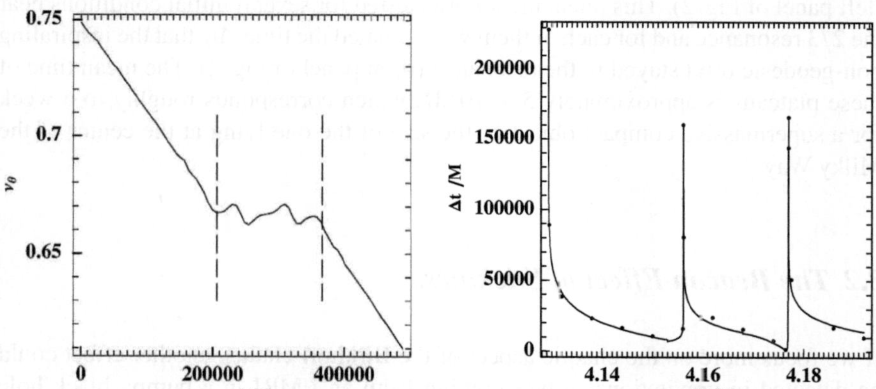

Fig. 2 The *left panel* shows the evolution of the ratio ν_θ as a function of the coordinate time t for a non-geodesic orbit. The vertical *dashed lines* demarcate the time intervals that the non-geodesic orbit spends in the interior of the 2/3-resonance. The *right panel* shows the time Δt_r needed by non-geodesic orbits to cross the chain of islands belonging to the 2/3-resonance as a function of their initial conditions $\rho(0)$ along the line $\dot{\rho} = 0$, $z = 0$. The parameters used are $\mu/M = 8 \times 10^{-5}$, $q = 0.95$, $\chi = 0.9$, $E(0) = 0.95$, $L_z(0) = 3\ M$, where $E(0)$, $L_z(0)$ are the initial values of E and L_z respectively

hand the anomalous quadrupole moment q to the χ^2 terms. Furthermore, we assumed a constant rate of energy and angular momentum loss due to gravitational radiation.

We applied the aforementioned scheme for a mass ratio $\mu/M = 8 \times 10^{-5}$, to evolve initial conditions near the 2/3 resonance, and found that the plateau also appears when we calculate the rotation number as a function of the coordinate time

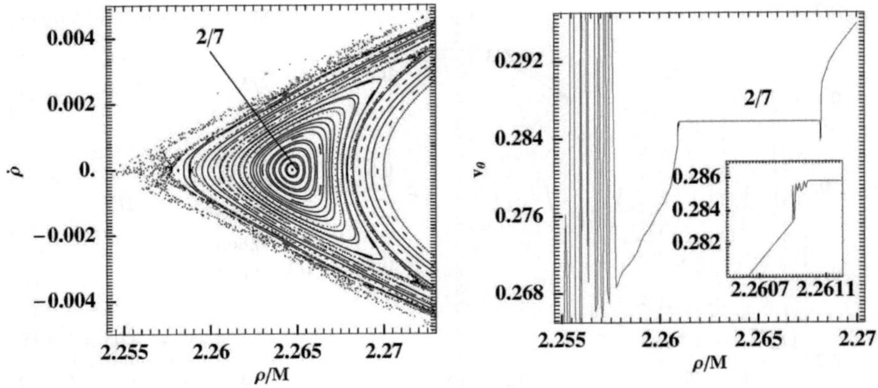

Fig. 3 The *left panel* shows a detail of the surface of section near an island 2/7. The *right panel* shows the *rotation curve* along the $\dot{\rho} = 0$ line of the surface of section presented in the *left panel*. Embedded in the *right panel* is the irregular variation of the rotation number just outside the left side of the 2/7-plateau. This irregular behavior is due to the chaotic layer surrounding the corresponding island. The parameters used are $E = 0.95$, $Lz = 2.995\ M$, $\chi = 0.9$, $q = 0.95$

(left panel of Fig. 2). This phenomenon was tested for several initial conditions near the 2/3 resonance and for each of them we estimated the time Δt_r that the inspiraling non-geodesic orbit stayed in the resonance (right panel of Fig. 2). The mean time of these plateaus is approximately $5 \times 10^4 M$, which corresponds roughly to a week for a supermassive compact object of the size of the one lying at the center of the Milky Way.

3.2 The Beacon Effect of Stickiness

If we focus more on the chaotic aspect of the Birkhoff chains, another effect could be detected in gravitational waves coming from an EMRI in a bumpy black hole spacetime background. This effect is connected with the phenomenon of stickiness [15]. The stickiness phenomenon concerns chaotic orbits which for various reasons stick for a long time interval in a region close to regular orbits. Therefore, their behavior in the frequency spectrum might resemble that of the regular orbits they are close to, before they depart from that region.

In the left panel of Fig. 3 we see a detail of the surface of section near the resonance 2/7. The stickiness appears in the region where chaotic orbits (scattered points on the surface of section) are confined by regular orbits. Even though their true character is detected by the rotation number, since ν_θ varies widely in the corresponding regions (right panel of Fig. 3), the phenomenology might be more complicated in the frequency spectrum. Namely, while a chaotic orbit stays near a regular orbit we might get a signal, i.e., distinct characteristic frequencies; when the orbit moves to a more prominent chaotic layer the frequency peaks in the signal will dissolve leaving only noise instead of a signal; later on when the orbit returns near a regular orbit the signal shall reappear, and so on. This effect resembles a beacon, where the signal appears and disappears.

4 Conclusions

The resonance and the stickiness effect are generic characteristics of the geodesic motion in any non-integrable Hamiltonian system describing a stationary and axisymmetric spacetime background like that of an axially symmetric perturbation of the Kerr spacetime. Therefore, they should be in principle detectable in the gravitational wave signal coming from an EMRI into a non-Kerr metric.

Acknowledgments G. L-G is supported by the DFG grant SFB/Transregio 7.

References

1. Ryan, F.: Gravitational waves from the inspiral of a compact object into a massive, axisymmetric body with arbitrary multipole moments. Phys. Rev. D **52**, 5707 (1995). doi:10.1103/PhysRevD.52.5707
2. Ryan, F.: Accuracy of estimating the multipole moments of a massive body from the gravitational waves of a binary inspiral. Phys. Rev. D **56**, 1845 (1997). doi:10.1103/PhysRevD.56.1845
3. Collins, N., Hughes, S.: Towards a formalism for mapping the spacetimes of massive compact objects: bumpy black holes and their orbits. Phys. Rev. D **69**, 124022 (2004). doi:10.1103/PhysRevD.69.124022
4. Amaro-Seoane, P., Aoudia, S., Babak, S., et al.: Low-frequency gravitational-wave science with eLISA/NGO. Class. Quantum Grav. **29**, 124016 (2012). doi:10.1088/0264-9381/29/12/124016
5. Bambi, C.: Testing the Kerr black hole hypothesis. Mod. Phys. Lett. A **26**, 2453 (2011). doi:10.1142/S0217732311036929
6. Johannsen, T.: Testing the no-hair theorem with Sgr A*. Adv. Astron. **2012**, 486750 (2012). doi:10.1155/2012/486750
7. Carter, B.: Global Structure of the Kerr family of gravitational fields. Phys. Rev. **174**, 1559 (1968). doi:10.1103/PhysRev.174.1559
8. Markakis, C.: Constants of motion in stationary axisymmetric gravitational fields. ArXiv e-prints arXiv:1202.5228 [astro-ph.SR] (2012)
9. Brink, J.: Formal solution of the fourth order Killing equations for stationary axisymmetric vacuum spacetimes. Phys. Rev. D **84**, 104015 (2011). doi:10.1103/PhysRevD.84.104015
10. Apostolatos, T., Lukes-Gerakopoulos, G., Contopoulos, G.: How to observe a non-Kerr spacetime using gravitational waves. Phys. Rev. Lett. **103**, 111101 (2009). doi:10.1103/PhysRevLett.103.111101
11. Lukes-Gerakopoulos, G., Apostolatos, T., Contopoulos, G.: Observable signature of a background deviating from the Kerr metric. Phys. Rev. D **81**, 124005 (2010). doi:10.1103/PhysRevD.81.124005

12. Contopoulos, G., Lukes-Gerakopoulos, G., Apostolatos, T.: Orbits in a non-Kerr dynamical system. Int. J. Bifurcat. Chaos **21**, 2261 (2011). doi:10.1142/S0218127411029768
13. Manko, V., Novikov, I.: Generalizations of the Kerr and Kerr-Newman metrics possessing an arbitrary set of mass-multipole moments. Class. Quantum Grav. **9**, 2477 (1992). doi:10.1088/0264-9381/9/11/013
14. Gair, J., Li, C., Mandel, I.: Observable properties of orbits in exact bumpy spacetimes. Phys. Rev. D **77**, 024035 (2008). doi:10.1103/PhysRevD.77.024035
15. G. Contopoulos, Order and chaos in dynamical astronomy. Astronomy and astrophysics library (Springer, 2002).
16. Gair, J.R., Glampedakis, K.: Improved approximate inspirals of test bodies into Kerr black holes. Phys. Rev. D **73**(6), 064037 (2006). doi:10.1103/PhysRevD.73.064037

The Conformal Einstein Field Equations for Trace-free Perfect Fluids

Christian Lübbe and Juan A. Valiente Kroon

Abstract A nonlinear stability analysis is carried out for the trace-free (radiation) perfect fluid Friedmann-Lemaître-Robertson-Walker models with a de Sitter-like cosmological constant. It is shown that the solutions close to the above FLRW spacetimes exist globally towards the future and are future geodesically complete. For this analysis we formulate the conformal Einstein field equations for a trace-free (radiation) perfect fluid in terms of the Levi-Civita connection of a conformally rescaled metric.

1 Introduction

During his time in Prague, 100 years ago, Einstein started his work on a general theory of relativity [1]. Since then many exact solutions have been found and analysed in detail—see e.g. [2, 3] for an overview. Typically these solutions lead to models which approximate certain features of our universe by making assumptions on the matter model or spacetime symmetries. As a result of this, a cosmological model is a representation of the universe at a particular averaging scale. This leads to the following important question: "How sensitive are the predictions derived from these models to perturbations". The answer is of interest for the following reasons. Firstly, the universe does not match an idealised model on all scales. Secondly, any observation of the universe (or a subsystem of it) gives rise to data that includes a certain margin of error. Thirdly, numerical calculations and simulations have made marked

C. Lübbe (✉)
Department of Mathematics, University College London,
Gower Street, London WC1E 6BT, UK
e-mail: c.luebbe@ucl.ac.uk

J. A. Valiente Kroon
School of Mathematical Sciences, Queen Mary University of London,
Mile End Road, London E1 4NS, UK

J. Bičák and T. Ledvinka (eds.), *Relativity and Gravitation*,
Springer Proceedings in Physics 157, DOI: 10.1007/978-3-319-06761-2_17,

progress in recent years. However, any simulation is limited by the finite precision of the individual computations and hence has to repeatedly deal with numerical errors. In all three scenarios one hopes that the unavoidable deviation from reality has negligible consequences for the predictions, as long as this deviation is sufficiently small. However, the concern about stability is not at all limited to cosmological scales. It equally applies to stars and other compact objects. One of the main open problems concerns the behaviour of black holes under perturbations—see e.g. [4].

Due to the nonlinear nature of the Einstein field equations, the question of stability is not straightforward. Seminal work by Friedrich [5, 6], using conformal methods, and Christodoulos and Klainermann [7], using a detailed analysis of the structure of the underlying evolution equations, have provided an essential starting point for this problem. In recent years the topic has seen several papers addressing the stability problem for a range of matter models, dimensions and types of cosmological constants.

The problem of nonlinear stability of the Euler-Einstein system for de Sitter-like spacetimes has been analysed in [8, 9]. It was shown that the Friedmann-Lemaître-Robertson-Walker (FLRW) solutions with a barotropic equation of state of the form $\tilde{p} = (\gamma - 1)\tilde{\rho}$, where $1 < \gamma < \frac{4}{3}$, are future asymptotically stable under small perturbations. The case of a pure radiation perfect fluid ($\gamma = \frac{4}{3}$) was not covered by this analysis. This motivated the authors to investigate this case using conformal methods [10]. The details and the results are outlined in this article.

We address the question of nonlinear stability for small perturbations of the FLRW solutions, which describe a trace-free perfect fluid with de Sitter-like cosmological constant. The main result is given by the following theorem:

Theorem 1 *Suppose one is given Cauchy initial data for the Einstein-Euler system with a de Sitter-like cosmological constant and equation of state for pure radiation $\tilde{p} = \frac{1}{3}\tilde{\rho}$. If the initial data is sufficiently close to data for a FLRW cosmological model with the same equation of state, value of the cosmological constant and spatial curvature $k = 1$, then the development exists globally towards the future, is future geodesically complete and remains close to the FLRW solution.*

2 Methodology and Analysis

For a spacetime with a pure radiation perfect fluid the problem of nonlinear stability is addressed using the conformal methods of [5, 11, 12]. In this approach, which is summarised below, the physical spacetime $(\tilde{\mathcal{M}}, \tilde{g}_{\mu\nu})$ is conformally embedded into a manifold $(\mathcal{M}, g_{\mu\nu})$, referred to as the unphysical spacetime here, where the metrics are related by

$$g_{\mu\nu} = \theta^2 \tilde{g}_{\mu\nu}. \tag{1}$$

The physical Einstein field equations

$$\tilde{R}_{\mu\nu} - \tfrac{1}{2}\tilde{R}\tilde{g}_{\mu\nu} + \lambda\tilde{g}_{\mu\nu} = \tilde{T}_{\mu\nu} \tag{2}$$

are reformulated in terms of the geometry of $(\mathscr{M}, g_{\mu\nu})$ and some rescaled matter variables, leading to a new set of equations referred to as the conformal Einstein field equations (CEFE). A key advantage of this approach is that one can study global problems in $(\tilde{\mathscr{M}}, \tilde{g}_{\mu\nu})$ in terms of a local analysis near conformal infinity in $(\mathscr{M}, g_{\mu\nu})$. For this analysis one has to show that the CEFE associated to the chosen matter model form a regular system of PDEs on $(\mathscr{M}, g_{\mu\nu})$. One then proceeds by formualting an evolution problem in the form of a first order symmetric hyperbolic (FOSH) system to deduce existence and uniqueness results. In order to derive the desired stability results, one shows that a chosen reference spacetime is a regular solution to the CEFE, respectively the associated FOSH system, and invokes Kato's stability theorem [13].

The CEFE have inherited degrees of freedom. One needs to fix a set of coordinates x^μ, a frame $\{e_k\}$ and the conformal factor θ. In the following, we use frame components with respect to the chosen frame $\{e_k\}$.

There are two main methods to fix these gauge freedoms. One method uses gauge source functions [5, 6, 11, 14] to evolve the coordinates, the frame and the conformal factor. In the other method, a congruence of conformal geodesics [15–17] or conformal curves [10] is used to construct a conformal Gaussian coordinate system. The congruence also induces a general Weyl connection, which is used to propagated the frame $\{e_k\}$, and a canonical choice of the conformal factor. Here we employ the first method. In particular, we fix the conformal factor θ locally by setting the unphysical Ricci scalar to $R = -\frac{1}{6}$ and work in the Levi-Civita connection induced by $g_{\mu\nu}$ in (1). Moreover, the frame $\{e_k\}$ will be g-orthonormal.

The general approach for trace-free matter models has been discussed in [6]. In [10] it was shown how to address the case of a pure radiation perfect fluid ($\gamma = \frac{4}{3}$). The energy momentum tensor is given by

$$\tilde{T}_{ij} = \frac{4}{3}\tilde{\rho}\tilde{u}_i\tilde{u}_j - \frac{1}{3}\tilde{\rho}\tilde{g}_{ij}, \tag{3}$$

where $\tilde{u}^\mu$ is the fluid flow velocity. As shown in [6], if one defines $T_{ij} = \theta^{-2}\tilde{T}_{ij}$, then

$$\nabla^i \tilde{T}_{ij} = 0 \quad \Leftrightarrow \quad \nabla^i T_{ij} = 0. \tag{4}$$

We thus define new variables $\rho = \theta^{-4}\tilde{\rho}$ and $u_i = \theta\tilde{u}_i$ so that

$$\theta^{-2}\tilde{T}_{ij} = T_{ij} = \frac{4}{3}\rho u_i u_j - \frac{1}{3}\rho g_{ij}. \tag{5}$$

It was shown in [18] how to obtain a FOSH system for (3) if $\tilde{\rho} \neq 0$ and $\tilde{u}_0 \neq 0$, where e_0 is timelike vector. The same method works for (5) using (4). Moreover the approach can be adapted to derive a FOSH system for $\rho_i = \nabla_i \rho$ and $u_{ij} = \nabla_i u_j$. It is necessary to introduce these new variables in order to close the overall FOSH system derived from the CEFE.

3 Existence and Uniqueness

Overall it can be shown that for $\rho \neq 0$ and $u_0 \neq 0$ the CEFE for a radiation fluid form a regular FOSH system. In particular this system is regular at conformal inifinity $\mathscr{I}$, where $\theta = 0$.

Given sufficiently smooth initial data, the CEFE have a unique solution $(\mathscr{M}, g_{\mu\nu})$, which implies a solution $(\tilde{\mathscr{M}}, \tilde{g}_{\mu\nu})$ to the Einstein field equations for a radiation fluid by setting $\tilde{\mathscr{M}} = \mathscr{M}|_{\{\theta>0\}}$, $\tilde{g}_{\mu\nu} = \theta^{-2} g_{\mu\nu}$. Thus, if the initial data is given at or near conformal infinity, then one can construct a spacetime $(\tilde{\mathscr{M}}, \tilde{g}_{\mu\nu})$ with a radiation fluid for which a part reaches all the way to conformal infinity $\mathscr{I}$.

4 The Reference Spacetimes

The family of FLRW metrics will be considered as the reference spacetime (sometimes also referred to as the background solution) against which the stability analysis is carried out. The FLRW metric may be written in the form (see e.g. [3], page 471):

$$ds^2_{FLRW} = dt^2 - \frac{a(t)^2}{(1+\frac{1}{4}kr^2)^2}(dr^2 + r^2(d\theta^2 + \sin^2\theta d\phi^2)). \tag{6}$$

As is well known, the FLRW spacetimes are conformally flat and they can be suitably rescaled to conformally embed them into the Einstein cosmos, whose metric is given by:

$$ds^2_{EC} = d\tau^2 - d\sigma^2_{\mathbb{S}^3} = d\tau^2 - (d\psi^2 + \sin^2\psi(d\theta^2 + \sin^2\theta d\phi^2)).$$

In this article we shall focus on FLRW cosmologies whose spatial sections have positive curvature ($k = 1$). Changing coordinates we have

$$ds^2_{FLRW} = a(t)^2 \left[d\tau^2 - d\sigma^2_{\mathbb{S}^3} \right] = a(t)^2 ds^2_{EC}. \tag{7}$$

Thus, the conformal factor is given by $\theta = 1/a(t)$. Our analysis is restricted to trace-free perfect fluids ($\gamma = \frac{4}{3}$) with $\lambda < 0$ in (2). There exists a static solution amongst the remaining FLRW solutions for which

$$a(t) = a_0 = \text{constant}, \qquad \lambda = \lambda_0 \equiv \tfrac{3}{2}a_0^{-2}.$$

For the dynamical FLRW solutions in our family we can state the following:

Proposition 1 *For a FLRW cosmology with* $k = 1$, $\gamma = \frac{4}{3}$, $\dot{a}(t_0) > 0$ *and* $\lambda < 0$, $\lambda \neq \lambda_0$, *the scale factor,* $a(t)$, *is a smooth, non-vanishing and monotonically increasing function for* $t \in [t_0, \infty)$, *with* $t = t_0 > 0$ *and* $a_0 = a(t_0) > 0$. *Furthermore,*

$$\tau_\infty := \int_{t_0}^{\infty} \frac{ds}{a(s)} < \infty,$$

and one has the limits

$$a \to \infty, \quad \dot{a}/a \to \sqrt{-\tfrac{1}{3}\lambda}, \quad \ddot{a}/a \to -\tfrac{1}{3}\lambda,$$

as $t \to \infty$. *The density for these models is given by*

$$\tilde{\rho} = \tilde{\rho}_0 a_0^4 / a^4,$$

where $\tilde{\rho}_0 = \tilde{\rho}(t_0)$. *In particular, one has that* $\tilde{\rho} \to 0$ *as* $t \to \infty$.

The proof of this proposition follows from direct inspection of the explicit solutions—see e.g. [3, p. 78]. Moreover, the spacetime is de Sitter-like for late times, that is conformal infinity is given by a spacelike hypersurface.

It is straightforward to check directly that the Einstein cosmos is a regular solution of the CEFE for a radiation fluid and satisfies our gauge choices (in particular $R[g_{EC}] = -\frac{1}{6}$).

5 Stability

Note that the spatial slices of the Einstein cosmos and our FLRW solutions are 3-spheres. Hence for the stability analysis we consider an initial surface $\mathscr{S}$ with the topology of $\mathbb{S}^3$ and $\tilde{g}$-unit normal $\tilde{n}$.

In the sequel, it will be assumed that one has a solution $(\mathscr{S}, \tilde{h}_{c\beta}, \tilde{K}_{\alpha\beta}, \tilde{\rho}, \tilde{u}^\alpha)$ to the (physical) $\lambda < 0$ Einstein-Euler perfect fluid constraint equations

$$\begin{aligned} \tilde{r} + \tilde{K}^2 - \tilde{K}_{\alpha\beta}\tilde{K}^{\alpha\beta} &= 2(\lambda - \tfrac{1}{3}\tilde{\rho}(4\tilde{u}_\parallel - 1)), \\ \tilde{D}^\alpha \tilde{K}_{\alpha\beta} - \tilde{D}_\beta \tilde{K} &= \tfrac{4}{3}\tilde{\rho}\tilde{u}_\parallel \tilde{u}_\beta, \end{aligned}$$

where $\tilde{D}_\beta$ and $\tilde{r}$ denote the Levi-Civita covariant derivative and the Ricci scalar of the intrinsic 3-metric $\tilde{h}_{\alpha\beta}$ of $\mathscr{S}$. $\tilde{K}_{\alpha\beta}$ is a symmetric 3-dimensional tensor corresponding to the extrinsic curvature of $\mathscr{S}$ with respect to the $\tilde{g}$–unit normal $\tilde{n}_\mu$. In addition, $\tilde{u}_\parallel \equiv \tilde{u}^\mu \tilde{n}_\mu$. Thus, one has Cauchy initial data for the Einstein-Euler system with a de Sitter-like cosmological constant and equation of state for pure radiation, $\tilde{p} = \frac{1}{3}\tilde{\rho}$.

From this data one can obtain a solution to the corresponding constraints induced by the CEFE.

For the stability analysis we make use of the Hilbert space $H^m(\mathbb{S}^3, \mathbb{R}^N)$ with $m \geq 4$—for more details see e.g. [5]. Using an extension of the general existence and stability Theorem by Kato [13] provided in [12] we are then able to prove the following more detailed version of Theorem 1.

Theorem 2 *Let* $\mathbf{w}_0$ *be initial data on* $\mathbb{S}^3$ *for CEFE for radiation fluids with* $\lambda < 0$ *such that* $\mathbf{w}_0$ *is sufficiently close to* $\mathring{\mathbf{w}}_0$ *(FLRW data with* $\lambda < 0$ *and* $k = 1$*). Then*

(i) a solution $\mathbf{w}$ *to the CEFE exists on* $[0, T] \times \mathbb{S}^3$ *with* $T > \tau_\infty$*,*
(ii) $\mathbf{w}$ *implies a* C^{m-2} *solution of the Einstein equations for a radiation fluid on* $\tilde{M} = \{p \in [0, T] \times \mathbb{S}^3 : \theta(p) > 0\}$*,*
(iii) the development exists globally towards the future,
(iv) $\tilde{\mathscr{M}}$ *is future geodesically complete and* $\mathscr{I}^+$ *is a space-like hypersurface.*
(v) $\mathbf{w}$ *remains close to the FLRW solution, and hence, it is nonlinearly stable.*

6 Discussion

As originally envisaged, the above theorem provides a proof for the case $\gamma = \frac{4}{3}$ missing in [9] using the conformal method. It should be noted that in the mean time the results in [9] have been extended to cover $\gamma = \frac{4}{3}$ as well [19]. In a next step one might consider spacetimes with null dust, such as the Robinson-Trautman and the Vaidya spacetimes, which describe radiating black holes. It remains to be seen whether the conformal method can in turn be generalised to cover more general perfect fluids with $\gamma \neq \frac{4}{3}$. This requires a formulation of the CEFE for matter models that have non-vanishing trace, which is an open problem in itself.

It is of interest, whether the CEFE for radiation fluid can be formulated using the alternative method with a congruence of conformal curves [10]. This method has the advantage that some of the PDEs reduce to transport equations along the congruence. Moreover, it allows one to prescribe/pretedict the location of conformal infinity, since the conformal factor is known a priori in terms of the specified initial data. This can simplify the explicit analysis of conformal infinity for the spacetime one is dealing with.

The use of the CEFE for the stability analysis presented here is restricted to regions near conformal infinity. The region near the initial singularity cannot be analysed as the CEFE are not regular any longer in the form given here. For an analysis of polytropic perfect fluid spacetimes near an initial singularity we refer the reader to [20]. The approach in [20] is based on conformal methods as well. However instead of the conformal factor vanishing, as it does at $\mathscr{I}$, it diverges to ∞ at the initial singularity. Nevertheless, the conformal geometry is perfectly regular in this region. It would be interesting to know whether a variant of the CEFE exists which is regular at the singularity and corresponds to the analysis in [20].

References

1. Einstein, A., Die Grundlage der allgemeinen Relativitätstheorie. Ann. Phys. (Leipzig) **49**, 769 (1916). doi:10.1002/andp.19163540702
2. Stephani, H., Kramer, D., MacCallum, M., Hoenselaers, C., Herlt, E.: Exact solutions of Einstein's field equations (2nd edn). Cambridge Monographs on Mathematical Physics. Cambridge University Press, New York (2003)
3. Griffiths, J., Podolský, J.: Exact space-times in Einstein's general relativity. Cambridge Monographs on Mathematical Physics. Cambridge University Press, New York (2009)
4. Dafermos, M., Rodnianski, I.: Lectures on black holes and linear waves, ArXiv e-prints arXiv:0811.0354 [gr-qc] (2008)
5. Friedrich, H.: On the existence of n-geodesically complete or future complete solutions of Einstein's field equations with smooth asymptotic structure. Commun. Math. Phys. **107**, 587 (1986). doi:10.1007/BF01205488
6. Friedrich, H.: On the global existence and the asymptotic behaviour of solutions to the Einstein-Maxwell-Yang-Mills equations. J. Differ. Geom. **34**, 275 (1991)
7. Christodoulou, D., Klainerman, S.: The Global Nonlinear Stability of the Minkowski Space, Princeton Mathematical Series, vol. 41. Princeton University Press, Princeton (1993)
8. Rodnianski, I., Speck, J.: The Stability of the irrotational Euler-Einstein system with a positive cosmological constant, ArXiv e-prints arXiv:0911.5501 [math-ph] (2009)
9. Speck, J.: The nonlinear future stability of the FLRW family of solutions to the Euler-Einstein system with a positive cosmological constant. Selecta Mathematica **18**, 633 (2012). doi:10.1007/s00029-012-0090-6
10. Lübbe, C., Valiente Kroon, J.A.: The extended conformal Einstein field equations with matter: the Einstein-Maxwell field. J. Geom. Phys. **62**, 1548 (2012). doi:10.1016/j.geomphys.2012.01.009
11. Friedrich, H.: Cauchy problems for the conformal vacuum field equations in general relativity. Commun. Math. Phys. **91**, 445 (1983). doi:10.1007/BF01206015
12. Friedrich, H.: On purely radiative space-times. Commun. Math. Phys. **103**, 35 (1986). doi:10.1007/BF01464281
13. Kato, T.: The Cauchy problem for quasi-linear symmetric hyperbolic systems. Arch. Ration. Mech. Anal. **58**, 181 (1975). doi:10.1007/BF00280740
14. Friedrich, H.: On the hyperbolicity of Einstein's and other gauge field equations. Commun. Math. Phys. **100**, 525 (1985). doi:10.1007/BF01217728
15. Friedrich, H.: Einstein equations and conformal structure: existence of anti-de Sitter-type space-times. J. Geom. Phys. **17**, 125 (1995). doi:10.1016/0393-0440(94)00042-3
16. Friedrich, H.: Conformal Einstein Evolution. In: Frauendiener, J., Friedrich, H. (eds.) The Conformal Structure of Space-Time: Geometry, Analysis, Numerics. Lecture Notes in Physics, pp. 1–102. Springer, New York (2002)
17. Lübbe, C., Valiente Kroon J.: A stability result for purely radiative spacetimes. J. Hyperbol. Differ. Equations **7**, 545 (2010). doi:10.1142/S0219891610002220
18. Choquet-Bruhat, Y.: General Relativity and the Einstein Equations. Oxford Mathematical Monographs. Oxford University Press, New York (2009)
19. Speck J.: Private communication
20. Anguige, K., Tod, K.: Isotropic cosmological singularities: I. Polytropic perfect fluid spacetimes. Ann. Phys. (N.Y.) **276**, 257 (1999). doi:10.1006/aphy.1999.5946

Canonical Gravity, Non-Inertial Frames, Relativistic Metrology and Dark Matter

Luca Lusanna

Abstract Clock synchronization leads to the definition of instantaneous 3-spaces (to be used as Cauchy surfaces) in non-inertial frames, the only ones allowed by the equivalence principle. ADM canonical tetrad gravity in asymptotically Minkowskian space-times can be described in this framework. This allows to find the York canonical basis in which the inertial (gauge) and tidal (physical) degrees of freedom of the gravitational field can be identified. A Post-Minkowskian linearization with respect to the asymptotic Minkowski metric (asymptotic background) allows to solve the Dirac constraints in non-harmonic 3-orthogonal gauges and to find non-harmonic TT gravitational waves. The inertial gauge variable York time (the trace of the extrinsic curvature of the 3-space) describes the general relativistic freedom in clock synchronization. After a digression on the gauge problem in general relativity and its connection with relativistic metrology, it is shown that dark matter, whose experimental signatures are the rotation curves and the mass of galaxies, may be described (at least partially) as an inertial relativistic effect (absent in Newtonian gravity) connected with the York time, namely with the non-Euclidean nature of 3-spaces as 3-sub-manifolds of space-time.

While in special relativity (SR) the use of non-inertial frames is optional, in general relativity (GR) only these are allowed by the equivalence principle, forbidding the existence of global inertial frames. In both cases the Lorentz signature of the space-time implies that there is no notion of instantaneous 3-space: the only intrinsic structure is the conformal one, i.e. the light-cone as the locus of incoming and outgoing radiation. *A convention on the synchronization of clocks is needed to define an instantaneous 3-space*, where one has to give the Cauchy data for the relevant wave equations. For instance the *1-way velocity of light* from one observer A to an observer B has a meaning only after choosing such a convention. In SR Einstein

L. Lusanna (✉)
Sezione INFN di Firenze, Polo Scientifico, Via Sansone 1, 50019 Sesto Fiorentino, FI, Italy
e-mail: lusanna@fi.infn.it

J. Bičák and T. Ledvinka (eds.), *Relativity and Gravitation*,
Springer Proceedings in Physics 157, DOI: 10.1007/978-3-319-06761-2_18,

convention for the synchronization of clocks in Minkowski space-time uses the 2-way (or round trip) velocity of light to identify the Euclidean 3-spaces of the inertial frames centered on an inertial observer A by means of only one clock. It is this velocity which is isotropic and constant in SR and replaces the standard of length in relativistic metrology [1]. Only in the inertial frames of SR the 1-way and the 2-way velocities coincide.

Therefore in References [2–4] a general theory of global non-inertial frames in Minkowski space-time was developed by using the $3+1$ point of view in which, besides the world-line of a time-like observer, one also gives a global, nice foliation of the space-time with instantaneous 3-spaces. In this way one avoids the coordinate singularities of the $1+3$ description (both those of Fermi coordinates and of the rotating disk). The time-like observer carries a standard atomic clock and τ is an arbitrary monotonically increasing function of the proper time of this clock. The space-like instantaneous 3-spaces Σ_τ are the mathematical idealization of a protocol for clock synchronization: all the clocks in the points of Σ_τ show the same time of the atomic clock of the observer. On each 3-space Σ_τ one chooses curvilinear 3-coordinates σ^r having the observer as origin. The Lorentz-scalar and observer-dependent coordinates $\sigma^A = (\tau, \sigma^r)$ are named *radar 4-coordinates*. The coordinate transformation $\sigma^A \mapsto x^\mu = z^\mu(\tau, \sigma^r)$ to the Cartesian coordinates x^μ defines the embedding $z^\mu(\tau, \sigma^r)$ of the 3-spaces Σ_τ into Minkowski space-time. The induced 4-metric on Σ_τ is the following functional of the embedding ${}^4g_{AB}(\tau, \sigma^r) = [z^\mu_A\, \eta_{\mu\nu}\, z^\nu_B](\tau, \sigma^r)$, where $z^\mu_A = \partial\, z^\mu / \partial\, \sigma^A$.

The $3+1$ point of view has allowed to get the description of arbitrary isolated systems (particles, strings, fluids, fields) admitting a Lagrangian formulation in arbitrary non-inertial frames by means of *parametrized Minkowski theories* [2–6]. In them the Lagrangian is coupled to an external gravitational field and then the gravitational 4-metric is replaced with the 4-metric ${}^3g_{AB}(\tau, \sigma^r)$ induced by an admissible $3+1$ splitting of Minkowski space-time. The new Lagrangian, a function of the matter and of the embedding, is invariant under frame-preserving diffeomorphisms and this type of general covariance implies that the embeddings are *gauge variables*, so that the transition among non-inertial frames is described as a *gauge transformation*: only the appearances change, not the physics. The metric ${}^3g_{AB}(\tau, \sigma^r)$ and the extrinsic curvature tensor ${}^3K_{rs}(\tau, \sigma^u)$ play the role of inertial potentials.

This framework allows us to define the *inertial and non-inertial rest frames* of the isolated systems, and to develop the rest-frame instant form of the dynamics and to build the explicit form of the Lorentz boosts for interacting systems. While the inertial rest frames have their Euclidean 3-spaces defined as space-like 3-manifolds of Minkowski space-time orthogonal to the conserved 4-momentum of the isolated system, the non-inertial rest frames are admissible non-inertial frames whose 3-spaces tend to those of some inertial rest frame at spatial infinity, where the 3-space becomes orthogonal to the conserved 4-momentum. This setting allows to study the problem of the relativistic center of mass with the associated external and internal (i.e. inside the 3-space) realizations of the Poincaré algebra in a way compatible with relativistic bound states [7–11], and to get a new Wigner-covariant formulation of relativistic quantum mechanics [12], with a solution of all the known problems introduced by SR.

This metrology-oriented solution of the problem of clock synchronization used in SR can be extended to GR, if Einstein space-times are restricted to the class of globally hyperbolic, topologically trivial, asymptotically Minkowskian space-times without super-translations and without Killing symmetries, which include the Christodoulou-Klainerman space-times [13, 14]. In these space-times one can define global non-inertial frames by using the same admissible $3+1$ splittings, centered on a time-like observer, to define the 3-spaces Σ_τ and the observer-dependent radar 4-coordinates $\sigma^A = (\tau; \sigma^r)$ employed in SR. This will allow to separate the *inertial* (gauge) degrees of freedom of the gravitational field (playing the role of inertial potentials) from the dynamical *tidal* ones at the Hamiltonian level.

In GR the gradients $z^\mu_A(\tau, \sigma^r)$ of the embeddings $x^\mu = z^\mu(\tau, \sigma^r)$, defining the admissible $3+1$ splittings of space-time, give the transition coefficients from radar to world 4-coordinates.

The components ${}^4g_{AB}(\tau, \sigma^r) = z^\mu_A(\tau, \sigma^r)\, z^\nu_B(\tau, \sigma^r)\, {}^4g_{\mu\nu}(z(\tau, \sigma^r))$ of the 4-metric will be the dynamical fields in the ADM action [15], written in the basis of radar 4-coordinates. Like in SR the 4-vectors $z^\mu_u(\tau, \sigma^r)$, tangent to the 3-spaces Σ_τ, are used to define the unit normal $l^\mu(\tau, \sigma^r) = z^\mu_A(\tau, \sigma^r)\, l^A(\tau, \sigma^r)$ to Σ_τ, while the 4-vector $z^\mu_\tau(\tau, \sigma^r)$ has the lapse function as component along the unit normal and the shift functions as components along the tangent vectors.

While in SR time and 3-space are absolute notions, in GR the space-time is a dynamical object [16–18]. Each solution (i.e. an Einstein 4-geometry) of Einstein's equations (or of the associated ADM Hamilton equations) dynamically selects a preferred $3+1$ splitting of the space-time, namely in GR the instantaneous 3-spaces are dynamically determined in the chosen world coordinate system, modulo the choice of the 3-coordinates into the 3-space and modulo the trace of the extrinsic curvature of the 3-space as a space-like sub-manifold of the space-time [19]. In GR the gravitational field is described by ten dynamical fields ${}^4g_{\mu\nu}(x)$, which also determine the *chrono-geometrical structure of space-time* through the line element $ds^2 = {}^4g_{\mu\nu}\, dx^\mu\, dx^\nu$. Therefore the 4-metric teaches relativistic causality to the other fields: it says to massless particles like photons and gluons which are the allowed world-lines at each point of space-time.

As shown in the first paper of References [20, 21], in the chosen class of space-times the 4-metric ${}^4g_{\mu\nu}(x)$ tends, in a suitable way, to the flat Minkowski 4-metric ${}^4\eta_{\mu\nu}$ at spatial infinity (to be used as an *asymptotic background* at spatial infinity in the linearization of the theory), where there are asymptotic inertial observers whose spatial axes may be identified by means of the fixed stars of star catalogues (the fixed stars can be considered as an empirical definition of spatial infinity of the observable universe). In absence of super-translations the asymptotic symmetries reduce to the asymptotic ADM Poincaré group. The ten *strong* asymptotic ADM Poincaré generators P^A_{ADM}, J^{AB}_{ADM} (they are fluxes through a 2-surface at spatial infinity) are well defined functionals of the 4-metric fixed by the boundary conditions at spatial infinity. Moreover in that paper it is also shown that the boundary conditions on the 4-metric required by the absence of super-translations imply that the only admissible $3+1$ splittings of space-time (i.e. the allowed global non-inertial frames) are the *non-inertial rest frames*: their 3-spaces are asymptotically orthogonal to the weak ADM 4-momentum. Therefore one gets $\hat{P}^r_{ADM} \approx 0$ as the rest-frame condition of the 3-universe with a mass and a rest spin fixed by the boundary conditions.

Finally, in the limit of vanishing Newton's constant ($G = 0$) the asymptotic ADM Poincaré generators become the generators of the special relativistic Poincaré group describing the matter present in the space-time, allowing the inclusion into GR of the classical version of the standard model of particle physics, whose properties are all connected with the representations of this group in the inertial frames of Minkowski space-time.

To define the canonical formalism the Einstein-Hilbert action for metric gravity (depending on the second derivative of the metric) must be replaced with the ADM action (the two actions differ by a surface term at spatial infinity). As shown in the first paper of References [20, 21], the Legendre transform and the definition of a consistent canonical Hamiltonian require the introduction of the DeWitt surface term at spatial infinity: the final canonical Hamiltonian turns out to be the *strong* ADM energy (a flux through a 2-surface at spatial infinity), which is equal to the *weak* ADM energy (expressed as a volume integral over the 3-space) plus constraints. Therefore there is not a frozen picture like in the "spatially compact space-times without boundaries" used in loop quantum gravity (where the canonical Hamiltonian vanishes), but an evolution generated by a Dirac Hamiltonian equal to the weak ADM energy plus a linear combination of the first class constraints. Also the other strong ADM Poincaré generators are replaced by their weakly equivalent weak form $\hat{P}^A_{ADM}$, $\hat{J}^{AB}_{ADM}$.

To take into account the fermion fields present in the standard particle model one must extend ADM gravity to ADM tetrad gravity. Since our class of space-times admits orthonormal tetrad and a spinor structure [22], the extension can be done by simply replacing the 4-metric in the ADM action with its expression in terms of cotetrad fields $E^{(\alpha)}_A(\tau, \sigma^r)$,

$$
{}^4g_{AB}(\tau, \sigma^r) = E^{(\alpha)}_A(\tau, \sigma^r)\, {}^4\eta_{(\alpha)(\beta)}\, E^{(\beta)}_B(\tau, \sigma^r),
$$

(α) are flat indices and ${}^4\eta_{(\alpha)(\beta)}$ the flat metric; by convention a sum on repeated indices is assumed. The cotetrad fields $E^{(\alpha)}_A$, considered as the basic 16 configurational variables in the ADM action, are the inverse of the tetrad fields $E^A_{(\alpha)}$, which are connected to the world tetrad fields by $E^\mu_{(\alpha)}(x) = z^\mu_A(\tau, \sigma^r)\, E^A_{(\alpha)}(z(\tau, \sigma^r))$. The cotetrads $E^{(\alpha)}_A(\tau, \sigma^r)$ are connected to cotetrads ${}^4\overset{\circ}{E}{}^{(\alpha)}_A(\tau, \sigma^r)$ adapted to the $3+1$ splitting of space-time, namely such that the inverse adapted time-like tetrad ${}^4\overset{\circ}{E}{}^A_{(o)}(\tau, \sigma^r)$ is the unit normal to the 3-space Σ_τ, by a standard Wigner boosts for time-like Poincaré orbits with parameters $\varphi_{(a)}(\tau, \sigma^r)$, $a = 1, 2, 3$.

This leads to an interpretation of gravity based on a congruence of time-like observers endowed with orthonormal tetrads: at each point of space-time the time-like axis is the unit 4-velocity of the observer, while the spatial axes are a (gauge) convention for observer's gyroscopes. This framework was developed in the second and third paper of References [20, 21].

Even if the action of ADM tetrad gravity depends upon 16 fields, the counting of the physical degrees of freedom of the gravitational field does not change, because this action is invariant not only under the group of 4-diffeomorphisms but also under the O(3,1) gauge group of the Newman-Penrose approach [23] (the extra gauge freedom acting on the tetrads in the tangent space of each point of space-time).

After having introduced the kinematical framework for the description of non-inertial frames in GR, we must study the dynamical aspects of the gravitational field to understand which variables are dynamically determined and which are the inertial effects hidden in the general covariance of the theory. Since at the Lagrangian level it is not possible to identify which components of the 4-metric tensor are connected with the gauge freedom in the choice of the 4-coordinates and which ones describe the dynamical degrees of freedom of the gravitational field, one must restrict oneself to the quoted class of globally hyperbolic, asymptotically Minkowskian space-times allowing a Hamiltonian description of ADM gravity. In canonical ADM gravity one can use Dirac theory of constraints to describe the Hamiltonian gauge group, whose generators are the first-class constraints of the model. The basic tool of this approach is the possibility to find so-called Shanmugadhasan canonical transformations [24, 25], which identify special canonical bases adapted to the first-class constraints (and also to the second-class ones when present). In these special canonical bases the vanishing of certain momenta (or of certain configurational coordinates) corresponds to the vanishing of well defined Abelianized combinations of the first-class constraints (Abelianized because the new constraints have exactly zero Poisson brackets even if the original constraints were not in strong involution). As a consequence, the variables conjugate to these Abelianized constraints are *inertial Hamiltonian gauge variables describing the Hamiltonian gauge freedom*.

Therefore, starting from the ADM action for tetrad gravity one defines the Hamiltonian formalism in a phase space containing 16 configurational field variables and 16 conjugate moments. One identifies the 14 first-class constraints of the system. The existence of these 14 first-class constraints implies that 14 components of the tetrads (or of the conjugate momenta) are Hamiltonian gauge variables describing the *inertial* aspects of the gravitational field (6 of these inertial variables describe the extra gauge freedom in the choice of the tetrads and in their transport along world-lines). Therefore there are only $2+2$ degrees of freedom for the description of the *tidal* dynamical aspects of the gravitational field (the two polarizations of gravitational waves in the linearized theory). The asymptotic ADM Poincaré generators can be evaluated explicitly. Till now the type of matter studied in this framework [26–28] consists of the electromagnetic field and of N charged scalar particles, whose signs of the energy and electric charges are Grassmann-valued to regularize both the gravitational and electromagnetic self-energies (it is both a ultraviolet and an infrared regularization).

If one would be able to include all the constraints in the Shanmugadhasan canonical basis, the $2+2$ tidal variables would be the *Dirac observables* of the gravitational field, invariant under the Hamiltonian gauge transformations. However such Dirac observables are not known: one only has statements about their existence. Moreover, in general they are not 4-scalar observables. The problem of the connection between the 4-diffeomorphism group and the Hamiltonian gauge group was studied in References [29, 30] by means of the inverse Legendre transformation and of the notion of dynamical symmetry. The conclusion is that on the space of solutions of Einstein equations there is an overlap of the two types of observables: there should exist special Shanmugadhasan canonical bases in which the $2+2$ Dirac observables become 4-scalars when restricted to the space of solutions of the Einstein equations (i.e. on-shell). In any case the identification of the inertial gauge components of the 4-metric is what is needed to make a fixation of 4-coordinates as required by relativistic metrology.

The best which can be done till now is the explicit identification of a Shanmugadhasan canonical transformation [19] (implementing the so-called York map and diagonalizing the York-Lichnerowicz approach) to a so-called York canonical basis adapted to 10 of the 14 first-class constraints. Only the super-Hamiltonian and supermomentum constraints, whose general solution is not known, are not included in the basis, but it is clarified which variables are to be determined by their solution, namely the 3-volume element (the determinant of the 3-metric) of the 3-space Σ_τ and three momenta conjugated to 3-coordinates on Σ_τ. The 14 inertial gauge variables turn out to be: (a) the six configurational variables $\varphi_{(a)}$ and $\alpha_{(a)}$ of the tetrads describing their O(3,1) gauge freedom; (b) the lapse and shift functions; (c) the 3-coordinates on the 3-space (their fixation implies the determination of the shift functions); (d) the York time ${}^3K$, i.e. the trace of the extrinsic curvature of the 3-spaces as 3-manifolds embedded into the space-time (its fixation implies the determination of the lapse function). It is the only gauge variable which is a momentum in the York canonical basis (instead in Yang-Mills theory all the gauge variables are configurational): this is due to the Lorentz signature of space-time, because the York time and three other inertial gauge variables can be used as 4-coordinates of the space-time. In this way an identification of the inertial gauge variables to be fixed to get a 4-coordinate system in relativistic metrology was found. While in SR all the components of the tetrads and their conjugate momenta are inertial gauge variables, in GR the two eigenvalues of the 3-metric with determinant one and their conjugate momenta describe the physical tidal degrees of freedom of the gravitational field. In the first paper of References [26–28] there is the expression of the Hamilton equations for all the variables of the York canonical basis.

An important remark is that in the framework of the York canonical basis the natural family of gauges is not the harmonic one, but the family of 3-orthogonal Schwinger time gauges in which the 3-metric in the 3-spaces is diagonal [26–28].

In conclusion, while the gauge group of the Lagrangian formulation of Einstein GR, the diffeomorphism group, implies that the 4-coordinates of the space-time are *gauge variables*, the Hamiltonian gauge group replaces them with the *inertial gauge variables*, York time and 3-coordinates on the instantaneous 3-space Σ_τ. In both cases one would like to re-express physical properties in terms either of 4-scalars or of Dirac observables becoming 4-scalars on-shell. However, on one side it is not yet known how to implement this program and on the other side this is not the praxis of experimental physics.

Inside the Solar System the experimental localization of macroscopic classical objects is unavoidably done by choosing some *convention* for the local 4-coordinates of space-time. Atomic physicists, NASA engineers and astronomers have chosen a series of reference frames and standards of time and length suitable for the existing technology [1, 31–33]. These conventions determine certain Post-Minkowskian (PM) 4-coordinate systems (in harmonic gauges) of an asymptotically Minkowskian space-time, in which the instantaneous 3-spaces are not strictly Euclidean. Then these reference frames are seen as a local approximation of a celestial reference frame (ICRS), where however the space-time has become a cosmological Friedman-Robertson-Walker (FRW) one, which is only conformally asymptotically Minkowskian at spatial infinity and therefore does not admit a Hamiltonian description. A search of a consistent patching of the 4-coordinates from inside the Solar System to the rest of the universe will start when the data from the future GAIA mission [34] for the cartography of the Milky Way will be available. This will allow a PM definition of a Galactic Reference System containing at least our Galaxy. Let us remark that notwithstanding the FRW instantaneous 3-spaces are not strictly Euclidean, all the books on galactic dynamics describe the galaxies by means of the Kepler theory in Galilei space-time.

Both in SR and GR an admissible $3+1$ splitting of space-time has two associated congruences of time-like observers [2–4], geometrically defined and not to be confused with the congruence of the world-lines of fluid elements, when relativistic fluids are added as matter in GR [35]. One of the two congruences, with zero vorticity, is the congruence of the Eulerian observers, whose 4-velocity field is the field of unit normals to the 3-spaces. This congruence allows us to re-express the non-vanishing momenta of the York canonical basis in terms of the expansion ($\theta = -{}^3K$) and of the shear of the Eulerian observers. This allows us to compare the Hamilton equations of ADM canonical gravity with the usual first-order non-Hamiltonian ADM equations deducible from Einstein equations given a $3+1$ splitting of space-time but without using the Hamiltonian formalism. As a consequence, one can extend our Hamiltonian identification of the inertial and tidal variables of the gravitational field to the Lagrangian framework and use it in the cosmological (conformally asymptotically flat) space-times: in them it is not possible to formulate the Hamiltonian formalism but the standard ADM equations are well defined. The time inertial gauge variable needed for relativistic metrology is now the expansion of the Eulerian observers of the given $3+1$ splitting of the globally hyperbolic cosmological space-time. It is this inertial gauge variable which has to be fixed in this way to reproduce experimental astronomical data and their astrophysical interpretation.

In conclusion we now have a framework for non-inertial frames in GR and an identification of the inertial gauge variables in asymptotically Minkowskian and also in cosmological space-times. See the third paper of References [26–28] for the possibility that the three main signatures of dark matter (rotation curves of galaxies; mass of galaxy clusters from the virial theorem and weak gravitational lensing; see the review in References [36, 37]) can be explained as only a relativistic inertial effect induced by the inertial gauge variable ${}^3K$ (the York time): a suitable choice of the 3-space in the celestial reference frame could simulate the effects explained by dark matter. Since in the PM Hamiltonian linearization of canonical tetrad gravity [26–28] the lapse function is $n = n_1 + \partial_\tau\, {}^3K + \ldots$ with n_1 describing the Newtonian potential in the non-relativistic limit, one can identify $\partial_\tau\, {}^3K$ with the Yukawa-like potential used in $f(R)$ gravity to simulate dark matter [38].

See Reference [39] for an extended and complete review of the approach. In this paper it is also shown (at a preliminary level) that the York time is connected also with dark energy in inhomogeneous cosmological space-times [40].

References

1. Lusanna, L.: Relativistic metrology: from earth to astrophysics. In: Cocco, L. (ed.) Modern Metrology Concerns (InTech, 2012), chap. 15. doi:10.5772/37508
2. Alba, D., Lusanna, L.: Charged particles and the electro-magnetic field in noninertial frames of Minkowski space-time I: admissible 3 + 1 splittings of Minkowski space-time and the noninertial rest frames. Int. J. Geom. Meth. Mod. Phys. **7**, 33 (2010). doi:10.1142/S021988781000394X
3. Alba, D., Lusanna, L.: Charged particles and the electro-magnetic field in noninertial frames of Minkowski space-time II: applications—rotating frames, Sagnac effect, Faraday rotation, wrap-up effect. Int. J. Geom. Meth. Mod. Phys. **7**, 185 (2010). doi:10.1142/S0219887810004051
4. Alba, D., Lusanna, L.: Generalized radar 4-coordinates and equal-time Cauchy surfaces for arbitrary accelerated observers. Int. J. Mod. Phys. D **16**, 1149 (2007). doi:10.1142/S021827180701064X
5. Lusanna, L.: The N- and 1-time classical descriptions of N-body relativistic kinematics and the electromagnetic interaction. Int. J. Mod. Phys. A **12**, 645 (1997). doi:10.1142/S0217751X9700058X
6. Lusanna, L.: The chrono-geometrical structure of special and general relativity: a re-visitation of canonical geometrodynamics. Int. J. Geom. Meth. Mod. Phys. **04**, 79 (2007). doi:10.1142/S0219887807001874
7. Alba, D., Lusanna, L., Pauri, M.: Centers of mass and rotational kinematics for the relativistic N-body problem in the rest-frame instant form. J. Math. Phys. **43**, 1677 (2002). doi:10.1063/1.1435424
8. Crater, H., Lusanna, L.; The rest-frame Darwin potential from the Lienard-Wiechert solution in the radiation gauge. Ann. Phys. (N.Y.) **289**, 87 (2001). doi:10.1006/aphy.2000.6129
9. Alba, D., Crater, H., Lusanna, L.: Towards relativistic atomic physics. Part 1. The rest-frame instant form of dynamics and a canonical transformation for a system of charged particles plus the electro-magnetic field. Can. J. Phys. **88**, 379 (2010). doi:10.1139/P09-037
10. Alba, D., Crater, H., Lusanna, L.: Towards relativistic atomic physics. Part II. Collective and relative relativistic variables for a system of charged articles plus the electromagnetic field. Can. J. Phys. **88**, 425 (2010). doi:10.1139/P09-038
11. Alba, D., Crater, H., Lusanna, L.: Hamiltonian relativistic two-body problem: center of mass and orbit reconstruction. J. Phys. A Math. Theory **40**, 9585 (2007). doi:10.1088/1751-8113/40/31/029
12. Alba, D., Crater, H., Lusanna, L.: Relativistic quantum mechanics and relativistic entanglement in the rest-frame instant form of dynamics. J. Math. Phys. **52**, 062301 (2011). doi:10.1063/1.3591131
13. Beig, R., Ó Murchadha, N.: The Poincaré group as the symmetry group of canonical general relativity. Ann. Phys. (N.Y.) **174**, 463 (1987). doi:10.1016/0003-4916(87)90037-6

14. Christodoulou, D., Klainerman, S.: The Global Nonlinear Stability of the Minkowski Space. Princeton Mathematical Series, vol. 41. Princeton University Press, Princeton (1993)
15. Arnowitt, R., Deser, S., Misner, C.: The dynamics of general relativity. In: Witten, L. (ed.) Gravitation: An Introduction to Current Research, pp. 227–265. Wiley, New York (1962). doi:10.1007/s10714-008-0661-1
16. Lusanna, L., Pauri, M.: Explaining Leibniz equivalence as difference of non-inertial appearances: dis-solution of the Hole Argument and physical individuation of point-events. Stud. Hist. Phil. Mod. Phys. **37**, 692 (2006). doi:10.1016/j.shpsb.2006.03.006
17. Lusanna, L., Pauri, M.: The physical role of gravitational and gauge degrees of freedom in general relativity—I: dynamical synchronization and generalized inertial effects. Gen. Relativ. Gravit. **38**, 187 (2006). doi:10.1007/s10714-005-0217-6
18. Lusanna, L., Pauri, M.: The physical role of gravitational and gauge degrees of freedom in general relativity—II: Dirac versus Bergmann observables and the objectivity of space-time. Gen. Relativ. Gravit. **38**, 229 (2006). doi:10.1007/s10714-005-0218-5
19. Alba, D., Lusanna, L.: The York map as a Shanmugadhasan canonical transformation in tetrad gravity and the role of non-inertial frames in the geometrical view of the gravitational field. Gen. Relativ. Gravit. **39**, 2149 (2007). doi:10.1007/s10714-007-0507-2
20. Lusanna, L.: The rest-frame instant form of metric gravity. Gen. Relativ. Gravit. **33**, 1579 (2001). doi:10.1023/A:1012297028267
21. Lusanna, L., Russo, S.: A new parametrization for tetrad gravity. Gen. Relativ. Gravit. **34**, 189 (2002). doi:10.1023/A:1015387310306
22. Geroch, R.: Spinor structure of space-times in general relativity. I. J. Math. Phys. **9**, 1739 (1968). doi:10.1063/1.1664507
23. Stewart, J.: Advanced General Relativity. Cambridge Monographs on Mathematical Physics. Cambridge University Press, Cambridge (1991)
24. Shanmugadhasan, S.: Canonical formalism for degenerate Lagrangians. J. Math. Phys. **14**, 677 (1973). doi:10.1063/1.1666377
25. Lusanna, L.: The Shanmugadhasan canonical transformation, function groups and the extended second Noether theorem. Int. J. Mod. Phys. A **8**, 4193 (1993). doi:10.1142/S0217751X93001727
26. Alba, D., Lusanna, L.: The Einstein-Maxwell-particle system in the York canonical basis of ADM tetrad gravity: I. The equations of motion in arbitrary Schwinger time gauges, ArXiv e-prints arXiv:0907.4087 [gr-qc] (2009)
27. Alba, D., Lusanna, L.: The Einstein-Maxwell-particle system: II. The weak field approximation in the non-harmonic 3-orthogonal gauges and Hamiltonian post-Minkowskian gravity: the N-body problem and gravitational waves with asymptotic background, ArXiv e-prints arXiv:1003.5143 [gr-qc] (2010)
28. Alba, D., Lusanna, L.: The Einstein-Maxwell-particle system in the York canonical basis of ADM tetrad gravity: III. The post-Minkowskian N-body problem, its post-Newtonian limit in non-harmonic 3-orthogonal gauges and dark matter as an inertial effect, ArXiv e-prints arXiv:1009.1794 [gr-qc] (2010)
29. Moncrief, V.: Spacetime symmetries and linearization stability of the Einstein equations. I. J. Math. Phys. **16**, 493 (1975). doi:10.1063/1.522572
30. Moncrief, V.: Space-time symmetries and linearization stability of the Einstein equations. II. J. Math. Phys. **17**, 1893 (1976). doi:10.1063/1.522814
31. Soffel, M., Klioner, S., Petit, G., et al.: The IAU 2000 resolutions for astrometry, celestial mechanics, and metrology in the relativistic framework: explanatory supplement. Astron. J. **126**, 2687 (2003). doi:10.1086/378162
32. McCarthy, D., Petit, G. (eds.): IERS Conventions (2003). No. IERS TN 32 in IERS Technical Notes (Verlag des BKG, Frankfurt am Main, 2004). http://www.iers.org/IERS/EN/Publications/TechnicalNotes/tn32.html
33. Kaplan, G.: The IAU resolutions on astronomical reference systems, time scales, and Earth rotation models: explanation and implementation. No. 179 in USNO Circular. U.S. Naval Observatory, Washington, DC, 2005. http://aa.usno.navy.mil/publications/docs/Circular_179.php

34. Jordan, S.: The GAIA project: technique, performance and status. Astron. Nachr. **329**, 875 (2008). doi:10.1002/asna.200811065
35. Alba, D., Lusanna, L.: Dust in the York canonical basis of ADM tetrad gravity: the problem of vorticity, ArXiv e-prints arXiv:1106.0403 [gr-qc] (2011)
36. Bartelmann, M.: The dark universe. Rev. Mod. Phys. **82**, 331 (2010). doi:10.1103/RevModPhys.82.331
37. Bean, R.: TASI lectures on cosmic acceleration (2010). ArXiv e-prints arXiv:1003.4468 [astro-ph.CO]
38. Capozziello, S., De Laurentis, M.: The dark matter problem from f(R) gravity viewpoint. Ann. Phys. (Berlin) **524**, 545 (2012). doi:10.1002/andp.201200109
39. Lusanna, L.: From clock synchronization to dark matter as a relativistic inertial effect, ArXiv e-prints arXiv:1205.2481 [gr-qc][(2012). Lecture at BOSS2011, Frascati, Italy, Accessed 9–13 May 2011
40. Plebański, J., Krasiński, A.: An Introduction to General Relativity and Cosmology. Cambridge University Press, Cambridge (2006)

Gravomagnetic Solenoids

Donald Lynden-Bell and Joseph Katz

Abstract We introduce strong field gravomagnetism and illustrate its use by constructing exact rolling toroidal solutions of Einstein's equations.

1 Introduction

The 1966 edition of the Classical theory of Fields by Landau and Lifshitz [1] gives Einstein's equations for general stationary metrics in a form that has strong analogies with Maxwell's electrodynamics. The technique identifies the points of space that lie along the time-like Killing vector, so it does not extend continuously inside ergospheres where the Killing vector becomes space-like. We write the metric in the form

$$ds^2 = \xi^2(dt - \mathscr{A}_k dx^k)^2 - \gamma_{kl} dx^k dx^l = g_{\mu\nu} dx^\mu dx^\nu. \quad (1)$$

We work in the positive definite three dimensional metric of space, γ_{kl}, where k and l run from 1 to 3. It is not a cross-section of the four metric by any surface, nevertheless we may define its Christoffel symbols λ^m_{kl} and the corresponding three-dimensional Ricci tensor of this gamma space, P^{kl}. We use commas to denote ordinary derivatives and semicolons to denote covariant derivatives in gamma space. The Ricci tensor of space-time will be denoted by $R_{\mu\nu}$. The divergence and curl are defined in gamma-space by

D. Lynden-Bell (✉) · J. Katz
Institute of Astronomy, Madingley Rd, Cambridge, UK
e-mail: dlb@ast.cam.ac.uk

J. Katz
The Racah Institute of Physics, Edmond Safra Campus, Givat Ram, 91904 Jerusalem, Israel
e-mail: jkatz@phys.huji.ac.il

J. Bičák and T. Ledvinka (eds.), *Relativity and Gravitation*, Springer Proceedings in Physics 157, DOI: 10.1007/978-3-319-06761-2_19,

$$\mathbf{div}\, \boldsymbol{E} = \gamma^{-\frac{1}{2}} \partial_k(\sqrt{\gamma} E^k); \quad (\mathbf{curl}\, \boldsymbol{E})^i = \gamma^{-\frac{1}{2}} \epsilon^{ijk} \partial_j E_k. \tag{2}$$

We define the gravomagnetic induction $\mathscr{B}$ by

$$\mathscr{B} = \mathbf{curl}\, \mathscr{A}, \tag{3}$$

where $\mathscr{A}_k$ is the vector potential defined in the metric (1). Clearly $\mathbf{div}\, \mathscr{B} = 0$ so $\mathscr{B}$ carries the gravomagnetic flux. Landau and Lifshitz rewrite Einstein's equations in gamma space; rewriting their equations in our notation we have with $\kappa = 8\pi G/c^4$,

$$\xi\, \mathbf{div}\, \mathbf{grad}\, \xi + \frac{1}{2} \xi^4 \mathscr{B}^2 = R_{00} = \kappa (T_{00} - \frac{1}{2} g_{00} T). \tag{4}$$

Henceforth we use units with $c = 1$ and $G = 1$. If we now define a field intensity vector $\mathscr{H} = \xi^3 \mathscr{B}$ then their second equation reads

$$(\mathbf{curl}\, \mathscr{H})^k = -2\kappa \xi T_0^k = -2\kappa J^k. \tag{5}$$

Notice a strong resemblance of this strong field equation to Maxwell's electrodynamic equation $\mathbf{curl}\, \mathbf{H} = 4\pi \mathbf{j}$. In both cases the current has no divergence however in general $\mathscr{H}$ has a divergence while $\mathscr{B}$ does not. Clearly $\mathscr{H}$ is the gradient of a scalar whenever $\mathbf{j}$ is zero. The $\mathscr{H}^k$ are the spatial components of the twist vector $\eta^{\mu\nu\sigma\tau} \xi_\nu D_\tau \xi_\sigma$ where D is the covariant derivative in the space-time $g_{\mu\nu}$. The last Einstein equation is

$$P^{kl} + \frac{1}{2} \xi^2 (\gamma^{kl} \mathscr{B}^2 - \mathscr{B}^k \mathscr{B}^l) - \xi^{-1} \xi^{;k;l} = R^{kl} = \kappa (T^{kl} - \frac{1}{2} g^{kl} T). \tag{6}$$

To illustrate the power of the analogy between electromagnetism and gravomagnetism we consider first a straight solenoid of length L with a wire wound n times around it carrying a current I; the magnetic flux inside is $F = 4\pi n I$. This flux emerges from the pole at one end of the solenoid and spreads over an area of order L^2 before returning to the pole at the other end. Even if n increases linearly with the the length L the equatorial field outside the solenoid F/L^2 decreases like $1/L$ so, as the length of the solenoid increases, the field outside tends to zero.

The analogous gravitational case is a rotating cylindrical shell and the case that is solved in General Relativity is the infinitely long cylinder. In such a case the external gravomagnetic field will be zero as the returning flux is sent to infinite distances. Thus it is not surprising that the external metric is static rather than stationary.

Our second, less trivial problem is the toroidal solenoid for which the electrical case is illustrated below. Here the magnetic field is confined within the solenoid and there are no poles from which it emerges. Indeed even if the torus carries a charge as well as a current the external field is purely electrical without any magnetism. The analogous gravitational problem is a massive toroidal shell that rolls around the circumference of its small cross-section so that the equator closest to the global

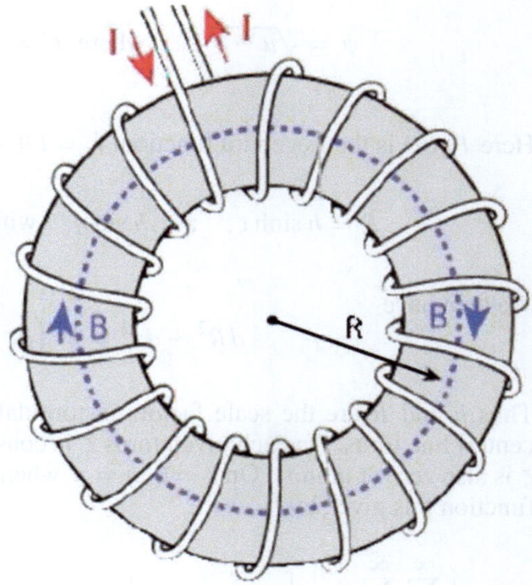

Fig. 1 In electromagnetism a wire carrying a current I tightly wound n times around a torus produces a field confined within the torus. In the mathematics the wire is replaced by a continuum whose total current is called I (not nI)

axis moves up while the equator furthest from the axis moves down. We find exact solutions to this problem in General Relativity in which the metric is static outside the torus but stationary and non-static inside the torus (Fig. 1).

2 The General Static Weyl Metric in Toroidal Coordinates

Weyl takes the metric in the form

$$ds^2 = e^{-2\psi}dt^2 - e^{2\psi}\left[e^{2k}\left(dz^2 + dR^2\right) + R^2 d\varphi^2\right]. \tag{7}$$

Then, in empty axially symmetric spaces Einstein's equations give $\nabla^2\psi = 0$, where ∇^2 is the flat space operator. Also setting

$$D = \partial_R - i\partial_z, \tag{8}$$

we have the Weyl equations

$$Dk = \tfrac{1}{4}Re^{4\psi}De^{-2\psi}De^{-2\psi}. \quad \text{So} \quad DkD\ln R = (D\psi)^2. \tag{9}$$

The general symmetric solution to these equations in toroidal coordinates is given by [2] whose solution does not include that for k given below,

$$\psi = \sqrt{u - v}\,U, \quad \text{where } U = \sum_{l=0}^{\infty} a_l P_L(u)\cos(l\eta). \tag{10}$$

Here $P_L(u)$ is the Legendre function, $L = l + \frac{1}{2}$; $u = \cosh\zeta$, $v = \cos\eta$,

$$R = h\sinh\zeta, \quad z = h\sin\eta, \quad \text{with } h = \frac{a}{\cosh\zeta - \cos\eta}. \tag{11}$$

We then have

$$dR^2 + dz^2 = h^2\left(d\zeta^2 + d\eta^2\right). \tag{12}$$

Thus h and R are the scale factors in toroidal coordinates, a is the radius of the central line torus. On each given torus ζ is constant. On the axis $R = 0$, $\zeta = 0$ and ζ is also zero at infinity. On $z = 0$, $\eta = \pi$ when $R < a$ and $\eta = 0$ when $R > a$. The function k is given by,

$$\begin{aligned} k =& \tfrac{1}{8}\sum_{l=0}^{\infty}\sum_{m=0}^{\infty} a_l a_m \left[c(l+m+1)k^1_{l,m} + c(l+m)k^0_{l,m} + c(l+m-1)k^{-1}_{l,m}\right] \\ &+ \tfrac{1}{8}\sum_{l=0}^{\infty}\sum_{m=0}^{\infty} a_l a_m \left[c(l-m+1)k^1_{l,-m} + c(l-m)k^0_{l,-m} + c(l-m-1)k^{-1}_{l,-m}\right], \end{aligned} \tag{13}$$

where $k^n_{l,m}$ are known terms of Legendre functions themselves and their derivatives and $c(m)$ is short for $\cos(m\eta)$.

3 A Toroidal Solenoid's Metric and Junction Conditions

Inside our torus we need a solution with a toroidal gravomagnetic field. For this we take Bonnor's nice solution [3] for the metric external to a light beam characterised as null dust. This clearly must have a toroidal gravomagnetic field. We generalise the metric by incorporating a constant conicity $\bar{k}$. Since the symmetry axis is no longer included in the part of the solution used within our torus, this does not generate a singularity and is necessary to accomplish the fitting to the external metric. Bonnor's metric is then

$$ds^2 = F[dt - (1 - F^{-1})d\overline{z}]^2 - (e^{2\bar{k}}d\overline{R}^2 + \overline{R}^2 d\varphi^2 + F^{-1}d\overline{z}^2), \tag{14}$$

where $F = 8I\,\ln(\overline{R}/a) + \overline{C}$, $\overline{C}$ is a constant and I the total current that generates the gravomagnetic field; in Bonnor's application it is the current caused by the light ray; in our application it is the total current that runs around our torus by the short way. Our task is now reduced to fitting this metric for the inside of our torus to our general

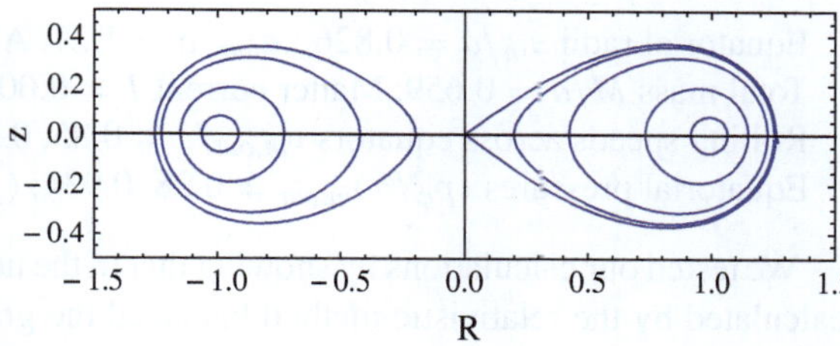

Fig. 2 Equipotentials of the Bach-Weyl static toroids

Weyl metric for the outside. Before attempting this full problem we solved the easy case of a static equipotential shell with a Bach-Weyl [4] exterior metric. The Riemann tensor inside the equipotential toroid is zero but the metric there though locally flat is actually "conical", so it is not globally flat. $\overline{k}$ is a negative constant not zero. It was solving this simpler problem that alerted us to the necessity of incorporating such a conicity into Bonnor's metric in the general problem. The shapes of the Bach-Weyl equipotential surfaces are shown in Fig. 2. Inside the limiting one that intersects the axis, the conicity, $\overline{k}$, is zero so for it the internal space is globally flat. A massive shell may be placed on any one of these equipotential surfaces leaving the shapes of the external equipotentials unchanged but of course changing the values of their potentials. Inside such a shell the potential is constant. We determined the limits to these masses so that the pressure components in the shell do not exceed the limits set by the energy conditions.

For our rolling toroid the complete set of Einstein's equations are given in (4), (5) and (6). None of these equations involve $\mathscr{A}$ itself as opposed to $\mathscr{B} = \mathbf{curl}\,\mathscr{A}$, so the boundary conditions for all such problems can be expressed in terms of $\mathscr{B}$ rather than $\mathscr{A}$. Four of them are strikingly similar to the boundary conditions in electrodynamics. These are: that the potential ψ is continuous, that $\mathscr{B}.\boldsymbol{n}$ is continuous, that $\mathscr{H} \times \boldsymbol{n} = 16\pi\mathbf{J}$ (the gravomagnetic field being zero outside) and that from integrating (1.4) the discontinuity in the potential gradient along the normal is related to an appropriate surface density. We also need the gamma metrics on the surfaces to be identical and from integrating (6) across the surface we require the discontinuities of the external curvatures of the gamma space 3-metrics to be related to the stresses in the toroid. While this procedure is equivalent to Israel's it is easier to carry out since it does not involve $\mathscr{A}$ and fits the surfaces in space rather than space-time. We give below an example of one of our strongly relativistic rolling toroids for which we carried out this procedure to ensure that the surface stresses obeyed the dominant energy conditions.

4 A Strongly Relativistic Example

The dominant energy condition limits how much mass our torus can withstand at a given rolling rate. Our example is as follows:

Equatorial radii $a_{in}/a = 0.826 : a_{out}/a = 1.21$. Axial ratio $b/a = 0.297$.
Total mass $M/a = 0.659$. Matter current $I = 0.008$
Rolling speeds across equators $v_{in,out}/c = 0.85, 0.09$.
Equatorial pressures $(p_\phi/\sigma)_{in,out} = 0.28, 0.934$; $(p_\eta/\sigma)_{in,out} = 0.1, -0.57$.

We tested our calculations by showing that in the non-relativistic limit the stresses calculated by the relativistic method balanced the gravity and the centrifugal force due to rolling. More details of this work are given in [5].

Acknowledgments We thank Jiří Bičák for his interest and advice during the work and for inviting us to his excellent conference "Relativity and Gravitation—100 Years after Einstein in Prague".

References

1. Landau, L., Lifshitz, E. The Classical Theory of Fields, pp. 354–355. Mir, Moscow (1966)
2. Frolov, V.P., Israel, W., Unruh, W.G.: The gravitational field of straight and circular cosmic strings: relation between gravitational mass, angular deficit, and internal structure. Phys. Rev. D **39**, 1084 (1989). doi:10.1103/PhysRevD.39.1084
3. Bonnor, W.B.: J. Math. Mech. **6**, 203 (1957)
4. Bach, R., Weyl, H.: Neue Lösungen der Einsteinschen Gravitationsgleichungen. Math. Z. **13**, 134 (1922)
5. Lynden-Bell, D., Katz, J.: Toroidal metrics: gravitational solenoids and static shells. Class. Quantum Grav. **29**, 5010 (2012). doi:10.1088/0264-9381/29/11/115010

Exact Dynamical AdS Black Holes and Wormholes with a Klein-Gordon Field

Hideki Maeda

Abstract We present an exact solution with spherical, plane, or hyperbolic symmetry in the Einstein-Klein-Gordon system with negative Λ in arbitrary dimensions. In the coordinate system we adopt, the scalar field is homogeneous and the spacetime represents an asymptotically locally AdS dynamical black hole or wormhole. In three dimensions, the scalar field becomes trivial and the solution reduces to the BTZ (Bañados-Teitelboim-Zanelli) black hole.

1 Motivation and Summary

The motivation of this study is twofold. Firstly to provide an exact AdS black hole which can be applied to the study of AdS/CFT duality in the dynamical context [1]. And secondly to find a possible final state of the recently-found nonlinear instability of the AdS vacuum [2].

The solution presented below may be a good model for further investigations to shed light on dynamical properties of AdS black holes. Interesting subjects are thermodynamical properties, dynamical stability, or Hawking radiation. This paper is based on [3].

2 System

We consider the Einstein-Klein-Gordon-Λ system in arbitrary $n(\geq 3)$ dimensions. The field equations are $G_{\mu\nu} + \Lambda g_{\mu\nu} = \kappa_n^2 T_{\mu\nu}$ and $\Box\phi = 0$, where

H. Maeda (✉)
Centro de Estudios Científicos (CECs), Arturo Prat 514, Valdivia, Chile
e-mail: hideki@cecs.cl

J. Bičák and T. Ledvinka (eds.), *Relativity and Gravitation*,
Springer Proceedings in Physics 157, DOI: 10.1007/978-3-319-06761-2_20,

$$T_{\mu\nu} = (\nabla_\mu\phi)(\nabla_\nu\phi) - \left(\frac{1}{2}\right) g_{\mu\nu}(\nabla\phi)^2. \tag{1}$$

In the present paper, we consider n-dimensional warped product spacetimes $(g_{\mu\nu}, \mathscr{M}^n) \approx (g_{AB}, M^2) \times (\gamma_{ij}, K^{n-2})$ with the line element

$$\begin{aligned} ds^2 &= g_{\mu\nu}\, dx^\mu dx^\nu \\ &= g_{AB}(y)\, dy^A dy^B + R(y)^2 \gamma_{ij}(z)\, dz^i dz^j, \end{aligned} \tag{2}$$

where g_{AB} is a Lorentzian metric on M^2 and R is a scalar on M^2. K^{n-2} is an $(n-2)$-dimensional unit space of constant curvature, where k denotes its curvature taking the values 1, 0, and -1, and γ_{ij} is the metric on K^{n-2}.

The generalized Misner-Sharp quasi-local mass is a scalar on M^2 defined by

$$m := \frac{(n-2)V_{n-2}^{(k)}}{2\kappa_n^2} R^{n-3}\Big(-\tilde{\Lambda}R^2 + k - (DR)^2\Big), \tag{3}$$

where $\tilde{\Lambda} := 2\Lambda/[(n-1)(n-2)]$, $(DR)^2 := g^{AB}(D_A R)(D_B R)$, and D_A is the covariant derivative on M^2 [4–7]. $V_{n-2}^{(k)}$ denotes the volume of K^{n-2} if it is compact and otherwise arbitrary. m has the monotonicity and positivity properties for arbitrary (positive) $V_{n-2}^{(k)}$ and is constant in vacuum [6, 7]. In the asymptotically flat or AdS case, that coefficient is fixed in such a way that it converges to the global mass such as the Arnowitt-Deser-Misner mass [8] or Abbott-Deser mass [9].

3 Exact Asymptotically Locally AdS Solutions

Here we particularly consider $\Lambda < 0$ and the metric in the following form:

$$ds^2 = H(\rho)^{-2}\Big(-dt^2 + d\rho^2 + S(t)\gamma_{ij}(z)\, dz^i dz^j\Big), \tag{4}$$

$$H(\rho) := \sqrt{-\tilde{\Lambda}} \sin\rho. \tag{5}$$

The domain of ρ is given by $N\pi < \rho < (N+1)\pi$, $(N \in \mathbb{Z})$, since $\rho = N\pi$ corresponds to the AdS infinity, where

$$\lim_{\rho\to N\pi} R^{\mu\nu}{}_{\rho\sigma} = \tilde{\Lambda}(\delta^\mu_\rho\delta^\nu_\sigma - \delta^\mu_\sigma\delta^\nu_\rho) \tag{6}$$

is satisfied. The function $S(t)$ is obtained by solving Einstein equations and the domain of t is determined by $S(t) > 0$. The areal radius is given by $R = (\varepsilon H)^{-1} S^{1/2}$, where $\varepsilon = \pm 1$ is chosen such that R is non-negative.

The Einstein equations require that the Klein-Gordon field is homogeneous $\phi = \phi(t)$. Then, the energy-momentum tensor has the form of $T^{\mu}{}_{\nu} = \mathrm{diag}(-\mu, \mu, \ldots, \mu)$, where $\mu := (1/2)H^2\dot{\phi}^2$ is the energy density of the scalar field. Finally, the system reduces to the following master equation for $X(t) := S^{(n-2)/2}$:

$$E = \frac{1}{2}\dot{X}^2 + V_{(k)}(X), \tag{7}$$

$$V_{(k)}(X) := \frac{(n-2)^2}{2}\left(kX^{2(n-3)/(n-2)} + X^2\right), \tag{8}$$

where a dot denotes the derivative with respect to t and E is an integration constant.

This class of solutions has been investigated with a stiff fluid because it is equivalent to a massless Klein-Gordon field if the gradient of the scalar field is timelike. The first solutions were obtained by Lake [10] and independently obtained by other authors [11–13] in the spherical case in four dimensions ($k = 1$ and $n = 4$). The four-dimensional solutions with general k were obtained in [14, 15].

The master Eq. (7) is solved analytically in three and four dimensions for any k but only for $k = 0$ in higher dimensions. In four dimensions, S is given by

$$S(t) = \frac{1}{2}(-k + 2C_1 \sin 2t), \tag{9}$$

where C_1 is a constant relating to E. The energy density of the scalar field μ and the generalized Misner-Sharp mass m are given by

$$\mu = \frac{(4C_1^2 - k^2)H^2}{4\kappa_4^2 S^2}, \quad m = \frac{V_2^{(k)}(4C_1^2 - k^2)}{4\kappa_4^2 \varepsilon H S^{1/2}}. \tag{10}$$

The scalar field with positive μ (namely $4C_1^2 > k^2$) is given by

$$\pm(\phi - \phi_0) = \begin{cases} \sqrt{\dfrac{1}{2\kappa_4^2}} \ln\left|\dfrac{\sqrt{4C_1^2 - k^2} + (-k\tan t + 2C_1)}{\sqrt{4C_1^2 - k^2} - (-k\tan t + 2C_1)}\right| & [k = 1, -1], \\ \sqrt{\dfrac{1}{2\kappa_4^2}} \ln\left|\dfrac{1 - \cos 2t}{\sin 2t}\right| & [k = 0], \end{cases} \tag{11}$$

while the scalar field with negative μ is

$$\pm(\phi - \phi_0) = i\sqrt{\frac{2}{\kappa_4^2}} \arctan\left(\frac{-k\tan t + 2C_1}{\sqrt{k^2 - 4C_1^2}}\right), \tag{12}$$

where $i^2 = -1$ and ϕ_0 is a constant. In arbitrary dimensions with $k = 0$, S and ϕ are given by

$$S(t) = C_1[\sin(n-2)t]^{2/(n-2)}, \tag{13}$$

$$\pm(\phi - \phi_0) = \sqrt{\frac{n-3}{(n-2)\kappa_n^2}} \ln\left|\frac{1-\cos(n-2)t}{\sin(n-2)t}\right|. \tag{14}$$

Even in other cases, the qualitative property of the solution is easily understood because the master equation represents a simple one-dimensional potential problem for the variable $X(t) (\geq 0)$. In general, ϕ, μ, and m are given by

$$\phi = \pm\sqrt{\frac{2(n-3)E}{(n-2)\kappa_n^2}} \int^t \frac{d\bar{t}}{S(\bar{t})^{(n-2)/2}}, \tag{15}$$

$$\mu = \frac{(n-3)EH^2}{(n-2)\kappa_n^2 S^{n-2}}, \quad m = \frac{EV_{n-2}^{(k)}}{(n-2)\kappa_n^2(\varepsilon H)^{n-3} S^{(n-3)/2}}, \tag{16}$$

where $S(t)$ is determined by the master Eq. (7). μ and m are positive (negative) for $E > (<) 0$ and then the scalar field is real (pure imaginary, namely ghost). In three dimensions ($n = 3$), the scalar field becomes trivial and we have $\mu = 0$ and $m =$constant. The spacetime is then locally (A)dS and represents the BTZ (Bañados-Teitelboim-Zanelli) black hole in the non-standard coordinates [16, 17].

Equation (16) shows that the spacetime is vacuum at AdS infinity ($H = 0$), but m blows up there. This means that the fall-off rate to the AdS infinity is slower than the Henneaux-Teitelboim condition and hence the spacetime is asymptotically AdS only locally.

4 Physical Interpretations

We clarify the causal structure of the spacetime (4) and give its physical interpretations. For this purpose, the following three facts are important; (i) $\rho = N\pi$ ($N \in \mathbb{Z}$) is AdS infinity, which is timelike, (ii) $S(t) = 0$ corresponds to curvature singularity, which is spacelike, and (iii) a light ray runs along a 45° straight line in the (ρ, t)-plane since the metric on M^2 in the solution (4) is conformally flat.

The present coordinate system covers the maximally extended spacetime and the Penrose diagrams for the spacetime (4) are presented in Fig. 1. (See Table 1.) The spacetime represents a dynamical black hole or wormhole depending on the parameters.

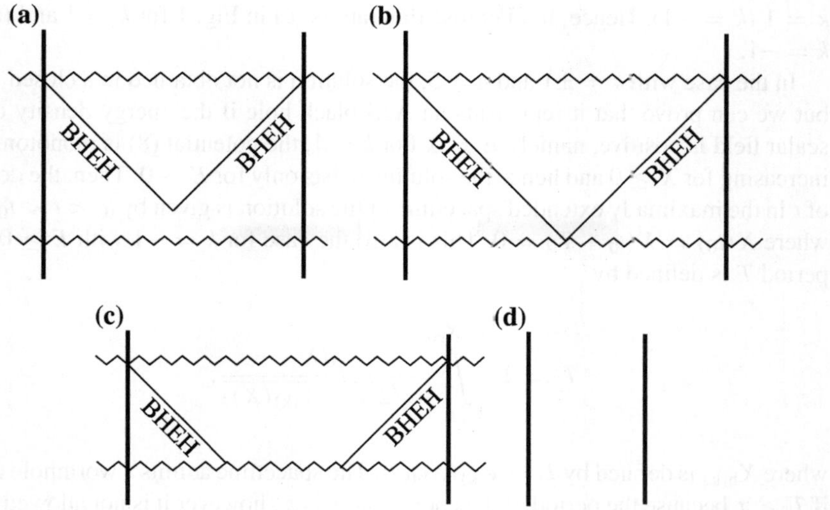

Fig. 1 The Penrose diagrams for the solution. A zigzag and a *thick line* correspond to a curvature singularity and AdS infinity, respectively. BHEH stands for the *black-hole event horizon*. **a–c** represent a *black hole*, while **d** represents a wormhole

Table 1 The Penrose diagrams of the solution with positive energy density and $n \geq 4$

	$n = 4$	$n \geq 5$
$k = 0$	Figure 1b	Figure 1c
$k = 1$	Figure 1c	Figure 1a, b, or c
$k = -1$	Figure 1a	Figure 1a, b, or c

In the case of $k = -1$ with negative energy density, the Penrose diagram is Fig. 1d

4.1 Dynamical AdS Black Holes

If the scalar field is real (and equivalently the energy density is positive), there are spacelike curvature singularities given by $S(t) = 0$ in the (ρ, t)-plane. As a result, the solution represents a dynamical black hole. Since both H and S are periodic, the (ρ, t) plane is divided by singularities and AdS infinities ($\rho = N\pi$) into an infinite number of portions. All the portions with positive S are equivalent.

First let us see the case with $k = 0$. Without loss of generality, we assume $C_1 > 0$ in (13) and consider a physical portion defined by $t = (0, \pi/(n-2))$ and $\rho = (0, \pi)$, which covers the maximally extended spacetime. The event horizon in this portion is given by $t = \rho - (n-3)/(n-2)\pi$ and $t = -\rho + \pi/(n-2)$ and the Penrose diagram is (b) in Fig. 1 for $n = 4$ and (c) for $n \geq 5$.

On the other hand, in the case with $k = 1, -1$ in four dimensions, the period of t in a physical portion is different. The period is shorter (longer) than $\pi/2$ for

$k = 1$ ($k = -1$). Hence, the Penrose diagram is (c) in Fig. 1 for $k = 1$ and (a) for $k = -1$.

In the case with $k = \pm 1$ and $n \geq 5$, the solution is not obtained in a closed form, but we can prove that it represents an AdS black hole if the energy density of the scalar field is positive, namely $E > 0$. For $k = 1$, the potential (8) is monotonically increasing for $X \geq 0$ and hence the solution exists only for $E > 0$. Then, the domain of t in the maximally extended spacetime of the solution is given by $t_0 < t < t_0 + T$, where $X(t_0) = X(t_0 + T) = 0$. This is also the case for $k = -1$ with $E \geq 0$. The period T is defined by

$$T := 2 \int_{X=0}^{X=X_{\mathrm{b(k)}}} \frac{dX}{\sqrt{2(E - V_{(\mathrm{k})}(X))}}, \tag{17}$$

where $X_{\mathrm{b(k)}}$ is defined by $E = V_{(\mathrm{k})}(X_{\mathrm{b(k)}})$. The spacetime admits a wormhole throat if $T \geq \pi$ because the period of the coordinate ρ is π, however it is not allowed if the scalar field has positive energy density. (See Appendix C in [3] for the proof.) Since $t = t_0$ and $t = t_0 + T$ are both spacelike curvature singularities, the corresponding Penrose diagram is Fig. 1a, b, and c for $\pi/2 < T < \pi$, $T = \pi/2$, and $0 < T < \pi/2$, respectively. Although the diagrams are different depending on the value of T, the solution represents a dynamical AdS black hole.

4.2 Dynamical AdS Wormholes

In the case of $k = -1$ in four dimensions, if $4C_1^2 < k^2$, then the energy density is negative and S is positive definite for $-\infty < t < \infty$. (There is no physical solution for $k = 1$ because S is negative definite if $4C_1^2 < k^2$.) The Klein-Gordon field then becomes ghost and there is no curvature singularity in the spacetime. As a result, the spacetime is a dynamical AdS wormhole described by the Penrose diagram (d) in Fig. 1.

It is shown that an AdS wormhole is realized also for $k = -1$ and $n \geq 5$ if $E < 0$; namely the energy density is negative (and equivalently the scalar field is ghost). In the case of $k = -1$, the potential (8) in the master equation has a negative minimum $V_{(-1)} = V_{\mathrm{ex}}$, where

$$V_{\mathrm{ex}} := -\frac{n-2}{2}\left(\frac{n-3}{n-2}\right)^{n-3} \quad (<0). \tag{18}$$

As a result, for the solution with E satisfying $V_{\mathrm{ex}} < E < 0$, the value of X (and hence S) oscillates and never becomes 0. Hence, the corresponding Penrose diagram is Fig. 1d and the spacetime describes a dynamical AdS wormhole.

References

1. Kinoshita, S., Mukohyama, S., Nakamura, S., Oda, K.: A holographic dual of Bjorken flow. Prog. Theor. Phys. **121**, 121 (2009). doi:10.1143/PTP.121.121
2. Bizoń, P., Rostworowski, A.: Weakly turbulent instability of anti-de Sitter spacetime. Phys. Rev. Lett. **107**, 031102 (2011). doi:10.1103/PhysRevLett.107.031102
3. Maeda, H.: Exact dynamical AdS black holes and wormholes with a Klein-Gordon field. Phys. Rev. D **86**, 044016 (2012). doi:10.1103/PhysRevD.86.044016
4. Misner, C., Sharp, D.: Relativistic equations for adiabatic, spherically symmetric gravitational collapse. Phys. Rev. **136**, 571 (1964). doi:10.1103/PhysRev.136.B571
5. Nakao, K.: On a quasi-local energy outside the cosmological horizon, ArXiv e-prints. [arXiv:gr-qc/9507022] (1995).
6. Hayward, S.: Gravitational energy in spherical symmetry. Phys. Rev D. **53**, 1938 (1996). doi:10.1103/PhysRevD.53.1938
7. Maeda, H., Nozawa, M.: Generalized Misner-Sharp quasilocal mass in Einstein-Gauss-Bonnet gravity. Phys. Rev. D **77**, 064031 (2008). doi:10.1103/PhysRevD.77.064031
8. Arnowitt, R., Deser, S., Misner, C.: The dynamics of general relativity. In: Witten, L. (ed.) Gravitation: an Introduction to Current Research, chap. 7, pp. 227–265. Wiley, New York (1962). doi:10.1007/s10714-008-0661-1
9. Abbott, L., Deser, S.: Stability of gravity with a cosmological constant. Nucl. Phys. B **195**, 76 (1982). doi:10.1016/0550-3213(82)90049-9
10. Lake, K.: Remark concerning spherically symmetric nonstatic solutions to the Einstein equations in the comoving frame. Gen. Relativ. Gravit. **15**, 357 (1983). doi:10.1007/BF00759164
11. Hajj-Boutros, J.: On spherically symmetric perfect fluid solutions. J. Math. Phys. **26**, 771 (1985). doi:10.1063/1.526565
12. Herrara, L., Ponce de León, J.: Perfect fluid spheres admitting a one-parameter group of conformal motions. J. Math. Phys. **26**, 778 (1985). doi:10.1063/1.526567
13. Van den Bergh, N., Wils, P.: Exact solutions for nonstatic perfect fluid spheres with shear and an equation of state. Gen. Relativ. Gravit. **17**, 223 (1985). doi:10.1007/BF00760245
14. Collins, C., Lang, J.: A class of self-similar perfect-fluid spacetimes, and a generalisation. Class. Quant. Grav. **4**, 61 (1987). doi:10.1088/0264-9381/4/1/009
15. Shaver, E., Lake, K.: Singularities in separable metrics with spherical, plane, and hyperbolic symmetry. Gen. Relativ. Gravit. **20**, 1007 (1988). doi:10.1007/BF00759022
16. Bañados, M., Teitelboim, C., Zanelli, J.: Black hole in three-dimensional spacetime. Phys. Rev. Lett. **69**, 1849 (1992). doi:10.1103/PhysRevLett.69.1849
17. Bañados, M., Henneaux, M., Teitelboim, C., Zanelli, J.: Geometry of the $2+1$ black hole. Phys. Rev. D **48**, 1506 (1993). doi:10.1103/PhysRevD.48.1506

2.5PN Kick from Black-Hole Binaries in Circular Orbit: Nonspinning Case

Chandra Kant Mishra, K. G. Arun and Bala R. Iyer

Abstract Using the Multipolar post-Minskowskian formalism, we compute the linear momentum flux from black-hole binaries in circular orbits and having no spins. The total linear momentum flux contains various types of *instantaneous* (which are functions of the retarded time) and *hereditary* (which depend on the dynamics of the binary in the past) terms both of which are analytically computed. In addition to the inspiral contribution, we use a simple model of plunge to compute the kick or recoil accumulated during this phase.

1 Introduction

In classical radiation theory, any form of radiation that has a preferred direction (anisotropic emission) results in the *recoil* of the system in the opposite direction. Such a recoil has important consequences in astrophysics like the pulsar acceleration due to the radiation asymmetry [1]. This effect can be understood in terms of multipole moments of the system, more specifically interference between different

C. K. Mishra · B. R. Iyer (✉)
Raman Research Institute, Sadashivanagar, Bangalore 560080, India
e-mail: bri@rri.res.in

B. R. Iyer
International Centre for Theoretical Sciences (ICTS), Bangalore 560012, India

C. K. Mishra
Indian Institute of Science, Bangalore 560012, India
e-mail: cmishra@iisertvm.ac.in

C. K. Mishra
IISER, Thiruvananthapuram, India

K. G. Arun
Chennai Mathematical Institute, Siruseri, Kelambakkam 603103, India
e-mail: kgarun@cmi.ac.in

J. Bičák and T. Ledvinka (eds.), *Relativity and Gravitation*,
Springer Proceedings in Physics 157, DOI: 10.1007/978-3-319-06761-2_21,

multipoles [2]. It was argued in Ref. [2], on general grounds, that the leading recoil effect in the case of electromagnetic radiation could come from the interference between electric dipole and electric quadrupole or magnetic dipole. (See Ref. [3] for a more detailed discussion.) A similar effect can happen in the case of gravitational radiation [2]. Since the leading emission in the case of Gravitational Waves (GWs) in GR is quadrupolar, the lowest order recoil effect in GR arises from the interference of mass quadrupole and mass octupole or current quadrupole.[1] This GW recoil can be seen as a consequence of linear momentum flux emission by the radiating system.

Let us now focus on a particular type of GW source known as coalescing compact binaries (CCBs). As the name indicates, these are systems consisting of neutron stars (NS) or black holes (BH) which go around each other in a bound orbit. The orbit keeps on shrinking due to gravitational radiation reaction until the two objects merge to form a compact object, most likely a BH. As is well known, the GW emission is not isotropic (due to the quadrupolar nature) and thus the emission is beamed in some direction. As the two bodies move in their respective orbits, this direction also keeps changing. If the orbit is closed, then over an orbit, the binary's centre of mass would return to the same point where it started, making the net recoil zero. However, if the orbit is not closed, as is the case for CCBs, there can be an accumulation of recoil over these orbits until the two bodies coalesce into a single BH, eventually manifesting as a kick to this newly formed BH. This is the GW recoil in the context of CCBs.

Gravitational recoil has important implications for astrophysics, especially models of black hole formation and growth (see e.g. [4]). If the recoil velocity of a merging compact binary system is greater than the escape velocity of the galaxy, it can result in the ejection of the newly formed black hole from the galaxy. Hence very accurate estimation of the recoil velocity of the merging compact binaries is important to understand the astrophysics of black hole formation and growth.

Evolution of CCBs can be divided into three phases: adiabatic *inspiral*, nonlinear *merger* and perturbative *ringdown*. The inspiral part and ringdown phases can be accurately modelled analytically using post-Newtonian (PN) approximation to general relativity [5] and BH perturbation theory [6], respectively. Due to recent successes in numerical relativity (see Refs. [7, 8] for reviews), the highly nonlinear merger phase can be modelled quite accurately by numerically solving Einstein's equations.

In this work, we use the PN approximation to calculate the linear momentum loss and resulting GW recoil from a black-hole binary[2] moving in circular orbits up to 2.5PN order extending the earlier works of Fittchett (Newtonian order) [9], Wiseman (1PN order [10]) and Blanchet et al. (through 2PN [11]). A more detailed account of the contents of this paper can be found in Ref. [12].

---

[1] Both mass octupole and current quadrupole have the same parity.

[2] Note that the expressions for the linear momentum flux (2) and that for the recoil velocity (7) are even applicable to compact binary systems involving NSs as components. However, the plunge computations presented here assume that both the components of the binary are BHs.

This article is organized in the following way. In Sect. 2, we write down the multipolar expansion of the linear momentum flux and the final expression for the flux in terms of source multipole moments. Section 3 derives the 2.5PN expression for the recoil and Sect. 4 discusses the numerical estimates of recoil. Conclusions are given in Sect. 5.

2 Multipole Expansion for Linear Momentum Flux

The general formula for linear momentum flux in the far-zone of the source in terms of symmetric trace-free (STF) radiative multipole moments is given in [13] and at relative 2.5PN order it takes the following form (see e.g. (4.20') of Ref. [13]):

$$\begin{aligned}\mathscr{F}^i_P(U) = \frac{G}{c^7}\Bigg\{&\left[\frac{2}{63}U^{(1)}_{ijk}U^{(1)}_{jk} + \frac{16}{45}\varepsilon_{ijk}U^{(1)}_{ja}V^{(1)}_{ka}\right]\\ &+\frac{1}{c^2}\left[\frac{1}{1134}U^{(1)}_{ijkl}U^{(1)}_{jkl} + \frac{1}{126}\varepsilon_{ijk}U^{(1)}_{jab}V^{(1)}_{kab} + \frac{4}{63}V^{(1)}_{ijk}V^{(1)}_{jk}\right]\\ &+\frac{1}{c^4}\left[\frac{1}{59400}U^{(1)}_{ijklm}U^{(1)}_{jklm} + \frac{2}{14175}\varepsilon_{ijk}U^{(1)}_{jabc}V^{(1)}_{kabc} + \frac{2}{945}V^{(1)}_{ijkl}V^{(1)}_{jkl}\right] + \mathscr{O}\left(\frac{1}{c^6}\right)\Bigg\}.\end{aligned} \tag{1}$$

In the equation above U_L and V_L denote the mass and current type radiative multipoles of the source (suffix L captures the multi-index structure of the U and V moments) and $U_L^{(n)}$ and $V_L^{(n)}$ denote the nth time derivatives of the moments: $\mathscr{O}(1/c^6)$ denotes the omission of terms at 3PN and at higher PN orders w.r.t the leading term. The moments which appear in the above formula are functions of retarded time $T - \frac{R}{c}$, where R denotes the distance of the source relative to the observer and T the time of observation, both in radiative coordinates. In the MPM formalism U_L and V_L are related to canonical moments M_L and S_L ((5.4)–(5.8) of [14]) which in turn are related to source moments $\{I_L, J_L, X_L, W_L, Y_L, Z_L\}$ [14, (5.9)–(5.11)]. Using these inputs, one can re-express the radiative multipole moments in terms of source moments. Then one notices that the total linear momentum flux consists of two parts: one type of terms are functions of the retarded time called *instantaneous terms* and the other which are sensitive to the dynamics of the source in its entire past are called *hereditary terms*. Explicit expressions for the two types of terms are given in (2.3)–(2.5) of Ref. [12]. Further, the explicit expressions for various source multipole moments are given in (3.1)–(3.4) of Ref. [12]. The only other input we require is the 2.5PN accurate equations of motion which can be found in [14, 15]. Using these inputs, and working in *harmonic coordinates*, we obtain the total linear momentum flux in terms of gauge independent variable $x = (m\omega)^{2/3}$ as

$$\mathcal{F}_P^i = -\frac{464}{105}\frac{c^4}{G}\sqrt{1-4\nu}\,x^{11/2}\,\nu^2\Bigg\{\Bigg[1+x\left(-\frac{452}{87}-\frac{1139}{522}\nu\right)+\frac{309}{58}\pi\,x^{3/2}$$
$$+x^2\left(-\frac{71345}{22968}+\frac{36761}{2088}\nu+\frac{147101}{68904}\nu^2\right)+x^{5/2}\left(-\frac{2663}{116}\pi-\frac{2185}{87}\pi\,\nu\right)\Bigg]\hat{\lambda}^i$$
$$+x^{5/2}\Bigg[-\frac{106187}{50460}+\frac{32835}{841}\log 2-\frac{77625}{3364}\log 3$$
$$+\left(\frac{10126}{4205}-\frac{109740}{841}\log 2+\frac{66645}{841}\log 3\right)\nu\Bigg]\hat{n}_i+\mathcal{O}\left(\frac{1}{c^6}\right)\Bigg\}. \quad (2)$$

In the equation above $\hat{\mathbf{n}}$ and $\hat{\boldsymbol{\lambda}}$ are related to the phase angle ψ as

$$\hat{\mathbf{n}} = \cos\psi\,\hat{\mathbf{e}}_x + \sin\psi\,\hat{\mathbf{e}}_y, \quad (3)$$
$$\hat{\boldsymbol{\lambda}} = -\sin\psi\,\hat{\mathbf{e}}_x + \cos\psi\,\hat{\mathbf{e}}_y. \quad (4)$$

3 Computation of the Recoil

Having computed the linear momentum flux for compact binaries in circular orbits, one can obtain the recoil velocity by integrating the momentum balance equation

$$\frac{dP^i}{dt} = -\mathcal{F}_P^i\,, \quad (5)$$

to get

$$\Delta P^i = -\int_{-\infty}^{t} dt'\,\mathcal{F}_P^i\,. \quad (6)$$

Performing the integration we find,

$$V^i_{\rm recoil} = \frac{464}{105}\,c\sqrt{1-4\nu}\,x^4\,\nu^2\Bigg\{\Bigg[1+x\left(-\frac{452}{87}-\frac{1139}{522}\nu\right)+\frac{309}{58}\pi\,x^{3/2}$$
$$+x^2\left(-\frac{71345}{22968}+\frac{36761}{2088}\nu+\frac{147101}{68904}\nu^2\right)+x^{5/2}\Bigg(-\frac{2663}{116}\pi$$
$$-\frac{2185}{87}\pi\,\nu\Bigg)\Bigg]\hat{n}_i+x^{5/2}\Bigg[\frac{106187}{50460}-\frac{32835}{841}\log 2+\frac{77625}{3364}\log 3$$
$$+\left(\frac{41034}{841}+\frac{109740}{841}\log 2-\frac{66645}{841}\log 3\right)\nu\Bigg]\hat{\lambda}_i+\mathcal{O}\left(\frac{1}{c^6}\right)\Bigg\}. \quad (7)$$

4 Numerical Estimates of the Recoil Velocity Including Plunge Contribution

As is evident from (7), the recoil velocity depends on the (symmetric) mass ratio ν and does *not* depend on the total mass. This is consistent with our understanding that the origin of the recoil is in the mass asymmetry. We finally want to numerically estimate the recoil velocities as a function of ν. Our calculation based on 2.5PN approximation yields a maximum kick velocity of $\sim$4 km/s as opposed to $\sim$22 km/s of the 2PN model given in Ref. [11]. Thus 2.5PN estimates predict a smaller kick velocity than the 2PN model. This result is obtained by integrating the linear momentum flux till the Innermost Stable Circular Orbit (ISCO), up till which PN approximation is considered to be valid.

Since the dominant contribution to the recoil comes towards the late stages of the binary evolution, we incorporate the contribution from the 'plunge' phase of the evolution beyond the last stable orbit, following and extending a model proposed in Ref. [11]. The method may be considered to be less sophisticated version of the Effective One Body approach [16, 17]. In this model, the plunge can be viewed as that of a test particle moving in the fixed Schwarzshild geometry of mass m. The contribution from the plunge phase is estimated using the PN formulae assuming they are valid even beyond ISCO. Since the PN representation is usually not reliable beyond ISCO, this should be a source of error and in general this computation is only a crude estimate. Further, the recoil velocity accumulated during the two phases (inspiral and the plunge) can be obtained by taking a vector sum of the two estimates. This is achieved by matching the circular orbit at the ISCO to a suitable plunge orbit. The final results of the numerical estimates of the recoil velocity are presented in Fig. 1. We compare the recoil velocity based on our 2.5PN inspiral + plunge model with that of 2PN inspiral + plunge model of Ref. [11]. (Specifications are the same as those of Fig. 1 of [11].) As is evident, our 2.5PN inspiral + plunge model predicts smaller recoil velocity for almost all values of ν. The maximum of the curve drops from $\sim$243 km/s of [11] to $\sim$180 km/s. This may be attributed to the oscillatory convergence of the PN series observed in various contexts (see e.g. [18]). While our estimates are within the error bars given by [11], one should keep in mind that there are also error bars associated with our estimates due to systematic effects such as the omission of higher order PN contributions.

5 Conclusion

Using the MPM formalism, we computed the linear momentum flux due to GW emission from inspiralling compact binaries moving in circular orbit up to 2.5PN order. This complements the computations of energy flux [19], waveform and

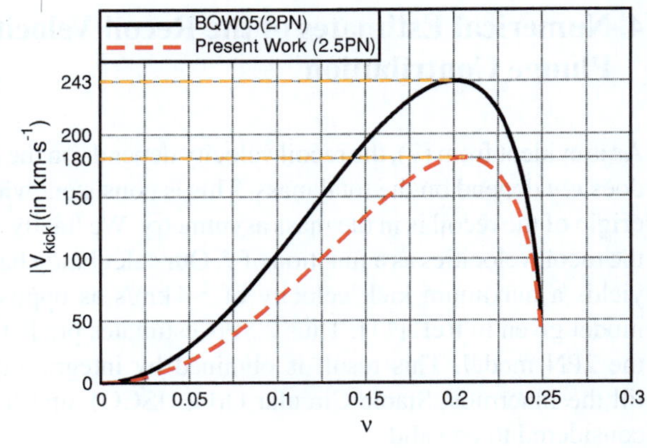

Fig. 1 Kick velocity imparted to the remnant of the compact binary coalescence as a function of symmetric mass ratio. This figure incorporates the contribution from the plunge phase of the binary evolution. The results of 2.5PN + plunge is compared against that of Ref. [11] where the estimation was done with a 2PN + plunge model

polarizations [20, 21] at 2.5PN order for circular orbits. Using the PN linear momentum flux and an analytical model of plunge [11], we have estimated the contributions to the recoil from the inspiral and plunge phase as a function of symmetric mass ratio.

References

1. Harrison, E.R., Tademaru, E.: Acceleration of pulsars by asymmetric radiation. Astrophys. J. **201**, 447 (1975). doi:10.1086/153907
2. Peres, A.: Classical radiation recoil. Phys. Rev. **128**, 2471 (1962). doi:10.1103/PhysRev.128.2471
3. Hughes, S.A., Favata, M., Holz, D.E.: How black holes get their kicks: radiation recoil in binary black hole mergers. In: Merloni, A., Nayakshin, S., Sunyaev, R.A. (eds.) Growing Black Holes: Accretion in a Cosmological Context, pp. 333–339 (2005). doi:10.1007/11403913_64
4. Merritt, D., Milosavljević, M., Favata, M., Hughes, S.A., Holz, D.E.: Consequences of gravitational radiation recoil. Astrophys. J. Lett. **607**, L9 (2004). doi:10.1086/421551
5. Blanchet, L.: Gravitational radiation from post-Newtonian sources and inspiralling compact binaries. Living Rev. Relativ. **5**, lrr-2002-3 (2002). http://www.livingreviews.org/lrr-2002-3
6. Sasaki, M., Tagoshi, H.: Analytic black hole perturbation approach to gravitational radiation. Living Rev. Relativ. **6**, lrr-2003-6 (2003). http://www.livingreviews.org/lrr-2003-6
7. Pretorius, F.: Binary Black Hole Coalescence. In: Colpi, M., et al. (eds.) Physics of Relativistic Objects in Compact Binaries: From Birth to Coalescense, Astrophysics and Space Science Library, vol. 359, pp. 305–370. Springer, Canopus Publishing Limited, Dordrecht, Bristol (2009)
8. Centrella, J., Baker, J.G., Kelly, B.J., van Meter, J.R.: Black-hole binaries, gravitational waves, and numerical relativity. Rev. Mod. Phys. **82**, 3069 (2010). doi:10.1103/RevModPhys.82.3069
9. Fitchett, M.J.: The influence of gravitational wave momentum losses on the centre of mass motion of a Newtonian binary system. Mon. Not. R. Astron. Soc. **203**, 1049 (1983)
10. Wiseman, A.G.: Coalescing binary systems of compact objects to (post)5/2 Newtonian order. 2. Higher order wave forms and radiation recoil. Phys. Rev. D **46**, 1517 (1992). doi:10.1103/PhysRevD.46.1517
11. Blanchet, L., Qusailah, M.S.S., Will, C.M.: Gravitational recoil of inspiraling black hole binaries to second post-Newtonian order. Astrophys. J. **635**, 508 (2005). doi:10.1086/497332

12. Mishra, C.K., Arun, K.G., Iyer, B.R.: 2.5PN linear momentum flux from inspiralling compact binaries in quasicircular orbits and associated recoil: nonspinning case. Phys. Rev. D **85**(4), 044021 (2012). doi:10.1103/PhysRevD.85.044021
13. Thorne, K.: Multipole expansions of gravitational radiation. Rev. Mod. Phys. **52**, 299 (1980). doi:10.1103/RevModPhys.52.299
14. Blanchet, L., Faye, G., Iyer, B.R., Sinha, S.: The third post-Newtonian gravitational wave polarizations and associated spherical harmonic modes for inspiralling compact binaries in quasi-circular orbits. Class. Quantum Grav. **25**(16), 165003 (2008). doi:10.1088/0264-9381/25/16/165003
15. Kidder, L.: Using full information when computing modes of post-Newtonian waveforms from inspiralling compact binaries in circular orbit. Phys. Rev. D **77**, 044016 (2008). doi:10.1103/PhysRevD.77.044016
16. Buonanno, A., Damour, T.: Effective one-body approach to general relativistic two-body dynamics. Phys. Rev. D **59**, 084006 (1999). doi:10.1103/PhysRevD.59 084006
17. Buonanno, A., Damour, T.: Transition from inspiral to plunge in binary black hole coalescences. Phys. Rev. D **62**, 064015 (2000). doi:10.1103/PhysRevD.62.064015
18. Damour, T., Iyer, B.R., Sathyaprakash, B.S.: Comparison of search templates for gravitational waves from binary inspiral. Phys. Rev. D **63**(4), 044023 (2001/2005). doi:10.1103/PhysRevD.63.044023. Erratum-ibid. **D** 72 029902 (2005)
19. Blanchet, L.: Energy losses by gravitational radiation in inspiraling compact binaries to 5/2 post-Newtonian order. Phys Rev. D **54**(1417), (1996). doi:10.1103/PhysRevD.54.1417. Erratum-ibid. **71**, 129904(E) (2005)
20. Arun, K.G., Blanchet, L., Iyer, B.R., Qusailah, M.S.S.: The 2.5PN gravitational wave polarizations from inspiralling compact binaries in circular orbits. Class. Quantum Grav. **21**(3771), (2004). doi:10.1088/0264-9381/21/15/010. Erratum-ibid. **22**, 3115 (2005)
21. Kidder, L.E., Blanchet, L., Iyer, B.R.: A note on the radiation reaction in the 2.5PN waveform from inspiralling binaries in quasi-circular orbits. Class. Quantum Grav. **24**, 5307 (2007). doi:10.1088/0264-9381/24/20/N01

A Reference for the Covariant Hamiltonian Boundary Term

James M. Nester, Chiang-Mei Chen, Jian-Liang Liu and Gang Sun

Abstract The Hamiltonian for dynamic geometry generates the evolution of a spatial region along a vector field. It includes a boundary term which determines both the value of the Hamiltonian and the boundary conditions. The value gives the quasi-local quantities: energy-momentum, angular-momentum/center-of-mass. The boundary term depends not only on the dynamical variables but also on their reference values; the latter determine the ground state (having vanishing quasi-local quantities). For our preferred boundary term for Einstein's GR we propose 4D isometric matching and extremizing the energy to determine the reference metric and connection values.

1 Introduction

Energy-momentum is the source of gravity. Gravitating bodies can exchange energy-momentum with gravity—*locally*—yet there is no well defined energy-momentum

---

J. M. Nester (✉) · C.-M. Chen · J.-L. Liu · G. Sun
Department of Physics and Center for Mathematics and Theoretical Physics,
National Central University, Chungli 320, Taiwan
e-mail: nester@phy.ncu.edu.tw

C.-M. Chen
e-mail: cmchen@phy.ncu.edu.tw

J.-L. Liu
e-mail: liujl@phy.ncu.edu.tw

G. Sun
e-mail: liquideal@gmail.com

J. M. Nester
Graduate Institute of Astronomy, National Central University,
Chungli 320, Taiwan

J. Bičák and T. Ledvinka (eds.), *Relativity and Gravitation*,
Springer Proceedings in Physics 157, DOI: 10.1007/978-3-319-06761-2_22,

density for gravity itself. This inescapable conclusion can be understood as a consequence of the equivalence principle (for a discussion see Misner et al. [1], Sect. 20.4).

2 Quasi-Local Energy-Momentum

The standard approaches aimed at identifying an energy-momentum density for gravitating systems always led to various non-covariant, reference frame dependent, energy-momentum complexes (such expressions are generally referred to as pseudotensors). There are two types of ambiguity. First, there was no unique expression, but rather many that were found by various investigators—including Einstein [2], Papapetrou [3], Landau-Lifshitz [4], Bergmann-Thompson [5],break Møller [6], Goldberg [7], and Weinberg [8]—so which expression should be used? And second—in view of the fact all of these expressions are inherently reference frame dependent—for a chosen expression which reference frame should be used to give the proper physical energy-momentum localization?

The more modern idea is *quasi-local*, i.e., energy-momentum should be associated not with a local density but rather with a closed 2-surface; for a comprehensive review, see Szabados [9].

One approach to energy-momentum is via the *Hamiltonian* (the generator of time evolution). It turns out that this actually includes all the classical pseudotensors as special cases, while taming their ambiguities—providing clear physical/geometric meaning [10, 11].

3 The Covariant Hamiltonian Formulation Results

We have developed a covariant Hamiltonian formalism that is applicable to a large class of geometric gravity theories [10–16]. For such theories the Hamiltonian 3-form $\mathscr{H}(N)$ is both a conserved Noether current,

$$d\mathscr{H}(N) \propto \text{field eqns} \simeq 0, \tag{1}$$

as well as the generator of the evolution of a spatial region along a space-time displacement vector field. It has the general form

$$\mathscr{H}(N) = N^{\mu}\mathscr{H}_{\mu} + d\mathscr{B}(N), \tag{2}$$

where $N^{\mu}\mathscr{H}_{\mu}$—which generates the evolution equations—is itself proportional to certain field equations (initial value constraints) and thus vanishes "on shell". Consequently the *value* of the Hamiltonian is determined by the total differential (boundary) term:

$$E(N, \Sigma) := \int_{\Sigma} \mathscr{H}(N) = \oint_{\partial \Sigma} \mathscr{B}(N). \tag{3}$$

Thus, the value is *quasi-local*. From this boundary term, with suitable choices of the vector field on the boundary, one can determine the quasi-local energy-momentum and angular momentum/center-of-mass.

It should be noted that the boundary 2-form $\mathscr{B}(N)$ can be modified—*by hand*—in any way without destroying the conservation property. (This is a particular case of the usual Noether conserved current ambiguity.) With this freedom one can arrange for almost any conserved quasi-local values. Fortunately the Hamiltonian's role in generating evolution equations tames that freedom.

4 Boundary Variation Principle, Reference Values

One must look to the boundary term in the *variation of the Hamiltonian* (see Lanczos [17–19]). Requiring it to vanish yields the *boundary conditions*. The Hamiltonian is *functionally differentiable* on the phase space of fields *satisfying these boundary conditions*. Modifying the boundary term changes the boundary conditions. (The different classical pseudotensors are each associated with a specific "superpotential" which can serve as the Hamiltonian boundary term, thus they correspond to Hamiltonians with different boundary conditions [10].)

In order to accommodate suitable boundary conditions one must, in general, also introduce certain *reference values* which represent the ground state of the field—the "vacuum" (or background field) values. To this end for any quantity α we let $\Delta\alpha := \alpha - \bar{\alpha}$ where $\bar{\alpha}$ is the reference value.

5 Preferred Boundary Term for GR

Some time ago we identified for GR two *covariant-symplectic* boundary terms [12]; one, which was also found[1] at about the same time by Katz, Bičák and Lynden-Bell [20, 21], has become our preferred choice[2]:

$$\mathscr{B}(N) = \frac{1}{2\kappa}(\Delta\Gamma^{\alpha}{}_{\beta} \wedge i_N \eta_{\alpha}{}^{\beta} + \bar{D}_{\beta} N^{\alpha} \Delta\eta_{\alpha}{}^{\beta}), \tag{4}$$

This choice corresponds to fixing the orthonormal coframe ϑ^{μ} (equivalently the metric) on the boundary:

[1] Via a different route, using a Noether type argument with a global reference.

[2] Here $\Gamma^{\alpha}{}_{\beta}$ is the connection one-form, $\eta^{\alpha\beta\cdots} := *(\vartheta^{\alpha} \wedge \vartheta^{\beta} \wedge \cdots)$ and i_N denotes the interior product (aka contraction) with the vector field N.

$$\delta\mathcal{H}(N) \sim di_N(\Delta\Gamma^{\alpha}{}_{\beta} \wedge \delta\eta_{\alpha}{}^{\beta}). \tag{5}$$

Like other choices, at spatial infinity it gives the ADM [22], MTW [1], Regge-Teitelboim [18], Beig-Ó Murchadha [23], Szabados [24, 25] energy, momentum, angular-momentum, center-of-mass.

Its special virtues include (i) at null infinity it directly gives the Bondi-Trautman energy and the Bondi energy flux [15], (ii) it is "covariant", (iii) it has a positive energy property, (iv) for small spheres it gives a positive multiple of the Bel-Robinson tensor, (v) it yields the first law of thermodynamics for black holes [13], (vi) for spherically symmetric solutions it has the hoop property [26].

6 The Reference and the Quasi-Local Quantities

For all other fields it is appropriate to choose vanishing reference values as the reference ground state—the vacuum. But for geometric gravity the standard ground state is the non-vanishing Minkowski metric. Thus a non-trivial reference is essential.

Using standard Minkowski coordinates y^i, a Killing field of the reference has the form $N^k = N_0^k + \lambda^k_{0l} y^l$, where the translation parameters N_0^k and the boost-rotation parameters $\lambda_0^{kl} = \lambda_0^{[kl]}$ are constants. The 2-surface integral of the Hamiltonian boundary term then gives a value of the form

$$\oint_S \mathcal{B}(N) = -N_0^k p_k(S) + \frac{1}{2}\lambda_0^{kl} J_{kl}(S), \tag{6}$$

which yields not only a quasi-local energy-momentum but also a quasi-local angular momentum/center-of-mass. The integrals $p_k(S)$, $J_{kl}(S)$ in the spatial asymptotic limit agree with accepted expressions for these quantities [1, 18, 23–25].

7 The Reference

For energy-momentum one takes N to be a translational Killing field of the Minkowski reference. Then the second term in our quasi-local boundary expression (4) vanishes. Let us note in passing that holonomically (with vanishing reference connection coefficients) the first term in (4) reduces to Freud's 1939 superpotential [27]. Thus we are in effect here making a proposal for good coordinates for the Einstein pseudotensor.

To construct a reference, choose, in a neighborhood of the desired spacelike boundary 2-surface S, four smooth functions y^i, $i = 0, 1, 2, 3$ with $dy^0 \wedge dy^1 \wedge dy^2 \wedge dy^3 \neq 0$; they define a Minkowski reference by $\bar{g} = -(dy^0)^2 + (dy^1)^2 + (dy^2)^2 + (dy^3)^2$. This is equivalent to finding a diffeomorphism for a neighborhood of the 2-surface into Minkowski space. The reference connection can now be obtained from the pullback of the flat Minkowski connection.

With constant N^k our quasi-local expression now takes the form

$$\mathscr{B}(N) = N^k x^\mu{}_k (\Gamma^\alpha{}_\beta - x^\alpha{}_j \, dy^j{}_\beta) \wedge \eta_{\mu\alpha}{}^\beta , \tag{7}$$

where $dy^k = y^k{}_\alpha dx^\alpha$ has the inverse $dx^\alpha = x^\alpha{}_k dy^k$.

8 Isometric Matching of the 2-Surface

The reference metric on the dynamical space has the components

$$\bar{g}_{\mu\nu} = \bar{g}_{ij} y^i{}_\mu y^j{}_\nu . \tag{8}$$

Consider the usual embedding restriction: isometric matching of the 2-surface S. This can be expressed quite simply in terms of quasi-spherical foliation adapted coordinates t, r, θ, φ as

$$g_{AB} \doteq \bar{g}_{AB} = \bar{g}_{ij} y^i_A y^j_B = -y^0_A y^0_B + \delta_{ij} y^i_A y^j_B , \tag{9}$$

where S is given by constant values of t, r, and A, B range over θ, φ. We use $\doteq$ to indicate a relation which holds only on the 2-surface S.

From a classic closed 2-surface into $\mathbb{R}^3$ embedding theorem—as long as one restricts S and $y^0(x^\mu)$ so that on S

$$g'_{AB} := g_{AB} + y^0_A y^0_B \tag{10}$$

is convex—one has a unique embedding. Wang and Yau have discussed in detail this type of embedding of a 2-surface into Minkowski controlled by one function in their recent quasi-local work [28, 29].

9 Complete 4D Isometric Matching

Our "new" proposal is: complete 4D isometric matching on S. (We remark that this was already suggested by Szabados back in 2000,[3] and he has since extensively explored this idea [30] in unpublished work.)

Complete 4D isometric matching imposes 10 constraints,

$$g_{\mu\nu}|_S \doteq \bar{g}_{\mu\nu}|_S \doteq \bar{g}_{ij} y^i{}_\mu y^j{}_\nu|_S ,$$

[3] At a workshop in Hsinchu, Taiwan.

on the 16 $y^i_\alpha(t_0, r_0, \theta, \varphi)$. On the 2-surface S these 16 quantities are actually determined by 12 independent embedding functions: $y^i, {y^i}_t, {y^i}_r$ (since from y^i on S one can get ${y^i}_\theta, {y^i}_\varphi$). There remain $2 = 12$–10 degrees of freedom in choosing the reference.

One could as an alternative use orthonormal frames. Then the 4D isometric matching can be represented by $\vartheta^\alpha \doteq \bar{\vartheta}^\alpha$. But the reference coframe has the form $\bar{\vartheta}^\alpha = dy^\alpha$. Thus one should Lorentz transform the coframe ϑ^α to match dy^α on the 2-surface S. This leads to an integrability condition: the 2-forms $d\vartheta^\alpha$ should vanish when restricted to the 2-surface:

$$d\vartheta^\alpha|_S \doteq 0. \tag{11}$$

This is 4 conditions restricting the 6 parameter local Lorentz gauge freedom. Which again reveals that after 4D isometric matching there remains $2 = 6$–4° of freedom in choosing our reference. By the way, this orthonormal frame formulation shows that our procedure can alternatively be viewed as finding a good frame for the "teleparallel gauge current" [31].

10 The Best Matched Reference Geometry

There are 12 embedding variables subject to 10 4D isometric matching conditions, or equivalently, 6 local Lorentz gauge parameters subject to 4 frame embedding conditions. To fix the remaining 2, one can regard the quasi-local value as a measure of the difference between the dynamical and the reference boundary values. So we propose taking the optimal "best matched" embedding as the one which gives the extreme value to the associated invariant mass $m^2 = -p_i p_j \bar{g}^{ij}$. This is reasonable, as one expects the quasi-local energy to be non-negative and to vanish only for Minkowski space.

More precisely, we note two different situations:

I: Given a 2-surface S find the critical points of m^2. This should determine the reference up to Poincaré transformations.

II: Given a 2-surface S and a vector field N, then one can look to the choices of the embedding variables that are a critical point of $E(N, S)$. (Afterward one could extremize over the choice of N.)

Based on some physical and practical computational arguments it seems reasonable to expect a unique solution in general.

For spherically symmetric systems (both static and dynamic), using this and some other related strategies we have found reasonable quasi-local energy results [32–35].

Acknowledgments We much benefited from discussing some of this material with M.-F. Wu. This work was supported by the National Science Council of the R.O.C. under the grants NSC-100-2119-M-008-018, NSC-101-2112-M-008-006 (JMN) and NSC 99-2112-M-008-005-MY3 (CMC) and in part by the National Center of Theoretical Sciences (NCTS).

References

1. Misner, C., Thorne, K., Wheeler, J.: Gravitation. W.H. Freeman, San Francisco (1973)
2. Trautman, A.: Conservation laws in general relativity. In: Witten, L. (ed.) Gravitation: An Introduction to Current Research, pp. 169–198. Wiley, New York (1962)
3. Papapetrou, A.: Einstein's theory of gravitation and flat space. Proc. R. Irish Acad. A **52**, 11 (1958)
4. Landau, L., Lifshitz, E.: The Classical Theory of Fields, 2nd edn. Pergamon Press; Addison-Wesley, Oxford; Reading, MA (1962)
5. Bergmann, P., Thompson, R.: Spin and angular momentum in general relativity. Phys. Rev. **89**, 400 (1953). doi:10.1103/PhysRev.89.400
6. Møller, C.: On the localization of the energy of a physical system in general theory of relativity. Ann. Phys. (N.Y.) **4**, 347 (1958). doi:10.1016/0003-4916(58)90053-8
7. Goldberg, J.: Conservation laws in general relativity, Phys. Rev. **111**, 315 (1958). doi:10.1103/PhysRev.111.315
8. Weinberg, S.: Gravitation and Cosmology: Principles and Applications of the General Theory of Relativity. Wiley, New York (1972)
9. Szabados, L.: Quasi-local energy-momentum and angular momentum in general relativity. Living Rev. Relativity **12**(4), lrr-2009-4 (2009). http://www.livingreviews.org/lrr-2009-4
10. Chang, C., Nester, J., Chen, C.M.: Pseudotensors and quasi-local energy-momentum, Phys. Rev. Lett. **83**, 1897 (1999). doi:10.1103/PhysRevLett.83.1897
11. Nester, J.: General pseudotensors and quasilocal quantities, Class Quantum Grav. **21**, S261 (2004). doi:10.1088/0264-9381/21/3/016
12. Chen, C.M., Nester, J., Tung, R.: Quasilocal energy-momentum for geometric gravity theories, Phys. Lett. A **203**, 5 (1995). doi:10.1016/0375-9601(95)92844-T
13. Chen, C.M., Nester, J.: Quasilocal quantities for general relativity and other gravity theories, Class. Quantum Grav. **16**, 1279 (1999). doi:10.1088/0264-9381/16/4/018
14. Chen, C.M., Nester, J.: A symplectic Hamiltonian derivation of quasilocal energy-momentum for GR. Grav. Cosmol. **6**, 257 (2000)
15. Chen, C.M., Nester, J., R.S. Tung, Hamiltonian boundary term and quasilocal energy flux, Phys. Rev. D **72**, 104020 (2005). doi:10.1103/PhysRevD.72.104020
16. Nester, J.: A manifestly covariant Hamiltonian formalism for dynamical geometry, Prog. Theor. Phys. Suppl. **172**, 30 (2008). doi:10.1143/PTPS.172.30
17. Lanczos, C.: The Variational Principles of Mechanics, Mathematical Expositions, vol. 4, 1st edn. University of Toronto Press, London (1949)
18. Regge, T., Teitelboim, C.: Role of surface integrals in the Hamiltonian formulation of general relativity. Ann. Phys. (N.Y.) **88**, 286 (1974). doi:10.1016/0003-4916(74)90404-7
19. Kijowski, J., Tulczyjew, W.: A Symplectic Framework for Field Theories, Lecture Notes in Physics, vol. 107. Springer, Berlin (1979). doi:10.1007/3-540-09538-1
20. Lynden-Bell, D., Katz, J., Bičák, J.: Mach's principle from the relativistic constraint equations, Mon. Not. R. Astron. Soc. **272**, 150 (1995). Erratum: ibid. 277, 1600 (1995)
21. Katz, J., Bičák, J., Lynden-Bell, D.: Relativistic conservation laws and integral constraints for large cosmological perturbations. Phys. Rev. D **55**, 5957 (1997). doi:10.1103/PhysRevD.55.5957
22. Arnowitt, R., Deser, S., Misner, C.: Energy and the criteria for radiation in general relativity. Phys. Rev. **118**, 1100 (1960). doi:10.1103/PhysRev.118.1100
23. Beig, R., Ó Murchadha, N.: The Poincaré group as the symmetry group of canonical general relativity. Ann. Phys. (N.Y.) **174**, 463 (1987). doi:10.1016/0003-4916(87)90037-6
24. Szabados, L.: On the roots of the Poincaré structure of asymptotically flat spacetimes, Class. Quantum Grav. **20**, 2627 (2003). doi:10.1088/0264-9381/20/13/312
25. Szabados, L.: The Poincaré structure and the centre-of-mass of asymptotically flat spacetimes. In: Frauendiener, J., Giulini, D., Perlick, V. (eds.) Analytical and Numerical Approaches to Mathematical Relativity vol. 692, pp. 157–184 (2006). doi:10.1007/3-540-33484-X8

26. Ó Murchadha, N., Tung, R.S., Xie, N., Malec, E.: Brown-York mass and the Thorne hoop conjecture, Phys. Rev. Lett. **104**, 041101 (2010). doi:10.1103/PhysRevLett.104.041101
27. Freud, P.: Uber die Ausdrücke der Gesamtenergie und des Gesamtimpulses eines materiellen Systems in der allgemeinen Relativitätstheorie, Ann. Math. **40**, 417 (1939). doi:10.2307/1968929
28. Wang, M.T., Yau, S.T.: Isometric embeddings into the Minkowski space and new quasi-local mass, Commun. Math. Phys. **288**, 919 (2009). doi:10.1007/s00220-009-0745-0
29. Wang, M.T., Yau, S.T.: Quasilocal mass in general relativity. Phys. Rev. Lett. **102**, 021101 (2009). doi:10.1103/PhysRevLett.102.021101
30. Szabados, L.: Quasi-Local Energy-Momentum and Angular Momentum in GR: The Covariant Lagrangian Approach (2005). Unpublished draft.
31. de Andrade, V., Guillen, L., Pereira, J.: Gravitational energy-momentum density in teleparallel gravity, Phys. Rev. Lett. **84**, 4533 (2000). doi:10.1103/PhysRevLett.84.4533
32. Wu, M.F., Chen, C.M., Liu, J.L., Nester, J.: Optimal choices of reference for a quasi-local energy, Phys. Lett. A **374**, 3599 (2010). doi:10.1016/j.physleta.2010.07.004
33. Liu, J.L., Chen, C.M., Nester, J.: Quasi-local energy and the choice of reference. Class. Quantum Grav. **28**, 195019 (2012). doi:10.1088/0264-9381/28/19/195019
34. Wu, M.F., Chen, C.M., Liu, J.L., Nester, J.: Optimal choices of reference for a quasilocal energy: Spherically symmetric spacetimes, Phys. Rev. D **84**, 084047 (2011). doi:10.1103/PhysRevD.84.084047
35. Wu, M.F., Chen, C.M., Liu, J.L., Nester, J.: Quasi-local energy for spherically symmetric spacetimes, Gen. Relativ. Gravit. **44**, 2401 (2012). doi:10.1007/s10714-012-1399-3

On a Five-Dimensional Version of the Goldberg-Sachs Theorem

Marcello Ortaggio, Vojtěch Pravda, Alena Pravdová and Harvey S. Reall

Abstract The recently developed generalization of the Goldberg-Sachs theorem to five-dimensional Einstein spacetimes is summarized. This generalization involves two steps. First it has been proven that in arbitrary dimension an Eistein spacetime admitting a multiple WAND admits also a multiple *geodetic* WAND. Second, in five dimensions, the 3×3 optical matrix of such geodetic multiple WAND can be cast to one of three canonical forms, each determined by two free parameters.

1 Introduction

Recently a generalization of the Petrov classification of the Weyl tensor to arbitrary dimension was developed [1] (see [2] for a recent review). It is thus natural to ask whether the four dimensional Goldberg-Sachs (GS) theorem can be in some form extended to higher dimensions.

M. Ortaggio (✉) · V. Pravda · A. Pravdová
Institute of Mathematics, Academy of Sciences, Žitná 25, 115 67 Prague 1, Czech Republic
e-mail: ortaggio@math.cas.cz

V. Pravda
e-mail: pravda@math.cas.cz

A. Pravdová
e-mail: pravdova@math.cas.cz

H. S. Reall
DAMTP, Centre for Mathematical Sciences, University of Cambridge, Wilberforce Road, Cambridge CB3 0WA, UK
e-mail: hsr1000@cam.ac.uk

J. Bičák and T. Ledvinka (eds.), *Relativity and Gravitation*,
Springer Proceedings in Physics 157, DOI: 10.1007/978-3-319-06761-2_23,

In four dimensions GS theorem proved to be a very useful tool for studying and constructing new, algebraically special solutions of the Einstein equations, e.g. the Kerr solution [3]. It states that *in a (non-conformally flat) Einstein spacetime,*[1] *a null vector field is a repeated principal null direction (of the Weyl tensor) if, and only if, it is geodetic and shear-free*.

A higher dimensional analogue of the principal null direction (PND) of the Weyl tensor is a so called Weyl Aligned Null Direction (WAND) which in four dimensions coincides with PND. We will say that a higher dimensional spacetime is algebraically special if it admits a *multiple* WAND (an analogue of a repeated PND).

In contrast with the four-dimensional case, higher dimensional Einstein spacetimes may admit (only in type D spacetimes) *non-geodetic* multiple WANDs. However, it has been shown in [4] that such spacetimes also always admit *geodetic* multiple WANDs. The higher dimensional generalization of the "geodetic" part of the GS theorem thus reads [4]:

Proposition 1 *An Einstein spacetime admits a multiple WAND if, and only if, it admits a geodetic multiple WAND.*

Therefore when considering algebraically special Einstein spacetimes without loss of generality one can always choose a *geodetic* multiple WAND.

For the formulation of the "shearfree" part of the GS theorem we need to introduce the $(d-2)\times(d-2)$ *optical matrix* ρ_{ij}

$$\rho_{ij} \equiv m^{\mu}_{(i)} m^{\nu}_{(j)} \nabla_{\nu} \ell_{\mu}, \tag{1}$$

corresponding to a geodetic null vector field $\boldsymbol{\ell}$ coinciding with a multiple WAND and with $m_{(i)}$ being orthonormal spacelike vectors orthogonal to $\boldsymbol{\ell}$.

In contrast with the four-dimensional case, algebraically special Einstein spacetimes may admit multiple WANDs with non-vanishing shear[2] and in fact in the generic case shear is non-vanishing. Therefore a generalization of *necessary*[3] conditions on ρ_{ij} following from the existence of multiple WAND will not be straightworward and in the next section we will thus limit ourselves to the case of five dimensions.

Let us conclude this section with mentioning some *special* classes of spacetimes where the *necessary* conditions on ρ_{ij} following from the multiple WAND condition have been known in any dimensions.

---

[1] An Einstein spacetime is a solution of the vacuum Einstein equation, possibly with a cosmological constant, i.e. with the Ricci tensor $R_{ab} = (R/d)g_{ab}$ in d dimensions.

[2] Shear is defined as traceless symmetric part of the optical matrix.

[3] *Sufficient* conditions on ρ_{ij} for $\boldsymbol{\ell}$ to be a multiple WAND are not in full generality known, but it has been shown [5, 6] that $\rho_{ij} = 0$ (Kundt class) and $\rho_{ij} \propto \delta_{ij}$ (Robinson-Trautman class) are examples of such sufficient conditions.

1.1 Necessary Conditions on ρ_{ij} for Various Special Classes of Spacetimes

1.1.1 Types N and III

The multiple WAND in vacuum spacetimes of type N must be geodetic, the optical matrix must have rank 2 and it can be put into a form [7]

$$\rho = b \,\mathrm{diag}\left(\begin{bmatrix} 1 & a \\ -a & 1 \end{bmatrix}, 0, \ldots, 0\right). \tag{2}$$

The same form of $\boldsymbol{\rho}$ also applies to type III Ricci flat spacetimes [7] that either

(i) are five-dimensional,
(ii) satisfy a certain genericity condition [7],
(iii) have a *non-twisting* multiple WAND (with vanishing rotation, i.e., $a = 0$).

Generalization of the above type N and III results from the Ricci-flat case to the Einstein case is straightforward, see [8]. Note that ρ_{ij} of the form (2) is shearfree for $d = 4$ but not for $d > 4$ (for $b \neq 0$).

1.1.2 Kerr-Schild Spacetimes

It can be shown [9, 10] that for Kerr-Schild (KS) spacetimes,[4]

$$g_{\mu\nu} = \bar{g}_{\mu\nu} + H k_\mu k_\nu, \tag{3}$$

with vanishing $T_{00} \equiv T_{ab}k^a k^b$ component of the energy–momentum tensor, the KS vector $\boldsymbol{k}$ is a geodetic multiple WAND and the optical matrix can be put into a block diagonal form

$$\rho = \alpha \mathrm{diag}\left(1, \ldots 1, \frac{1}{1+\alpha^2 b_1^2}\begin{bmatrix} 1 & -\alpha b_1 \\ \alpha b_1 & 1 \end{bmatrix}, \ldots, \frac{1}{1+\alpha^2 b_\nu^2}\begin{bmatrix} 1 & -\alpha b_\nu \\ \alpha b_\nu & 1 \end{bmatrix}, 0, \ldots, 0\right). \tag{4}$$

1.1.3 Asymptotically Flat Type II Spacetimes

Similar result as above holds also for asymptotically flat type II spacetimes [11].

It can be shown that in this case $\boldsymbol{\rho}$ obeys the *optical constraint*

$$\rho_{ik}\rho_{jk} \propto \rho_{(ij)}, \tag{5}$$

[4] $\bar{g}_{\mu\nu}$ is a metric of constant curvature and KS vector $\boldsymbol{k}$ is null with respect to $\bar{g}_{\mu\nu}$ and thus also with respect to $g_{\mu\nu}$.

which in fact holds if and only if $\boldsymbol{\rho}$ can be put in the canonical form (4) by appropriately choosing the frame. Note that the optical constraint also holds in the type N and III cases discussed above. In fact it turns out that in arbitrary dimension such form of $\boldsymbol{\rho}$ is "preferred" and holds for generic algebraically special spacetimes (see [12] for more precise formulation). Explicit examples of Einstein spacetimes with all multiple WANDs violating the optical constraint are however also known [2].

2 GS Theorem in Five Dimensions

Let us now summarize main results of [12], where the necessary conditions for $\boldsymbol{\ell}$ to be a multiple WAND in a five-dimensional algebraically special Einstein spacetime have been found:

Theorem 1 *In a five-dimensional algebraically special Einstein spacetime that is not conformally flat, there exists a geodetic multiple WAND $\boldsymbol{\ell}$ and its optical matrix can be put in one of the forms*

$$\text{(i)}\quad \mathrm{b}\begin{pmatrix} 1 & a & 0 \\ -a & 1 & 0 \\ 0 & 0 & 1+a^2 \end{pmatrix}, \tag{6}$$

$$\text{(ii)}\quad \mathrm{b}\begin{pmatrix} 1 & a & 0 \\ -a & 1 & 0 \\ 0 & 0 & 0 \end{pmatrix}, \tag{7}$$

$$\text{(iii)}\quad \mathrm{b}\begin{pmatrix} 1 & a & 0 \\ -a & -a^2 & 0 \\ 0 & 0 & 0 \end{pmatrix}. \tag{8}$$

If the spacetime is of type III or type N then the form must be (ii).

Note that for $b \neq 0$ matrices (i), (ii), (iii) have rank 3, 2, 1, respectively, and for $b = 0$ it is the Kundt spacetime. Only the case iii) with $a \neq 0 \neq b$ does not satisfy the optical constraint (5). However, it has been proven recently that this case cannot occur for genuine type II spacetimes [13] (see also [14]) and for type D one can show [12] the following:

Proposition 2 *A five-dimensional type D Einstein spacetime admits a geodetic multiple WAND violating the optical constraint if, and only if, it admits a non-geodetic multiple WAND.*

Note that all 5d Einstein spacetimes admitting a non-geodetic multiple WAND are explicitly known [4]. It turns out that these spacetimes always admit another geodetic multiple WAND satisfying optical constraint and one can thus conclude:

Proposition 3 *A five-dimensional algebraically special Einstein spacetime always admits a geodetic multiple WAND obeying the optical constraint.*

Due to Theorem 1 the number of independent components of the optical matrix ρ is reduced from 6 to 2 free parameters. This will lead to a considerable simplification of the GHP equations [8] and hopefully also to a discovery of new higher dimensional algebraically special Einstein spacetimes.

Let us conclude with some known examples of Einstein spacetimes belonging to the cases (i)–(iii) of the Theorem 1 [12]:

- case (i)
  $a \neq 0$: the Myers-Perry [15] black hole solution (cf. [16]) and in fact all non-degenerate (i.e. $\det\rho \neq 0$) Einstein Kerr-Schild metrics with Minkowski or (A)dS background [9, 10], 5d Kaluza-Klein bubble obtained by analytic continuation of a singly spinning Myers-Perry solution [12, 17];
  $a = 0$: this case corresponds to the Robinson-Trautman class [6] and in five dimensions it reduces to the Schwarzschild-Tangherlini metric (possibly with a cosmological constant).
- case (ii)
  $a \neq 0$: product of a 4d Ricci-flat algebraically special twisting solution with a flat 5th direction, e.g., the Kerr black string (i.e. the product of the 4d Kerr solution with a flat direction), or more generally warped product of a 4d algebraically special Einstein spacetime with a fifth direction (with non-vanishing cosmological constant);
  $a = 0$: a direct or warped product of any 4d Einstein type II Robinson-Trautman metric [18], e.g. the Schwarzschild black string solution.
- case (iii)
  Spacetimes belonging to this case are of type D ([13], see also [14]) and admit a non-geodetic multiple WAND. All such metrics were determined in [4]—the direct products $dS_3 \times S^2$ and $AdS_3 \times H^2$ or the analytical continuation of the 5d Schwarzschild solution [19] (generalized to include a cosmological constant and planar or hyperbolic symmetry);
  $a \neq 0$: a specific twisting geodetic multiple WAND, e.g., in $dS_3 \times S^2$, see [12];
  $a = 0$: a non-twisting, expanding and shearing geodetic multiple WAND in $dS_3 \times S^2$ or in the Kaluza-Klein bubble solution, see [12].

Acknowledgments The authors acknowledge support from research plan RVO: 67985840 and research grant no P203/10/0749.

References

1. Coley, A., Milson, R., Pravda, V., Pravdová, A.: Classification of the Weyl tensor in higher dimensions. Class. Quantum Grav. **21**, L35 (2004). doi:10.1088/0264-9381/21/7/L01
2. Ortaggio, M., Pravda, V., Pravdová, A.: Algebraic classification of higher dimensional spacetimes based on null alignment. Class. Quantum Grav. **30**, 013001 (2013). doi:10.1088/0264-9381/30/1/013001
3. Kerr, R.P.: Gravitational field of a spinning mass as an example of algebraically special metrics. Phys. Rev. Lett. **11**, 237 (1963). doi:10.1103/PhysRevLett.11.237

4. Durkee, M., Reall, H.: A higher-dimensional generalization of the geodesic part of the Goldberg-Sachs theorem. Class. Quantum Grav. **26**, 245005 (2009). doi:10.1088/0264-9381/26/24/245005
5. Ortaggio, M., Pravda, V., Pravdová, A.: Ricci identities in higher dimensions. Class. Quantum Grav. **24**, 1657 (2007). doi:10.1088/0264-9381/24/6/018
6. Podolský, J., Ortaggio, M.: Robinson-Trautman spacetimes in higher dimensions. Class. Quantum Grav. **23**, 5785 (2006). doi:10.1088/0264-9381/23/20/002
7. Pravda, V., Pravdová, A., Coley, A., Milson, R.: Bianchi identities in higher dimensions, Class. Quantum Grav. **21**, 2873 (2004). doi:10.1088/0264-9381/21/12/007. Corrigendum: ibid. **24**, 1691 (2007)
8. Durkee, M., Pravda, V., Pravdová, A., Reall, H.: Generalization of the Geroch-Held-Penrose formalism to higher dimensions. Class. Quantum Grav. **27**, 215010 (2010). doi:10.1088/0264-9381/27/21/215010
9. Ortaggio, M., Pravda, V., Pravdová, A.: Higher dimensional Kerr-Schild spacetimes. Class. Quantum Grav. **26**, 025008 (2009). doi:10.1088/0264-9381/26/2/025008
10. Málek, T., Pravda, V.: Kerr-Schild spacetimes with (A)dS background. Class. Quantum Grav. **28**, 125011 (2011). doi:10.1088/0264-9381/28/12/125011
11. Ortaggio, M., Pravda, V., Pravdová, A.: Asymptotically flat, algebraically special spacetimes in higher dimensions. Phys. Rev. D **80**, 084041 (2009). doi:10.1103/PhysRevD.80.084041
12. Ortaggio, M., Pravda, V., Pravdová, A., Reall, H.: On a five-dimensional version of the Goldberg-Sachs theorem. Class. Quantum Grav. **29**, 205002 (2012). doi:10.1088/0264-9381/29/20/205002
13. Wylleman, L.: (2014). In preparation
14. Reall, H.S., Graham, A.A.H., Turner, C.P.: On algebraically special vacuum spacetimes in five dimensions. Class. Quantum Grav. **30**(5), 055004 (2013). doi:10.1088/0264-9381/30/5/055004
15. Myers, R., Perry, M.: Black holes in higher dimensional space-times. Ann. Phys. (N.Y.) **172**, 304 (1986). doi:10.1016/0003-4916(86)90186-7
16. Pravda, V., Pravdová, A., Ortaggio, M.: Type D Einstein spacetimes in higher dimensions. Class. Quantum Grav. **24**, 4407 (2007). doi:10.1088/0264-9381/24/17/009
17. Dowker, F., Gauntlett, J., Gibbons, G., Horowitz, G.: Decay of magnetic fields in Kaluza-Klein theory. Phys. Rev. D **52**, 6929 (1995). doi:10.1103/PhysRevD.52.6929
18. Ortaggio, M., Pravda, V., Pravdová, A.: On higher dimensional Einstein spacetimes with a warped extra dimension. Class. Quantum Grav. **28**, 105006 (2011). doi:10.1088/0264-9381/28/10/105006
19. Witten, E.: Instability of the Kaluza-Klein vacuum. Nucl. Phys. B **195**, 481 (1982). doi:10.1016/0550-3213(82)90007-4

Gravitomagnetism: From Einstein's 1912 Paper to the Satellites LAGEOS and Gravity Probe B

Herbert Pfister

Abstract The first concrete calculations of (linear) gravitomagnetic effects were performed by Einstein in 1912–1913. Einstein also directly and decisively contributed to the "famous" papers by Thirring (and Lense) from 1918. Generalizations to strong fields were performed not earlier than in 1966 by Brill and Cohen. Extensions to higher orders of the angular velocity ω by Pfister and Braun (1985–1989) led to a solution of the centrifugal force problem and to a quasiglobal principle of equivalence. The difficulties but also the recent successes to measure gravitomagnetic effects are reviewed, and cosmological and Machian aspects of gravitomagnetism are discussed.

1 Einstein's Papers on Gravitomagnetism from 1912 and 1913

Einstein's paper "Is there a gravitational action analogous to electromagnetic induction?" [1] from July 1912 (presumably his last work in Prague) is exceptional in many ways: It is published in a journal for forensic medicine (as a birthday present for his friend Heinrich Zangger), and it is very short (4 pages in the original setting, equivalent to less than 1.5 pages in today's Physical Review). It introduces audacious new concepts: the model of a spherical mass shell with mass M and radius R (which is useful until today in general relativity, because it is the optimal substitute for Newton's mass point, and because it allows to treat systems with matter by solving only the vacuum equations of general relativity), moreover a new gravitomagnetic "force", and the first calculation of a dragging effect: If the mass shell is linearly accelerated with Γ, Einstein calculates that a test mass m at the center of the shell is dragged with acceleration $\gamma = \frac{3}{2}(M/R)\Gamma$ (in units with $G = c = 1$).

H. Pfister (✉)
Institute of Theoretical Physics, University of Tübingen, Auf der Morgenstelle 14, 72076 Tübingen, Germany
e-mail: herbert.pfister@uni-tuebingen.de

J. Bičák and T. Ledvinka (eds.), *Relativity and Gravitation*, Springer Proceedings in Physics 157, DOI: 10.1007/978-3-319-06761-2_24,

On the other hand, from today's perspective of general relativity, most details of the paper are wrong or inconsistent: The calculated mass increase of the test mass $m \rightarrow m' = m + mM/R$ is only a coordinate effect in general relativity [2]; in the calculated linear dragging acceleration γ the prefactor $\frac{3}{2}$ has to be substituted by $\frac{4}{3}$ [3], and, most importantly, a scalar relativistic gravity theory (which was the basis of Einstein's paper) can never produce a vectorial gravitomagnetic induction.

But the central new physical ideas of this paper (dragging and gravitomagnetism) kept command over Einstein when in 1913 (now in Zürich, with Grossman) he formulated the tensorial Entwurf theory. In the so-called Einstein-Besso manuscript [4] of June 1913 they calculated within this theory, besides the main topic of perihelion advance of Mercury, also a new value for the linear dragging acceleration ($\gamma = 2(M/R)\Gamma$), a Coriolis force inside a rotating spherical mass shell, and therefrom a rotational dragging of test masses (half the value in final general relativity), and a motion of the nodes of planets in the field of the rotating sun (1/4 of the value in general relativity). It is quite interesting which parts of this manuscript Einstein presented in his great and brilliant speech at the Naturforscher-Versammlung in Vienna in September 1913 [5], and which parts he omitted. When Einstein had finished general relativity in November 1915, he did not immediately come back to the questions of dragging and gravitomagnetism, because there were more urgent new problems (gravitational waves, cosmology, gravitational field energy, ...), and because he presumably imagined that the results on dragging and gravitomagnetism in general relativity would be similar to his results in the Entwurf theory.

2 The Papers of Thirring (and Lense) on Gravitomagnetism from 1918

It is well known that questions of dragging and gravitomagnetism in general relativity were first taken up in 1917–1918 by Hans Thirring (and Lense). Not so well known is that these papers owe nearly all their interesting and correct results to the direct interference of Einstein. Thirring had started his work in April 1917 (see Thirring [6] and Pfister [7]), with (partly wrong) calculations of centrifugal effects exerted by rotating mass shells and full bodies, and he did not realize that these effects, being of second order in the angular velocity ω, are ridiculously small for all laboratory and solar systems. In a letter of July 17 [8], Thirring informed Einstein about his work, together with some questions. Einstein's answer of August 2 [8] is short and polite, but admirably clear and concise. He stresses that much more important and realistic than centrifugal effects are Coriolis effects of first order of ω; he explains to Thirring the resulting dragging phenomena and the effects on the planets and moons in the solar system, and tells him that he has calculated all these effects (in the Entwurf theory), a fact which should have been known to Thirring from Einstein's speech [5] in Vienna in 1913. Only after this eye-opening lesson from Einstein is Thirring able to produce his two "famous" papers [9, 10] of 1918. Still these papers have severe deficits: For the rotating mass shell Thirring calculates in the weak field approximation the dragging acceleration of test masses of velocity v: $a = 2d\, v \times \omega$, with $d_{Th} = 4M/3R$, but only near the center of the shell (for $r \ll R$), and for the rotating full body he calculates the so-called Lense-Thirring effect $\boldsymbol{a} = 2v \times H$, with the gravitomagnetic dipole field $H = \frac{2MR^2}{5r^3}[\omega - 3(\omega r)r/r^2]$ only for $r \gg R$, which does not apply to the modern satellite experiments LAGEOS and Gravity Probe B. (See Sect. 5.) The centrifugal results of order ω^2 in [9] contain many errors: an integration error observed by Laue and Pauli in 1920, the error (observed by Lanczos [11]) that Thirring modelled the mass shell as dust, and did therefore not correctly solve the Einstein equations, and the result of an axial component of his centrifugal "force", for which he gave a wrong physical explanation. The contributions of Lense to [10] are anyhow only of minor, technical character: The transformation of Thirring's results from Cartesian coordinates to the orbital elements used in astronomy, and their evaluation for some planets and moons of the solar system.

In my judgement a more original and valuable (but seldom quoted) paper by Thirring is his [12] where he as the first person (and correctly) formulates the analogies between electromagnetism and the Einstein equations in linear approximation, discusses the different signs and a factor 4 of the basic equations of gravitomagnetism in comparison to electromagnetism, and here he even mentions the preliminary discussion of gravitomagnetism by Einstein in his Vienna speech [5] of 1913. (For a modern and more extended treatment of gravitomagnetism see Ciufolini and Wheeler [13].)

3 Generalizations to Strong Fields and Higher Orders of ω. Solution of the Centrifugal Force Problem

Considerable progress and extension of the work of Einstein and Thirring happened only in 1966 by the work of Brill and Cohen [14] who performed a first order rotational perturbation not of Minkowski spacetime but of the Schwarzschild solution, with the result for the dragging factor $d_{BC} = 4\alpha(2-\alpha)/((1+\alpha)(3-\alpha))$, with $\alpha = M/2R$, where R is the shell radius in isotropic coordinates. The important new physical result is that in the collapse limit $\alpha \to 1$ the factor d_{BC} attains the value 1: total dragging, and herewith a complete realization of the Machian postulate of relativity of rotation: in this limit the interior of the shell cuts itself off as a type of separate universe, and interior test particles are dragged along with the full angular velocity ω of the shell. As far as I know, Brill and Cohen were also the first to make clear that the interior Coriolis field applies to all $r < R$, and the exterior dipole field to all $r > R$. (The latter follows simply from symmetry arguments: a first order rotational perturbation of a spherical system produces quite generally a pure dipole field proportional to r^{-3}.)

An extension of this work to higher orders of ω, and in particular the problem of the notoriously wrong "centrifugal force" inside a rotating mass shell had to wait for another 19 years to be solved in [15]. The solution is based on two "new" observations which could and should have been made already in Thirring's time, but which were overlooked by all authors before 1985:

(a) Any physically realistic rotating body will suffer a centrifugal deformation in orders ω^2 and higher, and cannot be expected to keep its spherical shape.
(b) If we aim to realize inside the rotating mass shell quasi-Newtonian conditions with correct Coriolis and centrifugal forces—and no other forces!—, the interior of the mass shell obviously has to be a flat piece of spacetime. In the first order of ω this flatness is more or less trivial; however, in order ω^2 it is by no means trivial, and is indeed violated for Thirring's solution, due to the axial component of his "centrifugal force".

These observations lead to the mathematical question whether it is possible to connect a rotating flat metric through a mass shell (with, to begin with, unknown geometrical and material properties) to the non-flat but asymptotically flat exterior metric of a rotating body. In [15–17] we could show that this problem has (for given M, R, and $\omega \ll 1/R$) a unique solution in every order ω^n, and that the resulting mass shell has non-spherical (surprisingly oblate) geometry, non-spherical mass distribution, and differential rotation. Only in the collapse limit $R \to M/2$ the shell is again spherical and rigidly rotating, as was already deduced by de la Cruz und Israel [18].

4 A Quasi-Global Principle of Equivalence

The success with this "matter-induced centrifugal force" guided me to the following hypothesis of a "quasi-global equivalence principle in general relativity" [15]. In short: "Every acceleration field can be understood as a gravitational field." In more detail: If some finite laboratory (a flat region of spacetime) is in arbitrary accelerated motion relative to the fixed stars, then all motions of free particles and all physical laws, measured from laboratory axes, are modified by inertial forces. It is argued that exactly the same modified motions and laws can be induced (at least for some time) at all places of a laboratory at rest relative to the fixed stars, by suitable and suitably accelerated masses outside the laboratory, e.g., in a mass shell. After formulating this hypothesis in 1985, I found that similar ideas arose already in the years 1912–1913 in discussions of Einstein with Ehrenfest [19] and Mie [5]. But at that time these people were quite sceptical about such a "macroequivalence". Today there are good arguments for the validity of the hypothesis at least for small accelerations because for small rotations (in Pfister and Braun [15]) and small linear accelerations (in Pfister et al. [3]) the hypothesis has been explicitly proven, and because arbitrary accelerations can (at least in principle) be combined from linear and rotational accelerations.

5 Measuring Gravitomagnetism

I should like to comment on the difficulties but also successes to measure the new "force" gravitomagnetism. For laboratories on earth and for satellites we have on one hand a factor $M_{\mathrm{Earth}}/R_{\mathrm{Earth}} \approx 10^{-9}$ for any deviations from Newtonian gravity. For rotational effects there comes another factor $\omega_{\mathrm{Earth}} R_{\mathrm{Earth}}/c \approx 10^{-6}$, therefore a factor 10^{-15} for any gravitomagnetic field, in comparison to Newtonian gravity. (Already Einstein in his letter to Thirring from 1917 stated that "the effects stay far below the measurement error".) Since there exist no gravitomagnetic materials in nature, there comes typically another factor $v/c \leq 10^{-5}$ from the velocity v of the rotating parts of the measuring device (except where these are photons or neutrinos). The resulting demand of a total precision of 10^{-20} can presumably not be fulfilled by any laboratory experiment in the foreseeable future, why I judge all pertaining recent proposals as questionable, even if they use Bose-Einstein condensates as in [20]. For neutron stars, pulsars, and black holes the above numbers are of course much more favourable. But in these astrophysical systems there exist many competing, partly unknown or poorly understood processes so that it is again questionable whether they lead to a clear measurement of gravitomagnetism [21].

In contrast, already soon after the start of the first earth satellites (in 1957) there appeared proposals (e.g., by Ginzburg and Schiff) to use these for tests of general relativity, because in space there is automatically high vacuum and low temperature, and because such tests can accumulate data over long time (years). In an admirable effort over 40 years (and with expenses of 700 million US$) the Stanford Gravity Probe B project (a satellite with $r/R \approx 1.10$) has finally confirmed the Lense-Thirring or rather Schiff effect (precession of a gyroscope axis) with 19 % precision, much less than the originally expected precision of 1 % (Everitt et al. [22]). (The accompanying geodetic precession is not a gravitomagnetic effect, because the "gravitomagnetic invariant" $^{*}R \cdot R = \frac{1}{2}\epsilon^{\alpha\beta\gamma\delta} R_{\alpha\beta\mu\nu} R^{\mu\nu}_{\gamma\delta}$ is zero for this effect.) A somewhat better (10 %) confirmation of the Lense-Thirring effect was, however, performed already some years earlier by Ciufolini and Pavlis [23] by a (in principle) much simpler satellite experiment: the careful measurement of the orbits of the passive satellites LAGEOS I and II (with $r/R \approx 1.92$) over 11 years, together with a precise measurement of the earth multipole moments $J_2, J_4, \ldots$ by the satellites CHAMP and GRACE. An ingenious proposal by Ciufolini [24] to start LAGEOS II with orbital elements "complementary" to LAGEOS I, and hereby cancelling the multipole contributions, was unfortunately never realized. But the newly launched satellite LARES gives hope to confirm a gravitomagnetic effect soon with 1 % precision.

If gravitational waves can be analyzed in detail in the future, this will also be an indirect test for gravitomagnetism, because, similar to electromagnetism, gravitational waves have in equal parts gravitoelectric and gravitomagnetic contributions.

6 Cosmological Remarks

Although the dragging results of Einstein, Thirring, Brill-Cohen et al. with their asymptotically flat solutions do not really meet the Machian demand for a cosmological origin of inertia, it was proven by my PhD-student Klein [25], and by Bičák et al. [26], and by Schmid [27] that rotational perturbations of standard FRW cosmologies provide similar dragging results. Concerning the observational confirmation of the (non-causal!) determination of the local inertial frames by the cosmos as a whole, I should like to quote from the MTW-book [28]: "Consider a bit of solid ground near the geographic pole, and a support erected there, and from it hanging a pendulum. Though the sky is cloudy, the observer watches the track of the Foucault pendulum as it slowly turns through 360°. Then the sky clears and, **miracle of miracles**, the pendulum is found to be swinging all the time on an arc fixed relative to the far-away stars." The presently best measurement of this "non-rotation" (smaller than 10^{-9} of the earth angular velocity) comes from the terrestrial reference system realized by VLBI and GPS [29].

References

1. Einstein, A.: Gibt es eine Gravitationswirkung, die der elektrodynamischen Induktionswirkung analog ist? Vierteljahrsschr. Gerichtl. Med. öffentl. Sanitätswesen **44**, 37 (1912)
2. Brans, C.: Mach's principle and the locally measured gravitational constant in general relativity. Phys. Rev. **125**, 388 (1962). doi:10.1103/PhysRev.125.388
3. Pfister, H., Frauendiener, J., Hengge, S.: A model for linear dragging. Class. Quantum Grav. **22**, 4743 (2005). doi:10.1088/0264-9381/22/22/007
4. Klein, M., Kox, A., Renn, J., Schulman, R. (eds.): The Collected Papers of Albert Einstein, Vol. 4: The Swiss Years, Writings, 1912–1914, pp. 334–473. Princeton University Press, Princeton (1995)
5. Einstein, A.: Zum gegenwärtigen Stande des Gravitationsproblems, Phys. Z. **14**, 1249 (1913) (Reprinted in 'The Collected Papers of Albert Einstein', Vol. 4. Doc. **17**, 486–503)
6. Thirring, H.: Notebook 'Die Wirkung Rotierender Massen' (1917). http://phaidra.univie.ac.at/o:141725
7. Pfister, H.: On the history of the so-called Lense-Thirring effect. Gen. Rel. Grav **39**, 1735 (2007). doi:10.1007/s10714-007-0521-4
8. Thirring, H., Einstein, A.: 361: From Hans Thirring, 369: To Hans Thirring. In: Schulman, R., Kox, A., Janssen, M., Illy, J. (eds.) The Collected Papers of Albert Einstein, Vol. 8: The Berlin Years: Correspondence, 1914–1918, p. Docs. 361 and 369. Princeton University Press, Princeton (1998)
9. Thirring, H.: Über die Wirkung rotierender ferner Massen in der Einsteinschen Gravitationstheorie. Phys. Zs. **19**, 33 (1918)
10. Lense, J., Thirring, H.: Über den Einfluß der Eigenrotation der Zentralkörper auf die Bewegung der Planeten und Monde nach der Einsteinschen Gravitationstheorie. Phys. Z. **19**, 156 (1918)
11. Lanczos, K.: Zum Rotationsproblem der allgemeinen Relativitätstheorie. Zeits. Phys. **14**, 204 (1923). doi:10.1007/BF01340041
12. Thirring, H.: Über die formale Analogie zwischen den elektromagnetischen Grundgleichungen und den Einsteinschen Gravitationsgleichungen erster Näherung. Phys. Z. **19**, 204 (1918). Translated and commented on in Gen. Rel. Grav. **44**, 3217, 3225(2012). doi:10.1007/s10714-012-1451-3

13. Ciufolini, I., Wheeler, J.: The Gravitomagnetic Field and its Measurement in Gravitation and Inertia. Princeton University Press, Princeton, (1995) (Chap. 6. Princeton Series in Physics)
14. Brill, D., Cohen, J.: Rotating masses and their effect on inertial frames. Phys. Rev. **143**, 1011 (1966). doi:10.1103/PhysRev.143.1011
15. Pfister, H., Braun, K.: Induction of correct centrifugal force in a rotating mass shell. Class. Quantum Grav. **2**, 909 (1985). doi:10.1088/0264-9381/2/6/015
16. Pfister, H., Braun, K.: A mass shell with flat interior cannot rotate rigidly. Class. Quantum Grav. **3**, 335 (1986). doi:10.1088/0264-9381/3/3/008
17. Pfister, H.: Rotating mass shells with flat interiors. Class. Quantum Grav. **6**, 487 (1989). doi:10.1088/0264-9381/6/4/009
18. De La Cruz, V., Israel, W.: Spinning shell as a source of the Kerr metric. Phys. Rev. **170**, 1187 (1968). doi:10.1103/PhysRev.170.1187
19. Einstein, A., Ehrenfest, P.: 409: To Paul Ehrenfest, 411: From Paul Ehrenfest. In: Klein, M., Kox, A., Schulman, R. (eds.) The Collected Papers of Albert Einstein, Vol. 5: The Swiss Years: Correspondence, 1902–1914, p. Docs. 409 and 411. Princeton University Press, Princeton (1993)
20. van Zoest, T., Gaaloul, N., Singh, Y., et al.: Bose-Einstein condensation in microgravity. Science **328**, 1540 (2010). doi:10.1126/science.1189164
21. Stella, L., Possenti, A.: Lense-Thirring precession in the astrophysical context. Space. Sci. Rev. **148**, 105 (2009). doi:10.1007/s11214-009-9627-1
22. Everitt, C., DeBra, D., Parkinson, B., et al.: Gravity probe B: final results of a space experiment to test general relativity. Phys. Rev. Lett. **106**, 221101 (2011). doi:10.1103/PhysRevLett.106.221101
23. Ciufolini, I., Pavlis, E.: A confirmation of the general relativistic prediction of the Lense-Thirring effect. Nature **431**, 958 (2004). doi:10.1038/nature03007
24. Ciufolini, I.: Measurement of the Lense-Thirring drag on high-altitude, laser-ranged artificial satellites. Phys. Rev. Lett. **56**, 278 (1986). doi:10.1103/PhysRevLett.56.278
25. Klein, C.: Rotational perturbations and frame dragging in a Friedmann universe. Class. Quantum Grav. **10**, 1619 (1993). doi:10.1088/0264-9381/10/8/019
26. Bičák, J., Lynden-Bell, D., Katz, J.: Do rotations beyond the cosmological horizon affect the local inertial frame? Phys. Rev. D **69**, 064011 (2004). doi:10.1103/PhysRevD.69.064011
27. Schmid, C.: Cosmological gravitomagnetism and Mach's principle. Phys. Rev. D **74**, 044031 (2006). doi:10.1103/PhysRevD.74.044031
28. Misner, C., Thorne, K., Wheeler, J.: Gravitation. W.H. Freeman, San Francisco (1973)
29. Kovalevsky, J., Mueller, I., Kołaczek, B. (eds.): Reference Frames in Astronomy and Geophysics, Astrophysics and Space Science Library, vol. 154. Kluwer, Dordrecht (1989)

Evolution of the Einstein Equations to Future Null Infinity

Oliver Rinne and Vincent Moncrief

Abstract We describe recent progress with a formulation of the Einstein equations on constant mean curvature surfaces extending to future null infinity. Long-time stable numerical evolutions of an axisymmetric gravitationally perturbed Schwarzschild black hole have been obtained. Here we show how matter can be included in our formulation. We study late-time tails for the spherically symmetric Einstein–Yang–Mills equations both for initial data that disperse and that collapse to a black hole.

1 Introduction

The standard approach to numerical simulations of asymptotically flat spacetimes is to adopt the Cauchy formulation of general relativity and truncate the spatial slices at a finite distance, where boundary conditions must be imposed. Apart from leading to a well-posed initial-boundary value problem, such boundary conditions should also be *absorbing*, i.e. they should be consistent with the solution on the unbounded domain. The problem is that the correct boundary conditions are not known at a finite distance. At best one may appeal to linearised theory. Bad choices of boundary conditions are known to destroy relevant features of the solution. A far more elegant approach is to include future null infinity in the numerical domain, which is the true physical boundary of spacetime. In order to do this, we follow

O. Rinne (✉)
Max Planck Institute for Gravitational Physics (Albert Einstein Institute),
Am Mühlenberg 1, 14476 Potsdam, Germany
e-mail: oliver.rinne@aei.mpg.de

V. Moncrief
Department of Mathematics and Department of Physics, Yale University,
New Haven, CT 06520, USA
e-mail: vincent.moncrief@yale.edu

J. Bičák and T. Ledvinka (eds.), *Relativity and Gravitation*, Springer Proceedings in Physics 157, DOI: 10.1007/978-3-319-06761-2_25,

Penrose's approach and apply a conformal transformation to the spacetime metric, combined with a compactifying coordinate transformation. Rather than Friedrich's regular conformal field equations [1], we work directly with the Einstein equations in an ADM-like formulation on constant mean curvature (CMC) slices [2]. This formulation is reviewed in Sect. 2 and extended here to include matter sources. In Sect. 3 we review a first numerical implementation of this system, which achieved long-time stable evolution of a perturbed Schwarzschild black hole for the vacuum Einstein equations in axisymmetry. In Sect. 4 we include matter in the form of a Yang–Mills field, and we perform numerical simulations of the late-time decay of this field, restricted to spherical symmetry. Our evolutions include cases that form a black hole from regular initial data.

2 General Formulation

We decompose the spacetime metric ${}^{(4)}g_{\mu\nu}$ in ADM form,

$$ {}^{(4)}g = -N^2 dt^2 + g_{ij}(dx^i + X^i dt)(dx^j + X^j dt), \tag{1} $$

where g_{ij} is the induced metric on the $t = \text{const}$ slices, N is the lapse function and X the shift vector. The conformal spacetime metric ${}^{(4)}\gamma_{\mu\nu} = \Omega^{2(4)}g_{\mu\nu}$ is decomposed in a similar way,

$$ {}^{(4)}\gamma = -\tilde{N}^2 dt^2 + \gamma_{ij}(dx^i + X^i dt)(dx^j + X^j dt), \tag{2} $$

where we identify $\gamma_{ij} = \Omega^2 g_{ij}$ and $\tilde{N} = \Omega N$. The unit timelike normals of the physical and conformal spacetimes are related via $n^\mu = \Omega \tilde{n}^\mu$. The extrinsic curvature of the slices is defined as

$$ K_{ij} = -\tfrac{1}{2}\mathscr{L}_n g_{ij}, \tag{3} $$

where $\mathscr{L}$ denotes the Lie derivative. We require constant mean curvature,

$$ g^{ij} K_{ij} \equiv -K = \text{const}, \tag{4} $$

with $K > 0$ so that the slices approach future null infinity $\mathscr{I}^+$. Our fundamental evolution variable is the traceless part of the ADM momentum

$$ \pi^{\text{tr}\, ij} = -\mu_g \left(g^{ik} g^{jl} - \tfrac{1}{3} g^{ij} g^{kl} \right) K_{kl}, \tag{5} $$

where $\mu_g = \sqrt{\det(g_{ij})}$.

Before continuing, we show how matter can be included in our formalism. We restrict ourselves to tracefree energy-momentum tensors, $g^{\mu\nu}T_{\mu\nu} = 0$. Examples of matter models satisfying this condition include Maxwell and Yang–Mills fields and the conformally coupled scalar field. The tracefree condition insures that the matter evolution equations are conformally invariant (and hence are regular at $\mathscr{I}^+$), in particular,

$$ {}^{(4)}\gamma^{\mu\nu\,(4)}\tilde{\nabla}_\mu \tilde{T}_{\nu\rho} = \Omega^{-4\,(4)}g^{\mu\nu\,(4)}\nabla_\mu T_{\nu\rho} = 0, \tag{6} $$

where we have introduced a conformally rescaled energy-momentum tensor $\tilde{T}_{\mu\nu} \equiv \Omega^{-2}T_{\mu\nu}$. For this conformal energy-momentum tensor we introduce the usual projections

$$ \tilde{\rho} \equiv \tilde{n}^\mu \tilde{n}^\nu \tilde{T}_{\mu\nu}, \quad \tilde{J}^i \equiv -\gamma^{i\mu}\tilde{n}^\nu \tilde{T}_{\mu\nu}, \quad \tilde{S}_{ij} \equiv \gamma_i{}^\mu \gamma_j{}^\nu \tilde{T}_{\mu\nu}, \quad \tilde{S} \equiv \gamma^{ij}\tilde{S}_{ij}. \tag{7} $$

We are now ready to write down the ADM equations. Let $\tilde{\nabla}$ denote the Levi–Civita connection of γ, $\tilde{R}_{ij}$ its Ricci tensor and $\tilde{R}$ the Ricci scalar. The (not generally constant) mean curvature of the slices in the *conformal* spacetime is denoted by $\tilde{K}$. The evolution equations are

$$ \begin{aligned} \mathscr{L}_{\tilde{n}}\gamma_{ij} &= 2\mu_\gamma^{-1}\gamma_{ik}\gamma_{jl}\pi^{\mathrm{tr}\,kl} - \tfrac{2}{3}\gamma_{ij}\tilde{K}, & (8)\\ \mathscr{L}_{\tilde{n}}\pi^{\mathrm{tr}\,ij} &= -2\mu_\gamma^{-1}\gamma_{kl}\pi^{\mathrm{tr}\,ik}\pi^{\mathrm{tr}\,jl} - \tfrac{2}{3}\Omega^{-1}K\pi^{\mathrm{tr}\,ij} & \\ &\quad + \mu_\gamma\left[\tilde{N}^{-1}\tilde{\nabla}^i\tilde{\nabla}^j\tilde{N} - \tilde{R}^{ij} - 2\Omega^{-1}\tilde{\nabla}^i\tilde{\nabla}^j\Omega + \kappa\Omega^2\tilde{S}^{ij}\right]^{\mathrm{tr}}. & (9) \end{aligned} $$

The Hamiltonian and momentum constraints read

$$ \begin{aligned} 0 &= -4\Omega\tilde{\nabla}^i\tilde{\nabla}_i\Omega + 6\gamma^{ij}\Omega_{,i}\Omega_{,j} - \Omega^2\tilde{R} - \tfrac{2}{3}K^2 & \\ &\quad + \Omega^2\mu_\gamma^{-2}\gamma_{ik}\gamma_{jl}\pi^{\mathrm{tr}\,ij}\pi^{\mathrm{tr}\,kl} + 2\kappa\Omega^4\tilde{\rho}, & (10)\\ 0 &= \tilde{\nabla}_j(\Omega^{-2}\pi^{\mathrm{tr}\,ij}) + \kappa\mu_\gamma\tilde{J}^i. & (11) \end{aligned} $$

We also have an elliptic equation for the lapse arising from the constant mean curvature condition (4),

$$ \begin{aligned} 0 &= -\Omega^2\tilde{\nabla}^i\tilde{\nabla}_i\tilde{N} + 3\Omega\gamma^{ij}\tilde{N}_{,i}\Omega_{,j} - \tfrac{3}{2}\tilde{N}\gamma^{ij}\Omega_{,i}\Omega_{,j} + \tfrac{1}{6}\tilde{N}K^2 & \\ &\quad - \tfrac{1}{4}\tilde{N}\Omega^2\tilde{R} + \tfrac{5}{4}\tilde{N}\Omega^2\mu_\gamma^{-2}\gamma_{ik}\gamma_{jl}\pi^{\mathrm{tr}\,ij}\pi^{\mathrm{tr}\,kl} + \tfrac{1}{2}\kappa\tilde{N}\Omega^4(\tilde{S} + 2\tilde{\rho}). & (12) \end{aligned} $$

In [2] we fixed the spatial coordinates by imposing a (spatial) harmonic gauge condition. However, other choices are possible; for example, in Sects. 3 and 4 we use coordinates adapted to the symmetry. There is also a residual conformal gauge freedom inherent in the decomposition $\gamma_{\mu\nu} = \Omega^2 g_{\mu\nu}$. In [2] we fixed this by requiring the conformal scalar curvature $\tilde{R}$ to be constant. For the explicit forms of the conformal metrics used in Sects. 3 and 4 there is no remaining conformal gauge freedom.

The evolution equation (9) is formally singular at $\mathscr{I}^+$, where $\Omega = 0$. However in [2] we showed how the offending terms can in fact be evaluated at $\mathscr{I}^+$ in a regular way. This makes use of the fact that the constraints (10)–(11) and the CMC slicing condition (12) are also degenerate at $\mathscr{I}^+$. On a given spatial slice, we choose spherical polar coordinates such that the cut of the slice with $\mathscr{I}^+$ corresponds to $r = r_+ =$ const. We expand the fields in finite Taylor series in r about r_+ and substitute them in the degenerate elliptic equations. Thus we obtain the first three radial derivatives of Ω and the zeroth and first radial derivative of $\pi^{\mathrm{tr}\,ri}$ at $\mathscr{I}^+$. With this information we can evaluate the formally singular terms in the evolution equation (9) explicitly, provided that necessary conditions for smoothness of $\mathscr{I}^+$ are satisfied. These include the condition that $\mathscr{I}^+$ be shear-free and were obtained earlier in [3]. We show that these regularity conditions are preserved under the time evolution. While our analysis in [2] assumed vacuum, it is easy to see that it is unaffected by the addition of the matter sources, as will be shown in a forthcoming paper [4].

3 Axisymmetric Vacuum Gravity

The first numerical implementation [5] of the scheme presented in Sect. 2 assumed vacuum and axisymmetry. The spatial conformal metric is written in quasi-isotropic coordinates as

$$\gamma = e^{2\eta \sin\theta}(dr^2 + r^2 d\theta^2) + r^2 \sin^2\theta \, d\phi^2, \tag{13}$$

where η is a function of t, r and θ only, $\partial/\partial\phi$ being the Killing vector. Preservation of this form of the metric under the time evolution implies a first-order elliptic system for the shift vector similar to the Cauchy–Riemann equations.

The numerical implementation is based on fourth-order finite differences on a logically Cartesian grid in r and θ. The grid is allowed to be non-uniform in r in order to better resolve the steep gradients occurring near the horizon of the black hole spacetimes we consider. We use black hole excision, i.e. the inner boundary is placed just inside the horizon. This boundary is spacelike, so evolution equations do not require any boundary conditions there. The outer boundary is placed at $\mathscr{I}^+$, where the regularised form of the evolution equations is used, as outlined at the end of the previous section. One-sided derivatives are used at both boundaries. The constraint equations, CMC slicing condition and spatial gauge condition are solved at each time step using a nonlinear multigrid solver. The evolution equations are integrated in time using the method of lines with a fourth-order Runge–Kutta method.

As a first test problem, we evolve Schwarzschild spacetime. We use the Schwarzschild metric in constant-mean-curvature coordinates derived in [6] with parameters $M = 1$, $K = \frac{1}{2}$ and $C = 2$. The Schwarzschild solution has a flat spatial conformal metric, $\eta = 0$ in (13). We were able to evolve this solution for times as long as $10^3 M$ (and potentially longer) without any signs of instability, with approximate fourth-order convergence as expected.

Next, we include a gravitational wave perturbation. For this we choose η to be a Gaussian centred at $r = 0.5$ with width $\sigma = 0.05$ and amplitude $A = 10^{-4}$, initially at rest. For comparison, the black hole horizon is at $r = 0.0635$ and $\mathscr{I}^+$ is at $r = 1$.

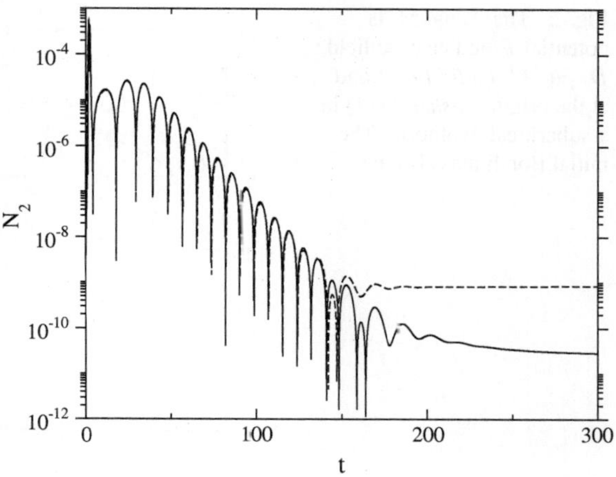

Fig. 1 $\ell = 2$ contribution to the Bondi news function for a gravitationally perturbed Schwarzschild black hole (mass $M = 1$) as a function of time. Numerical results for two different resolutions are shown, $(N_r, N_\theta) = (64, 8)$ (*dashed line*) and $(128, 16)$ (*solid line*)

We stress that this perturbation is evolved using the full nonlinear Einstein equations rather than linearised theory.

We extract the gravitational radiation emitted by the system by evaluating the Bondi news function [7] at $\mathscr{I}^+$ in Fig. 1. The quasi-normal mode ringing phase is clearly visible. The decay rate and frequency are consistent with the analytical result from linearised theory. At later times the numerical solution has not yet converged for the resolutions used here so we are currently unable to resolve the expected power-law tail.

4 Spherically Symmetric Einstein–Yang–Mills

In this section, we include matter in the form of a Yang–Mills field. This is conformally invariant; we choose to work in the conformal spacetime here. The energy-momentum tensor is

$$\tilde{T}_{\mu\nu} = \tilde{F}^{(a)}_{\mu\rho}\tilde{F}_{\nu}{}^{\rho(a)} - \tfrac{1}{4}{}^{(4)}\gamma_{\mu\nu}\tilde{F}^{(a)}_{\rho\sigma}\tilde{F}^{\rho\sigma(a)}, \tag{14}$$

where the field-strength tensor $\tilde{F}^{(a)}_{\mu\nu}$ is given in terms of the connection $\tilde{A}^{(a)}_{\mu}$ by

$$\tilde{F}^{(a)}_{\mu\nu} = \partial_\mu \tilde{A}^{(a)}_{\nu} - \partial_\nu \tilde{A}^{(a)}_{\mu} + f_{abc}\tilde{A}^{(b)}_{\mu}\tilde{A}^{(c)}_{\nu}. \tag{15}$$

Greek indices refer to the internal Yang–Mills gauge group, and the symbol f_{abc} is totally antisymmetric. Here we choose the gauge group to be SU(2), so Greek indices range over 1, 2, 3 and we may write $f_{abc} = g[abc]$, where g is the Yang–Mills coupling constant (taken to be $g = -2$ in the following) and $[abc]$ is totally antisymmetric with $[123] = 1$.

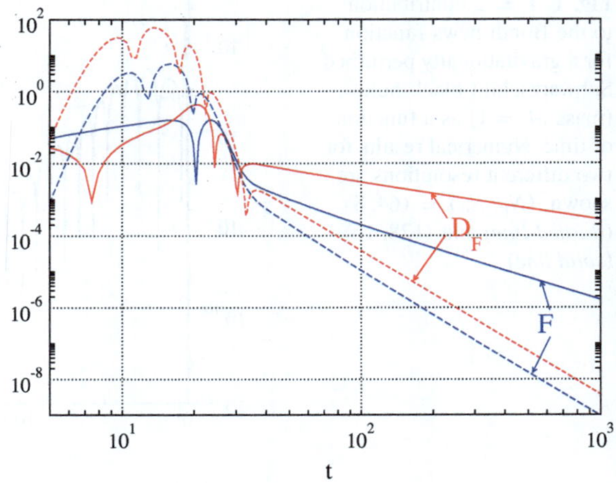

Fig. 2 The Yang–Mills potential F and electric field D_F at $\mathscr{I}^+$ (*solid lines*) and at the origin (*dashed lines*) in a subcritical evolution. The initial Bondi mass is 0.63

We now restrict ourselves to spherical symmetry. In this case we may always choose isotropic coordinates such that the spatial conformal metric is flat. This implies a first-order ordinary differential equation for the shift vector, which now has a radial component only. For the Yang–Mills connection we make the ansatz

$$\tilde{A}_i^{(a)} = [aij]x^j F(t, r), \quad \tilde{A}_0^{(a)} = 0. \tag{16}$$

A more general spherically symmetric ansatz will be considered in [4]. Energy-momentum conservation implies a nonlinear wave equation for F.

The numerical method used for this system is similar to the one described in Sect. 3. One difference is that in spherical symmetry, the traceless momentum $\pi^{\mathrm{tr}\,ij}$ only has one independent component, and we choose to solve the momentum constraint for it rather than its evolution equation. Hence the system is fully constrained and the only evolution equation used is the one for the Yang–Mills field. Also, our implementation allows for both regular and excised centres so that we may start from regular initial data until a black hole forms, which is then excised.

The Yang–Mills field F is taken to be a Gaussian centred at $r = 0.5$ with $\sigma = 0.05$ (again, $\mathscr{I}^+$ is at $r = 1$) and variable amplitude. The time derivative of F is chosen such that the pulse is approximately ingoing initially.

First we take the amplitude to be sufficiently small such that the field disperses. Figure 2 shows F at the origin and at $\mathscr{I}^+$ as a function of coordinate time. For the higher numerical resolution used in this $(1+1)$-dimensional simulation ($N_r = 4000$) the tail is now well resolved. At the origin (and in fact at any finite distance) the decay is approximately $F \sim t^{-4}$ whereas at $\mathscr{I}^+$, we find $F \sim t^{-2}$. This agrees with the results in [8], and the same decay exponents were found in the test field approximation [9].

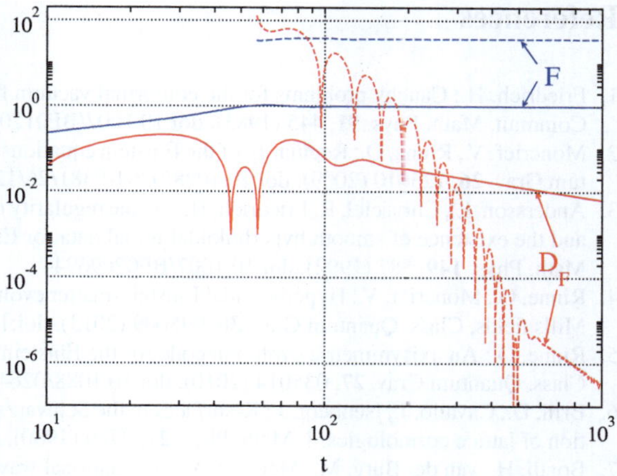

Fig. 3 The Yang–Mills potential F and electric field D_F at $\mathscr{I}^+$ (*solid lines*) and at the horizon (from when it forms, *dashed lines*) in a supercritical evolution. The initial Bondi mass is 3.0 and the final Bondi mass (which agrees with the final black hole mass) is 2.5

Let us also evaluate the electric field

$$\tilde{\mathscr{D}}^{i(a)} = \sqrt{-^{(4)}\gamma}\,\tilde{F}^{0i(a)} \equiv [aij]x^j D_F(t,r), \tag{17}$$

also shown in Fig. 2. While this decays at the same rate as F at the origin, it decays more slowly at $\mathscr{I}^+$, $D_F \sim t^{-1}$. This may seem surprising at first but can be explained by looking at the evolution equation for F,

$$F_{,t} = X^r F_{,r} + 2r^{-1} X^r F - \tilde{N} D_F. \tag{18}$$

At $\mathscr{I}^+$ the r-derivative of F appearing on the right-hand side must decay more slowly than F itself because F decays faster away from $\mathscr{I}^+$. From (18) we infer that D_F must also decay at the slower rate.

For sufficiently high amplitudes, the field collapses and a black hole forms (Fig. 3). Interestingly, the Yang–Mills potential F does not decay to zero in this case but approaches $F = 2/(gr^2)$, which is another vacuum state (the field strength tensor vanishes). The electric field shows the same power-law decay as in the subcritical evolution.

Acknowledgments O.R. gratefully acknowledges support from the German Research Foundation through a Heisenberg Fellowship and research grant RI 2246/2. V.M. was supported by NSF grant PHY-0963869 to Yale University.

References

1. Friedrich, H.: Cauchy problems for the conformal vacuum field equations in general relativity. Commun. Math. Phys. **91**, 445 (1983). doi:10.1007/BF01206015
2. Moncrief, V., Rinne, O.: Regularity of the Einstein equations at future null infinity. Class. Quantum Grav. **26**, 125010 (2009). doi:10.1088/0264-9381/26/12/125010
3. Andersson, L., Chruściel, P., Friedrich, H.: On the regularity of solutions to the Yamabe equation and the existence of smooth hyperboloidal initial data for Einstein's field equations. Commun. Math. Phys. **149**, 587 (1992). doi:10.1007/BF02096944
4. Rinne, O., Moncrief, V.: Hyperboloidal Einstein-matter evolution and tails for scalar and Yang–Mills fields, Class. Quantum Grav. **30**, 095009 (2013). doi:10.1088/0264-9381/30/9/095009
5. Rinne, O.: An axisymmetric evolution code for the Einstein equations on hyperboloidal slices. Class. Quantum Grav. **27**, 035014 (2010). doi:10.1088/0264-9381/27/3/035014
6. Brill, D., Cavallo, J., Isenberg, J.: K-surfaces in the Schwarzschild space–time and the construction of lattice cosmologies. J. Math. Phys. **21**, 2789 (1980). doi:10.1063/1.524400
7. Bondi, H., van der Burg, M., Metzner, A.: Gravitational waves in general relativity. VII. Waves from axi-symmetric isolated systems. Proc. R. Soc. London, Ser. A **269**, 21 (1962). doi:10.1098/rspa.1962.0161
8. Pürrer, M., Aichelburg, P.: Tails for the Einstein–Yang–Mills system. Class. Quantum Grav. **26**, 035004 (2009). doi:10.1088/0264-9381/26/3/035004
9. Zenginoğlu, A.: A hyperboloidal study of tail decay rates for scalar and Yang–Mills fields. Class. Quantum Grav. **25**, 175013 (2008). doi:10.1088/0264-9381/25/17/175013

Increase of Black Hole Entropy in Lanczos-Lovelock Gravity

Sudipta Sarkar

Abstract The striking similarity of the laws of black hole mechanics with thermodynamics was first established in case of general relativity (GR). A natural question is to ask whether this analogy is a peculiar property of GR or a robust feature of any generally covariant theory of gravity. We study this question in the context of Lanczos-Lovelock gravity and provide a proof of classical quasi-stationary second law.

1 Introduction

General relativity (GR), being quantum mechanically non-renormalizable, may make sense as a Wilsonian effective theory working perturbatively in powers of the dimensionless small parameter $G\,(Energy)^{D-2}$, where G is the D-dimensional Newton's constant. Then the Einstein-Hilbert Lagrangian is the lowest order term (other than the cosmological constant) in a derivative expansion of generally covariant actions for a metric theory, and the presence of higher curvature terms is presumably inevitable. In general, the specific form of these terms will depend on the detailed features of the quantum gravity model. Still, from a purely classical point of view, a natural modification of the Einstein-Hilbert action is to include terms preserving the diffeomorphism invariance and still leading to an equation of motion containing no more than second order time derivatives. Interestingly, this generalization is unique, [1, 2] and goes by the name of Lanczos-Lovelock gravity. Lanczos-Lovelock gravity is free from perturbative ghost [3] and leads to a well-defined initial value formalism [4]. The lowest order Lanczos-Lovelock correction term in space time dimensions $D > 4$, namely the Gauss-Bonnet term, also appears as a low energy α' correction in case of heterotic string theory [3, 5]. Hence, it is interesting to pursue various

S. Sarkar (✉)
The Institute of Mathematical Sciences, Chennai, India
e-mail: sudiptas@imsc.res.in

J. Bičák and T. Ledvinka (eds.), *Relativity and Gravitation*,
Springer Proceedings in Physics 157, DOI: 10.1007/978-3-319-06761-2_26,

classical and semi-classical properties of Lanczos-Lovelock gravity. For example, the striking similarity of the laws of black hole mechanics with thermodynamics was first established in case of general relativity [6] and a natural question is to ask whether this analogy is a peculiar property of GR or a robust feature of any generally covariant theory of gravity. Studying the properties of black holes in a general Lanczos-Lovelock theory may provide a partial answer to this important question.

The equilibrium state version of first law for black holes was established by Wald and collaborators [7, 8] for any arbitrary diffeomorphism invariant theory of gravity. The entropy of the black hole can be expressed as a local geometric quantity integrated over a space-like cross section of the horizon and is associated with the Noether charge of Killing isometry that generates the horizon.

Implicit in the investigations which uses the Wald entropy in these theories is the assumption that the entropy associated with a horizon behaves like ordinary thermodynamic entropy. But, the equilibrium state version of first law for black holes, established by Wald and collaborators [7, 8] requires the existence of a stationary black hole with regular bifurcation surface. As a result, from the equilibrium state version of first law, it is not immediately clear whether the Wald entropy always increases under physical processes, except for black holes in GR, in which the "area theorem" asserts that area of a black hole can not decrease in any process provided null energy condition holds for the matter fields [9]. The area theorem, in turn, follows from Raychaudhuri equation and crucially depends on the contracted Einstein's equation $R_{ab}k^a k^b = 8\pi\, T_{ab}k^a k^b$ where k^a is the tangent to the horizon. Since the entropy of black holes is no longer proportional to area in Lanczos-Lovelock models of gravity, there is no obvious assurance that the entropy still obeys an increase theorem. As a result, the question of validity of the second law of black hole thermodynamics for arbitrary theory of gravity remains an unresolved issue. Except for the case of $f(R)$-gravity [10], there is no proof of the analog of Hawking's area theorem beyond GR. In the quasi-stationary case, an argument for second law valid for all diffeomorphism invariant gravity theories was given in [10]; but it is based on the assumption that the stationary comparison version of the first law implies the physical process version for quasi-stationary processes.

For the thermodynamic interpretation to be valid, we would expect horizon entropy to increase when a black hole in the Lanczos-Lovelock model participates in some physical process, like, e.g., accretion of matter. Recently, a direct proof of the physical process version of first law is proposed for Einstein-Gauss-Bonnet (EGB) gravity [11] which establishes that the net change of black hole entropy during a physical process is positive as long as matter satisfies null energy condition.

Here, we investigate this question for general Lanczos-Lovelock models and show that during a physical process, the Wald entropy of stationary black holes in general Lanczos-Lovelock gravity monotonically increases provided the matter stress energy tensor obeys null energy condition. As a result, not only the net change of the entropy is positive, but the entropy is increasing at every cross section of the horizon. In this paper, we will present the essential idea and main steps of the calculations. For more details of the derivation, see Kolekar et al. [12].

Let us start with a brief review of the properties of stationary, non-extremal, Killing horizons. (We adopt the metric signature $(-,+,+,+,\dots)$ and our sign conventions are the same as those of [13].) In a D-dimensional spacetime, the event horizon is a null hypersurface $\mathscr{H}$ parametrized by an affine parameter λ. The vector field $k^a = (\partial_\lambda)^a$ is tangent to the horizon and obeys geodesic equation. All $\lambda =$ constant slices are space-like and foliate the horizon. Any point p on such slices has coordinates $\{\lambda, x^A\}$ where x^A, $(A = 2, \dots, D)$ are the coordinates of a point on $\lambda = 0$ slice connected with p by a horizon generator. We can construct a basis with the vector fields, $\{k^a, l^a, e^a_A\}$ where l^a is a second null vector such that $l^a k_a = -1$. The induced metric on any slice is $\gamma_{ab} = g_{ab} + 2k_{(a}l_{b)}$ and $k^a\gamma_{ab} = 0 = l^a\gamma_{ab}$. The change of the induced metric from one slice to another can be obtained from the metric evolution equation [13],

$$\mathscr{L}_k\gamma_{ab} = 2\left(\sigma_{ab} + \frac{\theta}{(D-2)}\gamma_{ab}\right), \tag{1}$$

where σ_{ab} is the shear and θ is the expansion of the horizon. If the event horizon is also a Killing horizon,[1] i.e. the horizon generators are the orbits of a Killing field $\xi^a = (\partial/\partial v)^a$, which is null on the horizon, then the surface gravity κ of the horizon is defined as $\xi^a\nabla_a\xi^b = \kappa\,\xi^b$. For stationary spacetimes with a Killing horizon, both the expansion and shear vanish and using Raychaudhuri equation and the evolution equation for shear, we obtain [13, 14] that on the horizon,

$$\xi^a\xi^c\gamma_i^b\gamma_k^d R_{abcd} = 0 = R_{ab}\xi^a\xi^b = 0, \tag{2}$$

and

$$\xi^a\gamma_i^b\gamma_j^c\gamma_k^d R_{abcd} = 0. \tag{3}$$

We would like to emphasize that in order to derive these relationships, we have only used the fact that the horizon is a Killing horizon with zero expansion and shear without assuming any further symmetry.

We would like to consider the situation when a stationary black hole is perturbed by a weak matter stress energy tensor and ultimately settles down to a stationary state in the asymptotic future. Since the black hole is stationary in the asymptotic future, the vector field ξ^a is an exact Killing vector at late times. The accretion process is assumed to be slow such that all changes of the dynamical fields are first order in some suitable bookkeeping parameter ϵ and that we can neglect all viscous effects. More specifically, we assume that, $\theta \sim \sigma_{ab} \sim \mathscr{O}(\epsilon)$.

In GR, a concrete example of such a physical process is a black hole of mass M slowly accreting matter for a finite time and ultimately settling down to a stationary

---

[1] Here we make an assumption, that the event horizon of a stationary black hole is also a Killing horizon with regular bifurcation surface. Although this is certainly true for GR, we are not aware of any proof for Lanczos-Lovelock gravity.

state. Then a linearized version of the Raychaudhuri equation gives,

$$\frac{d\theta}{d\lambda} \approx -R_{ab}k^a k^b = -8\,\pi\, T_{ab}k^a k^b, \tag{4}$$

where we have used Einstein's equation to get the second equality. If the matter stress tensor satisfies null energy condition, i. e. $T_{ab}k^a k^b \geq 0$, the rate of change of the expansion is negative on any slice prior to the asymptotic future. Since the expansion vanishes in the future, the generators must have positive expansion during the accretion process. As a result, the area is monotonically increasing in the physical process. Note that, the result is crucially dependent on the field equation. As a result, the monotonicity of the horizon area is only valid in case of GR. Our aim is to prove a same statement for the Wald entropy during a dynamical change of the black holes in Lanczos-Lovelock gravity.

We shall now turn our attention to the features of Lanczos-Lovelock gravity. As discussed before, a natural generalization of the Einstein-Hilbert Lagrangian is provided by the Lanczos-Lovelock Lagrangian, which is the sum of dimensionally extended Euler densities,

$$\mathscr{L}^D = \sum_{m=0}^{[D-1)/2]} \alpha_m \mathscr{L}_m^D, \tag{5}$$

where the α_m are arbitrary constants and $\mathscr{L}_m^D$ is the mth order Lanczos-Lovelock term given by,

$$\mathscr{L}_m^D = \frac{1}{16\pi} \sum_{m=0}^{[D-1)/2]} \frac{1}{2^m} \delta^{a_1 b_1 \ldots a_m b_m}_{c_1 d_1 \ldots c_m d_m} R^{c_1 d_1}_{\ \ a_1 b_1} \cdots R^{c_m d_m}_{\ \ a_m b_m}, \tag{6}$$

where R^{cd}_{ab} is the D dimensional curvature tensor and the generalized alternating tensor $\delta^{\cdots}_{\cdots}$ is totally anti-symmetric in both set of indices. The Einstein-Hilbert Lagrangian is a special case of (6) when $m = 1$. The field equation of Lanczos-Lovelock theory is, $G_{ab}/(16\pi) + \alpha_m E_{(m)ab} = (1/2)T_{ab}$ where,

$$E^i_{(m)j} = -\frac{1}{16\pi}\frac{1}{2^{m+1}} \delta^{i a_1 b_1 \ldots a_m b_m}_{j c_1 d_1 \ldots c_m d_m} R^{c_1 d_1}_{\ \ a_1 b_1} \cdots R^{c_m d_m}_{\ \ a_m b_m}, \tag{7}$$

and $m \geq 2$. For convenience, we have written the GR part (i.e. for $m = 1$) separately so that the GR limit can be easily verified by setting all α_m's to zero. Spherically symmetric black hole solutions in Lanczos-Lovelock gravity was derived in [15, 16] and the Wald entropy associated with a stationary Killing horizon is [17–19],

$$S = \frac{1}{4}\int \rho\,\sqrt{\gamma}\,dA, \tag{8}$$

where the entropy density

$$\rho = \left(1 + \sum_{m=2}^{[D-1)/2]} 16\pi m \alpha_m \,{}^{(D-2)}L_{(m-1)}\right). \tag{9}$$

The integration is over $(D-2)$-dimensional space-like cross-section of the horizon and ${}^{(D-2)}L_{(m-1)}$ is the intrinsic $(m-1)$th Lanczos-Lovelock scalar of the horizon cross-section. We would like to prove that this entropy always increases when a black hole is perturbed by a weak matter stress energy tensor of $\mathscr{O}(\epsilon)$ provided the matter obeys null energy condition.

The change in entropy is [10],

$$\Delta S = \frac{1}{4}\int_{\mathscr{H}} \left(\frac{d\rho}{d\lambda} + \theta\,\rho\right) d\lambda \,\sqrt{\gamma}\, dA. \tag{10}$$

We define a quantity Θ as

$$\Theta = \left(\frac{d\rho}{d\lambda} + \theta\,\rho\right). \tag{11}$$

In case of GR, Θ is equal to the expansion parameter of the null generators. But, in case of a general gravity theory, Θ is the rate of change of the entropy associated with a infinitesimal portion of horizon (see Jacobson et al. [10] for similar construction in $f(R)$ gravity). We would like to prove that given null energy condition holds, Θ is positive on any slice in a physical process.

In order to proceed, we would like to study the rate of change of Θ along the congruence using Raychaudhuri equation and the evolution equation of shear [13]. We are only interested in quantities first order in perturbation over a background stationary spacetime. Therefore, when we encounter a product of two quantities X and Y, to extract the part linear in perturbation, we will always express such a product as,

$$XY \approx X^{(B)}\,Y^{(P)} + X^{(P)}\,Y^{(B)}, \tag{12}$$

where $X^{(B)}$ is the value of the quantity X evaluated on the stationary background and $X^{(P)}$ is the perturbed value of X linear in perturbation. Note that, on the stationary background, Raychaudhuri equation demands $R^{(B)}_{ab}k^a k^b = 0$ and since $T^{(B)}_{ab}k^a k^b = 0$, we have $E^{(B)}_{(m)ab}k^a k^b = 0$. Also, to simplify the calculation, we use diffeomorphism freedom to make the null geodesic generators of the event horizon of the perturbed black hole coincide with the null geodesic generators of the background stationary black hole [20].

Using the perturbation scheme mentioned above and the evolution equation of θ and σ_{ab} to linear order as $d\theta/d\lambda \approx -R^{(P)}_{ab}k^a k^b$ and $d\sigma_{ab}/d\lambda \approx C^{(P)}_{acdb}k^c k^d$ and further using conditions (2) and (3) on the stationary background, the evolution equation of Θ to linear order in perturbation can be written as [12]

$$\frac{d\Theta}{d\lambda} = -8\pi\, T_{ab}k^a k^b + \mathcal{O}(\epsilon^2). \tag{13}$$

Equation (13) shows that if the null energy condition holds, the rate of change of Θ is always negative during a slow classical dynamical process (i.e. ignoring the terms which are higher order in the perturbation) which perturbs the black hole and leads to a new stationary state. Since the final state is assumed to be stationary, both θ and σ, and as a consequence, Θ vanish in the asymptotic future. Hence, we can use the same argument as with the expansion parameter in case of GR to conclude that Θ must be positive at every slice during the physical process. As a result, we conclude that the horizon entropy of black holes in Lanczos-Lovelock gravity is a monotonically increasing function during any quasi-stationary physical process, i.e.

$$\frac{dS}{d\lambda} \geq 0, \tag{14}$$

which is what we set out to prove.

In case of a dynamical scenario, it is possible to write down several candidates for the black hole entropy beyond GR [21], such that all the expressions have the same stationary limit. We have actually chosen a particular expression and the validity of (14) favors such a choice. In fact, in Ref. [8], a local and geometrical prescription for the entropy of dynamical black holes is proposed. This proposal is based on a boost invariant construction and agrees with the Wald's Noether charge formula for stationary black holes and their perturbations. Interestingly, for Lanczos-Lovelock gravity, the entropy expression used in this work matches the expression obtained from the boost invariant construction. Consequently, our result provides a strong justification in favor of the prescription for dynamical entropy as proposed in Ref. [8]. This may also be important to decide the right candidate for the entropy of non-stationary black holes for non-Lanczos-Lovelock gravity models. Also, one would like to relax the quasi-stationarity physical process assumption and calculate the full change of the Wald entropy along the horizon to understand the validity of classical second law for the Lanczos-Lovelock gravity.

The last point is related to the special status enjoyed by Lanczos-Lovelock models. The derivation presented here used identities which are very specific to Lanczos-Lovelock models and do not generalize to an arbitrary theory of gravity. Therefore, it would be worthwhile to find a general approach which can answer whether classical second law holds in a physical process for any diffeomorphism invariant gravity theory or applies to a special class of action functionals. This may be useful as a criterion to select a sub-class of diffeomorphism invariant actions as preferred theories where a consistent formulation of black hole thermodynamics is possible.

References

1. Lanczos, C.: A remarkable property of the Riemann-Christoffel tensor in four dimensions. Ann. Math. **39**, 842 (1938). doi:10.2307/1968467
2. Lovelock, D.: The Einstein tensor and its generalizations. J. Math. Phys. **12**, 498 (1971). doi:10.1063/1.1665613
3. Zwiebach, B.: Curvature squared terms and string theories. Phys. Lett. B **156**, 315 (1985). doi:10.1016/0370-2693(85)91616-8
4. Choquet-Bruhat, Y.: The Cauchy problem for stringy gravity. J. Math. Phys. **29**, 1891 (1988). doi:10.1063/1.527841
5. Sen, A.: Black hole entropy function, attractors and precision counting of microstates. Gen. Relativ. Gravit. **40**, 2249 (2008). doi:10.1007/s10714-008-0626-4
6. Bardeen, J., Carter, B., Hawking, S.: The four laws of black hole mechanics. Commun. Math. Phys. **31**, 161 (1973). doi:10.1007/BF01645742
7. Wald, R.: Black hole entropy is the Noether charge. Phys. Rev. D **48**, 3427 (1993). doi:10.1103/PhysRevD.48.R3427
8. Iyer, V., Wald, R.: Some properties of Noether charge and a proposal for dynamical black hole entropy. Phys. Rev. D **50**, 846 (1994). doi:10.1103/PhysRevD.50.846
9. Hawking, S.: Black holes in general relativity. Commun. Math. Phys. **25**, 152 (1972). doi:10.1007/BF01877517
10. Jacobson, T., Kang, G., Myers, R.: Increase of black hole entropy in higher curvature gravity. Phys. Rev. D **52**, 3518 (1995). doi:10.1103/PhysRevD.52.3518
11. Chatterjee, A., Sarkar, S.: Physical process first law and increase of horizon entropy for black holes in Einstein-Gauss-Bonnet gravity. Phys. Rev. Lett. **108**, 091301 (2012). doi:10.1103/PhysRevLett.108.091301
12. Kolekar, S., Padmanabhan, T., Sarkar, S.: Entropy increase during physical processes for black holes in Lanczos-Lovelock gravity. Phys. Rev. D **86**, 021501 (2012). doi:10.1103/PhysRevD.86.021501
13. Wald, R.: General Relativity. University of Chicago Press, Chicago (1984)
14. Vega, I., Poisson, E., Massey, R.: Intrinsic and extrinsic geometries of a tidally deformed black hole. Class. Quantum Gravit. **28**, 175006 (2011). doi:10.1088/0264-9381/28/17/175006
15. Boulware, D., Deser, S.: String generated gravity models. Phys. Rev. Lett. **55**, 2656 (1985). doi:10.1103/PhysRevLett.55.2656
16. Boulware, D., Deser, S.: Effective gravity theories with dilations. Phys. Lett B **175**, 409 (1986). doi:10.1016/0370-2693(86)90614-3
17. Visser, M.: Dirty black holes: entropy versus area. Phys. Rev. D **48**, 583 (1993). doi:10.1103/PhysRevD.48.583
18. Jacobson, T., Myers, R.: Black hole entropy and higher curvature interactions. Phys. Rev. Lett. **70**, 3684 (1993). doi:10.1103/PhysRevLett.70.3684
19. Visser, M.: Dirty black holes: entropy as a surface term. Phys. Rev. D **48**, 5697 (1993). doi:10.1103/PhysRevD.48.5697
20. Gao, S., Wald, R.: The "Physical process" version of the first law and the generalized second law for charged and rotating black holes. Phys. Rev. D **64**, 084020 (2001). doi:10.1103/PhysRevD.64.084020
21. Jacobson, T., Kang, G., Myers, R.: On black hole entropy. Phys. Rev. D **49**, 6587 (1994). doi:10.1103/PhysRevD.49.6587

On the Stability Operator for MOTS and the 'Core' of Black Holes

José M. M. Senovilla

Abstract Small deformations of marginally (outer) trapped surfaces are considered by using their stability operator. In the case of spherical symmetry, one can use these deformations on any marginally trapped round sphere to prove several interesting results. The concept of 'core' of a black hole is introduced: it is a minimal region that one should remove from the spacetime in order to get rid of all possible closed trapped surfaces. In spherical symmetry one can prove that the spherical marginally trapped tube is the boundary of a core. By using a novel formula for the principal eigenvalue of the stability operator, I will argue how to pursue similar results in general black-hole spacetimes.

1 Introduction: Basic Concepts and Notation

Let S denote a closed marginally outer trapped surface (MOTS) in the spacetime $(\mathscr{V}, g)$. This means that the outer null expansion vanishes $\theta_{\mathbf{k}} = 0$, where the two future-pointing null vector fields orthogonal to S are denoted by ℓ and $\mathbf{k}$, the latter is declared to be outer, and we set $\ell^\mu k_\mu = -1$ as a convenient normalization. If in addition the other null expansion is non-positive ($\theta_\ell \leq 0$), then S is called a marginally trapped surface (MTS). I will also use the concept of outer trapped surface (OTS) when just $\theta_k < 0$ and of future trapped surface (TS) if both expansions are negative: $\theta_k < 0$ and $\theta_\ell < 0$. A hypersurface foliated by M(O)TS is called a marginally (outer) trapped tube, abbreviated to M(O)TT. For further explanations check [1–5].

J. M. M. Senovilla (✉)
Física Teórica, Universidad del País Vasco, Apartado 644, 48080 Bilbao, Spain
e-mail: josemm.senovilla@ehu.es

J. Bičák and T. Ledvinka (eds.), *Relativity and Gravitation*,
Springer Proceedings in Physics 157, DOI: 10.1007/978-3-319-06761-2_27,

1.1 Stability Operator for MOTS

As proven in [6, 7], the variation of the vanishing expansion $\delta_{f\mathbf{n}}\theta_{\mathbf{k}}$ along any normal direction $f\mathbf{n}$ such that $k_\mu n^\mu = 1$ reads

$$\delta_{f\mathbf{n}}\theta_{\mathbf{k}} = -\Delta_S f + 2s^B \overline{\nabla}_B f + f\left(K_S - s^B s_B + \overline{\nabla}_B s^B - G_{\mu\nu}k^\mu \ell^\nu\big|_S - \frac{n^\rho n_\rho}{2} W\right), \tag{1}$$

where K_S is the Gaussian curvature on S, Δ_S its Laplacian, $G_{\mu\nu}$ the Einstein tensor, $\overline{\nabla}$ the covariant derivative on S, $s_B = k_\mu e^\sigma_B \nabla_\sigma \ell^\rho$ (with $\mathbf{e}_B$ the tangent vector fields on S), and

$$W \equiv G_{\mu\nu}k^\mu k^\nu\big|_S + \sigma^2,$$

with σ^2 the shear scalar of $\mathbf{k}$ at S. Obviously $W \geq 0$ whenever $G_{\mu\nu}k^\mu k^\nu\big|_S \geq 0$ (for instance if the null convergence condition holds [8]). Under this hypothesis, $W = 0$ can only happen if $G_{\mu\nu}k^\mu k^\nu\big|_S = \sigma^2 = 0$. This leads to Isolated Horizons [2], and I shall assume $W > 0$ throughout.

Note that the direction $\mathbf{n}$ is selected by fixing its norm:

$$\mathbf{n} = -\ell + \frac{n_\mu n^\mu}{2}\mathbf{k}, \tag{2}$$

and observe also that the causal character of $\mathbf{n}$ is totally unrestricted.

The right-hand side in formula (1) defines a differential operator $L_{\mathbf{n}}$ acting (linearly) on the function f: $\delta_{f\mathbf{n}}\theta_{\mathbf{k}} \equiv L_{\mathbf{n}} f$. $L_{\mathbf{n}}$ is an elliptic operator on S, called *the stability operator* for the MOTS S in the normal direction $\mathbf{n}$. $L_{\mathbf{n}}$ is not self-adjoint in general, however it has a real principal eigenvalue $\lambda_{\mathbf{n}}$, and the corresponding (real) eigenfunction $\phi_{\mathbf{n}}$ can be chosen to be positive on S [6, 7]. The (strict) stability of the MOTS S is ruled by the (positivity) non-negativity of the principal eigenvalue $\lambda_{\mathbf{n}}$ [6, 7].

2 Spherically Symmetric Spacetimes

In advanced coordinates, spherically symmetric spacetimes have the line-element

$$ds^2 = -e^{2\alpha}\left(1 - \frac{2m}{r}\right)dv^2 + 2e^\alpha dv dr + r^2 d\Omega^2,$$

where α and m are functions of v and r. For each round sphere defined by $\{r, v\} =$consts., its future null normals are

$$\ell = -e^{-\alpha}\partial_r, \quad \mathbf{k} = \partial_v + \frac{1}{2}\left(1 - \frac{2m}{r}\right)e^\alpha \partial_r,$$

so that their null expansions are:

$$\theta_{\mathbf{k}}^{sph} = \frac{e^{\alpha}}{r}\left(1 - \frac{2m}{r}\right), \quad \theta_{\ell}^{sph} = -\frac{2e^{-\alpha}}{r}.$$

The apparent 3-horizon A3H : $r - 2m(r, v) = 0$ ($\Leftrightarrow \theta_{\mathbf{k}}^{sph} = 0$) is an MTT. A3H is actually the only *spherically symmetric* MTT: the only spherically symmetric hypersurface foliated by MTSs—be they round spheres or not [9].

The round spheres are untrapped if $r > 2m$, and trapped if $r < 2m$. One can further prove [9] that any closed trapped surface cannot be fully contained in a region with $r \geq 2m$, so that all of them must intersect the region $\{r < 2m\}$. However, how much must a TS penetrate into $\{r < 2m\}$?

Let $\varsigma \subset$ A3H be any MT round sphere (i.e., $\theta_{\mathbf{k}}^{sph} = 0$) defined by $r = r_{\varsigma} =$const. The variation $\delta_{f\mathbf{n}}\theta_{\mathbf{k}}^{sph}$ along normal directions simplifies drastically in this case, because $\sigma^2 = 0$ ($\mathbf{k}$ is shear-free) and $s_B = 0$. In other words, most of the terms in the variation formula vanish and the variation simplifies to

$$\delta_{f\mathbf{n}}\theta_{\mathbf{k}}^{sph} = -\Delta_{\varsigma} f + f\left(\frac{1}{r_{\varsigma}^2} - G_{\mu\nu}k^{\mu}\ell^{\nu} - \frac{1}{2}n_{\rho}n^{\rho}G_{\mu\nu}k^{\mu}k^{\nu}\right).$$

Selecting $f =$constant, the vector $\mathbf{n}$ such that the expression enclosed in brackets vanishes produces no variation on $\theta_{\mathbf{k}}^{sph}$, meaning that $\mathbf{n}$ is tangent to the A3H simply leading to other marginally trapped round spheres on A3H. Let us call such a vector field $\mathbf{m}$, so that $\mathbf{m} = -\ell + \frac{m_{\mu}m^{\mu}}{2}\mathbf{k}$, with $\frac{1}{r_{\varsigma}^2} - G_{\mu\nu}k^{\mu}\ell^{\nu}\big|_{\varsigma} - \frac{m_{\rho}m^{\rho}}{2}G_{\mu\nu}k^{\mu}k^{\nu}\Big|_{\varsigma} = 0$, characterizes A3H.

Consider now the parts of A3H with $G_{\mu\nu}k^{\mu}k^{\nu} > 0$ (i.e. $W > 0$). From the properties of $\mathbf{m}$ one deduces that the perturbation along $f\mathbf{n}$ will enter into the region with trapped round spheres (that is, $\{r < 2m\}$) at points with $f(n_{\mu}n^{\mu} - m_{\mu}m^{\mu}) > 0$. Note that

$$\left(G_{\rho\sigma}k^{\rho}k^{\sigma}|_{\varsigma}\right) f\left(n_{\mu}n^{\mu} - m_{\mu}m^{\mu}\right) = -2\left(\Delta_{\varsigma} f + \delta_{f\mathbf{n}}\theta_{\mathbf{k}}^{sph}\right). \quad (3)$$

In order to construct examples of TSs which lie partly in $\{r > 2m\}$, consider the case $n_{\mu}n^{\mu} - m_{\mu}m^{\mu} > 0$. For this choice the deformed surface enters the region $\{r < 2m\}$ at points with $f > 0$. Setting $f \equiv a_0 + \tilde{f}$ for some as yet undetermined function $\tilde{f}$ and a constant a_0, (3) can be split into two parts

$$\left(G_{\rho\sigma}k^{\rho}k^{\sigma}|_{\varsigma}\right) a_0\left(n_{\mu}n^{\mu} - m_{\mu}m^{\mu}\right) + 2\delta_{f\mathbf{n}}\theta_{\mathbf{k}}^{sph} = 0,$$

$$\frac{1}{2}\left(G_{\rho\sigma}k^{\rho}k^{\sigma}|_{\varsigma}\right)\left(n_{\mu}n^{\mu} - m_{\mu}m^{\mu}\right) = -\frac{\Delta_{\varsigma}\tilde{f}}{\tilde{f}} > 0.$$

By our assumptions the first of these implies that $\delta_{f\mathbf{n}}\theta_{\mathbf{k}}^{sph} < 0$ if $a_0 > 0$, so that the deformed surface will be trapped. The second, in turn, is a mild restriction on the function $\tilde{f}$. A simple solution is to choose $\tilde{f}$ to be an eigenfunction of the Laplacian Δ_ς, say $\tilde{f} = c_l P_l$ for a fixed $l \in \mathbb{N}$ and constant c_l, where P_l are the Legendre polynomials.

Even more interestingly, we are ready to answer the question of how small the fraction of any closed TS that extends outside $\{r < 2m\}$ can be made. The aim is to produce a C^2 function $\tilde{f}$ defined on the sphere (i) obeying the inequality $-\dfrac{\Delta_\varsigma \tilde{f}}{\tilde{f}} > 0$, and (ii) positive only in a region that we can make arbitrarily small. By choosing a sufficiently small constant a_0 requirement (ii) implies that the part of the surface extending outside $\{r > 2m\}$ can be made arbitrarily small. To find $\tilde{f}$ explicitly, introduce stereographic coordinates $\{\rho, \varphi\}$ on the sphere, so that the Laplacian takes the form $\Delta_\varsigma = \Omega^{-1}\left(\partial_\rho^2 + \frac{1}{\rho}\partial_\rho + \frac{1}{\rho^2}\partial_\varphi^2\right)$, $\Omega = \frac{4r_\varsigma^2}{(1+\rho^2)^2}$. Then, a solution for $\tilde{f}$ is the axially symmetric function

$$\tilde{f}(\rho) = \begin{cases} c_1\left(e^{\frac{1}{2a}(2a-\rho^2)} - 1\right) & \rho^2 < 4a \\ \dfrac{8c_1 a}{e}\dfrac{1}{\rho^2} - c_1(1+e^{-1}) & \rho^2 > 4a. \end{cases} \tag{4}$$

This function is C^2 (and can be further smoothed if necessary), and it is positive only if $\rho^2 < 2a$, that is on a disk surrounding the origin (the pole) whose size can be chosen at will. It obeys

$$-\frac{\Delta_\varsigma \tilde{f}}{\tilde{f}} = \begin{cases} \dfrac{\Omega^{-1}}{a^2}\dfrac{2a-\rho^2}{1-e^{-\frac{1}{2a}(2c-\rho^2)}} & \rho^2 < 4a \\ \dfrac{32a\Omega^{-1}}{\rho^4}\dfrac{\rho^2}{(e+1)\rho^2 - 8a}, & \rho^2 > 4a, \end{cases}$$

which is always larger than zero. Thus we have proven the following important and perhaps surprising result [9].

Theorem 1 (Bengtsson and JMMS 2011) *In spherically symmetric spacetimes, there are closed f-trapped surfaces (topological spheres) penetrating both sides of the (non-isolated part of the) apparent 3-horizon* $A3H \backslash A3H^{iso}$ with arbitrarily small portions *outside the region* $\{r > 2m\}$.

3 Cores

The (future)-trapped region $\mathscr{T}$ of a spacetime is defined as the set of points $x \in \mathscr{V}$ such that x lies on a closed (future) TS [9]. This is a space–time concept, not to be confused with the outer trapped region within spacelike hypersurfaces, which is defined as the union of the interiors of all (bounding) OTS in the given hypersurface [6, 10]. I denote by $\mathscr{B}$ the boundary of the future trapped region $\mathscr{T}$: $\mathscr{B} \equiv \partial \mathscr{T}$.

Closed TSs are clairvoyant, highly non-local objects [2, 9]. They cross MTTs and even enter flat portions of the space–time [9, 11, 12]. In conjunction with the non-uniqueness of MTTs [1, 9], this poses a fundamental puzzle for the physics of black holes. Although several solutions can be pursued, a popular one is trying to define a preferred MTT. Hitherto, though, there has been no good definition for that. We have put forward a novel strategy [9]. The idea is based on the simple question: *what part of the spacetime is absolutely indispensable for the existence of the black hole?*

Definition 1 (*Cores of Black Holes*) A region $\mathscr{Z}$ is called the *core* of the f-trapped region $\mathscr{T}$ if it is a minimal closed connected set that needs to be removed from the spacetime in order to get rid of all closed f-trapped surfaces in $\mathscr{T}$, and such that any point on the boundary $\partial \mathscr{Z}$ is connected to $\mathscr{B} = \partial \mathscr{T}$ in the closure of the remainder.

- Here, "minimal" means that there is no other set $\mathscr{Z}'$ with the same properties and properly contained in $\mathscr{Z}$.
- The final technical condition states that the excised space–time $(\mathscr{V} \backslash \mathscr{Z}, g)$ has the property that $\forall x \in \mathscr{V} \backslash \mathscr{Z} \cup \partial \mathscr{Z}$ there is continuous curve $\gamma \subset \mathscr{V} \backslash \mathscr{Z} \cup \partial \mathscr{Z}$ joining x and $\mathscr{B}$ (γ can have zero length if $\mathscr{B} \cap \partial \mathscr{Z} \neq \emptyset$). The reason why this is needed is explained in [9].

In spherically symmetric spacetimes one can prove that the region $\mathscr{Z} \equiv \{r \leq 2m\}$ is a core [9]. The proof is founded on the previous Theorem 1. It should be observed that this is an interesting and maybe deep result, for the concept of core is global and requires full knowledge of the future while A3H is quasi-local. It is thus surprising that A3H $= \partial \mathscr{Z}$.

Actually, one can further prove that in spherically symmetric spacetimes, $\mathscr{Z} = \{r \leq 2m\}$ are the only spherically symmetric cores of $\mathscr{T}$. Therefore, $\partial \mathscr{Z} =$ A3H are the only spherically symmetric boundaries of a core. Nevertheless, there exist non-spherically symmetric cores of the f-trapped region in spherically symmetric spacetimes. This implies the non-uniqueness of cores, and of their boundaries [9]. Still, the identified core $\mathscr{Z} = \{r \leq 2m\}$ might be unique in the sense that its boundary $\partial \mathscr{Z} =$ A3H is a MTT: we do not know whether other cores share this property or not [9].

To study whether or not Theorem 1 can be generalized to general situations, thereby providing the possibility of selecting a unique MTT as the boundary of a selected core, consider the family of operators, parameterized by a function $z \in C^\infty(S)$, with a similar structure as that of $L_{\mathbf{n}}$: $L_z f = -\Delta_S f + 2 s^B \overline{\nabla}_B f + z f$. Each L_z has a principal *real* eigenvalue λ_z—which depends on z—and the corresponding eigenfunction $\phi_z > 0$. For any given z one easily gets

$$\oint_S L_z f = \oint_S \left(2 s^B \overline{\nabla}_B f + z f\right) = \oint_S \left(z - 2 \overline{\nabla}_B s^B\right) f,$$

in particular for the principal eigenfunction

$$\lambda_z \oint_S \phi_z = \oint_S \left(z - 2\overline{\nabla}_B s^B\right) \phi_z.$$

This provides

1. a formula for the principal eigenvalue

$$\lambda_z = \frac{\oint_S \left(z - 2\overline{\nabla}_B s^B\right) \phi_z}{\oint_S \phi_z}. \tag{5}$$

2. bounds for λ_z

$$\min_S \left(z - 2\overline{\nabla}_B s^B\right) \leq \lambda_z \leq \max_S \left(z - 2\overline{\nabla}_B s^B\right). \tag{6}$$

3. and that $\lambda_z - \left(z - 2\overline{\nabla}_B s^B\right)$ must vanish somewhere on S for all z.

On any MOTS, varying $\theta_{\mathbf{k}} = 0$ along the direction $\phi_z \mathbf{n}$ one derives

$$\frac{L_{\mathbf{n}}\phi_z}{\phi_z} = \lambda_z - z + K_S - s^B s_B + \overline{\nabla}_B s^B - G_{\mu\nu}k^\mu \ell^\nu\big|_S - \frac{n^\rho n_\rho}{2} W.$$

Thus, *whenever* $W \neq 0$ on S, one can choose for any z a variation vector $\mathbf{m}_z = -\ell + M_z \mathbf{k}$ such that the righthand side vanishes

$$M_z = \frac{m_z^\rho m_{z\rho}}{2} = \frac{1}{W}\left(\lambda_z - z + K_S - s^B s_B + \overline{\nabla}_B s^B - G_{\mu\nu}k^\mu \ell^\nu\big|_S\right), \tag{7}$$

hence $\delta_{\phi_z \mathbf{m}_z} \theta_{\mathbf{k}} = 0$. Observe that this $\mathbf{m}_z$ depends on the chosen function z. The general variation of $\theta_{\mathbf{k}}$ along $\mathbf{m}_z$ reads

$$\delta_{f \mathbf{m}_z} \theta_{\mathbf{k}} = -\Delta_S f + 2s^B \overline{\nabla}_B f + f(z - \lambda_z) = (L_z - \lambda_z) f, \tag{8}$$

so that the stability operator $L_{\mathbf{m}_z}$ of S along $\mathbf{m}_z$ is simply $L_z - \lambda_z$ which obviously has a vanishing principal eigenvalue. The directions $\mathbf{m}_z$ define locally MOTTs including any given *stable* MOTS S [6, 7]. These MOTTs will generically be different for different z. In fact, given that $\forall z_1, z_2 \in C^\infty(S)$, $\mathbf{m}_{z_1} - \mathbf{m}_{z_2} = \frac{1}{W}\left(\lambda_{z_1} - z_1 - \lambda_{z_2} + z_2\right) \mathbf{k}$, one can easily prove that

$$\mathbf{m}_{z_1} = \mathbf{m}_{z_2} \Longleftrightarrow z_1 - z_2 = \text{const.}$$

Now, for any given z rewrite $\delta_{f\mathbf{n}} \theta_{\mathbf{k}} = L_{\mathbf{n}} f$ using (7) so that

$$\frac{W}{2} f\left(n^\rho n_\rho - m_z^\rho m_{z\rho}\right) = (L_z - \lambda_z) f - \delta_{f\mathbf{n}} \theta_{\mathbf{k}}. \tag{9}$$

Consider the particular function $z = 2\overline{\nabla}_B s^B$. This may be the natural generalization of the spherically symmetric MTT shown above. Observe that, for such a choice of z, and letting $L \equiv L_{2\overline{\nabla}_B s^B}$, its principal eigenvalue (say μ) vanishes, as follows immediately from either (5) or (6). Moreover,

$$Lf = -\Delta_S f + 2\overline{\nabla}_B \left(f s^B\right) = -\overline{\nabla}_B \left(\overline{\nabla}^B f - 2 f s^B\right),$$

so that L is a divergence and thus $\oint_S Lf = 0, \ \forall f$. Moreover, (9) reduces to

$$\frac{W}{2} f\left(n^\rho n_\rho - m^\rho m_\rho\right) = Lf - \delta_{f\mathbf{n}} \theta_{\mathbf{k}}, \tag{10}$$

where now the vector $\mathbf{m} = -\ell + \frac{m^\rho m_\rho}{2}\mathbf{k}$ is defined by

$$\frac{m^\rho m_\rho}{2} = \frac{1}{W}\left(K_S - \overline{\nabla}_B s^B - s^B s_B - G_{\mu\nu} k^\mu \ell^\nu \big|_S\right),$$

as follows from (7). For any other direction $\mathbf{m}_z$ defining a local M(O)TT

$$\frac{W}{2}\left(m_z^\rho m_{z\rho} - m^\rho m_\rho\right) = \lambda_z - (z - 2\overline{\nabla}_B s^B),$$

and therefore point (3) above leads to

Result 1 *The local M(O)TT defined by the direction* $\mathbf{m}$ *is such that any other nearby local M(O)TT must interweave it with non-trivial intersections to both of its sides, that is to say, the vector* $\mathbf{m}_z - \mathbf{m}$ *changes causal character on any of its M(O)TSs.*

Concerning cores, I try to follow the same steps as in spherical symmetry, and thus I start with a function $f = a_0\phi + \tilde{f}$ for a constant $a_0 > 0$ and $\phi > 0$ is the principal eigenfunction of L. Then (10) becomes

$$\frac{W}{2}(a_0\phi + \tilde{f})\left(n^\rho n_\rho - m^\rho m_\rho\right) = L\tilde{f} - \delta_{f\mathbf{n}} \theta_{\mathbf{k}}.$$

that can be split into two parts:

$$\frac{W}{2} a_0 \phi\left(n^\rho n_\rho - m^\rho m_\rho\right) = -\delta_{f\mathbf{n}} \theta_{\mathbf{k}}, \tag{11}$$

$$\frac{W}{2} \tilde{f}\left(n^\rho n_\rho - m^\rho m_\rho\right) = L\tilde{f}. \tag{12}$$

Equation (11) tells us that $\delta_{f\mathbf{n}} \theta_{\mathbf{k}} < 0$ whenever $\mathbf{n}$ points "above" $\mathbf{m}$ if $a_0 > 0$ is chosen. Therefore, using (12) the problem one needs to solve can be reformulated as follows: *Is there a function $\tilde{f}$ on S such that (i) $L\tilde{f}/\tilde{f} \geq \varepsilon > 0$, (ii) $\tilde{f}$ changes sign on S, (iii) $\tilde{f}$ is positive in a region as small as desired?* To prove that there are OTSs penetrating both sides of the MOTT it is enough to comply with points (i) and (ii) only. This does happen if L has more real eigenvalues, for any real eigenvalue is strictly positive (as $\mu = 0$), hence the corresponding eigenfunction must change sign on S, because integration of $L\psi = \lambda\psi$ on S implies $\oint \psi = 0$. However, even if there are no other real eigenvalues the result might still hold in general. In any case, the above leads to the analysis of the condition $L\tilde{f}/\tilde{f} > 0$ for functions $\tilde{f}$.

Acknowledgments Supported by grants FIS2010-15492 (MICINN), GIU06/37 (UPV/EHU) and P09-FQM- 4496 (J. Andalucía–FEDER) and UFI 11/55 (UPV/EHU).

References

1. Ashtekar, A., Galloway, G.: Some uniqueness results for dynamical horizons. Adv. Theor. Math. Phys. **9**, 1 (2005)
2. Ashtekar, A., Krishnan, B.: Isolated and dynamical horizons and their applications. Living Rev. Relativ. **7**(10), lrr-2004-10 (2004). http://www.livingreviews.org/lrr-2004-10
3. Senovilla, J.: Classification of spacelike surfaces in spacetime. Class. Quantum Grav. **24**, 3091 (2007). doi:10.1088/0264-9381/24/11/020
4. Senovilla, J.: Trapped surfaces. Int. J. Mod. Phys. D **20**, 2139 (2011). doi:10.1142/S0218271811020354
5. Wald, R.: General Relativity. University of Chicago Press, Chicago (1984)
6. Andersson, L., Mars, M., Simon, W.: Local existence of dynamical and trapping horizons. Phys. Rev. Lett. **95**, 111102 (2005). doi:10.1103/PhysRevLett.95.111102
7. Andersson, L., Mars, M., Simon, W.: Stability of marginally outer trapped surfaces and existence of marginally outer trapped tubes. Adv. Theor. Math. Phys. **12**, 853 (2008)
8. Hawking, S., Ellis, G.: The Large Scale Structure of Space–Time. Cambridge Monographs on Mathematical Physics (Cambridge University Press, Cambridge (1973)
9. Bengtsson, I., Senovilla, J.: Region with trapped surfaces in spherical symmetry, its core, and their boundaries. Phys. Rev. D **83**, 044012 (2011). doi:10.1103/PhysRevD.83.044012
10. Andersson, L., Metzger, J.: The area of horizons and the trapped region. Commun. Math. Phys. **290**, 941 (2009). doi:10.1007/s00220-008-0723-y
11. Åman, J., Bengtsson I., Senovilla J.: Where are the trapped surfaces? In: Lazkoz, R., Vera, R. (eds.) Gravitation in the Large, J. Phys.: Conf. Ser., vol. 229, p. 012004 (IOP, 2010). doi:10.1088/1742-6596/229/1/012004
12. Bengtsson, I., Senovilla, J.: Note on trapped surfaces in the Vaidya solution. Phys. Rev. D **79**, 024027 (2009). doi:10.1103/PhysRevD.79.024027

The Twin Paradox in Static Spacetimes and Jacobi Fields

Leszek M. Sokołowski

Abstract The twin paradox of special relativity formulated in the geometrical setting of general relativity gives rise to the problem of determining the longest timelike curve between a given pair of points. As a first step one solves the local problem for a bundle of nearby curves (geodesics) in terms of Jacobi fields and conjugate points. These provide important information about geometrical properties of the given spacetime. The second step, to determine the globally maximal length curve in the set of all timelike curves with common endpoints, is harder and may be effectively performed only in spacetimes with high symmetries.

1 Formulation of the Problem

The twin paradox in special relativity may be considered on two levels of comprehending. The first, lowest level of understanding deals with the problem of why the effect is asymmetric at all: why one twin turns out younger than the other whereas the time dilation effect for two clocks in uniform relative motion is actually symmetric. Most textbooks on special relativity explain the problem on this level by discussing the simplest (traditional) version of the paradox in which one twin remains all the time at rest in one inertial frame and the other uniformly moves to a distant star, then suddenly turns back and returns again at a constant velocity. By considering the hyperplanes of simultaneity of the astronaut during his travel back and forth one shows that the twin at rest must be older at the reunion than the astronaut; however such calculation can be effectively performed only in this simplest case of twins' motions. In consequence it does not provide a deeper understanding of the paradox. In fact, what does occur if both the twins move at non-uniform velocities?

L. M. Sokołowski (✉)
Astronomical Observatory, Jagiellonian University, Orla 171, 30-244 Krakow, Poland
e-mail: lech.sokolowski@uj.edu.pl

J. Bičák and T. Ledvinka (eds.), *Relativity and Gravitation*,
Springer Proceedings in Physics 157, DOI: 10.1007/978-3-319-06761-2_28,

The generic form of the paradox may be elucidated only on the second level which requires to go beyond the elementary algebra of the special Lorentz transformation. The solution is well known to the expert on relativity (though it is rather infrequently stressed in textbooks) and is based on the identification of physical time measured by a moving clock = proper time = length of the moving clock's worldline. The problem is then reduced to a purely geometrical one of calculating worldline lengths in Minkowski spacetime. Let the twins A and B travel at arbitrary velocities measured in some inertial frame from point (event) P to Q, then their common age at P will increase at Q by

$$s_A = \int_P^Q ds_A = c \int_{t_1}^{t_2} \sqrt{1 - \left(\frac{\mathbf{v}_A}{c}\right)^2}\, dt,$$

$$s_B = \int_P^Q ds_B = c \int_{t_1}^{t_2} \sqrt{1 - \left(\frac{\mathbf{v}_B}{c}\right)^2}\, dt \neq s_A.$$

respectively. Here t_1 and t_2 are the time coordinates of the points in the inertial frame. In the geometrical setting the effect of different ageing of the twins is obvious. Coming back to the simplest version of the paradox, what is less obvious and is rather surprising at first glance as it stands in contradiction to our experience in Euclidean geometry, is that the twin at rest gets older than the twin moving on a curved (accelerated) worldline. This is due to the reverse triangle inequality which gives rise to the theorem that in Minkowski spacetime the timelike straight line is the longest timelike curve between any pair of its points. Yet a curved timelike line joining two points may be arbitrarily small and it makes no sense to ask of how to move from P to (chronologically related) Q in order to use as little as possible of proper time—the interval $s(P, Q)$ may be arbitrarily close to zero. The properly posed problem is which timelike worldline from P to Q has the largest length and in flat spacetime is clear.

The geometrical twin paradox becomes much more interesting in curved spacetimes since the variety of possible physically relevant motions is much greater than in the flat case. Besides comparing lengths of concrete worldlines in a given spacetime one may ask if there are whole classes of worldlines which are longer than curves in other classes and, first of all, which timelike curves attain the maximal length. The first question is relevant in static spacetimes: what makes one twin younger than the other—velocity (with respect to a static observer) or acceleration (acting e.g. on the twin at rest)? Special cases investigated in first works on the subject do not allow one to infer general statements. For instance, Abramowicz et al. [1], Abramowicz and Bajtlik [2] investigated static worldlines and circular geodesics in Schwarzschild spacetime and concluded that 'in all situations in which the absolute motion may be defined in terms of some invariant global properties of the spacetime, the twin who moves faster with respect to the global standard of rest is younger at the reunion, irrespectively to twins' accelerations'. This conclusion is, however, false already in Schwarzschild world since introducing a third twin moving on a radial timelike geodesic, first upwards and then downwards, one can show that his worldline is longer than those of the other twins. Usually in this and many other spacetimes two points may be connected by different geodesics and a worldline of the static observer and the multitude of possibilities concerning their lengths precludes generic conclusions such as that above.

Generically one can only seek for timelike curves having maximal length and this is the problem we are dealing with here. To the best of our knowledge the first who gave the correct but imprecise answer to the problem was Feynman (whilst at Princeton in the 1940): the longest worldline is a timelike geodesic. The answer may be deduced by analogy with the flat spacetime (geodesics are straight lines), but is insufficient if there are two or more timelike geodesics with common endpoints, as in the example above.

The general rigorous solution of the problem is achieved in two steps. The first step is contained in Hawking and Ellis' book [3] and we summarize it here in the form of three propositions.

Proposition 1 *In any convex normal neighbourhood, if p and q can be joined by a timelike curve, then the unique timelike geodesic connecting them has length strictly greater than that of any other piecewise smooth timelike curve between the points.*[1]

The existence of a convex normal neighbourhood is crucial here. If q does not lie in this neighbourhood of p then there are several timelike geodesics from p to q with different lengths, as in the Schwarzschild case outside the event horizon. But how to recognize whether given p and q can be connected by a *unique* timelike geodesic (i.e. that their neighbourhood is normal)? The first step described in [3] deals with bundles of nearby geodesics. Here the key notion is that of conjugate points.

Let $Z^\mu(s)$ be a geodesic deviation vector field on a timelike geodesic γ with tangent unit vector $u^\alpha(s)$. If Z^μ is chosen orthogonal to u_μ, then it satisfies the geodesic deviation equation

$$\frac{D^2}{ds^2} Z^\mu = R^\mu{}_{\alpha\beta\nu}\, u^\alpha\, u^\beta\, Z^\nu.$$

Any solution Z^μ of the equation is called a Jacobi field on γ. Points p and q on γ are said to be *conjugate* if there is Jacobi field $Z^\mu \neq 0$ such that $Z^\mu(q) = 0$ iff $Z^\mu(p) = 0$. If p and q are conjugate then one or more nearby geodesics intersect γ at p and q (or pass infinitesimally close to γ at these points) and they have different lengths. More precisely, if a geodesic γ joining points p_1 and p_2 has a point q conjugate to p_1 belonging to the segment $p_1 p_2$, then there exists a nearby timelike curve (not necessarily a geodesic) with endpoints p_1 and p_2 which is longer than γ. If there are no conjugate points, γ is the longest curve in the set of nearby curves.

---

[1] Proposition 4.5.3 in [3].

Proposition 2 *A timelike geodesic attains the local maximum of length (i.e. among nearby curves) from p_1 to p_2 iff there is no point conjugate to p_1 on the segment $p_1 p_2$.*[2]

The very existence of conjugate points (but not their localization) is determined by

Proposition 3 *If $R_{\alpha\beta}\, u^\alpha\, u^\beta \geq 0$ on a timelike geodesic γ and if the tidal force $R_{\mu\alpha\nu\beta}\, u^\alpha\, u^\beta \neq 0$ at some point p_0 on γ, there will be a pair of conjugate points somewhere on γ (providing that the geodesic can be extended sufficiently far).*[3]

Returning to the twin paradox one concludes that the problem of which twin will be older at the reunion has a general (i.e. without computing the lengths of concrete worldlines) solution only if one of the twins' worldlines is a timelike geodesic free of conjugate points. From this conclusion one gets, e. g. that in Schwarzschild spacetime the radial geodesic (flight up and down), being free of conjugate points, is longer than the circular geodesic with the same endpoints since the latter has a conjugate point in the middle of its segment; this outcome is hard to derive from the analytic expressions for their lengths and and the two expressions can be compared only numerically.

This is, however, not the full solution of the problem we are interested in. Conjugate points allow one to find locally maximal curves but not the globally maximal ones. We are now entering the second step in solving the problem. The radial geodesic in Schwarzschild spacetime is longest in the bundle of nearby curves and is longer than the distant circular geodesic, but a priori there might exist a timelike geodesic, distant from the radial one and longer than that. We are looking for globally maximal length geodesics, i.e. curves whose length between their points attains maximal value. This maximal length is named *Lorentzian distance function* $d(p, q)$. Properties of maximal timelike geodesics, or curves realizing the distance function, are investigated in global Lorentzian geometry [4]. In [5] we quote six theorems from the book [4] which in our opinion are most relevant for searching globally maximal geodesics. Without glancing at them one may expect that these are mathematical 'existence theorems' stating the presence of some global properties of geodesics if some global conditions hold. They cannot be and are not 'constructive'. In fact, in the search for locally maximal curves one investigates a bundle of close geodesics and their properties may be expressed in terms of the geodesic deviation equation. Yet in the problem of global maximality one compares the lengths of curves which besides their common endpoints are distant from each other. This means that there is no local analytic tool such as a differential equation (which expresses local properties of a mathematical object) allowing one to seek for the globally maximal curve. To find out which curve has the length equal to the distance function one has to study all timelike curves joining the given pair of points. In other terms there is no effective algorithmic procedure which provides in a *finite* number of steps the maximal

[2] Proposition 4.5.8 in [3].

[3] Proposition 4.4.2 in [3].

(unique or not) timelike curve for the given endpoints. The search is not hopeless if one considers a spacetime with global symmetries (isometries). At least in the case of static spherically symmetric spacetimes it is possible to show that some class of timelike geodesics (the radial ones) consists of curves which are maximal on some segments.

In this conference contribution we wish to make only a general introduction to the problem of maximal curves in Lorentzian spacetimes, which has evolved from the twin paradox in special relativity. Investigations of locally and globally maximal timelike curves are still at an initial stage and some results have been found in the simplest spacetimes of general relativity. These results reveal a multitude of possibilities concerning Jacobi fields and conjugate points. For details we refer the interested reader to author's recent work [6] and to his forthcoming papers.

References

1. Abramowicz, M., Bajtlik, S., Kluźniak, W.: The twin paradox on the photon sphere. Phys. Rev. A **75**, 044101 (2007). doi:10.1103/PhysRevA.75.044101
2. Abramowicz, M., Bajtlik, S.: Adding to the paradox: the accelerated twin is older (2009). ArXiv e-prints 0905.2428[physics.class-ph]
3. Hawking, S., Ellis, G.: The Large Scale Structure of Space–Time. In: Cambridge Monographs on Mathematical Physics. Cambridge University Press, Cambridge (1973)
4. Beem, J., Ehrlich, P., Easley, K.: Global Lorentzian Geometry. In: Monographs and Textbooks in Pure and Applied Mathematics, vol. 202, 2nd edn. Marcel Dekker, New York (1996)
5. Sokołowski, L. M., Golda, Z. A.: Jacobi fields, conjugate points and cut points on timelike geodesics in special spacetimes. ArXiv e-prints arxiv:1402.3976[gr-qc]
6. Sokołowski, L.: On the twin paradox in static spacetimes: I. Schwarzschild metric. Gen. Relativ. Gravit. **44**, 1267 (2012). doi:10.1007/s10714-012-1337-4

Geodesic Deviation in Kundt Spacetimes of any Dimension

Robert Švarc and Jiří Podolský

Abstract Using the invariant form of the equation of geodesic deviation, which describes relative motion of free test particles, we investigate a general family of D-dimensional Kundt spacetimes. We demonstrate that local influence of the gravitational field can be naturally decomposed into Newton-type tidal effects typical for type II spacetimes, longitudinal deformations mainly present in spacetimes of algebraic type III, and type N purely transverse effects corresponding to gravitational waves with $\frac{1}{2}D(D-3)$ independent polarization states. We explicitly study the most important examples, namely exact pp-waves, gyratons, and VSI spacetimes. This analysis helps us to clarify the geometrical and physical interpretation of the Kundt class of nonexpanding, nontwisting and shearfree geometries.

1 Geometry of Kundt Spacetimes

The scalars Θ (expansion), A^2 (twist) and σ^2 (shear) characterizing optical properties of an affinely parameterized geodesic null congruence k^a are

$$\Theta = \frac{1}{D-2} k^a{}_{;a}, \quad A^2 = -k_{[a;b]}k^{a;b}, \quad \sigma^2 = k_{(a;b)}k^{a;b} - \frac{1}{D-2}(k^a{}_{;a})^2. \qquad (1)$$

Purely geometric definition of the Kundt family of spacetimes, namely that it admits nonexpanding ($\Theta = 0$), nontwisting ($A = 0$) and shearfree ($\sigma = 0$) such a congruence, implies that there exist suitable coordinates in which the line element of any Kundt spacetime can be written as [1–5]

R. Švarc (✉) · J. Podolský
Institute of Theoretical Physics, Charles University in Prague, V Holešovičkách 2, 18000 Praha 8, Czech Republic
e-mail: robert.svarc@mff.cuni.cz

J. Podolský
e-mail: podolsky@mbox.troja.mff.cuni.cz

J. Bičák and T. Ledvinka (eds.), *Relativity and Gravitation*,
Springer Proceedings in Physics 157, DOI: 10.1007/978-3-319-06761-2_29,

$$ds^2 = g_{ij}(u,x)\,dx^i dx^j + 2\,g_{ui}(r,u,x)\,dx^i du - 2\,du\,dr + g_{uu}(r,u,x)\,du^2. \quad (2)$$

The coordinate r is the affine parameter along the congruence $k^a = \partial_r$, $u =$ const. are null (wave)surfaces, and $x \equiv (x^2, x^3, \ldots, x^{D-1})$ are $D-2$ spatial coordinates in the transverse Riemannian space. Notice that the spatial part g_{ij} of the metric must be independent of r, all other metric components g_{ui} and g_{uu} can, in principle, be functions of all the coordinates (r, u, x). No specific Einstein field equations have been employed yet.

For such most general Kundt line element (2) a lengthy calculation gives the following components of the Riemann curvature tensor

$$\begin{aligned}
R_{rprq} &= 0, & (3)\\
R_{rpru} &= -\tfrac{1}{2} g_{up,rr}, & (4)\\
R_{ruru} &= -\tfrac{1}{2} g_{uu,rr} + \tfrac{1}{4} g^{ij} g_{ui,r} g_{uj,r}, & (5)\\
R_{rpkq} &= 0, & (6)\\
R_{rpuq} &= \tfrac{1}{2} g_{up,rq} + \tfrac{1}{4} g_{up,r} g_{uq,r} - \tfrac{1}{4} g^{ij} g_{ui,r}\left(2g_{j(p,q)} - g_{pq,j}\right), & (7)\\
R_{rupq} &= g_{u[p,q],r}, & (8)\\
R_{ruup} &= g_{u[u,p],r} + \tfrac{1}{4} g^{ri} g_{up,r} g_{ui,r} - \tfrac{1}{4} g^{ij} g_{ui,r}\left(2g_{j(u,p)} - g_{up,j}\right), & (9)\\
R_{kplq} &= {}^{S}R_{kplq}, & (10)\\
R_{upkq} &= g_{p[k,q],u} - g_{u[k,q],p}\\
&\quad + \tfrac{1}{4}\left[g_{uk,r}\left(g_{pq,u} - 2g_{u(p,q)}\right) - g_{uq,r}\left(g_{pk,u} - 2g_{u(p,k)}\right)\right]\\
&\quad + \tfrac{1}{4} g^{ri}\left[g_{uk,r}\left(2g_{i(p,q)} - g_{pq,i}\right) - g_{uq,r}\left(2g_{i(p,k)} - g_{pk,i}\right)\right]\\
&\quad + \tfrac{1}{4} g^{ij}\left(2g_{j(u,q)} - g_{uq,j}\right)\left(2g_{i(p,k)} - g_{pk,i}\right)\\
&\quad - \tfrac{1}{4} g^{ij}\left(2g_{j(u,k)} - g_{uk,j}\right)\left(2g_{i(p,q)} - g_{pq,i}\right), & (11)\\
R_{upuq} &= g_{u(p,q),u} - \tfrac{1}{2}\left(g_{pq,uu} + g_{uu,pq}\right) + \tfrac{1}{4} g^{rr} g_{up,r} g_{uq,r}\\
&\quad - \tfrac{1}{4} g_{uu,r}\left[2g_{u(p,q)} - g_{pq,u} - g^{ri}\left(2g_{i(p,q)} - g_{pq,i}\right)\right]\\
&\quad + \tfrac{1}{4} g_{up,r}\left[g_{uu,q} - g^{ri}\left(2g_{i(u,q)} - g_{uq,i}\right)\right]\\
&\quad + \tfrac{1}{4} g_{uq,r}\left[g_{uu,p} - g^{ri}\left(2g_{i(u,p)} - g_{up,i}\right)\right]\\
&\quad + \tfrac{1}{4} g^{ij}\left(2g_{j(u,p)} - g_{up,j}\right)\left(2g_{i(u,q)} - g_{uq,i}\right)\\
&\quad - \tfrac{1}{4} g^{ij}\left(2g_{uj,u} - g_{uu,j}\right)\left(2g_{i(p,q)} - g_{pq,i}\right), & (12)
\end{aligned}$$

where i, j, k, l, p, q denote the spatial components (and derivatives w.r.t.) x. The superscript "^{S}" labels tensor quantities corresponding to the spatial metric g_{ij}, with derivatives taken only with respect to the coordinates x. The components of the Ricci tensor are

$$\begin{aligned}
R_{rr} &= 0, & (13)\\
R_{rk} &= -\tfrac{1}{2} g_{uk,rr}, & (14)
\end{aligned}$$

$$
\begin{aligned}
R_{ru} = & -\tfrac{1}{2} g_{uu,rr} + \tfrac{1}{2} g^{ri} g_{ui,rr} + \tfrac{1}{2} g^{pq} g_{up,rq} \\
& + \tfrac{1}{2} g^{pq} g_{up,r} g_{uq,r} - \tfrac{1}{4} g^{pq} g^{ij} g_{ui,r} \left(2 g_{jp,q} - g_{pq,j}\right),
\end{aligned} \tag{15}
$$

$$
R_{pq} = {}^S R_{pq} - g_{u(p,q),r} - \tfrac{1}{2} g_{up,r} g_{uq,r} + \tfrac{1}{2} g^{kl} g_{uk,r} \left(2 g_{l(p,q)} - g_{pq,l}\right), \tag{16}
$$

$$
\begin{aligned}
R_{uu} = & -\tfrac{1}{2} g^{rr} g_{uu,rr} - 2 g^{ri} g_{u[u,i],r} + \tfrac{1}{2} g^{pq} \left(2 g_{up,uq} - g_{pq,uu} - g_{uu,pq}\right) \\
& -\tfrac{1}{2} g^{rp} g^{rq} g_{up,r} g_{uq,r} + \tfrac{1}{2} g^{rr} g^{pq} g_{up,r} g_{uq,r} + \tfrac{1}{2} g^{pq} g^{ri} g_{up,r} (2 g_{q(u,i)} - g_{ui,q}) \\
& -\tfrac{1}{4} g^{pq} g_{uu,r} \left[2 g_{up,q} - g_{pq,u} - g^{ri} \left(2 g_{ip,q} - g_{pq,i}\right)\right] \\
& +\tfrac{1}{2} g^{pq} g_{up,r} \left[g_{uu,q} - g^{ri} \left(2 g_{i(u,q)} - g_{uq,i}\right)\right] \\
& +\tfrac{1}{4} g^{pq} g^{ij} \left(2 g_{j(u,p)} - g_{up,j}\right) \left(2 g_{i(u,q)} - g_{uq,i}\right) \\
& -\tfrac{1}{4} g^{pq} g^{ij} \left(2 g_{uj,u} - g_{uu,j}\right) \left(2 g_{ip,q} - g_{pq,i}\right),
\end{aligned} \tag{17}
$$

$$
\begin{aligned}
R_{uk} = & -\tfrac{1}{2} g^{rr} g_{uk,rr} - g_{u[u,k],r} + g^{ri} (g_{u[i,k],r} - g_{k[u,i],r}) \\
& + g^{pq} (g_{p[k,q],u} - g_{u[k,q],p}) - \tfrac{1}{2} g^{ri} g_{uk,r} g_{ui,r} \\
& +\tfrac{1}{4} g^{pq} g^{ri} \left[4 g_{uq,r} g_{k[p,i]} + g_{uk,r} (2 g_{i(p,q)} - g_{pq,i})\right] \\
& +\tfrac{1}{4} g^{pq} \left[2 g_{up,r} g_{uq,k} - g_{uk,r} (2 g_{up,q} - g_{pq,u})\right] \\
& +\tfrac{1}{4} g^{pq} g^{ij} \left(2 g_{j(u,q)} - g_{uq,j}\right) \left(2 g_{i(p,k)} - g_{pk,i}\right) \\
& -\tfrac{1}{4} g^{pq} g^{ij} \left(2 g_{j(u,k)} - g_{uk,j}\right) \left(2 g_{ip,q} - g_{pq,i}\right),
\end{aligned} \tag{18}
$$

and the Ricci scalar curvature of the Kundt spacetime (2) is given by

$$
\begin{aligned}
R = & {}^S R + g_{uu,rr} - 2 g^{ri} g_{ui,rr} - 2 g^{pq} g_{up,rq} \\
& -\tfrac{3}{2} g^{pq} g_{up,r} g_{uq,r} + g^{pq} g^{kl} g_{uk,r} (2 g_{lp,q} - g_{pq,l}).
\end{aligned} \tag{19}
$$

2 Applying the Field Equations

So far we have not specified the matter content of the spacetimes. Now, following the approach presented in [4], we can determine the r-dependence of the metric (2) using the Einstein field equations $R_{ab} - \frac{1}{2} R\, g_{ab} + \Lambda g_{ab} = 8\pi T_{ab}$. Since $R_{rr} = 0$ and $g_{rr} = 0$, there is an obvious restriction on the energy-momentum tensor allowed in the Kundt family, namely $T_{rr} = 0$. *Assuming* $T_{rk} = 0$, we can directly integrate the Einstein equation $R_{rk} = 0$ with (14), yielding g_{uk} *linear* in r. Using the field equation $R_{ru} + \frac{1}{2} R - \Lambda = 8\pi T_{ru}$, this implies that the component T_{ru} must be independent of r. Taking the trace of Einstein's equations we can also determine the r-dependence of g_{uu}: if the trace T of energy-momentum tensor T_{ab} does not depend on the coordinate r, the metric function g_{uu} can only be (at most) *quadratic* in r, see (19). Under these conditions

$$
\mathrm{d}s^2 = g_{ij}\, \mathrm{d}x^i \mathrm{d}x^j + 2\,(e_i + f_i\, r)\, \mathrm{d}x^i \mathrm{d}u - 2\, \mathrm{d}u\, \mathrm{d}r + (a\, r^2 + b\, r + c)\, \mathrm{d}u^2, \tag{20}
$$

where all the functions g_{ij}, e_i, f_i, a, b and c are independent of r, and are constrained by the specific Einstein equations [4]. In particular, any *vacuum* Kundt metric, possibly with a cosmological constant Λ and/or aligned electromagnetic field, can be written in the form (20).

3 Geodesic Deviation in an Arbitrary Spacetime

In our recent work [6] we demonstrated that the equation of geodesic deviation, which describes relative motion of nearby free test particles, can in *any* D-dimensional spacetime be expressed in the invariant form

$$\begin{aligned}
\ddot{Z}^{(1)} &= \frac{2\Lambda}{(D-2)(D-1)}\, Z^{(1)} + \Psi_{2S}\, Z^{(1)} + \frac{1}{\sqrt{2}}\,(\Psi_{1T^j} - \Psi_{3T^j})\, Z^{(j)} \\
&\quad + \frac{8\pi}{D-2}\Big[T_{(1)(1)}\, Z^{(1)} + T_{(1)(j)}\, Z^{(j)} - \Big(T_{(0)(0)} + \frac{2\,T}{D-1}\Big) Z^{(1)} \Big], \qquad (21)\\
\ddot{Z}^{(i)} &= \frac{2\Lambda}{(D-2)(D-1)}\, Z^{(i)} - \Psi_{2T^{(ij)}}\, Z^{(j)} + \frac{1}{\sqrt{2}}\,(\Psi_{1T^i} - \Psi_{3T^i})\, Z^{(1)} \\
&\quad - \frac{1}{2}\,(\Psi_{0^{ij}} + \Psi_{4^{ij}})\, Z^{(j)} + \frac{8\pi}{D-2}\Big[T_{(i)(1)}\, Z^{(1)} \\
&\quad + T_{(i)(j)}\, Z^{(j)} - \Big(T_{(0)(0)} + \frac{2\,T}{D-1}\Big) Z^{(i)} \Big], \qquad (22)
\end{aligned}$$

with $i, j = 2, \ldots, D-1$. Here $Z^{(1)}, Z^{(2)}, \ldots, Z^{(D-1)}$ are spatial components of the separation vector $\boldsymbol{Z} = Z^a\, \boldsymbol{e}_a$ between the test particles in a natural interpretation orthonormal frame $\{\boldsymbol{e}_a\}$ where $\boldsymbol{e}_{(0)} = \boldsymbol{u}$ is the velocity vector of the fiducial test particle ($\boldsymbol{e}_a \cdot \boldsymbol{e}_b = \eta_{ab}$), $\ddot{Z}^{(1)}, \ddot{Z}^{(2)}, \ldots, \ddot{Z}^{(D-1)}$ are the corresponding relative accelerations, T_{ab} are frame components of the energy-momentum tensor, and the scalars $\Psi_{A^{\ldots}}$ defined as

$$\begin{aligned}
\Psi_{0^{ij}} &= C_{abcd}\, k^a\, m_i^b\, k^c\, m_j^d, \\
\Psi_{1T^i} &= C_{abcd}\, k^a\, l^b\, k^c\, m_i^d, & \Psi_{1^{ijk}} &= C_{abcd}\, k^a\, m_i^b\, m_j^c\, m_k^d, \\
\Psi_{2S} &= C_{abcd}\, k^a\, l^b\, l^c\, k^d, & \Psi_{2^{ijkl}} &= C_{abcd}\, m_i^a\, m_j^b\, m_k^c\, m_l^d, \\
\Psi_{2T^{ij}} &= C_{abcd}\, k^a\, m_i^b\, l^c\, m_j^d, & \Psi_{2^{ij}} &= C_{abcd}\, k^a\, l^b\, m_i^c\, m_j^d, \\
\Psi_{3T^i} &= C_{abcd}\, l^a\, k^b\, l^c\, m_i^d, & \Psi_{3^{ijk}} &= C_{abcd}\, l^a\, m_i^b\, m_j^c\, m_k^d, \\
\Psi_{4^{ij}} &= C_{abcd}\, l^a\, m_i^b\, l^c\, m_j^d, & & \qquad (23)
\end{aligned}$$

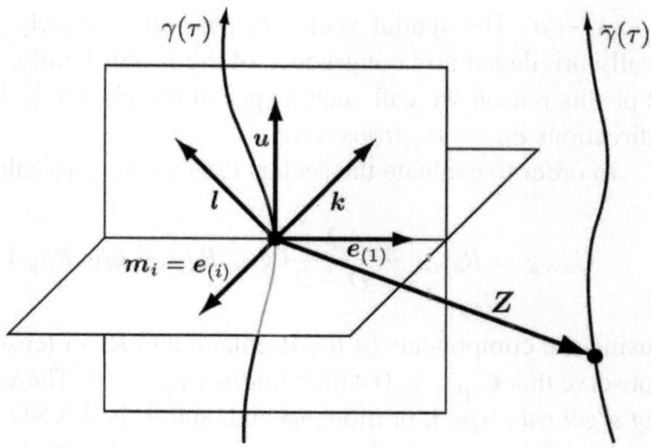

Fig. 1 Evolution of the separation vector $\boldsymbol{Z}$ that connects particles moving along geodesics $\gamma(\tau)$, $\bar{\gamma}(\tau)$ is given by the equation of geodesic deviation (21) and (22). Its components are expressed in the orthonormal frame $\{\boldsymbol{e}_a\}$, $\boldsymbol{e}_{(0)} = \boldsymbol{u}$. The associated null frame $\{\boldsymbol{k}, \boldsymbol{l}, \boldsymbol{m}_i\}$ is also indicated

$i, j, k, l = 2, \ldots, D-1$, are components of the Weyl tensor with respect to the null frame $\{\boldsymbol{k}, \boldsymbol{l}, \boldsymbol{m}_i\}$ associated with $\{\boldsymbol{e}_a\}$ via the relations $\boldsymbol{k} = \frac{1}{\sqrt{2}}(\boldsymbol{u} + \boldsymbol{e}_{(1)})$, $\boldsymbol{l} = \frac{1}{\sqrt{2}}(\boldsymbol{u} - \boldsymbol{e}_{(1)})$, $\boldsymbol{m}_i = \boldsymbol{e}_{(i)}$, see Fig. 1.

Components of the Weyl tensor (23) are listed by their boost weight and directly generalize the standard Newman–Penrose complex scalars Ψ_A known from the $D = 4$ case [6, 7]. In equations (21), (22), only the "electric part" of the Weyl tensor represented by the scalars in the left column of (23) occurs, and there are various constraints and symmetries, for example

$$\begin{aligned}
&\Psi_{1T^i} = \Psi_{1^k{}_{ki}}\,, \; \Psi_{2S} = \tfrac{1}{2}\Psi_{2kl}{}^{kl}\,, \; \Psi_{2T^{(ij)}} = \tfrac{1}{2}\Psi_{2ikj}{}^{k}\,, \; \Psi_{3T^i} = \Psi_{3^k{}_{ki}}\,, \\
&\Psi_{0^{ij}} = \Psi_{0^{(ij)}}\,, \quad \Psi_{0^k{}_k} = 0\,, \qquad\qquad \Psi_{4^{ij}} = \Psi_{4^{(ij)}}\,, \quad \Psi_{4^k{}_k} = 0\,.
\end{aligned} \tag{24}$$

4 Geodesic Deviation in Kundt Spacetimes

For the general Kundt spacetime (2), the null interpretation frame adapted to an arbitrary observer moving along a timelike geodesic $\gamma(\tau)$ with velocity $\boldsymbol{u} = \dot{r}\,\partial_r + \dot{u}\,\partial_u + \dot{x}^i\,\partial_i$ takes the form

$$\begin{aligned}
\boldsymbol{k} &= \frac{1}{\sqrt{2}\,\dot{u}}\,\partial_r\,, \quad \boldsymbol{l} = \Big(\sqrt{2}\,\dot{r} - \frac{1}{\sqrt{2}\,\dot{u}}\Big)\partial_r + \sqrt{2}\,\dot{u}\,\partial_u + \sqrt{2}\,\dot{x}^i\,\partial_i\,, \\
\boldsymbol{m}_i &= \frac{1}{\dot{u}}\,m_i^j\,(g_{ju}\,\dot{u} + g_{jk}\,\dot{x}^k)\,\partial_r + m_i^j\,\partial_j\,,
\end{aligned} \tag{25}$$

where m_i^j satisfy $g_{jl}\,m_i^j\,m_k^l = \delta_{ik}$ to fulfil $\boldsymbol{m}_i \cdot \boldsymbol{m}_k = \delta_{ik}$, $\boldsymbol{k} \cdot \boldsymbol{l} = -1$. Vector $\boldsymbol{k}$ is oriented along the nonexpanding, nontwisting and shearfree null congruence $k^a = \partial_r$ defining the Kundt family. Moreover, $\boldsymbol{u} = \frac{1}{\sqrt{2}}(\boldsymbol{k} + \boldsymbol{l})$ and $\boldsymbol{e}_{(1)} = \frac{1}{\sqrt{2}}(\boldsymbol{k} - \boldsymbol{l}) =$

$\sqrt{2}\,\boldsymbol{k} - \boldsymbol{u}$. The spatial vector $\boldsymbol{e}_{(1)}$ is thus uniquely determined by the geometrically privileged null congruence of the Kundt family, and the observer's velocity $\boldsymbol{u}$. For this reason we call such a special direction $\boldsymbol{e}_{(1)}$ *longitudinal*, while the $D-2$ directions $\boldsymbol{e}_{(i)} = \boldsymbol{m}_i$ *transverse*.

In order to evaluate the scalars (23) we need to calculate the Weyl tensor

$$C_{abcd} = R_{abcd} - \frac{2}{D-2}(\,g_{a[c}\,R_{d]b} - g_{b[c}\,R_{d]a}) + \frac{2\,R\,g_{a[c}\,g_{d]b}}{(D-1)(D-2)}, \tag{26}$$

using the components of the Riemann and Ricci tensors (3)–(19). We immediately observe that $C_{rprq} = 0$ which implies $\Psi_{0^{ij}} = 0$. Therefore, *all Kundt spacetimes are of algebraic type* I, or more special, and ∂_r is WAND.

Restricting now to the important subfamily (20) for which

$$g_{ui} = e_i(u,x) + f_i(u,x)\,r, \quad g_{uu} = a(u,x)\,r^2 + b(u,x)\,r + c(u,x), \tag{27}$$

we obtain $R_{rpru} = 0$, $R_{rp} = 0$ which implies $C_{rpru} = 0$, $C_{rpkq} = 0$ so that $\Psi_{1T^i} = 0$, $\Psi_{1^{ijk}} = 0$. Since all Weyl scalars of boost weights 2 and 1 vanish, the metric (20) represents Kundt spacetimes of algebraic type II (or more special). Equations (21), (22) for the geodesic deviation (omitting the frame components of T_{ab} encoding the direct influence of matter) in the case of the Kundt class of spacetimes (20) thus simplify considerably to

$$\begin{aligned}\ddot{Z}^{(1)} &= \frac{2\Lambda}{(D-2)(D-1)}\,Z^{(1)} + \Psi_{2S}\,Z^{(1)} - \frac{1}{\sqrt{2}}\,\Psi_{3T^j}\,Z^{(j)}, \\ \ddot{Z}^{(i)} &= \frac{2\Lambda}{(D-2)(D-1)}\,Z^{(i)} - \Psi_{2T^{(ij)}}\,Z^{(j)} - \frac{1}{\sqrt{2}}\,\Psi_{3T^i}\,Z^{(1)} - \frac{1}{2}\,\Psi_{4^{ij}}\,Z^{(j)},\end{aligned} \tag{28}$$

where the only nonvanishing Weyl scalars are

$$\begin{aligned}\Psi_{2S} &= -R_{ruru} + \tfrac{2}{D-2}\,R_{ru} + \tfrac{1}{(D-1)(D-2)}\,R, \\ \Psi_{2T^{ij}} &= m_i^p m_j^q \big[R_{rpuq} - \tfrac{1}{D-2}\,\big(g_{pq}\,R_{ru} - R_{pq}\big) - \tfrac{1}{(D-1)(D-2)}\,R\,g_{pq}\big], \\ \Psi_{3T^j} &= \sqrt{2}\,m_j^p \Big\{\dot{x}^k \big[R_{ruru}\,g_{kp} - R_{rkup} - R_{rukp} - \tfrac{1}{D-2}\,\big(g_{kp}\,R_{ru} + R_{kp}\big)\big] \\ &\quad + \dot{u}\,\big[R_{ruru}\,g_{up} - R_{ruup} - \tfrac{1}{D-2}\,\big(g_{up}\,R_{ru} + R_{up}\big)\big]\Big\}, \\ \Psi_{4^{ij}} &= 2\,m_{(i}^p m_{j)}^q \Big\{\dot{x}^k \dot{x}^l \big[R_{rpuq}\,g_{kl} - g_{pk}(2R_{rluq} - g_{lq}\,R_{ruru} + 2R_{rulq}) \\ &\quad + R_{kplq} - \tfrac{1}{D-2} g_{pq}\,(g_{kl}\,R_{ru} + R_{kl})\big] \\ &\quad + 2\dot{u}\dot{x}^k \big[R_{rpuq}\,g_{uk} - g_{up}(R_{rkuq} - R_{ruru}\,g_{qk} + R_{rukq}) \\ &\quad - R_{ruup}\,g_{qk} + R_{upkq} - \tfrac{1}{D-2} g_{pq}(g_{uk}\,R_{ru} + R_{uk})\big] \\ &\quad + \dot{u}^2 \big[R_{rpuq}\,g_{uu} - g_{uq}\,\big(2R_{ruup} - g_{up}\,R_{ruru}\big) + R_{upuq}\end{aligned}$$

$$-\frac{1}{D-2}g_{pq}\left(g_{uu}R_{ru}+R_{uu}\right)\Big]\Big\}, \tag{29}$$

and the components R_{abcd} are explicitly given by (5)–(12), R_{ab} by (15)–(18), and the Ricci scalar curvature R is given by (19).

The relative motion of free test particles in any Kundt spacetime (20) is thus composed of the *isotropic influence* of the cosmological constant Λ, *Newton-like tidal deformations* represented by Ψ_{2S}, $\Psi_{2T(ij)}$, *longitudinal* accelerations associated with the direction $+\mathbf{e}_{(1)}$ given by Ψ_{3Tj}, and *transverse gravitational waves* propagating along $+\mathbf{e}_{(1)}$ encoded in the symmetric traceless matrix Ψ_{4ij}, see (24). The invariant amplitudes (29) combine the curvature of the Kundt spacetime with kinematics of the specific geodesic motion. In contrast to longitudinal and transverse wave effects, the Newton-like deformations caused by Ψ_{2S} and $\Psi_{2T(ij)}$ are independent of the observer's velocity components $\dot{x}^i$ and $\dot{u}$.

More details can be found in our recent publications [8, 9].

5 Discussion of Particular Subfamilies

The Kundt class involves several physically interesting subfamilies, for example pp-waves including gyratons and VSI spacetimes.

The *pp-waves* are defined by admitting a covariantly constant null vector field k^a [2, 3]. They thus belong to the Kundt class with all metric functions independent of r, which is the metric (20) with $f_i = 0$, $a = 0 = b$:

$$\mathrm{d}s^2 = g_{ij}(u,x)\,\mathrm{d}x^i\mathrm{d}x^j + 2\,e_i(u,x)\,\mathrm{d}x^i\mathrm{d}u - 2\,\mathrm{d}u\,\mathrm{d}r + c(u,x)\,\mathrm{d}u^2. \tag{30}$$

The components e_i encode the possible presence of gyratonic matter.

The *VSI spacetimes* have the property that their scalar curvature invariants of all orders vanish identically. As shown in [10], these spacetimes must be of the form (20) with flat transverse space $g_{ij} = \delta_{ij}$:

$$\mathrm{d}s^2 = \delta_{ij}\,\mathrm{d}x^i\mathrm{d}x^j + 2\,(e_i + f_i\,r)\,\mathrm{d}x^i\mathrm{d}u - 2\,\mathrm{d}u\,\mathrm{d}r + (a\,r^2 + b\,r + c)\,\mathrm{d}u^2. \tag{31}$$

It is straightforward to apply our general results (28) to these particular subcases by evaluating the corresponding Weyl scalars (29) and discussing their specific influence on test particles. We have to restrict ourselves only to the simplest case here,[1] to *vacuum VSI pp-waves without gyratons*:

$$\mathrm{d}s^2 = \delta_{ij}\,\mathrm{d}x^i\mathrm{d}x^j - 2\,\mathrm{d}u\,\mathrm{d}r + c(u,x)\,\mathrm{d}u^2. \tag{32}$$

Since $\Lambda = 0$, $\Psi_{2S} = 0 = \Psi_{2Tij}$, $\Psi_{3Tj} = 0$, the geodesic deviation reduces to

[1] A thorough discussion of other cases can be found in our subsequent papers [3, 9].

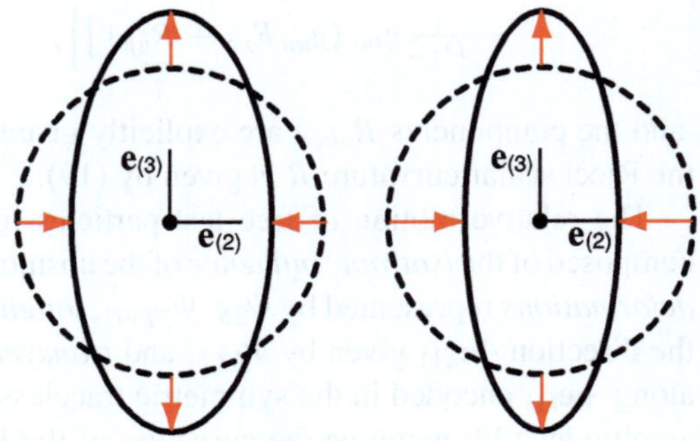

Fig. 2 Standard (*left*) and one of peculiar deformations of a detector indicating extension of the gravitation wave into higher dimensions (*right*)

$$\ddot{Z}^{(1)} = 0, \qquad \ddot{Z}^{(i)} = -\tfrac{1}{2}\,\Psi_{4ij}\,Z^{(j)}. \tag{33}$$

This clearly represents *gravitational waves* propagating along the spatial direction (1), with the test particles influenced only in the *transverse* directions $(i) = (2), (3), \ldots, (D-1)$. The elements of the *symmetric and traceless* $(D-2)\times(D-2)$ matrix $\Psi_{4ij} = -\dot{u}^2\,c_{,ij}$ (where $\dot{u}$ is a constant) directly encode the corresponding wave amplitudes. Obviously, there are $\frac{1}{2}D(D-3)$ independent *polarization states*.

Taking, e.g., a quadratic function $c(x) \equiv \sum_{i=2}^{D-1} \mathscr{A}_i\,(x^i)^2$ where the constants must satisfy $\sum_{i=2}^{D-1} \mathscr{A}_i = 0$, the wave-amplitude matrix becomes $\Psi_{4ij} = -2\dot{u}^2$ diag$(\mathscr{A}_2, \mathscr{A}_3, \ldots)$. Relative motion of (initially static) particles given by (33) can be explicitly integrated: in the spatial directions with *positive eigenvalues* $\mathscr{A}_i > 0$ they *recede* as $Z^{(i)}(\tau) = Z_0^{(i)} \cosh\left(\sqrt{\mathscr{A}_i}\,|\dot{u}|\,\tau\right)$, while with *negative eigenvalues* $\mathscr{A}_i < 0$ they *converge* as $Z^{(i)}(\tau) = Z_0^{(i)} \cos\left(\sqrt{\mathscr{A}_i}\,|\dot{u}|\,\tau\right)$, and in the directions where $\mathscr{A}_i = 0$ the particles stay fixed, $Z^{(i)}(\tau) = Z_0^{(i)}$.

In principle, the presence of higher-dimensional components of gravitational waves could be observed by detectors in our (1+3)-dimensional universe as the *violation of the standard TT-property*. Indeed, taking the simplest case $D = 5$, the matrix reads $\Psi_{4ij} = -2\dot{u}^2\,\mathrm{diag}(\mathscr{A}_2, \mathscr{A}_3, \mathscr{A}_4)$ where $\mathscr{A}_2 = -(\mathscr{A}_3 + \mathscr{A}_4)$. In the absence of the higher-dimensional component, $\mathscr{A}_4 = 0$, an interferometer in our space detects usual deformations shown in the left part of Fig. 2. But if $\mathscr{A}_4 \neq 0$ then $\mathscr{A}_2 \neq -\mathscr{A}_3$, and a peculiar deformation, such as on the right part of Fig. 2, would be observed.

Acknowledgments R.Š. was supported by the grants GAČR 202/09/0772 and SVV-267301, and J.P. by the grants GAČR P203/12/0118 and MSM0021620860.

References

1. Kundt, W.: Exact solutions of the field equations: twist-free pure radiation fields. Proc. R. Soc. London, Ser. A **270**, 328 (1962). doi:10.1098/rspa.1962.0224
2. Stephani, H., Kramer, D., MacCallum, M., Hoenselaers, C., Herlt, E.: Exact solutions of Einstein's field equations, 2nd edn. Cambridge Monographs on Mathematical Physics. Cambridge University Press, Cambridge (2003)
3. Griffiths, J., Podolský, J.: Exact space-times in Einstein's general relativity. Cambridge Monographs on Mathematical Physics. Cambridge University Press, Cambridge (2012)
4. Podolský, J., Žofka, M.: General Kundt spacetimes in higher dimensions. Class. Quantum Gravity **26**, 105008 (2009). doi:10.1088/0264-9381/26/10/105008
5. Coley, A., Hervik, S., Papadopoulos, G., Pelavas, N.: Kundt spacetimes. Class. Quantum Gravity **26**, 105016 (2009). doi:10.1088/0264-9381/26/10/105016
6. Podolský, J., Švarc, R.: Interpreting spacetimes of any dimension using geodesic deviation. Phys. Rev. D **85**, 044057 (2012). doi:10.1103/PhysRevD.85.044057
7. Krtouš, P., Podolský, J.: Asymptotic structure of radiation in higher dimensions. Class. Quantum Gravity **23**, 1603 (2006). doi:10.1088/0264-9381/23/5/011
8. Podolský, J., Švarc, R.: Explicit algebraic classification of Kundt geometries in any dimension. Class. Quantum Gravity **30**, 125007 (2013). doi:10.1088/0264-9381/30/12/125007
9. Podolský, J., Švarc, R.: Physical interpretation of Kundt spacetimes using geodesic deviation. Class. Quantum Gravity **30**, 205016 (2013). doi:10.1088/0264-9381/30/20/205016
10. Coley, A., Fuster, A., Hervik, S., Pelavas, N.: Higher dimensional VSI spacetimes, Class. Quantum Gravity **23**, 7431 (2006). doi:10.1088/0264-9381/23/24/014

A Class of Conformal Curves on Spherically Symmetric Spacetimes

Juan A. Valiente Kroon and Christian Lübbe

Abstract A class of curves with special conformal properties (conformal curves) is studied on the Reissner–Nordström spacetime. It is shown that initial data for the conformal curves can be prescribed so that the resulting congruence of curves does not contain any conjugate points in the domain of outer communication of the spacetime and extend smoothly to future and past null infinity. The results of this analysis are expected to be of relevance for the discussion of the Reissner–Nordström and other spherically symmetric spacetimes as solutions to the conformal field equations and for the numerical global evaluation of static black hole spacetimes.

1 Introduction

Conformal methods constitute a powerful tool for the discussion of global properties of spacetimes—in particular those representing black holes. The conformal structure of static electrovacuum black hole spacetimes is, to some extent, well understood—see e.g. [1, 2]. However, the constructions involved often require several changes of variables and the introduction of some type of null coordinates. This choice of coordinates may not be the most convenient to undertake an analysis of global or asymptotic properties of a spacetime by means of the *conformal Einstein field equations*—see e.g. the discussion in [3]. A key issue in this respect, is how to construct in a systematic or canonical fashion a conformal extension of the spacetime which, in addition, eases the analysis of the underlying conformal field equations—for a

J. A. V. Kroon (✉)
School of Mathematical Sciences, Queen Mary University of London,
Mile End Road, London E1 4NS, UK
e-mail: j.a.valiente-kroon@qmul.ac.uk

C. Lübbe
Department of Mathematics, University College London, Gower Street,
London WC1E 6BT, UK

J. Bičák and T. Ledvinka (eds.), *Relativity and Gravitation*,
Springer Proceedings in Physics 157, DOI: 10.1007/978-3-319-06761-2_30,

review of the conformal equations and the issues involved in their analysis see e.g. [4]. In the case of vacuum spacetimes, gauges based on the use of *conformal geodesics* offer such a systematic approach—see e.g. [5, 6]. Conformal geodesics are invariants of the conformal structure: a conformal transformations map conformal geodesics into a conformal geodesics—this is not the case with standard geodesics unless they are null.

One of the main advantages of the use of conformal geodesics in the construction of gauge (and coordinate) systems in a vacuum spacetime is that they provide an a priori conformal factor which can be read off directly from the data one has specified to generate the congruence of conformal geodesics. Hence, one has a canonical procedure to generate a conformal extension of the spacetime in question. In addition, gauge systems based on conformal geodesics give rise to a fairly straightforward hyperbolic reduction of the conformal Einstein field equations in which most of the evolution equations are, in fact, transport equations—see e.g. [4, 7].

The useful property of having an a priori conformal factor is lost when one considers conformal geodesics in non-vacuum spacetimes. Nevertheless, in [8] it has been shown that this property can be recovered if one considers a more general class of curves—the *conformal curves*. These conformal curves satisfy equations similar to the conformal geodesic equations, but with a different coupling to the curvature of the spacetime. In the vacuum case they coincide with the conformal geodesic equations. Gauges based on this class of curves have been used in [8] to revisit the stability proofs for the Minkowski and the de Sitter spacetimes given in [9] and to obtain a stability result for purely radiative electrovacuum spacetimes. An important remark concerning these results is that they deal with spacetimes, or are restricted to regions of spacetimes, that do not contain singularities.

2 Conformal Curves

Given an interval $I \subseteq \mathbb{R}$, let $\boldsymbol{x}(\tau)$, with $\tau \in I$, denote a curve in a spacetime $(\tilde{\mathcal{M}}, \tilde{\boldsymbol{g}})$ and let $\boldsymbol{b}(\tau)$ denote a 1-form along $\boldsymbol{x}(\tau)$. Furthermore, let $\dot{\boldsymbol{x}} \equiv \mathrm{d}\boldsymbol{x}/\mathrm{d}\tau$ denote the tangent vector field of the curve $\boldsymbol{x}(\tau)$. The pairing between the vector $\dot{\boldsymbol{x}}$ and the 1-form $\boldsymbol{b}$ is denoted by $\langle \boldsymbol{b}, \dot{\boldsymbol{x}} \rangle$. In [8] the following equations have been introduced for the pair $(\boldsymbol{x}(\tau), \boldsymbol{b}(\tau))$:

$$\tilde{\nabla}_{\dot{x}}\dot{\boldsymbol{x}} = -2\langle \boldsymbol{b}, \dot{\boldsymbol{x}} \rangle \dot{\boldsymbol{x}} + \tilde{\boldsymbol{g}}(\dot{\boldsymbol{x}}, \dot{\boldsymbol{x}})\boldsymbol{b}^{\sharp}, \tag{1a}$$

$$\tilde{\nabla}_{\dot{x}}\boldsymbol{b} = \langle \boldsymbol{b}, \dot{\boldsymbol{x}} \rangle \boldsymbol{b} - \tfrac{1}{2}\tilde{\boldsymbol{g}}^{\sharp}(\boldsymbol{b}, \boldsymbol{b})\dot{\boldsymbol{x}}^{\flat} + \tilde{\boldsymbol{H}}(\dot{\boldsymbol{x}}, \cdot), \tag{1b}$$

where $\tilde{\nabla}_{\dot{x}}$ denotes the directional derivative of the Levi–Civita connection of the metric $\tilde{\boldsymbol{g}}$, while $\tilde{\boldsymbol{H}}$ denotes a rank 2 covariant tensor which upon the conformal transformation $\boldsymbol{g} = \Theta^2 \tilde{\boldsymbol{g}}$ transforms as:

$$\tilde{H}_{\mu\nu} - H_{\mu\nu} = \nabla_\mu \Upsilon_\nu + \Upsilon_\mu \Upsilon_\nu - \tfrac{1}{2} g^{\lambda\rho} \Upsilon_\lambda \Upsilon_\rho \, g_{\mu\nu},$$

where $\Upsilon_\mu \equiv \Theta^{-1}\nabla_\mu\Theta$. This transformation law is formally identical to that of the *Schouten tensor*

$$\tilde{L}_{\mu\nu} \equiv \tfrac{1}{2}(\tilde{R}_{\mu\nu} - \tfrac{1}{6}\tilde{R}\,\tilde{g}_{\mu\nu}).$$

The equation (1a)–(1b) are called the *conformal curve equations*. In [8] the choice $\tilde{\boldsymbol{H}} = \lambda\tilde{\boldsymbol{g}}$, has been adopted so that equation (1b) reduces to

$$\tilde{\nabla}_{\dot{x}}\boldsymbol{b} = \langle \boldsymbol{b}, \dot{\boldsymbol{x}}\rangle \boldsymbol{b} - \tfrac{1}{2}\tilde{\boldsymbol{g}}^\sharp(\boldsymbol{b},\boldsymbol{b})\dot{\boldsymbol{x}}^\flat + \lambda\dot{\boldsymbol{x}}^\flat. \tag{2}$$

In the following analysis we fix the initial conditions for the conformal curve equations such that the curves of the congruence are orthogonal to the initial hypersurface $\mathscr{S}$, while the initial value of the 1-form $\boldsymbol{b}$ is given by $\Omega^{-1}\mathrm{d}\Omega$ where Ω is a conformal factor inducing a 1-point compactification of an asymptotically Euclidean end.

3 Conformal Curves in the Reissner–Nordström Spacetime

A natural question to be asked is whether conformal geodesics, and more generally, the class of conformal curves introduced in [10] can be used to analyse global aspects of black hole spacetimes. In this respect, in [5] it has been shown that the maximal extension of the Schwarzschild spacetime, the so-called Schwarzschild-Kruskal spacetime [11], can be covered with a congruence of conformal geodesics which has no conjugate points. The conformal Gaussian gauge system obtained using this congruence offers a vantage perspective for the study of conformal properties of the Schwarzschild spacetime and for its global evaluation by means of numerical methods—see e.g. [12].

In the present contribution we briefly review the main results of [10] where the question has been raised to what extent a similar construction can be performed for the Reissner–Nordström spacetime. The consideration of the Reissner–Nordström spacetime is, for several reasons, natural. The inclusion of the electromagnetic field provides a model of angular momentum—see e.g. [13, 14]. We expect our analysis to provide insights into more general (i.e. less symmetric) situations—e.g. the Kerr and Kerr–Newman spacetimes. In addition, there is the expectation that black hole spacetimes with timelike singularities could be more tractable from the point of view of the conformal geometry than black holes with spacelike singularities. This expectation is based on the analysis of the structure of spatial infinity of the Schwarzschild spacetime. In this case the well understood divergence of the Weyl tensor at spatial infinity can also be regarded as the timelike singularity of a negative mass Schwarzschild spacetime—see e.g. [15] for a conformal diagram of this.

The main result of the analysis in [10] is the following:

Theorem 1 *The domains of outer communication of the extremal and the non-extremal Reissner–Nordström spacetime can be covered with a timelike congruence of conformal curves which contains no conjugate points. The congruence of conformal curves extends smoothly to null infinity.*

Numerical solutions of the conformal curve equations show, in fact, that conjugate points do arise after the curves have crossed the horizon and entered the black hole region. From the perspective of the Cauchy problem for the Einstein field equations, these conjugate points are not a major concern as one is mainly interested in the behaviour of the spacetime in the domain of outer communication and at the horizon. This is, in particular, the case in the problem of the so-called *non-linear stability of black hole spacetimes*—see e.g. [16].

Our main result provides a suitable conformal gauge to analyse the properties of the Reissner–Nordström spacetime by means of the conformal Einstein field equations. In particular, it opens the possibility of global numerical evaluations of the spacetime [17] similar to the ones carried out in [12] for the Schwarzschild spacetime.

4 Conformal Curves in Other Spacetimes

The analysis of [5] and [10] suggest the possibility of obtaining analogues of Theorem 1 for the Schwarzschild–de Sitter and Schwarzschild–anti de Sitter spacetimes. More generally, one could also consider the Reissner–Nordström–de Sitter/anti-de Sitter spacetimes. These spacetimes are static. In order to consider dynamical situations, while at the same time retaining the assumption of spherical symmetry, one has to bring into play other matter models. A suitable test case is given by the Vaidya solution. Alternatively, one could consider spacetimes arising from the (spherically symmetric) Einstein-scalar field and Einstein–Maxwell-scalar field system.

In general terms, it is conjectured that for spacetimes with a Maxwell field it should be possible to obtain a result similar to that of Theorem 1 where at least the outer domain of communication of the spacetime can be covered without having conjugate points. If no Maxwell field is present then the situation should resemble that of the analysis of the Schwarzschild spacetime in [5] where even the regions inside the black hole can be covered by the congruence.

5 Qualitative Behaviour of the Curves

For our choice of initial data, the analysis of [10] identified three different types of possible behaviours for the conformal curves. First of all, one has curves starting in the asymptotic region of an initial hypersurface $\mathscr{S}$ which extend up to (and including) future null infinity. Null infinity is reached for a finite value of a conformally

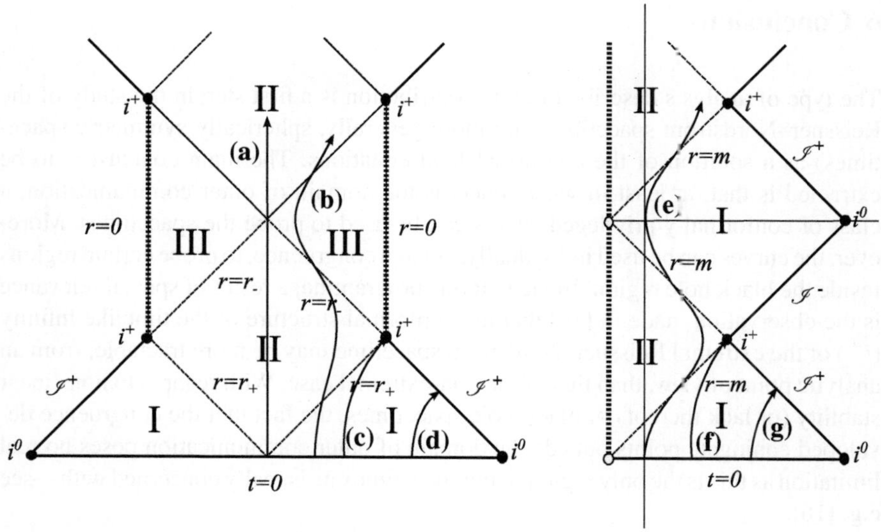

Fig. 1 Schematic illustration of the behaviour of conformal curves. To the *left* the non-extremal case: **a** the curve starting at $r_* = r_+$; **b** a curve with $r_+ < r_* < r_\circledast$; **c** the critical curve $r_* = r_\circledast$; **d** a curve with $r_* > r_\circledast$. To the *right* the extremal case: **e** a curve with $r_* < r_\circledast$; **f** the critical curve $r_* = r_\circledast$; **g** a curve with $r_* > r_\circledast$. *The curves are not depicted to scale*

privileged parameter τ. A different type of behaviour is given by conformal curves starting closer to the horizon. These curves exhibit a periodic behaviour on the conformal diagram of the spacetime, reaching the horizon in a finite value of τ and then entering the so-called regions II and III (in the non-extremal case) and the region II (in the extremal case) before re-emerging in a domain of outer communication. An extreme case of this behaviour is that of the conformal curve starting exactly at the bifurcation sphere of the non-extremal case—this corresponds to the curve (a) of Fig. 1. A curious property of these curves is that they cannot get arbitrarily close to the singularity. Hence, there are regions inside the black hole region which cannot be probed by means of the curves. Separating the behaviour of curves entering the black hole and those escaping to null infinity, one has *critical curves* which exactly hit timelike infinity (i^+) in a finite amount of the parameter τ. These curves are of special relevance for disentangling the conformal structure of i^+. In particular, the analysis of [10] shows that in the non-extremal case, the critical curves become null at i^+. A similar behaviour had been identified in the Schwarzschild spacetime [4]. Remarkably, for the extremal case the critical conformal curves remain timelike even at i^+.

6 Conclusions

The type of analysis described in this contribution is a first step in the study of the Reissner–Nordström spacetime (and more generally, spherically symmetric spacetimes) as a solution of the conformal field equations. The main conclusion to be extracted is that, at least in what concerns the domain of outer communication, a class of conformally privileged curves can be used to probe the spacetimes. Moreover, the curves can be used individually, not as a congruence, to probe certain regions inside the black hole region. In view of this programme, a result of special relevance is the observation made in [10] that the conformal structure of the timelike infinity (i^+) of the extremal Reissner–Nordström spacetime may be more tractable, from an analytic point of view, than that of the non-extremal case. With regards to non-linear stability (or lack thereof) of black-hole spacetimes, the fact that the congruence developed conjugate points outside the domain of outer communication poses no real limitation as this is the only region of the spacetime one is really concerned with—see e.g. [16].

Comparing the conformal diagrams of the Reissner–Nordström spacetime and the the Kerr spacetime, it is natural to wonder how much of the structure observed in the present analysis has an analogue in the Kerr solution. For example, it is natural to conjecture that the domain of outer communication of the Kerr spacetime can be covered by means of a non-singular congruence of conformal geodesics reaching beyond null infinity. It is likely that this congruence will degenerate after it has crossed the event horizon and that the curves will have some type of singularity avoiding properties so that there may exists regions in the black hole region which can not be probed by this congruence. A more tantalising possibility is that, as in the case of the extremal Reissner–Nordström spacetime, the extreme Kerr may have a more tractable structure at i^+. In any case, the analysis of conformal geodesics in the Kerr spacetime is bound to be much more complicated as the warped product structure of the line element is lost.

References

1. Griffiths, J., Podolský, J.: Exact Space–Times in Einstein's General Relativity. Cambridge Monographs on Mathematical Physics (Cambridge University Press, Cambridge, New York (2009)
2. Hawking, S., Ellis, G.: The Large Scale Structure of Space–Time. Cambridge Monographs on Mathematical Physics (Cambridge University Press, Cambridge (1973)
3. Friedrich, H.: Gravitational fields near space-like and null infinity, J. Geom. Phys. **24**, 83 (1998). doi:10.1016/S0393-0440(97)82168-7
4. Friedrich, H.: Conformal Einstein evolution. In: Friedrich, H., Frauendiener, J. (eds.) The Conformal Structure of Space–Time: Geometry, Analysis, Numerics. Lecture Notes in Physics, vol. 604, pp. 1–50. Springer, Berlin, New York (2002)
5. Friedrich, H.: Conformal geodesics on vacuum spacetimes. Comm. Math. Phys. **235**, 513 (2003). doi:10.1007/s00220-003-0794-8

6. Friedrich, H., Schmidt, B.: Conformal geodesics in general relativity. Proc. R. Soc. London, Ser. A **414**, 171 (1987)
7. Friedrich, H.: Smoothness at null infinity and the structure of initial data. In: Chruściel, P., Friedrich, H. (eds.) The Einstein Equations and the Large Scale Behavior of Gravitational Fields: 50 Years of the Cauchy Problem in General Relativity, pp. 121–203. Birkhäuser, Basel (2004)
8. Lübbe, C., Valiente Kroon, J.: The extended conformal Einstein field equations with matter: the Einstein–Maxwell field. J. Geom. Phys. **62**, 1548 (2012). doi:10.1016/j.geomphys.2012.01.009
9. Friedrich, H.: On the global existence and the asymptotic behaviour of solutions to the Einstein–Maxwell–Yang–Mills equations. J. Differ. Geom. **34**, 275 (1991)
10. Lübbe, C., Valiente Kroon, J.: A class of conformal curves in the Reissner–Nordström spacetime (2013). ArXiv e-prints arXiv:1301.5458 [gr-qc]
11. Kruskal, M.: Maximal extension of Schwarzschild metric. Phys. Rev. D **119**, 1743 (1960). doi:10.1103/PhysRev.119.1743
12. Zenginoğlu, A.: A conformal approach to numerical calculations of asymptotically flat spacetimes. Ph.D. thesis, Max Planck Institute for Gravitational Physics (AEI) and University of Potsdam, Potsdam (2006)
13. Dafermos, M.: Stability and instability of the Cauchy horizon for the spherically symmetric Einstein–Maxwell-scalar field equations. Ann. Math. **158**, 875 (2003). doi:10.4007/annals.2003.158.875
14. Dafermos, M.: The interior of charged black holes and the problem of uniqueness in general relativity. Comm. Pure Appl. Math. **LVIII**, 0445 (2005). doi:10.1002/cpa.20071
15. Schmidt, B., Walker, M.: Analytic conformal extensions of asymptotically flat spacetimes. J. Phys. A: Math. Gen. **16**, 2187 (1983). doi:10.1088/0305-4470/16/10/015
16. Dafermos, M., Rodnianski, I.: Lectures on black holes and linear waves (2008). ArXiv e-prints 0811.0354[gr-qc]
17. Valiente Kroon, J.: Global evaluations of static black hole spacetimes (2012). (In preparation)

Black Hole Collisions in Asymptotically de Sitter Spacetimes

Miguel Zilhão , Vitor Cardoso, Leonardo Gualtieri, Carlos Herdeiro, Ulrich Sperhake and Helvi Witek

Abstract We report on the first dynamical evolutions of black holes in asymptotically de Sitter spacetimes. We focus on the head-on collision of equal mass binaries and compare analytical and perturbative methods with full blown nonlin-

M. Zilhão (✉)
Centro de Física do Porto, Departamento de Física e Astronomia, Faculdade de Ciências da Universidade do Porto, Rua do Campo Alegre, 4169-007 Porto, Portugal
e-mail: mzilhao@fc.up.pt

M. Zilhão
Center for Computational Relativity and Gravitation, Rochester Institute of Technology, Rochester, NY 14623, USA

M. Zilhão · C. Herdeiro
Departamento de Física da Universidade de Aveiro, Campus de Santiago, 3810-183 Aveiro, Portugal

V. Cardoso
Department of Physics and Astronomy, University of Mississippi, Oxford, MS 38677-1848, USA

L. Gualtieri
Dipartimento di Física, "Sapienza" Università di Roma, Piazzale Aldo Moro 5, 00185 Roma, RM, Italy

L. Gualtieri
Sezione Roma1, INFN, Piazzale Aldo Moro 5, 00185 Roma, RM, Italy

U. Sperhake
Institut de Ciències de l'Espai (CSIC-IEEC), Facultat de Ciències Campus UAB, E-08193 Bellaterra, Spain

U. Sperhake
California Institute of Technology, Pasadena, CA 91125, USA

V. Cardoso · U. Sperhake · H. Witek
Centro Multidisciplinar de Astrofísica—CENTRA, Departamento de Física, Instituto Superior Técnico–IST, Universidade de Lisboa, Av. Rovisco Pais 1, 1049-001 Lisboa, Portugal

U. Sperhake · H. Witek
DAMTP, Centre for Mathematical Sciences, Wilberforce Road, Cambridge CB3 0WA, UK

J. Bičák and T. Ledvinka (eds.), *Relativity and Gravitation*,
Springer Proceedings in Physics 157, DOI: 10.1007/978-3-319-06761-2_31,

ear simulations. Our results include an accurate determination of the merger/scatter transition (consequence of an expanding background) for small mass binaries and a test of the Cosmic Censorship conjecture, for large mass binaries. We observe that, even starting from small separations, black holes in large mass binaries eventually lose causal contact, in agreement with the conjecture.

1 Introduction

Nonlinear dynamics in cosmological backgrounds has the potential to teach us immensely about our universe, and also to serve as prototype for nonlinear processes in generic curved spacetimes. *de Sitter* spacetime is the simplest accelerating universe—a maximally symmetric solution of Einstein's equations with a positive cosmological constant—which seems to model quite well the present cosmological acceleration [1].

Key questions concerning the evolution towards a de Sitter, spatially homogeneous universe are how inhomogeneities develop in time and, in particular, if they are washed away by the cosmological expansion [2]. Answering them requires controlling the imprint of the gravitational interaction between localised objects on the large-scale expansion. Conversely, the cosmological dynamics should leave imprints in strong gravitational phenomena like primordial black hole formation [3] or the gravitational radiation emitted in a black hole binary coalescence, which carry signatures of the cosmological acceleration as it travels across the universe. Identifying these signatures is not only of conceptual interest but also phenomenologically relevant, in view of the ongoing efforts to directly detect gravitational radiation.

Finally, dynamics in asymptotically de Sitter spacetimes could also teach us about more fundamental questions such as cosmic censorship: two black holes of sufficiently large mass in de Sitter spacetime would, upon merger, give rise to a too large black hole to fit in its cosmological horizon. In this case the end state would be a naked singularity. This possibility begs for a time evolution of such a configuration. Does the time evolution of non-singular data containing two black holes result in a naked singularity, or are potentially offending black holes simply driven away from each other by the cosmological expansion?

Following [4], we here report on numerical evolutions of black hole binaries in an asymptotically de Sitter geometry. Even though we consider a range of values for the cosmological constant far larger than those which are phenomenologically viable, these results provide useful insight on the general features of dynamical black hole processes in spacetimes with a cosmological constant, which can improve our understanding of our universe.

1.1 Schwarzschild-de Sitter

The Schwarzschild-de Sitter spacetime, written in static coordinates reads

$$ds^2 = -f(R)dT^2 + f(R)^{-1}dR^2 + R^2 d\Omega_2. \quad (1)$$

The solution is characterised by two parameters: the black hole mass m and the Hubble parameter H,

$$f(R) = 1 - 2m/R - H^2R^2, \quad H \equiv \sqrt{\Lambda/3}. \quad (2)$$

$f(R)$ has two zeros, at $R = R_\pm$, $R_- < R_+$, if

$$0 < mH < mH_{\text{crit}}, \quad mH_{\text{crit}} \equiv \sqrt{1/27}. \quad (3)$$

These zeros are the location of the black hole event horizon (R_-) and of a cosmological horizon (R_+). If $H = 0$, then $R_- = 2m$; if $m = 0$, then $R_+ = 1/H$. If $H, m \neq 0$, then $R_- > 2m$ and $R_+ < 1/H$.

The basic dynamics in this spacetime may be inferred by looking at radial timelike geodesics. They obey the equation $(dR/d\tau)^2 = E^2 - f(R)$, where τ is the proper time and E is the conserved quantity associated to the Killing vector field $\partial/\partial T$. In the static patch ($R_- < R < R_+$), E can be regarded as energy. From this equation we see that $f(R)$ is an effective potential. This potential has a maximum at

$$R_{\max} = (m/H^2)^{1/3}. \quad (4)$$

Geodesics starting from rest (i.e. $dR/d\tau(\tau = \tau_0) = 0$) will fall into the black hole if $R_- < R < R_{\max}$ or move away from the black hole if $R_{\max} < R < R_+$.

As we will discuss in the next section, the initial data for an evolution in the de Sitter universe can be computed in a similar manner as has been done in asymptotically flat space as long as one chooses a foliation with extrinsic curvature K_{ij} having only a trace part [5, 6]. Such a coordinate system is known for Schwarzschild-de Sitter: *McVittie coordinates* [7]. These are obtained from static coordinates by the transformation $(T, R) \to (t, r)$ given by

$$R = (1 + \xi)^2 a(t) r, \quad T = t + H \int \frac{R\,dR}{f(R)\sqrt{1 - 2m/R}}, \quad (5)$$

where $a(t) = \exp(Ht)$ and $\xi \equiv \frac{m}{2a(t)r}$. One obtains McVittie's form for Schwarzschild-de Sitter:

$$ds^2 = -\left(\frac{1-\xi}{1+\xi}\right)^2 dt^2 + a(t)^2(1+\xi)^4(dr^2 + r^2 d\Omega_2). \quad (6)$$

By setting $m = 0$ in McVittie coordinates one recovers an FRW cosmological model with flat spatial curvature and an exponentially growing scale factor. The cosmological horizon $\mathcal{H}_C$ discussed above, located at $R = 1/H$, stands at $r_{\mathcal{H}_C} = 1/(He^{Ht})$. The spatial sections of $\mathcal{H}_C$ seem to be shrinking down in this coordinate system. What happens, in fact, is that the exponentially fast expansion is taking any observer to the outside of $\mathscr{H}_C$. This is a well known phenomenon in studies of inflation and has important consequences for the numerical evolution.

1.2 Numerical Setup

For the numerical implementation we make use of the Baumgarte, Shapiro, Shibata and Nakamura (BSSN) formulation, *e.g.* [8, 9] with the *moving puncture* technique [10, 11]. In terms of the BSSN variables $\chi, \tilde{\gamma}_{ij}, \tilde{A}_{ij}, \tilde{\Gamma}^k$ (cf. App. A in [9]), the evolution equations are the standard ones except for the one of the trace of the extrinsic curvature which becomes $\left(\partial_t - \mathscr{L}_\beta\right) K = [\ldots] - \alpha\Lambda$, where $[\ldots]$ denotes the standard right-hand side of the BSSN equations in the absence of source terms (see *e.g.* [8]). Moreover, a new variable $\bar{\chi} = \exp(2Ht)\chi$ was evolved instead of χ [2].

In references [5, 6] it was observed that imposing a spacetime slicing obeying $K^i{}_j = -H\delta^i_j$, and a spatial metric of the form $\mathrm{d}l^2 = \psi^4\tilde{\gamma}_{ij}\mathrm{d}x^i\mathrm{d}x^j$, the equations to be solved in order to obtain initial data are equivalent to those in vacuum. In particular, for a system of N black holes momentarily at rest (with respect to the given spatial coordinate patch), the conformal factor ψ takes the form

$$\psi = 1 + \sum_{i=1}^{N} \frac{m_i}{2|r - r_{(i)}|}. \tag{7}$$

There are $N + 1$ asymptotically de Sitter regions, as $|r - r_{(i)}| \to 0, +\infty$; the total mass for observers in the common asymptotic region ($|r - r_{(i)}| \to +\infty$) is $\sum_i m_i$ [6].

1.3 Numerical Results

For binary black hole initial data, we start by reproducing the results of Nakao et. al [6], where the critical distance between two black holes for the existence of a common Black hole Apparent Horizon (BAH) already at $t = 0$ was studied. We thus prepare initial data (7) with $m_1 = m_2$ and take all quantities in units of the total mass $m = m_1 + m_2$. The two punctures are set initially at symmetric positions along the z axis. The critical value for the cosmological constant, for which the black hole and cosmological horizon coincide is now $mH_{\text{crit}} = 1/\sqrt{27}$. We call *small (large) mass binaries* those, for which $H < H_{\text{crit}}$ ($H > H_{\text{crit}}$). Our results for the critical separation in small mass binaries, at $t = 0$, as function of the Hubble parameter are shown in Fig. 1. The line (diamond symbols) agrees, after a necessary normalisation, with Fig. 14 of [6].

We now consider head-on collisions of two black holes with no initial momentum, i.e. the time evolution of these data. For subcritical Hubble constant $H < H_{\text{crit}} =$

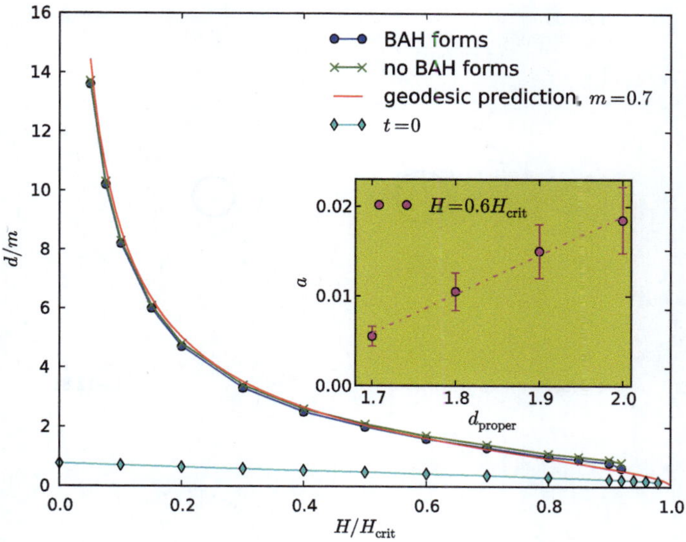

Fig. 1 Critical coordinate distance for small mass binaries as a function of $H/H_{\rm crit}$. We obtain this estimate from the coordinate distance to the horizon, (4), for a particular value of m. The $t = 0$ line refers to the critical separation between having or not having a common BAH in the initial data. The inset shows details of the approach to the critical line for $H = 0.6H_{\rm crit}$, where a is an acceleration parameter

$1/(\sqrt{27}m)$, we monitor the evolution of the areal radius of the BAHs and that of the Cosmological Apparent Horizons (CAHs) of an observer at $z = 0$. For instance, for $H = 0.9H_{\rm crit}$ and proper (initial) separation $3.69m$ we find that the areal radii of the BAH and CAH are approximately constant and equal to $R_{\rm BAH} \simeq 2.36m$ and $R_{\rm CAH} \simeq 4.16m$, respectively. As expected the two initial BAHs, as well as the final horizon, are inside the CAH. As a comparison, a Schwarzschild-de Sitter spacetime with the same H has $R_{\rm BAH} \simeq 2.43m$ and $R_{\rm CAH} \simeq 4.16m$. This suggests that the interaction effects (binding energy and emission of gravitational radiation) are of the order of a few per cent for this configuration.

By performing a large set of simulations for various cosmological parameters H and initial distance d, we have bracketed the critical distance for the merger/scatter region as a function of the Hubble parameter H for the "dynamical" case, i.e., the initial *coordinate* distance between the black holes such that no common BAH forms. The results are displayed in Fig. 1 (circles and × symbols).

As expected the critical distance becomes larger as compared to the initial data value ("$t = 0$" line): there are configurations for which a common BAH is absent in the initial data but appears during the evolution (just as in asymptotically flat spacetime). The numerical results can be qualitatively well approximated by a point particle prediction—from (4). To do such comparison a transformation to McVittie coordinates needs to be done; we have performed such transformation at McVittie time $t = 0$. Intriguingly, for a particular value of $m \simeq 0.7$, the point particle approx-

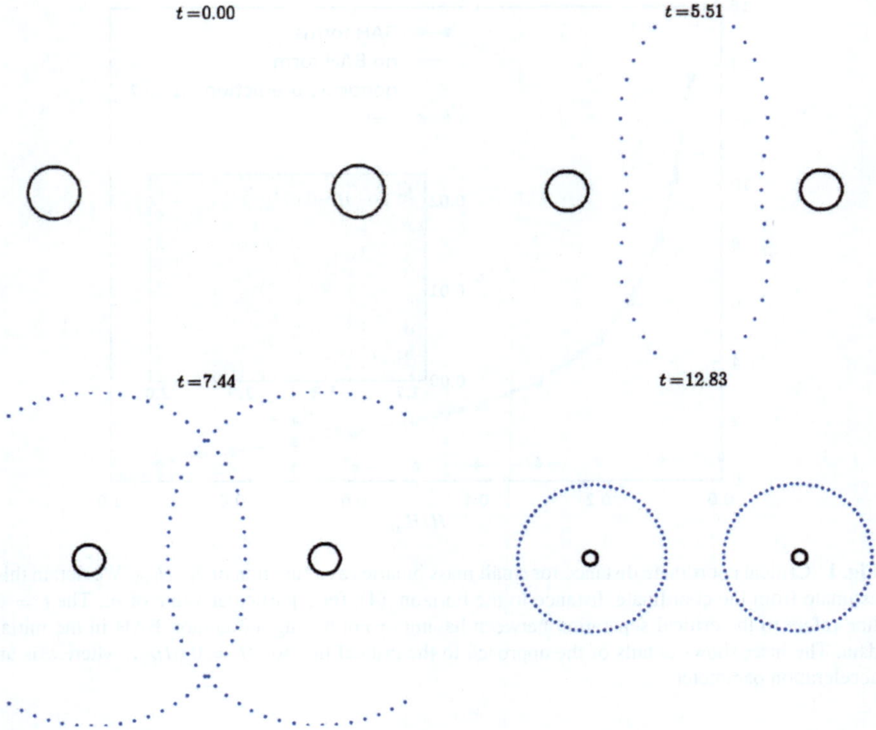

Fig. 2 Snapshots at different times of a simulation with $H = 1.05H_{\rm crit}$, and an initial coordinate distance $d/m = 1.5002$. The dotted *blue line* denotes the CAHs (for the observers moving with the *black holes*). The full animation for this simulation can be found in [12]

imation matches quantitatively very well the numerical result; the curve obtained from the geodesic prediction in Fig. 1 is barely distinguishable from the numerical results.

A further interesting feature concerns the approach to the critical line. For an initially static binary close to the critical initial separation, the coordinate distance d scales as $d = d_0 + at^2$. In general the acceleration parameter scales as $\log a = C + \Gamma \log(d - d_0)$, where $\Gamma = 1$ in the geodesic approximation. A fit to our numerical results for $H = 0.6H_{\rm crit}$ (dashed curve in the inset of Fig. 1) for example yields $C = -3.1$, $\Gamma = 0.9$ in rough agreement with this expectation. Details of this regime are given in the inset of Fig. 1.

Finally, we have performed evolutions with $H > H_{\rm crit}$. On the assumption of weak gravitational wave release, such evolutions can test the cosmic censorship conjecture since the observation of a merger in such case would reveal a violation of the conjecture [13]. From general arguments and from the simulations with $H < H_{\rm crit}$, we know the cosmological repulsion will dominate for sufficiently large initial distance and in that case we can even expect that a CAH for the observer at $z = 0$ will not encompass the BAHs. This indicates the black holes are no longer in causal contact and therefore can never merge. Our numerical results confirm this overall picture. To test the potentially dangerous configurations, we focus on the regime in which the black holes are initially very close. A typical example is depicted in Fig. 2, for a supercritical cosmological constant $H = 1.05 H_{\text{crit}}$, and an initial coordinate distance $d/m = 1.5002$. Even though the initial separation is very small, we find that the holes move *away* from each other, with a proper separation increasing as the simulation progresses. In fact, further into the evolution, a distorted CAH appears, and remains for as long as the simulation lasts. The evolution therefore indicates that the spacetime becomes, to an excellent approximation, empty de Sitter space for the observer at $z = 0$ and that the black holes are not in causal contact. Observe that qualitatively similar evolutions can be found in small mass binaries when the initial distance is larger than the critical value

1.4 Final Remarks

We have presented evidence that the numerical evolution of black hole spacetimes in de Sitter universes is under control. Our results open the door to new studies of strong field gravity in cosmologically interesting scenarios. In closing, we would like to mention that our results are compatible with cosmic censorship in cosmological backgrounds. However, an analytic solution with multiple (charged and extremal) black holes in asymptotically de Sitter spacetime is known, and has been used to study cosmic censorship violations [14]. In *collapsing* universes a potential violation of the conjecture has been reported, although the conclusion relied on singular initial data. To clarify this issue, it would be of great interest to perform numerical evolution of large mass black hole binaries, analogous to those performed herein, but in collapsing universes. This will require adaptations of our setup, since the "expanding" behaviour discussed of the coordinate system will turn into a "collapsing" one, which raises new numerical challenges.

References

1. Komatsu, E., Smith, K., Dunkley, J., et al.: Seven-year Wilkinson microwave anisotropy probe (WMAP) observations: cosmological interpretation. Astrophys. J. Suppl. Ser. **192**, 18 (2011). doi:10.1088/0067-0049/192/2/18
2. Shibata, M., Nakao, K.i., Nakamura, T., Maeda, K.i.: Dynamical evolution of gravitational waves in the asymptotically de Sitter space-time. Phys. Rev. D 50, **708** (1994). doi:10.1103/PhysRevD.50.708
3. Shibata, M., Sasaki, M.: Black hole formation in the Friedmann universe: formulation and computation in numerical relativity. Phys. Rev. D **60**, 084002 (1999). doi:10.1103/PhysRevD.60.084002
4. Zilhão, M., Cardoso, V., Gualtieri, L., et al.: Dynamics of black holes in de Sitter spacetimes. Phys. Rev. D **85**, 104039 (2012). doi:10.1103/PhysRevD.85.104039

5. Nakao, K.I., Nakamura, T., Oohara, K.I., Maeda, K.I.: Numerical study of cosmic no-hair conjecture: formalism and linear analysis. Phys. Rev. D **43**, 1788 (1991). doi:10.1103/PhysRevD.43.1788
6. Nakao, K.I., Yamamoto, K., Maeda, K.I.: Apparent horizons of an N-black-hole system in a space-time with a cosmological constant. Phys. Rev. D **47**, 3203 (1993). doi:10.1103/PhysRevD.47.3203
7. McVittie, G.: The mass-particle in an expanding universe. Mon. Not. R. Astron. Soc. **93**, 325 (1933)
8. Alcubierre, M.: Introduction to 3+1 numerical relativity, International Series of Monographs on Physics, vol. 140. Oxford University Press, Oxford (2008)
9. Sperhake, U.: Binary black-hole evolutions of excision and puncture data. Phys. Rev. D **76**, 104015 (2007). doi:10.1103/PhysRevD.76.104015
10. Campanelli, M., Lousto, C., Marronetti, P., Zlochower, Y.: Accurate evolutions of orbiting black-hole binaries without excision. Phys. Rev. Lett. **96**, 111101 (2006). doi:10.1103/PhysRevLett.96.111101
11. Baker, J.G., Centrella, J., Choi, D.I., Koppitz, M., van Meter, J.: Gravitational wave extraction from an inspiraling configuration of merging black holes. Phys. Rev. Lett. **96**, 111102 (2006). doi:10.1103/PhysRevLett.96.111102
12. Black hole dynamics beyond general relativity (containing animations of black hole collisions). http://blackholes.ist.utl.pt/?page=Files#h-4
13. Hayward, S., Shiromizu, T., Nakao, K.I.: A cosmological constant limits the size of black holes. Phys. Rev. D **49**, 5080 (1994). doi:10.1103/PhysRevD.49.5080
14. Brill, D., Horowitz, G., Kastor, D., Traschen, J.: Testing cosmic censorship with black hole collisions. Phys. Rev. D **49**, 840 (1994). doi:10.1103/PhysRevD.49.840

On the Effects of Rotating Gravitational Waves

Jiří Bičák, Joseph Katz, Tomáš Ledvinka and Donald Lynden-Bell

Abstract We study effects of gravitational waves which in the first order form a time-symmetric ingoing and then outgoing pulse of rotating waves. The influence of the angular momentum of these waves on the rotation of local inertial frames with respect to the frames at great distances is analyzed by solving the relevant Einstein equation to second order. Also the apparent motions of the fixed stars on the celestial sphere as seen through rotating waves from the local inertial frame in the centre are calculated and displayed.

1 Introduction

It was just 100 years ago in Prague when Einstein wrote the paper [1] in which he, for the first time, expressed his understanding of Mach's Principle. Within his pre-General Relativity theory in which there was only one metric function he considered a mass point inside a shell accelerated "upwards" and found that the mass-point is dragged along by the shell.

Many formulations and studies of Mach's Principle appeared during the last 100 years, most of them were analyzed in the Tübingen conference in 1993 which led to the remarkable volume [2] containing lectures as well as valuable discussions. We

J. Bičák · T. Ledvinka (✉)
Institute of Theoretical Physics, Charles University,
180 00 Prague 8, Czech Republic
e-mail: tomas.ledvinka@mff.cuni.cz

J. Bičák · D. Lynden-Bell
Institute of Astronomy, Madingley Road, Cambridge CB3 0HA, UK

J. Bičák
Albert Einstein Institute, Am Mühlenberg 1, 14476 Golm, Germany

J. Katz
The Racah Institute of Physics, Givat Ram, 91904 Jerusalem, Israel

J. Bičák and T. Ledvinka (eds.), *Relativity and Gravitation*,
Springer Proceedings in Physics 157, DOI: 10.1007/978-3-319-06761-2_32,

studied Machian effects in various contexts, both in asymptotically flat spacetimes and within cosmological perturbation theory—see, e.g., [3], and number of references therein; later, cf. Schmid [4].

More recently, we investigated a subtle question whether dragging of inertial frames should be attributed also to gravitational waves. After the discovery of binary pulsars losing energy and angular momentum as a consequence of emitting gravitational radiation it would be surprising if gravitational waves did not have an influence on local inertial frames. However, there are still doubts uttered about the status of gravitational stress-energy as compared with stress-energy tensor $T_{\mu\nu}$ of matter in relation to Machian ideas (see, e.g., [2], p. 83). In [5, 6] we analyzed dragging by cylindrical gravitational waves.

In the present work based on Ref. [7] we investigate the effects of rotating gravitational waves in a more general, asymptotically flat setting. We again start out from linearized theory and construct an ingoing rotating pulse of radiation which later transforms into an outgoing pulse. While in the cylindrical case our waves were characterized by just one harmonic index m governing the number of wave crests in φ, now the situation becomes considerably richer involving both spherical harmonic indices l and m.

Near the origin the first-order metric of our waves behaves as r^l, so the region around the origin will be very nearly flat for l sufficiently large. When, however, a local inertial frame is introduced at the origin, we find that its axes rotate with the angular velocity ω_0 with respect to the lines $\varphi = \text{const}$ of the global frame, i.e., with respect to stars at infinity. The situation thus indeed resembles the interior of a collapsing slowly rotating shell—see [8] where the vorticity of the lines $\varphi = \text{const}$ is given in covariant form.

2 Rotating Scalar and Gravitational Waves

We first construct a solution of the scalar wave equation in a form of the rotating wave pulse written in spherical coordinates t, r, θ, ϕ

$$\psi_{lm}(t, r, \theta, \phi) = Q_l(t, r) Y_{lm}(\theta, \phi) = B_l 2^l l! \, \frac{(r/a)^l Y_{lm}(\theta, \phi)}{\left[\left((a + it)^2 + r^2\right)/a^2\right]^{l+1}}, \tag{1}$$

where l, m are harmonic indices, a is typical wave pulse width and B_l is the amplitude. Here both numerator and denominator are complex functions so the actual wave profile given by its real part has plenty of features as can be seen in Fig. 1. Among them the rotating character of the wave, its regularity and the fact that for high values of l the wave is concentrated near a shell with radius $r^2 = a^2 + t^2$ are most important.

In the construction of rotating gravitational waves within linearized Einstein theory we then use the Regge-Wheeler equation for odd-parity waves [9] which on the flat Minkowski background simplifies to a usual wave equation $\Box\psi = 0$. We decom-

Fig. 1 The snapshots of the rotating waves ψ_{lm} in times $t/a = -1, -\frac{1}{2}, 0, \frac{1}{2}, 1$ (from *left* to *right*) for $l = 17, m = 11$. Waves are localized in the radial direction so that they resemble a falling and rotating shell

pose the metric perturbations into tensor harmonics [9–11] and consider only the odd-parity waves with the Regge-Wheeler gauge condition ($(i) = (1), (2)$ denotes the first- and second-order perturbations)

$$h_{\mu\nu}^{(i)} = \sum_{lm} \frac{\sqrt{2l(l+1)}}{r}\Big[- h_{0lm}^{(i)}(t,r) c_{0lm\,\mu\nu} + i h_{1lm}^{(i)}(t,r) c_{lm\,\mu\nu} \Big], \qquad (2)$$

where c_{0lm}, c_{lm} are the odd-parity harmonics [11]. The first order radial functions $h_{0lm}^{(1)}(t,r) = -\partial_r(r^2 Q_l)/(l^2+l-2)$ and $h_{1lm}^{(1)}(t,r) = -\partial_t(r^2 Q_l)/(l^2+l-2)$ of the odd-parity metric perturbations we directly obtain from the radial part $Q_l(t,r)$ of the scalar field ψ_{lm}.

In [7] we also analyze and relate the energy and angular momentum densities of scalar and gravitational waves.

3 Second Order Perturbations

To determine the influence of gravitational waves $h_{\mu\nu}^{(1)}$ on the rotation of local inertial frames at the axis of symmetry due to the second-order metric perturbations $h_{\mu\nu}^{(2)}$ we solve the equations

$$R_{\mu\nu}^{(1)}[h^{(2)}] = - <R_{\mu\nu}^{(2)}[h^{(1)}, h^{(1)}]>, \qquad (3)$$

where we introduced the averaging symbol $<>$. We expand both sides in tensor spherical harmonics. For general l the l.h.s. yields a hyperbolic set of equations for radial functions $h_0^{(2)}$ and $h_1^{(2)}$ indicating non-instantaneous effects, but the inertial frames at the origin will be influenced primarily by the *dipole* perturbations and it is well known that for $l = 1$ one can fix $h_1^{(2)} = 0$ by an appropriate gauge transformation [9, 12] and arrive thus at *elliptic* equation for $h_0^{(2)}$. In this equation the axially symmetric component $<R_{t\varphi}^{(2)}[h^{(1)}, h^{(1)}]>$ on the r.h.s. of (3) appears as the only source for the dipole second-order perturbations

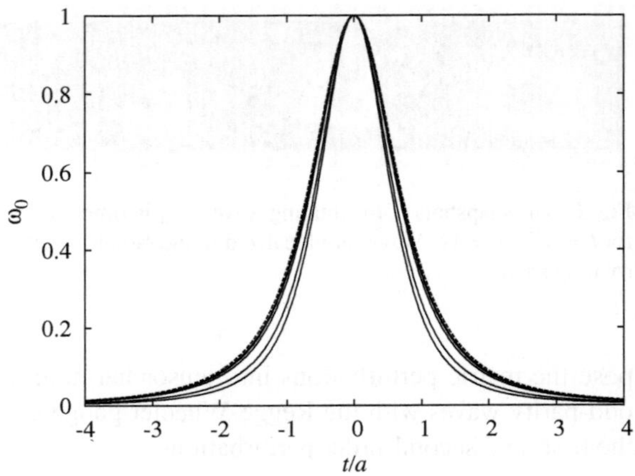

Fig. 2 The dependence of normalized angular velocity of the central inertial frame $\omega_0(l,1;t)/\omega_0(l,1;t=0)$ on the parameter $l = 2, 3, 10, 20, 30$ (from inside to out). The *dashed line* indicates the limit for large l

$$\partial_{rr} h^{(2)}_{\ 0}(t,r) - \frac{2}{r^2} h^{(2)}_{\ 0}(t,r) = \frac{4\pi}{\sqrt{2l(l+1)}r} \int_0^\pi <R^{(2)}_{t\varphi}> \partial_\theta Y_{l0}\, d\theta. \tag{4}$$

This equation can be solved by variation of constants. While the Coriolis and centrifugal accelerations are higher order in the angular velocity, ω_0 of the rotation of an inertial frame (of a gyroscope) located near the origin, entering $g^{(2)}_{t\varphi} = -\omega_0\, r^2 \sin^2\theta$, is determined by

$$\omega_0 = \frac{1}{4\pi} \iint_{0\,S^2}^{\infty} R^{(2)}_{t\varphi}[h^{(1)}, h^{(1)}]\, d\Omega\, \frac{dr}{r}. \tag{5}$$

Although $R^{(2)}_{t\varphi}[h^{(1)}, h^{(1)}]$ has a complicated structure, we obtain the angular velocity ω_0 in the closed, although quite lengthy, form. The profiles of $\omega_0(t)$ are in Fig. 2.

4 Observing Stars Through Gravitational Waves

We evaluate the first-order effects of the waves on the propagation of photons which apparently change the position of distant stars in the sky as seen by an observer fixed in the flat region at the origin. We found that the change of apparent star's celestial coordinates $\delta\varphi, \delta\theta$ can be computed as a perturbation of the ingoing radial null geodesic. This change of the direction of a momentum of the photon registered at time T can be written using integrals along the unperturbed ray. They yield quite simple formula for trajectory of the star with initial coordinates φ, θ in a form of conformal mapping of a straight line in a complex plane by function z^{-l-2}:

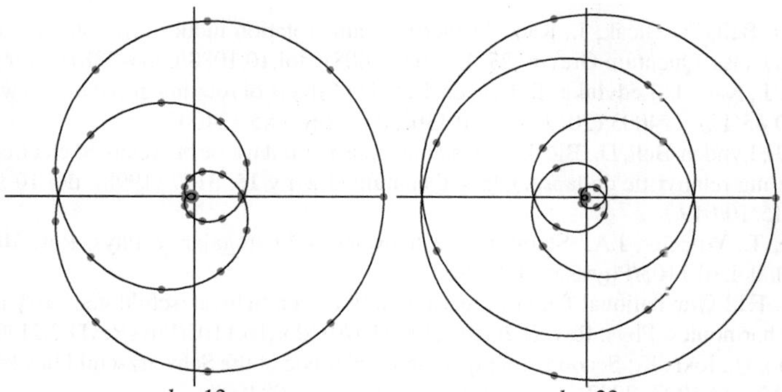

Fig. 3 Since light from distant stars is influenced by the gravitational waves the observed positions of the stars change. An observer at the origin can record the apparent position of the stars on the celestial sphere on a photographic plate. When appropriately scaled and rotated, the trajectories of all stars are the same. A star starts at the origin of the plate ($x = y = 0$ in the planes above) and moves along closed trajectories the structure of which becomes more complicated with increasing l. The trajectory of a star with celestial longitude φ fits in an ellipse with semi-axes $\Delta\theta$, $\Delta\varphi$ rotated on the celestial sphere by the angle $l\pi/2 + m\varphi$ (see Eq. (6))

$$\frac{\delta\varphi(T)}{\Delta\varphi} + i\frac{\delta\theta(T)}{\Delta\theta} = \frac{i^l\, e^{im\varphi}}{\left(1 + i\frac{T}{a}\right)^{l+2}}, \tag{6}$$

where $\Delta\theta = \hat{B}_l m P_l^m(\cos\theta)/\sin\theta$, $\Delta\varphi = -\hat{B}_l P_l'^m(\cos\theta)$, and $\hat{E}_l = N_l^m 2(l-1)!$ See Fig. 3 for examples of star image trajectories.

Acknowledgments J. B. and T. L. would like to acknowledge the support from the Czech Science Foundation by grant GAČR No. 14-37086G.

References

1. Einstein, A.: Gibt es eine Gravitationswirkung, die der elektrodynamischen Induktionswirkung analog ist? Vierteljahrsschrift für gerichtliche Medizin und öffentliches Sanitätswesen **44**, 37 (1912)
2. Barbour, J., Pfister, H. (eds.): Mach's Principle: from Newton's Bucket to Quantum Gravity, Einstein Studies, vol. 6. Birkhäuser, Boston (1995)
3. Bičák, J., Katz, J., Lynden-Bell, D.: Cosmological perturbation theory, instantaneous gauges, and local inertial frames. Phys. Rev. D **76**(6), 063501 (2007). doi:10.1103/PhysRevD.76.063501
4. Schmid, C.: Mach's principle: exact frame-dragging via gravitomagnetism in perturbed Friedmann-Robertson-Walker universes with $K = (\pm 1, 0)$. Phys. Rev. D **79**, 064007 (2009). doi:10.1103/PhysRevD.79.064007
5. Bičák, J., Katz, J., Lynden-Bell, D.: Gravitational waves and dragging effects. Class. Quantum Gravity **25**, 165017 (2008). doi:10.1088/0264-9381/25/16/165017

6. Lynden-Bell, D., Bičák, J., Katz, J.: Inertial frame rotation induced by rotating gravitational waves. Class. Quantum Gravity **25**, 165018 (2008). doi:10.1088/0264-9381/25/16/165018
7. Bičák, J., Katz, J., Ledvinka, T., Lynden-Bell, D.: Effects of rotating gravitational waves. Phys. Rev. D **85**(12), 124003 (2012). doi:10.1103/PhysRevD.85.124003
8. Katz, J., Lynden-Bell, D., Bičák, J.: Instantaneous inertial frame but retarded electromagnetism in rotating relativistic collapse. Class. Quantum Gravity **15**, 3177 (1998). doi:10.1088/0264-9381/15/10/019
9. Regge, T., Wheeler, J.A.: Stability of a Schwarzschild singularity. Phys. Rev. **108**(4), 1063 (1957). doi:10.1103/PhysRev.108.1063
10. Zerilli, F.J.: Gravitational field of a particle falling in a Schwarzschild geometry analyzed in tensor harmonics. Phys. Rev. D **2**(10), 2141 (1970). doi:10.1103/PhysRevD.2.2141
11. Nakano, H., Ioka, K.: Second order quasi-normal mode of the Schwarzschild black hole. Phys. Rev. D **76**, 084007 (2007). doi:10.1103/PhysRevD.76.084007
12. Bičák, J.: On the theories of the interacting perturbations of theReissner-Nordström black holes. Czech. J. Phys. B **29**, 945 (1979)

Variations on Spacetimes with Boost-Rotation Symmetry

Jiří Bičák and David Kofroň

Abstract Some new results on the boost-rotation symmetric spacetimes representing pairs of rotating charged objects accelerated in opposite directions are summarized. A particular attention is paid to (a) the Newtonian limit analyzed using the Ehlers frame theory and (b) the special-relativistic limit of the C-metric. Starting from the new, simpler form of the rotating charged C-metric we also show how to remove nodal singularities and obtain a rotating charged black hole freely falling in an external electromagnetic field.

1 Boost-Rotation Symmetric Spacetimes

Boost-rotation/axial symmetric spacetimes are important explicit examples of exact solutions of Einstein field equations describing non-trivially moving sources of gravitational and electromagnetic field [1]. The only "initial" assumption that is made is the existence of two Killing vectors: boost Killing vector ξ_B^μ whose orbits are hyperbolas and axial Killing vector ξ_ϕ^μ with closed circular orbits. The metric of a general electrovacuum rotating boost-rotation symmetric spacetime in global coordinates reads:

$$\mathrm{d}s^2 = -\frac{e^{\mu}\left(z\mathrm{d}t - t\mathrm{d}z + \Omega(z^2-t^2)\,\mathrm{d}\phi\right)^2 - e^{\nu}\left(z\mathrm{d}z - t\mathrm{d}t\right)^2}{z^2-t^2} + e^{\nu}\mathrm{d}\rho^2 + e^{-\mu}\rho r^2\mathrm{d}\phi^2\,. \tag{1}$$

Functions μ, ν and Ω depend on $a = z^2 - t^2$ and $b = r^2$ only and are determined by the Ernst equations (basic consequence of the Einstein field equations under these

---

J. Bičák (✉) · D. Kofroň
Institute of Theoretical Physics, Charles University in Prague, Prague, Czech Republic
e-mail: jiri.bicak@mff.cuni.cz

D. Kofroň
e-mail: d.kofron@gmail.com

J. Bičák and T. Ledvinka (eds.), *Relativity and Gravitation*,
Springer Proceedings in Physics 157, DOI: 10.1007/978-3-319-06761-2_33,

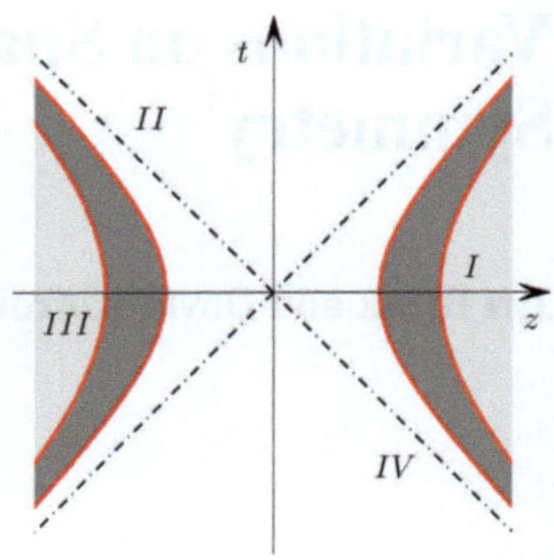

Fig. 1 Space-time diagram of boost-rotation symmetric solution. This diagram schematically indicates the *world lines* of sources (*thick lines*) and different conicity regions of the axis (*grey*), or world-sheets of the black-holes horizons

symmetries) and the character of sources. The nonlinear Ernst equations are difficult to solve but if rotation Ω and electromagnetic field vanish, then the basic Einstein field equation reduces to the wave equation on an auxiliary flat spacetime $\Box\mu = 0$, while ν can be determined by quadrature.

The main features of the boost-rotation symmetric spacetimes are:

(a) they describe uniformly accelerated sources;
(b) are asymptotically flat at null infinity except, in general, at two its generators;
(c) the hypersurfaces $z^2 = t^2$, where the boost Killing vector ξ_B^μ is null, invariantly divide the spacetime into four quadrants (see Fig. 1),

  (c1) below the roof (reg. I and III)—locally Weyl metrics,
  (c2) above the roof (reg. II and IV)—locally Einstein-Rosen, or Gowdy metrics;

(d) are of algebraic type I, in general; the C-metric, describing accelerated black holes, is of type D;
(e) are radiative with a non-vanishing news function;
(f) along the axis of symmetry there are conical singularities in general—they can be interpreted as strings or struts that cause the acceleration.

2 Newtonian Limit

Newtonian limit of a relativistic spacetime greatly corroborates its physical interpretation. We perform the limit within the framework of the Ehlers frame theory (see [2] for more details and references). The key point is the causality constant $\lambda = c^{-2}$ which becomes zero in the limit. To do the limit we have to choose a suitable set of observers and a naturally adapted coordinate system.

Using (1) with functions μ and ν known, we first introduce a shift of the coordinate origin by putting $z = \zeta + \lambda^{-1}g^{-1}$; otherwise the particles would "disappear" to infinity (see Fig. 2).

Our procedure then results in a classical point particle undergoing uniform acceleration, $z = \frac{1}{2}gt^2$, which generates classical field described by the Newtonian gravi-

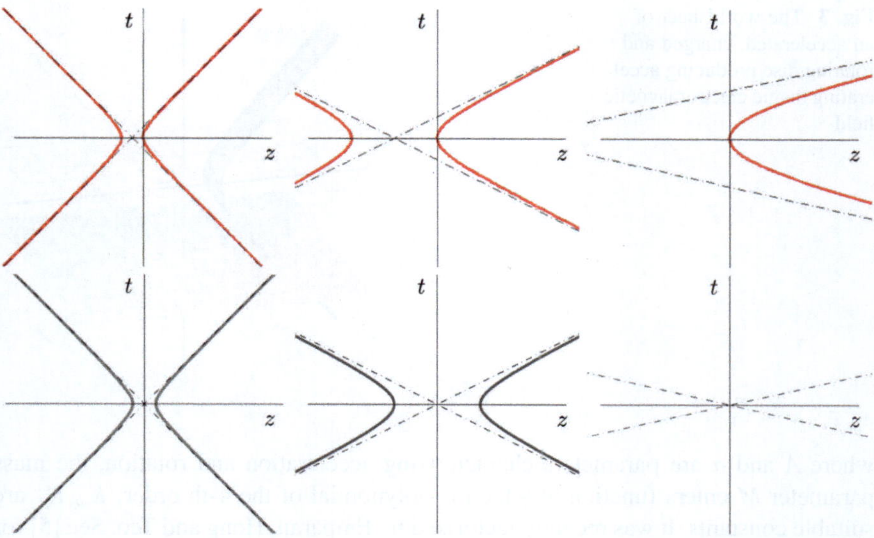

Fig. 2 The sequence of spacetime diagrams with decreasing $\lambda = c^{-2}$ when making the coordinate shift (*upper part*) and without it (*lower part*). The *worldlines* of the sources are shown

tational potential $\Phi = m/\sqrt{r^2 + (z - \frac{1}{2}gt^2)^2}$. This follows from the limit of explicit examples of these spacetimes [2].

Our results thus strongly support the physical significance of the boost-rotation symmetric spacetimes (in contrast to some previous conclusions [3] which made the limit in the regions *II* and *IV*).

3 The Rotating Charged C-Metric

The rotating charged C-metric is a special case of the boost-rotation symmetric spacetimes. It describes two rotating charged black holes. It can be written in the form

$$
\begin{aligned}
ds^2 = \frac{1}{A^2(x-y)^2} \Bigg\{ & \frac{\mathcal{G}(y)}{1+(aAxy)^2} \left[\left(1+(aAx)^2\right) K_\tau d\tau + aA\left(1-x^2\right) K_\phi d\phi \right]^2 \\
& - \frac{1+(aAxy)^2}{\mathcal{G}(y)} dy^2 + \frac{1+(aAxy)^2}{\mathcal{G}(x)} dx^2 + \frac{\mathcal{G}(x)}{1+(aAxy)^2} \\
& \left[\left(1+(aAy)^2\right) K_\phi d\phi - aA\left(1-y^2\right) K_\tau d\tau \right]^2 \Bigg\}, \qquad (2)
\end{aligned}
$$

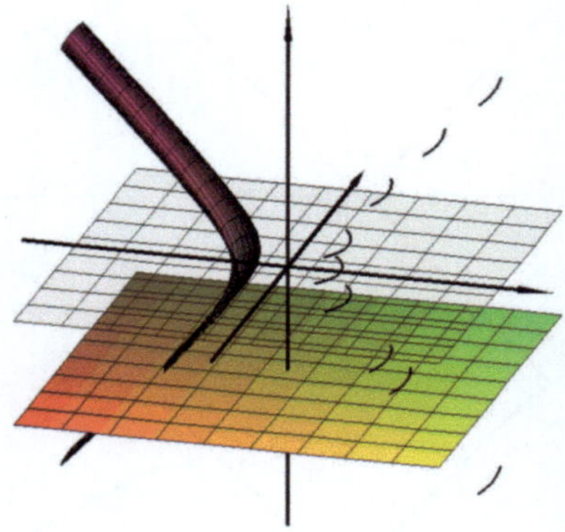

Fig. 3 The worldsheet of an accelerated, charged and rotating disc producing accelerating magic electromagnetic field

where A and a are parameters characterizing acceleration and rotation, the mass parameter M enters function $\mathscr{G}$ which is polynomial of the 4-th order; K_ϕ, K_τ are suitable constants. It was recently factorized by Emparan, Hong and Teo. See [5] for details and references. The transformation between the forms (2) and (1) is explicitly known.

The flat spacetime limit (i.e. $G \to 0$) of the charged rotating C-metric can be shown to lead to an electromagnetic field of two counter-rotating bent charged discs undergoing uniform acceleration (see Fig. 3, and [4] for details).

4 Removing the Conical Singularities

The charged C-metric represents a pair of uniformly accelerated black holes with mass m, charge q and acceleration A. These can be "immersed" in an external electric field E using the appropriate generating technique by Ernst. (The field breaks the asymptotic flatness.) For a suitably chosen value of E the axis becomes regular everywhere. This is because by adding the external field we include the "physical" source of the acceleration in the solution—the electromagnetic force. Utilizing the new factorized form of the C-metric, Ernst's simple "equilibrium condition" $mA = qE$ valid for small accelerations is generalized to an arbitrary A. See [5] where rotation is also included.

Acknowledgments We would like to acknowledge the support from the Czech Science Foundation by grant GAČR No. 14-37086G.

References

1. Bičák, J., Schmidt, B.: Asymptotically flat radiative space-times with boost-rotation symmetry: the general structure. Phys. Rev. D **40**, 1827 (1989). doi:10.1103/PhysRevD.40.1827
2. Bičák, J., Kofroň, D.: The Newtonian limit of spacetimes for accelerated particles and black holes. Gen. Relativ. Gravit. **41**, 153 (2009). doi:10.1007/s10714-008-0662-0
3. Lazkoz, R., Valiente Kroon, J.A.: The Newtonian limit of space-times describing uniformly accelerated particles. Proc. R. Soc. London, Ser. A **460**, 995 (2004). doi:10.1098/rspa.2003.1172
4. Bičák, J., Kofroň, D.: Accelerating electromagnetic magic field from the C-metric. Gen. Relativ. Gravit. **41**, 1981 (2009). doi:10.1007/s10714-009-0816-8
5. Bičák, J., Kofroň, D.: Rotating charged black holes accelerated by an electric field. Phys. Rev. D **82**, 024006 (2010). doi:10.1103/PhysRevD.82.024006

On the Existence and Properties of Helically Symmetric Systems

Jiří Bičák, Martin Scholtz and Paul Tod

Abstract By an argument similar to that of Gibbons and Stewart, but in a different coordinate system and less restrictive gauge, we show that any weakly asymptotically simple, analytic vacuum or electro-vacuum spacetime which is periodic in time is necessarily stationary. We generalized this theorem to the presence of scalar fields and, among other results, derived new expressions for the Bondi mass in this case. Here we summarize these results and also briefly discuss some new considerations concerning the periodic solutions within linearized theory of gravity.

1 Introduction

The inspiral and coalescence of binary black holes or neutron stars appears to be the most promising source for the detectors of gravitational waves, so that there has been much effort going into the development of numerical codes and analytic approximation methods to find the corresponding solutions of Einstein's equations. One of the recent approaches [1–3], assumes the existence of a helical Killing vector k. In a co-rotating frame k generates time translations but it becomes null on the light cylinder and is spacelike outside. Hence, the spacetime is not stationary but it is still periodic in the region where k is spacelike. Helical symmetry implies equal amounts of outgoing and ingoing radiation present for all the time, so the spacetime is not expected to be asymptotically flat.

J. Bičák
Faculty of Mathematics and Physics, Institute of Theoretical Physics, Charles University, V Holešovičkách 2, Prague, Czech Republic

M. Scholtz (✉)
Department of Applied Mathematics, Faculty of Transportation Sciences, Czech Technical University, Na Florenci 5, Prague, Czech Republic
e-mail: scholtz@fd.cvut.cz

P. Tod
Mathematical Institute, Oxford OX1 3LB, UK

J. Bičák and T. Ledvinka (eds.), *Relativity and Gravitation*, Springer Proceedings in Physics 157, DOI: 10.1007/978-3-319-06761-2_34,

2 Non-existence of Asymptotically Flat Solutions Periodic in Time

In [4] the authors used the spin-coefficient formalism to prove that any asymptotically flat vacuum periodic solution of Einstein's equations is necessarily stationary. They have proved the existence of Killing vector ∂_u in the neighbourhood of null infinity $\mathscr{I}$ which, however, is null by construction everywhere and hence does not imply stationarity. In fact, even in the flat spacetime (stationary!) there is no Killing vector which is everywhere null and extends to a translation on $\mathscr{I}$. Therefore, the Minkowski spacetime is not stationary according to the definition given in [4]. In [5] we introduced a different coordinate system in which we were able to prove the existence of Killing vector K^a which is null on $\mathscr{I}$ but timelike in its neighbourhood, see Fig. 1. Moreover, in [5, 6] we generalized the proof for the presence of electromagnetic fields and scalar fields and derived new expressions for the Bondi mass of two kinds of scalar fields, massless Klein–Gordon and conformal scalar field.[1] Our main results are summarized in the following theorems and corollary.

Theorem 1 *A weakly asymptotically simple vacuum or electro-vacuum spacetime which is periodic in time and analytic in a neighbourhood of $\mathscr{I}$ necessarily has a Killing vector which is timelike in the interior and extends to a translation on $\mathscr{I}$. The same holds for spacetimes with massless Klein–Gordon fields*[2] *and for spacetimes with conformally invariant scalar fields.*

Corollary 1 *In any weakly asymptotically simple, stationary electro-vacuum spacetime which is analytic in a neighbourhood of $\mathscr{I}$, the electromagnetic field is also stationary. The same holds for spacetimes with massless Klein–Gordon fields.*

Theorem 2 *The Bondi mass of the spacetime which is a weakly asymptotically simple solution to Einstein-massless-Klein–Gordon equations is given in terms of the standard Newman–Penrose coefficients by*

$$M_B = -\frac{1}{2\sqrt{\pi}} \oint \left(\Psi_2^0 + \tfrac{1}{3}\,\partial_u(\phi^0\bar{\phi}^0) + \sigma^0\,\dot{\bar{\sigma}}^0 \right) \mathrm{d}S, \tag{1}$$

where ϕ is the complex scalar field. The corresponding mass-loss formula reads

$$\dot{M}_B = -\frac{1}{2\sqrt{\pi}} \oint \left(\dot{\sigma}^0\dot{\bar{\sigma}}^0 + 2\,\dot{\phi}^0\,\dot{\bar{\phi}}^0 \right) \mathrm{d}S,$$

where the dot means the derivative with respect to time u.

[1] Massless Klein–Gordon field is a solution to the standard wave equation $\nabla_a\nabla^a\phi = 0$ while the conformal scalar field satisfies the conformally invariant equation $(\nabla_a\nabla^a + R/6)\phi = 0$.

[2] In fact, this remains to be true also for a self-interacting scalar fields for which potential $V(\phi)$ is appropriately decaying at $\mathscr{I}$ (see [6]).

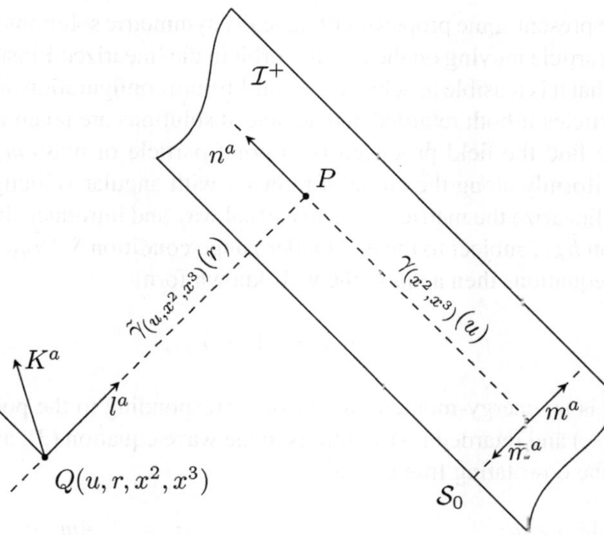

Fig. 1 The construction of the coordinate system and the Newman–Penrose null tetrad: the Killing vector K^a is null on $\mathscr{I}^+$ but timelike in its neighbourhood. Coordinate u is an affine parameter along null generators $\gamma_{(x^2,x^3)}(\mathrm{u})$ of $\mathscr{I}^+$, r is an affine parameter along null generators $\tilde{\gamma}_{(u,x^2,x^3)}(r)$ of hypersurfaces intersecting $\mathscr{I}$ at cuts u = constant. On these two-sphere cuts, the coordinates x^2 and x^3 are introduced in an arbitrary way

The Bondi mass of the spacetime with conformal scalar sources is given by the standard Newman–Penrose expression identical with (1) *for* $\phi^0 = 0$. *The Bondi mass-loss formula acquires the form*

$$\dot{M}_B = -\frac{1}{2\sqrt{\pi}} \oint \left(\dot{\sigma}^0 \dot{\bar{\sigma}}^0 + 2\,(\dot{\phi}^0)^2 - \phi^0\,\ddot{\phi}^0 \right) \mathrm{d}S, \tag{2}$$

where ϕ *is the real scalar field. Expression* (2) *is not negative semidefinite; a consequence of the null energy condition violation for conformally coupled scalar fields.*

3 Helical Symmetry in Linearized Gravity

In this last section we turn to another aspect of the helical symmetry. Although no exact solutions of the Einstein equations possessing this symmetry are known, such solutions have been constructed in various "toy models" as, for example, in scalar gravity [7] or Nordström theory [8]. In [9] the authors have found solutions describing five dimensional asymptotically AdS black holes with scalar field. These spacetimes have only one Killing vector (and hence are not stationary and axisymmteric) of the form $K = \partial_t + \omega\partial_\psi$, which is tangent to the null generators of the horizon and can be asymptotically timelike, null or spacelike, depending on the parameters of the solution. Thus, these solutions exhibit a kind of helical symmetry.

Here we present some properties of helically symmetric solutions representing the fields of a particle moving on the circular orbit in the linearized Einstein's theory. We also show that it is feasible to achieve the equilibrium configuration of a binary system of such particles if both retarded and advanced solutions are taken into account.

First we find the field produced by a point particle of mass m_A ("particle A") moving uniformly along the circle of radius a with angular velocity ω in the plane $z = 0$. We linearize the metric tensor in a usual way and introduce the trace-reversed perturbation $\bar{h}_{\mu\nu}$, subject to the de Donder gauge condition $\nabla^\mu \bar{h}_{\mu\nu} = 0$. Linearized Einstein's equations then acquire the well-known form

$$\Box \bar{h}_{\mu\nu} = -16\pi\, T_{\mu\nu}, \tag{3}$$

where $T_{\mu\nu}$ is an energy-momentum tensor corresponding to the point particle. The advanced ($-$) and retarded ($+$) solutions to the wave equation (3), after the transformation to the co-rotating frame, read

$$\begin{aligned}
\tilde{\bar{h}}^{\pm}_{00} &= -\frac{4m_A\gamma}{\rho_\pm}\left(1-\omega^2 ar\cos\theta_\pm\right)^2, & \tilde{\bar{h}}^{\pm}_{11} &= -\frac{4m_A\gamma}{\rho_\pm}\omega^2 a^2 \sin^2\theta_\pm,\\
\tilde{\bar{h}}^{\pm}_{01} &= -\frac{4m_A\gamma}{\rho_\pm}\left(1-\omega^2 ar\cos\theta_\pm\right)\omega a\sin\theta_\pm, & &\\
\tilde{\bar{h}}^{\pm}_{02} &= \frac{4m_A\gamma}{\rho_\pm}\left(1-\omega^2 ar\cos\theta_\pm\right)\omega ar\cos\theta_\pm, & &\\
\tilde{\bar{h}}^{\pm}_{12} &= \frac{4m_A\gamma}{\rho_\pm}\,\omega^2 a^2 r\sin\theta_\pm\cos\theta_\pm, & \tilde{\bar{h}}^{\pm}_{22} &= -\frac{4m_A\gamma}{\rho_\pm}\omega^2 a^2 r^2\cos^2\theta_\pm,
\end{aligned} \tag{4}$$

where the functions $\theta_\pm$ and $\rho_\pm$ are given implicitly by

$$\begin{aligned}
\theta_\pm &= \mp\omega R_\pm + \phi_0 - \hat{\phi}, \quad \rho_\pm = R_\pm \pm \omega ar\sin\theta_\pm,\\
R_\pm &= \sqrt{a^2 + r^2 + z^2 - 2ar\cos\theta_\pm}.
\end{aligned} \tag{5}$$

In terms of inertial coordinates, $\theta_\pm = \omega t_\pm + \phi_0 - \phi$, where $t_\pm = t \mp R_\pm$.

Due to the well-known inconsistency of the linearized theory, the solutions (4) do not obey the gauge condition imposed. However, we have $\nabla^\mu \bar{h}^{\pm}_{\mu\nu} = \mathcal{O}\left(\alpha^2\right)$, where $\alpha = \omega a/c$ (we set $c = 1$), so our calculations are consistent up to order $\mathcal{O}(\alpha)$. In particular, the Ricci tensor does not vanish outside the world line of particle A, but $R_{\mu\nu} = \mathcal{O}\left(\alpha^2\right)$.

Gravitational field given by the advanced solution in (4) does not display usual peeling properties near future null infinity $\mathscr{I}^+$. In particular, the Newman–Penrose components of the Weyl spinor Ψ_m, $m = 0, \ldots 4$, decay as $f(r)r^{-5+m}$, where $f(r)$ is an oscillating function of r.

The leading term in the asymptotic expansion of Ψ_4 for both advanced and retarded fields is

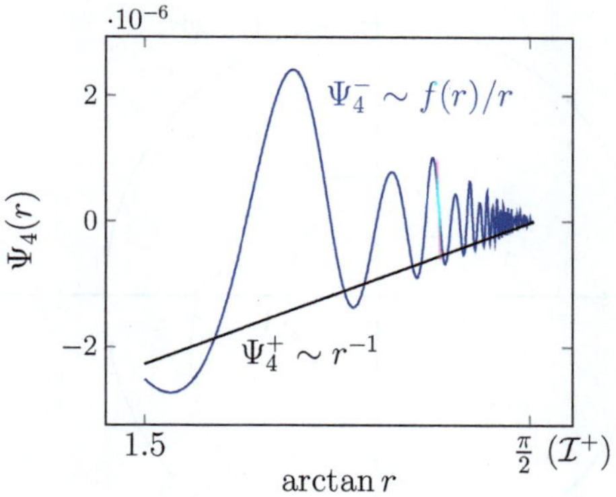

Fig. 2 The comparison of behaviour of Ψ_4 scalar in the neighbourhood of $\mathscr{I}^+$ in the case of the retarded solution (Ψ_4^+) and the advanced solution (Ψ_4^-). The value of $\Psi_4^\pm$ is plotted against rescaled coordinate arctan r, so that the point $\pi/2$ on horizontal axis corresponds to $\mathscr{I}^+$. The retarded solution Ψ_4^+ exhibits usual r^{-1} decay while the advanced solution oscillates with increasing frequency

$$\Psi_4^\pm = \frac{m\,\alpha^4\,\gamma}{2r(1+\alpha\sin\theta_\pm)}\left(10\alpha^2 + 15\alpha\sin\theta_\pm - \sin 3\theta_\pm + (2\alpha^2 - 8)\cos 2\theta_\pm\right), \tag{6}$$

where θ_+ is r-independent function given implicitly by (cf. [5])

$$\theta_+ = \alpha\,\cos\theta_+ + \alpha\,u,$$

while θ_- is an oscillating function of r,

$$\theta_- = \omega(u+r) + \omega\sqrt{a^2 + r^2 - 2ar\cos\theta_-}\,.$$

Thus, the retarded field decays in a usual way near $\mathscr{I}^+$ but the advanced field decays in an oscillatory manner in such a way that the limit of the rescaled field does not exist at $\mathscr{I}^+$, see Fig. 2. Similar behaviour has been observed in the case of helically symmetric electromagnetic field, see [10].

Next we consider the motion of the test particle B in the spacetime with metric (4). For purely retarded (advanced) field of particle A, particle B is expected to move along the spiral with decreasing (increasing) radius as an inspection of retarded (advanced) effects suggests. This expectation is confirmed by the numerical solution of geodesic equation the results of which are shown in Fig. 3. In order to achieve a circular motion of particle B, it is necessary to take the solution in the time-symmetric form

$$\bar{h}_{\mu\nu} = \tfrac{1}{2}\left(\bar{h}^+_{\mu\nu} + \bar{h}^-_{\mu\nu}\right).$$

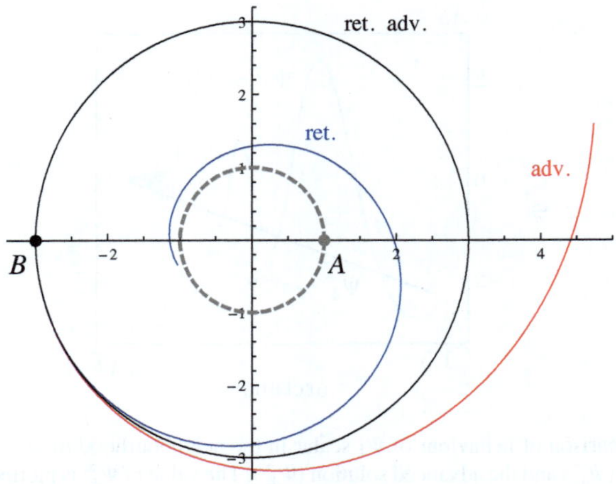

Fig. 3 The motion of test particle B in the field of particle A with mass $m = 0.44041$ (satisfying the equilibrium condition for the time-symmetric field), and with velocity $\alpha = 0.1$. The trajectories are plotted for purely retarded, purely advanced, and for the time-symmetric field produced by particle A

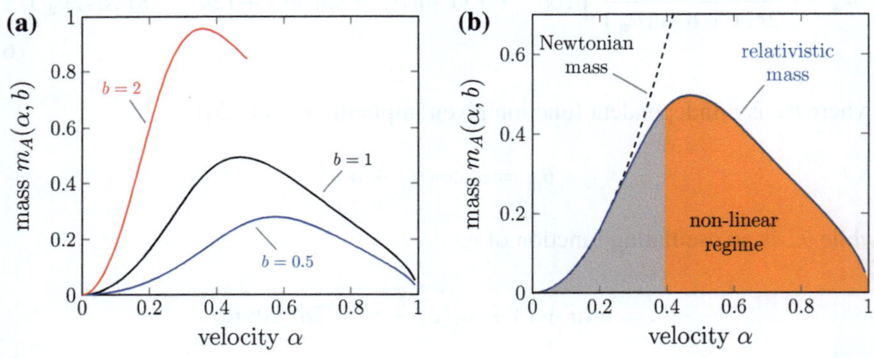

Fig. 4 **a** The equilibrium value of the mass of particle A as a function of its velocity for selected values of radius b along which particle B orbits with velocity αb. **b** Comparison with the Newtonian value of equilibrium mass for $b = 1$ when the circular orbits of both particles have the same radius. The *orange region* corresponds to the case when $|\bar{h}_{\mu\nu}| > |\eta_{\mu\nu}|$ so that the linearized theory breaks down

With this choice, the angular acceleration $\ddot{\phi}$ becomes zero and the only remaining condition of equilibrium comes from the radial component of the geodesic equation, $\ddot{r} = 0$. For a given velocity, α, of particle A and for a given radius b of the orbit of particle B, the condition of equilibrium can be solved explicitly with respect to the mass m_A of particle A. In Fig. 4a we plot the mass $m_A = m_A(\alpha, b)$ as a function of the velocity α for selected values of parameter b (we use units in which $a = 1$); in Fig. 4b the relativistic results are compared with the Newtonian results for $b = 1$. Thus, if the mass $m_A(\alpha, b)$ is chosen so as to satisfy the equilibrium condition,

$\ddot{r} = 0$, particle B will move along the circle of radius b with the velocity αb in the (time-symmetric) field of particle B.

A complete discussion of the equilibrium of a binary system on circular orbits will be given in [11].

Acknowledgments This work was supported by the grant GAČR 202/09/00772 of the Czech Republic.

References

1. Blackburn, J.K., Detweiler, S.: Close black-hole binary systems. Phys. Rev. D **46**, 2318 (1992). doi:10.1103/PhysRevD.46.2318
2. Detweiler, S.: Periodic solutions of the Einstein equations for binary systems. Phys. Rev. D **50**, 4929 (1994). doi:10.1103/PhysRevD.50.4929
3. Klein, C.: Binary black hole spacetimes with a helical Killing vector. Phys. Rev. D **70**(12), 124026 (2004). doi:10.1103/PhysRevD.70.124026
4. Gibbons, G.W., Stewart, J.M.: Absence of asymptotically flat solutions of Einstein's equations which are periodic and empty near infinity. In: Bonnor, W.B., Islam, J.N., MacCallum, M.A.H. (eds.) Classical General Relativity. pp. 77–94. Cambridge University Press, Cambridge (1984)
5. Bičák, J., Scholtz, M., Tod, P.: On asymptotically flat solutions of Einstein's equations periodic in time: I. Vacuum and electrovacuum solutions, Class. Quantum Gravity **27**(5), 055007 (2010). doi:10.1088/0264-9381/27/5/055007
6. Bičák, J., Scholtz, M., Tod, P.: On asymptotically flat solutions of Einstein's equations periodic in time: II. Spacetimes with scalar-field sources. Class. Quantum Gravity **27**(17), 175011 (2010). doi:10.1088/0264-9381/27/17/175011
7. Beig, R., Heinzle, J.M., Schmidt, B.G.: Helically symmetric N-particle solutions in scalar gravity. Phys. Rev. Lett. **98**(12), 121102 (2007). doi:10.1103/PhysRevLett.98.121102
8. Bruneton, J.P.: The two body problem: analytical results in a toy model of relativistic gravity. In: Alimi, J.M., Füzfa, A. (eds.) Albert Einstein Century International Conference, American Institute of Physics Conference Series, vol. 861, pp. 558–565 (2006). doi: 10.1063/1.2399624
9. Dias, Ó.J.C., Horowitz, G.T., Santos, J.E.: Black holes with only one Killing field. J. High Energy Phys. **7**, 115 (2011). doi:10.1007/JHEP07(2011)115
10. Bičák, J., Schmidt, B.G.: Helical symmetry in linear systems. Phys. Rev. D **76**(10), 104040 (2007). doi:10.1103/PhysRevD.76.104040
11. Bičák, J., Bohata, M., Scholtz, M.: Helical symmetry in linearized Einstein's gravity (in preparation)

Probing the Spacetime Structure Through Dynamics

Felipe T. Falciano

Abstract We shall show that the self-interaction of a field can be geometrized together with its perturbations in the sense that both dynamics are controlled by the same metric.

1 Introduction

The main idea in the analogue models program is to simulate gravitational configurations in condense matter physics or using electromagnetic fields in non-linear medium (see [1, 2]). Hence, to study some of the features of general relativity theory by using controlled laboratory systems. The connection can be made once we notice that the evolution of the perturbations of a given field can be described in a geometrical language. Even though the majority of works in the literature concern only the perturbative regime, it has been shown in Ref. [3] that one can also geometrize the dynamics of the background field simultaneously with its perturbations.

The geometrization of the perturbation dynamics alone has only a kinematical value. It is a change of description valid only for the perturbations. Thus, the effective metric that defines the evolution of the perturbations cannot be interpreted as a real geometry as it is done in general relativity.

For a scalar nonlinear theory with $L(\varphi, w)$, where $w \equiv \gamma^{\mu\nu}\partial_\mu\varphi\,\partial_\nu\varphi$ is the kinetic term, the equation of motion is a quasi-linear second order partial differential equation for φ. Its principal part that defines the effective metric

$$\hat{g}^{\mu\nu} \equiv L_w\gamma^{\mu\nu} + 2L_{ww}\partial^\mu\varphi\partial^\nu\varphi,$$

F. T. Falciano (✉)
ICRA-Centro Brasileiro de Pesquisas Físicas—CBPF,
Xavier Sigaud st., 150, urca, Rio de Janeiro 22290-180, Brazil
e-mail: ftovar@cbpf.br

J. Bičák and T. Ledvinka (eds.), *Relativity and Gravitation*,
Springer Proceedings in Physics 157, DOI: 10.1007/978-3-319-06761-2_35,

determines the causal structure of the theory (see [4] for details). Furthermore, by constructing a Riemannian affine structure such that $\hat{g}_{\alpha\beta||\nu} = 0$, the rays describing the perturbations of the scalar field follow null geodesics in the effective metric $\hat{g}_{\mu\nu}$. Therefore, the effective metric determines the causal structure and controls the propagation of the field's excitations in the geometrical optics limit.

1.1 Geometrization of Field Dynamics

The general approach in analogue models deals simultaneously with two metrics, one for the perturbations and another for the background. The geometrization scheme is valid only for the perturbations while the background field works only as a medium which defines the effective metric. This is probably the main difficulty in trying to consider the effective metric as an emergent metric with the status of a Riemannian metric as in general relativity.

A considerable improvement in this program is to include the background field in the geometric description. Thus, we want to define an emergent metric that simultaneously encodes the dynamics of the background field and its perturbations.

It has been shown in Ref. [3] that any scalar non-linear theory described by the Lagrangian $L(w, \varphi)$ is equivalent to the field φ propagating in an emergent spacetime with metric $\hat{h}_{\mu\nu}(\varphi, \partial\varphi)$ and a suitable source $j(\varphi, \partial\varphi)$. The emergent metric $\hat{h}_{\mu\nu}$ and the source field j are both constructed explicitly in terms of the field and its derivatives. In addition, in the optical limit, the wave vectors associated with perturbations follow null geodesics in the same $\hat{h}_{\mu\nu}$ metric. Therefore, there is an emergent spacetime "generated" by the non-linearity of the scalar field dynamics which dictates the propagation of the scalar field.

The equation of motion for the scalar field can be written as

$$\frac{1}{\sqrt{-\gamma}}\partial_\mu\left(\sqrt{-\gamma}\, L_w, \partial_\nu\varphi\, \gamma^{\mu\nu}\right) = \frac{1}{2}L_\varphi. \tag{1}$$

We shall define the emergent metric and its inverse as

$$\hat{h}_{\mu\nu} \equiv \frac{L_w}{\sqrt{1+\beta w}}\left(\gamma_{\mu\nu} - \frac{\beta}{1+\beta w}\varphi_{,\mu}\varphi_{,\nu}\right), \quad \text{with} \quad \beta \equiv 2L_{ww}/L_w,$$

$$\hat{h}^{\mu\nu} \equiv \frac{\sqrt{1+\beta w}}{L_w}\left(\gamma^{\mu\nu} + \beta\varphi^{,\mu}\varphi^{,\nu}\right).$$

To show that the above equation can be written as a Klein-Gordon in the $\hat{h}_{\mu\nu}$ metric, we need to calculate its determinant which amounts to $\sqrt{-\hat{h}} = \frac{L_w^2}{(1+\beta w)^{3/2}}\sqrt{-\gamma}$. Therefore, a straightforward calculation shows that

$$\sqrt{-\hat{h}}\hat{h}^{\mu\nu}\partial_\nu\varphi = \sqrt{-\hat{\gamma}}L_w\gamma^{\mu\nu}\partial_\nu\varphi. \tag{2}$$

Comparing the above equation with (1), we see that

$$\Box_{\hat{h}}\varphi = j(\varphi, \partial\varphi), \quad \text{with} \quad j(\varphi, \partial\varphi) \equiv \frac{L_\varphi}{2L_w^2}(1+\beta w)^{3/2}. \tag{3}$$

The last but very important point is to show that the effective metric associated with the perturbations is the same as defined above. This is straightforward once we realize that $\hat{h}_{\mu\nu}$ and $\hat{g}_{\mu\nu}$ are conformally related, i.e. $\hat{h}^{\mu\nu} = \frac{\sqrt{1-\beta w}}{L_w^2}\hat{g}^{\mu\nu}$. Thus, the rays also propagate as null geodesics in $\hat{h}^{\mu\nu}$.

The emergent metric $\hat{h}_{\mu\nu}$ encodes simultaneously the dynamics of the nonlinear field and its perturbations. This result goes further in the analogue program inasmuch it includes the dynamics of the background field. Note, however, that it is imperative that the non-linearity should be in the kinetic term. Algebraic non-linearities such as $L(w, \varphi) = w + V(\varphi)$ with $V(\varphi)$ any function of the scalar field, trivialize the effective metric in the form of the Minkowski metric. Thus, it is the non-linearity in w that is essential to generate the curved emergent spacetime. Additionally, in the other sense, in the particular case of a theory where the Lagrangian does not depend explicitly on φ, i.e. $L(w)$, equation (3) reduces to a "free" wave propagating in a curved spacetime generated by itself

$$\Box_{\hat{h}}\varphi = 0. \tag{4}$$

We have used the term "free" field above but one should keep in mind that the emergent metric depends non-trivially on the scalar field φ. Thus, the above "free" Klein-Gordon equation is actually a complicated non-linear equation for φ. Notwithstanding, as we have discussed in Sect. 1, there is a similar situation in general relativity when we use the term free scalar field. There, the metric appearing in the Klein-Gordon equation brings information from Einstein's equation that also depends on φ, hence making the coupled system a very non-linear and involved system of equations (see [3] for details).

2 General Remarks

The early proposal of general theory of relativity to encode gravitational interactions within the metrical structure of the spacetime is strongly supported by the universality of the interaction and the equality between inertial and gravitational masses. In this manner, once this identification is done, the evolution of a particle in flat spacetime subjugated to gravitational forces is equivalent to a Riemannian curved spacetime free of any force. In this scenario there is a unique metric that works as a background structure where the dynamics of all fields are defined.

We have argued that the non-linearities in the kinetic term of the scalar field also allow us to define a curved spacetime which incorporates its dynamics. The main difference from general relativity is that the metric thus defined depends explicitly on the scalar field while in GR the spacetime metric depends only implicitly through Einstein's equations. In addition, we do not have any guarantee that this metric is universal in the sense that all fields would incorporate it in their dynamics. On the contrary, this result seems to show that the geometrization of each non-linear theory would define different metrics. Nevertheless, the possibility of defining a single metric for interacting non-linear fields is not excluded. If we consider two interacting non-linear fields it might be possible to define a single Riemannian metric that encodes both non-linearities. Notwithstanding, the inclusion of interaction is a non-trivial step that deserves careful analysis in the future.

Acknowledgments We would like to thank E. Goulart, M. Novello and J. D. Toniato for helpful comments and useful discussions.

References

1. Gordon, W.: Zur Lichtfortpflanzung nach der Relativitätstheorie. Ann. Phys. (Leipzig) **72**, 421 (1923). doi:10.1002/andp.19233772202
2. Novello, M., Visser, M., Volovik, G. (eds.): Artificial Black Holes. World Scientific, Singapore (2002)
3. Goulart, E., Novello, M., Falciano, F., Toniato, J.: Hidden geometries in nonlinear theories: a novel aspect of analogue gravity. Class. Quantum Gravity **28**, 245008 (2011). doi:10.1088/0264-9381/28/24/245008
4. Jeffrey, A., Taniuti, T.: Non-linear wave propagation, with applications to physics and magnetohydrodynamics, mathematics in science and engineering, vol. 9. Academic Press, London (1964)

Analytical Conformal Compactification of Schwarzschild Spacetime

Jakub Haláček and Tomáš Ledvinka

Abstract We discuss a construction of the coordinates simultaneously covering the complete Schwarzschild manifold as well as its conformal extension beyond $\mathscr{I}^{\pm}$. We provide an example of such coordinates and show they are analytical both at horizon and at null infinity. We also show, that having such analytical compactification can improve convergence in certain numerical applications.

The Penrose–Carter diagrams became a standard way to visualize various aspects of geometrical objects and physical processes in black hole spacetimes. The most widely used prescription to compactify the Schwarzschild coordinates comes from the textbook [1]. It is known it depicts the regions near infinity dissimilarly to those of compactified Minkowski spacetime and thus more recent textbooks present modified transformations from Kruskal to compactified coordinates [2, 3]. Unfortunately, even these transformations do not provide analytical coordinates at null infinites, $\mathscr{I}^{\pm}$.

Spacetimes which at distant regions resemble the Minkowski spacetime form a class of *asymptotically flat spacetimes* (*AFS*) [4, 5]. For such spacetimes coordinates and appropriate conformal factor Ω exist that make the conformally related metric

$$\widetilde{ds}^2 = \Omega^2\, ds^2 \tag{1}$$

regular at null infinity, where the conformal factor must vanish at infinity and the leading terms of its expansion near $\mathscr{I}$ are prescribed as follows:

$$\Omega(\mathscr{I}^{\pm}) = 0, \qquad \widetilde{\nabla}_{\mu}\Omega(\mathscr{I}^{\pm}) \neq 0, \tag{2}$$

J. Haláček (✉) · T. Ledvinka
Faculty of Mathematics and Physics, Institute of Theoretical Physics,
Charles University, V Holešovičkách 2, 180 00 Praha 8, Czech Republic
e-mail: j.halacek@gmail.com

T. Ledvinka
e-mail: ledvinka@gmail.com

J. Bičák and T. Ledvinka (eds.), *Relativity and Gravitation*,
Springer Proceedings in Physics 157, DOI: 10.1007/978-3-319-06761-2_36,

$$\Omega(i^0) = 0, \qquad \widetilde{\nabla}_\mu \Omega(i^0) = 0, \qquad \widetilde{\nabla}_\mu \widetilde{\nabla}_\nu \Omega(i^0) = 2\widetilde{g}_{\mu\nu}(i^0). \tag{3}$$

We construct an explicit transformation and a conformal factor satisfying these conditions for the Schwarzschild spacetime with

$$ds^2 = -\left(1 - \frac{2M}{r}\right) dt^2 + \frac{dr^2}{1 - \frac{2M}{r}} + r^2 d\omega^2. \tag{4}$$

We generalize the coordinates smoothly covering $\mathscr{I}^\pm$ given in [4] in the form of a direct transformation between Schwarzschild coordinates t, r and the compactified null coordinates u, v

$$f(r(u,v)) = h(v) + h(-u), \tag{5}$$

$$t(u,v) = h(v) - h(-u). \tag{6}$$

Here $f(r) = r + 2M \ln[r/(2M) - 1]$ denotes Regge–Wheeler tortoise coordinate. We put the horizon at $u = 0$ and $v = 0$, the past null infinity $\mathscr{I}^-$ at $u = -\pi/2$, and $\mathscr{I}^+$ at $v = \pi/2$. If the function h is written as a combination of two analytic functions $h(z) = \alpha(z) + 2M \ln \beta(z)$, then using the usual conformal factor $\Omega \sim \cos u \cos v$, we can show that the conformally related metric (1) obtained by such transformation is analytic both at the horizon and at $\mathscr{I}^\pm$ if certain behavior of the functions α and β at $z = 0$ and $z = \pm\pi/2$ is satisfied.

A careful analysis shows that the function

$$h(x) = \frac{M}{\cos x} + 2M \ln \frac{1 - \cos x}{\sin x \, \cos x} \tag{7}$$

leads to the transformation (5–6) which provides analytic coverage of all regions of Kruskal's complete manifold as well as of the regions beyond null infinites of the conformally related manifold. The compactified line element (1) can then be simplified into

$$\widetilde{\mathrm{d}s}^2 = \frac{1 - \dfrac{2M}{r(u,v)}}{4M^2 \sin u \sin v} \, \mathrm{d}u \, \mathrm{d}v + \Omega^2 r(u,v)^2 \mathrm{d}\omega^2, \tag{8}$$

where we have absorbed the derivatives of h into the conformal factor

$$\Omega(u,v) = \frac{\cos u \cos v}{4M^2 \sqrt{(1+\cos u)^2 - 2\cos^3 u} \sqrt{(1+\cos v)^2 - 2\cos^3 v}}. \tag{9}$$

Note that no factors similar to $\exp(-r/2M)$, which in Kruskal's coordinates spoil the behavior at null infinities, appear in (8).

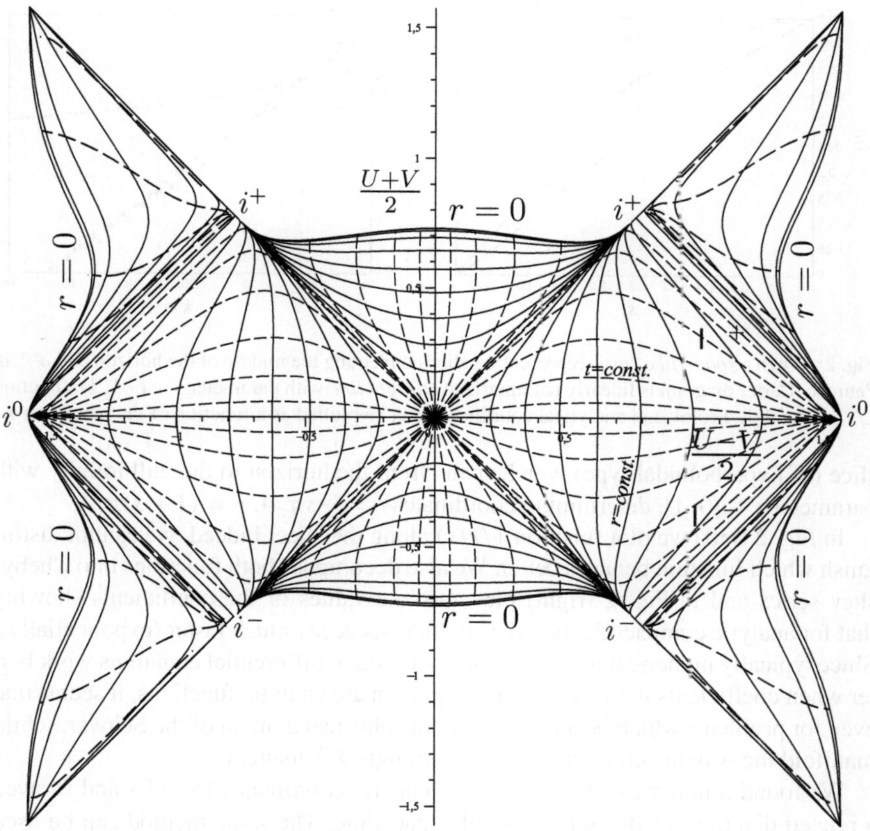

Fig. 1 The Carter-Penrose diagram of the Schwarzschild spacetime covering its analytical extension beyond $\mathscr{I}^{\pm}$

In Fig. 1 we illustrate the way the coordinates t and r cover the Penrose–Carter diagram of compactified Schwarzschild spacetime, namely, we can see how these coordinates behave near spatial infinity i^0, the way singularity $r = 0$ and event horizon $r = 2M$ meet at future time-like infinity i^+, and how a region of negative r appears behind $\mathscr{I}^{\pm}$. The fact that the conformal metric is an analytic function of compactified coordinates has to be be proven from mathematical properties of functions which appear in (5)–(9)—the most complicated is the proof of analyticity at $\mathscr{I}^{\pm}$, where we have to use either the implicit function theorem or theorem on properties of solution of ordinary differential equations.

To emphasize the importance of analytic coordinates, we compare the coordinates (5–6) with those suggested in [2] in a simple test. We assume a situation when a problem is formulated as a differential equation which is numerically solved on the compactified Schwarzschild spacetime. Coefficients in this equation reflect the curved geometry of the spacetime and typically contain function $1/r = \Omega\ \tilde{g}_{\theta\theta}^{-1/2}$. This function inherits analytic properties of $\tilde{g}_{\mu\nu}$ and Ω. In this test we consider a

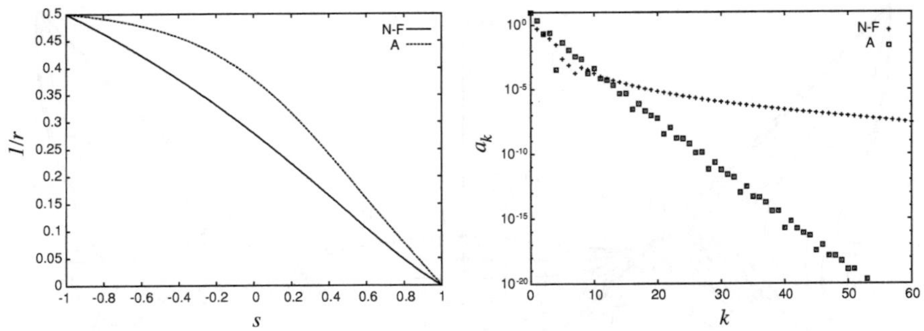

Fig. 2 When a *space-like curve* representing slice connecting the middle of the horizon and $\mathscr{I}^+$ in Penrose–Carter diagram is linearly parametrized *straight line* with parameter $s \in (-1, 1)$, function $1/r$ in Novikov-Frolov and analytical coordinates can be plotted as a function of the parameter s

slice (of hyperboloidal type) which spans from the horizon to the null infinity with parameter $s \in [0, 1]$, determining coordinates $u = -s\pi/4, v = (1 + s)\pi/4$.

In Fig. 2 (left) we plot function $1/r(s)$ along the slice. Indeed, we cannot distinguish which function behaves better. We also decompose both functions into Chebyshev series and in Fig. 2 (right) plot absolute values of the coefficients showing that for analytic compactification the coefficients decay much faster (exponentially). Since typically numerical methods of the solution of differential equations work better when coefficients in the differential equation are analytic functions, it seems that even for problems which occur entirely in the physical domain of the Schwarzschild manifold the way the coordinates pierce through $\mathscr{I}^\pm$ matters.

We found a new way of constructing analytic coordinates for *AFS* and showed a related diagram for the Schwarzschild spacetime. The same method can be used for other vacuum *AFSs*, e.g. Reissner–Nordström or extreme Reissner–Nordström solutions. Advantages of using this coordinate system for numerical methods in general relativity were demonstrated.

Acknowledgments We are grateful for support from the grants GAUK 606412, GAČR 14-37086G, SVV-267301 (J.H.), and GAČR 202/09/0772 (T.L.).

References

1. Misner, C.W., Thorne, K.S., Wheeler, J.A.: Gravitation, reprinted edn. W. H. Freeman and Company, New York (2013) (First publication 1970; 25th printing)
2. Frolov, V., Novikov, I.: Black Hole Physics: Basic Concepts and New Developments. Fundamental Theories of Physics. Springer, New York (1998)
3. Grøn, Ø., Hervik, S.: Einstein's General Theory of Relativity: With Modern Applications in Cosmology. Springer, New York (2007)
4. Ashtekar, A., Hansen, R.: A unified treatment of null and spatial infinity in general relativity. I-Universal structure, asymptotic symmetries, and conserved quantities at spatial infinity. J. Math. Phys. **19**, 1542 (1978). doi:10.1063/1.523863
5. Penrose, R.: Conformal treatment of infinity. In: DeWitt, C., DeWitt, B. (eds.) Relativity, Groups and Topology, pp. 563–584. Gordon and Breach, New York (1964)

Solutions in the 2 + 1 Null Surface Formulation

Tina A. Harriott and J. G. Williams

Abstract The null surface formulation of general relativity (NSF) differs from the standard approach by featuring a function Z, describing families of null surfaces, as the prominent variable, rather than the metric tensor. It is possible to reproduce the metric, to within a conformal factor, by using Z (entering through its third derivative, which is denoted by Λ) and an auxiliary function Ω. The functions Λ and Ω depend upon the spacetime coordinates, which are usually introduced in a manner that is convenient for the null surfaces, and also upon an additional angular variable. A brief summary of the $(2+1)$-dimensional null surface formulation is presented, together with the NSF field equations for Λ and Ω. A few special solutions are found and the properties of one of them are explored in detail.

1 Introduction

Frittelli et al. [1–3] have introduced an alternative approach to general relativity called the null surface formulation (NSF). In this approach, it is not the metric g_{ab} that plays a primary role, but a function Z, which is used to specify families of null surfaces. If needed, a metric can be constructed up to a conformal factor from a knowledge of Z and an auxiliary function Ω. A $(2+1)$-dimensional version of the NSF has been developed by Forniet al. [4, 5], Tanimoto [6] and Silva-Ortigoza [7].

T. A. Harriott (✉)
Department of Mathematics and Computer Science,
Mount Saint Vincent University, Halifax, NS B3M 2J6, Canada
e-mail: tina.harriott@msvu.ca

J. G. Williams
Department of Mathematics and Computer Science,
Winnipeg Institute for Theoretical Physics, Brandon University,
Brandon, MB R7A 6A9, Canada
e-mail: williams@brandonu.ca

J. Bičák and T. Ledvinka (eds.), *Relativity and Gravitation*,
Springer Proceedings in Physics 157, DOI: 10.1007/978-3-319-06761-2_37,

Central to the NSF in 2+1 dimensions is a third-order ordinary differential equation,

$$u''' = \Lambda(u, u', u'', \varphi),$$

where the prime denotes differentiation with respect to the angular variable $\varphi \in S^1$. Solutions are written $u = Z(x^a; \varphi)$ with x^a $(a = 0, 1, 2)$ representing three constants of integration which are to be identified with coordinates in (2 + 1)-dimensional spacetime.

The NSF uses *intrinsic* coordinates [2],

$$\begin{aligned} u \equiv \theta^0 &:= Z(x^a; \varphi), \\ \omega \equiv \theta^1 &:= u' \equiv \eth u \equiv \eth Z(x^a; \varphi), \\ \rho \equiv \theta^2 &:= u'' \equiv \eth^2 u \equiv \eth^2 Z(x^a; \varphi), \end{aligned}$$

(where $\eth := \partial/\partial\varphi$ denotes the derivative with respect to φ when x^a is held fixed) to derive field equations that are consistent with general relativity,

$$2[\eth(\partial_\rho\Lambda) - \partial_\omega\Lambda - \tfrac{2}{9}(\partial_\rho\Lambda)^2]\partial_\rho\Lambda - \eth^2(\partial_\rho\Lambda) + 3\,\eth(\partial_\omega\Lambda) - 6\,\partial_u\Lambda = 0,$$

$$3\,\eth\Omega = \Omega\,\partial_\rho\Lambda, \quad \partial_\rho^2\Omega = \kappa T_{\rho\rho}\Omega.$$

2 Nontrivial Solution

In the present paper, instead of using our previous light cone cut approach [8], we find a nontrivial solution directly by making the simplifying assumption that Λ and Ω depend only upon ρ: $\Lambda = \Lambda(\rho)$ and $\Omega = \Omega(\rho)$. This implies $\Omega = \Lambda^{1/3}$. For further simplicity, assume that Λ takes the particular form $\Lambda = (a+\rho)^k$ where a and k are constants. This leads to the quadratic, $(2/9)\,k^2 - k + 1 = 0$, which has solutions $k = 3$ and $k = 3/2$. Ignoring the choice $k = 3$ (which leads to empty space), we choose $k = 3/2$. This gives the solution

$$\Lambda = (a+\rho)^{3/2}, \quad \Omega = (a+\rho)^{1/2},$$

with a nonzero source term,

$$T_{\rho\rho} = -\frac{1}{4\kappa\,(a+\rho)^2},$$

and corresponds to the metric

$$ds^2 = (a+\rho)^{-1}\left[\,\tfrac{1}{4}(a+\rho)\,du^2 + (a+\rho)^{1/2}\,dud\omega - 2\,dud\rho + d\omega^2\,\right].$$

The three independent curvature scalars of 2 + 1 dimensions are found to be

$$R = \frac{1}{32}, \quad R_{ab}R^{ab} = \frac{3}{1024}, \quad \frac{\det \|R_{ab}\|}{\det \|g_{ab}\|} = -\left(\frac{1}{32}\right)^3,$$

and the components of the Einstein tensor are

$$G_{uu} = -\frac{3}{256}, \quad G_{u\omega} = -\frac{3}{128}(a+\rho)^{-1/2}, \quad G_{u\rho} = \frac{3}{64}(a+\rho)^{-1},$$
$$G_{\omega\omega} = -\frac{3}{64}(a+\rho)^{-1}, \quad G_{\omega\rho} = \frac{1}{8}(a+\rho)^{-3/2}, \quad G_{\rho\rho} = -\frac{1}{4}(a+\rho)^{-2}.$$

The null surface formulation of general relativity does not distinguish between conformally related spacetimes, and so a conformally flat spacetime would be an uninteresting example. The Cotton-York tensor C_{ab} is nonzero for the above solution, indicating that the spacetime is not conformally flat:

$$C_{uu} = -\frac{1}{256}, \quad C_{u\omega} = -\frac{1}{128}(a+\rho)^{-1/2}, \quad C_{u\rho} = \frac{1}{64}(a+\rho)^{-1},$$

$$C_{\omega\omega} = -\frac{1}{64}(a+\rho)^{-1}, \quad C_{\omega\rho} = \frac{3}{64}(a+\rho)^{-3/2}, \quad C_{\rho\rho} = -\frac{3}{32}(a+\rho)^{-2}.$$

In 2+1 dimensions, the Einstein equations, $G_{ab} = \kappa\, T_{ab}$, are sometimes replaced by the Einstein-Cotton field equations of topologically massive gravity (thereby allowing gravitational excitations):

$$G_{ab} + \lambda\, g_{ab} + \frac{1}{m} C_{ab} = \kappa\, T_{ab}.$$

The constant m can take either sign. (In fact, in 2 + 1 dimensions, this is also true for κ). It is straightforward to show that the metric under consideration satisfies the field equations of topologically massive gravity for a perfect fluid source, $T_{ab} = (\mu + p) U_a U_b + p\, g_{ab}$, with velocity U_a given by

$$U_u = 0, \quad U_\omega = (a+\rho)^{-1/2}, \quad U_\rho = -2(a+\rho)^{-1},$$

and with constant μ and p. Specifically:

$$m = -3/8, \quad \mu = -p, \quad p = \frac{1}{\kappa}\left(\lambda - \frac{1}{192}\right).$$

The most interesting case comes from choosing $\lambda = 1/192$. This gives a topologically massive gravity solution analogous to the regular de Sitter solution: a vacuum solution with nonzero cosmological constant and nonzero expansion θ.

Acknowledgments This work was supported by the Mount Saint Vincent University Dean of Arts and Science Travel Fund. Discussions with Dr. Ted Newman and Dr. Simonetta Frittelli during the authors' visits to Pittsburgh are gratefully acknowledged.

References

1. Frittelli, S., Kozameh, C., Newman, E.: Lorentzian metrics from characteristic surfaces. J. Math. Phys. **36**, 4975 (1995). doi:10.1063/1.531209
2. Frittelli, S., Kozameh, C., Newman, E.: GR via characteristic surfaces. J. Math. Phys. **36**, 4984 (1995). doi:10.1063/1.531210
3. Frittelli, S., Kozameh, C., Newman, E.: Linearized Einstein theory via null surfaces. J. Math. Phys. **36**, 5005 (1995). doi:10.1063/1.531211
4. Forni, D., Iriondo, M., Kozameh, C.: Null surfaces formulation in three dimensions. J. Math. Phys. **41**, 5517 (2000). doi:10.1063/1.533422
5. Forni, D., Iriondo, M., Kozameh, C., Parisi, M.: Understanding singularities in Cartan's and null surface formulation geometric structures. J. Math. Phys. **43**, 1584 (2002). doi:10.1063/1.1408282
6. Tanimoto, M. : On the null surface formalism—formulation in three dimensions and gauge freedom, ArXiv:e-prints arXiv:gr-qc/9703003 (1997)
7. Silva-Ortigoza, G.: Null surfaces and their singularities in three-dimensional Minkowski spacetime. Gen. Relativ. Gravit. **32**, 2243 (2000). doi:10.1023/A:1001943407824
8. Harriott, T., Williams, J.: Light cone cut solution in the 2+1 null surface formulation. In: Damour, T., Jantzen, R., Ruffini, R. (eds.) Proceedings of the Twelfth Marcel Grossmann Meeting on General Relativity, vol. 12, pp. 1896–1898. World Scientific, Singapore; Hackensack, NJ (2012). doi:10.1142/9789814374552_0355

Electric and Magnetic Weyl Tensors in Higher Dimensions

S. Hervik, M. Ortaggio and L. Wylleman

Abstract Recent results on purely electric (PE) or magnetic (PM) spacetimes in n dimensions are summarized. These include: Weyl types; diagonalizability; conditions under which direct (or warped) products are PE/PM.

1 Definition and General Properties

The standard decomposition of the Maxwell tensor F_{ab} into its electric and magnetic parts **E** and **B** with respect to (wrt) an observer (i.e., a unit time-like vector u) can be extended to any tensor in an n-dimensional spacetime [1–3]. Here we summarize the results of [3] about the Weyl tensor, and the connection with the null alignment classification [4, 5].

Consider the u-orthogonal projector $h_{ab} = g_{ab} + u_a u_b$. The "electric" and "magnetic" parts of C_{abcd} can be defined, respectively, as [3]

$$(C_+)^{ab}{}_{cd} = h^{ae} h^{bf} h_c{}^g h_d{}^h C_{efgh} + 4u^{[a} u_{[c} C^{b]e}{}_{d]f} u_e u^f, \tag{1}$$

$$(C_-)^{ab}{}_{cd} = 2h^{ae} h^{bf} C_{efk[c} u_{d]} u^k + 2u_k u^{[a} C^{b]kef} h_{ce} h_{df}. \tag{2}$$

S. Hervik
Faculty of Science and Technology, University of Stavanger,
4036 Stavanger, Norway
e-mail: sigbjorn.hervik@uis.no

M. Ortaggio (✉)
Institute of Mathematics, Academy of Sciences of the Czech Republic,
Žitná 25, 115 67 Prague 1, Czech Republic
e-mail: ortaggio@math.cas.cz

L. Wylleman
Faculty of Applied Sciences TW16, Ghent University, Galglaan 2,
9000 Ghent, Belgium
e-mail: lode.wylleman@ugent.be

J. Bičák and T. Ledvinka (eds.), *Relativity and Gravitation*,
Springer Proceedings in Physics 157, DOI: 10.1007/978-3-319-06761-2_38,

These extend the well-known 4D definitions [6, 7]. In any orthonormal frame adapted to u the electric [magnetic] part accounts for the Weyl components with an even [odd] number of indices u. *At a spacetime point (or region) the Weyl tensor is called purely electric [magnetic] (from now on, PE [PM]) wrt u if* $C_- = 0\,[C_+ = 0]$. The corresponding spacetime is also called PE [PM]. Several conditions on PE/PM Weyl tensors follow.

Proposition 1 (Bel-Debever-like criteria [3]). *A Weyl tensor* C_{abcd} *is: (i) PE wrt u iff* $u_a g^{ab} C_{bc[de} u_{f]} = 0$; *(ii) PM wrt u iff* $u_{[a} C_{bc][de} u_{f]} = 0$.

Proposition 2 (Eigenvalues [3]). *A PE [PM] Weyl operator*[1] *is diagonalizable, and possesses only real [purely imaginary] eigenvalues. Moreover, a PM Weyl operator has at least* $\frac{(n-1)(n-4)}{2}$ *zero eigenvalues.*

Proposition 3 (Algebraic type [3]). *A Weyl tensor which is PE/PM wrt a certain u can only be of type G,* I_i*, D or O. In the type* I_i *and D cases, the second null direction of the timelike plane spanned by u and any WAND is also a WAND (with the same multiplicity). Furthermore, a type D Weyl tensor is PE iff it is type D(d), and PM iff it is type D(abc).*

Proposition 4 (Uniqueness of u [3]). *A PE [PM] Weyl tensor is PE [PM] wrt: (i) a unique u (up to sign) in the type* I_i *and G cases; (ii) any u belonging to the space spanned by all double WANDs (and only wrt such us) in the type D case (noting also that if there are more than two double WANDs the Weyl tensor is necessarily PE* (*type D*(*d*)) [10]).

2 PE Spacetimes

Proposition 5 ([3]). *All spacetimes admitting a shearfree, twistfree, unit timelike vector field u are PE wrt u. In coordinates such that* $u = V^{-1}\partial_t$*, the line-element reads*

$$ds^2 = -V(t,x)^2 dt^2 + P(t,x)^2 \xi_{\alpha\beta}(x) dx^\alpha dx^\beta. \tag{3}$$

The above metrics include, in particular, direct, warped and doubly warped products with a one-dimensional timelike factor, and thus all *static* spacetimes (see also [11]). For a warped spacetime (M, g) with $M = M^{(n_1)} \times M^{(n_2)}$, one has $g = e^{2(f_1+f_2)}\left(g^{(n_1)} \oplus g^{(n_2)}\right)$, where $g^{(n_i)}$ is a metric on the factor space $M^{(n_i)}$ $(i = 1, 2)$ and f_i are functions on $M^{(n_i)}$ ($M^{(n_i)}$ has dimension n_i, $n = n_1 + n_2$, and $M^{(n_1)}$ is Lorentzian).

Proposition 6 (Warps with $n_1 = 2$ [3, 11]). *A (doubly) warped spacetime with* $n_1 = 2$ *is either type O, or type D(d) and PE wrt any u living in* $M^{(n_1)}$*; the uplifts of the null directions of the tangent space to* $(M^{(n_1)}, g^{(n_1)})$ *are double WANDs of*

[1] In the sense of the Weyl operator approach of [8] (see also [9]).

(M,g). If $(M^{(n_2)},g^{(n_2)})$ is Einstein the type specializes to D(bd), and if it is of constant curvature to D(bcd).

In particular, all spherically, hyperbolically or plane symmetric spacetimes belong to the latter special case.

Proposition 7 (Warps with $n_1 = 3$ [3, 11]). *A (doubly) warped spacetime with $(M^{(n_1)},g^{(n_1)})$ Einstein and $n_1 = 3$ is of type D(d) or O. The uplift of* any *null direction of the tangent space to $(M^{(n_1)},g^{(n_1)})$ is a double WAND of (M,g), which is PE wrt any u living in $M^{(n_1)}$.*

Proposition 8 (Warps with $n_1 > 3$ [3, 11]). *In a (doubly) warped spacetime*

(i) *if $(M^{(n_1)},g^{(n_1)})$ is an Einstein spacetime of type D, (M,g) can be only of type D (or O) and the uplift of a double WAND of $(M^{(n_1)},g^{(n_1)})$ is a double WAND of (M,g)*

(ii) *if $(M^{(n_1)},g^{(n_1)})$ is of constant curvature, (M,g) is of type D(d) (or O) and the uplifts of* any *null direction of the tangent space to $(M^{(n_1)},g^{(n_1)})$ is a double WAND of (M,g); (M,g) is PE wrt any u living in $M^{(n_1)}$.*

Proposition 9 (PE direct products [3]). *A direct product spacetime $M^{(n)} = M^{(n_1)} \times M^{(n_2)}$ is PE wrt a u that lives in $M^{(n_1)}$ iff u is an eigenvector of $R^{(n_1)}_{ab}$, and $M^{(n_1)}$ is PE wrt u. (u is then also an eigenvector of the Ricci tensor R_{ab} of $M^{(n)}$, i.e., $R_{ui} = 0$.)*

A conformal transformation (e.g., to a (doubly) warped space) will not, of course, affect the above conclusions about the Weyl tensor. There exist also direct products which are PE wrt a vector u *not* living in $M^{(n_1)}$ [3].

Also the presence of certain (Weyl) isotropies (e.g., $SO(n-2)$ for $n > 4$) implies that the spacetime is PE, see [3, 8] for details and examples.

3 PM Spacetimes

Proposition 10 (PM direct products [3]). *A direct product spacetime $M^{(n)} = M^{(n_1)} \times M^{(n_2)}$ is PM wrt a u that lives in $M^{(n_1)}$ iff all the following conditions hold (where $R_{(n_i)}$ is the Ricci scalar of $M^{(n_i)}$):*

(i) *$M^{(n_1)}$ is PM wrt u and has a Ricci tensor of the form $R^{(n_1)}_{ab} = \frac{R_{(n_1)}}{n_1} g^{(n_1)}_{ab} + u_{(a}q_{b)}$ (with $u^a q_a = 0$).*

(ii) *$M^{(n_2)}$ is of constant curvature and $n_2(n_2-1)R_{(n_1)} + n_1(n_1-1)R_{(n_2)} = 0$.*

Further, $M^{(n)}$ is PM Einstein iff $M^{(n_1)}$ is PM Ricci-flat and $M^{(n_2)}$ is flat.

See [3] for explicit (non-Einstein) examples. However, in general PM spacetimes are most elusive. For example,

Proposition 11 ([3]). *PM Einstein spacetimes of type D do not exist.*

In [3] also several results for PE/PM Ricci and Riemann tensors have been worked out, along with corresponding examples. In general, we observe that PE/PM tensors provide examples of *minimal tensors* [12]. Thanks to the *alignment theorem* [13], the latter are of special interest since they are precisely the *tensors characterized by their invariants* [13] (cf. also [3]). This in turn sheds new light on the classification of the Weyl tensor [5], providing a further invariant characterization that distinguishes the (minimal) types G/I/D from the (non-minimal) types II/III/N.

Acknowledgments M.O. acknowledges support from research plan RVO: 67985840 and research grant no P203/10/0749.

References

1. Senovilla, J.: Super-energy tensors. Class. Quantum Grav. **17**, 2799 (2000). doi:10.1088/0264-9381/17/14/313
2. Senovilla, J.: General electric-magnetic decomposition of fields, positivity and Rainich-like conditions. In: Pascual-Sánchez, J., Floría, L., San Miguel, A., Vicente, F. (eds.) Reference Frames and Gravitomagnetism, pp. 145–164. World Sicentific, Singapore (2001)
3. Hervik, S., Ortaggio, M., Wylleman, L.: Minimal tensors and purely electric or magnetic spacetimes of arbitrary dimension, ArXiv e-prints 1203.3563 [gr-qc] (2012)
4. Milson, R., Coley, A., Pravda, V., Pravdová, A.: Alignment and algebraically special tensors in Lorentzian geometry. Int. J. Geom. Meth. Mod. Phys. **2**, 41 (2005). doi:10.1142/S0219887805000491
5. Coley, A., Milson, R., Pravda, V., Pravdová, A.: Classification of the Weyl tensor in higher dimensions. Class. Quantum Grav. **21**, L35 (2004). doi:10.1088/0264-9381/21/7/L01
6. Matte, A.: Sur de nouvelles solutions oscillatoires de équations de la gravitation. Can. J. Math. **5**, 1 (1953)
7. Stephani, H., Kramer, D., MacCallum, M., Hoenselaers, C., Herlt, E.: Exact solutions of Einstein's field equations, 2nd edn. Cambridge Monographs on Mathematical Physics. Cambridge University Press, Cambridge (2003)
8. Coley, A., Hervik, S.: Higher dimensional bivectors and classification of the Weyl operator. Class. Quantum Grav. **27**, 015002 (2010). doi:10.1088/0264-9381/27/1/015002
9. Coley, A., Hervik, S., Ortaggio, M., Wylleman, L.: Refinements of the Weyl tensor classification in five dimensions. Class. Quantum Grav. **29**, 155016 (2012). doi:10.1088/0264-9381/29/15/155016
10. Wylleman, L.: On Weyl type II or more special spacetimes in higher dimensions (in preparation)
11. Pravda, V., Pravdová, A., Ortaggio, M.: Type D Einstein spacetimes in higher dimensions. Class. Quantum Grav. **24**, 4407 (2007). doi:10.1088/0264-9381/24/17/009
12. Richardson, R., Slodowy, P.: Minimum Vectors for real reductive algebraic groups. J. London Math. Soc. **42**, 409 (1990). doi:10.1112/jlms/s2-42.3.409
13. Hervik, S.: A spacetime not characterized by its invariants is of aligned type II, Class. Quantum Grav. **28**, 215009 (2011). doi:10.1088/0264-9381/28/21/215009

Phase Structure of Five Dimensional Black Di-ring

Hideo Iguchi

Abstract We investigate the phase structure of black di-ring in five-dimensional asymptotically flat vacuum gravity. We numerically plot the points of black di-rings in the phase diagram to study the region covered by black di-ring. The distribution of black di-ring shows that the area of black di-ring is always less than the maximum value of black ring. The plot indicates that there are black di-ring configurations whose area parameters are arbitrarily close to zero.

1 Introduction

In five dimensions, in addition to the solutions with a single horizon, there exist solutions with disconnected event horizons. Black Saturn which is a spherical black hole surrounded by a black ring was constructed by the inverse scattering method [1]. It was shown that the black rings can be superposed concentrically by using the Bäcklund transformation [2]. This black di-ring solution also can be constructed by the inverse scattering method [3].

The existence of multi-black hole configurations implies continuous non-uniqueness of five-dimensional black holes. The phase diagram of the black Saturn was investigated in [1, 4]. The plot of random sets of points in the phase diagram showed that the black Saturn covers the wide region of the phase diagram. The phases of black Saturn were investigated by the method based on the thin and long ring approximation: the black Saturn can be modeled as a simple superposition of a Myers-Perry black hole and a very thin black ring [1]. It was argued that the configurations that approach maximal entropy for fixed mass and angular momentum are black Saturns with a nearly static black hole and a very thin black ring.

H. Iguchi (✉)
College of Science and Technology, Nihon University, Narashinodai, Funabashi, Chiba, Japan
e-mail: iguchi.h@phys.ge.cst.nihon-u.ac.jp

J. Bičák and T. Ledvinka (eds.), *Relativity and Gravitation*, Springer Proceedings in Physics 157, DOI: 10.1007/978-3-319-06761-2_39,

The black di-ring also indicates "an infinite non-uniqueness" [2]. It was confirmed that there are infinite number of black di-rings for the same mass and angular momentum. The phase structure of a black di-ring in thermodynamic equilibrium was investigated in [5, 6]. Distributions of black di-rings in the phase diagram have not been fully investigated. When we approximate the black di-ring as a simple superposition of two concentric black rings, we can roughly estimate the region covered by the black di-ring in the phase diagram. The maximum of the area would be smaller than the one of black Saturn for the same mass and angular momentum. Because of the strong non-linearity, however, we need rigorous analysis for the distributions of black di-ring in the phase diagram for a decisive conclusion.

2 Rod Structure of Black Di-ring

The rod structure analysis is a very useful tool to understand a higher dimensional black hole solution. The rod structure of black di-ring is composed by two semi-infinite rods and four finite rods. Two of four finite rods are timelike and the other two finite and two semi-infinite rods are spacelike. The finite timelike rods have direction vectors $(1, \Omega_i, 0)$ if the solution has angular momentum along S^1 direction. The rod structure of black di-ring rotating along S^1 direction is described in Fig. 1 (left). This solution was constructed from the seed solution, whose rod structure is described in Fig. 1 (right), by the inverse scattering method [3]. The seed solution has six finite rods. We define six parameters by using the lengths of these finite rods as in Fig. 1. Physical variables of black di-ring are expressed by using these six parameters.

3 Phase Structure of Black Di-ring

The physical variables of black di-ring are calculated from the exact expressions of the solution. Following [1], we normalize the ADM angular momentum and the area of horizons as

$$j^2 = \frac{27\pi}{32G}\frac{J^2}{M^3}, \quad a_h = \frac{3}{16}\sqrt{\frac{3}{\pi}}\frac{A_h}{(GM)^{3/2}}, \tag{1}$$

to compare the physical properties of black objects with the same ADM mass.

We fix the scaling freedom by $d_1 + d_2 + d_3 + d_4 = 1$. The balance conditions impose two constraints on the parameters. As a result, the balanced black di-ring has three dimensionless parameters. In the analysis, we choose d_2, d_3 and d_4 as the three parameters for the balanced black di-ring. The parameter d_1 is determined by the scaling. The parameters p and q are determined by solving the balance conditions.

To investigate the region of the phase diagram covered by black di-ring, we plot the point (j^2, a_h) corresponding to the sets of parameters (d_2, d_3, d_4). The result is shown in Fig. 2 (left). Figure 2 (right) is a similar plot for black Saturns.

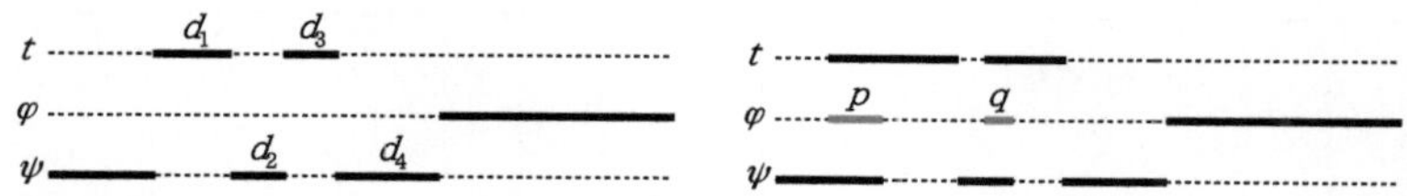

Fig. 1 *Left* Rod structure of black di-ring. *Right* Rod structure of seed solution of black di-ring with S^1 rotation

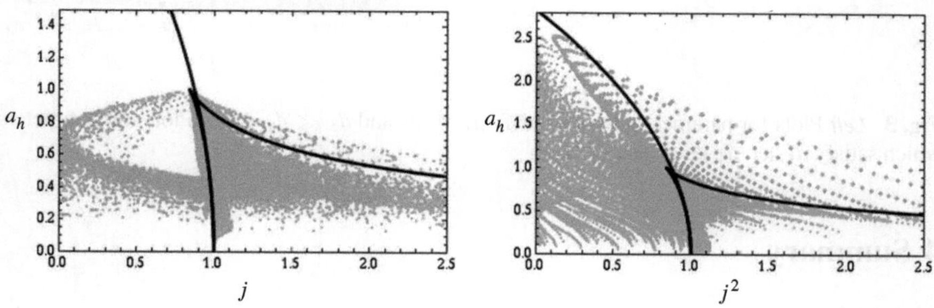

Fig. 2 *Left* Distribution of black di-rings in the phase diagram. The *black bold curves* are the phases of the Myers-Perry black hole and the black ring. *Right* Distribution of black Saturns in the phase diagram. The *black bold curves* are the phases of the Myers-Perry black hole and the black ring

The total area of black di-ring can not become larger than the maximum of black ring $a_h = 1$. There are black di-ring configurations with total area a_h greater than the black ring with the same j^2. It can be confirmed that the black di-ring solution with $j = 0$ is possible while maintaining balance similar to black Saturn. In the plot of Fig. 2 (left) the low entropy black di-ring $a_h \lesssim 0.2$ is scarcely distributed except around $j^2 = 1$. When $d_1 = d_3 = 0$ the area of black di-ring becomes exactly zero. If we simply set $d_1 = d_3 = 0$, it can be easily shown that the balance condition for spacelike rod d_4 is violated. Therefore we have to choose parameters such that p and q become very small in addition to d_1 and d_3 for the small area black di-ring.

The low entropy black di-ring would be constructed by two different configurations. One is a double thin ring and the other is a combination of a nearly extremal fat ring with a large thin ring. The double thin ring configuration will be constructed by choosing the rod parameters as $d_1 \ll d_2$ and $d_3 \ll d_4$. The corresponding plot of the phase diagram becomes like Fig. 3 (left). The second configuration will be constructed by $d_1, d_3, d_4 \ll d_2$ and $d_3 \gg d_4$. The corresponding plot is given in Fig. 3 (right). Both plots show that the black di-rings can exist in the region $0 < a < 0.2$ of the phase diagram.

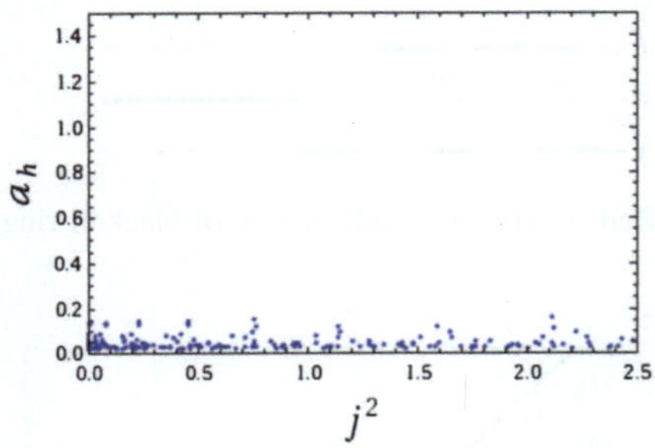

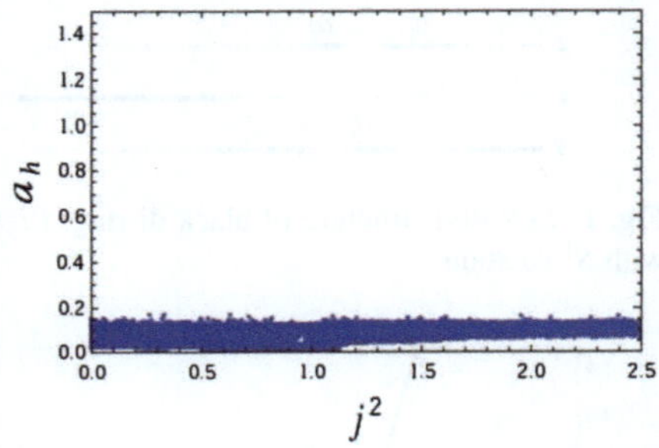

Fig. 3 *Left* Plots for parameters which satisfy $d_1 \ll d_2$ and $d_3 \ll d_4$. *Right* Plots for parameters which satisfy $d_1, d_3, d_4 \ll d_2$ and $d_3 \gg d_4$

4 Summary

We analyzed the phase structure of the black di-ring. The distribution of black di-rings in the phase diagram shows infinite non-uniqueness of the black di-ring. The configurations of black di-ring span the open strip $0 < a_h < 1$ and $j^2 > 0$.

Acknowledgments This work is partially supported by Grant-in-Aid for Scientific Research (C) (No. 23540319) from Japanese Ministry of Education, Science, Sports, and Culture.

References

1. Elvang, H., Figueras, P.: Black Saturn. J. High Energy Phys. **2007**(05), 050 (2007). doi:10.1088/1126-6708/2007/05/050
2. Iguchi, H., Mishima, T.: Black di-ring and infinite nonuniqueness. Phys. Rev. D **75**, 064018 (2007). doi:10.1103/PhysRevD.75.064018. [Erratum: ibid. **78**, 069903 (2008)]
3. Evslin, J., Krishnan, C.: The black di-Ring: an inverse scattering construction. Class. Quant. Grav. **26**, 125018 (2009). doi:10.1088/0264-9381/26/12/125018
4. Elvang, H., Emparan, R., Figueras, P.: Phases of five-dimensional black holes. J. High Energy Phys. **2007**(05), 056 (2007). doi:10.1088/1126-6708/2007/05/056
5. Iguchi, H., Mishima, T.: Thermodynamic black di-rings. Phys. Rev. D **82**, 084009 (2010). doi:10.1103/PhysRevD.82.084009
6. Emparan, R., Figueras, P.: Multi-black rings and the phase diagram of higher-dimensional black holes. J. High Energy Phys. **2010**(11), 022 (2010). doi:10.1007/JHEP11(2010)022

The Null Geodesics in the Black Saturn Spacetime

Alicja Konieczny

Abstract We calculate numerically null geodesics in the Black Saturn spacetime. Our analysis is restricted to the rotation axis in the domain of outer communication. The geodesics are calculated for two different Black Saturn configurations with the same ADM mass and angular momentum.

1 Introduction

In 2007 Henriette Elvang and Pau Figueras presented the single spinning, uncharged Black Saturn metric, which can be defined as "a Black Ring balanced by rotation around a concentric spherical black hole in an asymptotically flat spacetime" [1]. This solution was constructed by inverse scattering method. It is an exact, stationary, asymptotically flat $4+1$ dimensional vacuum solution of Einstein's equations, where angular momentum keeps the configuration in equilibrium. This solution is really interesting, because of the 2-fold continuous non-uniqueness for fixed ADM mass and ADM angular momentum. The configuration with zero angular momentum measured at infinity makes the Schwarzschild-Tangherlini solution non-unique. The solution permits two rotational planes and also the possibility of charged Black Saturn, what is not investigated here. Such a complicated solution does not allow us to check algebraically if Einstein's vacuum equations are satisfied, but in [1] authors describe numerical tests, which show that all components of the Ricci tensor vanish.

A. Konieczny (✉)
Astronomical Observatory, Jagiellonian University,
ul. Orla 171, 30-244 Krakow, Poland
e-mail: alicja.konieczny@uj.edu.pl

J. Bičák and T. Ledvinka (eds.), *Relativity and Gravitation*,
Springer Proceedings in Physics 157, DOI: 10.1007/978-3-319-06761-2_40,

1.1 The Metric

The metric is rather complicated, and it might be written as follows [1]

$$ds^2 = -\frac{H_y}{H_x}\left[dt + \left(\frac{\omega_\psi}{H_y} + q\right)d\psi\right]^2 + H_x\left[k^2 P\left(d\rho^2 + dz^2\right) + \frac{G_y}{H_y}d\psi^2 + \frac{G_x}{H_x}d\phi^2\right], \tag{1}$$

where H, G, ω, P are functions of ρ and z coordinates, q is a constant included in order to ensure asymptotic flatness and k is an integration constant. The explicit formula of these functions can be found in [1].

2 Geodesics

If the space-time possesses n Killing vectors ξ_μ: $\nabla_\mu \xi_\nu + \nabla_\nu \xi_\mu = 0$, there are n conserved quantities connected with these vectors. The conjugate momenta are given by $p_\mu = g_{\mu\nu}\frac{dx^\nu}{d\lambda}$ where $\mu, \nu = 1, 2, \ldots, d$, for d dimensional space-time. If the Killing field ∂_μ exists, p_μ is conserved and can be used to simplify the equations. Another equation is obtained from the null condition $g_{\mu\nu}\frac{dx^\mu}{d\lambda}\frac{dx^\nu}{d\lambda} = 0$. All that allows us to determine the path of photons for the given metric. From the form of the metric it is clear that there are three Killing vectors $\partial_t, \partial_\psi, \partial_\phi$ which generate three conserved quantities along the geodesics:

$$-e = g_{tt}\frac{dt}{d\lambda} + g_{t\psi}\frac{d\psi}{d\lambda} = -\frac{H_y}{H_x}\frac{dt}{d\lambda} - \frac{H_y q + \omega_\psi}{H_x}\frac{d\psi}{d\lambda}, \tag{2}$$

$$l_1 = g_{\psi\psi}\frac{d\psi}{d\lambda} + g_{t\psi}\frac{dt}{d\lambda} = -\frac{H_y q + \omega_\psi}{H_x}\frac{dt}{d\lambda} + \left(\frac{G_y H_x}{H_y} - \frac{H_y\left(q + \frac{\omega_\psi}{H_y}\right)}{H_x}\right)\frac{d\psi}{d\lambda}, \tag{3}$$

$$l_2 = g_{\phi\phi}\frac{d\phi}{d\lambda} = G_x\frac{d\phi}{d\lambda}. \tag{4}$$

We set $l_1 = l_2 = 0$, which means that the angular momenta are equal to zero, e might be interpreted as just the scale of affine parameter and is chosen to be equal to one. Now, taking Taylor expansion of the metric functions we calculate them in the limit of $\rho \to 0$ and take into account only the leading terms. These calculations are based on the code used in [2]. The null condition takes the form (assuming that $d\rho/d\lambda = 0$)

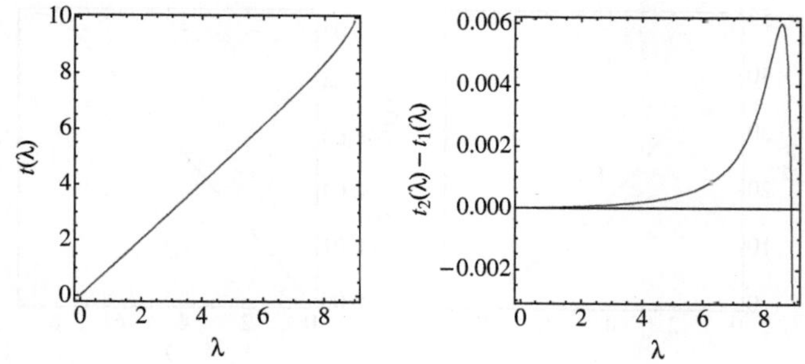

Fig. 1 *Left panel*—function $t(\lambda)$ for different Black Saturn configurations and the same ADM mass $M = 1$ and ADM angular momentum $J = 0.5$. *Right panel*—plot of $t_2(\lambda) - t_1(\lambda)$ difference between $t(\lambda)$ functions

$$\frac{e^2(-G_y H_x^2 + (H_y q + \omega_\psi)^2) + G_y H_x^2 H_y k^2 P\left(\frac{dz}{d\lambda}\right)^2}{G_y H_x H_y} = 0. \tag{5}$$

3 The Results

We have considered the null geodesic equations on the axis of rotation, in the region of outer communication, reduced them to the simplest form and then solved numerically in a particular case. In one of the parameterizations, the Black Saturn is defined by free parameters a_5, a_4, a_3, which are related to the position and shape of the horizons of the central black hole and the surrounding black ring [1]. In this paper, two different sets of these parameters are investigated. It is impossible to calculate the geodesics substituting the $\rho = 0$ into the metric functions, because this limit is non-trivial.

The metric functions are analytic near $\rho = 0$ in ρ for $z > a_2$. Thus, we use Taylor expansion in ρ, about $\rho = 0$. In this work we present the numerical solutions for two Black Saturn configurations, such that the ADM mass M and ADM angular momentum J are the same. If we solve the equations (3.30) and (3.31) from [1] for fixed values of ADM mass $M = 1$ and angular momentum $J = 0.5$, it is now possible to calculate the a_3 and a_5 parameters for two parameter values $a_4 = \frac{1}{2}, \frac{1}{4}$. The following sets of parameters and Black Saturnconfigurations were investigated:

1. $a_5 = 0.180344, a_4 = \frac{1}{4}, a_3 = 0.81897$;
2. $a_5 = 0.45903, a_4 = \frac{1}{2}, a_3 = 0.85904$.

Using Wolfram Mathematica, the equations for $t(\lambda)$ and $z(\lambda)$ were numerically solved for this set of parameters, and the results are presented below. The functions $t(\lambda)$ seem to be the same, see Fig. 1 (left panel), but the difference is visible in a plot of $t_1(\lambda) - t_2(\lambda)$ see Fig. 1 (right panel). Almost the same situation occurs for functions $z(\lambda)$, see Fig. 2 (left panel) and (right panel). The Fig. 3 (left panel) shows the $t(z)$ dependence, where the t coordinate became the proper time for a distant,

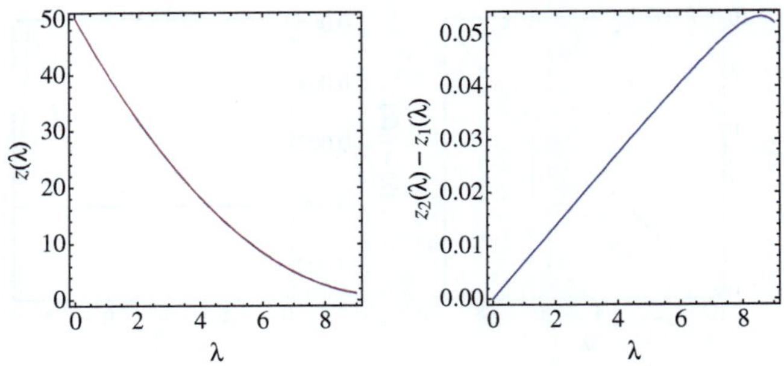

Fig. 2 *Left panel*—Function $z(\lambda)$ for different Black Saturn configurations and the same ADM mass $M = 1$ and ADM angular momentum $J = 0.5$. *Right panel*—plot of $z_2(\lambda) - z_1(\lambda)$ difference between $z(\lambda)$ functions

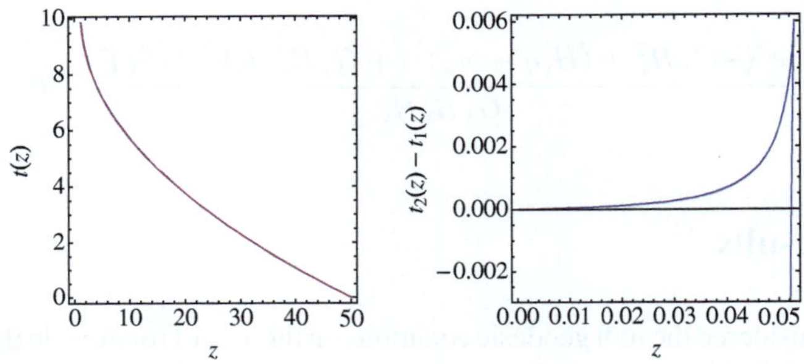

Fig. 3 *Left panel*—The parametric plot of $t(z)$ geodesics for fixed Black Saturn ADM mass $M = 1$ and ADM angular momentum $J = 0.5$. *Right panel*—plot of difference $t_2(z_2) - t_1(z_1)$ of $t(z)$

stationary observer. For increasing t, z tends to $a_2 = 1$, and it is now visible, that there is a horizon of the central black hole.

References

1. Elvang, H., Figueras, P.: Black saturn. J. High Energy Phys. **2007**(05), 050 (2007). doi:10.1088/1126-6708/2007/05/050
2. Chruściel, P., Eckstein, M., Szybka, S.: On smoothness of black saturns. J. High Energy Phys. **2010**(11), 048 (2010). doi:10.1007/JHEP11(2010)048

Conformal Symmetries on the Horizon and Black Hole Entropy in Generic Dimensions

Jianwei Mei

Abstract In this short note I briefly summarize some recent results regarding conformal symmetries on the horizon of general stationary and axisymmetric black holes in generic dimensions.

1 Introduction

In an essay in 2007, Steve Carlip used the term "universality" to reflect the fact that there exist a plethora of different ideas or calculations which (more or less) all lead to the same correct result for black hole entropy [1]. Since the discovery of the AdS/CFT correspondence it is expected that black hole entropy is related to some conformal symmetries (either at the spatial infinity or on the horizon) of the black holes [2–4]. But the term "universality" suggests a slightly more intriguing possibility, namely that one might be able to learn general lessons about the black hole entropy by studying the conformal symmetries alone, but without having to know the full detail of the underlying theory of quantum gravity. This point, if true, can be helpful in understanding the entropy of black holes in those dimensions where a consistent theory of quantum gravity (e.g. string theory) is not currently available. So it is of great importance to understand possible conformal symmetries related to black holes in detail. Ideally, one would also like to know the full detail of the dual conformal field theory (CFT) in the end.

In this short note, I briefly summarize some new results obtained during the past few years, focusing on the more detailed understanding of conformal symmetries on the horizon of general stationary and axisymmetric black holes in arbitrary dimensions.

J. Mei (✉)
Max Planck Institute for Gravitational Physics (Albert Einstein Institute),
Am Mühlenberg 1, 14476 Golm, Germany
e-mail: jwmei@aei.mpg.de

J. Bičák and T. Ledvinka (eds.), *Relativity and Gravitation*,
Springer Proceedings in Physics 157, DOI: 10.1007/978-3-319-06761-2_41,

2 Extremal Black Holes and the Kerr/CFT Correspondence

The new development started with the Kerr/CFT conjecture [5]. Back in 1997, Strominger already noticed that for any black hole having an $AdS(3)$ factor on its horizon, the entropy can always be calculated in the same fashion as that for the BTZ black hole [3]. In [5], the authors noticed that the near horizon limit of the extremal Kerr (NHEK) metric at a fixed polar angle is a quotient of some warped AdS_3. Since the BTZ black holes are quotients of $AdS(3)$, this close analogy indicates that what has been done for the BTZ black hole may also be done for the NHEK. Indeed, by using the techniques of Brown and Henneaux [2], the authors of [5] found that quantum gravity on the NHEK background could be dual to a chiral CFT with a central charge $c_L = 12J$ and a temperature $T_L = \frac{1}{2\pi}$, where J is the angular momentum of the Kerr black hole. Cardy's formula then reproduces exactly the Bekenstein-Hawking entropy.

Among many works that followed, one particularly important observation was made in [6]. It was noticed that the near horizon limit of all extremal black holes shares the same form of the metric, and this allows one to explain the entropy of all extremal black holes through the Kerr/CFT conjecture once and for all [7]. In a slightly different notation than [6], the common near-horizon metric is given by [7]

$$ds^2 = \frac{2f_r^0}{\Delta_0''}\Big[-(1+r^2)dt^2 + \frac{dr^2}{1+r^2}\Big] + g_{ij}^0 d\theta^i d\theta^j \\ + g_{ab}^0(d\phi^a + k^a r dt)(d\phi^b + k^b r dt), \tag{1}$$

where t is the time, r is the radius, θ^i are the longitudinal angles and ϕ^a are the azimuthal angles. For the functions, f_r^0, g_{ij}^0 and g_{ab}^0 depend on θ^i, while Δ_0'' and k^a are constants. The constants k^a are related to an interesting observation [8]: for each non-vanishing angular momentum there is an independent copy of Virasoro algebra and each copy is equally good in reproducing the Bekenstein-Hawking entropy! In more detail, related to each direction ϕ^a, the corresponding central charge and the temperature of the CFT are [6, 7]

$$c^a = \frac{3k^a}{2\pi}\mathscr{A}_{rea}\,, \quad T^a = \frac{1}{2\pi k^a}\,, \tag{2}$$

where $\mathscr{A}_{rea}$ is the area of the black hole horizon. Cardy formula then yields

$$S = \frac{1}{3}\pi^2 c^a T^a = \frac{\mathscr{A}_{rea}}{4}\,. \tag{3}$$

Although one often prefers to understand the Kerr/CFT correspondence in relation to string theory, it also fits our theme discussed in the introduction. In particular, the near-horizon metric (1) is a blow-up of the thin region extremely close to the horizon of the black hole. So the conformal symmetries found in the background of the near-horizon metric must also reside on/near the horizon.

3 Conformal Symmetries for Non-extremal Black Holes

The Kerr/CFT correspondence, as it was originally conjectured in [5], only works for extremal black holes. This is because it is impossible to take the near horizon limit for the metrics of non-extremal black holes.

However, as Carlip has pointed out recently, there is no need to take the near-horizon limit in order to expose the conformal symmetries [9]. Just like in [4], one can impose boundary conditions on the horizon and also recover the conformal symmetries. In [10], this was carried out for general stationary and axisymmetric black holes in generic dimensions. Similarly to the extremal case, this was made possible by identifying the common form of all the relevant metrics,

$$ds^2 = f\left[-\frac{\Delta}{\nu^2}dt^2 + \frac{dr^2}{\Delta}\right] + g_{ij}d\theta^i d\theta^j + g_{ab}(d\phi^a - w^a dt)(d\phi^b - w^b dt), \quad (4)$$

where the coordinates are the same as in (1), but the functions f, ν, g_{ij}, g_{ab} and w^a now depend on both r and θ^i, while Δ only depends on r. More details of the metric can be found in [10]. Note that (1) is easily recovered by taking the near horizon limit of the above metric for extremal black holes. One of our main result is that, just like for extremal black holes [7, 8], in the non-extremal case there is also an independent copy of Virasoro algebra for each non-vanishing angular momentum, and each Virasoro algebra is equally good in reproducing the Bekenstein-Hawking entropy. But for non-extremal black holes, the result is off by a factor of 2 [10]. We wish to resolve this problem in the near future.

4 Conformal Symmetries Without Boundary Conditions

There are possible issues related to the boundary conditions used in the Kerr/CFT conjecture [11, 12]. But this does not necessarily harm the conformal symmetries, because there are other ways to expose them. For example, one can study the dynamics of a probe scalar field near the black hole horizon, and as a result people have suggested "hidden conformal symmetry" related to the Kerr black hole [13]. Further evidence of the hidden conformal symmetry was also obtained through the "subtracted geometry" [14]. In all these works the presence of the conformal symmetry is inferred from the existence of approximate $SL(2, R)$ symmetries in the radial wave equation of a probing scalar field. Inspired by these results, we then notice that for a stationary and axisymmetric black hole the on-shell Einstein-Hilbert action has a rigid $SL(m, R)$ symmetry, where m is the number of ignorable coordinates (t and ϕ^a)

in the system. Corresponding to each rotation (ϕ^a) one has an $SL(2, R)$ subgroup from the $SL(m, R)$. We then find that each of this $SL(2, R)$ is enhanced to a Virasoro algebra on the black hole horizon [15]. (See also [16] for an earlier indication.) So again one has the nice result that there is an independent copy of Virasoro algebra for each non-vanishing angular momentum.

5 Outlook

As said in [14] there has been little concrete progress towards a CFT interpretation of general non-extremal black holes, despite some optimistic conventional wisdom.

From the recent results of the Kerr/CFT correspondence, many fingers seem to be pointing at the existence of conformal symmetries on the black hole horizon. But much still needs to be done to expose the conformal symmetries and to explicitly show how they can account for the black hole entropy. For us, we are still off by a factor of 2 in the Carlip-like treatment, and we still have not been able to abstract any physical information from the conformal symmetries identified from the on-shell Einstein-Hilbert action on the horizon.

But if we are eventually successful with this procedure, it is likely to expose some common features in the quantum nature of black holes and such features are likely to be shared by all spacetime dimensions.

References

1. Carlip, S.: Symmetries, horizons, and black hole entropy. Gen. Relativ. Gravit. **39**, 1519 (2007). doi:10.1007/s10714-007-0467-6
2. Brown, J., Henneaux, M.: Central charges in the canonical realization of asymptotic symmetries: an example from three-dimensional gravity. Commun. Math. Phys. **104**, 207 (1986)
3. Strominger, A.: Black hole entropy from near-horizon microstates. J. High Energy Phys. **1998**(02), 009 (1998). doi:10.1088/1126-6708/1998/02/009
4. Carlip, S.: Black hole entropy from conformal field theory in any dimension. Phys. Rev. Lett. **82**, 2828 (1999). doi:10.1103/PhysRevLett.82.2828
5. Guica, M., Hartman, T., Song, W., Strominger, A.: The Kerr/CFT correspondence. Phys. Rev. D **80**, 124008 (2009). doi:10.1103/PhysRevD.80.124008
6. Chow, D., Cvetič, M., Lu, H., Pope, C.: Extremal black hole/CFT correspondence in (gauged) supergravities. Phys. Rev. D **79**, 084018 (2009). doi:10.1103/PhysRevD.79.084018
7. Mei, J.: The entropy for general extremal black holes. J. High Energy Phys. **2010**(04), 005 (2010). doi:10.1007/JHEP04(2010)005
8. Lü, H., Mei, J., Pope, C.: Kerr-AdS/CFT correspondence in diverse dimensions. J. High Energy Phys. **2009**(04), 054 (2009). doi:10.1088/1126-6708/2009/04/054
9. Carlip, S.: Extremal and nonextremal Kerr/CFT correspondences. J. High Energy Phys. **2011**(04), 076 (2011). doi:10.1007/JHEP04(2011)076
10. Mei, J.: On the general Kerr/CFT correspondence in arbitrary dimensions. J. High Energy Phys. **2012**(04), 113 (2012). doi:10.1007/JHEP04(2012)113
11. Amsel, A., Horowitz, G., Marolf, D., Roberts, M.: No dynamics in the extremal Kerr throat. J. High Energy Phys. **2009**(09), 044 (2009). doi:10.1088/1126-6708/2009/09/044

12. Dias, O., Reall, H., Santos, J.: Kerr-CFT and gravitational perturbations. J. High Energy Phys. **2009**(08), 101 (2009). doi:10.1088/1126-6708/2009/08/101
13. Castro, A., Maloney, A., Strominger, A.: Hidden conformal symmetry of the Kerr black hole. Phys. Rev. D **82**, 024008 (2010). doi:10.1103/PhysRevD.82.024008
14. Cvetic, M., Larsen, F.: Conformal symmetry for general black holes. J. High Energy Phys. **2012**(02), 122 (2012). doi:10.1007/JHEP02(2012)122
15. Mei, J.: Conformal symmetries of the Einstein-Hilbert action on horizons of stationary and axisymmetric black holes. Class. Quantum Grav. **29**, 095020 (2012)
16. Carlip, S.: Near horizon conformal symmetry and black hole entropy. Phys. Rev. Lett. **88**, 241301 (2002). doi:10.1103/PhysRevLett.88.241301

Finsler Spacetimes and Gravity

Christian Pfeifer and Mattias Wohlfarth

Abstract We consider the geometry of spacetime based on a non-metric, Finslerian, length measure, which, in terms of physics, represents a generalized clock. Our definition of Finsler spacetimes ensure a well defined notion of causality, a precise description of observers and a geometric background for field theories. Moreover we present our Finsler geometric extension of the Einstein equations, which determine the geometry of Finsler spacetimes dynamically.

1 Introduction

For hundred years Lorentzian manifolds serve as geometric background for physics. Equipped with the standard model of particle physics this led to the explanation of a huge amount of observations. However, on this basis we have to conclude that 96 % of the universe is unknown; called dark matter and dark energy [1]. Today most explanation attempts for this fact come from modifications of the standard model of particle physics; but possibly a well controlled extension of the geometric background for physics is able to shed light on the dark universe.

Here we present Finsler spacetimes which are capable to serve as generalized geometric background for physics providing:

- a precise well-defined notion of causality,
- a notion of observers and their measurements,
- a geometric background for field theories,
- and gravitational dynamics consistent with general relativity.

Further details beyond this invitation can be found in our articles [2, 3].

C. Pfeifer (✉) · M. Wohlfarth
II. Institut für Theoretische Physik und Zentrum für Mathematische Physik,
Universität Hamburg Luruper Chaussee 149, 22761 Hamburg, Germany
e-mail: christian.pfeifer@desy.de

J. Bičák and T. Ledvinka (eds.), *Relativity and Gravitation*,
Springer Proceedings in Physics 157, DOI: 10.1007/978-3-319-06761-2_42,

2 Finsler Geometry and its Causal Structure

One of the fundamental measurements in physics is the measurement of time. Its theoretical description is given by Einstein's clock postulate: The time that passes for an observer between two events is given by the length of the observers worldline connecting the events. In case the geometry of spacetime is fundamentally determined by a metric this length is given by

$$S[x] = \int d\tau \sqrt{g_{ab}(x)\dot{x}^a\dot{x}^b}\,. \tag{1}$$

The key idea for Finsler spacetimes is a more general description of the measurement of time which still realizes the weak equivalence principle:

$$S[x] = \int d\tau F(x, \dot{x})\,. \tag{2}$$

It is based on a one-homogeneous function F on the tangent bundle which determines the geometry of spacetime. This so called Finsler geometry is a well known mathematical framework which extends Riemannian metric geometry [4]. However this standard Finsler geometry breaks down as soon as F has a non-trivial null-structure $N_x = \{y \in T_xM | F(x, y) = 0\}$. For generalizations of the Lorentzian metric length measures we introduce our definition of Finsler spacetimes which ensure the existence of a precise notion of causality and the existence of a well-defined geometry.

The description of Finsler spacetimes requires the tangent bundle TM of the spacetime manifold M. We consider the tangent bundle in manifold induced coordinates $(x, y) = Z \in TM$, $Z = y^a\partial_{a|x}$ and its tangent spaces $T_{(x,y)}TM$ in the coordinate basis $\{\partial_a = \frac{\partial}{\partial x^a}, \bar{\partial}_a = \frac{\partial}{\partial y^a}\}$.

A Finsler spacetime (M, L, F) is a smooth manifold M equipped with a continuous function $L : TM \mapsto \mathbb{R}$ such that

- L is smooth on the tangent bundle without the zero section $TM\setminus\{0\}$,
- L is reversible $|L(x, -y)| = |L(x, y)|$,
- L is positively homogeneous of degree $r \geq 2$: $L(x, \lambda y) = \lambda^r L(x, y)$,
- ${g^L}_{ab} = \frac{1}{2}\bar{\partial}_a\bar{\partial}_b L$ is non-degenerate on $TM\setminus A$, $A \subset TM$ measure zero,
- $\forall x \in M$ there exists a non-empty closed connected set $S_x \in T_xM$ where: $|L(x, y)| = 1$ and $sign({g^L}_{ab}) = (\varepsilon, -\varepsilon, -\varepsilon, -\varepsilon)$ with $\varepsilon = \frac{|L(x,y)|}{L(x,y)}$.

The Finsler function F, which defines the geometric clock is a derived object and defined as $F = |L|^{\frac{1}{r}}$; the Finsler metric is ${g^F}_{ab} = \frac{1}{2}\bar{\partial}_a\bar{\partial}_b F^2$.

Our definition of Finsler spacetimes guarantees a causal structure in each tangent space: S_x is the shell of unit timelike vectors which defines a cone of timelike directions with null boundary, as displayed in Fig. 1.

The geometry of Finsler spacetimes is solely derived from derivatives of L in terms of the unique Cartan non-linear connection coefficients: $N^a{}_b = \frac{1}{4}\bar{\partial}_b(g^{Laq}$

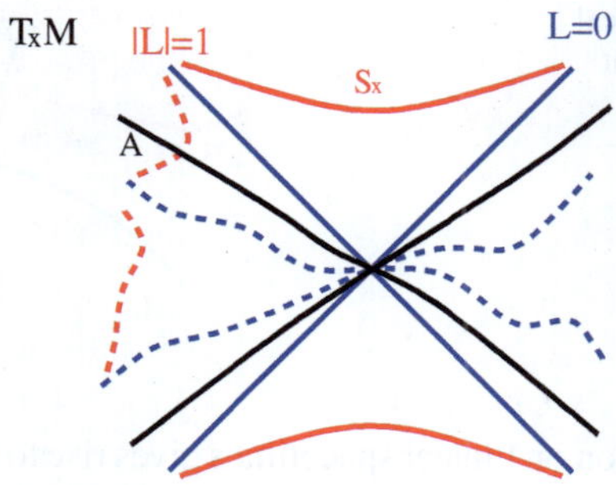

Fig. 1 Causal structure of Finsler spacetime

$(y^m \partial_m \bar{\partial}_q L - \partial_q L))$. The connection between our definition of Finsler spacetimes and standard Finsler geometry is given by the following theorem: *Wherever L and F are both differentiable they encode the same geometry, i.e.* $N[L] = N[F^2]$.

3 Observers, Matter Fields, and Gravity

The nonlinear connection coefficients split TTM and T*TM into horizontal and vertical space by $\{\delta_a = \partial_a - N^b{}_a \bar{\partial}_b, \bar{\partial}_a\}$ and $\{dx^a, \delta y^a = dy^a + N^a{}_b dx^b\}$, as displayed in Fig. 2. The horizontal (co-)tangent space is identified with the (co-)tangent space along the manifold directions.

Timelike observers move on worldlines $x(\tau) \in M$ with trajectory $(x, \dot{x}) \in TM$ and $\dot{x}$ in the cone of timelike vectors. A horizontal orthonormal frame defines their time and space directions along the manifold $\{E_a\} = \{E_0 = \dot{x}^a \delta_a, E_\alpha\}$; $g^F_{(x,\dot{x})}(E_\mu, E_\nu) = -\eta_{\mu\nu}$. Measurable quantities are components of horizontal tensors evaluated in this frame at the observers' *TM* position.

The geometry of Finsler spacetimes is built from tensors on *TM*; hence physical fields coupling to this geometry will be of the same kind. Lagrange densities on *TM* require the canonical Sasaki-type *TM*-metric $G = -g^F{}_{ab}(dx^a dx^b + F^{-2} \delta y^a \delta y^b)$, which allows us to couple field theories to Finsler spacetime geometry as follows: Choose an action for a p-form $\phi(x)$ on $(M, g) : S[\phi, g] = \int_M \sqrt{g} \mathscr{L}(g, \phi, d\phi)$, use the Lagrangian for a zero homogenous p-form field $\Phi(x, y)$ on (TM, G), introduce Lagrange multipliers to restrict the p-form field to be horizontal, integrate over the unit tangent bundle $\Sigma = \{(x, y) \in TM | F(x, y) = 1\}$ to obtain the p-form field action $S_m[\Phi, L, \lambda] = \int_\Sigma (\sqrt{g^F h^F} \mathscr{L}(G, \Phi, d\Phi) + \lambda(1 - P^H)\Phi)_{|\Sigma}$. Variation yields the equations of motion, the vanishing of all non horizontal components on shell and the source term of the gravitational dynamics $T_{|\Sigma}$. Our coupling principle ensures that in case the Finsler spacetime is metric, field theories and gravitational dynamics equal those of general relativity.

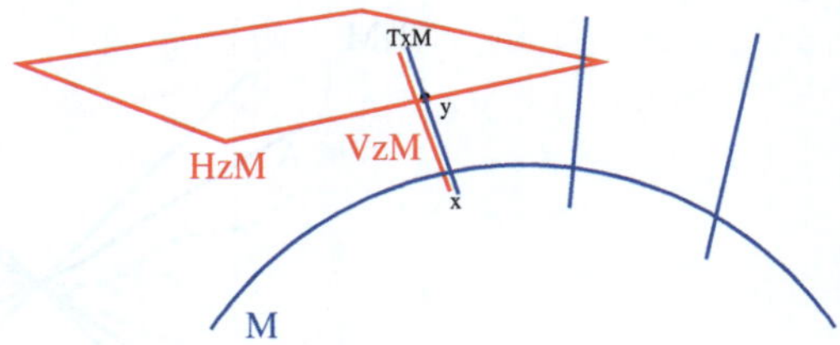

Fig. 2 Horizontal and vertical tangent space to the tangent bundle

The geodesic deviation on Finsler spacetimes gives rise to a tensor causing relative gravitational acceleration $\nabla_{\dot{x}}\nabla_{\dot{x}}V^a = R^a{}_{bc}(x,\dot{x})\dot{x}^b V^c$. This non-linear curvature given by $R^a{}_{bc} = \delta_{[b}N^a{}_{c]}$ leads to the curvature scalar $\mathscr{R}^{\mathscr{F}} = R^a{}_{ab}y^b$. No further dependence on L or its derivatives appears, thus we choose $\mathscr{R}^{\mathscr{F}}$ as Lagrangian for our Finsler gravity action $S[L,\Phi] = \int_\Sigma(\sqrt{g^F h^F}\mathscr{R}^{\mathscr{F}})_{|\Sigma} + S_m[L,\Phi]$. Variation with respect to the L yields the Finsler gravity field equation

$$g^{Fab}\bar{\partial}_a\bar{\partial}_b R^F - \frac{6}{F^2}R^F + 2g^F_{ab}\big(\nabla_a S_b + S_a S_b + \bar{\partial}_a(y^q\nabla_q S_b)\big) = -\kappa T_{|\Sigma}\,. \quad (3)$$

It contains the curvature scalar, a measure of the departure from metric geometry S, and a Finsler version of the Levi-Civita derivative. In case the function L is the metric length measure the Finsler gravity equation is equivalent to the Einstein equations.

4 Conclusion

We constructed a theory of gravity for spacetimes equipped with a general Finsler length measure. In case the Finsler length equals the metric length our theory becomes general relativity, hence all solutions of the Einstein equations are solutions to our Finsler gravity equation. The implications of Finsler spacetime gravity on the dark universe can be studied by spherically symmetric and cosmological solutions that go beyond metric geometry. A perturbative first order Finsler solution around the Schwarzschild and Friedmann-Robertson-Walker metric is work in progress.

References

1. Spergel, D., Bean, R., Doré, O., et al.: Three-year Wilkinson microwave anisotropy probe (WMAP) observations: implications for cosmology. Astrophys. J. Suppl. **170**, 377 (2007). doi:10.1086/513700
2. Pfeifer, C., Wohlfarth, M.: Causal structure and electrodynamics on Finsler spacetimes. Phys. Rev. D **84**, 044039 (2011). doi:10.1103/PhysRevD.84.044039
3. Pfeifer, C., Wohlfarth, M.: Finsler geometric extension of Einstein gravity. Phys. Rev. D **85**, 064009 (2012). doi:10.1103/PhysRevD.85.064009
4. Bao, D., Shern, S.S., Chen, Z.: An introduction to Riemann-Finsler geometry. In: Graduate Texts in Mathematics, vol. 200. Springer, Berlin (2000)

Lagrangian Analysis of 'Trivial' Symmetries in Models of Gravity

Debraj Roy

Abstract We study the differences between Poincaré and canonical Hamiltonian symmetries in models of gravity through the corresponding Noether identities and show that they are equivalent modulo trivial gauge symmetries.

1 Introduction

Poincaré symmetry is a fundamental symmetry of nature and a gauge theory of the Poincaré group can be used to model theories of gravity. This Poincaré gauge theory (PGT) was developed by Utiyama [1], Kibble [2], Sciama [3] and later on by various authors [4]. PGT is built on a global manifold with local orthonormal frames glued to each spacetime point by frame fields or triads (in 3D). The triads ${b^i}_\mu$ are used to translate between the global (index: Greek) and local (index: Latin) frames. To construct a gauge theory, connections ${\omega^i}_\mu$ are introduced replacing partial derivatives by corresponding covariant derivatives. The corresponding field strengths give rise to the gravitational fields of curvature ${R^i}_{\mu\nu}$ and torsion ${T^i}_{\mu\nu}$

$$ {R^i}_{\mu\nu} = \partial_\mu {\omega^i}_\nu - \partial_\mu {\omega^i}_\nu + {\varepsilon^i}_{jk}\, {\omega^j}_\mu {\omega^k}_\nu, \tag{1} $$

$$ {T^i}_{\mu\nu} = \nabla_\mu {b^i}_\nu - \nabla_\nu {b^i}_\mu . \tag{2} $$

These fields can now be used to write actions describing gravity in Riemann-Cartan spacetime. Imposition of a condition on torsion through equations of motion (in vacuum) may lead one to a spacetime with only curvature and no torsion—the usual Einstein GR on Riemannian manifold.

D. Roy (✉)
S. N. Bose National Centre for Basic Sciences, Block-JD, Sector III, Salt Lake, Kolkata 700098, India
e-mail: debraj@bose.res.in

J. Bičák and T. Ledvinka (eds.), *Relativity and Gravitation*,
Springer Proceedings in Physics 157, DOI: 10.1007/978-3-319-06761-2_43,

As gauge theories of the Poincaré group, Poincaré symmetries are already inbuilt. A Dirac canonical analysis of symmetries on the other hand also yields a set of gauge symmetries for the same models. By a gauge symmetry here we mean any continuous symmetry of the basic fields that leave the action invariant. The total number of independent gauge symmetries are however limited by the number of independent, primary first class constraints [5]. So it transpires that there is a discrepancy with established results in the apparent off-shell in-equivalence between the Poincaré and canonical Hamiltonian symmetries. Here we study and resolve this from a lagrangian point of view.

2 Noether Identities and Trivial Symmetries

For specifics of discussion, we take up the Mielke-Baekler model [6] describing a cosmologically topological model of gravity with torsion. The action for the model is

$$S=\int \mathrm{d}^3\mathrm{x}\, \varepsilon^{\mu\nu\rho}\Big[ab^i{}_\mu R_{i\nu\rho}-\frac{\Lambda}{3}\varepsilon_{ijk}b^i{}_\mu b^j{}_\nu b^k{}_\rho+\alpha_3\Big(\omega^i{}_\mu\partial_\nu\omega_{i\rho} +\frac{1}{3}\varepsilon_{ijk}\,\omega^i{}_\mu\omega^j{}_\nu\omega^k{}_\rho\Big)+\frac{\alpha_4}{2}b^i{}_\mu T_{i\nu\rho}\Big], \tag{3}$$

where the terms are the Einstein-Cartan term, cosmological term, Chern-Simons term (in connection) and the torsion term, respectively. The Euler derivatives corresponding to the independent canonical fields are:

$$\begin{aligned}\frac{\delta S}{\delta b^i{}_\mu}&=\varepsilon^{\mu\nu\rho}\Big[a\,R_{i\nu\rho}+\alpha_4\,T_{i\nu\rho}-\Lambda\,\varepsilon_{ijk}b^j{}_\nu b^k{}_\rho\Big],\\ \frac{\delta S}{\delta\omega^i{}_\mu}&=\varepsilon^{\mu\nu\rho}\Big[\alpha_3\,R_{i\nu\rho}+a\,T_{i\nu\rho}+\alpha_4\,\varepsilon_{ijk}b^j{}_\nu b^k{}_\rho\Big].\end{aligned}\tag{4}$$

The model independent Poincaré symmetries (subscript 'P') are [7]

$$\begin{aligned}\delta_P b^i{}_\mu&=-\varepsilon^i{}_{jk}b^j{}_\mu\theta^k-\partial_\mu\xi^\rho\, b^i{}_\rho-\xi^\rho\,\partial_\rho b^i{}_\mu,\\ \delta_P\omega^i{}_\mu&=-\partial_\mu\theta^i-\varepsilon^i{}_{jk}\omega^j{}_\mu\theta^k-\partial_\mu\xi^\rho\,\omega^i{}_\rho-\xi^\rho\,\partial_\rho\omega^i{}_\mu,\end{aligned}\tag{5}$$

while the canonical symmetries generated by the first-class gauge generator constructed through an off-shell algorithm [5, 8–10] are

$$\begin{aligned}\delta_H b^i{}_\mu&=\nabla_\mu\varepsilon^i-p\,\varepsilon^i{}_{jk}\,b^j{}_\mu\varepsilon^k+\varepsilon^i{}_{jk}\,b^j{}_\mu\tau^k,\\ \delta_H\omega^i{}_\mu&=\nabla_\mu\tau^i-q\,\varepsilon^i{}_{jk}\,b^j{}_\mu\varepsilon^k.\end{aligned}\tag{6}$$

An inspection of the two symmetries (5) and (6) reveal that the canonical symmetries are structurally dependent on the form of the action while the Poincaré symmetries are independent of particular action. Also, to compare the two symmetries, we have to first find a suitable mapping between the different sets of gauge parameters. To find this, we take recourse to the Noether identities corresponding to the symmetries [11].

A Noether identity corresponds to each continuous gauge symmetry of an action, marked by an independent gauge parameter. In fact, the identity is a direct consequence of the invariance of the action. To see this, let us consider a generic gauge symmetry expressed as terms proportional to the gauge parameter (ε^{μ}) and its derivative

$$\delta q_i = R_{i\mu}\varepsilon^{\mu} + \tilde{R}^{\nu}_{i\mu}\,(\partial_{\nu}\varepsilon^{\mu}). \tag{7}$$

The invariance of the action, step by step, leads to

$$\begin{aligned} \delta S &= \int \frac{\delta \mathscr{L}}{\delta q_i}\delta q_i = \int \frac{\delta \mathscr{L}}{\delta q_i}\left(R_{i\mu}\varepsilon^{\mu} + \tilde{R}_{i\mu}{}^{\nu}\,\partial_{\nu}\varepsilon^{\mu}\right) \\ &= \int \left[\frac{\delta \mathscr{L}}{\delta q_i}R_{i\mu} - \partial_{\nu}\left(\frac{\delta \mathscr{L}}{\delta q_i}\tilde{R}_{i\mu}{}^{\nu}\right)\right]\varepsilon^{\mu} = 0, \end{aligned} \tag{8}$$

where the quantity within braces forms the Noether identity due to the arbitrary nature of each of the gauge parameters.

The Noether identities corresponding to PGT symmetries are

$$\begin{aligned} P_k &= \frac{\delta S}{\delta b^i{}_{\mu}}\varepsilon^i{}_{jk}b^j{}_{\mu} + \frac{\delta S}{\delta \omega^i{}_{\mu}}\varepsilon^i{}_{jk}\omega^j{}_{\mu} - \partial_{\mu}\left(\frac{\delta S}{\delta \omega^k{}_{\mu}}\right) = 0, \\ R_{\rho} &= \frac{\delta S}{\delta b^i{}_{\mu}}\partial_{\rho}b^i{}_{\mu} + \frac{\delta S}{\delta \omega^i{}_{\mu}}\partial_{\rho}\omega^i{}_{\mu} - \partial_{\mu}\left(b^i{}_{\rho}\frac{\delta S}{\delta b^i{}_{\mu}} + \omega^i{}_{\rho}\frac{\delta S}{\delta \omega^i{}_{\mu}}\right) = 0, \end{aligned} \tag{9}$$

and that corresponding to canonical Hamiltonian symmetries of the Mielke-Baekler action are

$$\begin{aligned} A_k &= \frac{\delta S}{\delta b^i{}_{\mu}}\varepsilon^i{}_{jk}b^j{}_{\mu} + \frac{\delta S}{\delta \omega^i{}_{\mu}}\varepsilon^i{}_{jk}\omega^j{}_{\mu} - \partial_{\mu}\left(\frac{\delta S}{\delta \omega^k{}_{\mu}}\right) = 0 \\ B_k &= -\partial_{\mu}\left(\frac{\delta S}{\delta b^k{}_{\mu}}\right) + \frac{\delta S}{\delta b^i{}_{\mu}}\varepsilon^i{}_{jk}\omega^j{}_{\mu} - p\frac{\delta S}{\delta b^i{}_{\mu}}\varepsilon^i{}_{jk}b^j{}_{\mu} - q\frac{\delta S}{\delta \omega^i{}_{\mu}}\varepsilon^i{}_{jk}b^j{}_{\mu} = 0 \end{aligned} \tag{10}$$

A comparison between (9) and (10) immediately shows that one identity from each pair is already equivalent: $P_k = A_k$. Comparing the nature of the other identities it is seen that the term $-\omega^k{}_{\rho}A_k - b^k{}_{\rho}B_k$ gives

$$-R_\rho + \frac{\delta S}{\delta b^i{}_\mu}\left(\frac{\alpha_3}{\Delta}\,\eta^{ij}\,\varepsilon_{\mu\nu\rho}\right)\frac{\delta S}{\delta b^j{}_\nu} + \frac{\delta S}{\delta b^i{}_\mu}\left(\frac{-a}{\Delta}\,\eta^{ij}\,\varepsilon_{\mu\nu\rho}\right)\frac{\delta S}{\delta \omega^j{}_\nu}$$
$$+\frac{\delta S}{\delta \omega^i{}_\mu}\left(\frac{-a}{\Delta}\,\eta^{ij}\,\varepsilon_{\mu\nu\rho}\right)\frac{\delta S}{\delta b^j{}_\nu} + \frac{\delta S}{\delta \omega^i{}_\mu}\left(\frac{\alpha_4}{\Delta}\,\eta^{ij}\,\varepsilon_{\mu\nu\rho}\right)\frac{\delta S}{\delta b^j{}_\nu} = 0,$$

where $\Delta = 2\,(\alpha_3\alpha_4 - a^2)$. The terms proportional to square of Euler derivatives are antisymmetric in their coefficients and as such drop out *without* having to use the equations of motion, i.e. without having to set the Euler derivatives to zero. Thus we get back the Poincaré Noether identities from the canonical hamiltonian Noether identities, their difference being just 'trivial' gauge identities [12]. Thus substituting $R_\rho = -b^k{}_\rho B_k - \omega^k{}_\rho A_k$ and $P_k = -A_k$ in $\delta S = \int \left(\theta^k P_k + \xi^\rho R_\rho\right) = 0$ gives

$$\int \left[(-\theta^k - \xi^\rho \omega^k{}_\rho)\,A_k \,+\, (-b^k{}_\rho \xi^\rho)\,B_k\right] = 0. \tag{11}$$

Comparing this with $\delta S = \int \left(\varepsilon^k A_k + \tau^k B_k\right) = 0$ gives us the required map between the two sets of gauge parameters.

$$\varepsilon^i = -\xi^\rho b^i{}_\rho \quad \text{and} \quad \tau^i = -\theta^i - \xi^\rho \omega^i{}_\rho. \tag{12}$$

So the Noether identities help us to generate the required map between different sets of gauge parameters and show the equivalence of the two symmetries as their difference is just 'trivial'!

References

1. Utiyama, R.: Invariant theoretical interpretation of interactions. Phys. Rev. **101**, 1597 (1956). doi:10.1103/PhysRev.101.1597
2. Kibble, T.: Lorentz invariance and the gravitational field. J. Math. Phys. **2**, 212 (1961). doi:10.1063/1.1703702
3. Sciama, D.: On the analogy between charge and spin in general relativity. In: Recent Developments in General Relativity, pp. 415–439. Pergamon Press/PWN, New York (1962)
4. Hehl, F., von der Heyde, P., Kerlick, G., Nester, J.: General relativity with spin and torsion: foundation and prospects. Rev. Mod. Phys. **48**, 393 (1976). doi:10.1103/RevModPhys.48.393
5. Henneaux, M., Teitelboim, C., Zanelli, J.: Gauge invariance and degree of freedom count. Nucl. Phys. B **332**, 169 (1990). doi:10.1016/0550-3213(90)90034-B
6. Mielke, E., Baekler, P.: Topological gauge model of gravity with torsion. Phys. Lett. A **A156**, 399 (1991). doi:10.1016/0375-9601(91)90715-K
7. Blagojevic, M., Cvetkovic, B.: Canonical structure of 3-D gravity with torsion. In: Benton, C.V. (ed.) Trends in General Relativty and Quantum Cosmology, Horizons in World Physics, vol. 253, pp. 103–123. Nova Science Publishers, New York (2006)
8. Banerjee, R., Rothe, H., Rothe, K.: Hamiltonian approach to Lagrangian gauge symmetries. Phys. Lett. B **463**, 248 (1999). doi:10.1016/S0370-2693(99)00977-6
9. Banerjee, R., Rothe, H., Rothe, K.: Master equation for Lagrangian gauge symmetries. Phys. Lett. B **479**, 429 (2000). doi:10.1016/S0370-2693(00)00323-3

10. Banerjee, R., Gangopadhyay, S., Mukherjee, P., Roy, D.: Symmetries of topological gravity with torsion in the hamiltonian and lagrangian formalisms. J. High Energy Phys. **2010**(02), 075 (2010). doi:10.1007/JHEP02(2010)075
11. Banerjee, R., Roy, D.: Poincare gauge symmetries, Hamiltonian symmetries and trivial gauge transformations. Phys. Rev. D **84**, 124034 (2011). doi:10.1103/PhysRevD.84.124034
12. Henneaux, M., Teitelboim, C.: Quantization of Gauge Systems. Princeton University Press, Princeton (1992)

Quasi-normal Frequencies, Horizon Area Spectra and Multi-horizon Spacetimes

Jozef Skákala

Abstract We analyse the behavior of the asymptotic frequencies of the spherically symmetric multi-horizon spacetimes (in particular Reissner-Nordström, Schwarzschild-deSitter, Reissner-Nordström-deSitter) and provide some suggestions for how to interpret the results following the spirit of the modified Hod's conjecture. The interpretation suggested is in some sense analogical to the Schwarzschild case, but has some new specific features.

1 Introduction

This short paper refers to work done over a longer period of time contained in papers [1–3] and also to some extent in [4]. Black holes, when perturbed, show certain characteristic damped oscillations which are called quasi-normal modes (QNMs). The low damped black hole quasi-normal modes are of potential astrophysical interest, as they carry information about the black hole parameters. More than a decade ago it was conjectured that also the highly damped (asymptotic) modes might have physical importance, as they might carry information about quantum black holes [5]. The original conjecture by Hod [5] was later modified by Maggiore [6]. The conjecture of [6] was used in case of Schwarzschild black hole and also in cases of other black holes to derive the area spectra of the black hole horizon. What is still missing is a (unconstrained) application of the conjecture to the most immediate generalizations of Schwarzschild spacetime, to the spherically symmetric multi-horizon spacetimes: Reissner-Nordström (R-N), Schwarzschild-deSitter (S-dS) and Reissner-Nordström-deSitter (R-N-dS) spacetime. This work is trying to fill the gap.

J. Skákala (✉)
UFABC, Santo André, São Paulo, Brazil
e-mail: jozef.skakala@ufabc.edu.br

J. Bičák and T. Ledvinka (eds.), *Relativity and Gravitation*,
Springer Proceedings in Physics 157, DOI: 10.1007/978-3-319-06761-2_44,

2 Spherically Symmetric Multi-horizon Black Hole Spacetimes

The *modified* Hod's (or Maggiore's) conjecture relates asymptotic QNM frequencies ($\omega = \omega_R + i \cdot \omega_I$) to the black hole ADM mass quantum transition as:

$$\Delta M = \lim_{n\to\infty} (\omega_{nI} - \omega_{n-1I}) \doteq \lim_{n\to\infty} \Delta_{(n,n-1)}\omega_{nI}. \tag{1}$$

(For a detailed reasoning see [6]. Also everywhere in the paper we use Planck units and n labels the QNM frequencies monotonically with respect to the imaginary part.) For the multi-horizon spherically symmetric black hole spacetimes (R-N, S-dS, R-N-dS) and the scalar, vector, tensor perturbations the asymptotic QNM frequencies obey the following equation:

$$\sum_{A=1}^{M} C_A \exp\left(\sum_{i=1}^{N} Z_{Ai}\frac{2\pi\omega}{|\kappa_i|}\right) = 0. \tag{2}$$

Z_{Ai} takes one of the values $Z_{Ai} = 0, 1, 2$, furthermore N is the number of horizons and κ_i are the surface gravities of the different horizons. The analytical solutions of formulas (2) are not known and this prevented many people of using the modified Hod's conjecture in R-N, S-dS, R-N-dS cases. In [3] we analysed the behaviour of the solutions of (2) with the following results: If the ratio of all of the surface gravities is a rational number then all the frequencies split in a *finite* number of equispaced families (labeled by a) of the form:

$$\omega_{an} = (\text{offset})_a + in \cdot \text{lcm}(|\kappa_1|, |\kappa_2|, \ldots, |\kappa_N|). \tag{3}$$

Here lcm is the least common multiple of the numbers in the bracket, hence $\text{lcm}(|\kappa_1|, \ldots, |\kappa_N|) = p_1|\kappa_1| = \cdots = p_N|\kappa_N|$, where $\{p_1, \ldots, p_N\}$ is a set of relatively prime integers. If the ratio of arbitrary two of the surface gravities is irrational, then there does *not* exist an equispaced subsequence in the sequence of asymptotic QNM frequencies. Moreover one can prove [1] for the R-N black hole (but one expects it to hold for all the three cases) that, in case the ratio of the surface gravities is irrational, the $n \to \infty$ limit for $\Delta_{n,n-1}\omega_{nI}$ does *not* exist. Also for the rational ratio of the surface gravities and the R-N black hole the only case in which the limit $n \to \infty$ $\Delta_{n,n-1}\omega_{nI}$ exists is if *all* the frequencies are given by families with the same $(\text{offset})_I$. But this cannot be the case when the ratio of the surface gravities is given by two relatively prime integers whose product is an odd number [1].

The previous considerations seem to suggest, that the modified Hod's conjecture has very little chance to survive the multi-horizon case. However the significantly different behaviour for the cases of rational/irrational ratios of surface gravities and the general splitting of frequencies into families seem to indicate something important. Moreover, it was already observed that surface gravities rational ratios have significant consequences for the multi-horizon spacetime thermodynamics [7]. Based on this observations let us pick the R-N case where the thermodynamical interpretation is straightforward and consider the following: Let us presuppose that both of the horizons in the R-N spacetime, the outer horizon with the area A_+ and the inner Cauchy horizon with the area A_- have equispaced area spectra given as[1] $A_\pm = 8\pi l_p^2 \gamma \cdot n_\pm$.

The perturbations are supposed to carry *no* charge, so one expects that only the ADM mass of the black hole will be changed. Thus [2] one can write the change of the areas of the black hole horizons as:

$$\Delta A_\pm = \frac{8\pi \Delta M}{\kappa_\pm}. \tag{4}$$

But $\Delta A_\pm$ can be given only as $\Delta A_\pm = 8\pi\gamma m_\pm$, which implies $\Delta M = \gamma\kappa_\pm m_\pm$. Furthermore, this implies $m_+\kappa_+ = m_-\kappa_-$ and thus $\kappa_+/\kappa_- = m_-/m_+$. This means that if the single ADM mass transitions have to be allowed the surface gravities ratio must be rational. Furthermore if one wants the emitted mass quantum to be as small as possible, such that it is still compatible with the quantization of the two horizon's areas one obtains:

$$\Delta M = \gamma \cdot \mathrm{lcm}(\kappa_+, |\kappa_-|). \tag{5}$$

Then modified Hod's conjecture suggests that

$$\lim_{n\to\infty} \Delta_{n,n-1}\omega_{nI} = \gamma \cdot \mathrm{lcm}(\kappa_+, |\kappa_-|). \tag{6}$$

This is indeed true if one takes the following interpretation of the frequencies (slight modification of Maggiore's conjecture): the straightforward extension of Maggiore's conjecture to the multi-horizon case is misleading, in fact only the equispaced families carry information about the quantum black hole mass transitions. (Every frequency belongs to one of the families.) Thus one has to first identify the equispaced families and then take the limit in the spacing in the imaginary part of the frequencies within each of the families. Such interpretation then fixes together with the formula (3) the parameter γ to be $\gamma = 1$. This means the area spectra of both of the horizons are given as $8\pi n$. Let us remind here, that the same analysis can be repeated for both S-dS and R-N-dS spacetimes: Assuming that all the horizons have the same equispaced area spectra, the single M parameter transitions lead to the surface gravities rational ratio condition and the QNM frequencies given by the formula fix the spectra of all the horizons to be $8\pi n$, (after one considers our generalization/modification of Maggiore's conjecture).

[1] This type of spectrum with $\gamma = 1$ is the one originally suggested by Bekenstein for the black hole horizon area and represents the far most popular choice in the current literature. Let us also mention that by different, but not too different lines of thinking as we present, it was already speculated that both of the horizons in the R-N spacetime have the same area spectra of the Bekenstein form.

3 Conclusions

To summarize: We suppose that also in the multi-horizon case the modified Hod's conjecture provides information about the spacetime horizons spectra, only the way the information is encoded is more tricky than in the single horizon case. (This is hardly anything surprising as the quantization of more than one horizon might play a role in the game.) The QNM frequencies are consistent with each of the horizons being quantized with spectra given as $8\pi n$ (in Planck units). If these conclusions are accepted then still many open questions remain, for example, if similar interpretation could survive in case of charged black holes that occur after a collapse of matter. (In such case the black hole interior is very different to the extremely idealized R-N, R-N-dS cases. For example a weak mass-inflation singularity occurs at the inner horizon, but some results indicate that possibility of crossing the inner horizon might still be considered.)

References

1. Skákala, J.: Non-extremal Reissner-Nordström black hole: do asymptotic quasi-normal modes carry information about the quantum properties of the black hole? J. High Energy Phys. **2012**(01), 144 (2012). doi:10.1007/JHEP01(2012)144
2. Skakala, J.: Quasinormal modes, area spectra and multi-horizon spacetimes. J. High Energy Phys. **2012**(06), 094 (2012). doi:10.1007/JHEP06(2012)094
3. Skakala, J., Visser, M.: Generic master equations for quasi-normal frequencies. J. High Energy Phys. **2010**(11), 070 (2010). doi:10.1007/JHEP11(2010)070
4. Skakala, J., Visser, M.: Semi-analytic results for quasi-normal frequencies. J. High Energy Phys. **2010**(08), 061 (2010). doi:10.1007/JHEP08(2010)061
5. Hod, S.: Bohr's correspondence principle and the area spectrum of quantum black holes. Phys. Rev. Lett. **81**, 4293 (1998). doi:10.1103/PhysRevLett.81.4293
6. Maggiore, M.: The Physical interpretation of the spectrum of black hole quasinormal modes. Phys. Rev. Lett. **100**, 141301 (2008). doi:10.1103/PhysRevLett.100.141301
7. Choudhury, T., Padmanabhan, T.: Concept of temperature in multi-horizon spacetimes: analysis of Schwarzschild-de Sitter metric. Gen. Relativ. Gravit. **39**, 1789 (2007). doi:10.1007/s10714-007-0489-0

Asymptotically AdS Spacetimes and Isometric Embeddings

Steven Willison

Abstract An algebraic global isometric embedding of the nonrotating BTZ black hole is presented. The ambient spacetime is $M^{2,3}$, the 3+2 dimensional flat spacetime. We also present the analogous embedding for the Euclidean BTZ spacetime and by performing a kind of double analytic continuation construct a 1-parameter family of embeddings of cosmological AdS spacetime into $M^{2,3}$ which coincide asymptotically with the embedded BTZ manifold of the appropriate mass. Finally we note that the family of embeddings of cosmological AdS$_n$ into $M^{2,n}$ generalises to higher dimensions.

1 Some Differential Geometric Preliminaries

Anti de Sitter space $\widetilde{\text{AdS}}$ is a Lorentzian manifold and as such all of the geometry can be understood in terms of intrinsically defined properties. However, to manifest more symmetries one often considers the following model: Let $M^{p,n}$ denote flat spacetime of signature $-p+n$. The submanifold:

$$\mathscr{A} := \{X^\mu \in \mathbb{M}^{2,n-1} | X^\mu X^\nu \eta_{\mu\nu}^{(2,n-1)} = -1\}, \tag{1}$$

has the same intrinsic geometry as Anti de Stter space $\widetilde{\text{AdS}}^n$. By this we mean that the submanifold so defined is diffeomorphic to $S \times \mathbb{R}^3$ and the induced metric is that of a maximally symmetric Lorentzian spacetime with sectional curvatures -1. By representing the spacetime as a submanifold certain geometrical facts become clear: The intersections with hyperplanes $\mathscr{P}$ through the origin are geodesics, specifically

S. Willison (✉)
CENTRA, Departamento de Física,
Instituto Superior Técnico, Universidade Técnica de Lisboa,
Av. Rovisco Pais 1, 1049-001 Lisboa, Portugal
e-mail: steven.willison@ist.utl.pt

J. Bičák and T. Ledvinka (eds.), *Relativity and Gravitation*,
Springer Proceedings in Physics 157, DOI: 10.1007/978-3-319-06761-2_45,

$\mathscr{A} \cap \mathscr{P}^{2,n-2}$ is a timelike geodesic and $\mathscr{A} \cap \mathscr{P}^{1,n-1}$ is a pair of disconnected spacelike geodesics; $\mathscr{A}$ contains closed timelike curves; the global isometry group is $SO(n,1)$. This embedding of $\widetilde{\mathrm{AdS}}^n$ is well known. Two spacetimes of special interest which are locally isometric to $\widetilde{\mathrm{AdS}}$ are the cosmological AdS spacetime and the BTZ [1] spacetime in 2+1 dimensions. The former is the universal covering space obtained by unwrapping the timelike geodesics. The latter is a quotient space. Therefore the local physics can be described as a constrained theory in flat spacetime using $\mathscr{A}$ described above. However, to address global issues it can be helpful to have a simple global isometric embedding. Below we present such embeddings for the norotating BTZ spacetime (obtained in [2]) and also for AdS^n.

A smooth *embedding* $\phi : M \to N$ is an injective map such that $\phi(M)$ is homeomorphic to M and ϕ_* is injective. Note $dim(N) \geq dim(M)$. Let (M,h) and (N,g) be pseudo-Riemannian manifolds. A smooth embedding $\phi : M \to N$ is an *isometric embedding* if $\phi^* g = h$. If M is the entire maximally extended spacetime we call this a *global isometric embedding*.

2 Embedding the BTZ and Cosmological AdS Spacetimes

In 2+1 dimensional gravity with negative cosmological constant, the Einstein equation in vacuum is equivalent to $R^{\mu\nu}{}_{\kappa\lambda} = -\frac{1}{l^2}(\delta^\mu_\kappa \delta^\nu_\lambda - \delta^\nu_\kappa \delta^\mu_\lambda)$. We set $l = 1$. The spherically symmetric solution of mass a^2 has the static form [1]

$$ds^2 = (r^2 - a^2)d\tau^2 + \frac{dr^2}{r^2 - a^2} + r^2 d\phi^2$$

outside of the event horizon ($r = a$). Since we are interested in global embeddings, we introduce the Kruskal type coordinate system:

$$ds^2 = 4\frac{-dt^2 + dx^2}{(1 + t^2 - x^2)^2} + a^2 \frac{(1 - t^2 + x^2)^2}{(1 + t^2 - x^2)^2} d\phi^2 .$$

The domain of the coordinates is $-1 < -t^2 + x^2 < 1$, $\phi \sim \phi + 2\pi$. This covers the maximally extended space-time. The event horizons and bifurcation surface are $x = \pm t$ and $x = t = 0$ respectively. Singularities ($t^2 - x^2 = 1$) and conformal infinity ($x^2 - t^2 = 1$) are not considered part of the spacetime for our purposes.

Lemma 1 [2] *The nonrotating BTZ black hole spacetime can be globally isometrically embedded into the region* $X^0 > 0$ *of* $\mathbb{M}^{2,3}$. *The image is the intersection of quadric hypersurfaces:*

$$(X^1)^2 + (X^2)^2 = \frac{a^2}{1 + a^2}(X^0)^2, \quad (X^3)^2 - (X^4)^2 = -1 + \frac{1}{1 + a^2}(X^0)^2.$$

Table 1 Various embeddings relevant to three dimensional gravity. The left and right colums are related by $X_4 \leftrightarrow iX_4$

Embedding in $\mathbb{M}_{3,2}$	Embedding in $\mathbb{M}_{4,1}$
BTZ (mass = a^2)	Euclidean BTZ (mass = a^2)
$(X^1)^2 + (X^2)^2 = \frac{a^2}{1+a^2}(X^0)^2$,	$(X^1)^2 + (X^2)^2 = \frac{1}{a^2+1}(X^0)^2$,
$(X^3)^2 - (X^4)^2 = \frac{1}{a^2+1}(X^0)^2 - 1$,	$(X^3)^2 + (X^4)^2 = \frac{a^2}{1+a^2}(X^0)^2 - 1$,
$X^0 > 0$ ($X^0 = 0$ singular).	$X^0 > 0$.
AdS_3	"Thermal AdS_3" (mass = $1/a^2$)
$(X^1)^2 + (X^2)^2 = \frac{a^2}{1+a^2}(X^0)^2 - 1$,	$(X^1)^2 + (X^2)^2 = \frac{1}{a^2+1}(X^0)^2 - 1$,
$(X^3)^2 - (X^4)^2 = \frac{1}{a^2+1}(X^0)^2$,	$(X^3)^2 + (X^4)^2 = \frac{a^2}{1+a^2}(X^0)^2$,
$X^0 > 0$ (2 copies of AdS).	$X^0 > 0$.

The past and future singularities are located at the intersection of the two constraint surfaces with the hyperplane $X^0 = 0$.

The proof was given in Ref. [2]. Combining the constraint equations we have $X^\mu X^\nu \eta^{(2,3)}_{\mu\nu} = -1$ therefore a global embedding into $\widetilde{AdS_4}$ exists. By lifting the restriction $X^0 > 0$ we obtain two copies of BTZ joined at the singularity, but it is not a true embedding at $X^0 = 0$: the tangent space map is not injective (the central singularity is a conical singularity).

We may make an analytic continuation $X^4 \to iX^4$, whence we obtain an embedding of the Euclidean black hole into $\mathbb{M}^{(1,4)}$ [1]. In fact there is another embedding, of an Euclidean black hole with mass parameter $1/a$, which has the same asymptotic form for large X_0. They are related by $(X^1, X^2) \leftrightarrow (X^3, X^4)$. We shall refer to this as "thermal AdS." The reason for this apparently arbitrary distinction is that upon making the analytic continuation $X^4 \to -iX^4$ we obtain now a global embedding of (two copies of) the cosmological AdS_3 which coincide asymptotically with the exterior regions of the black hole. All of this is summarised in Table 1. In the case of the embedding of AdS_3, the parameter a has no intrinsic geometrical meaning, and therefore no direct physical meaning. We call $2\pi a$ the *extrinsic temperature*[2] in this context since it is the temperature of the embedded BTZ spacetime to which it is asymptotic.

The complete picture contained in Table 1 is peculiar to three dimensions and depends on the fact that the black hole and AdS are related by a double Wick rotation which exchanges the role of the angular coordinate with that of the Euclidean time. However, we are able to present here the following result pertaining to higher dimensions:

---

[1] This belongs to a class of immersions of H^3 into H^4 obtained in Ref. [3]

[2] Another kind of extrinsic notion of temperature, based on a local embedding modeled on $\mathscr{A}^3$, was introduced in Ref. [4].

Lemma 2 *Let $\mathbb{M}_{n,2}$ be pseudo-Euclidean space with two time directions and standard coordinates $X^A = (T, X^1, \ldots, X^n, S)$ and α be a positive real number. Then the submanifold $\{X^A \in \mathbb{M}_{n,2} | X^A X^B \eta_{AB} = -1;\ (X^n)^2 = S^2 + \frac{\alpha^2}{1+\alpha^2} T^2;\ T, X^n > 0\}$ is homeomorphic to $\mathbb{R}^n$ and globally isometric to AdS_n.*

Proof We introduce angular coordinates (θ^i), $i = 1, \ldots, n-2$ on the unit sphere $\sigma^a \sigma^a = 1$ in $\mathbb{R}^{n-1}$. We then consider a cylindrical polar system of coordinates (τ, r, θ^i) on $\mathbb{R}^n$. For convenience set $r = \sinh \chi$, $\chi \geq 0$. Then

$$X^a = \sinh \chi\, \sigma^a(\theta^i), \qquad a = 1, \ldots, n-1, \tag{2}$$

$$X^n = \alpha \cosh \chi \cosh(\tau/\alpha), \tag{3}$$

$$S = \alpha \cosh \chi \sinh(\tau/\alpha), \tag{4}$$

$$T = \sqrt{1+\alpha^2} \cosh \chi, \tag{5}$$

can be verified to extend to a global embedding ($\chi = 0$ is purely a coordinate singularity—the image is a smooth submanifold at $X^a = 0$). The pullback of the Minkowski metric w.r.t. this embedding is

$$ds^2 = -\cosh^2 \chi d\tau^2 + d\chi^2 + \sinh^2 \chi d\Omega^2_{n-2}, \tag{6}$$

$d\Omega^2_{n-2}$ being the metric of the unit sphere. This is the metric of AdS_n. □

Finally we note that it follows from Lemma 2 that there is a one-parameter family of global isometric embeddings of AdS_n into $\widetilde{\text{AdS}}_{n+1}$.

Acknowledgments The research leading to these results has received funding from the European Union Seventh Framework Programme (FP7/2007-2013) under grant agreement N PCOFUND-GA-2009-246542 and from the Foundation for Science and Technology of Portugal.

References

1. Bañados, M., Teitelboim, C., Zanelli, J.: Black hole in three-dimensional spacetime. Phys. Rev. Lett. **69**, 1849 (1992). doi:10.1103/PhysRevLett.69.1849
2. Willison, S.: The Banados, Teitelboim, and Zanelli spacetime as an algebraic embedding. J. Math. Phys. **52**, 042503 (2011). doi:10.1063/1.3579486
3. Nomizu, K.: Isometric immersions of the hyperbolic plane into the hyperbolic space. Math. Ann. **205**, 181 (1973). doi:10.1007/BF01349228
4. Deser, S., Levin, O.: Mapping Hawking into Unruh thermal properties. Phys. Rev. D **59**, 064004 (1999). doi:10.1103/PhysRevD.59.064004

Part II
Cosmology and Relativistic Astrophysics

A Cosmological Concordance Model with Particle Creation

J. S. Alcaniz, H. A. Borges, S. Carneiro, J. C. Fabris, C. Pigozzo and W. Zimdahl

Abstract A constant-rate creation of dark particles in the late-time FLRW spacetime provides a cosmological model in accordance with precise observational tests. The matter creation backreaction implies in this context a vacuum energy density scaling linearly with the Hubble parameter, which is consistent with the vacuum expectation value of the QCD condensate in a low-energy expanding spacetime. Both the cosmological constant and coincidence problems are alleviated in this scenario. We discuss the cosmological model that arises in this context and present a joint analysis of observations of the first acoustic peak in the cosmic microwave background (CMB) anisotropy spectrum, the Hubble diagram for supernovas of type Ia (SNIa), the distance scale of baryonic acoustic oscillations (BAO) and the distribution of large scale structures (LSS). We show that a good concordance is obtained, albeit with a higher value of the present matter abundance than in the standard model.

The understanding of the gravitational role of vacuum fluctuations is in general a difficult problem, since their energy density usually depends on the renormalization method used and on an adequate definition of the vacuum state in the curved background. In the case of free massless fields in de Sitter spacetime, the renormalized vacuum density is $\Lambda \approx H^4$ [1–4], which in a low-energy universe leads to a too tiny cosmological term.

In the case we consider the vacuum energy of interacting fields, it has been suggested that in a low energy, approximately de Sitter background the vacuum conden-

J. S. Alcaniz
Observatório Nacional, Rio de Janeiro-RJ, Brazil

H. A. Borges · C. Pigozzo · S. Carneiro (✉)
Instituto de Física, Universidade Federal da Bahia,
Salvador, BA, Brazil
e-mail: saulo.carneiro.ufba@gmail.com

J. C. Fabris · W. Zimdahl
Departamento de Física, Universidade Federal do Espírito Santo,
Vitória-ES, Brazil

J. Bičák and T. Ledvinka (eds.), *Relativity and Gravitation*,
Springer Proceedings in Physics 157, DOI: 10.1007/978-3-319-06761-2_46,

sate originated from the QCD phase transition leads to $\Lambda \approx m^3 H$, where $m \approx 150$ MeV is the energy scale of the transition [5–11]. These results are in fact intuitive. In a de Sitter background the energy per observable degree of freedom is given by the temperature of the horizon, $E \approx H$. For a massless free field this energy is distributed in a volume $1/H^3$, leading to a density $\Lambda \approx H^4$, as above. For a strongly interacting field in a low energy space-time, on the other hand, the occupied volume is $1/\mathrm{m}^3$, owing to confinement, and the expected density is $\Lambda \approx m^3 H$.

Such a late-time variation law for the vacuum term can also be derived as a back-reaction of the creation of non-relativistic dark particles in the expanding spacetime [12]. The Boltzmann equation for this process is

$$\frac{1}{a^3}\frac{d}{dt}\left(a^3 n\right) = \Gamma n, \tag{1}$$

where n is the particle number density and Γ is a constant creation rate. By taking $\rho_m = nM$, it can also be written as

$$\dot{\rho}_m + 3H\rho_m = \Gamma\rho_m, \tag{2}$$

where M is the mass of the created particle. Let us take, in addition to (2), the Friedmann equation

$$\rho_m + \Lambda = 3H^2, \tag{3}$$

with the vacuum term satisfying the equation of state $p_\Lambda = -\Lambda$. Using (2) and (3) we obtain the conservation equation for the total energy,

$$\dot{\rho} + 3H(\rho + p) = 0, \tag{4}$$

provided we take[1]

$$\Lambda = 2\Gamma H. \tag{5}$$

This is the time-variation law predicted for the vacuum density of the QCD condensate, with $\Gamma \approx m^3$. Dividing it by $3H^2$, we obtain

$$\Gamma = \frac{3}{2}\left(1 - \Omega_m\right) H, \tag{6}$$

where $\Omega_m = 1 - \Omega_\Lambda \equiv \rho_m/(3H^2)$ is the relative matter density (for simplicity, we are considering only the spatially flat case). In the de Sitter limit ($\Omega_m = 0$), we have $\Gamma = 3H/2$, that is, the creation rate is equal (apart from a numerical factor) to the thermal bath temperature predicted by Gibbons and Hawking in the de Sitter spacetime [13]. It also means that the scale of the future de Sitter horizon

[1] Strictly speaking, this result is only exact if we neglect the conserved baryons in the balance equations. Since baryons represent only about 5 % of the total energy content, this can be considered a good approximation.

is determined, through Γ, by the energy scale of the QCD phase transition, the last cosmological transition we have. For the present time we have, from (6) (with $\Omega_m \approx 1/3$), $H_0 \approx \Gamma \approx m^3$, and hence $\Lambda \approx m^6$, where H_0 is the current Hubble parameter. The former result is an expression of the Eddington-Dirac large number coincidence [14]. The later—also known as Zeldovich's relation [15]—gives the correct order of magnitude for Λ.

The corresponding cosmological model has a simple analytical solution, which reduces to the CDM model for early times and to a de Sitter universe for $t \to \infty$ [16]. It has the same free parameters of the standard model and presents good concordance when tested against type Ia supernovas, baryonic acoustic oscillations, the position of the first peak of CMB and the matter power spectrum [12, 17–21]. Furthermore, the coincidence problem is alleviated, because the matter density contrast is suppressed in the asymptotic future, owing to the matter production [12, 20].

With $\Lambda = 2\Gamma H$ we obtain, from the Friedmann equations, the solution [16–19]

$$\frac{H}{H_0} \approx 1 - \Omega_{m0} + \Omega_{m0}(1+z)^{3/2} , \tag{7}$$

where here, for simplicity, we have not added radiation. For high redshifts the matter density scales as $\rho_m(z) = 3H_0^2\Omega_{m0}^2 z^3$. The extra factor Ω_{m0}—as compared to the ΛCDM model—is owing to the late-time process of matter production. In order to have nowadays the same amount of matter, we need less matter in the past. Or, in other words, if we have the same amount of matter in the past (say, at the time of matter-radiation equality), this will lead to more matter today. We can also see from (7) that, in the asymptotic limit $z \to -1$, the solution tends to the de Sitter solution. Note that, like the ΛCDM model, the above model has only two free parameters, namely Ω_{m0} and H_0. On the other hand, it can not be reduced to the ΛCDM case except for $z \to -1$. In this sense, it is falsifiable, that is, it may be ruled out by observations.

The Hubble function (7) can be used to test the model against background observations like SNIa, BAO and the position of the first peak in the CMB spectrum [17–19]. The analysis of the matter power spectrum was performed in [20], where, for simplicity, baryons were not included and the cosmological term was not perturbed. In a subsequent publication a gauge-invariant analysis, explicitly considering the presence of late-time non-adiabatic perturbations, has shown that the vacuum perturbations are indeed negligible, except for scales near the horizon [21].

We show in Table 1 the best-fit results for Ω_{m0} (with H_0 marginalized) with three samples of supernovas: the SDSS and Constitution compilations calibrated with the MLCS2k2 fitter, and the Union2 sample. For the sake of comparison, we also show the best-fit results for the spatially flat ΛCDM model. We should have in mind that the Union2 dataset is calibrated with the Salt2 fitter, which makes use of a fiducial ΛCDM model for including high-z supernovas in the calibration. Therefore, that sample is not model-independent and, in the case of the standard model, the test should be viewed as rather a test of consistence. From the table we can see that for the model with particle creation the concordance is quite good. For the samples

Table 1 2σ limits to Ω_{m0} (SNe + CMB + BAO + LSS)

Test	$\Lambda(t)$CDM		ΛCDM	
	Ω_{m0}	χ^2_{min}/ν	Ω_{m0}	χ^2_{min}/ν
Union2 (SALT2)	$0.420^{+0.009}_{-0.010}$	1.063	0.235 ± 0.011	1.027
SDSS (MLCS2k2)	$0.450^{+0.014}_{-0.010}$	0.842	$0.260^{+0.013}_{-0.016}$	1.231
Constitution (MLCS2k2-17)	$0.450^{+0.008}_{-0.014}$	1.057	0.270 ± 0.013	1.384

calibrated with the MLCS2k2 fitter it is actually better than in the ΛCDM case. As anticipated above, the present matter density is higher than in the standard case, with $\Omega_{m0} \approx 0.45$.

With the concordance values of Ω_{m0} in hand, we can obtain the age parameter of the Universe, as well as the redshift of transition between the decelerated and accelerated phases. They are given, respectively, by [16–19]

$$H_0 t_0 = \frac{2 \ln \Omega_{m0}}{3(\Omega_{m0} - 1)}, \tag{8}$$

$$z_T = \left[2\left(\frac{1}{\Omega_{m0}} - 1\right)\right]^{2/3} - 1. \tag{9}$$

In the case of the SDSS and Constitution samples, this leads to $H_0 t_0 = 0.97$ and $z_T = 0.81$, in good agreement with standard predictions and astronomical limits [22]. For $H_0 \approx 70\,\text{km/(s.Mpc)}$, we have $t_0 \approx 13.5\,\text{Gyr}$.

Particle creation is something expected in expanding spacetimes [23]. In spite of the difficulty in deriving the production rate and backreaction in general, this phenomenon may in principle be related with inflation [24] and with the present cosmic acceleration, a possibility already considered in different ways by some authors [25, 26] . We have shown that a constant-rate creation of non-relativistic dark particles at late times leads indeed to a viable concordance model.

References

1. Ford, L.: Quantum vacuum energy in general relativity. Phys. Rev. D **11**, 3370 (1975). doi:10.1103/PhysRevD.11.3370
2. Dowker, J., Critchley, R.: Effective Lagrangian and energy momentum tensor in de Sitter space. Phys. Rev. D **13**, 3224 (1976). doi:10.1103/PhysRevD.13.3224
3. Davies, P.: Singularity avoidance and quantum conformal anomalies. Phys. Lett. B **68**, 402 (1977). doi:10.1016/0370-2693(77)90504-4
4. Starobinsky, A.: A new type of isotropic cosmological models without singularity. Phys. Lett. B **91**, 99 (1980). doi:10.1016/0370-2693(80)90670-X

5. Schützhold, R.: Small cosmological constant from the QCD trace anomaly? Phys. Rev. Lett. **89**, 081302 (2002). doi:10.1103/PhysRevLett.89.081302
6. Klinkhamer, F., Volovik, G.: Gluonic vacuum, q-theory, and the cosmological constant. Phys. Rev. D **79**, 063527 (2009). doi:10.1103/PhysRevD.79.063527
7. Urban, F., Zhitnitsky, A.: The cosmological constant from the ghost: A toy model. Phys. Rev. D **80**, 063001 (2009). doi:10.1103/PhysRevD.80.063001
8. Urban, F., Zhitnitsky, A.: The cosmological constant from the QCD Veneziano ghost. Phys. Lett. B **688**, 9 (2010). doi:10.1016/j.physletb.2010.03.080
9. Urban, F., Zhitnitsky, A.: The QCD nature of dark energy. Nucl. Phys. B **835**, 135 (2010). doi:10.1016/j.nuclphysb.2010.04.001
10. Ohta, N.: Dark energy and QCD ghost. Phys. Lett. B **695**, 41 (2011). doi:10.1016/j.physletb.2010.11.044
11. Holdom, B.: From confinement to dark energy. Phys. Lett. B **697**, 351 (2011). doi:10.1016/j.physletb.2011.02.024
12. Alcaniz, J., Borges, H., Carneiro, S., et al.: A cosmological concordance model with dynamical vacuum term. Phys. Lett. B **716**, 165 (2012). doi:10.1016/j.physletb.2012.08.014
13. Gibbons, G., Hawking, S.: Cosmological event horizons, thermodynamics, and particle creation. Phys. Rev. D **15**, 2738 (1977). doi:10.1103/PhysRevD.15.2738
14. Mena Marugán, G., Carneiro, S.: Holography and the large number hypothesis. Phys. Rev. D **65**, 087303 (2002). doi:10.1103/PhysRevD.65.087303
15. Bjorken, J.: Emergent photons and gravitons: the problem of vacuum structure. In: Kostelecký, V. (ed.) CPT and Lorentz Symmetry, pp. 1–5. World Scientific Publishing, Singapore, Hackensack, NJ (2011). doi:10.1142/9789814327688_0001
16. Borges, H., Carneiro, S.: Friedmann cosmology with decaying vacuum density. Gen. Relativ. Gravit. **37**, 1385 (2005). doi:10.1007/s10714-005-0122-z
17. Pigozzo, C., Dantas, M., Carneiro, S., Alcaniz, J.: Observational tests for Λ(t)CDM cosmology. J. Cosmol. Astropart. Phys. **2011**(08), 022 (2011). doi:10.1088/1475-7516/2011/08/022
18. Carneiro, S., Dantas, M., Pigozzo, C., Alcaniz, J.: Observational constraints on late-time $\Lambda(t)$ cosmology. Phys. Rev. D **77**, 083504 (2008). doi:10.1103/PhysRevD.77.083504
19. Carneiro, S., Pigozzo, C., Borges, H., Alcaniz, J.: Supernova constraints on decaying vacuum cosmology. Phys. Rev. D **74**, 023532 (2006). doi:10.1103/PhysRevD.74.023532
20. Borges, H., Carneiro, S., Fabris, J., Pigozzo, C.: Evolution of density perturbations in decaying vacuum cosmology. Phys. Rev. D **77**, 043513 (2008). doi:10.1103/PhysRevD.77.043513
21. Zimdahl, W., Borges, H., Carneiro, S., Fabris, J., Hipolito-Ricaldi, W.: Non-adiabatic perturbations in decaying vacuum cosmology. J. Cosmol. Astropart. Phys. **2011**(04), 028 (2011). doi:10.1088/1475-7516/2011/04/028
22. Hansen, B., Brewer, J., Fahlman, G., et al.: The white dwarf cooling sequence of the globular cluster messier 4. Astrophys. J. **574**, L155 (2002). doi:10.1086/342528
23. Parker, L., Toms, D.: Quantum Field Theory in Curved Spacetime: Quantized Fields and Gravity. Cambridge Monographs on Mathematical Physics. Cambridge University Press, Cambridge, New York (2009)
24. Carneiro, S., Tavakol, R.: On vacuum density, the initial singularity and dark energy. Gen. Relativ. Gravit. **41**, 2287 (2009). doi:10.1007/s10714-009-0846-2
25. Lima, J., Germano, A., Abramo, L.: FRW type cosmologies with adiabatic matter creation. Phys. Rev. D **53**, 4287 (1996). doi:10.1103/PhysRevD.53.4287
26. Zimdahl, W., Schwarz, D., Balakin, A., Pavon, D.: Cosmic anti-friction and accelerated expansion. Phys. Rev. D **64**, 063501 (2001). doi:10.1103/PhysRevD.64.063501

From ‘Nothing’ to Inflation and Back Again

Vladimír Balek

Abstract A procedure for solving Wheeler-DeWitt equation in Euclidean region, following step by step the construction of tunneling wave function in nonrelativistic quantum mechanics by Banks, Bender and Wu, is proposed. Solutions for a universe satisfying no-boundary condition and a universe created from ‘nothing’ are compared to the corresponding solutions for a particle in a two-dimensional potential well, and effects of indefiniteness of metric and zero energy in Wheeler-DeWitt equation are discussed.

1 Introduction

The basic two minisuperspace solutions of Wheeler-DeWitt equation were proposed as long ago as at the beginning of 80’s. In 1983, Hartle and Hawking introduced their no-boundary condition and used it to construct an inflationary solution in which scalar field was decoupled from gravity and its effect on the expansion of the universe was mimicked by the cosmological constant [1]. Hawking then replaced the conformal coupling of the scalar field by the mass term and obtained a truly inflationary solution [2, 3]. Since then the model has been explored repeatedly. To mention just two remarkable results, the universe was shown to avoid Big Crunch by quantum bounce for all, not just fine-tuned, initial conditions [4]; and the model was used to demonstrate the collapse of the measurement of time with the help of quantum mechanical degrees of freedom in the period of maximal expansion [5]. Shortly before the no-boundary solution appeared, in 1982, Vilenkin proposed another solution describing creation of the universe from ‘nothing’ [6, 7]. Linde advocated the use of such solution ‘in those situations where the scale parameter a itself must be quantized’ [8], and proposed a heuristic derivation of it via inverse Wick rotation. Later there appeared

V. Balek (✉)
Department of Theoretical Physics, Comenius University, Bratislava, Slovakia
e-mail: balek@fmph.uniba.sk

J. Bičák and T. Ledvinka (eds.), *Relativity and Gravitation*, Springer Proceedings in Physics 157, DOI: 10.1007/978-3-319-06761-2_47,

a different approach to 'creationist' cosmology, adopting the no-boundary condition but making use of density matrix rather than wave function [9].

Both Hawking and Vilenkin solutions contain an Euclidean region in which the wave function describes tunneling of the universe, either from a finite radius to a point (Hawking solution) or from a point to a finite radius (Vilenkin solution). In nonrelativistic quantum mechanics, a complete WKB solution in the tunneling region in two dimensions was obtained by Banks et al. [10]. The solution refers to a particle escaping from a potential well, but with some effort can be modified to apply also to the opposite case when a particle tunnels from outside the barrier into the well. The cosmological problem is two-dimensional and allows for WKB approximation, therefore it would be natural to employ the same construction in it. However, as for now the solution by Banks, Bender and Wu was apparently not used in this way, although Vilenkin mentions in [7] that his tunneling solution is 'similar' to it.

In this note a procedure for computing the wave function of the universe in the Euclidean region is outlined. In Sect. 2 the construction by Banks, Bender and Wu is summarized, rewritten so that it works both ways, outwards as well as inwards; in Sect. 3 new features of the construction appearing when it is carried over to cosmology are discussed; and in Sect. 4 possible applications of the theory are suggested.

2 Tunneling in Quantum Mechanics

Consider a particle in two dimensions obeying Schrödinger equation

$$\left[-\partial_x^2 - \partial_y^2 + \tfrac{1}{4}(x^2 + y^2) - \tfrac{1}{4}\varepsilon(x^4 + y^4 + 2cx^2y^2) - E\right]\psi = 0, \tag{1}$$

where the parameters ε and c satisfy $0 < \varepsilon \ll 1$ and $|c| < 1$. The potential into which the particle is placed consists of a circularly symmetric well and a quatrefoil barrier around it. We are interested in two processes, tunneling of the particle from the well to the other side of the barrier and *vice versa*, both along the positive x axis. The particle tunneling outwards has a (quasi) discrete spectrum, which coincides to a great precision, if the energies are not too high, with the spectrum in a well extrapolated to infinity. From the requirement that only an outgoing wave appears behind the barrier it follows that the energies acquire a small imaginary part equal to $-\frac{1}{2}\times$ the decay rate of the state inside the well. The particle tunneling inwards, on the other hand, can have any energy. In what follows we assume that ψ separates in variables x, y inside the well and the energy E_y going to the y direction equals $\frac{1}{2}$. In particular, the particle tunneling outwards can be in the ground state with $E = 1$ and $E_x = E_y = \frac{1}{2}$.

Let us start with rewriting the potential near the x axis as

$$V = V_0 + \tfrac{1}{4}ky^2, \tag{2}$$

where

$$V_0 = \tfrac{1}{4}x^2(1 - \varepsilon x^2), \quad k = 1 - 2\varepsilon c x^2. \tag{3}$$

The idea of solving (1) in the tunneling region is to separate a WKB wave function in the longitudinal direction and a bell-shaped function with a variable width in the transversal direction out of ψ. Thus, we write

$$\psi = A p^{-1/2} \exp\left(\mp \int p dx - \tfrac{1}{4} f y^2\right), \quad p = \sqrt{V_0 - E_x}, \tag{4}$$

with f depending on x and A depending on both x and y. The upper and lower signs refer to a particle tunneling outwards and inwards respectively. Equation for f is obtained by collecting the terms in the original equation proportional to y^2 and putting them equal to zero. In this way we relate the function f, defining the width of the wave function, to the function k, defining the width of the valley the particle is tunneling through; and we obtain an equation for A that separates after y is replaced by an appropriate variable proportional to it.

In the leading order of WKB approximation equations for f and A lose the second derivative with respect to x and, in addition, the function p multiplying the first derivative loses the constant E_x under the square root sign. After rescaling $x \to \varepsilon^{-1/2} x$ we have

$$2p\frac{d}{dx} \to xw\frac{d}{dx} = -x^2\frac{d}{dw}, \quad w = \sqrt{1 - x^2},$$

and equation for f takes the form

$$\pm x^2 \frac{df}{dw} = f^2 - k, \quad k = 1 - 2cx^2. \tag{5}$$

The equation can be linearized by introducing an auxiliary function u such that

$$f = \mp \frac{x^2}{u}\frac{du}{dw}. \tag{6}$$

After inserting this into (5) we find that u obeys equation for associated Legendre function with the lower index given by

$$\nu(\nu + 1) = 2c \tag{7}$$

and the upper index equal to ± 1. Equation for A separates in the variables w and $s = y/u$, and if we write A as WS, where W is a function of w and S is a function of s, we find that S is either cosine or hyperbolic cosine or constant. Matching the tunneling solution to the solution inside the well singles out the third possibility.

Thus, A depends on w only, and after a little algebra one finds

$$A = const \left(\frac{1-w}{1+w}\right)^{\pm 1/4} u^{-1/2}. \tag{8}$$

Let us examine the behavior of f. To simplify the analysis, we return from the variable w to x with the help of the formula

$$x\frac{d}{dw} = -w\frac{d}{dx}.$$

First we find the asymptotic of f for $x \sim 0$. In order that the internal and external solutions match, f must equal 1 at $x = 0$; thus, we write f as $1 + \Delta$ and skip the Δ^2 term in (5). We obtain the equation

$$\mp x\frac{d\Delta}{dx} = 2(\Delta + cx^2), \tag{9}$$

whose solution is

$$\Delta = \begin{cases} -\frac{1}{2}cx^2 \\ 2cx^2(\log x + \mathscr{C}), \end{cases} \tag{10}$$

where the upper and lower expression refer to the upper and lower sign in (9) and $\mathscr{C}$ is integration constant. We can see that the width of the tunneling wave function is given uniquely for a particle tunneling outwards, but the stripes on which the wave function is nonzero form a one-parametric sequence for a particle tunneling inwards. The reason is presumably that the waves incident on the barrier along the positive x axis form a one-parametric sequence, too, and different waves tunnel along different stripes.

From the asymptotic of f we can determine u. For $w \sim 1$, or equivalently, $x \doteq \sqrt{2(1-w)} \sim 0$, the asymptotics of the associated Legendre functions of first and second kind are

$$P_\nu^{-1} = \tfrac{1}{2}x, \quad Q_\nu^1 = -x^{-1}.$$

(It suffices to consider one upper index in P_ν and Q_ν since the functions with opposite upper indices are proportional to each other.) To obtain $f = 1$ for $x = 0$ we must choose u equal to P_ν^{-1} and to Q_ν^1 + some coefficient $\times$ P_ν^{-1} for a particle tunneling out- and inwards. Moreover, by including the next-to-leading term into the expansion of Q_ν^1,

$$\Delta Q_\nu^1 = \tfrac{1}{4}\nu(\nu+1)\big[2\log\tfrac{x}{2} + \psi(\nu) + \psi(\nu+2) - \psi(2) + \gamma\big]x + \tfrac{1}{4}x,$$

where ψ is digamma function and γ is Euler-Mascheroni constant, we can find the coefficient in front of P_ν^{-1} in the latter case.

Finally, we can use u to compute f on the whole interval of x. Note that for $c > \frac{1}{2}$ the function k becomes negative before x reaches 1, so that the valley the particle is tunneling through turns to a slope. According to (5), this drives the derivative df/dx to more negative values and makes f to fall down faster for a particle tunneling outwards. Nevertheless, f stays positive and ψ stays suppressed in the transversal direction up to $x = 1$ for any c. If the particle is tunneling inwards, f increases near $x = 1$ for the values of c in question, and only with decreasing c it starts to decrease. For low enough c's, f passes through zero before x reaches 1; however, this holds only for $\mathscr{C} = 0$ and can be cured by choosing positive $\mathscr{C}$.

3 Tunneling in Cosmology

Consider a closed universe with a scalar field living in it, and denote the radius of the universe by a and the value of the field by ϕ. Suppose, furthermore, that the field is massive with the mass m and the theory includes cosmological constant Λ; thus, the field has quadratic potential shifted upwards by $\lambda = \frac{1}{3}\Lambda$. The wave function of the universe satisfies the Wheeler-DeWitt equation

$$\big[-\partial_a^2 + a^{-2}\partial_\phi^2 + a^2(1 - \lambda a^2 - m^2 a^2 \phi^2)\big]\psi = 0. \tag{11}$$

To account for the ambiguity due to operator ordering, the operator $-\partial_a^2$ has to be modified to $-a^{-\mu}\partial_a a^\mu \partial_a$ with an arbitrary μ. However, in the WKB approximation we are interested in, this affects only the preexponential factor in ψ, therefore we can put $\mu = 0$.

Equation (11) is almost identical to the equation we obtain from (1) by passing from rectangular to polar coordinates, restricting the polar angle to $|\varphi| \ll 1$ and replacing the operator $-r^{-1}\partial_r r \partial_r$, again with the reference to WKB approximation, by $-\partial_r^2$. The new equation reads

$$\big[-\partial_r^2 - r^{-2}\partial_\phi^2 + \tfrac{1}{4}r^2(1 - \varepsilon r^2 + 2\varepsilon\gamma r^2\varphi^2) - E\big]\psi = 0, \tag{12}$$

where $\gamma = 1 - c$. We will solve this equation in a similar way as (1) and discuss equation (11) later. First we express ψ as

$$\psi = Bq^{-1/2}\exp\Big(\mp\int q dr - \tfrac{1}{4}g\varphi^2\Big), \quad q = \sqrt{\mathscr{V}_0 - E}, \tag{13}$$

with $\mathscr{V}_0 = \frac{1}{4}r^2(1 - \varepsilon r^2)$ and g related to $\kappa = 2\varepsilon\gamma r^4$ in a similar way as f was related to k. After rescaling $r \to \varepsilon^{-1/2}r$, $\varphi \to \varepsilon^{1/2}\varphi$, $\kappa \to \varepsilon^{-1}\kappa$ and $g \to \varepsilon^{-1}g$ and introducing $\xi = \sqrt{1 - r^2}$, we have

$$\pm r^2\frac{dg}{d\xi} = r^{-2}g^2 - \kappa, \quad \kappa = 2\gamma r^4, \tag{14}$$

and after writing g as

$$g = \mp \frac{r^4}{v} \frac{dv}{d\xi}, \tag{15}$$

we find that v obeys equation for Gegenbauer function with the lower index given by

$$\alpha(\alpha + 3) = -2\gamma \tag{16}$$

and the upper index equal to $\frac{3}{2}$. Finally, for the function B we obtain

$$B = const\ v^{-1/2}. \tag{17}$$

To fix the combination of Gegenbauer functions in v we must explore the behavior of g for $r \sim 0$. When doing so we notice that now there are *two* terms proportional to y^2 in the exponential near the origin, the term $-\frac{1}{4}g\varphi^2 \doteq -\frac{1}{4}r^{-2}gy^2$ and the term $\mp\frac{1}{4}y^2$ coming from

$$\int q d\hat{r} \sim \int \tfrac{1}{2}\hat{r} d\hat{r} = \tfrac{1}{4}\hat{r}^2 = \tfrac{1}{4}(\hat{x}^2 + y^2),$$

where we have denoted the original, non-rescaled variables x and r by $\hat{x}$ and $\hat{r}$. These two terms must add to produce the term $-\frac{1}{4}y^2$ appearing in the solution inside the well. The resulting asymptotics of g are

$$g = \begin{cases} \frac{1}{2}\gamma r^4 \\ 2r^2[1 + cr^2(\log r + \mathscr{D})]. \end{cases} \tag{18}$$

This yields v equal to $C_\alpha^{3/2}$ and $D_\alpha^{3/2}$ + some coefficient × $C_\alpha^{3/2}$ for a particle tunneling out- and inwards respectively; and knowing v, we can determine the behavior of g on the whole interval of r.

Once we have found f we do not need to compute g from scratch. Instead, we can express g in terms of f. For that purpose we insert

$$\int q d\hat{r} = \int \tfrac{1}{2}\hat{r}\xi d\hat{r} = \varepsilon^{-1} \int \tfrac{1}{2} r\xi dr = -\varepsilon^{-1}\tfrac{1}{6}(\xi^3 - 1) \doteq \int p d\hat{x} + \tfrac{1}{4}\xi y^2$$

into (13) and compare the resulting expression to (4). We obtain

$$g = r^2(f \mp \xi), \tag{19}$$

where f is to be regarded as a function of r. This coincides with the function g constructed previously, if we express $C_\alpha^{3/2}$ and $D_\alpha^{3/2}$ in terms of P_ν^{-1} and Q_ν^1 and put $\mathscr{D} = \mathscr{C} - \frac{1}{4}$.

Equation (11) differs from (12) in that it has reversed sign of the kinetic term in ϕ direction and vanishing energy. Because of the former property the metric of the kinetic term is indefinite. The sign of the mass term is reversed or stays the same depending on whether m is real, which corresponds to a metastable vacuum, or imaginary, which corresponds to an unstable vacuum. However, it is the *relative* signs with respect to the kinetic terms in a and ϕ directions which matter; and no matter what the absolute sign, one of these signs stays the same while the other is reversed.

The two solutions of equation (11), with minus and plus sign in the exponential, describe tunneling of the universe outwards, from a point to a finite radius, and inwards, from a finite radius to a point. An immediate consequence of the indefinitness of metric is that the signs in the equation for g are switched. As a result, we obtain one-parametric class of g's for a universe tunneling outwards and a single g for a universe tunneling inwards. Having just one solution in the latter case is consistent with the observation that the wave function of the universe is determined completely by the no-boundary condition. Having infinitely many solutions in the former case can be explained by the fact that the energy is zero, which means that the imaginary part of energy is zero, which means that the outgoing probability current stays finite up to the zero radius of the universe. The corresponding outgoing waves must be put into the theory by hand and apparently form a one-parametric sequence, similarly as ingoing waves in the quantum mechanical problem with a particle tunneling inwards.

In the previous discussion we have assumed that m is real. However, as mentioned by Vilenkin in [7], when considering a universe tunneling outwards it is reasonable to pass to imaginary m. The point is that the tunneling path shortens, and the wave function becomes less suppressed behind the barrier, if the shift of the potential λ increases. Thus, the tunneling is most effective at the global maximum of the potential. Imaginary m and tunneling outwards in cosmology corresponds to $c > 1$ and tunneling inwards in quantum mechanics. For such tunneling we find, in addition to a one-parametric class of g's with the asymptotic given in the lower line of (18), one more g with the asymptotic

$$g = -\tfrac{1}{2}\gamma r^4. \tag{20}$$

This solution must be abandoned since the corresponding f equals -1 at $x = 0$, which means that the wave function explodes in the transversal direction. However, in the cosmological setting such g seems admissible, and even privileged because of its one-to-one correspondence with g appearing in the problem with a universe tunneling inwards.

4 Conclusion

We have shown how the procedure by Banks, Bender and Wu can be carried over to cosmology and used to construct the wave function of the universe in the Euclidean region. The way how to do that has been only sketched here, the details will be given elsewhere [11]. In particular, we have skipped the discussion of the behavior of the wave function 'inside the well', in the region where the radius of the universe is close to zero. This question is vital for the construction, because as long as the tunneling solution is not matched with the solution 'inside the well', it remains unjustified.

For a universe tunneling inwards, tunneling solution converts at the edges of Euclidean region into oscillatory one, describing time-symmetric evolution during which the universe repeatedly crosses that region. The crossings do not seem to change the course of the evolution, but can still have some imprint on it, and the knowledge of the exact form of tunneling solution can help to identify that imprint. For a universe tunneling outwards there apparently exists, in addition to 'regular' solution which behaves like that for a universe tunneling inwards, a one-parametric class of solutions with markedly different behavior. Such solutions can mediate tunneling not only to the maximum of potential but also to its minimum, therefore can be helpful when contemplating the possibility of creating a universe with scalar field in metastable state directly from 'nothing'.

References

1. Hartle, J., Hawking, S.: Wave function of the universe. Phys. Rev. D **28**, 2960 (1983). doi:10.1103/PhysRevD.28.2960
2. Hawking, S.: Quantum cosmology. In: DeWitt, B., Stora, R. (eds.) Relativity, Groups and Topology II, pp. 333–379. North-Holland, Amsterdam (1984)
3. Hawking, S.: The quantum state of the universe. Nucl. Phys. B **239**, 257 (1984). doi:10.1016/0550-3213(84)90093-2
4. Singh, P., Toporensky, A.: Big crunch avoidance in $k = 1$ semiclassical loop quantum cosmology. Phys. Rev. D **69**, 104008 (2004). doi:10.1103/PhysRevD.69.104008
5. Höhn, P., Kubalová, E., Tsobanjan, A.: Effective relational dynamics of a nonintegrable cosmological model. Phys. Rev. D **86**, 065014 (2012). doi:10.1103/PhysRevD.86.065014
6. Vilenkin, A.: Creation of universes from nothing. Phys. Lett. B **117**, 25 (1982). doi:10.1016/0370-2693(82)90866-8
7. Vilenkin, A.: Boundary conditions in quantum cosmology. Phys. Rev. D **33**, 3560 (1986). doi:10.1103/PhysRevD.33.3560
8. Linde, A.: Particle Physics and Inflationary Cosmology, Contemporary Concepts in Physics, vol. 5. Harwood, Chur (1990)
9. Barvinsky, A., Kamenshchik, A.: Cosmological landscape from nothing: some like it hot. J. Cosmol. Astropart. Phys. **2006**(09), 014 (2006). doi:10.1088/1475-7516/2006/09/014
10. Banks, T., Bender, C., Wu, T.: Coupled anharmonic oscillators: I. Equal-mass case. Phys. Rev. D **8**, 3346 (1973). doi:10.1103/PhysRevD.8.3346
11. Balek, V., Genzor, J.: (in prep)

Quasinormal Modes from a Naked Singularity

Cecilia Chirenti, Jozef Skákala and Alberto Saa

Abstract What should be the quasinormal modes associated with a spacetime that contains a naked singularity instead of a black hole? In the present work we address this problem by studying the scattering of scalar fields on a curved background described by a Reissner-Nordström spacetime with $q > m$. We show that there is a qualitative difference between cases with $1 < q^2/m^2 \lesssim 9/8$ and cases with $q^2/m^2 \gtrsim 9/8$. We discuss the necessary conditions for the well-posedness of the problem, and present results for the low l and large l limit.

1 Introduction

The naked Reissner-Nordström (R-N) singularity is a classical general relativistic solution in electrovacuum. The solution is expected to have a very limited meaning, due to the fact that such singularities cannot be created by a gravitational collapse, or by dropping a charge into the black hole. (By weak cosmic censorship conjecture general naked singularities should be prohibited in general theory of relativity, although there are indications that by including quantum effects the violations of the conjecture could be considered [1].) Moreover, a naked singularity created from some exotic initial data conditions should become quickly neutralized (classically, or via quantum pair production). Some results also indicate that if one considers electro-gravitational perturbations the R-N naked singularity becomes linearly unstable [2]. However it was discovered that the scalar field scattering problem on such a singular

C. Chirenti (✉) · J. Skákala
Centro de Matemática, Computação e Cognição, UFABC,
Santo André-SP 09210-170, Brazil
e-mail: cecilia.chirenti@ufabc.edu.br

A. Saa
Departamento de Matemática Aplicada, Universidade Estadual de Campinas,
Campinas-SP 13083-859, Brazil

J. Bičák and T. Ledvinka (eds.), *Relativity and Gravitation*,
Springer Proceedings in Physics 157, DOI: 10.1007/978-3-319-06761-2_48,

background can be still well defined [3–8], since the waves remain regular at the origin. Despite of this nice property of the scattering problem, the spacetime is not globally hyperbolic and the time evolution of the fields is not unique [9, 10]. This means one has to specify additional boundary condition at the singularity to obtain a fully unique time evolution. Another way of seeing the problem is through the language of operators: one can understand the spatial part of the wave operator to be a positive symmetric operator acting on a L^2 Hilbert space, and then obtain the scalar field dynamics through a suitable positive self-adjoint extension of such a symmetric operator [3, 4]. (One "preferred" way in which such a self-adjoint extension can be always realized is through the so called Friedrich's extension [3], which will also be the case of this paper.) Anyway, after uniquely specifying the dynamics, one should be able to characterize the scattering by a set of characteristic oscillations, the quasi-normal modes.

Low damped quasi-normal modes are in general used as a possible source of information about potential astrophysical objects (such as neutron stars, black holes), the highly damped modes are potentially interesting from the point of view of quantum gravity. Since a lot of work was devoted to the problem of quasinormal modes of the Reissner-Nordström black hole, it might be interesting to observe what happens if one transits from the R-N black hole case to the R-N naked singularity case (with a reflective boundary condition). Information about "what happens" shows how many features of the quasi-normal modes of the black hole spacetimes are specific to the black holes themselves and what features survive much more general conditions. Thus, briefly, we hope that despite of the fact that most likely the R-N naked singularity model does not correspond to a realistic physical situation, there are still many interesting things one can learn from such a model. One of them is a question that we want to answer in the present paper, in particular what will be the behaviour of the *low* damped quasinormal modes when departing from the R-N black hole to the R-N naked singularity.

2 The Time Evolution Problem for a Scalar Field in the R-N Naked Singularity

In this section we will follow the standard analysis of the scalar field evolution in a curved background. (As an example of such an analysis see the treatment of Schwarzschild black hole perturbations in [11, 12]. For a review that presents also such techniques, see for example [13].) Take the Klein-Gordon equation for the complex (charged) scalar field:

$$\frac{1}{\sqrt{-g}}\partial_\mu(\sqrt{-g}\, g^{\mu\nu}\partial_\nu \Psi) = 0, \tag{1}$$

with the metric line element given as

$$g_{\mu\nu}dx^{\mu}dx^{\nu} = -f(r)dt^2 + f(r)^{-1}dr^2 + r^2 d\Omega^2. \tag{2}$$

For the Reissner-Nordström (R-N) singularity the function $f(r)$ is in Planck units given as:

$$f(r) = 1 - \frac{2m}{r} + \frac{q^2}{r^2}, \quad (q^2 > m^2). \tag{3}$$

Take the decomposition of the field into the spherical harmonics

$$\Psi(t, r, \theta, \phi) = \sum_{l,m} \psi_l(t, r) Y_{ml}(\theta, \phi). \tag{4}$$

After we separate the variables we obtain the following reduced equation

$$\frac{d^2\psi_l(t,r)}{dt^2} = \frac{f(r)}{r^2}\frac{d}{dr}\left[r^2 f(r)\frac{d\psi_l(t,r)}{dr}\right] - \frac{l(l+1)f(r)}{r^2}\psi_l(t,r). \tag{5}$$

The unique solution of the given initial data problem is obtained if the condition of boundedness of the Green's function leads to a *unique* way to construct Green's function from the two linerly independent solutions U_{l1}, U_{l2} of the homogeneous equation associated to (5). Unfortunately for the case of R-N naked singularity both of the linearly independent solutions U_{l1}, U_{l2} are regular at 0 and the problem is underdetermined.

Is there any intuitive physical condition that we can further impose on the fields, that will uniquely select the appropriate Green's function? At least to get the geometrical optics continuous extension of the black hole case one can impose the condition that nothing falls in or out of the singularity. This means there is neither absorption nor superradiation in the scattering and the S-matrix of the K-G field is a unitary operator.

Further in the text we will employ the field vanishing condition at 0. (This boundary condition at the singularity corresponds to what is known as Friedrich's extension of a symmetric operator.) Thus the quasinormal modes will relate to the scattering problem following from the time evolution determined by the boundary condition $\psi(0, t) = 0$ [14].

3 The Scalar Wave Scattering on a Naked Singularity

Using ϕ_l defined as $\phi_l(r, t) = r\psi_l(r, t)$ and x the tortoise coordinate given by the condition:

$$\frac{dr}{dx} = f(r), \tag{6}$$

one can rewrite (5) into the following form

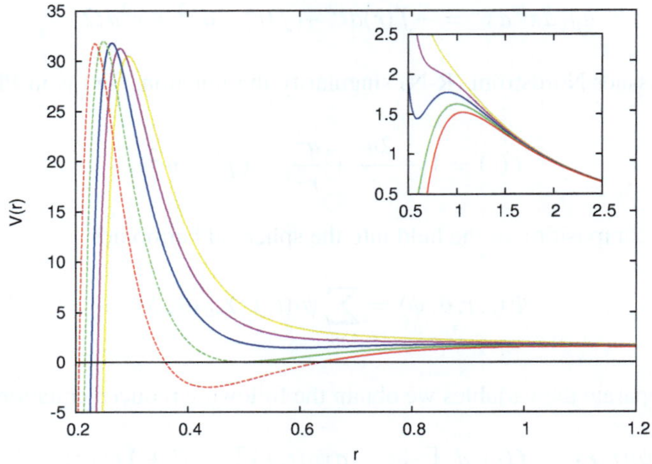

Fig. 1 Potential $V(r)$ given by (8) with $l = 2$, $m = 0.5$ for $q = 0.48, 0.5, 0.52, 0.54$ and 0.56. (The curves from *left* to *right* correspond to the increase of charge.) Note that the *dashed* part of the potential (for $q = 0.48$ and 0.5) is inside the black hole horizon

$$\frac{\partial^2 \phi_l(x,t)}{\partial t^2} - \frac{\partial^2 \phi_l(x,t)}{\partial x^2} = V(m,q,l,x)\phi_l(x,t), \tag{7}$$

with

$$V(m,q,l,x) = \left[\frac{l(l+1)}{r^2(x)} + \frac{2m}{r^3(x)} - \frac{2q^2}{r^4(x)}\right] f\big(r(x)\big). \tag{8}$$

And, for the normal modes $e^{-i\omega t}\phi_l(r)$, we can write

$$\frac{\partial^2 \phi_l(x)}{\partial x^2} + \left[\omega^2 - V(m,q,l,x)\right]\phi_l(x) = 0. \tag{9}$$

If $q > m$, we can see that $f(r)$ given by (3) has no zeros for real arguments, but (6) can be analytically integrated.

The potential (8) has for the ratio q^2/m^2 less than approximately $9/8$ and the relevant x (in the naked singularity case the domain of x is constrained) 3 extrema, one smaller "outer" maximum, one dominant "inner" maximum and minimum in the potential valley between them. (For $r \to 0$ the function $V(r) \to -\infty$.) For q^2/m^2 more than approximately $9/8$ the potential has only one maximum (thus only one peak). These features of the potential (8) can be seen in the Fig. 1.

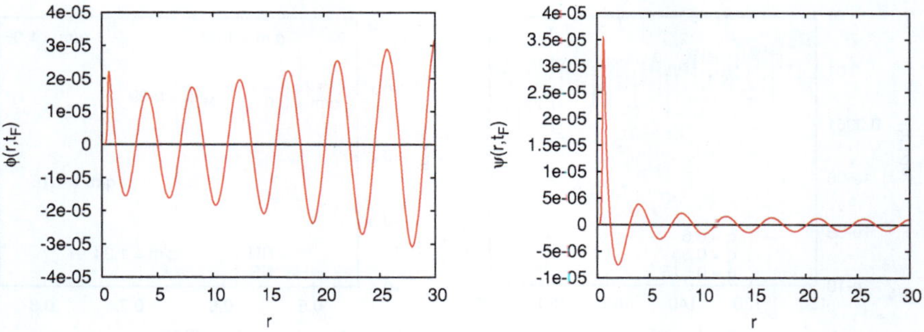

Fig. 2 *Left* Behavior of ϕ with $l = 2$ as a function of r near the center $r = 0$ for a late time $t_F = 350$, shown here in order to exemplify the effect of conditions (11) and (12) in the numerical integration, for a spacetime with $q = 0.5$ and $q = 0.52$. *Right* The same as in the left plot, but this time for the function $\psi = \phi/r$

4 The Naked Singularity for the Small Wave Mode Numbers: Numerical Results for the Frequencies

Our objective in this section is to solve (7) with potential (8) numerically, in the case where $q > m$ as described in the last section. To do this, we rewrite (7) in terms of the light-cone variables $u = t - x$ and $v = t + x$, where x corresponds to the tortoise coordinate (6), as

$$\frac{\partial^2\phi}{\partial t^2} - \frac{\partial^2\phi}{\partial x^2} = -4\frac{\partial^2\phi}{\partial u\partial v} = V(r)\phi \,, \tag{10}$$

that can be integrated with the boundary conditions

$$\phi(r = 0, t) = \phi(u, v = u + 2x_0) = 0 \,, \tag{11}$$

$$\phi(u = 0, v) = e^{-\frac{(v-v_c)^2}{2\sigma^2}} \,, \tag{12}$$

where condition (11) is a necessary condition on the field ϕ near the origin (see the discussion on Fig. 2 below) and condition (12) defines an "arbitrary" relevant initial signal to be propagated. We use the algorithm

$$\phi_N = \phi_W + \phi_E - \phi_S - \frac{\phi_W + \phi_E}{8} V \Delta_v \Delta_u \,, \tag{13}$$

where Δ_u and Δ_v are the integration steps in u and v, respectively. Note that here V is the potential (8) evaluated at the same r coordinate as ϕ_S (and ϕ_N).

As we can see in Fig. 2, the boundary conditions (11) and (12) ensure the necessary conditions on the fields ϕ and ψ near the center. As we discussed previously in Sect. 2, the physically correct boundary condition for ψ is $\psi(0, t) = 0$. From this we must have for $\phi(r, t) = r\psi(r, t)$ that $\phi(0, t) = 0$ and $\phi'(0, t) = 0$.

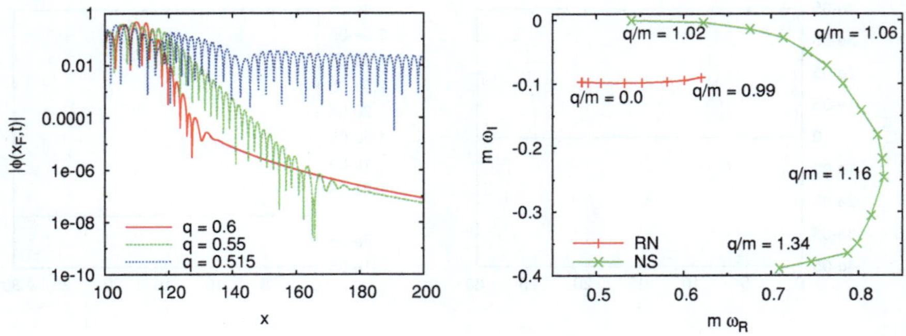

Fig. 3 *Left* $\phi(x_F, t)$ with $l = 2$ at $x_F = 100$ for a spacetime with $m = 0.5$ and different values of $q > m$. *Right* Frequencies of the fundamental mode with $l = 2$ in the $\omega_R \times \omega_I$ plane, parametrized by the q/m ration

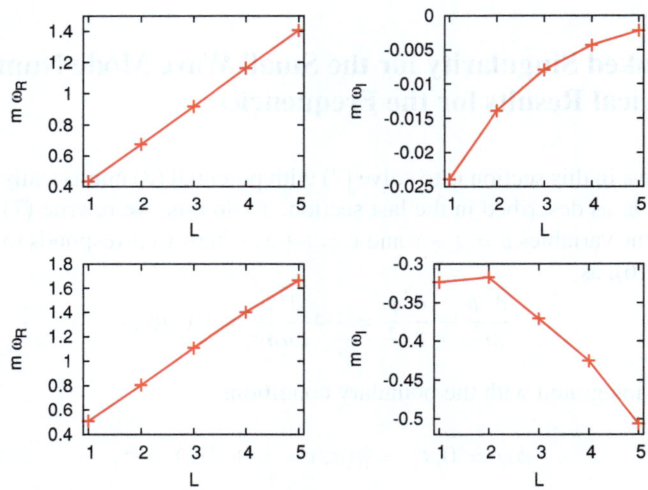

Fig. 4 *Above* frequencies of the fundamental mode as a function of l for $m = 0.5$ and $q = 0.52$ ($q^2/m^2 < 9/8$). *Below* same as above, but this time for $m = 0.5$ and $q = 0.6$ ($q^2/m^2 > 9/8$)

In the left plot of Fig. 3 we present some typical time evolutions of ϕ, for a $l = 2$ and different q/m ratios. In the right plot we present the obtained frequencies of the QNMs (fundamental mode) in the $\omega_R \times \omega_I$ plane. We can see a discontinuity in the frequencies as $q/m \to 1$, as was expected from the discussion of the potential $V(r)$ (see Fig. 1). We also point here that we see no significant changes, but rather a smooth behavior as $q^2/m^2 \to 9/8$ ($q/m \to 1.06$ in the plot). But we see a point of inflection in ω_R at $q/m \approx 1.16$, for which we did not find an analytical explanation.

Finally, in Fig. 4 we explore how the frequencies of the QNMs change with l. As usual in black hole scattering problems, we see that the oscillation frequency ω_R increases with l. But the qualitative behaviour of ω_I changes significantly with q/m.

In the upper plots ($q^2/m^2 \lesssim 9/8$), $|\omega_I|$ decreases with l, that is, the damping time is longer. In the lower plots ($q^2/m^2 \gtrsim 9/8$), we have the opposite tendency. This behaviour is connected to the potential $V(r)$. It might be also interesting to mention that in case $q^2/m^2 \gtrsim 9/8$, there is a qualitative similarity in the behaviour of the imaginary part of the frequencies as a function of l, between the case when l is small and the l large limit.

5 Conclusions

In this paper we analysed the problem of scalar field scattering on a R-N naked singularity background from the point of view of quasi-normal modes. The evolution on the R-N naked singularity is non-unique unless one specifies additional boundary condition representing a "hair" of the singularity. The quasi-normal modes then carry information about the "hair". We applied a particular boundary condition, that nothing comes out, or in from the singularity and analysed analytically, as well as numerically, the characteristic oscillations of the scalar field perturbations (low damped quasi-normal modes). In [14] we also analysed the eikonal $l \gg 1$ case via the analytical approach confirming the intuition obtained through the massless particle viewpoint, and showed that an approach based on analytical approximations can be useful also for the small l wave mode numbers. For the small l-s we calculated the frequencies numerically via the characteristic integration method.

The basic results can be summarized as follows: for the large l there is a continuous transition in the low damped QNM modes between the R-N black hole and the R-N naked singularity (see [14]). However, when the ratio q^2/m^2 becomes larger than approximately 9/8 then the picture becomes significantly different and the low damped modes do *not* exist for large l-s. (This is a very different picture from the BH based intuition.) For the small l numbers the modes face a discontinuous transition when transiting from the black hole to the naked singularity. Furthermore, the l dependence $|\omega_I|$ (for small l) changes as q^2/m^2 becomes larger than approximately 9/8: $|\omega_I|$ decreases for $q^2/m^2 \lesssim 9/8$ and increases for $q^2/m^2 \gtrsim 9/8$. It might be interesting to notice that for $q^2/m^2 \gtrsim 9/8$ the increase of $|\omega_I|$ as a function of l (for small l-s) matches the behaviour of $|\omega_I|$ for large l-s. In the case of large l-s and $q^2/m^2 \gtrsim 9/8$ we have shown that $|\omega_I|$ of the fundamental mode grows at least cubically with l and thus, as we already mentioned, the low damped modes do not exist [14].

References

1. Richartz, M., Saa, A.: Challenging the weak cosmic censorship conjecture with charged quantum particles. Phys. Rev. D **84**, 104021 (2011). doi:10.1103/PhysRevD.84.104021
2. Dotti, G., Gleiser, R., Pullin, J.: Instability of charged and rotating naked singularities. Phys. Lett. B **644**, 289 (2007). doi:10.1016/j.physletb.2006.12.004
3. Wald, R.: Dynamics in nonglobally hyperbolic, static space-times. J. Math. Phys. **21**, 2802 (1980). doi:10.1063/1.524403
4. Ishibashi, A., Wald, R.: Dynamics in non-globally-hyperbolic static space-times II: general analysis of prescriptions for dynamics. Class. Quantum Grav. **20**, 3815 (2003). doi:10.1088/0264-9381/20/16/318
5. Sandberg, V.: Light scattering properties of naked singularities. Phys. Rev. D **12**, 2226 (1975). doi:10.1103/PhysRevD.12.2226
6. Gibbons, G.: Quantized fields propagating in plane wave space-times. Commun. Math. Phys. **45**, 191 (1975). doi:10.1007/BF01629249
7. Horowitz, G., Marolf, D.: Quantum probes of space-time singularities. Phys. Rev. D **52**, 5670 (1995). doi:10.1103/PhysRevD.52.5670
8. Pitelli, J., Letelier, P.: Quantum singularities in static spacetimes. Int. J. Mod. Phys. D **20**, 729 (2011). doi:10.1142/S0218271811019062
9. Ishibashi, A., Hosoya, A.: Who's afraid of naked singularities? Probing timelike singularities with finite energy waves. Phys. Rev. D **60**, 104028 (1999). doi:10.1103/PhysRevD.60.104028
10. Martellini, M., Reina, C., Treves, A.: Klein-Gordon field in a naked singularity background. Phys. Rev. D **17**, 2573 (1978). doi:10.1103/PhysRevD.17.2573
11. Schmidt, B., Stewart, J.: The scalar wave equation in a Schwarzschild space-time. Proc. R. Soc. Lond. A Math. Phys. Sci. **367**(1731), 503 (1979)
12. Leaver, E.: Spectral decomposition of the perturbation response of the Schwarzschild geometry. Phys. Rev. D **34**, 384 (1986). doi:10.1103/PhysRevD.34.384
13. Nollert, H.P.: Quasinormal modes: the characteristic 'sound' of black holes and neutron stars. Class. Quantum Grav. **16**, R159 (1999). doi:10.1088/0264-9381/16/12/201
14. Chirenti, C., Saa, A., Skákala, J.: Quasinormal modes for the scattering on a naked Reissner-Nordström singularity. Phys. Rev. D **86**, 124008 (2012). doi:10.1103/PhysRevD.86.124008

Tracing a Relativistic Milky Way Within the RAMOD Measurement Protocol

Mariateresa Crosta

Abstract Advancement in astronomical observations and technical instrumentation implies taking into account the general relativistic effects due to the dynamical gravitational fields encountered by light while propagating from a star to the observer. Therefore, data exploitation for Gaia-like space astrometric mission (ESA, launch 2013) requires a fully relativistic interpretation of the inverse ray-tracing problem, namely the development of a highly accurate astrometric model, named RAMOD, in accordance with the geometrical environment affecting light propagation itself and the precepts of the theory of measurement. This could open a new rendition of the stellar distances and proper motions, or even an alternative detection perspective of many subtle relativistic effects suffered by light while it is propagating and subsequently recorded in the physical measurements.

1 Introduction

The role of astrometry has been revitalized thanks to the space mission Gaia [1] which will be launched by ESA not earlier than September 2013. The expected end-of-life astrometric performance, at the level of μ as accuracy, requires to take into account light deflections effects due to the Solar System bodies. This implies that any astrometric measurement has to be modelled in a way that stellar light propagation and detection should be both conceived in a general relativistic framework. As matter of fact, the trajectory of a photon is traced by solving the null geodesic in a curved space-time dictated by General Relativity (GR) and at the same time, the detection process usually takes place in a geometrical environment generated by a n-body distribution as it is that of our Solar System (SS).

M. Crosta (✉)
Istituto Nazionale di Astrofisica, Osservatorio Astronomico di Torino, via Osservatorio 20, 10025 Pino Torinese, Italy
e-mail: crosta@oato.inaf.it

J. Bičák and T. Ledvinka (eds.), *Relativity and Gravitation*, Springer Proceedings in Physics 157, DOI: 10.1007/978-3-319-06761-2_49,

Nowadays, a few approaches exist that model light propagation in a relativistic context. Among them, the post-Newtonian (pN) and the post-Minkowskian (pM) approximations are those mainly used ([2–4] and references therein). Inside the Consortium constitued for the Gaia data reduction (Gaia CU3, Core Processing, DPAC) two different formulations of relativistic light propagation have been developed to model astrometric observations of distant sources by an SS observer: (i) the GREM formulation [3], known as the Gaia baseline model IAU coordinate based [5], and (ii) the RAMOD model [6], an alternative approach fully compliant with the precepts of local measurement in a relativistic setting. Actually RAMOD is a family of astrometric models of increasing intrinsic accuracy conceived to solve the inverse ray-tracing problem in a general relativistic framework. Their theoretical equivalence to the 1-μ as accuracy level has been recently demonstrated ([7] and reference therein) and will be exploited, in a process, called in the Gaia jargon, Astrometric Verification Unit (AVU) by comparing the results of two fully independent astrometric reconstructions of the celestial sphere to assess all-sky scientific reliability on position, including parallax, and proper motions. The link between the models is crucial as far as the Gaia's goal is concerned: the unbiased measurements, i.e. independently from models, of the most fundamental astrophysical stellar parameters (absolute distance, angular position, velocity, and mass) for approximately 1 billion individual stars!

It can be inferred that the treatment of light propagation in time-dependent gravitational fields encompasses issues from fundamental astronomy to cosmology ([8–13] and references therein). The accurate measurement of the motions of stars in our Galaxy can also provide access to the cosmological signatures in the disk and halo, while astrometric experiments from within our Solar System can probe possible deviations from GR in an unrivaled way just one century after Einsteins's great discoveries. With the Gaia mission approaching launch, Relativistic Astrometry is about to trace the geometry of the visible Milky Way.

2 The Astrometric Problem

The astrometric problem consists, firstly, in solving the null geodesic for the single stellar photon, in order to trace back the light trajectory to the initial position of the emitting source and, then, determine its astrometric parameters through the astrometric observable, according to the chosen reference frames. Differently from the other approaches, RAMOD's full solution requires the integration of a set of coupled non-linear differential equations, called "master equations". The unknown of these equations is the local line-of-sight $\bar{\ell}$ as measured by the fiducial observer $\mathbf{u}$ at the point of observation in her/his rest-space. At the time of observation, $\bar{\ell}$ provides the boundary condition for uniquely solving the light path by means of the relativistic definition of the observable [14] and the satellite-observer frames [15]. The main purpose of the RAMOD approach is to express the null geodesic through all the physical quantities entering the process of measurement without any approximations, in order to entangle all the possible interactions of light with the background geometry. RAMOD uses a 3+1 characterization of space-time in order to measure physical phenomena along the proper time and on the rest-space of a set of fiducial observers according to the following measurement protocol [16]: (i) specify the phenomenon under investigation; (ii) identify the covariant equations which describe it; (iii) identify the observer who makes the measurements; (iv) choose a frame adapted to that observer allowing the space-time splitting into the observer's space and time; (v) understand the locality properties of the measurement under consideration (local or non-local with respect to the background curvature); (vi) identify the frame components of those quantities which are the observational targets; (vii) find a physical interpretation of the above components following a suitable criterion; (viii) verify the degree of the residual ambiguity in the interpretation of the measurements and decide the strategy to evaluate it (i.e. comparing what is already known).

Solving the astrometric problem in practice means to compile an astrometric catalog with the same order of accuracy as the measurements. To what extent, then, is the process of star coordinate "reconstruction" consistent with General Relativity and Theory of Measurements?

3 The Geometry of the Astrometric Problem

Gaia-like measurement takes place inside the Solar System, i.e. a weakly relativistic gravitationally bound system, described by the metric $g_{\alpha\beta} = \eta_{\alpha\beta} + h_{\alpha\beta} + O(h^2)$. Now, in order to gauge how much curvature can be considered local or not with respect to the measurement, let us resort to the virial theorem which requires an energy balance of the order of $|h_{\alpha\beta}| \leq U/c^2 \sim v^2/c^2$, where v is the characteristic relative velocity within the system.[1] Therefore the level of accuracy is fixed by the order of the small quantity $\varepsilon \sim (v/c)$. Since the system is weakly relativistic, the perturbation tensor $h_{\alpha\beta}$ contributes with even terms in ε to g_{00} and g_{ij} (lowest order ε^2) and with odd terms in ε to g_{0i} (lowest order ε^3, [16, 17]); its spatial variations are of the order of $|h_{\alpha\beta}|$, while its time variation is of the order of $\varepsilon|h_{\alpha\beta}|$. This means that at the order of ε^3, not only the time dependence of the background metric cannot be ignored any longer, but also the vorticity, which measures—in the process of foliation—how a world-line of an observer rotates around a neighboring one, can be neglected being proportional to the g_{0i} term of the metric (see details in [7]). Consequently, it is not possible to define a rest-space of a fiducial observer that covers the entire space-time. Any observer **u** can be considered at rest with respect to the coordinates x^i *only locally*, and for this reason **u** is called the *local barycentric observer*, as identified in [18]. The master equations satisfied by the vector field $\bar{\ell}$ up to the ε^3 order of accuracy are

[1] For a typical velocity ~30 km/s, $(v/c)^2$ ~1 milli-arcsec.

$$\frac{d\bar{\ell}^0}{d\sigma} = \bar{\ell}^i \bar{\ell}^j h_{0j,i} + \frac{1}{2} h_{00,0}, \tag{1}$$

$$\frac{d\bar{\ell}^k}{d\sigma} = \frac{1}{2} \bar{\ell}^k \bar{\ell}^i \bar{\ell}^j h_{ij,0} - \bar{\ell}^i \bar{\ell}^j \left(h_{kj,i} - \frac{1}{2} h_{ij,k} \right) - \frac{1}{2} \bar{\ell}^k \bar{\ell}^i h_{00,i}$$
$$- \bar{\ell}^i \left(h_{k0,i} + h_{ki,0} - h_{0i,k} \right) + \frac{1}{2} h_{00,k} + \bar{\ell}^k \bar{\ell}^i h_{0i,0} - h_{k0,0}, \tag{2}$$

named "RAMOD4 master equations" in the dynamical case [7, 18], σ being the parameter of the null geodesic. Note that there is a differential equation also for the $\bar{\ell}^0$ component, which represents an opportunity to better decipher light propagation in future developments.

The ε^2 regime, instead, is referred to as the "static case", or "static space-time", i.e. a stationary space-time in which a time-like Killing vector field $\mathbf{u}$ has vanishing vorticity [6]. In this case the parameter σ on u is the proper time of the physical observers who transport the spatial coordinates without shift. Any hypersurface $t(x, y, z) = constant$, at each different coordinate time t, can be considered the rest space *everywhere* of the observer $\mathbf{u}$ and the geometry that each photon feels is, then, identified with the weak relativistic metric where $g_{0i} = 0$. In these circumstances we can define a one-parameter local diffeomorphism which maps each point of the null geodesic to the point on the slice at the time of observation, say $S(t_o)$ [6]:

$$\frac{d\bar{\ell}^k}{d\sigma} = -\bar{\ell}^k \left(\frac{1}{2} \bar{\ell}^i h_{00,i} \right) - \delta^{ks} \left(h_{sj,i} - \frac{1}{2} h_{ij,s} \right) \bar{\ell}^i \bar{\ell}^j + \frac{1}{2} \delta^{ks} h_{00,s}. \tag{3}$$

Equation (3) determines light propagation in the static case, and are called "RAMOD3 master equations" [6]. Nevertheless, (2) can be integrated taking into account also the expansion of the congruence $\mathbf{u}$ [7] and, then, velocity of a uniform source can be included from $\partial_0 h_{00}$. Only a vorticity-free space-time allows to parametrize simultaneously the mapped trajectory with respect to the Center of Mass on $S(t_o)$; if the Euclidean scalar product is applied, the RAMOD procedure for the parametrization generalizes the one used in [2] or [3, 7].

The fact that light tracing is different with or without the vorticity term makes evident how the RAMOD recipe, based on a measurement protocol, differs from a direct "coordinate" approach which, instead, does not need to discriminate the accuracy of the geometry to be involved.

4 Matching Physics and Coordinates at High Accuracy

The quantity $\bar{\ell}$ is the unitary four-vector representing the *local line-of-sight* of the incoming photon as measured by the local observer $\mathbf{u}$ in his/her gravitational environment; it represents a physical quantity in any case, with or without vorticity. By implementing its coordinate expressions straightaway, (2), i.e. those for the spatial components, are converted into the coordinate ones derived in [2] at the first pM approximation of the null geodesic [7]. This result was expected, since both models are deduced from the null geodesic in a weak field regime. Then, once such an equivalence is obtained, one could solve the master equation in the RAMOD framework by applying the same procedure as adopted in [2]. However, consistently with the reasoning of the previous section, only RAMOD3 master equation can be transformed into the solution given by [2], since the parametrization in RAMOD is possible only in a vorticity-free space-time. In fact, if one assumes a constant light direction and a perturbed straight line trajectory, the equivalence of the two parametrizations implies a change of coordinates which transforms (2) into the same parametrized equation (36) used in [2]. Nevertheless, the integration of the null geodesic in [2] intends to consider the gravitomagnetic effects. In addition, the metric coefficients $h_{\alpha\beta}$ depend on the retarded distance $r_{(a)}$ as discussed in [18]. This means that one has to compute the spatial coordinate distance $r_{(a)}$ from the points on the photon trajectory to the a-th gravity source at the appropriate retarded time and up to the required accuracy. Hence, if we wish our model to be accurate to ε^3, it suffices that the retarded distance r contributes to the gravitational potentials—which we recall are at least of order ε^2—with terms of the order of ε. Instead, to the order of ε^2 (static geometry), the contribution of the relative velocities of the gravitating sources can be neglected. Indeed, in the stationary case, with expansion, one can choose to further expand the retarded distance in order to keep the terms depending on the source's velocity up to the desired accuracy. Obviously the effects due to the bodies' velocity cannot be related to a gravity-magnetic effect, at least up to the scale where the vorticity can be neglected. Actually, the positions of the bodies can be recorded as subsequent snapshots onto the mapped trajectories and deduced as "postponed" corrections in the reconstruction of the photon's path.

The importance of the measurement protocol in setting the correct role of the coordinates, and thus avoiding misinterpretations of parallel but different quantities, is also discussed in [14], where, within the context of the Gaia mission (ESA, [1]), a first comparison between RAMOD and GREM (Gaia RElativistic Model, [3]) was carried out via the extrapolation of the aberrational term in the *local* light direction. Differences, that already exist at the level of the aberration effect, suggest particular care in the interpretation of the final catalog. Another example which shows how the accurate inclusion of the geometry redraws a standard measurement, is given by the formula for the Doppler shift in [13]. The spectroscopic and astrometric data that will be provided by the new generation of satellites can be complemented with one another, thus leading to a general-relativistic Doppler which is exact up to and including the ε^3 terms. It is also showed that a previously proposed Doppler-shift formula is definitely not adequate to this task, since it misses relevant relativistic corrections already at ε^2.

5 Conclusions

Modeling light propagation is intrinsically connected to the identification of the geometry where photons move. The different conception of RAMOD provides a method to exploit highly accurate observations to their full extent, as could be the case for the astrometric data coming from the ESA mission Gaia, possibly a new beginning in the field of Relativistic Astrometry. The comparison between different light modeling approaches is extremely important since Gaia will "change" our scientific vision and we are implementing new methods using real data. By comparing different formulations of a null geodesic we have the opportunity to exploit the advantages of the different methods and improve on our understanding of light propagation. As far as RAMOD method is concerned, the geometrical distinction between the master equations introduces a criterion to disentangle coordinate and physical effects.

In RAMOD the vorticity term cannot be neglected at the order of ε^3: ignoring it locally is valid only in a small neighborhood compared to the scale of vorticity itself. When the vorticity term is needed the light trajectory cannot be laid out on a unique rest-space of simultaneity from the observer to the star, wherever the latter could be located. Without vorticity RAMOD allows a parametrization of the light trajectory and sets the level of reciprocal consistency with the existing approaches. Only master equations of RAMOD4, i.e. the case of a dynamical space-time, fully preserve the active content of gravity. This solution is accurate enough to implement Relativistic Astrometry beyond Gaia. Considering the number of objects that can be observed in high accuracy regime, the local line-of-sight, as a physical entity, can be also used in the future for an "inverse parameter problem" approach, able to statistically determine the metric also outside the Solar System [19].

References

1. Turon, C., O'Flaherty, K., Perryman, M.: The three-dimensional universe with Gaia, p. 732 (ESA Special Publication SP-576, 2005). http://sci.esa.int/science-e/www/object/index.cfm?fobjectid=37100
2. Kopeikin, S., Mashhoon, B.: Gravitomagnetic effects in the propagation of electromagnetic waves in variable gravitational fields of arbitrary-moving and spinning bodies. Phys. Rev. D **65**(6), 064025 (2002). doi:10.1103/PhysRevD.65.064025
3. Klioner, S.: A practical relativistic model for microarcsecond astrometry in space. Astron. J. **125**, 1580 (2003). doi:10.1086/367593
4. Teyssandier, P., Le Poncin-Lafitte, C.: General post-Minkowskian expansion of time transfer functions. Class. Quantum Grav. **25**, 145020 (2008). doi:10.1088/0264-9381/25/14/145020
5. Soffel, M., Klioner, S., Petit, G., et al.: The IAU 2000 resolutions for astrometry, celestial mechanics, and metrology in the relativistic framework: explanatory supplement. Astron. J. **126**, 2687 (2003). doi:10.1086/378162
6. de Felice, F., Crosta, M., Vecchiato, A., Lattanzi, M., Bucciarelli, B.: A general relativistic model of light propagation in the gravitational field of the solar system: the static case. Astrophys. J. **607**, 580 (2004). doi:10.1086/383244

7. Crosta, M.: Tracing light propagation to the intrinsic accuracy of spacetime geometry. Class. Quantum Grav. **28**(23), 235013 (2011). doi:10.1088/0264-9381/28/23/235013
8. Will, C.M.: The confrontation between general relativity and experiment. Living Rev. Relativ. **9**(3) (2006). http://www.livingreviews.org/lrr-2006-3
9. Turyshev, S.: Experimental tests of general relativity. Annu. Rev. Nucl. Part. Sci. **58**, 207 (2008). doi:10.1146/annurev.nucl.58.020807.111839
10. Crosta, M., Mignard, F.: Microarcsecond light bending by Jupiter. Class. Quantum Grav. **23**, 4853 (2006). doi:10.1088/0264-9381/23/15/006
11. Kopeikin, S., Gwinn, C., Johnston, K.: Sub-microarcsecond astrometry and new horizons in relativistic gravitational physics. In: Johnston, K., McCarthy, D., Luzum, B., Kaplan, G. (eds.) Towards Models and Constants for Sub-Microarcsecond Astrometry, pp. 303–307. International Astronomical Union, Paris (2000)
12. Damour, T., Donoghue, J.: Equivalence principle violations and couplings of a light dilaton. Phys. Rev. D **82**(8), 084033 (2010). doi:10.1103/PhysRevD.82.084033
13. de Felice, F., Preti, G., Crosta, M., Vecchiato, A.: Relativistic satellite astrometry: the stellar radial velocity. Astron. Astrophys. **528**, A23 (2011). doi:10.1051/0004-6361/201015375
14. Crosta, M., Vecchiato, A.: Gaia relativistic astrometric models. I. Proper stellar direction and aberration. Astron. Astrophys. **509**, A37 (2010). doi:10.1051/0004-6361/200912691
15. Bini, D., Crosta, M., de Felice, F.: Orbiting frames and satellite attitudes in relativistic astrometry. Class. Quantum Grav. **20**, 4695 (2003). doi:10.1088/0264-9381/20/21/009
16. de Felice, F., Bini, D.: Classical Measurements in Curved Space-Times. Cambridge Monographs on Mathematical Physics. Cambridge University Press, Cambridge (2010)
17. Misner, C., Thorne, K., Wheeler, J.: Gravitation. W.H. Freeman, San Francisco (1973)
18. de Felice, F., Vecchiato, A., Crosta, M., Bucciarelli, B., Lattanzi, M.: A general relativistic model of light propagation in the gravitational field of the solar system: the dynamical case. Astrophys. J. **653**, 1552 (2006). doi:10.1086/508701
19. Tarantola, A.: Inverse Problem Theory: Methods for Data Fitting and Model Parameter Estimation. Elsevier Science, Amsterdam (1987)

Is There a Flatness Problem in Classical Cosmology?

Phillip Helbig

Abstract I discuss various definitions of the flatness problem and previous claims that it does not exist. I also present a new quantitative argument which shows that it does not exist in cosmological models which collapse in the future.

1 Introduction

Questioning the existence of the flatness problem might seem to some like questioning the existence of the expansion of the universe. The flatness problem (e.g. [1]) and the fact that inflation can solve it (e.g. [2]) have become part of standard cosmology, at least for many definitions of 'standard'. How can something so fundamental not exist? My own view is that the emphasis has been so much on the *solution* of the flatness problem through inflation that the flatness problem itself has been rather neglected and its existence just assumed without being investigated in detail.

2 Basic Cosmology

I assume that, at the level of detail necessary, the universe can be described by the Friedmann–Lemaître equation

$$\dot{R}^2 = \frac{8\pi G\rho R^2}{3} + \frac{\Lambda R^2}{3} - kc^2, \tag{1}$$

P. Helbig (✉)
Thomas-Mann-Str. 9, 63477 Maintal, Germany
e-mail: helbig@astro.multivax.de

J. Bičák and T. Ledvinka (eds.), *Relativity and Gravitation*,
Springer Proceedings in Physics 157, DOI: 10.1007/978-3-319-06761-2_50,

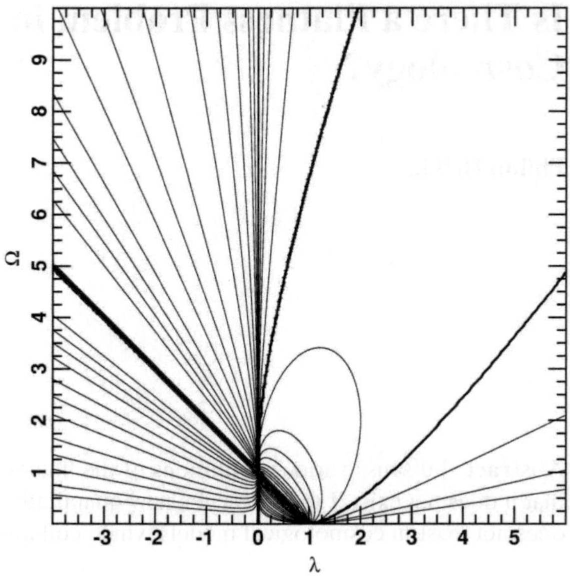

Fig. 1 Evolutionary trajectories in the λ-Ω plane

where the symbols have their usual meaning (e.g. [3]). It can be useful to express (1) with the values of the dimensionless parameters as observed now, denoted by the suffix 0. This leads to

$$\dot{R}^2 = \dot{R}_0^2 \left(\frac{\Omega_0 R_0}{R} + \frac{\lambda_0 R^2}{R_0^2} - K_0 \right). \tag{2}$$

Note that, at any time,

$$R = \frac{c}{H} \frac{\text{sign}(K)}{\sqrt{|K|}} . \tag{3}$$

In general, H, λ and Ω all change with time. (See [4] for an excellent discussion of the evolution of λ and Ω.) The change in λ with time is due entirely to the change in H with time, since Λ is constant; the variation in Ω is due both to variation in H and to the decrease in density as the universe expands. For the present discussion, the basic information needed can be seen in Fig. 1, referring at the moment only to the thick lines and curves. The vertical line corresponds to $\lambda = 0$; the diagonal line corresponds to $k = 0$ with $k = -1$ below it and $k = +1$ above it. The curve near the vertical line (corresponding to the A1 curve in [4]) separates models which will collapse (to the left) from those which will expand forever (to the right). Models on the curve start arbitrarily close to the Einstein– de Sitter model (like all non-empty big-bang models) and asymptotically approach the static Einstein model which has $\lambda = \Omega = \infty$ (since $H = 0$; Λ and ρ have finite values). The other curve (corresponding to the A2 curve in [4]) separates big-bang models (to the left) from non–big-bang models (to the right); the latter contract from an infinite to a finite size, then expand forever. Models on the curve start at the static Einstein model and asymptotically approach the de Sitter model (the latter feature is shared with all models which expand forever and have $\lambda > 0$). The thin curves show some sample trajectories in the λ-Ω parameter space. (Note that all the thick lines, curves and points of intersection in Fig. 1 are also trajectories.) Also, note that the trajectories do not cross; this means that the history of a cosmological model (i.e. the way λ and Ω change with time) is completely determined by the values at any point on it (in practice, by measuring the values at the present time, λ_0 and Ω_0).

3 A Very Brief History of the Flatness Problem

The flatness problem appears in two forms. One states that if $\Omega \approx 1$ today, then in the early universe it was arbitrarily close to 1; the assumption is that some 'mechanism' is needed to explain this 'fine-tuning' (e.g. [2]). (It is usually not stated but almost always assumed that no fine-tuning would be necessary if Ω were not ≈ 1 today.) The other states that if Ω changes with time, then we should be surprised that Ω is (still) ≈ 1 today [5].[1] Coles and Ellis [6] discuss three 'solutions' to the flatness problem—$\Omega \equiv 1$ (and $\lambda \equiv 0$), $k = 0$, anthropically selected special time—which, however, are ultimately unsatisfactory. Are there any satisfactory ones?

The flatness problem is often presented as a fine-tuning problem (e.g. [2]): if Ω is near 1 to day, then at some time t_{fine} in the past it must have been 1 to a very high accuracy. I refer to this sense of the flatness problem as the 'qualitative flatness problem'. This argument is completely bogus, as has been pointed out by many authors [5, 6]: *all* non-empty models begin their evolution at the Einstein– de Sitter model, so of course the further back in time one goes, the 'more finely tuned' Ω is. The point is, within the context of classical cosmology, there is nothing special about a time t_{fine} chosen so that Ω is very close to 1 at that point.

Evrard and Coles [7] (see also Coles and Ellis [6]) also point out that the assumption implicit in the qualitative flatness problem, namely that some wide range of Ω values are a priori equally likely at some early time, constitutes a prior which is incompatible with the assumption of minimal information. This can be regarded as a quantitative solution to the qualitative flatness problem (or, perhaps, an argument against its existence).

The qualitative flatness problem thus does not exist; it is merely a consequence of the way in which a universe, described by the Friedmann–Lemaître equation, evolves and how dimensionless observable quantities such as Ω are defined. Nevertheless,

[1] Historically, the flatness problem was first discussed during a time when λ was thought to be zero. If λ is not constrained to be zero, then the flatness problem should be re-phrased as the Einstein–de Sitter problem, i.e. the question is why the universe is (in some sense) close to the Einstein–de Sitter model (which is an unstable fixed point and a repulsor) today when $|\lambda|$ and Ω can take on values between 0 and ∞. However, for brevity I will continue to use the term 'flatness problem' even for the more general case and sometimes mention only the change in Ω with time.

even if it is not a puzzle why $\Omega = 1$ at early times, one can still ask whether we should be surprised that $\Omega \approx 1$ today. The rest of this article is concerned mainly with the second form: should we be surprised that $\Omega \approx 1$ today? This 'quantitative flatness problem' is more subtle, but also has solutions within the context of classical cosmology.

4 Cosmological Models Which Collapse in the Future

All cosmological models (assumed to be expanding now) with $\lambda < 0$ will collapse in the future: $\ddot{R}$ is negative for all values of R and for large R is proportional to R. Models with $\lambda = 0$ will collapse for $\Omega > 1$. In addition, models with $\lambda > 0$ will collapse provided that $\Omega > 1$ (which in this case implies $K > 0$, i.e. $k = +1$), $q > 0$ and $\alpha < 1$, where

$$\alpha = \text{sign}(K)\frac{27\Omega^2\lambda}{4K^3} \tag{4}$$

[4, 5]. (The A1 and A2 curves mentioned above have $\alpha = 1$.) In Fig. 1, these are in the area between $\lambda = 0$ and the A1 curve. Empty big-bang models start arbitrarily close to the Milne model with (λ,Ω) values of (0, 0); non-empty big-bang models start arbitrarily close to the Einstein–de Sitter model with (λ,Ω) values of (0, 1). The evolution of λ and Ω can be viewed as trajectories in the parameter space: λ and Ω evolve from the starting point to infinity and return along the same trajectory. (For the definitive discussion, see [4]; a very useful visualization can be found in [8].) The interesting question with regard to the flatness problem is the amount of time spent in various parts of parameter space. To quantify this, I have calculated the quotient of the age of the universe now and at the time of maximum expansion as a function of λ and Ω. The age of the universe is given by

$$t = \int_0^{R(t)} \frac{\mathrm{d}R}{\sqrt{\dot{R}_0^2\left(\frac{\Omega_0 R_0}{R} + \frac{\lambda_0 R^2}{R_0^2} - K_0\right)}}, \tag{5}$$

which follows from (2). For the current age, the upper limit is given by (3); at the time of maximum expansion it is found by calculating the (smallest) zero of $\dot{R}^2$ (since $\dot{R}^2$ cannot be negative). This is shown in Fig. 2. It is clear that large values of λ and Ω occur only during a relatively short time in the history of the universe, near the time of maximum expansion (at the precise time of maximum expansion, λ and Ω are infinite since $H = 0$). Note that this argument is completely independent of H_0.

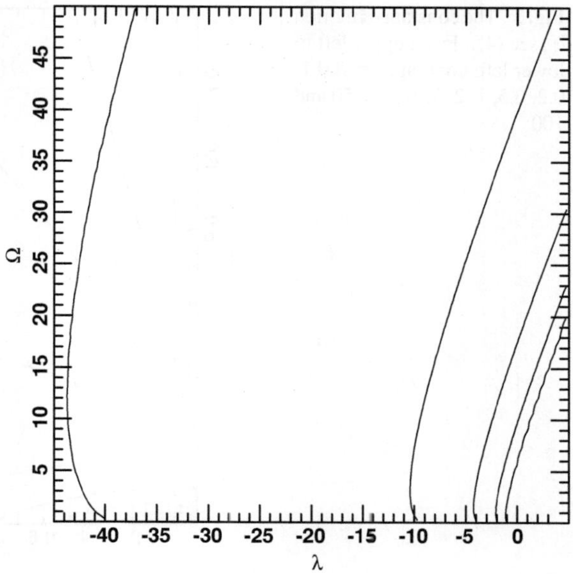

Fig. 2 The age of the universe as a fraction of the time between the big bang and maximum expansion. Contours, from right to left, are at 0.5, 0.6, 0.7, 0.8 and 0.9

5 Cosmological Models Which Expand Forever: Reversing the Fine-tuning Problem

Lake [5] has presented a solution which solves the flatness problem as well for models with $k = +1$ which will expand forever. (For non-collapsing models, large values of λ and Ω are possible only for $k = +1$.) Trajectories in the λ-Ω plane have a constant of motion given by (4). It seems natural to distinguish cosmological models on the basis of their value of α. Large values of λ and Ω are possible only for $\alpha \lesssim 1$. This is shown in Fig. 3. (Note that, for clarity, only $\Omega > 1$ is shown!) It is obvious that $\alpha \leq 1$ is a necessary condition for having infinitely large values of λ or Ω. Already for $\alpha = 2$ the maximum value of λ is just 2 (for $\Omega = 2$) and the maximum value of Ω is ≈ 3.5 (for $\lambda = 1.25$).

In this case, the fine-tuning argument is reversed; only in the case of fine-tuning do λ and Ω become arbitrarily large. This demonstrates quantitatively that there is no quantitative flatness problem regarding arbitrarily large values of λ or Ω for models which expand forever. This argument is also independent of the value of H_0. However, all non-empty models which expand forever asymptotically approach the de Sitter model at $(\lambda, \Omega) = (1, 0)$. Thus, one final aspect of the quantitative flatness problem remains: Ω can become arbitrarily small. This is investigated in the next section.

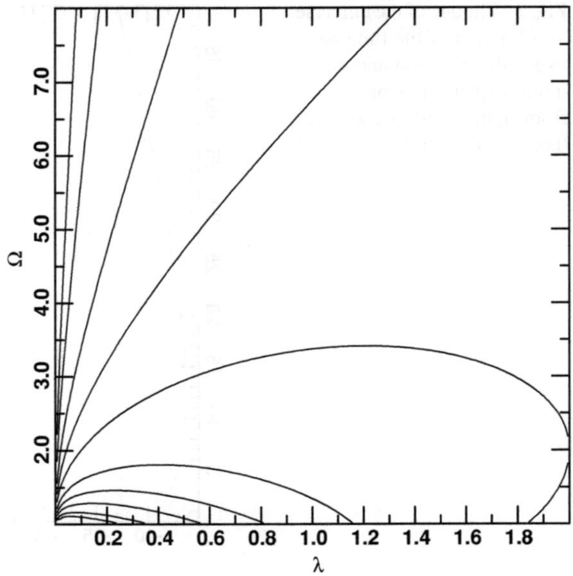

Fig. 3 The constant of motion α (see (4)). From upper left to lower left, contours are at 0.1, 0.2, 0.5, 1, 2, 5, 10, 20, 50 and 100

6 Cosmological Models Which Expand Forever: the Weak Anthropic Principle

I have now covered the entire λ-Ω parameter space except for big-bang models with (a) $q < 0$ (which implies $\lambda > 0$) and (b) Ω less than ≈ 2 (all three values of k are possible) and shown that in all cases there is no flatness problem. What about this remaining portion of parameter space? Models here all have $K \approx 0$ and approach the de Sitter model asymptotically. This means that there is no flatness problem in the restricted sense, as pointed out by Lake [5]. However, Ω becomes arbitrarily small (and λ arbitrarily close to 1). Thus, there is still a problem in that we do not observe such values, even though they exist for almost the entire (infinite) lifetime of the universe. This is essentially the question 'if the universe lasts forever, then why are we near the beginning?' Note that this question could be asked at any time. One could leave it at that and say that since any finite age is arbitrarily close to the beginning, there is nothing special about our time and thus no flatness problem in the time-scale sense (i.e. the quantitative flatness problem, why is Ω not arbitrarily small today). This is discussed in more detail in [9].

7 Summary

The qualitative flatness problem, i.e. the puzzle why the universe was arbitrarily close to the Einstein–de Sitter model (or, for an empty universe, the Milne or de Sitter model) at early times, does not exist. It is merely a consequence of the way λ and Ω are defined. Neither does the quantitative flatness problem exist: although the cosmological parameters in general evolve with time, it is not puzzling that we don't observe extreme values for them today. In the case of models which will collapse in the future this is because large (absolute) values of λ and Ω occur only during a relatively short time in the lifetime of such a universe, namely near the time of maximum expansion. λ and Ω can become large only when H becomes small, and this happens only during the time when the universe is at or near its maximum size. (Arbitrarily small (absolute) values, if they occur at all, also occur for only a relatively short time.) For models which will expand forever, large values are possible only for $k = +1$. However, this occurs only for $\alpha \approx 1$. In this case, the fine-tuning argument is reversed; only in the case of fine-tuning do λ and Ω become arbitrarily large. Since all models which will expand forever asymptotically approach $\Omega = 0$, arbitrarily small values of Ω can occur. Those with $\lambda = 0$ (and hence $k = -1$) approach the Milne model with $\Omega = 0$; models with $\lambda > 0$, whatever the value of k, approach the de Sitter model with $\lambda = 1$ (the Milne and de Sitter models themselves are of course stationary points). (If $\lambda = 0$ at any time then $\lambda = 0$ at all times. Otherwise, arbitrarily small values of λ, if they occur at all, occur only for a relatively short time.) However, if H_0 has a value similar to or smaller than the observed value, small values of Ω will occur only in the far future when anthropic arguments probably make the observation of such a low value of Ω unlikely. While (for $\lambda > 0$) a higher value of H_0 would allow a low value of Ω even for an age near the observed age, such a universe would have spent only a very short time during which Ω was not very small, so structure formation would have been strongly suppressed.

References

1. Dicke, R., Peebles, P.J.E.: The Big Bang cosmology-Enigmas and Nostrums. In: Hawking, S., Israel, W. (eds.) General Relativity: An Einstein Centenary Survey, pp. 504–517. Cambridge University Press, Cambridge (1979)
2. Guth, A.: Inflationary universe: a possible solution to the horizon and flatness problems. Phys. Rev. D **23**(2), 347 (1981). doi:10.1103/PhysRevD.23.347
3. Kayser, R., Helbig, P., Schramm, T.: A general and practical method for calculating cosmological distances. Astron. Astrophys. **318**(3), 680 (1997)
4. Stabell, R., Refsdal, S.: Classification of general relativistic world models. Mon. Not. R. Astron. Soc. **132**(3), 379 (1966)
5. Lake, K.: The flatness problem and Λ. Phys. Rev. Lett. **94**(20), 201102 (2005). doi:10.1103/PhysRevLett.94.201102
6. Coles, P., Ellis, G.: Is the Universe Open or Closed?, Cambridge Lecture Notes in Physics, vol. 7. Cambridge University Press, Cambridge (1997)
7. Evrard, G., Coles, P.: Getting the measure of the flatness problem. Class. Quantum Grav. **12**(10), L93 (1995). doi:10.1088/0264-9381/12/10/001
8. Leahy, J.: Solutions of the Friedman equation (2003). http://www.jb.man.ac.uk/~jpl/cosmo/friedman.html
9. Helbig, P.: Is there a flatness problem in classical cosmology? Mon. Not. R. Astron. Soc. **421**, 561 (2012). doi:10.1111/j.1365-2966.2011.20334.x

Cosmology in $f(R)$ Exponential Gravity

Luisa Jaime, Marcelo Salgado and Leonardo Patiño

Abstract Using an approach that treats the Ricci scalar itself as a degree of freedom, we analyze the cosmological evolution within an $f(R)$ model that has been proposed recently (exponential gravity) and that can be viable for explaining the accelerated expansion and other features of the Universe. This approach differs from the usual scalar–tensor method and, among other things, it spares us from dealing with unnecessary discussions about *frames*. It also leads to a simple system of equations which is particularly suited for a numerical analysis.

1 Introduction

Since the discovery of the accelerated expansion of the Universe from the observations of supernovae Ia [1–3] and its interpretation using the ΛCDM model of standard cosmology, a large amount of investigations have been devoted to explain the same phenomenon but using dark energy substances different from the cosmological constant Λ. One can rank the dark energy models from the most "conservative" to the most "radical" ones. Among the former we can mention those which do not introduce new fields or modifications to general relativity but which consider that inhomogeneities in the Universe could be enough to account for the observations [4]. There are also the models that introduce new fields and perfect fluids with exotic

L. Jaime · M. Salgado (✉)
Instituto de Ciencias Nucleares, Universidad Nacional Autónoma de México,
A.P. 70-543, 04510 México, D.F., México
e-mail: marcelo@nucleares.unam.mx

L. Jaime · L. Patiño
Facultad de Ciencias, Universidad Nacional Autónoma de México, A.P. 50-542,
04510 México, D.F., México

J. Bičák and T. Ledvinka (eds.), *Relativity and Gravitation*,
Springer Proceedings in Physics 157, DOI: 10.1007/978-3-319-06761-2_51,

equations of state within the framework of general relativity (GR) in order to avoid the "problems" associated with Λ [5–8]. For instance, *quintessence*, *k-essence*, and Chaplygin gas are some of the most popular models of this kind. The most radical attempts to explain the accelerated expansion are perhaps those which propose to modify GR while keeping the hypothesis of homogeneity and isotropy as a first approximation. Many alternate theories of gravity have been proposed to explain this phenomenon as well as those related with the dark matter (e.g. rotation curves of galaxies). The modified $f(R)$ metric gravity is just one of such theories and maybe the most analyzed one in the last ten years, where the geometry takes care of mimicking the dark energy. This alternative is certainly radical since GR has been thoroughly tested for almost one hundred years and it has not only supported all the tests but in addition most of its predictions have been confirmed as well. Thus, the challenge of modified gravity is both to be consistent with GR tests and also to explain the phenomena they were called for. This is not a trivial task and many $f(R)$ models have failed in the attempt. The story concerning dark substances does not end with the accelerated expansion. The measurements of the angular distribution of cosmic background radiation anisotropies in the sky can also be explained by the ΛCDM model, and therefore, the task for the alternative models, theories or dark energy substances is even more demanding.

As we mentioned, $f(R)$ theories have been studied in detail in a recent past and it is out of the scope of the present article to discuss all the properties, problems and features associated with some of the specific models proposed before (see [9–13] for a review).

Our aim is to report the results of a potentially viable candidate, termed *exponential gravity*, as a model for the accelerated expansion, but using an approach that has been proposed recently by us [14] and which avoids the identification with the scalar–tensor theories. The reason to follow this "unorthodox" method is because in some cases the scalar–tensor (ST) method can lead to ill-defined potentials, and moreover because we want to circumvent any possible discussion concerning the use of *frames* (Einstein vs. Jordan). Debates of this sort plague the subject, some of which have only led to create confusion instead of shedding light.

With our technique we propose to treat the Ricci scalar itself as a degree of freedom, instead of using $\phi = f_R$ as in the ST method (hereafter a subindex R indicates ∂/∂_R). Our approach also spare us of inverting all the quantities depending on R for treating them as functions of ϕ. Moreover, we have found that in several specific applications the field equations can be recasted in a rather friendly way that allows us to treat them numerically or even analytically [14, 15]. In the next section we present our method and apply it to the Friedmann-Roberson-Walker (FRW) spacetime within the scope of analyzing the cosmological evolution using the *exponential gravity* model. The analysis of other viable $f(R)$ models using the current approach can be seen in [14] and in references therein using other techniques.

2 $f(R)$ Theories, the Ricci Scalar Approach

The action in $f(R)$ gravity is given by:

$$S[g_{ab}, \psi] = \int \frac{f(R)}{2\kappa} \sqrt{-g}\, d^4x + S_{\rm matt}[g_{ab}, \psi]. \tag{1}$$

where $\kappa \equiv 8\pi G_0$ (we use units where $c = 1$), $f(R)$ is a sufficiently differentiable but otherwise *a priori* arbitrary function of the Ricci scalar R, and ψ represents schematically the matter fields. The field equation obtained from (1) is:

$$f_R R_{ab} - \frac{1}{2} f g_{ab} - (\nabla_a \nabla_b - g_{ab}\Box)\, f_R = \kappa T_{ab}. \tag{2}$$

where f_R indicates $\partial_R f$, $\Box = g^{ab}\nabla_a\nabla_b$ is the covariant D'Alambertian and T_{ab} is the energy-momentum tensor (EMT) of matter associated with the ψ fields. From this equation it is straightforward to obtain the following equation and its trace [14, 15]

$$G_{ab} = \frac{1}{f_R}\left[f_{RR}\nabla_a\nabla_b R + f_{RRR}(\nabla_a R)(\nabla_b R) - \frac{g_{ab}}{6}\Big(R f_R + f + 2\kappa T\Big) + \kappa T_{ab}\right], \tag{3}$$

$$\Box R = \frac{1}{3 f_{RR}}\left[\kappa T - 3 f_{RRR}(\nabla R)^2 + 2f - R f_R\right], \tag{4}$$

where $(\nabla R)^2 := g^{ab}(\nabla_a R)(\nabla_b R)$ and $T := T^a_{\ a}$.

The idea is then to solve simultaneously (3) and (4) for the metric g_{ab} and R as a system of coupled partial differential equations.

It is important to mention that the field equations imply that the EMT of matter alone is conserved, i.e., it satisfies $\nabla_a T^{ab} = 0$.

In this contribution we shall focus on the model $f(R) = R_*[\tilde{R} - \lambda(1 - e^{-\tilde{R}})]$, referred to as *exponential gravity*, where $\tilde{R} = R/R_*$, λ is a positive dimensionless constant and $R_* > 0$ is also a constant that fixes the built-in scale and which is of the order of the current Hubble parameter H_0^2. This kind of exponential models have been analyzed in the past by several authors using a different technique [16–21]. Other variants of this model have also been analyzed [22]. The scalar $f_R = 1 - \lambda e^{-\tilde{R}}$ is positive provided $R > R_* \ln\lambda$ and this condition ensures $G_{\rm eff} := G_0/f_R > 0$. This latter is always satisfied in the cosmological solutions given below. The possible de Sitter points correspond to trivial solutions $R = R_1 = const.$ of (4) in vacuum ($T = 0$) and give rise to an effective cosmological constant $\Lambda_{\rm eff} = R_1/4$ in (3) ($G_{ab} = -g_{ab}\Lambda_{\rm eff}$ in vacuum). Here $R_1 > 0$ is a critical point of the "potential" $V(R)$ such that $V_R(R_1) = 0$ with $V_R := (2f - Rf_R)/3 = R_*[\tilde{R}(1 + \lambda e^{-\tilde{R}}) - 2\lambda(1 - e^{-\tilde{R}})]/3$, and $V_{RR} = 1 - \lambda(1 + \tilde{R})e^{-\tilde{R}}$. The "potential" is given by

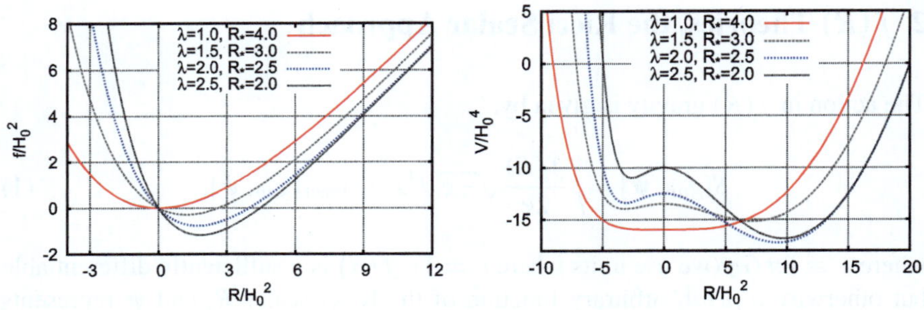

Fig. 1 $f(R)$ exponential gravity (*left panel*) and the potential $V(R)$ (*right panel*) for several values of λ

$V(R) = R_*^2[\tilde{R}(\tilde{R}-4\lambda) - 2\lambda(\tilde{R}+3)e^{-\tilde{R}})]/6$. For $0 < \lambda \leq 1$ one can easily see that $V(R)$ has just one critical point at $R = 0$ which is not a de Sitter point as in this case $\Lambda_{\rm eff} \equiv 0$. The point is a global minimum (c.f. $V_{RR}(0) = 1 - \lambda$). For $\lambda > 1$ there is a local maximum at $R = 0$ and a global minimum at $R = R_1 > 0$ which corresponds to the actual de Sitter point that the cosmological solution reaches asymptotically in the future. There is also a local minimum at $R = R_2 < 0$, but it is an anti de Sitter point which is never reached as the cosmological solutions take place only in the domain $R > 0$. The potential $V(R)$ is depicted in Fig. 1 (right panel) where one can appreciate the critical points just described. In the high curvature regime $\tilde{R} \gg 1$, we have $f(R) \approx R - \lambda R_*$, and thus the model acquires an effective cosmological constant $\Lambda_{\rm eff}^{\infty} := \lambda R_*/2$ (c.f. left panel of Fig. 1). From the figure we see that for λ sufficiently high, the de Sitter point verifies $\tilde{R}_1 \gg 1$, and thus $R_1 \approx 2\lambda R_*$ as it turns out from $V_R(R_1) = 0$. Therefore $\Lambda_{\rm eff} \approx \Lambda_{\rm eff}^{\infty}$. Finally, we stress that $f_{RR} = \lambda e^{-\tilde{R}}/R_* > 0$ which ensures that no singularities are found in the equations due to this scalar and moreover it guarantees that given $V_{RR}(R_1) > 0$ the effective mass $m^2 := V_{RR}/f_{RR} = (f_R - Rf_{RR})/(3f_{RR})|_{R_1}$ of the scalar mode is positive.

3 Cosmology in $f(R)$

We assume the spatially flat FRW metric given by:

$$ds^2 = -dt^2 + a^2(t)\left[dr^2 + r^2\left(d\theta^2 + \sin^2\theta d\varphi^2\right)\right]. \tag{5}$$

From (3) and (4) we have

$$\ddot{R} = -3H\dot{R} - \frac{1}{3f_{RR}}\left[3f_{RRR}\dot{R}^2 + 2f - f_R R + \kappa T\right], \tag{6}$$

$$H^2 = \frac{\kappa}{3}(\rho + \rho_X), \tag{7}$$

$$\dot{H} = -H^2 - \frac{\kappa}{6}\{\rho + \rho_X + 3(p_{\rm rad} + p_X)\}, \tag{8}$$

where a dot stands for d/dt and $H = \dot{a}/a$, is the Hubble expansion. In the above equations we have included the energy density ρ associated with matter (baryons, dark matter and radiation) as well as the geometric dark energy (GDE) density ρ_X and pressure p_X given respectively by Jaime et al. [14]

$$\rho_X = \frac{1}{\kappa f_R}\left\{\frac{1}{2}(f_R R - f) - 3f_{RR}H\dot{R} + \kappa\rho(1 - f_R)\right\}, \tag{9}$$

$$p_X = -\frac{1}{3\kappa f_R}\left\{\frac{1}{2}(f_R R + f) + 3f_{RR}H\dot{R} - \kappa(\rho - 3p_{\rm rad} f_R)\right\}. \tag{10}$$

Another differential equation that can be used to solve for H instead of (8) is given by $R = 6(\dot{H} + 2H^2)$. This latter is no other than the Ricci scalar computed directly from the metric (5). Equation (7) amounts to the modified Hamiltonian constraint which we use to set the initial data and also to monitor the accuracy of the numerical solutions at every integration step. At this regard, we stress that we shall not use the cosmic time t but instead $\alpha = \ln(a/a_0)$ as "time" parameter (see [14]), where a_0 is the present value of a. Notice that at the de Sitter point $R \to R_1 = const.$ where $2f(R_1) = R_1 f_R(R_1)$ and with $\rho \to 0$, $p_{\rm rad} \to 0$ (9) and (10), lead to $\rho_X \to \Lambda_{\rm eff}/\kappa$ and $p_X \to -\Lambda_{\rm eff}/\kappa$, respectively, and from (7) and (8), $H^2 \to H^2_{\rm vac} = \Lambda_{\rm eff}/3 = R_1/12$, and $q = -\ddot{a}/(aH^2) \to q_{\rm vac} = -1$. So the main idea behind all $f(R)$ models is that as the Universe evolves, $R \to R_1$, and thus the GDE dominates and mimics an effective cosmological constant that allows to explain the accelerated expansion required to account for the observations.

The matter variables obey the conservation equation $\dot{\rho}_i = -3H(\rho_i + p_i)$ for each fluid component (with $p_{\rm bar,DM} = 0$ and $p_{\rm rad} = \rho_{\rm rad}/3$) which integrates straightforwardly and gives rise to the usual expression for the energy density of matter plus radiation: $\rho = (\rho^0_{\rm bar} + \rho^0_{\rm DM})(a/a_0)^{-3} + \rho^0_{\rm rad}(a/a_0)^{-4}$, where the knotted quantities indicate their values today. The X–fluid variables (9) and (10) also satisfy a conservation equation similar to the one above, but with an EOS $\omega_X := p_X/\rho_X$ that evolves in cosmic time. Other possible inequivalent definitions of ρ_X, p_X and ω_X have been adopted in the past, but they suffer of several drawbacks (see [14] for a detailed discussion).

The total EOS is defined by $\omega_{\rm tot} = (p_{\rm rad} + p_X)/(\rho + \rho_X)$ which using (9) and (10) yields

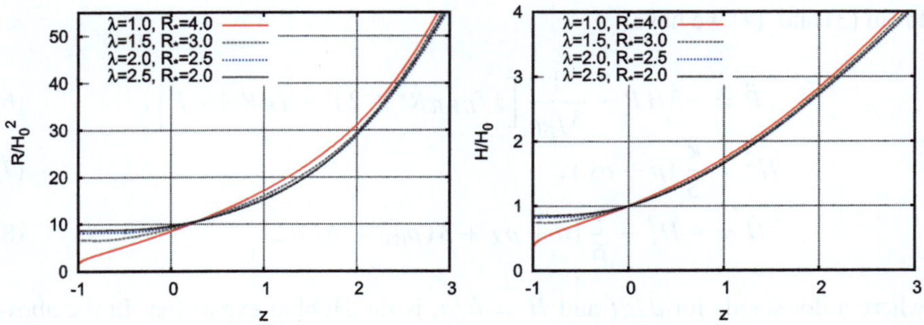

Fig. 2 Ricci scalar (*left panel*) and the Hubble expansion (*right panel*) for several values of λ and R_* (given in units of H_0^2)

$$\omega_{\text{tot}} = -\frac{1}{3}\left[\frac{\frac{1}{2}(f_R R + f) + 3 f_{RR} H \dot{R} - \kappa\rho}{\frac{1}{2}(f_R R - f) - 3 f_{RR} H \dot{R} + \kappa\rho}\right]. \tag{11}$$

This EOS allows us to track the epochs where the Universe is expanding in a decelerating or accelerating fashion. If $\omega_{\text{tot}} < -1/3$ then $\ddot{a} > 0$, while $\ddot{a} < 0$ if $\omega_{\text{tot}} > -1/3$.

4 Numerical Results and Discussion

We integrate the differential equations forward, from past to future, starting from a given $z = a_0/a - 1$, where matter dominates, to $z \to -1$ where the GDE prevails. The initial conditions are fixed as in [14]. Figure 2 shows the Hubble expansion and the Ricci scalar for several values of λ. In all the cases where $\lambda > 1$, R reaches the de Sitter point at the global minimum of $V(R)$. For $\lambda = 1$ because the potential is very flat around the global minimum at $R = 0$, and also due to the friction term, R varies very slowly as it approaches the minimum. This explains why in this case the model also mimics a cosmological constant. Figure 3 depicts the fraction of dimensionless densities $\Omega_i = \kappa\rho_i/(3H^2)$ which satisfy the constraint $\Omega_{\text{rad}} + \Omega_{\text{matt}} + \Omega_X = 1$, where $\Omega_{\text{matt}} := \Omega_{\text{bar}} + \Omega_{\text{DM}}$. The radiation contribution, although taken into account, is very small and cannot be appreciated from the plots. The current abundances at $z = 0$ (today) match reasonably well the predicted values of the ΛCDM model and the exponential models show an adequate matter domination era. The EOS of GDE is plotted in Fig. 4 (left panel), and like in other $f(R)$ models [14], it oscillates around the phantom divide value $\omega_\Lambda = -1$ before reaching its asymptotic value as $z \to -1$. The total EOS depicted in Fig. 4 (right panel) shows that at higher z the Universe is dominated by matter with $\omega_{\text{tot}} \sim 0$, and then interpolates to the value $\omega_{\text{tot}} = -1$ in the far future. At $z = 0$, ω_{tot} is similar to the value $\omega_{\text{tot}} \sim 0.75$ predicted

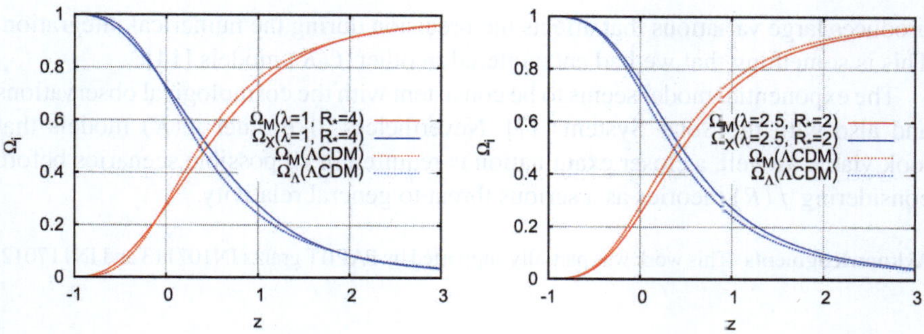

Fig. 3 Evolution of Ω_{matt} (*red solid line*) and Ω_X (*blue solid line*) for $\lambda = 1$ and $R_* = 4H_0^2$ (*left panel*) and $\lambda = 2.5$ and $R_* = 2H_0^2$ (*right panel*). For reference the corresponding quantities of the ΛCDM model are included in each panel (*dashed lines*)

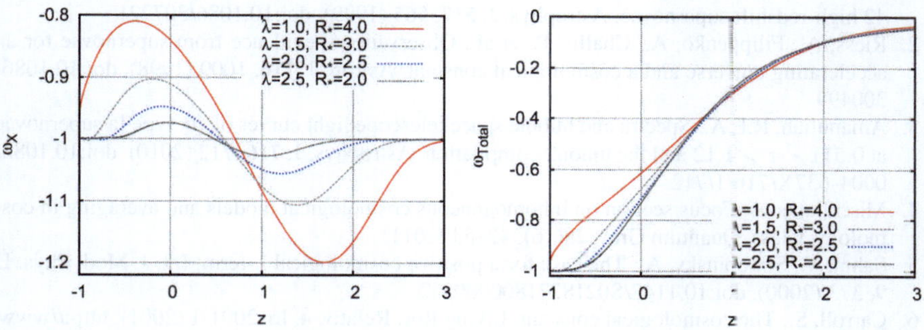

Fig. 4 The EOS ω_X (*left panel*) and the total EOS ω_{tot} (*right panel*)

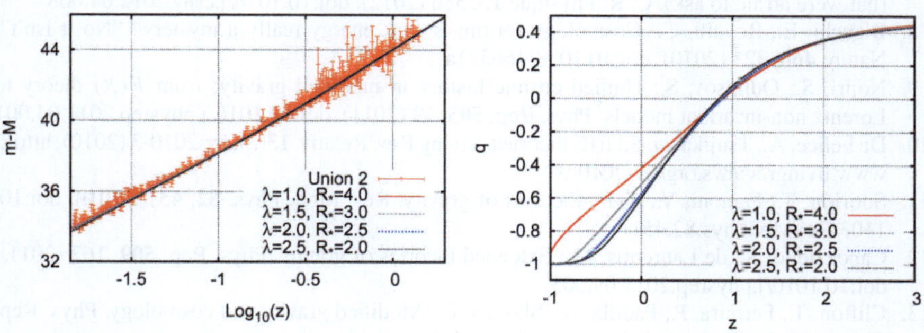

Fig. 5 Luminous distance (*left panel*) compared with the Union 2 data [3]. Deceleration parameter (*right panel*)

by the ΛCDM model. Figure 5 (left panel) shows the (modulus) luminous distance computed as in [14] and the deceleration parameter (right panel).

In these exponential models it is technically difficult to integrate far in the past since $f_{RR} \to 0$ exponentially. Since this quantity appears in the denominator of (6), it produces large variations that affects the precision during the numerical integration. This is something that we had encountered in other $f(R)$ models [14].

The exponential model seems to be consistent with the cosmological observations and also with the Solar System [17]. Nevertheless, like other $f(R)$ models that look viable as well, a closer examination is required in all possible scenarios before considering $f(R)$ theories as a serious threat to general relativity.

Acknowledgments This work was partially supported by PAPIIT grants IN107113 and IN117012.

References

1. Perlmutter, S., Aldering, G., Goldhaber, G., et al.: Measurements of omega and lambda from 42 high-redshift supernovae. Astrophys. J. **517**, 565 (1999). doi:10.1086/307221
2. Riess, A., Filippenko, A., Challis, P., et al.: Observational evidence from supernovae for an accelerating universe and a cosmological constant. Astron. J. **116**, 1009 (1998). doi:10.1086/300499
3. Amanullah, R.E.A.: Spectra and hubble space telescope light curves of six type Ia supernovae at $0.511 < z > 1.12$ and the union2 compilation. Astrophys. J. **716**, 712 (2010). doi:10.1088/0004-637X/716/1/712
4. Miscellaneous: Focus section on inhomogeneous cosmological models and averaging in cosmology. Class. Quantum Grav. **28**(16), 42–53 (2011)
5. Sahni, V., Starobinsky, A.: The case for a positive cosmological λ-term. Int. J. Mod. Phys. D **9**, 373 (2000). doi:10.1142/S0218271800000542
6. Carroll, S.: The cosmological constant. Living Rev. Relativ. **4**, lrr-2001-1 (2001). http://www.livingreviews.org/lrr-2001-1
7. Martin, J.: Everything you always wanted to know about the cosmological constant problem (but were afraid to ask). C. R. Physique **13**, 566 (2012). doi:10.1016/j.crhy.2012.04.008
8. Bianchi, E., Rovelli, C.: Cosmology forum: is dark energy really a mystery? "No it isn't". Nature **466**, 321 (2010). doi:10.1038/466321a
9. Nojiri, S., Odintsov, S.: Unified cosmic history in modified gravity: from $F(R)$ theory to Lorentz non-invariant models. Phys. Rep. **505**, 59 (2011). doi:10.1016/j.physrep.2011.04.001
10. De Felice, A., Tsujikawa, S.: f(R) theories. Living Rev. Relativ. **13**(3), lrr-2010-3 (2010). http://www.livingreviews.org/lrr-2010-3
11. Sotiriou, T., Faraoni, V.: $f(R)$ theories of gravity. Rev. Mod. Phys. **82**, 451 (2010). doi:10.1103/RevModPhys.82.451
12. Capozziello, S., de Laurentis, M.: Extended theories of gravity. Phys. Rep. **509**, 167 (2011). doi:10.1016/j.physrep.2011.09.003
13. Clifton, T., Ferreira, P., Padilla, A., Skordis, C.: Modified gravity and cosmology. Phys. Rep. **513**, 1 (2012). doi:10.1016/j.physrep.2012.01.001
14. Jaime, L., Patino, L., Salgado, M.: $f(R)$ cosmology revisited. ArXiv e-prints arXiv:1206.1642 [gr-qc] (2012)
15. Jaime, L., Patiño, L., Salgado, M.: Robust approach to $f(R)$ gravity. Phys. Rev. D **83**, 024039 (2011). doi:10.1103/PhysRevD.83.024039
16. Cognola, G., Elizalde, E., Nojiri, S., et al.: Class of viable modified $f(R)$ gravities describing inflation and the onset of accelerated expansion. Phys. Rev. D **77**, 046009 (2008). doi:10.1103/PhysRevD.77.046009
17. Linder, E.: Exponential gravity. Phys. Rev. D **80**, 123528 (2009). doi:10.1103/PhysRevD.80.123528
18. Yang, L., Lee, C.C., Luo, L.W., Geng, C.Q.: Observational constraints on exponential gravity. Phys. Rev. D **82**, 103515 (2010). doi:10.1103/PhysRevD.82.103515

19. Bamba, K., Geng, C.Q., Lee, C.C.: Cosmological evolution in exponential gravity. J. Cosmol. Astropart. Phys. **2010**(08), 021 (2010). doi:10.1088/1475-7516/2010/08/021
20. Elizalde, E., Nojiri, S., Odintsov, S., Sebastiani, L., Zerbini, S.: Nonsingular exponential gravity: a simple theory for early- and late-time accelerated expansion. Phys. Rev. D **83**, 086006 (2011). doi:10.1103/PhysRevD.83.086006
21. Elizalde, E., Odintsov, S., Sebastiani, L., Zerbini, S.: Oscillations of the F(R) dark energy in the accelerating universe. Eur. Phys. J. C **72**, 1843 (2012). doi:10.1140/epjc/s10052-011-1843-7
22. Zhang, P.: Testing gravity against the early time integrated Sachs-Wolfe effect. Phys. Rev. D **73**, 123504 (2006). doi:10.1103/PhysRevD.73.123504

Regular and Chaotic Motion in General Relativity: The Case of a Massive Magnetic Dipole

Ondřej Kopáček, Jiří Kovář, Vladimír Karas and Yasufumi Kojima

Abstract Circular motion of particles, dust grains and fluids in the vicinity of compact objects has been investigated as a model for accretion of gaseous and dusty environment. Here we further discuss, within the framework of general relativity, figures of equilibrium of matter under the influence of combined gravitational and large-scale magnetic fields, assuming that the accreted material acquires a small electric charge due to interplay of plasma processes and photoionization. In particular, we employ an exact solution describing the massive magnetic dipole and we identify the regions of stable motion. We also investigate situations when the particle dynamics exhibits the onset of chaos. In order to characterize the measure of chaoticness we employ techniques of Poincaré surfaces of section and of recurrence plots.

1 Introduction

This work represents a continuation of our steady effort [1–3] to understand dynamic properties of charged test particles being exposed to the simultaneous action of strong gravitational and electromagnetic fields surrounding compact objects – neutron stars and black holes. As we bear astrophysical motivation in our mind we choose such fields which could constitute a reasonable model of a real situation occurring in the vicinity of these objects. Survey of the test particle trajectories might be regarded as

O. Kopáček (✉) · V. Karas
Astronomical Institute, Academy of Sciences, Boční II 1401/1a, 14131 Prague, Czech Republic
e-mail: kopacek@ig.cas.cz

J. Kovář
Institute of Physics, Faculty of Philosophy and Science, Silesian University in Opava, Bezručovo nám. 13, 74601 Opava, Czech Republic

Y. Kojima
Department of Physics, Hiroshima University, Higashi-Hiroshima 739-8526, Japan

J. Bičák and T. Ledvinka (eds.), *Relativity and Gravitation*, Springer Proceedings in Physics 157, DOI: 10.1007/978-3-319-06761-2_52,

a single particle approximation to the complex dynamics of the astrophysical plasma which is applicable once the plasma is diluted sufficiently. Regions of diluted plasma are likely to be found above and below the main accretion body of astrophysical systems driven by compact objects.

In this contribution (which is based mainly on results previously published in [4]) we investigate the motion of the charged test particles around a massive magnetic dipole described by Bonnor's exact solution of coupled Einstein-Maxwell equations [5]. Such setup allows motion in the off-equatorial lobes if the parameters of the system are chosen carefully. We investigate motion in these lobes. We are particularly curious about the dynamic regime of motion (chaotic versus regular) and how does it change if we alter some of the parameters. Besides the standard technique of Poincaré surfaces of section we employ the recurrence analysis [6] and show that recurrence plots might serve as an alternative tool to the surfaces of section when analysing individual trajectories.

2 Massive Magnetic Dipole

Using spheroidal coordinates (t, r, θ, ϕ) and geometrized units $c = G = 1$ the line element of Bonnor's exact solution [5] describing the static spacetime around massive magnetic dipole and corresponding vector potential A_α are given as follows

$$ds^2 = -\left(\frac{P}{Y}\right)^2 \mathrm{d}t^2 + \frac{P^2Y^2}{Q^3Z}(\mathrm{d}r^2 + Z\mathrm{d}\theta^2) + \frac{Y^2Z\sin^2\theta}{P^2}\mathrm{d}\phi^2, \tag{1}$$

$$A_\alpha = \left(0, 0, 0, \frac{2abr\sin^2\theta}{P}\right), \tag{2}$$

where $P = r^2 - 2ar - b^2\cos^2\theta$, $Q = (r-a)^2 - (a^2+b^2)\cos^2\theta$, $Y = r^2 - b^2\cos^2\theta$ and $Z = r^2 - 2ar - b^2$.

The solution is characterized by two independent parameters a and b. Inspection of the asymptotic behaviour of the field reveals that these are related to the total mass of the source M as $M = 2a$ and to the magnetic dipole moment μ as $\mu = 2ab$. The solution has relatively complicated singular behviour at $P = 0$, $Q = 0$, $Z = 0$ and $Y = 0$. However, here we are interested in the regular part of the spacetime only. Therefore we restrict ourselves to $Z > 0$ which translates to the condition $r > r_{\mathrm{h}} \equiv a + \sqrt{a^2 + b^2}$. We investigate the test particle dynamics above the horizon r_{h} only.

The solution is asymptotically flat (for $a = 0$ exactly flat). The metric (1) actually represents a magnetostatic limit of a more general exact solution [7] suggested to describe the exterior field of a rotating neutron star. In the case of Bonnor's solution, the rotation is not considered and the value of a quadrupole mass moment is fixed by values of the parameters a and b. Setting $b = 0$ reduces the metric to Zipoy-Voorhees metric with $\delta = 2$ [8, 9].

Generalized Hamiltonian ("super Hamiltonian") describing the motion of the ionised test particle of charge $\tilde{q}$ is given as follows [10]:

$$\mathscr{H} = \frac{1}{2} g^{\mu\nu}(\pi_\mu - \tilde{q} A_\mu)(\pi_\nu - \tilde{q} A_\nu), \tag{3}$$

where π_μ is the generalized (canonical) momentum.

Hamiltonian equations of motion are given in a standard way:

$$\mathrm{d}x^\mu/\mathrm{d}\lambda = \partial\mathscr{H}/\partial\pi_\mu, \quad d\pi_\mu/d\lambda = -\partial\mathscr{H}/\partial x^\mu, \tag{4}$$

where $\lambda = \tau/m$ is the affine parameter, τ the proper time and m represents the rest mass of the particle.

The second Hamilton's equation ensures that the momenta

$$\pi_t = p_t + \tilde{q} A_t \equiv -\tilde{E}, \tag{5}$$

$$\pi_\phi = p_\phi + \tilde{q} A_\phi \equiv \tilde{L}, \tag{6}$$

represent constants of motion, reflecting stationarity and axial symmetry of the considered background.

Numerical integration of Hamilton's equations (4) is carried out using the multistep Adams-Bashforth-Moulton solver of variable order. In several cases when higher accuracy is demanded we employ 7-8th order Dormand-Prince method. Initial values of non-constant components of the canonical momentum $\pi_r(0)$ and $\pi_\theta(0)$ are obtained from $u^r(0)$ (which we set) and $u^\theta(0)$ which is calculated from the normalization condition $g^{\mu\nu}u_\mu u_\nu = -1$ where we always choose the non-negative root as a value of $u^\theta(0)$.

Two-dimensional (i.e. related to the motion in two coordinates, r and θ) effective potential may be expressed as follows:

$$V_{\mathrm{eff}}^2 = \frac{P^2}{Y^2}\left[1 + \frac{P^2}{Y^2 Z \sin^2\theta}\left(L - q\mu\frac{r \sin^2\theta}{P}\right)^2\right], \tag{7}$$

where we introduce specific angular momentum $L \equiv \tilde{L}/m$ and specific charge $q \equiv \tilde{q}/m$.

3 Dynamics of Test Particles

We perform a numerical survey of dynamics of test particles moving within potential wells formed along both the equatorial and off-equatorial circular orbits (so called halo orbits, illustrated by Fig. 1). In order to do so, we apply several complementary methods of investigation of nonlinear dynamic systems. First of all, we construct

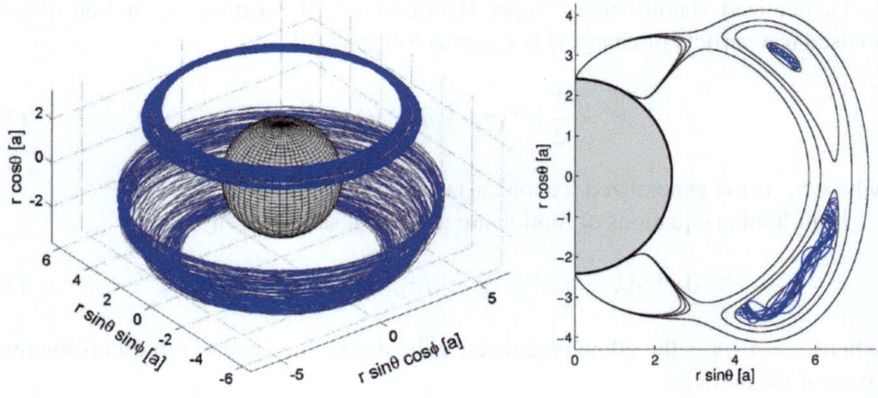

Fig. 1 Off-equatorial trajectories of charged test particle with $L = -2.356\,a$ and $q = 5.581$ in the Bonnor spacetime with $b = 1\,a$. In the left panel we present a stereometric projection of two trajectories: the upper one with $E = 0.8169$ shows ordered motion while with the higher energy the dynamics acquires properties of deterministic chaos (bottom trajectory with $E = 0.8182$). Poloidal projection of these trajectories along with several iso-contours of the effective potential is shown in the right panel. Both particles were launched at $r(0) = 6\,a$, $\theta(0) = \pi/3$ with $u^r(0) = 0$. Grey color marks $r = r_h$ surface in both plots

Poincaré surfaces of section which give an overall perspective of the phase space dynamics on a given energy hypersurface (for a given values of system parameters). For the inspection of individual trajectories, however, we prefer to analyse their recurrence plots [6], which proved to be very useful method in our previous work [1]. Besides other properties of recurrence plots (RPs), we highlight their ability to clearly distinguish between chaotic and regular dynamics on a short time scale, thus reducing the integration time needed for the analysis. We also investigate intrinsic frequencies of the orbits employing the rotation number ν [11] that allows us to detect and locate resonances of the system.

In order to compare dynamic properties of the system in all three cases (particle in a non-magnetized $b = 0$ spacetime, uncharged particle in a magnetized spacetime, and a general case $q \neq 0$, $b \neq 0$), we first investigate the motion in equatorial potential wells, since there are no circular halo orbits for $b = 0$. In Fig. 2, we present series of Poincaré surfaces of section with $\theta_{\rm sec} = \pi/2$ along with the corresponding plots of a rotation number as a function of initial value of radial coordinate.

The upper left panel of Fig. 2 shows the Poincaré surface of the test particles trajectories when the magnetic dipole is switched off by setting $b = 0$. We observe perfectly ordered motion with no traces of secondary fixed points nested in Birkhoff islands nor the chaotic orbits. Such a simple pattern on the section is characteristic for integrable systems. The integrability conjecture is also supported by the behaviour of rotation number, which is smooth and non-constant throughout the lobe. While we show roughly 50 trajectories on the section, the ν-plot is constructed from around 1000 of them, which allows much more detailed inspection for the possible presence of tiny chaotic domains or faint resonances.

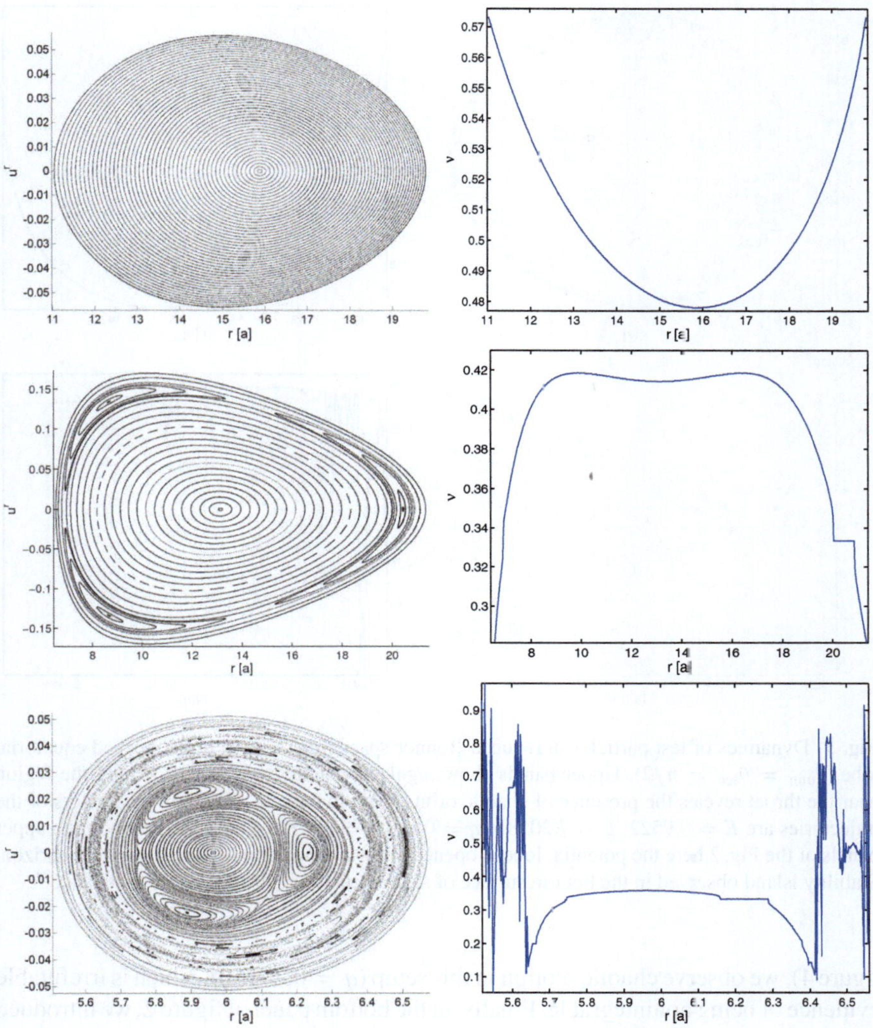

Fig. 2 Dynamics in equatorial lobes ($\theta_{\min} = \theta_{\text{sec}} = \pi/2$). Left panels show equatorial surfaces of section while the corresponding rotation curves are shown on the right. Common parameters of the trajectories in the top panels are $E = 0.951$, $L = 7.2058\,a$, $q = 0$ and $b = 0$ for which the potential minimum appears at $r_{\min} = 15\,a$ with $V_{\min} = 0.9494$. Middle panels show the situation for particles with $E = 0.94$, $L = 6.1076\,a$, $q = 0$ and $b = 2.8535\,a$ which brings the stable circular orbit to $r_{\min} = 10\,a$ with $E_{\min} = 0.9234$. Bottom panels are plotted for $E = 0.818$, $L = -2.7277\,a$, $q = 4.7181$ and $b = 1\,a$ ($V_{\min} = 0.8165$ at $r_{\min} = 6\,a$)

If we perturb the system by the magnetic field, however, the chain of Birkhoff islands develops. Although we know that there are some integrable system with resonant islands of single multiplicity [11], here its presence arouses suspicion of nonintegrability, since none were present for $b = 0$. Indeed, in the following (see

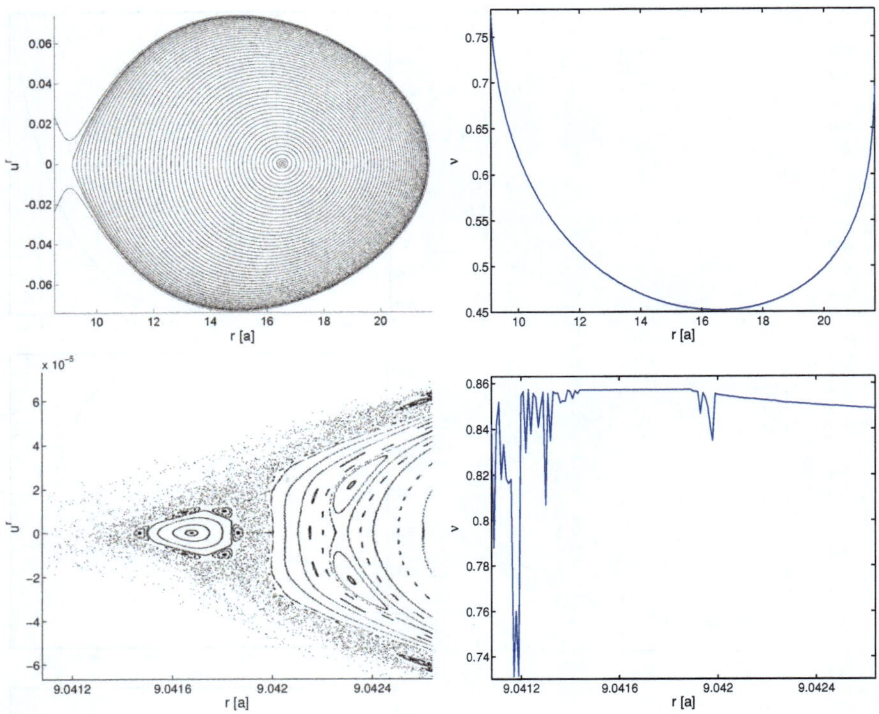

Fig. 3 Dynamics of test particles in reduced Bonnor spacetime ($b = 0$) in the opened equatorial lobe ($\theta_{\min} = \theta_{\text{sec}} = \pi/2$). Upper panels show regular dynamics, however, zooming the region near the throat reveals the presence of chaotic orbits (bottom panels). Common parameters of the trajectories are $E = 0.9522$, $L = 7.2058\,a$, $q = 0$ and $b = 0$. Unlike the case shown in the upper panels of the Fig. 2 here the potential lobe is opened allowing the particles to fall onto the horizon. Stability island observed in the bottom surface of section corresponds to $\nu = 6/7$ resonance

figure 4), we observe chaotic motion in this setup ($q = 0, b \neq 0$), which is irrefutable evidence of being nonintegrable. Finally, in the bottom panel of figure 2, we introduce the charge of the test particle. We choose such a combination of parameters that leads to the same value of ratio $r_{\min}/r_{\text{h}}$, as it acquired in the previous uncharged case. This makes the two cases better comparable and the effect of the newly introduced electromagnetic forces more distinguishable. In the last surface of section, we really observe much more complex patterns compared to those of uncharged particles. KAM curves of quasiperiodic orbits are present as well as several Birkhoff chains of islands corresponding to the resonances of intrinsic frequencies of the system. These are interwoven with pronounced chaotic layers. Such a picture is typical for a considerably perturbed system far from integrability.

However, further examination of the dynamics in opened potential lobes in the non-magnetized system presented in Fig. 3 reveals presence of narrow zones of chaotic orbits which proves the system nonintegrable. These chaotic orbits correspond to those particles which actually leave the potential well after certain amount

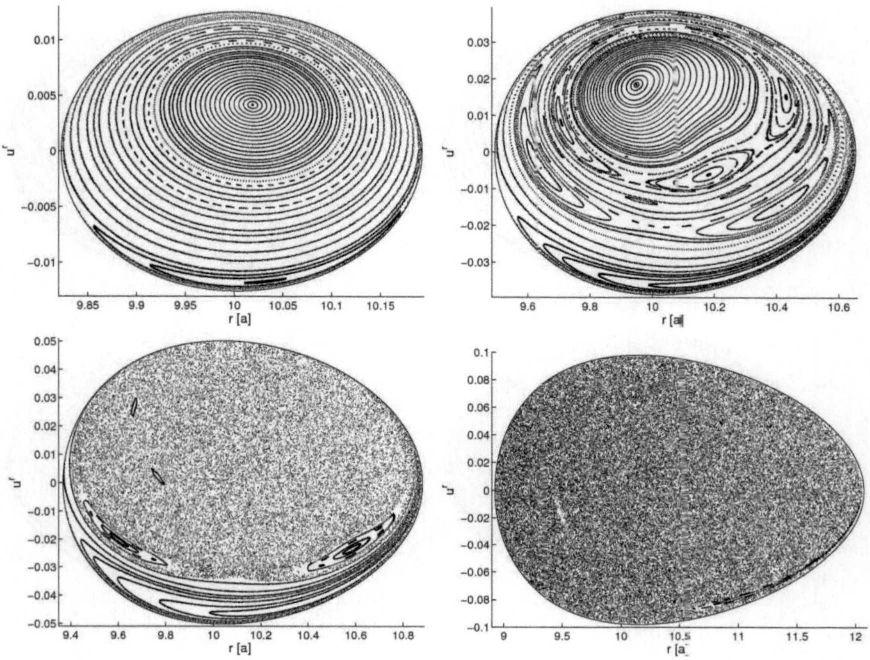

Fig. 4 Poincaré surfaces ($\theta_{\rm sec} = \pi/3$) for uncharged particles moving within halo lobes in the Bonnor spacetime with $b = 5.9771\,a$. Particles with $L = 3.6743\,a$ are launched from the vicinity of the potential minimum ($r_{\rm min} = 10\,a$, $\theta_{\rm min} = \theta_{\rm sec} = \pi/3$ and $V_{\rm min} = 0.8717$) with various values of energy. The *upper left panel* shows the section for the level $E = 0.8718$ (small halo lobe), in the *upper right* we set $E = 0.873$ (large halo lobe), $E = 0.8739$ produces cross-equatorial lobe which just emerged from symmetric halo lobes (*bottom left panel*), while with $E = 0.88$ we obtain the large cross-equatorial lobe which almost opens

of time. Coincidentally, the nonintegrability of motion in a general Zipoy-Voorhees spacetime was very recently shown by Lukes-Gerakopoulos in [12] where the issue was treated in detail. We summarize that although the dynamics in the non-magnetized Bonnor spacetime (i.e. Zipoy-Voorhees with $\delta = 2$) is typically regular (we actually found no chaotic orbit in closed lobes) the underlying system is not integrable.

In the following, we compare dynamics in off-equatorial potential wells for uncharged (Fig. 4) and charged particles (Fig. 5). For the sake of better comparability, these are both chosen to have $\theta_{\rm min} = \theta_{\rm sec} = \pi/3$ and equal value of $r_{\rm min}/r_h$. Both series show Poincaré surfaces of section of particles being launched from the vicinity of off-equatorial potential minima (in which the circular halo orbit resides) differing in energy E, which governs the size and shape of the lobe (E sets the level at which the effective potential surface is being intersected). Comparing Figs. 4 and 5 we conclude that charging the test particle makes it more prone to chaotic dynamics and we also observe that the energy of the particle triggers chaotic motion.

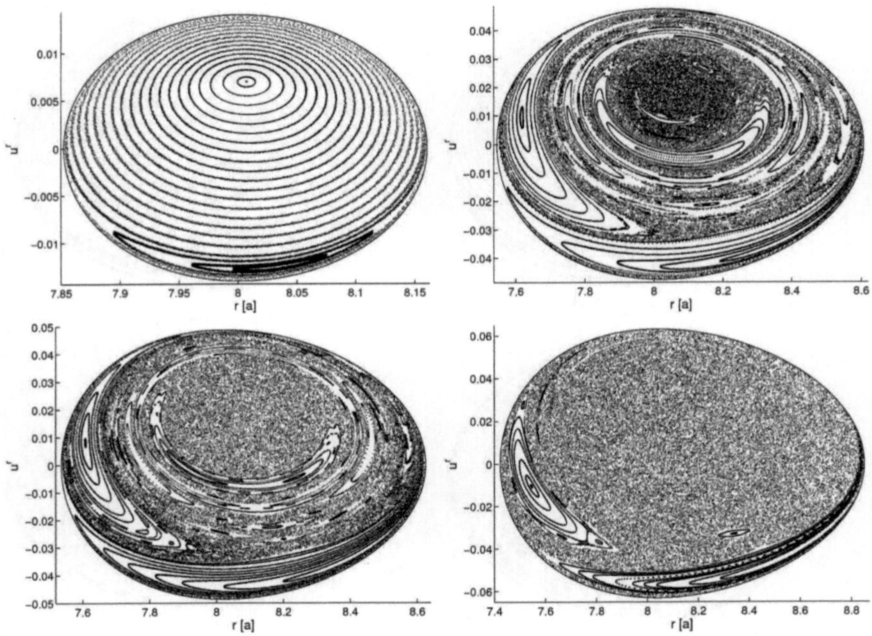

Fig. 5 Poincaré surfaces of section ($\theta_{\rm sec} = \pi/3$) for charged particles with $q = 0.1259$ moving within halo lobes in the Bonnor spacetime with $b = 4.5393\,a$. Particles with $L = -3.5486\,a$ are launched from a vicinity of the potential minimum ($r_{\rm min} = 8\,a$, $\theta_{\rm min} = \theta_{\rm sec} = \pi/3$ and $V_{\rm min} = 0.8475$) with various values of energy. The upper left panel shows the section for the level $E = 0.8477$ (small halo lobe), in the *upper right* we set $E = 0.8495$ (large halo lobe), $E = 0.8496$ produces cross-equatorial lobe which just emerged from symmetric halo lobes (*bottom left panel*) while with $E = 0.851$ we obtain the large cross-equatorial lobe

To illustrate the continuous transition from ordered to chaotic dynamics, we pick a particular trajectory of charged particle and plot series of Poincaré sections along with corresponding recurrence plots in Fig. 6. We launch the particle from the locus of the off-equatorial potential minimum with various values of energy E, while other parameters remain fixed. The sequence begins with the energy corresponding to a small halo lobe where we observe ordered motion manifested by narrow curves on the surface of section and, simple diagonal pattern of the recurrence plot (upper left panels of Fig. 6). Increasing the energy, however, gradually shifts the dynamics towards deterministic chaos—trajectory becomes more and more ergodic as it spans larger fraction of given energy hypersurface in the phase space.

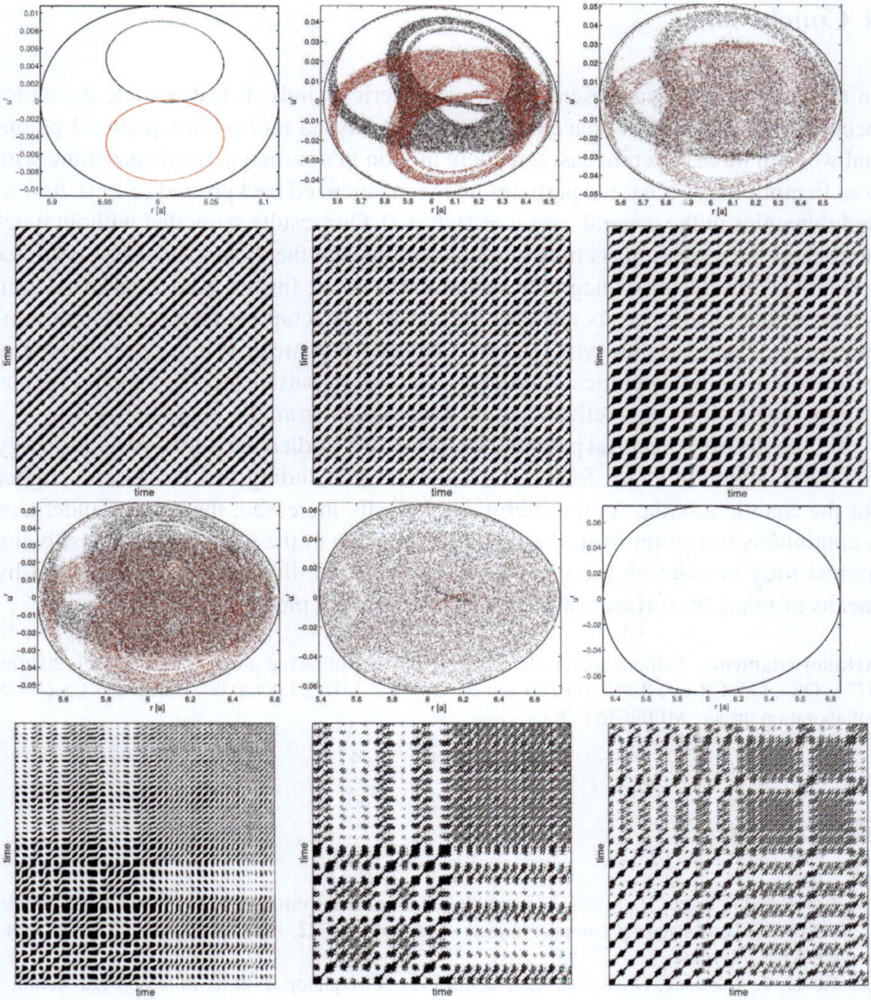

Fig. 6 Poincaré surfaces of section ($\theta_{\text{sec}} = \pi/3$) and corresponding recurrence plots for charged particle with $q = 5.581$ moving within halo lobes in the Bonnor spacetime with $b = 1\,a$. Particle with $L = -2.356\,a$ is launched just from the locus of the off-equatorial potential well minimum ($r_{\min} = 6\,a$, $\theta_{\min} = \theta_{\text{sec}} = \pi/3$ and $V_{\min} = 0.81675$) with various values of energy (from the upper left to the bottom right: $E = 0.8168$, $E = 0.818$, $E = 0.8182$, $E = 0.8183$, $E = 0.819$ and $E = 0.8198$). First three pairs of plots show the situation in the halo lobe, while bottom plots reveal the dynamics after merging of the lobes. A decisive surface of section cannot be constructed for the particle in an opened lobe ($E = 0.8198$) as it escapes after several intersections with the surface, while the corresponding RP shows the chaotic nature of the motion unambiguously. Unlike the previous figures, we plot all intersection points: downward crossings with $u^{\theta} \geq 0$ (*black dot*) as well as those resulting from upward crossings with $u^{\theta} < 0$ (*red dot*) on the Poincaré surfaces

4 Conclusions

In this contribution we presented a brief numerical study of test particle dynamics occurring in the Bonnor spacetime. First we analysed motion in equatorial potential wells in three different cases, namely motion in non-magnetized spacetime with $b = 0$, motion of uncharged particles on the magnetized background ($q = 0, b \neq 0$) and dynamics in the general case $q \neq 0$, $b \neq 0$. Our results show that without magnetic field the system hosts mostly regular orbits and the dynamics of test particles resembles closely fully integrable systems. However, further numerical inspection revealed that chaotic orbits are also present in this setup proving the system non-integrable. Then we observed that the magnetic parameter b introduces profound perturbation of the dynamics. Moreover, a charge of particle acts as an extra perturbation, which shifts magnetized system even farther from the integrability.

Within the off-equatorial potential wells we also studied the role of particle energy E on the degree of chaos found in the system, concluding that it acts as a trigger for the chaotic motion. As the energy is gradually increased, the system undergoes a continuous transition from the regular behaviour to the chaotic dynamics, being almost fully ergodic on the given hypersurface. We illustrated such transition by means of Poincaré surfaces of section and recurrence plots.

Acknowledgments Authors appreciate support from the following projects: GA ČR ref. 202/09/0772 (OK), GA ČR ref. P209/10/P190 and Synergy CZ.1.07/2.3.00/20.0071 (JK) and Czech-US collaboration project ME09036 (VK).

References

1. Kopáček, O., Karas, V., Kovář, J., Stuchlík, Z.: Transition from regular to chaotic circulation in magnetized coronae near compact objects. Astrophys. J. **722**, 1240 (2010). doi:10.1088/0004-637X/722/2/1240
2. Kovář, J., Kopáček, O., Karas, V., Stuchlík, Z.: Off-equatorial orbits in strong gravitational fields near compact objects II: halo motion around magnetic compact stars and magnetized black holes. Class. Quantum Grav. **27**, 135006 (2010). doi:10.1088/0264-9381/27/13/135006
3. Kovář, J., Stuchlík, Z., Karas, V.: Off-equatorial orbits in strong gravitational fields near compact objects. Class. Quantum Grav. **25**, 095011 (2008). doi:10.1088/0264-9381/25/9/095011
4. Kovář, J., Kopáček, O., Karas, V., Kojima, Y.: Regular and chaotic orbits near a massive magnetic dipole. Class. Quantum Grav. **30**, 025010 (2013). doi:10.1088/0264-9381/30/2/025010
5. Bonnor, W.: An exact solution of the Einstein-Maxwell equations referring to a magnetic dipole. Z. Phys. **190**, 444 (1966). doi:10.1007/BF01327262
6. Marwan, N., Carmen, M.: Romano, M. Thiel, J. Kurths, Recurrence plots for the analysis of complex systems. Phys. Rep. **438**, 237 (2007). doi:10.1016/j.physrep.2006.11.001
7. Pachón, L., Rueda, J., Sanabria-Gómez, J.: Realistic exact solution for the exterior field of a rotating neutron star. Phys. Rev. D **73**, 104038 (2006). doi:10.1103/PhysRevD.73.104038
8. Zipoy, D.: Topology of some spheroidal metrics. J. Math. Phys. **7**, 1137 (1966). doi:10.1063/1.1705005
9. Voorhees, B.: Static axially symmetric gravitational fields. Phys. Rev. D **2**, 2119 (1970). doi:10.1103/PhysRevD.2.2119
10. Misner, C., Thorne, K., Wheeler, J.: Gravitation. Freeman, San Francisco (1973)

11. Contopoulos, G.: Order and Chaos in Dynamical Astronomy. Astronomy and Astrophys Library. Springer, Berlin (2002)
12. Lukes-Gerakopoulos, G.: Nonintegrability of the Zipoy-Voorhees metric. Phys. Rev. D **86**, 044013 (2012). doi:10.1103/PhysRevD.86.044013

The Fitting Problem in a Lattice Universe

Julien Larena

Abstract We present a regular cubic lattice solution to Einstein field equations that is exact at second order in a small parameter. We show that this solution is kinematically equivalent to the Friedmann-Lemaître-Robertson-Walker (FLRW) solution with the same averaged energy density. This allows us to discuss the fitting problem in that framework: are observables along the past lightcone of observers equivalent to those in the analogue FLRW model obtained by smoothing spatially the distribution of matter? We find a criterion on the compacity of the objects that must be satisfied in order for the answer to this question to be positive and given by perturbative arguments. If this criterion is not met, the answer to this question must be addressed fully non perturbatively along the past lightcone, even though the spacetime geometry can be described perturbatively.

1 What is the Fitting Problem?

Cosmology is unique among physical sciences for a certain number of reasons. First, the Universe is given once and for all, and there is no possibility to compare it to another Universe. This can usually be overcome by supposing that the initial conditions for the Big Bang model must be generic in some reasonable sense, or that some mechanism (e.g. inflation) is responsible for making them generic. Second, we are the only available observers in the Universe; there might be other observers, but we do not have access to their observations. The only piece of information we have on the Universe comes from our past lightcone, and a few local (geological) measurements on our worldline. Therefore, in general, one cannot rely purely on observations to fully determine the nature and dynamics of the Universe: one has to introduce extra assumptions on the theory of gravitation, the geometry of the Universe on large scales

J. Larena (✉)
Department of Mathematics, Rhodes University, Grahamstown 6140, South Africa
e-mail: j.larena@ru.ac.za

J. Bičák and T. Ledvinka (eds.), *Relativity and Gravitation*,
Springer Proceedings in Physics 157, DOI: 10.1007/978-3-319-06761-2_53,

and the physical nature of its matter content. In the present paper, we will suppose throughout that gravity obeys the laws of General Relativity, and we will concentrate on the other two points: the geometry of the Universe and its matter content. In the standard model of cosmology, it is assumed that 'on average', on sufficiently large scales, the distribution of matter in the Universe is well described by a set of perfect fluids whose energy densities and pressures are locally homogeneous and isotropic; this results in Friedmann-Lemaître-Robertson-Walker (FLRW) Universes with spatial sections that are maximally symmetric, i.e. determined entirely by their constant Gaussian curvature. This assumption is based on the observed almost isotropy of the Cosmic Microwave Background around us, together with the Copernican Principle, and is usually called the Cosmological Principle. It is clear that it is an extrapolation outside our past lightcone, since the notion of average implicitly present in this principle tells us something about the spatial distribution of matter, starting from its distribution along our past lightcone. In a nutshell, the fitting problem [1] can be summarized by the question: does the effective FLRW model obtained by extrapolating the observed properties down our past lightcone coincide with the effective FLRW model obtained by smoothing the spatial distribution of matter? Of course, this question is not independent on the set of observers used to define the notion of spatial distribution: it makes use of a preferred set of observers, called fundamental in the standard model; usually, in the late-time Universe, the fundamental observers are supposed to be comoving with virialised objects such as galaxies, so as to include us among fundamental observers. In this paper, we will try and address the fitting problem by considering a special dynamical solution to the field equations consisting in a regular cubic lattice of initial cell size L and of objects of equal masses M. The solution is exact at second order in the small parameter $\sqrt{M/L}$ and we will see that it exhibits, on average, the same dynamical behaviour than the equivalent flat FLRW model with a non-relativistic fluid of density $\rho = M/L^3$, therefore showing that, at second order in $\sqrt{M/L}$, there is no backreaction in the model [2]. Then, because the solution for the metric is exact at this order, we will be able to solve the Sachs equations at second order in $\sqrt{M/L}$ in order to reconstruct observables such that the distance-redshift relation. We will see that this solution for observables presents some divergences linked with the compacity of the object: if the extension η of the objects is too small, the perturbative expansion of the solution of the Sachs equations is no longer valid, even though the perturbative expansion of the solution of the Einstein field equations remains stable. Namely, we will show that observables in this model remain very close to the observables calculated in the analogue FLRW model with energy density $\rho = M/L^3$, as long as the parameters of the lattice obey: $\frac{M}{L} \ll \mathcal{O}(1)\left(\frac{\eta}{L}\right)^4$. If this condition is not satisfied, then observables cannot be calculated perturbatively, even though the metric is well approximated by the perturbative expansion, and one must solve the full system of Sachs equations [3]. Results regarding this complete integration are hinted at in this paper. These results illustrate the importance of the fitting problem in cosmology: the kinematically averaged FLRW model and the FLRW reconstructed by fitting observations might differ significantly (if calculated perturbatively, at least), even in the absence of (kinematical) backreaction. We work in units $G = c = 1$.

2 A Lattice Universe: Kinematics and Observables

2.1 The Cubic Lattice Solution

Let us start by describing the lattice solution (see [2] for a complete derivation); we will only sketch the results and discuss their implications. We start with a cubic lattice of size L with identical masses M at the centre of each cell. If the masses on the lattice are to represent typical galaxies, we can choose, as our typical parameters $M \sim 10^{11} M_{\odot}$ and $L \sim 1$ Mpc, where $M_{\odot} \sim 10^{30}$ kg is the Solar mass, and L is of order of the intergalactic distances. Then, the natural parameter of the lattice is $R_S/L \sim 10^{-8} \ll 1$, where $R_S = 2M$ is the Schwarzschild radius of the masses. Therefore, we can look for a solution expanded into powers of $\sqrt{M/L}$ (in [2] we prove that there is no perturbative solution in powers of M/L); this will lead to linearised field equations that can be solved exactly. We choose coordinates that are comoving with the masses: $g_{00} = -1$, and spatial coordinates are Cartesian and adapted to the symmetries of the lattice. The distribution of matter is therefore a three dimensional Dirac comb with the masses located at $\mathbf{x_n} = L\mathbf{n}$, $\mathbf{n} \in \mathbb{Z}^3$; the energy momentum tensor is then: $T_{ab} = T_{00}\delta_a^0\delta_b^0$ such that:

$$T_{00} \propto M \sum_{\mathbf{n}\in\mathbb{Z}^3} \delta^{(3)}(\mathbf{x} - L\mathbf{n}).$$

Actually, the field equations do not have a solution for such a source term [2, 4], because the formal series solution presents a UV divergence coming from the point-like nature of the masses. Therefore, we introduce a UV cut-off by giving a small but finite extension to the masses, η and by replacing the Dirac deltas by their standard approximation: $\delta(x - nL) \sim \frac{1}{\eta\sqrt{\pi}} e^{-\frac{(x-nL)^2}{\eta^2}}$. Then, we write the source term in Fourier series, and we expand the field equations in powers of $\sqrt{M/L}$ and solve them order by order, to find the following solution at second order: $\forall i \in \{1, 2, 3\}$, $g_{0i} = 0$, and $\forall (i, j) \in \{1, 2, 3\}^2$:

$$g_{ij} = \delta_{ij}\left[1 + 2\varepsilon\sqrt{\frac{M}{L}}\sqrt{\frac{8\pi}{3}}\frac{t}{L} + \frac{2M}{L}\left(f_\eta(\mathbf{x}) + \frac{2\pi t^2}{3L^2}\right)\right] + \frac{M}{L}t^2\partial^2_{ij} f_\eta(\mathbf{x}), \quad (1)$$

where $\varepsilon = \pm 1$ and:

$$f_\eta(\mathbf{x}) = \frac{1}{\pi}\sum_{\mathbf{n}\in\mathbb{Z}^3_*} \frac{e^{-\frac{\pi^2|\mathbf{n}|^2\eta^2}{L^2}}}{|\mathbf{n}|^2} e^{\frac{2\pi}{L}i\mathbf{n}.\mathbf{x}}. \quad (2)$$

Let us insist on the fact that this solution is exact at order M/L. We can now calculate the rate of expansion between two masses of the lattice. For that, consider two masses on the x-axis (all the other axes are equivalent, by symmetry), separated by a coordinate distance NL, for N an integer. The physical distance between the two masses is given by $l(t) = \int_0^{NL} \sqrt{g_{xx}}dx$, and, expanding the square root to order M/L we find the effective scale factor of the lattice:

$$a(t) \equiv \frac{l(t)}{NL} = 1 + \varepsilon\sqrt{\frac{8\pi}{3}}\sqrt{\frac{M}{L^3}}t - \frac{2\pi Mt^2}{3L^3}. \tag{3}$$

The Hubble flow defined by $H(t) = \dot{a}(t)/a(t)$ is then found to be, at order M/L: $H(t) = \varepsilon\sqrt{\frac{8\pi}{3}}\sqrt{\frac{M}{L^3}} - \frac{4\pi Mt}{L^3}$. Thus defining the initial Hubble rate $H_0 = \varepsilon\sqrt{\frac{8\pi}{3}}\sqrt{\frac{M}{L^3}}$ and choosing the expanding solution, $\varepsilon = 1$, we get:

$$H(t) = H_0 - \frac{3}{2}H_0^2 t + \mathcal{O}(H_0^3), \tag{4}$$

and this corresponds exactly, at order $M/L \propto H_0^2$ to a flat FLRW model filled with non-relativistic dust. The result is actually valid at order $(M/L)^{3/2}$ [2]. Thus, the model with discrete masses on a cubic lattice, once smoothed, is identical to a FLRW model with dust, with the corresponding smeared energy density. This means that, from purely kinematical considerations, one cannot distinguish between the average, homogeneous fluid description of the lattice and the exact behaviour of this lattice: there is no backreaction (in the sense of [5]) associated with spatially smoothing the lattice.

2.2 Observables, Compacity and the Fitting Problem

Now that we have a solution of the field equations that does not display backreaction, we can try and address the fitting problem by comparing observables in the lattice with observables in the corresponding, smoothed FLRW Universe with the same kinematics: any discrepancy between the two will be a sign that there exists a fitting problem. In order to carry the comparison, we calculate the distance/redshift relation in the lattice. The 4-velocity of an observer in one of the objects of mass M is given by $u^a = (1, 0, 0, 0)$ according to our choice of coordinates, and we define λ, an affine parameter down light rays. We denote by O and S the locations of observer and the source respectively. Given the normalisation chosen in [3], the null vector of a past-directed light ray, k^a is such that $k_O^0 = 1$, so that we have, for the redshift: $1 + z(\lambda) = \frac{(k^a u_a)_S}{(k^a u_a)_O} = k^0(\lambda)$. Therefore, we can solve the 0 component of the null geodesic equation order by order in terms of $\sqrt{M/L}$. The distance is obtained similarly by solving Sachs equations [6] expanded at order M/L. The Sachs equations are actually exactly solvable in this perturbative scheme because the equations for the isotropic expansion and the shear decouple from each other at that order. Details of the calculations can be found in [3]. In terms of the past-directed affine parameter $\lambda < 0$ (and $\lambda = 0$ at the observer), we find that, at order M/L:

$$z(\lambda) = -\sqrt{\frac{M}{L}}\sqrt{\frac{8\pi}{3}}\frac{\lambda}{L} + \frac{M}{L}\left(\frac{14\pi\lambda^2}{3L^2} + \left[f_\eta(\mathbf{x}(\lambda)) - \lambda\partial_i f_\eta(\mathbf{x}(\lambda))v^i\right]_0^\lambda\right), \tag{5}$$

$$\begin{aligned} r_A(\lambda) = & -\lambda + \frac{2\pi}{3}\frac{M}{L}\frac{\lambda^3}{L^2}\left[1 + \sum_{(n,p,q)\in\mathscr{D}_\mathbf{v}} e^{-\frac{\pi^2(n^2+p^2+q^2)\eta^2}{L^2}}\right] \\ & + \frac{2}{\pi}\frac{M}{L}\sum_{\mathbf{n}\in\mathbb{N}_*^3\backslash\mathscr{D}_\mathbf{v}} e^{-\frac{\pi^2(n^2+p^2+q^2)\eta^2}{L^2}} \sum_{l=1}^{l=4}\left[-\lambda\frac{\cos\left(\frac{2\pi\lambda\mathbf{v}.\mathbf{u}_l}{L}\right)}{(\mathbf{v}.\mathbf{u}_l)^2} + \frac{L}{\pi}\frac{\sin\left(\frac{2\pi\lambda\mathbf{v}.\mathbf{u}_l}{L}\right)}{(\mathbf{v}.\mathbf{u}_l)^3}\right]. \end{aligned} \tag{6}$$

Here: $\mathbf{u}_1 = (n, p, q)$, $\mathbf{u}_2 = (n, -p, -q)$, $\mathbf{u}_3 = (n, p, -q)$, $\mathbf{u}_4 = (n, -p, q)$, and $\mathscr{D}_\mathbf{v} = \{(n, p, q) \in \mathbb{N}_*^3 : \exists l \in \{1, 2, 3, 4\}/\mathbf{u}_l.\mathbf{v} = 0\}$. This means that the first sum is over all the triplets that cancel one at least of the $\mathbf{u}_l.\mathbf{v}$, whereas the second sum is over all the other triplets. These expressions coincide with their FLRW counterparts for the FLRW model obtained by smoothing the distribution of masses of the lattice, up to *a priori* small corrections proportional to M/L (the parts that are non-polynomial in λ). Actually, it turns out that the additional terms in the expression for $r_A(\lambda)$ are not generally small, because some denominators in the second sum become extremely small and the corrections to the FLRW distance become of order $\sqrt{M/L}$, or even 1, instead of being of order M/L; see [3] for a detailed discussion of these effects. By carefully studying these additional terms, we arrive at the conclusion that the perturbative corrections to the FLRW distance/redshift relation remain small provided:

$$\frac{M}{L} \ll \mathscr{O}(1) \times \left(\frac{\eta}{L}\right)^4. \tag{7}$$

This relation between the mass of the object, M/L, and the compactness of the lattice, η/L, shows that if objects are too compact, the perturbative expansion breaks down, as far as the calculation of observables down a past lightcone is concerned, even though the perturbative calculations remain a good way of estimating the spacetime geometry (i.e. of solving the Einstein field equations). Similar problems were encountered in perturbations of an FLRW background in [7]. If this criterion is not satisfied, the perturbative calculations cannot be trusted, as second order terms (in $\sqrt{M/L}$) become of order $\mathscr{O}(1)$: one needs to integrate the system of Sachs equations without any perturbative expansion, thus retaining the coupling between isotropic expansion and shear. This is an ongoing work [8] and preliminary results indicate that when Eq. (7) is not satisfied, the contribution of the shear modifies significantly the FLRW observables: the corrections are usually smaller than the order 1 corrections predicted by the perturbative expansion presented here, but they are definitely significant to raise the issue of a fitting problem. For example, Fig. 1 shows the percent change $\delta r_A(z) = 100 \times \left(r_A(z) - r_A^{FLRW}(z)\right)/r_A^{FLRW}(z)$ in the angular distance between a

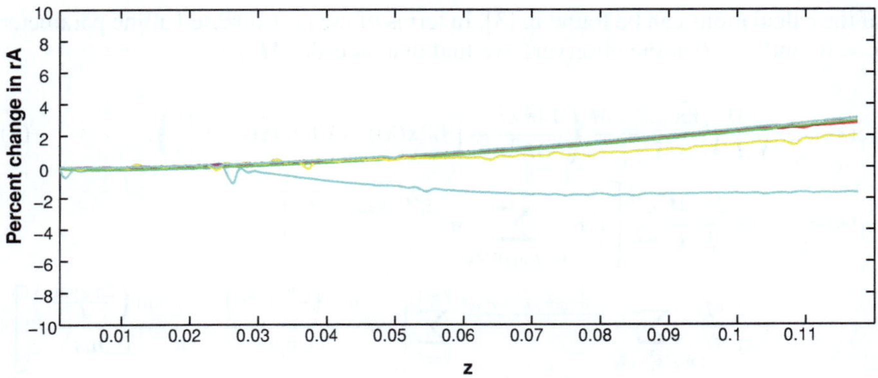

Fig. 1 Percent change $\delta r_A(z)$ in the angular distance between a lattice with $M = 10^{12} M_{\odot}$, $L = 1$ Mpc, $\eta = 0.01L$ and its FLRW counterpart, obtained from a complete integration of Sachs equations. The result is presented for 25 different directions on the celestial sphere of the observer located at the centre of one of the masses

Table 1 Some typical lattices and their characteristic parameters

Object	R_S (Mpc)	L (Mpc)	η (Mpc)	Criterion passed
Neutron star	10^{-19}	10^{-6}	10^{-18}	No
Galaxy (disk)	10^{-8}	1	10^{-2}	No (marginally)
Galaxy (disk+DM halo)	10^{-7}	1	0.05	Yes (marginally)
Galaxy cluster	10^{-4}	30	20	Yes

The choices are only indicative. The last column answers the question: does such a lattice satisfy the criterion (7)?

lattice with $M = 10^{12} M_{\odot}$, $L = 1$ Mpc, $\eta = 0.01L$ (lattice of galaxy-like objects) that does not satisfy the criterion (7) and the equivalent smoothed FLRW model, obtained from a complete integration of Sachs equations. We see that the 'divergence' problem encountered for such lattices when using perturbative methods is somehow 'cured' by solving the full system, even though, differences appear and seem to be systematically increasing with the redshift, irrespective of the direction on the sky.

Table 1 presents the typical values of M, L and η for some lattices of standard astrophysical objects and shows whether the criterion (7) is satisfied or not for such lattices. We see that a lattice of galaxies composed of their disk only marginally fails to pass the criterion, whereas when we include the Dark Matter halo, they pass the test marginally. A lattice of cluster-like objects passes the criterion easily, but $\eta/L \sim 1$, and one can hardly talk of a lattice of separated objects in that case: such objects could not really be considered as virialised, independent objects as we did in this work.

3 Discussion

We have presented a toy model of the Universe in the form of a regular cubic lattice of equal masses of typical size η whose kinematics is identical on large scales to the FLRW model obtained by smoothing the distribution of masses; this model does not exhibit any backreaction. We have seen that, despite the fact that a perturbative expansion in terms of R_S/L gives a very good approximation for the geometry of space-time, it is not suitable for the accurate calculation of observables in the model if the objects are too compact. Specifically, we have shown that a perturbative calculation of observables can be trusted only if the parameters of the lattice satisfy (7). If this is not the case, a non perturbative approach is needed to fully take into account the effect of the Weyl curvature sourcing the shear of bundles of null geodesics. Somehow, this was to be expected: the perturbative calculation decouples the shear from the isotropic expansion, making the observables independent on the Weyl curvature, but we know that in a mostly empty Universe (masses very compact), the behaviour of null ray bundles must be dominated by the Weyl curvature. The bound (7) gives a quantitative criterion to decide what 'too compact' means. This also illustrates the importance of the fitting problem: if (7) is not satisfied, the FLRW reconstructed by smoothing spatially the kinematics of the model differs systematically and sometimes significantly from the FLRW model fitting the observations on the past lightcone of observers. Choosing the order of magnitude of the parameters of the model to represent something 'realistic' is difficult, but it is interesting to note from Table 1 that galaxy-like objects and lattices are exactly at the transition between the lattices that pass the criterion (7) and those which do not. This might be extremely important in the precise characterisation of the properties of Dark Energy, and the detailed exploration of the consequences of this bound as well as the precise non perturbative estimates in the cases when it is not satisfied are the subject of an ongoing work [8].

References

1. Ellis, G.: Relativistic cosmology—Its nature, aims and problems. In: Bertotti B., de Felice F., Pascolini A. (eds.) General Relativity and Gravitation Conference, Fundamental Theories of Physics, vol. 9, pp. 215–288. Reidel, Dordrecht (1984)
2. Bruneton, J.P., Larena, J.: Dynamics of a lattice universe: the dust approximation in cosmology. Class. Quantum Grav. **29**, 155001 (2012). doi:10.1088/0264-9381/29/15/155001
3. Bruneton, J.P., Larena, J.: Observables in a lattice Universe: the cosmological fitting problem. Class. Quantum Grav. **30**, 025002 (2013). doi:10.1088/0264-9381/30/2/025002
4. Korotkin, D., Nicolai, H.: A periodic analog of the Schwarzschild solution, ArXiv e-prints [http://arxiv.org/abs/gr-qc/9403029arXiv:gr-qc/9403029] (1994)
5. Buchert, T.: On average properties of inhomogeneous fluids in general relativity. I: Dust cosmologies. Gen. Relativ. Gravit. **32**, 105 (2000). doi:10.1023/A:1001800617177
6. Sachs, R.: Gravitational waves in general relativity VI. The outgoing radiation condition. Proc. R. Soc. Lond. A **264**, 309 (1961). doi:10.1098/rspa.1961.0202

7. Clarkson, C., Umeh, O.: Is backreaction really small within concordance cosmology? Class. Quantum Grav. **28**, 164010 (2011). doi:10.1088/0264-9381/28/16/164010
8. Larena, J., Bruneton, J.P.: Observables in a lattice Universe: non perturbative treatment (2012) (In preparation)

Hair of Astrophysical Black Holes

Maxim Lyutikov

Abstract The "no hair" theorem is not formally applicable for black holes formed from collapse of a rotating neutron star. Rotating neutron stars can self-produce particles via vacuum breakdown forming a highly conducting plasma magnetosphere such that magnetic field lines are effectively "frozen-in" the star both before and during collapse. In the limit of no resistivity, this introduces a topological constraint which prohibits the magnetic field from sliding off the newly-formed event horizon. As a result, during collapse of a neutron star into a black hole, the latter conserves the number of magnetic flux tubes $N_B = e\Phi_\infty/(\pi c\hbar)$, where Φ_∞ is the initial magnetic flux through the hemispheres of the progenitor and out to infinity. The black hole's magnetosphere subsequently relaxes to the split monopole magnetic field geometry with self-generated currents outside the event horizon. The dissipation of the resulting equatorial current sheet leads to a slow loss of the anchored flux tubes, a process that makes the black hole bald on long resistive time scales rather than the short light-crossing time scales expected from the vacuum "no-hair" theorem.

1 Introduction

The "no hair" theorem [1] postulates that all black hole solutions of the Einstein-Maxwell equations of gravitation and electromagnetism in general relativity can be completely characterized by only three externally observable classical parameters: mass, electric charge, and angular momentum. The key point in the classical proof [2] is that the outside medium is a vacuum. In contrast, the surroundings of astrophysi-

M. Lyutikov (✉)
Department of Physics, Purdue University, 525 Northwestern Avenue,
West Lafayette, IN 47907-2036, USA
e-mail: lyutikov@purdue.edu

M. Lyutikov
INAF Osservatorio Astrofisico di Arcetri, Largo Enrico Fermi 5, 50125 Firenze, Italy

J. Bičák and T. Ledvinka (eds.), *Relativity and Gravitation*,
Springer Proceedings in Physics 157, DOI: 10.1007/978-3-319-06761-2_54,

cal high energy sources like pulsars and black holes can rarely be treated as vacuum [3–5]. The ubiquitous presence of magnetic fields combined with high (often relativistic) velocities produce inductive electric fields with electric potential drops high enough to break the vacuum via various radiative effects (curvature emission followed by a single photon pair production in magnetic field, or inverse Compton scattering followed by a two photon pair production). For example, in case of neutron stars the rotation of the magnetic field lines frozen into the crust generates an inductive electric field, which, due to the high conductivity of the neutron star interior, induces surface charges. The electric field of these induced surface charges has a component parallel to the dipolar magnetic field. These parallel electric fields accelerate charges to the energy $\mathscr{E} \sim eB_s R_s (\Omega R_0/c)^2$, where B_s and R_s are the surface magnetic field, radius of a neutron star and Ω is the angular rotation frequency. The resulting primary beam of leptons produces a dense secondary plasma via vacuum breakdown. Thus, in case of neutron stars the electric charges and currents are self-generated: no external source is needed. Rotating black holes can also lead to a similar vacuum break-down [4].

We demonstrated that contrary to the prediction of the "no hair" theorem, the collapse of a rotating neutron star into the black hole results in a formation of a long lived self-generated conducting BH magnetosphere. This results from the violation of the key assumption of the "no hair" theorem, that the outside is vacuum, and allows a black hole to preserve open magnetic flux tubes that initially connect to the neutron star surface.

2 The Black Hole Hair: The Conserved Poloidal Magnetic Flux

Consider collapse of a rotating neutron star into the BH. Before the onset of the collapse, the electric currents within the neutron star create poloidal magnetic field. Rotation of the poloidal magnetic field lines and the resulting inductive electric field lead to the creation, through vacuum breakdown, of the conducting plasma and poloidal electric currents. The presence of a conducting plasma then imposes a topological constraint, that the magnetic field lines which initially were connecting the neutron star surface to the infinity must connect the black hole horizon to the infinity.

During the collapse, as the surface of a neutron star approaches the horizon, the closed magnetic field lines will be quickly absorbed by the black hole, while the open field lines (those connecting to infinity) have to remain open by the frozen-in condition. Thus, a black hole can have only open fields lines, connecting its horizon to the infinity. There is a well known solution that satisfies this condition: an exact split monopolar solution for rotating magnetosphere due to [6]; it was generalized to Schwarzschild metrics by [4]. We recently found an exact non-linear *time-dependent* split monopole-type structure of magnetospheres driven by spinning and collapsing neutron star in Schwarzschild geometry [7]. We demonstrated that the collapsing neutron star enshrouded in a self-generated conducting magnetosphere does not allow a quick release of the magnetic fields to infinity.

Thus, if a collapsing black hole can self-sustain the plasma production in its magnetosphere, the magnetic field lines that were initially connecting the neutron star surface to infinity will connect the black hole horizon to the infinity. Each hemisphere then keeps the magnetic flux that was initially connected to the infinity. For a neutron star with the surface magnetic field B_{NS} and the initial pre-collapse radius R_{NS} and period P_{NS}, the magnetic flux through each hemisphere connecting to infinity is $\Phi_\infty \approx 2\pi^2 B_{NS} R_{NS}^3/(P_{NS}c)$ [3]. Using quantization of the magnetic flux [8], this corresponds to a conserved quantum number of magnetic flux tubes

$$N_B = e\Phi_\infty/(\pi c\hbar) = 2\pi B_{NS} e R_{NS}^3/(c^2 \hbar P_{NS}) = 10^{41} \frac{B_{NS}}{10^{12}\ \mathrm{G}} \frac{P_{NS}}{1\ \mathrm{msec}}. \quad (1)$$

This quantum number is the black hole "hair": an observer at infinity can measure the corresponding Poynting flux and infer the number N_B.

3 Numerical Simulations

We have performed numerical simulations that confirm the basic principle that the "no-hair" theorem and related time-dependent vacuum simulations are not applicable to a plasma-filled black hole magnetosphere. We do not model the process of vacuum breakdown and the subsequent formation of a plasma-filled magnetosphere. Instead, we assume the neutron star already created a plasma-filled magnetosphere (or that the black hole self-generates a plasma-filled magnetosphere), and we assume that the neutron star has already collapsed to a black hole. Only once an event horizon has formed would the magnetic field begin to slip-off the black hole in vacuum, so starting with an event horizon should be a strong enough test—one should not have to follow the collapse of the neutron star to a black hole as long as a plasma is present. The goal of the simulations is to measure the decay timescale of the magnetic flux threading the event horizon of the black hole: $\Phi_{\mathrm{EM}} = (1/2)\int_S dS|B^r|$ as integrated over the surface (S) of the black hole horizon (Fig 1). We show that the magnetic dipole decay seen in vacuum solutions is avoided or delayed by three effects: (1) presence of plasma and self-generation of toroidal currents; (2) black hole spin induced poloidal currents; and (3) plasma pressure support of current layers generated internally by dissipating currents. These effects cause the field to avoid vacuum-like decay of the dipole magnetic field and help support the newly-formed split-monopole magnetic field against magnetic reconnection (Fig. 2).

These GRMHD simulations use the fully conservative, shock-capturing GRMHD scheme called HARM [9] using Kerr-Schild coordinates in the Kerr metric for a sequence of spins.

We perform simulations that either use the force-free or use the fully energy-conserving MHD equations of motions. These approximate, respectively, the limits

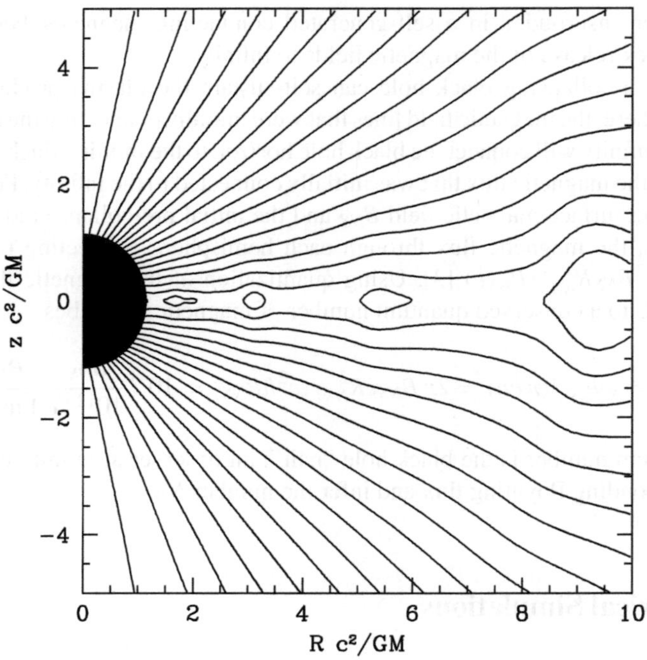

Fig. 1 A contour plot of the magnetic flux ($\Psi = RA_{\phi}$) showing the inner (cylindrical radius) $R < 10\,\mathrm{GM/c^2}$ for the MHD $a = 0.99$ model described in the text. The structure of the magnetosphere relaxes to monopolar-like solution, as predicted by [7]. Note also the development of the tearing modes and the formation of magnetic islands in the equatorial current sheet

of radiatively efficient emission and radiatively inefficient emission once the plasma has been generated. That is, if the electromagnetic field dominates the rest-mass and internal energy density over most of the volume outside current sheets, then the force-free limit corresponds to an instantaneous loss (such as radiation) of magnetic energy dissipated in current sheets, while the fully energy-conserving MHD limit without cooling corresponds to all dissipated energy going into internal + kinetic energy that remains in the system and sustains the current sheet against dissipation. A non-energy-conserving system of equations or simulation code would be unable to properly follow the energy conservation process of electromagnetic dissipation within the current sheet that leads to plasma formation there. The force-free electrodynamics equations of motion are not solely relied upon because they are undefined within current sheets and any particular resistive force-free electrodynamics equations [10–12] still leave some degree of ambiguity in how the resistivity would map onto the full magnetohydrodynamical (MHD) equations. For the MHD equations, an ideal $\gamma = 4/3$ gas equation of state is chosen, which can be considered as mimicking a radiatively inefficient high-energy particle distribution component generated by the dissipation of the currents within the reconnecting layer.

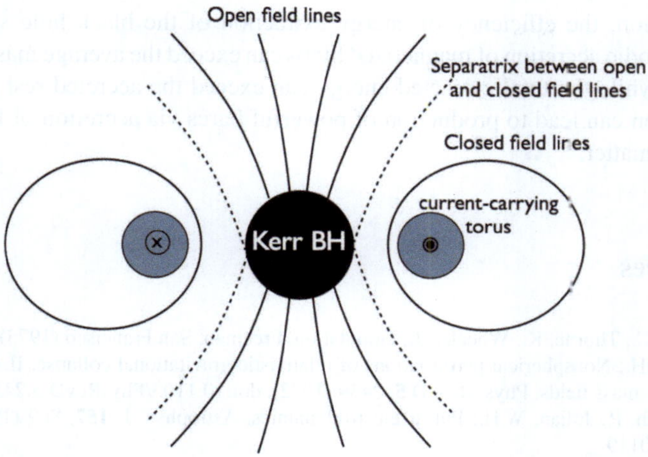

Fig. 2 Schematic presentations of magnetic flux surfaces in the BH-torus system. Toroidal electric current in the torus creates poloidal magnetic field. The field lines that intersect the BH are twisted by the rotation of the space-time (carry poloidal electric current) and open up to infinity. There are two types of magnetic field lines separated by a separatrix (*dashed lines*): closed field lines and open magnetic field lines that intersect the Kerr BH. (The section shows only the poloidal component of the magnetic field.) After the torus is accreted, the open magnetic field lines remain on the BH, relaxing to a twisted monopolar structure [4, 6]

4 Astrophysical Applications

The fact that isolated black holes formed in a collapse of rotating neutron stars can retain their open magnetic flux for times much longer than the collapse time implies that isolated BHs can spindown electromagnetically, converting the rotational energy in the electromagnetic wind. This can be important for short GRBs: in [13] we address the key problem of the neutron star merger paradigm in application to short GRBs, the presence of energetic prompt tails and flares at very long time scales, orders of magnitude longer than the active stage of the merger. We identify the prompt GRB spike as coming from the energy dissipation of the wind powered by a transient accretion torus surrounding the newly formed GRB. It's duration is limited by the life of the torus, tens to hundreds of milliseconds. The long extended emission comes from wind powered by the isolated rotating BH, that produces equatorially-collimated outflow. It's duration is limited by the retention time scale of the magnetic field, and it contains *more total energy than the prompt spike*.

Thus, the proposed model for short GRBs implies a different type of collimation of the outflow than the conventionally envisioned jet-like structure, at least in the prompt tail stage. An observer on the axis see only the axially collimated prompt emission generated by the BH-torus system, while an observer at medium polar angles sees both the prompt spike and the equatorially-collimated extended tail.

In addition, the efficiency of energy extraction of the black hole spin energy during episodic accretion of magnetized blobs can exceed the average mass accretion rate $\dot{M}c^2$, while the total extracted energy can exceed the accreted rest mass. This phenomenon can lead to production of powerful flares via accretion of fairly small amount of matter.

References

1. Misner, C., Thorne, K., Wheeler, J.: Gravitation. Freeman, San Francisco (1973)
2. Price, R.H.: Nonspherical perturbations of relativistic gravitational collapse. II. Integer-spin, zero-rest-mass fields. Phys. Rev. D **5**, 2439 (1972). doi:10.1103/PhysRevD.5.2439
3. Goldreich, P., Julian, W.H.: Pulsar electrodynamics. Astrophys. J. **157**, 869 (1969). doi:10.1086/150119
4. Blandford, R.D., Znajek, R.L.: Electromagnetic extraction of energy from Kerr black holes. Mon. Not. R. Astron. Soc. **179**, 433 (1977)
5. Muslimov, A.G., Tsygan, A.I.: General relativistic electric potential drops above pulsar polar caps. Mon. Not. R. Astron. Soc. **255**, 61 (1992)
6. Michel, F.C.: Rotating magnetosphere: a simple relativistic model. Astrophys. J. **180**, 207 (1973). doi:10.1086/151956
7. Lyutikov, M.: Electromagnetic power of merging and collapsing compact objects. Phys. Rev. D **83**(12), 124035 (2011). doi:10.1103/PhysRevD.83.124035
8. Landau, L.D., Lifshitz, E.M.: Fluid Mechanics. Course of Theoretical Physics. Pergamon Press, Oxford (1959)
9. McKinney, J.C.: General relativistic magnetohydrodynamic simulations of the jet formation and large-scale propagation from black hole accretion systems. Mon. Not. R. Astron. Soc. **368**, 1561 (2006). doi:10.1111/j.1365-2966.2006.10256.x
10. Lyutikov, M.: Explosive reconnection in magnetars. Mon. Not. R. Astron. Soc. **346**, 540 (2003). doi:10.1046/j.1365-2966.2003.07110.x
11. Gruzinov, A.: Strong-Field Electrodynamics, ArXiv e-prints [http://arxiv.org/abs/0802.1716arXiv:0802.1716] (2008)
12. Li, J., Spitkovsky, A., Tchekhovskoy, A.: Resistive solutions for pulsar magnetospheres. Astrophys. J. **746**, 60 (2012). doi:10.1088/0004-637X/746/1/60
13. Lyutikov, M.: The electromagnetic model of short GRBs, the nature of prompt tails, supernova-less long GRBs, and highly efficient episodic accretion. Astrophys. J. **768**, 63 (2013). doi:10.1088/0004-637X/768/1/63

Backreaction Effects on the Luminosity-Redshift Relation in Inhomogeneous Cosmology

Giovanni Marozzi

Abstract We recall a general gauge invariant formalism for defining cosmological averages that are relevant for observations based on light-like signals. Using such formalism, together with adapted "geodesic light-cone" coordinates, the effect of a stochastic background of cosmological perturbations on the luminosity-redshift relation is computed to second order. The resulting expressions are free from both ultraviolet and infrared divergences, implying that such perturbations cannot mimic a sizable fraction of dark energy. Different averages are estimated and depend on the particular function of the luminosity distance being averaged. The energy flux, being minimally affected by perturbations at large z, is proposed as the best choice for precision estimates of dark-energy parameters. Nonetheless, its irreducible (stochastic) variance induces statistical errors on $\Omega_\Lambda(z)$ typically lying in the few-percent range.

1 Introduction

Establishing the existence of dark energy and determining its parameters is one of the central issues in modern cosmology. Evidence for a sizable dark-energy component in the cosmic fluid comes from different sources: CMB anisotropies, models of large-scale-structure formation and, most directly, the luminosity redshift relation of Type Ia supernovae, used as standard candles.

In this latter case, on which we concentrate our attention, the analysis is usually made in the simplified context of a homogeneous and isotropic (FLRW) cosmology. The issue has then been raised about whether inhomogeneities may affect the

G. Marozzi (✉)
Collège de France, 11 Place M. Berthelot, 75005 Paris, France
e-mail: giovanni.marozzi@unige.ch

G. Marozzi
Département de Physique Théorique and CAP, Université de Genève,
24 quai Ernest-Ansermet, 1211 Genève 4, Switzerland

J. Bičák and T. Ledvinka (eds.), *Relativity and Gravitation*,
Springer Proceedings in Physics 157, DOI: 10.1007/978-3-319-06761-2_55,

conclusion of such a naive analysis. In particular, one should address this question in the presence of stochastically isotropic and homogeneous perturbations of the kind predicted by inflation. In such a context, the possibility that sub-horizon perturbations may simulate a substantial fraction of dark energy, or that they may at least play some role in the context of near-future precision cosmology, has to be seriously considered.

In order to address these issues we follow [1] and study the luminosity-redshift relation in a spatially-flat ΛCDM model. The luminosity distance d_L now depends on the redshift z as well as on the angular coordinates of the sources, and must be inserted in an appropriate light-cone and *ensemble* average [2, 3].

The paper is organized as follows. In Sect. 2 we briefly present the prescription used to average on null hypersurfaces. In Sect. 3 we discuss the effect of a stochastic background of inhomogeneities on different functions of the luminosity-redshift relation. Our conclusive remarks are then presented in Sect. 4.

2 Gauge Invariant Light-Cone Averaging

Let us here briefly present a general gauge invariant formalism for defining cosmological averages that are relevant for observations based on light-like signals. Following [2], we start with a spacetime integral where the four-dimensional integration region is bounded by two hypersurfaces, one spacelike and the other one null (corresponding e.g. to the past light cone of some observer). Let us choose, in particular, the region inside the past light cone of an observer bounded in the past by the hypersurface defined by $A(x) = A_0$: clearly a gauge invariant definition of the integral of a scalar $S(x)$ over such a hypervolume can be written as

$$I(S; -; A_0, V_0) = \int_{\mathcal{M}_4} d^4x\sqrt{-g}\,\Theta(V_0 - V)\Theta(A - A_0)S(x), \tag{1}$$

where $V(x)$ is a scalar satisfying $\partial_\mu V \partial^\mu V = 0$, and where the "−" symbol on the l.h.s. denotes the absence of delta-like window functions.

Starting with this hypervolume integral we can construct covariant and gauge invariant hypersurface and surface integrals considering the variation of the volume average along the flow lines n_μ normal to the reference hypersurface $\Sigma(A)$ defined by $A(x)$ equal to a constant.

Considering the variation of the hypervolume integral by shifting the light cone $V = V_0$ along n_μ, we obtain the integral on the past light cone itself starting from a given hypersurface in the past

$$I(S; V_0; A_0) = \int d^4x\sqrt{-g}\,\delta(V_0 - V)\Theta(A - A_0)\frac{|\partial_\mu V \partial^\mu A|}{\sqrt{-\partial_\nu A \partial^\nu A}}S(x). \tag{2}$$

While considering the variation of the hypervolume integral both by shifting the light cone $V = V_0$ and the hypersurface $A = A_0$ along n_μ, we obtain the integral on the 2-sphere embedded in the past light cone

$$I(S; V_0, A_0; -) = \int d^4x \sqrt{-g}\, \delta(V_0 - V)\delta(A - A_0)|\partial_\mu V \partial^\mu A| S(x) . \quad (3)$$

We note, finally, that averages of a scalar S over different (hyper)surfaces are trivially defined by:

$$\langle S \rangle_{V_0, A_0} = \frac{I(S; V_0, A_0; -)}{I(1; V_0, A_0; -)} ; \quad (4)$$

$$\langle S \rangle_{V_0}^{A_0} = \frac{I(S; V_0; A_0)}{I(1; V_0; A_0)} . \quad (5)$$

3 Backreaction on the Luminosity-Redshift Relation

Let us start by recalling the standard expression for the luminosity distance in an unperturbed flat ΛCDM model, with present fractions of critical density Ω_Λ and $\Omega_m = 1 - \Omega_\Lambda$:

$$d_L^{FLRW}(z) = \frac{1+z}{H_0} \int_0^z \frac{dz'}{\left[\Omega_\Lambda + \Omega_m (1+z')^3\right]^{1/2}} . \quad (6)$$

Consider now the expression for d_L in the corresponding perturbed geometry. Combining light-cone and ensemble averages (denoted, respectively, by brackets and over-bars), we can write the averaged result in the form:

$$\overline{\langle d_L \rangle}(z) = d_L^{FLRW} \left[1 + f_d(z)\right] , \quad (7)$$

where $f_d(z)$ represents the "backreaction" on d_L due to inhomogeneities. For consistency, d_L has to be computed (at least) up to the second perturbative order since *ensemble* averages of first-order quantities are vanishing for stochastic perturbations. A detailed computation of $f_d(z)$ would thus enable to extract the "true" value of the dark-energy parameters from the measurement of $\overline{\langle d_L \rangle}(z)$ after taking the correction into account.

However, as already stressed in [3], given the covariant (light-cone) average of a perturbed (inhomogeneous) observable S the average of a generic function of this observable differs, in general, from the function of its average, i.e. $\overline{\langle F(S) \rangle} \neq F(\overline{\langle S \rangle})$. Expanding the observable to second order as $S = S_0 + S_1 + S_2 + \cdots$, one finds:

$$\overline{\langle F(S) \rangle} = F(S_0) + F'(S_0)\overline{\langle S_1 + S_2 \rangle} + F''(S_0)\overline{\langle S_1^2/2 \rangle} , \quad (8)$$

where $\overline{\langle S_1 \rangle} \neq 0$ as a consequence of the "induced backreaction" terms (see [3]). Thus different functions of the luminosity distance are differently affected by the inhomogeneities, and require different "subtraction" procedures. Finding the function that minimizes the backreaction will help of course for a precision estimate of the cosmological parameters.

The average value of Φ, obviously controlled by the average of d_L^{-2}, has to be carried out on the past light cone of the observer, at a fixed redshift z, using the gauge-invariant prescription described. This is most conveniently done [2, 3] in the so-called geodesic light-cone gauge (GLC), where the metric depends on six arbitrary functions (Υ, U^a, γ_{ab}, $a, b = 1, 2$), and the line element takes the form (with $\widetilde{\theta}^1 = \widetilde{\theta}, \widetilde{\theta}^2 = \widetilde{\phi}$):

$$ds^2 = \Upsilon^2 dw^2 - 2\Upsilon dw d\tau + \gamma_{ab}(d\widetilde{\theta}^a - U^a dw)(d\widetilde{\theta}^b - U^b dw). \tag{9}$$

In the GLC gauge the past light cone is defined by the condition $w = w_0 =$ const, and the redshift is given by:

$$1 + z = \Upsilon(w_0, \tau_0, \widetilde{\theta}^a)/\Upsilon(w_0, \tau, \widetilde{\theta}^a). \tag{10}$$

Furthermore, the luminosity distance of the source is simply expressed as [3] $d_L = (1+z)^2 \gamma^{1/4} (\sin\widetilde{\theta})^{-1/2}$, yielding the following exact result [4]:

$$\langle d_L^{-2} \rangle (z, w_0) = \frac{4\pi (1+z)^{-4}}{\int d^2\widetilde{\theta}^a \sqrt{\gamma(w_0, \tau(z, \widetilde{\theta}^a), \widetilde{\theta}^b)}}, \tag{11}$$

where $\gamma = \det \gamma_{ab}$, and $\tau(z, \widetilde{\theta}^a)$ is obtained by solving (10). The above expression has a simple physical interpretation: the averaged flux, for a given z, is inversely proportional to the proper area of the surface lying on our past light-cone at the given value of z.

To compute this quantity in the perturbed geometry of our interest, we need to express it in a gauge where the stochastic background of cosmological perturbations is explicitly known up to second order. To this purpose, we can use the standard Poisson gauge where we include first and second-order scalar perturbations, neglecting their tensor and vector counterparts. Performing the relevant transformations to second order we arrive at the following analogue of (7):

$$\overline{\langle d_L^{-2} \rangle} = \left(d_L^{FLRW}\right)^{-2} \overline{(I_\Phi(z))^{-1}} \equiv \left(d_L^{FLRW}\right)^{-2} [1 + f_\Phi(z)], \tag{12}$$

where I_Φ has in general the following structure:

$$I_\Phi(z) = \int \frac{d\tilde{\phi} d\tilde{\theta} \sin\tilde{\theta}}{4\pi} \Big[1 + \mathscr{I}_1 + \mathscr{I}_{1,1} + \mathscr{I}_2\Big](\tilde{\theta}, \tilde{\phi}, z). \tag{13}$$

Here $\mathscr{I}_1$, $\mathscr{I}_{1,1}$, $\mathscr{I}_2$ are, respectively, the first-order, quadratic first-order, and genuine second-order contributions of our stochastic fluctuations. After solving the relevant perturbation equations [5] they can all be expressed in terms of the first-order Bardeen potential $\Psi(x, \eta)$. Using the stochastic properties of this perturbation, and expanding in Fourier modes $\Psi_k(\eta)$, we can then obtain an expression for $\overline{(I_\Phi)^{-1}}$ where the scalar perturbations only appear through the so-called dimensionless power spectrum, $\mathscr{P}(k, \eta) = (k^3/2\pi^2)|\Psi_k(\eta)|^2$.

Considering a ΛCDM model we have to proceed with an approximate numerical integration. The result can be then written in the form

$$f_\Phi(z) = \int_0^\infty \frac{dk}{k}\, \mathscr{P}(k)\Big[f_{1,1}(k, z) + f_2(k, z)\Big]. \tag{14}$$

At leading order the contribution, in the region of z relevant for dark-energy phenomenology, comes from terms of the type $f(k, z) \sim (k/\mathscr{H}_0)^2 \tilde{f}(z)$, and we can write, to a very good accuracy,

$$f_\Phi(z) \simeq \Big[\tilde{f}_{1,1}(z) + \tilde{f}_2(z)\Big] \int_0^\infty \frac{dk}{k} \left(\frac{k}{\mathscr{H}_0}\right)^2 \mathscr{P}(k). \tag{15}$$

The absolute value (and sign) of $f_\Phi(z)$ is illustrated in Fig. 1, showing that the backreaction of a realistic spectrum of stochastic perturbations induces negligible corrections to the averaged flux at large z. In addition, such corrections have the wrong z-dependence (in particular change sign at some z) to simulate even a tiny dark energy component. For the considered spectrum (behaving as $k^{n_s-5}\log^2 k$ at large k, see [6]) the spectral integral is convergent and very weakly sensitive to the chosen value of the UV cutoff [3] representing here the limit of validity of our perturbative approach.

The small value of $|f_\Phi|$ at large z leads us to conclude that the averaged flux is a particularly appropriate quantity for extracting from the observational data the "true" cosmological parameters. On the other hand, the situation is somewhat different for other functions of d_L.

Indeed, let's apply the general result (8) to the flux variable, $S = \Phi$, and consider two important examples: $F(\Phi) = \Phi^{-1/2} \sim d_L$, and $F(\Phi) = -2.5\log_{10}\Phi +$ const $\sim \mu$ (the distance modulus). For d_L, following the notations of (7) and using the general result (8), we obtain:

$$f_d = -(1/2) f_\Phi + (3/8)\overline{\langle(\Phi_1/\Phi_0)^2\rangle}. \tag{16}$$

Similarly, for the distance modulus we obtain:

$$\overline{\langle\mu\rangle} - \mu^{FLRW} = -1.25(\log_{10} e)\Big[2 f_\Phi - \overline{\langle(\Phi_1/\Phi_0)^2\rangle}\Big]. \tag{17}$$

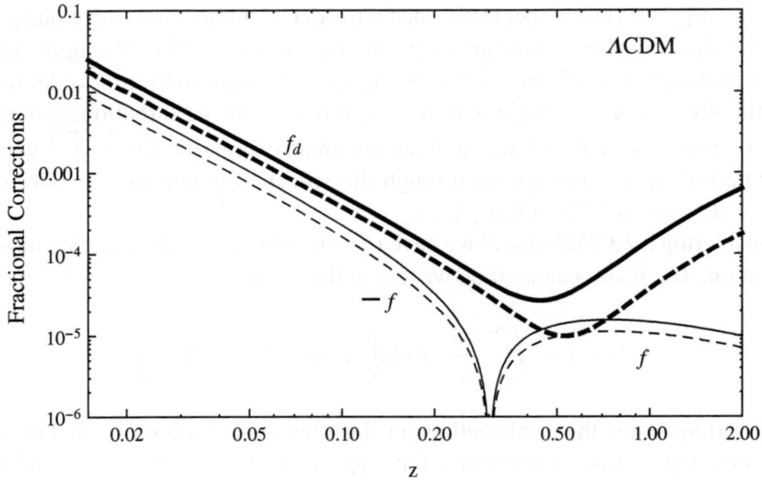

Fig. 1 The correction f_Φ of (12) (*thin curves*) is compared with the correction f_d of (16) (*thick curves*), for a ΛCDM model with $\Omega_\Lambda = 0.73$. We have used two different cutoff values: $k_{UV} = 0.1\,\mathrm{Mpc}^{-1}$ (*dashed curves*) and $k_{UV} = 1\,\mathrm{Mpc}^{-1}$ (*solid curves*). We have used for $\mathscr{P}(k)$ the inflationary scalar spectrum with the WMAP parameters [7] and the transfer function given in [6] (see also [3])

As clearly shown by the two above equations, the corrections to the averaged values of d_L and μ are qualitatively different from those of the flux, because of the extra contribution (inevitable for any non-linear function of the flux) proportional to the square of the first-order fluctuations. The averaged flux corrections have leading spectral contributions of the type $k^2\mathscr{P}(k)$; on the contrary, the new corrections to d_L and μ are due to the so-called "lensing effect", they dominate at large z, and have leading spectral contributions of the type $k^3\mathscr{P}(k)$ (as already discussed in [3]). The explicit numerical integration, reported in Fig. 1, confirms that $|f_\Phi| \ll f_d$ at large z. We stress that even the k^3-enhanced contributions are UV-finite for the case under consideration.

Let us now briefly discuss to what extent the enhanced corrections due to the squared first-order fluctuations can affect the determination of the dark-energy parameters if quantities other than the flux are used in the fits. To this purpose we consider the much used (average of the) distance modulus given in (17), referred as usual to the homogeneous Milne model with $\mu^M = 5\log_{10}[(2+z)z/(2H_0)]$. In Fig. 2 we compare the averaged value $\overline{\langle\mu\rangle} - \mu^M$ with the corresponding expression in a homogeneous ΛCDM model with different values of Ω_Λ. We also show the expected dispersion around the averaged result, represented by the square root of the variance [3]. The latter is given by:

$$\sqrt{\overline{\langle\mu^2\rangle} - \left(\overline{\langle\mu\rangle}\right)^2} = \pm 2.5(\log_{10} e)\sqrt{\overline{\langle(\Phi_1/\Phi_0)^2\rangle}}; \tag{18}$$

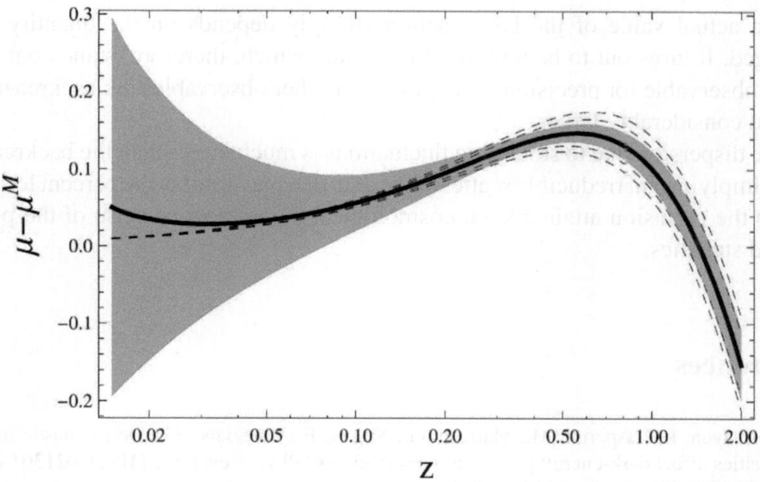

Fig. 2 The averaged distance modulus $\overline{\langle\mu\rangle} - \mu^M$ (*thick solid curve*), and its dispersion of (18) (*shaded region*) are computed for $\Omega_\Lambda = 0.73$ and compared with the homogeneous value for the unperturbed ΛCDM models with $\Omega_\Lambda = 0.69, 0.71, 0.73, 0.75, 0.77$ (*dashed curves*). We have used $k_{UV} = 1\text{Mpc}^{-1}$ and the same spectrum as in Fig. 1

while for the flux we simply find:

$$\sqrt{\overline{\langle(\Phi/\Phi_0)^2\rangle} - \left(\overline{\langle\Phi/\Phi_0\rangle}\right)^2} = \pm\sqrt{\overline{\langle(\Phi_1/\Phi_0)^2\rangle}} \,. \quad (19)$$

As illustrated in Fig. 2, we find that, even for the distance modulus, the effect of inhomogeneities on the average only affects the determination of Ω_Λ at the third decimal figure (see also Fig. 1), at least for the inflationary power spectrum with the ΛCDM transfer function of [6]: in that case, the curves for $\overline{\langle\mu\rangle}$ and μ^{FLRW} are practically coincident at large z.

4 Conclusions

The main results presented in this paper can be summarized as follows. We have recalled a covariant and gauge invariant formalism to average on null hypersurfaces and to analyze the effects of inhomogeneities on astrophysical observables related to light-like (massless) signals.

Then we have seen how the gauge invariant light-cone averaging of the luminosity-redshift relation leads to results which are automatically free from UV/IR divergences for *any* function of the luminosity distance, and, as a consequence, cannot simulate a substantial fraction of dark energy.

The actual value of the backreaction strongly depends on the quantity being averaged. It turns out to be minimal for the flux which, therefore, stands out as the safest observable for precision cosmology. For other observables the backreaction is instead considerably larger.

The dispersion due to stochastic fluctuations is much larger than the backreaction itself, implying an irreducible scatter of the data that may limit to the percent level (see Fig. 2) the precision attainable on cosmological parameters because of the present limited statistics.

References

1. Ben-Dayan, I., Gasperini, M., Marozzi, G., Nugier, F., Veneziano, G.: Do stochastic inhomogeneities affect dark-energy precision measurements? Phys. Rev. Lett. **110**(2), 021301 (2013). doi:10.1103/PhysRevLett.110.021301
2. Gasperini, M., Marozzi, G., Nugier, F., Veneziano, G.: Light-cone averaging in cosmology: formalism and applications, J. Cosmol. Astropart. Phys. **2011**(07), 008 (2011). doi:10.1088/1475-7516/2011/07/008
3. Ben-Dayan, I., Gasperini, M., Marozzi, G., Nugier, F., Veneziano, G.: Backreaction on the luminosity-redshift relation from gauge invariant light-cone averaging, J. Cosmol. Astropart. Phys. 2012(04), 036 (2012). doi:10.1088/1475-7516/2012/04/036
4. Ben-Dayan, I., Marozzi, G., Nugier, F., Veneziano, G.: The second-order luminosity-redshift relation in a generic inhomogeneous cosmology. J. Cosmol. Astropart. Phys. **2012**(11), 045 (2012). doi:10.1088/1475-7516/2012/11/045
5. Bartolo N., Matarrese S., Riotto A.: The full second-order radiation transfer function for large-scale CMB anisotropies. J. Cosmol. Astropart. Phys. **2006**(05), 010 (2006). doi:10.1088/1475-7516/2006/05/010
6. Eisenstein, D., Hu, W.: Baryonic features in the matter transfer function. Astrophys. J. **496**, 605 (1998). doi:10.1086/305424
7. Komatsu, E., Smith, K., Dunkley, J., et al.: Seven-year Wilkinson microwave anisotropy probe (WMAP) observations: cosmological interpretation. Astrophys. J. Suppl. Ser. **192**, 18 (2011). doi:10.1088/0067-0049/192/2/18

Scalar Averaging in Szekeres Models

Roberto A. Sussman

Abstract We introduce a formalism of scalar proper volume weighted averages (the "q-average") acting on compact comoving domains of quasi-spherical Szekeres models with a dust source. Although these models do not admit isometries, the resulting averaged scalars are spherically symmetric functionals that can be defined as local functions (the "q-scalars") by considering a varying averaging domain. The fluctuations of the density and Hubble scalar with respect to their corresponding q-scalars determine the Riemann, Weyl, electric Weyl and shear tensors through irreducible covariant algebraic expansions. The q-average of all invariant scalars formed by contractions of these tensors are directly related to statistical variance and covariance moments of the density and Hubble scalar with respect to their q-averages. The q-scalars and q-averages, together with their fluctuations, lead to complete systems of evolution equations and algebraic constraints that fully determine the dynamics of the models. However, these evolution equations lack the "back-reaction" correlation terms characteristic of Buchert's averaging scheme.

1 Introduction

The Szekeres dust models are a well known class of exact solution of Einstein's equations that admit (in general) no Killing vectors [1]. For this reason they are natural candidates to construct models of cosmological inhomogeneities that are much less idealized than spherically symmetric configurations that arise from the Lemaître–Tolman–Bondi (LTB) models [2] (their spherical limiting case). However, Szekeres models provide also an ideal theoretical framework for the study of generic properties of inhomogeneity, such as averaging. In this brief article we introduce a formalism of

R. A. Sussman (✉)
Instituto de Ciencias Nucleares, UNAM, Circuito Exterior S/N, C.U.,
04510 México, D.F., México
e-mail: sussman@nucleares.unam.mx

J. Bičák and T. Ledvinka (eds.), *Relativity and Gravitation*,
Springer Proceedings in Physics 157, DOI: 10.1007/978-3-319-06761-2_56,

proper volume weighted averages acting on compact comoving domains, and thus we consider only the "quasi-spherical" QS subclass of models for which such domains always exist [3].

2 Quasi-Spherical Szekeres Models

QS Szekeres dust models can be described by the following metric reminiscent of the spherically symmetric LTB metric:

$$\mathrm{d}s^2 = -\mathrm{d}t^2 + \frac{\mathscr{E}^2\, Y'^2}{1-K}\mathrm{d}r^2 + Y^2\left[\mathrm{d}x^2 + \mathrm{d}y^2\right], \tag{1}$$

where $Y = Y(t, r, x, y)$ and $\mathscr{E} = \mathscr{E}(r, x, y)$ are given by

$$Y = \frac{R}{\mathscr{E}}, \qquad \mathscr{E} = \frac{S}{2}\left[1 + \left(\frac{x-P}{S}\right)^2 + \left(\frac{x-Q}{S}\right)^2\right], \tag{2}$$

with $R = R(t, r)$ being the area distance that appears in LTB metrics and $S(r)$, $P(r)$, $Q(r)$ are the Szekeres characteristic free arbitrary functions, so that (1) becomes the LTB metric if they are constants. The metric function Y satisfies an analogous Friedman–like equation as R in LTB models:

$$\dot{Y}^2 = \frac{2\tilde{M}}{Y} - \tilde{K}, \qquad \tilde{M} = \frac{M(r)}{\mathscr{E}^3}, \qquad \tilde{K} = \frac{K(r)}{\mathscr{E}^2}. \tag{3}$$

The main covariant scalars are the density, ρ, the Hubble expansion scalar $\Theta = \nabla_a u^a$ and the Ricci scalar ${}^3\mathscr{R}$ of hypersurfaces ${}^3\mathscr{T}[t]$ orthogonal to u^a take also LTB–like form:

$$\frac{4\pi}{3}\rho = \frac{\tilde{M}'}{3Y^2Y'}, \qquad \mathscr{H} \equiv \frac{\Theta}{3} = \frac{(Y^2Y')^{\cdot}}{3Y^2Y'}, \qquad \mathscr{K} \equiv \frac{{}^3\mathscr{R}}{6} = \frac{(\tilde{K}Y)'}{3Y^2Y'}. \tag{4}$$

3 Quasi-Local Average and Quasi-Local Scalars

Let A be a scalar function defined along an arbitrary hypersurface ${}^3\mathscr{T}[t]$ whose proper volume element is $\mathrm{d}\mathscr{V} = \mathscr{F}^{-1}Y^2Y'\mathrm{d}r\mathrm{d}x\mathrm{d}y$, with $\mathscr{F} \equiv \sqrt{1-K}/\mathscr{E}$, the quasi-local scalar average of A for a compact comoving domain $\mathscr{D}[r_b]$ bounded by $r = r_b$ is the linear functional

$$\langle A\rangle_q[r_b] = \frac{\int_{\mathscr{D}} A\mathscr{F}\mathrm{d}\mathscr{V}}{\int_{\mathscr{D}} \mathscr{F}\mathrm{d}\mathscr{V}} = \frac{\int \mathrm{d}y \int \mathrm{d}x \int_0^{r_b} AY^2\, Y'\mathrm{d}r}{\int \mathrm{d}y \int \mathrm{d}x \int_0^{r_b} Y^2\, Y'\mathrm{d}r}, \tag{5}$$

which applied to $A = \rho, \mathscr{H}, \mathscr{K}$ yields [with the help form (3) and (4)] averaged quantities that do not depend on the "non-spherical" coordinates (x, y) (even if the A depend on all 4 coordinates, see [3, 4]):

$$\frac{4\pi}{3}\langle\rho\rangle_q[r_b] = \frac{\tilde{M}_b}{Y_b^3} = \frac{M_b}{R_b^3}, \qquad \langle\mathscr{K}\rangle_q[r_b] = \frac{\tilde{K}_b}{Y_b^2} = \frac{K_b}{R_b^2}, \tag{6}$$

$$\langle\mathscr{H}\rangle_q[r_b] = \frac{\dot{Y}_b}{Y_b} = \frac{\dot{R}_b}{R_b}, \qquad \langle\mathscr{H}\rangle_q^2[r_b] = \frac{8\pi}{3}\langle\rho\rangle_q[r_b] - \langle\mathscr{K}\rangle_q[r_b], \tag{7}$$

where the subindex $_b$ denotes evaluation at $r = r_b$. Since r_b is arbitrary, we can construct the following functions of (t, r) that evaluate locally (the "q-scalars") from the functionals (6)–(7) by considering domains with varying boundary:

$$\frac{4\pi}{3}\rho_q = \frac{M}{R^3}, \quad \mathscr{H}_q = \frac{\dot{R}}{R}, \quad \mathscr{K}_q = \frac{K}{R^2}, \qquad \mathscr{H}_q^2 = \frac{8\pi}{3}\rho_q - \mathscr{K}_q. \tag{8}$$

The relevant curvature and kinematic tensors of the models: the Riemann ($\mathscr{R}^{ab}_{cd}$), Ricci ($\mathscr{R}_{ab}$), Weyl (C^{acbd}) and electric Weyl ($E^{ab} = u_c u_d C^{acbd}$) tensors, as well as the shear tensor (σ_{ab}), are all expressible in terms of irreducible algebraic decompositions containing only the metric, the projection tensor $h_{ab} = g_{ab} + u_a u_b$ and a common divergence-less tensor $\mathbf{e}_{ab} = h_{ab} - 3\eta_a\eta_b$ ($\eta_a = \sqrt{h_{rr}}\delta^r_a$), with the coefficients given by ρ, $\mathscr{H}$ and their fluctuations with respect to ρ_q, $\mathscr{H}_q$:

$$\mathscr{R}^{ab}_{cd} = \frac{8\pi}{3}\rho\left(3\delta^{[a}_{[c}\delta^{b]}_{d]} + 6\delta^{[a}_{[c}u^{b]}u_{d]} - \delta^a_{[c}\delta^b_{d]}\right) + C^{ab}_{cd}, \tag{9}$$

$$\mathscr{R}^a_b = 4\pi\rho\left(h^a_b + u^a u_b\right), \quad C^{ab}_{cd} = \frac{4\pi}{3}\mathbf{D}_q(\rho)(h^{[a}_{[c} - 3u_{[c}u^{[a})\mathbf{e}^{b]}_{d]}), \tag{10}$$

$$\sigma_{ab} = -\mathbf{D}_q(\mathscr{H})\,\mathbf{e}_{ab}, \quad E_{ab} = \frac{4\pi}{3}\mathbf{D}_q(\rho)\mathbf{e}_{ab}, \tag{11}$$

where the fluctuations $\mathbf{D}_q(\rho)$ and $\mathbf{D}_q(\mathscr{H})$ are defined as

$$\frac{4\pi}{3}\mathbf{D}_q(\rho) = \frac{4\pi}{3}(\rho - \rho_q) = \frac{4\pi}{3}\frac{\rho_q'}{3Y'/Y} = \Psi_2, \tag{12}$$

$$\mathbf{D}_q(\mathscr{H}) = \mathscr{H} - \mathscr{H}_q = \frac{\mathscr{H}_q'}{3Y'/Y} = -\Sigma, \tag{13}$$

with Σ and the conformal invariant Ψ_2 (the eigenvalues of E_{ab} and σ_{ab} in terms of $\mathbf{e}_{ab}$) given by

$$\Sigma = \sigma_{ab}\mathbf{e}^{ab} = -\frac{1}{3}\left(\frac{\dot{Y}'}{Y'} - \frac{\dot{Y}}{Y}\right), \qquad \Psi_2 = E_{ab}\mathbf{e}^{ab} = \frac{\tilde{M}}{Y^3} - \frac{4\pi}{3}\rho. \tag{14}$$

4 Statistical Fluctuations and Invariant Scalars

The statistical fluctuation of a Szekeres scalar A with respect to its q-average $\langle A\rangle_q$ is given by

$$\mathbf{D}_q^{\mathrm{st}}(A) = A(t,r,x,y) - \langle A\rangle_q[r_b](t) \quad \Rightarrow \quad \langle \mathbf{D}^{\mathrm{st}}(A)\rangle_q[r_b] = 0, \tag{15}$$

and is a non-local quantity that depends in inner points of the domain and also on its boundary $r = r_b$. Evidently, the fluctuations $\mathbf{D}_q(\rho)$ and $\mathbf{D}_q(\mathscr{H})$ in (12) and (13) are not statistical fluctuations, as they are evaluated locally and thus $\langle \mathbf{D}_q(\rho)\rangle_q[r_b] \neq 0$ and $\langle \mathbf{D}_q(\mathscr{H})\rangle_q[r_b] \neq 0$ hold. However, as proven in [5, 6], the averages of local quadratic fluctuations coincides with the average of quadratic statistical fluctuations, and thus relates these averages with the variance and covariance statistical moments $\mathbf{Var}_q$ and $\mathbf{Cov}_q$[1]

$$\langle\, [\mathbf{D}_q(\rho)]^2\,\rangle_q = \langle\, [\mathbf{D}_q^{\mathrm{st}}(\rho)]^2\,\rangle_q = \langle \rho^2\rangle_q - \langle\rho\rangle_q^2 = \mathbf{Var}_q(\rho), \tag{16}$$

$$\langle\, [\mathbf{D}_q(\mathscr{H})]^2\,\rangle_q = \langle\, [\mathbf{D}_q^{\mathrm{st}}(\mathscr{H})]^2\,\rangle_q = \langle \mathscr{H}^2\rangle_q - \langle\mathscr{H}\rangle_q^2 = \mathbf{Var}_q(\mathscr{H}), \tag{17}$$

$$\begin{aligned}\langle\, \mathbf{D}_q(\rho)\,\mathbf{D}_q(\mathscr{H})\,\rangle_q &= \langle\, \mathbf{D}_q^{\mathrm{st}}(\rho)\,\mathbf{D}_q^{\mathrm{st}}(\mathscr{H})\,\rangle_q = \langle \rho\mathscr{H}\rangle_q - \langle\rho\rangle_q\langle\mathscr{H}\rangle_q \\ &= \mathbf{Cov}_q(\rho,\mathscr{H}),\end{aligned} \tag{18}$$

where we omitted the domain indicator $[r_b]$ to simplify the notation.

The relation between the local fluctuations $\mathbf{D}_q(\rho)$, $\mathbf{D}_q(\mathscr{H})$ and the covariant scalars Ψ_2 and Σ in (12) and (13) illustrates an interesting and appealing relation between the q-average and the characteristic tensors of the models through the properties (16), (17) and (18): the q-averages of all quadratic contractions of the curvature and shear tensors in (9)–(11) are directly expressible in terms of statistical moments of ρ and $\mathscr{H}$ with respect to $\langle\rho\rangle_q$ and $\langle\mathscr{H}\rangle_q$:

$$\langle \sigma_{ab}\sigma^{ab}\rangle_q = 6\langle\Sigma^2\rangle_q = 6\mathbf{Var}_q(\mathscr{H}), \tag{19}$$

$$\langle E_{ab}E^{ab}\rangle_q = 6\langle(\Psi_2)^2\rangle_q = \frac{32\pi^2}{3}\mathbf{Var}_q(\rho), \tag{20}$$

$$\langle \sigma_{ab}E^{ab}\rangle_q = 6\langle\Sigma\,\mathscr{E}\rangle_q = 8\pi\,\mathbf{Cov}_q(\rho,\mathscr{H}), \tag{21}$$

$$\langle C_{abcd}C^{abcd}\rangle_q = \frac{256\pi^2}{3}\,\mathbf{Var}_q(\rho) = \frac{4}{3}\,\mathbf{Var}_q(\mathscr{R}) = 8\langle E_{ab}E^{ab}\rangle_q, \tag{22}$$

$$\langle \mathscr{R}_{abcd}\mathscr{R}^{abcd}\rangle_q = \frac{256\pi^2}{3}\left[\mathbf{Var}_q(\rho) + \frac{5}{4}\langle\rho^2\rangle_q\right] = \frac{4}{3}\mathbf{Var}_q(\mathscr{R}) + \frac{5}{3}\langle\mathscr{R}^2\rangle_q, \tag{23}$$

[1] This result was proven for LTB models, but it is straightforward to prove that it also holds for the QS Szekeres models.

where we used the fact that the Ricci scalar is $\mathscr{R} = 8\pi\rho$, and thus: $\langle \mathscr{R} \rangle_q = 8\pi \langle \rho \rangle_q$ and $\mathscr{R}_{ab}\mathscr{R}^{ab} = \mathscr{R}^2$. We can express the ratio of Weyl versus Ricci curvature ($\Psi_2/\mathscr{R}$) and anisotropic vs isotropic expansion ($\Sigma/\mathscr{H}$) as ratios between the q-scalars ρ_q, $\mathscr{H}_q$ and their counterparts ρ, $\mathscr{H}$:

$$\frac{6\Psi_2}{\mathscr{R}} = 1 - \frac{\rho_q}{\rho}, \qquad \frac{\Sigma}{\mathscr{H}} = \frac{\mathscr{H}_q}{\mathscr{H}} - 1. \tag{24}$$

Also, the quadratic ratio of Weyl to Ricci curvatures is expressible as the ratio of the averages of $(\Psi_2)^2$ and $\mathscr{R}^2$, and as a sort of "standard deviation" of ρ with respect to $\langle\rho\rangle_q$:

$$\frac{6\langle E_{ab}E^{ab}\rangle_q}{\langle \mathscr{R}_{ab}\mathscr{R}^{ab}\rangle_q} = \frac{6\langle(\Psi_2)^2\rangle_q}{\langle(\mathscr{R})^2\rangle_q} = \frac{6\mathbf{Var}_q(\rho)}{\langle\rho^2\rangle_q} = \frac{\langle\rho^2\rangle_q - \langle\rho\rangle_q^2}{\langle\rho^2\rangle_q}. \tag{25}$$

A similar standard deviation of $\mathscr{H}$ with respect to $\langle\mathscr{H}\rangle_q$ follows as the quotient of averages of quadratic covariant scalars $\sigma_{ab}\sigma^{ab}$ and $\mathscr{H}^2$:

$$\frac{\langle\sigma_{ab}\sigma^{ab}\rangle_q}{6\langle\mathscr{H}^2\rangle_q} = \frac{\langle\Sigma^2\rangle_q}{\langle\mathscr{H}^2\rangle_q} = \frac{\mathbf{Var}_q(\mathscr{H})}{\langle\mathscr{H}^2\rangle_q} = \frac{\langle\mathscr{H}^2\rangle_q - \langle\mathscr{H}\rangle_q^2}{\langle\mathscr{H}^2\rangle_q}, \tag{26}$$

where we used (14).

5 Comparison with Buchert's Average

Buchert's scalar average is the standard proper volume average, $\langle A\rangle_p[r_b]$, hence it is defined by (5) with $\mathscr{F} = 1$:

$$\langle A\rangle_p[r_b] = \frac{\int_{\mathscr{D}} A\mathrm{d}\mathscr{V}}{\int_{\mathscr{D}} \mathrm{d}\mathscr{V}} = \frac{\int \mathrm{d}y \int \mathrm{d}x \int_0^{r_b} A\mathscr{F}^{-1}Y^2\,Y'\mathrm{d}r}{\int \mathrm{d}y \int \mathrm{d}x \int_0^{r_b} \mathscr{F}^{-1}Y^2\,Y'\mathrm{d}r}. \tag{27}$$

Evidently, the scalars Σ and Ψ_2 are not related to the the local fluctuations $\mathbf{D}_p(\rho)$, $\mathbf{D}_p(\mathscr{H})$ (analogous to $\mathbf{D}_p(\rho)$, $\mathbf{D}_p(\mathscr{H})$) through a closed and straightforward manner as (12)–(13). Therefore, the relations between the Buchert's averages of invariant quadratic scalars and the variance and covariance moments with respect to this average is much more complicated than the simple elegant relations (19)–(23). Likewise, we cannot express with this average the ratio of Weyl to Ricci curvature as in (25) and (26).

It is straightforward to show that the "back-reaction" correlations terms that appear when applying Buchert's average to evolution equations vanish if we apply the q-average (5). The Raychaudhuri equation for Szekeres models is

$$\dot{\mathscr{H}} = -\mathscr{H}^2 - \frac{\kappa}{6}\rho - 2\,\Sigma^2, \tag{28}$$

averaging on both sides, using (13) and the commutation rule (we omit the domain indicator)

$$\frac{\partial}{\partial t}\langle A\rangle_q - \left\langle\frac{\partial A}{\partial t}\right\rangle_q = \langle A\rangle\dot{}_q - \langle\dot{A}\rangle_q = 3\langle\mathscr{H}\,A\rangle_q - 3\langle\mathscr{H}\rangle_p\langle A\rangle_q, \tag{29}$$

for $A = \mathscr{H}$ we obtain

$$\langle\mathscr{H}\rangle\dot{}_q[r_b] = -\langle\mathscr{H}\rangle_q^2[r_b] - \frac{4\pi}{3}\langle\rho\rangle_q[r_b] + 2\mathscr{Q}_q[r_b], \tag{30}$$

where $\mathscr{Q}_q[r_b]$ is the back-reaction term :

$$\begin{aligned}\mathscr{Q}_q[r_b] &\equiv \langle\big(\mathscr{H} - \langle\mathscr{H}\rangle_q\big)^2\rangle_q - \langle\big(\mathscr{H} - \mathscr{H}_q\big)^2\rangle_q \\ &= \langle\Big[\mathbf{D}_q^{\text{st}}(\mathscr{H})\Big]^2\rangle_q[r_b] - \langle\big[\mathbf{D}_q(\mathscr{H})\big]^2\rangle_q[r_b] = 0,\end{aligned} \tag{31}$$

which vanishes identically for every domain as a consequence of (16) (see [5, 6]). Hence, (30) reduces exactly to a FLRW Raychaudhuri equation given in terms of q-averages. While Buchert's average satisfies the same commutation rule (29), the scalar Σ^2 is not directly related to a local fluctuation of $\mathscr{H}$ [i.e. as in the relation (13)], hence its application to the Raychaudhuri equation (28) yields a different result: equation (30) with the $_p$ average, but with nonzero back-reaction:

$$\begin{aligned}\mathscr{Q}_p[r_b] &\equiv \langle\big(\mathscr{H} - \langle\mathscr{H}\rangle_p\big)^2\rangle_p - \langle\big(\mathscr{H} - \mathscr{H}_q\big)^2\rangle_p \\ &= \langle\Big[\mathbf{D}_p^{\text{st}}(\mathscr{H})\Big]^2\rangle_p[r_b] - \langle\Big[\mathbf{D}_q^{\text{st}}(\mathscr{H})\Big]^2\rangle_p[r_b] \neq 0,\end{aligned} \tag{32}$$

where we used (17).

6 Evolution Equations

While the re-interpretation of the dynamics through the presence of extra back-reaction terms is not possible with the q-average, the latter yields evolution equations that are complete and self-consistent, as opposed to the evolution equations from Buchert's average that require making extra assumptions on the back-reaction terms in order to close the system. As shown in [3], if we define relative fluctuations (or "exact perturbations") as

$$\Delta^{(A)} \equiv \frac{\mathbf{D}_q(A)}{A_q} = \frac{A - A_q}{A_q} = \frac{A_q'/A_q}{3Y'/Y}, \qquad A = \rho,\ \mathscr{H},\ \mathscr{K}, \tag{33}$$

the dynamics of the models becomes completely determined by the following system of autonomous evolution equations:

$$\dot{\rho}_q = -3\rho_q \mathscr{H}_q, \tag{34}$$

$$\dot{\mathscr{H}}_q = -\mathscr{H}_q^2 - \frac{4\pi}{3}\rho_q, \tag{35}$$

$$\dot{\Delta}^{(\rho)} = -3(1+\Delta^{(\rho)})\,\mathscr{H}_q \Delta^{(\mathscr{H})}, \tag{36}$$

$$\dot{\Delta}^{(\mathscr{H})} = -(1+3\Delta^{(\mathscr{H})})\,\mathscr{H}_q \Delta^{(\mathscr{H})} + \frac{4\pi\rho_q}{3\mathscr{H}_q}(\Delta^{(\mathscr{H})} - \Delta^{(\rho)}), \tag{37}$$

subjected to the algebraic constraints

$$\mathscr{H}_q^2 = \frac{8\pi}{3}\rho_q - \mathscr{K}_q, \qquad 2\Delta^{(\mathscr{H})} = \Omega_q\,\Delta^{(\rho)} + (1-\Omega_q)\,\Delta^{(\mathscr{K})}, \tag{38}$$

where we have introduced the following q-scalar analogous to a FLRW Omega factor

$$\Omega_q \equiv \frac{8\pi\rho_q}{3\mathscr{H}_q^2}, \qquad \Omega_q - 1 = \frac{\mathscr{K}_q}{\mathscr{H}_q^2}. \tag{39}$$

If we consider instead non-local statistical relative fluctuations

$$\Delta_{\text{NL}}^{(A)} = \frac{\mathbf{D}_q^{\text{st}}(A)}{\langle A\rangle_q[r_b]}, \tag{40}$$

such that

$$1 + \Delta_{\text{NL}}^{(A)} = \frac{A_q(r)}{\langle A\rangle_q[r_b]}(1+\Delta^{(A)}), \tag{41}$$

we obtain the following system of evolution equations:

$$\langle\mathscr{H}\rangle\dot{}_q[r_b] = -\langle\mathscr{H}\rangle_q^2[r_b] - \frac{4\pi}{3}\langle\rho\rangle_q[r_b], \tag{42}$$

$$\langle\rho\rangle\dot{}_q[r_b] = -3\langle\mathscr{H}\rangle_q[r_b]\langle\rho\rangle_q[r_b], \tag{43}$$

$$\dot{\Delta}_{\text{NL}}^{(\rho)} = -3(1+\Delta_{\text{NL}}^{(\rho)})\,\langle\mathscr{H}\rangle_q[r_b]\,\Delta_{\text{NL}}^{(\mathscr{H})}, \tag{44}$$

$$\begin{aligned}\dot{\Delta}_{\text{NL}}^{(\mathscr{H})} &= -(1+3\Delta_{\text{NL}}^{(\mathscr{H})})\,\langle\mathscr{H}\rangle_q[r_b]\,\Delta_{\text{NL}}^{(\mathscr{H})} + \frac{4\pi\,\langle\rho\rangle_q[r_b]}{3\langle\mathscr{H}\rangle_q^2[r_b]}(\Delta_{\text{NL}}^{(\mathscr{H})} - \Delta_{\text{NL}}^{(\rho)}) \\ &\quad -2\langle\mathscr{H}\rangle_q[r_b]\left(1 - \frac{\mathscr{H}_q(r)}{\langle\mathscr{H}\rangle_q[r_b]}\right)^2 + 4(\mathscr{H}_q(r) - \langle\mathscr{H}\rangle_q[r_b])\,\Delta_{\text{NL}}^{(\mathscr{H})},\end{aligned} \tag{45}$$

which, in order to render a fully complete system, must be supplemented by the evolution equations (34) and (35) for ρ_q and $\mathscr{H}_q$. This system is subjected to the same algebraic constraints (38), but given in terms of q-averages and non-local relative fluctuations. Evidently, (42)–(45) is much more complicated than (34)–(37), and both systems coincide for comoving observers at the domain boundary $r = r_b$ where $\langle \mathscr{H} \rangle_q[r_b] = \mathscr{H}_q(r_b)$ holds for all t.

7 Conclusions

Szekeres models provide an ideal tool to explore the theoretical consequences of non-trivial inhomogeneity, and in particular, the relation between averaging and the geometric objects that characterize inhomogeneous spacetimes. We have shown how a suitable weighted scalar average (the q-average) allows us to relate the average of invariant scalars and statistical variance and covariance moments of the density and Hubble scalar. We have also shown that the dynamics of the models can be completely determined by evolution equations constructed with these averaged scalars (and functions constructed with them) and their fluctuations, which can be local or non-local (statistical). While the evolution equations of the q-averages lack "back-reaction" terms characteristic of Buchert's average (the q-average with unit weight factor), these evolution equations are complete and self-consistent systems that can be handled numerically in the multiple potential applications of the Szekeres quasi-spherical solution to model building in Cosmology and General Relativity.

Acknowledgments The author acknowledges financial support from grant CONACYT 132132.

References

1. Plebański, J., Krasiński, A.: An introduction to general relativity and cosmology. Cambridge University Press, Cambridge (2006)
2. Bolejko, K., Sussman, R.: Cosmic spherical void via coarse-graining and averaging non-spherical structures. Phys. Lett. B **697**, 265 (2011). doi:10.1016/j.physletb.2011.02.007
3. Sussman, R., Bolejko, K.: A novel approach to the dynamics of Szekeres dust models. Class. Quantum Grav. **29**, 065018 (2012). doi:10.1088/0264-9381/29/6/065018
4. Bolejko, K.: Volume averaging in the quasispherical Szekeres model. Gen. Relat. Gravit. **41**, 1585 (2009). doi:10.1007/s10714-008-0727-0
5. Sussman, R.A.: Back-reaction and effective acceleration in generic LTB dust models. Class. Quantum Gravit. **28**(23), 235002 (2011). doi:10.1088/0264-9381/28/23/235002
6. Sussman, R.A.: Weighed scalar averaging in LTB dust models: part I. Statistical fluctuations and gravitational entropy. Class. Quantum Gravit. **30**(6), 065015 (2013). doi:10.1088/0264-9381/30/6/065015

On the Interplay Between Radial and Angular Reflection Emissivity from the Black Hole Accretion Disc

Jiří Svoboda, Michal Dovčiak, René W. Goosmann and Vladimír Karas

Abstract Accretion processes around relativistic compact objects, such as black holes or neutron stars, can be well studied through X-ray spectroscopy. The disc reflection spectra detected in observations of several active galactic nuclei and X-ray binaries in our and nearby galaxies suggest high steepness of the radial emissivity. This can be primarily caused by compactness of the illuminating radiation. In our recent paper [1], we showed that the measurement of the steep radial emissivity index might also be over-estimated by ignoring the radial ionisation structure and the proper angular-emissivity law, which is non-trivial in the fully relativistic regime. In this paper, we demonstrate the interplay of the angular and radial emissivity. Employing an improper angular emissivity in the reflection models leads to over-estimated values for the black-hole angular momentum and the radial-emissivity index (by about 10 %).

1 Introduction

X-ray spectroscopy provides a unique way to study accretion physics in the strong-gravity field. X-ray reflection spectra have been used to measure angular momentum of black holes as well as the other geometrical parameters of the accretion discs, such as its inclination angle, radial emissivity, inner and outer disc radius (for a review, see

J. Svoboda (✉)
European Space Astronomy Centre of ESA, P.O. Box 78, Villanueva de la Cañada, 28691 Madrid, Spain
e-mail: jsvoboda@sciops.esa.int

J. Svoboda · M. Dovčiak · V. Karas
Astronomical Institute, Academy of Sciences, Boční II 1401, 14131 Prague, Czech Republic

R.W. Goosmann
Observatoire astronomique de Strasbourg, Equipe Hautes Energies 11 rue de l 'Université, 67000 Strasbourg, France

J. Bičák and T. Ledvinka (eds.), *Relativity and Gravitation*, Springer Proceedings in Physics 157, DOI: 10.1007/978-3-319-06761-2_57,

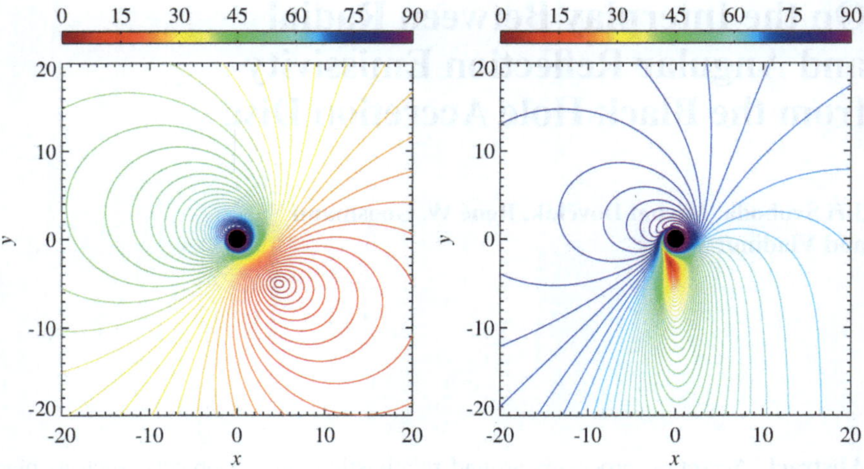

Fig. 1 Contours of the local emission angle, $\theta_e(r, \phi)$, near a maximally rotating black hole, $a = 1$, depicted in the equatorial plane (x, y). The black hole and the accretion disc rotate counter clockwise. The inner region is shown up to $r = 20$ gravitational radii from the black hole. A distant observer is located towards the *top* of the figure. The inclination angle is 30 (*left*) and 70° (*right*). Adopted from [2]

e.g. [3]). Current relativistic kernels that are applied to reflection models to include the relativistic effects on the spectral shape are based on a series of simplified assumptions and, in particular, they assume a single/broken power-law form of the radial reflection emissivity and an angular emissivity law defined by a simple analytical formula, most frequently employing a limb-darkening profile [4]. Due to the abberation effect and strongly curved photon trajectories, the emission angle is not identical with the inclination angle. It is particularly prominent at the innermost region where it is always very high (see Fig. 1). Such a complex structure implies a non-trivial link between the radial and the angular emissivity when fitting the data.

The intrinsic disc radial emissivity is naturally expected to decrease with increasing distance, i.e. the reflection emissivity is

$$\varepsilon(r) = r^{-q}, \tag{1}$$

where q is the emissivity index that can be constant over all radii or a varying quantity. The thermal dissipation of the disc decreases as r^{-3} [5]. Therefore, the simplest assumption is postulating the same dependence for the reflection. The most energetic photons are produced in the innermost regions, where by consequence the strongest irradiation of the disc occurs. In addition, assuming a point-like X-ray source at height h on the disc axis, the irradiation of the disc in the absence of any relativistic effect is proportional to $(r^2 + h^2)^{-3/2} \propto r^{-3}$, as shown e.g. by [6]. An emissivity profile with $q = 3$ is therefore considered as *standard*, while steeper/flatter indices may need to be explained.

Steep emissivity profiles have been measured in X-ray spectra of several active galactic nuclei, such as MCG-6-30-15 [7], 1H0707-495 [8], and IRAS 13224-3809 [9], as well as X-ray binaries with a black hole, such as XTE J1650-500 [10], GX 339-4 [11], and Cyg X-1 [12], or with a neutron star, e.g. Cyg X-2 [13]. The measured indices reach values up to $q \approx 7$.

Martocchia et al. [14], and Wilkins and Fabian [15] examined whether the required steep emissivity law as well as the predicted equivalent width of the cold reflection line of iron and the Compton reflection component can be reproduced in a phenomenological (lamp-post) model where the X-ray illuminating source is located on the common symmetry axis of the black hole and the equatorial accretion disc. They suggested that the radial emissivity function of the reflection component steepens when the height parameter of the primary irradiation source decreases.

In this paper, we discuss the effect of the angular emissivity in addition to the assumption of the compact centrally localised corona. The reader may also look at [2] and [1], where more geometrical cases are considered as well as the radially dependent ionisation structure (in the latter one). This paper is organised as follows. In Sect. 2, we describe our model set-up. Analytical approaches of the angular emissivity are compared with the numerical one in Sect. 3. Inter-dependence of the angular and radial emissivity is demonstrated in Sect. 4, and the main conclusions are drawn in Sect. 5.

2 Model Pre-requisites

The configuration of a very compact corona located on the rotational axis just above the black hole, known also as the lamp-post scheme, was studied as a simple disc-corona scenario by George and Fabian [16] and Matt et al. [17]. In this scenario, the irradiation far from the source radially decreases as r^{-3}. In the central region, the relativistic effects influence the disc illumination, thus shape the reflection spectra of black hole accretion discs [18]. As a result, the different parts of the disc are irradiated with different intensities, making the emissivity profiles in reflection models distinct from the standard value of $q = 3$. If the height of the source is sufficiently close to the black hole event horizon, light bending implies higher irradiation of the innermost region compared to the outer parts of the disc. The exact profile of the radial emissivity depends on the geometrical properties of the source. Different cases of axial, orbiting, jet, and extended sources were studied by Wilkins and Fabian [15]. The steepest profiles were obtained for point-like sources at small heights along the vertical axis.

The physical set-up of our model is a combination of the general-relativistic lamp-post scheme for an X-ray illuminated accretion disc near a rotating black hole [14] based on the KY package [19] and a self-consistent Monte Carlo scheme for the X-ray reprocessing within the disc environment [20]. We used local (re-) emission tables that were computed by the radiative transfer code NOAR [21] for the case of "cold" reflection (i.e. for neutral or weakly ionised matter). Photo-absorption, Compton scattering, and the fluorescent emission of the iron K line are considered.

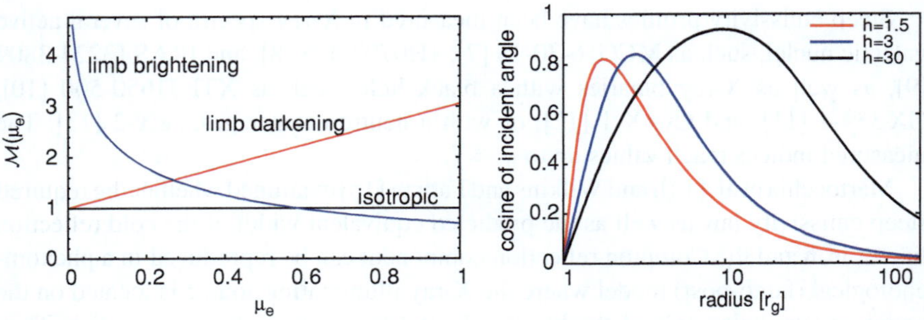

Fig. 2 *Left* Comparison of simple approximations of angular emissivity defined by (2). *Right* Radial dependence of the incident angle in the lamp-post geometry for different heights. The spin is $a = 1$ and the inclination angle is 30°

A stratified plane-parallel atmosphere irradiated by a power law with the photon index $\Gamma = 1.9$ is assumed. A large number of primary photons were sampled in the 2–300 keV energy range to ensure sufficiently high quality statistics. At all local emission angles, the Poissonian noise level is much smaller than any relevant spectral feature. The computations were done for various incident local emission angles, both polar and azimuthal.

3 Angular Emissivity: Simple Approximations Versus Numerical Results

The angular emissivity law, $\mathcal{M}(\mu_e, r_e, E_e)$, defines the distribution of the intrinsic intensity outgoing from each radius r_e of the disc surface with respect to the perpendicular direction. The emission angle $\theta_e = \arccos \mu_e$ is measured from the disc normal direction to the equatorial plane, in the disc co-moving frame. The general expression is usually simplified to a simple function of μ_e. Three cases are the most frequently considered:

$$\begin{array}{l} \text{Case 1:} \\ \text{Case 2:} \; \mathcal{M}(\mu_e) = \\ \text{Case 3:} \end{array} \begin{cases} 1 + 2.06\,\mu_e & \text{(limb-darkening) [4]} \\ 1 & \text{(locally isotropic emission)} \\ \ln(1 + \mu_e^{-1}) & \text{(limb-brightening) [22]} \end{cases} \tag{2}$$

Figure 2 (left panel) shows the comparison of these analytical approaches. Limb brightening and limb darkening are completely different. While the limb-darkening law diminishes radiation reflected into high emission angles the limb-brightening profile enhances such a reflected emission. Although limb darkening is the most frequently used approximation in the current spectroscopic analyses, several X-ray simulations of irradiated disc atmospheres suggest the presence of limb brightening (see e.g. [23–25]) instead.

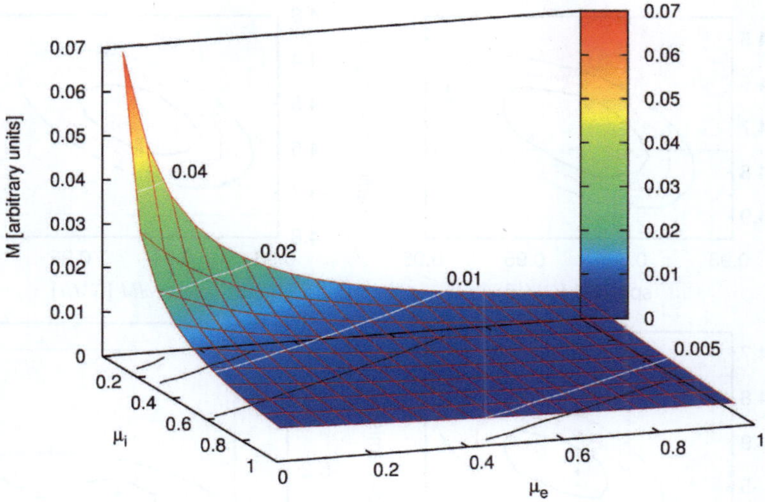

Fig. 3 Angular emissivity from our numerical model depending on the incident angle μ_i and the emission angle μ_e

Our results also exhibit the evidence for limb brightening, as shown in Fig. 3. The incident angle plays an important role. A strong limb-brightening effect occurs for grazing incident angles, while almost isotropic dependence is obtained for irradiation of the disc from above. We note that in the strongly curved space-time around a rapidly rotating black hole, the irradiation from "above" does not occur under the lamp, as it would be in the Newtonian physics. In the innermost region, both the incident and emission angles are grazing (see Fig. 1 and the right panel of Fig. 2). This is due to the combined effects of aberration and light bending which grow greatly near the inner rim of the disc.

4 Inter-dependence Between Angular and Radial Emissivity

The angular emissivity is linked with the radial emissivity owing to the significant radial dependence of the emission angle (see Fig. 1). The local emission angle spans the entire range, from 0 (perpendicular, face-on) to 90° (grazing, edge-on) degrees. The highest value is obtained in the innermost region, and only far from the centre the emission angle is consistent with the inclination angle, i.e. $\mu_e(r, \phi) \rightarrow \mu_{obs}$ for $r \rightarrow \infty$.

Figure 4 shows the contour plots between the black hole angular momentum and the radial emissivity index for different cases of angular emissivity. The simulated data were created with our numerical procedure. The details of the simulations are described in Sect. 2 of [1]. The radial-emissivity index is slightly under-estimated by

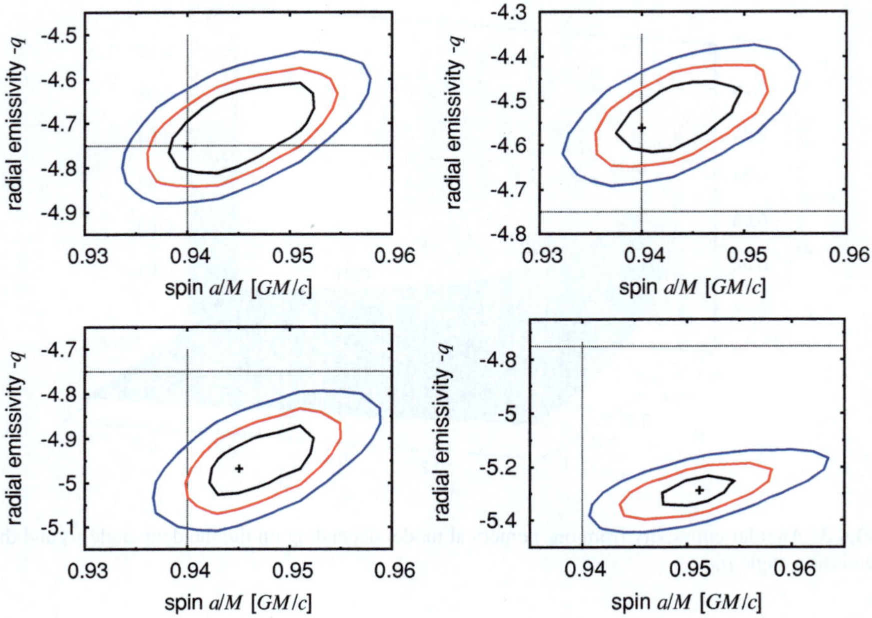

Fig. 4 Contour plots of the spin a and the radial emissivity parameter q. The data were generated with the lamp-post model with the height $h = 1.5\, r_{\rm g}$. The default value of the spin was $a = 0.94$, which is indicated by a *dashed line* in the graph. Different prescriptions for the angular emissivity were used: *Top left* Angular emissivity from numerical calculations. *Top right* Limb brightening. *Bottom left* Isotropic. *Bottom right* Limb darkening. The *contour lines* correspond to 1σ, 2σ, and 3σ levels. Adopted from [1]

the model with limb brightening and it is slightly over-estimated by the model with the isotropic angular emissivity. This illustrates the fact that our numerical results do exhibit limb-brightening effect but on average over all incident angles it is not as strong effect as limb brightening defined by the analytical formula in (2). The model with limb darkening clearly produces the worst fit, it over-estimates the radial emissivity index by $\approx 10\,\%$ and also over-estimates the spin-value.

5 Conclusions

The angular emissivity significantly modifies the total emission expected from the reflection on a black hole accretion disc. The non-trivial dependence of the emission angle on the radial and azimuthal coordinate in the disc implies that (i) there is a link between the radial and angular emissivity, and (ii) the black hole spin measurements via reflection models are affected by the employed angular-emissivity prescription. Our numerical model suggests the presence of limb brightening whose exact profile depends on the incident angle, and thus on the geometry of the primary-radiation source. The strongest limb-brightening effect is obtained for the highest value of the incident angle while almost isotropic dependence is present in the case of irradiation from above. The highest reflection efficiency occurs at the innermost region around a rapidly rotating black hole, where both the incident and emission angles are largest. Averaged over all incident angles, our limb-brightening effect is weaker than the analytical prescription of the limb-brightening law defined by Haardt [22].

References

1. Svoboda, J., Dovčiak, M., Goosmann, R., et al.: Origin of the X-ray disc-reflection steep radial emissivity. Astron. Astrophys. **545**, A106 (2012). doi:10.1051/0004-6361/201219701
2. Svoboda, J., Dovčiak, M., Goosmann, R., Karas, V.: Role of emission angular directionality in spin determination of accreting black holes with a broad iron line. Astron. Astrophys. **507**, 1 (2009). doi:10.1051/0004-6361/200911941
3. Reynolds, C., Nowak, M.: Fluorescent iron lines as a probe of astrophysical black hole systems. Phys. Rep. **377**, 389 (2003). doi:10.1016/S0370-1573(02)00584-7
4. Laor, A.: Line profiles from a disk around a rotating black hole. Astrophys. J. **376**, 90 (1991). doi:10.1086/170257
5. Novikov, I., Thorne, K.: Black hole astrophysics. In: DeWitt, C., DeWitt, B. (eds.) Black Holes, pp. 343–450. Gordon and Breach, New York (1973)
6. Reynolds, C., Begelman, M.: Iron fluorescence from within the innermost stable orbit of black hole accretion disks. Astrophys. J. **488**, 109 (1997). doi:10.1086/304703
7. Fabian, A., Vaughan, S., Nandra, K., et al.: A long hard look at MCG-6-30-15 with XMM-Newton. Mon. Not. R. Astron. Soc. **335**, L1 (2002). doi:10.1046/j.1365-8711.2002.05740.x
8. Fabian, A., Zoghbi, A., Ross, R., et al.: Broad line emission from iron K- and L-shell transitions in the active galaxy 1H0707-495. Nature **459**, 540 (2009). doi:10.1038/nature08007
9. Ponti, G., Gallo, L., Fabian, A., et al.: Relativistic disc reflection in the extreme NLS1 IRAS13224-3809. Mon. Not. R. Astron. Soc. **406**, 2591 (2010). doi:10.1111/j.1365-2966.2010.16852.x
10. Miniutti, G., Fabian, A., Miller, J.: The relativistic Fe emission line in XTE J1650–500 with BeppoSAX: Evidence for black hole spin and light-bending effects? Mon. Not. R. Astron. Soc. **351**, 466 (2004). doi:10.1111/j.1365-2966.2004.07794.x
11. Miller, J.: Relativistic X-Ray lines from the inner accretion disks around black holes. Astron. Astrophys. Rev. **45**, 441 (2007). doi:10.1146/annurev.astro.45.051806.110555
12. Fabian, A., Wilkins, D., Miller, J., et al.: On the determination of the spin of the black hole in Cyg X-1 from X-ray reflection spectra. Mon. Not. R. Astron. Soc. **424**, 217 (2012). doi:10.1111/j.1365-2966.2012.21185.x
13. Cackett, E., Miller, J., Ballantyne, D., et al.: Relativistic lines and reflection from the inner accretion disks around neutron stars. Astrophys. J. **720**, 205 (2010). doi:10.1088/0004-637X/720/1/205
14. Martocchia, A., Karas, V., Matt, G.: Effects of Kerr space-time on spectral features from X-ray illuminated accretion discs. Mon. Not. R. Astron. Soc. **312**, 817 (2000). doi:10.1046/j.1365-8711.2000.03205.x
15. Wilkins, D., Fabian, A.: Understanding X-ray reflection emissivity profiles in AGN: Locating the X-ray source. Mon. Not. R. Astron. Soc. **424**, 1284 (2012). doi:10.1111/j.1365-2966.2012.21308.x
16. George, I., Fabian, A.: X-ray reflection from cold matter in active galactic nuclei and X-ray binaries. Mon. Not. R. Astron. Soc. **249**, 352 (1991)

17. Matt, G., Perola, G., Piro, L.: The iron line and high energy bump as X-ray signatures of cold matter in Seyfert 1 galaxies. Astron. Astrophys. **247**, 25 (1991)
18. Miniutti, G., Fabian, A.: A light bending model for the X-ray temporal and spectral properties of accreting black holes. Mon. Not. R. Astron. Soc. **349**, 1435 (2004). doi:10.1111/j.1365-2966.2004.07611.x
19. Dovčiak, M., Karas, V., Yaqoob, T.: An extended scheme for fitting X-Ray data with accretion disk spectra in the strong gravity regime. Astrophys. J. Suppl. Ser. **153**, 205 (2004). doi:10.1086/421115
20. Goosmann, R., Czerny, B., Mouchet, M., et al.: Magnetic flares in active galactic nuclei: modeling the iron Kα line. Astron. Nachr. **327**, 977 (2006). doi:10.1002/asna.200610674
21. Dumont, A.M., Abrassart, A., Collin, S.: A code for optically thick and hot photoionized media. Astron. Astrophys. **357**, 823 (2000)
22. Haardt, F.: Anisotropic Comptonization in thermal plasmas—spectral distribution in plane-parallel geometry. Astrophys. J. **413**, 680 (1993). doi:10.1086/173036
23. Czerny, B., Różańska, A., Dovčiak, M., Karas, V., Dumont, A.M.: The structure and radiation spectra of illuminated accretion disks in AGN. II. Flare/spot model of X-ray variability. Astron. Astrophys. **420**, 1 (2004). doi:10.1051/0004-6361:20035741
24. Goosmann, R., Mouchet, M., Czerny, B., et al.: Iron lines from transient and persisting hot spots on AGN accretion disks. Astron. Astrophys. **475**, 155 (2007). doi:10.1051/0004-6361:20078273
25. Różańska, A., Madej, J., Konorski, P., Sadowski, A.: Iron lines in model disk spectra of Galactic black hole binaries. Astron. Astrophys. **527**, A47 (2011). doi:10.1051/0004-6361/201015626

Critical-Curve Topologies of Triple Gravitational Lenses

Kamil Daněk and David Heyrovský

Abstract An extrasolar analog of the Sun–Jupiter–Saturn system has been discovered recently by detecting its gravitational microlensing action on the flux from a background star [1]. More generally, however, gravitational lensing by a system of three bodies has not yet been satisfactorily analyzed theoretically. Correct interpretation of microlensing light curves requires an understanding of the geometry of the underlying lens caustic and critical curves. These curves correspond to source positions and image positions, respectively, with infinite point-source-flux amplification. Following the pioneering Erdl and Schneider analysis of the parameter dependence of binary lensing [2], we extend their approach to special cases of the triple lens. While the binary lens is characterized by two parameters, three more parameters are needed to describe the triple lens. We present here an example of a three-dimensional cut through the five-dimensional parameter space, identifying the boundaries of regions with different critical-curve topology.

1 Lens Equation and Critical-Curve Topology

An n-point-mass gravitational lens deflects light from a background source and forms multiple images, which can be found by solving the lens equation

$$\boldsymbol{\beta} = \boldsymbol{\theta} - \sum_{i=1}^{n} \frac{4GM_i}{c^2} \frac{D_{LS}}{D_L D_S} \frac{\boldsymbol{\theta} - \boldsymbol{\theta}_\mathbf{i}}{|\boldsymbol{\theta} - \boldsymbol{\theta}_\mathbf{i}|^2}, \tag{1}$$

where $\boldsymbol{\beta}$ is the angular source position, $\boldsymbol{\theta}$ angular image position, $\boldsymbol{\theta}_\mathbf{i}$ angular lens positions, M_i lens masses, G gravitational constant, c speed of light, distances D_L

K. Daněk (✉) · D. Heyrovský
Institute of Theoretical Physics, Faculty of Mathematics and Physics,
Charles University in Prague, Prague, Czech Republic
e-mail: Kamil.Danek@gmail.com

J. Bičák and T. Ledvinka (eds.), *Relativity and Gravitation*,
Springer Proceedings in Physics 157, DOI: 10.1007/978-3-319-06761-2_58,

from observer to lens, D_S observer to source, and D_{LS} lens to source. We can express all angular positions in units of the Einstein radius

$$\theta_E = \sqrt{\frac{4GM}{c^2}\frac{D_{LS}}{D_L D_S}} \tag{2}$$

corresponding to total mass M, and instead of vectors we use complex notation introduced by [3], so that $\boldsymbol{\beta} \to \zeta, \boldsymbol{\theta} \to z, \boldsymbol{\theta}_i \to z_i$.

For $n = 3$ with relative masses $\mu_i = M_i/M$ we then get the triple lens equation in the form

$$\zeta = z - \frac{\mu_1}{\bar{z} - \bar{z}_1} - \frac{\mu_2}{\bar{z} - \bar{z}_2} - \frac{\mu_3}{\bar{z} - \bar{z}_3}. \tag{3}$$

Its critical curve parameterized by phase $\varphi \in [0, \pi)$ is given by

$$\frac{\mu_1}{(z - z_1)^2} + \frac{\mu_2}{(z - z_2)^2} + \frac{\mu_3}{(z - z_3)^2} = e^{-2i\varphi}, \tag{4}$$

and the caustic $\zeta(z(\varphi))$ is obtained by plugging critical-curve points $z(\varphi)$ into (3). The critical curve generally consists of several separate loops that may merge (or split) only when the critical curve passes through a saddle point of the lens-equation Jacobian. These points can be found by solving

$$\frac{\mu_1}{(z - z_1)^3} + \frac{\mu_2}{(z - z_2)^3} + \frac{\mu_3}{(z - z_3)^3} = 0. \tag{5}$$

Lens parameters corresponding to these topology changes can be obtained using the Sylvester matrix method to find the conditions for the existence of common roots of polynomials obtained from (4) and (5).

For the binary lens it is possible to obtain the conditions in analytical form, but for the general triple lens the equations are prohibitively intricate. However, for various simple two-parameter triple-lens models the conditions for topology change can be found in the form of polynomial equations in terms of the lens parameters. Several such examples can be found in [4]. We also found an algorithm for obtaining the conditions numerically for three-parameter triple-lens models. We demonstrate here the results obtained for the general equal-mass triple lens.

2 Equal-Mass Triple Lens

In this model we treat the "size" and "shape" of the triangular lens configuration separately. The size is expressed by the perimeter p, while the shape is determined

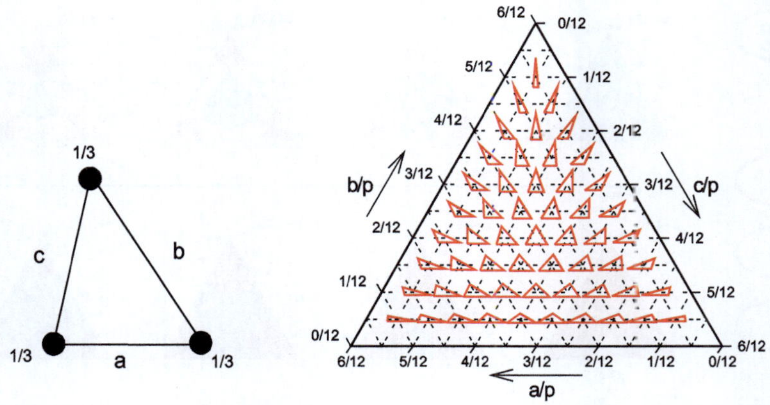

Fig. 1 *Left panel* Sketch of triple lens with equal-mass lenses at the vertices *Right panel* Relative side-length ternary plot showing the shapes of triangles corresponding to the grid-point at their centroid

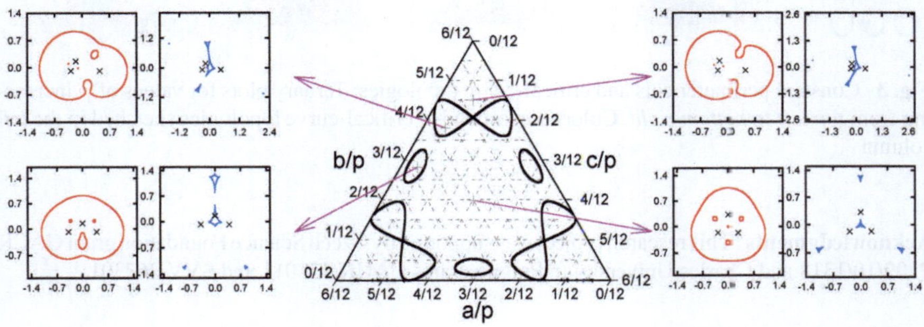

Fig. 2 *Center* Parameter space cut for perimeter $p = 1.68$ divided according to critical-curve topology. *Left and Right* Examples of critical curves (*red*) and caustics (*blue*) from each region, with lens positions marked by *black* crosses

by the position in the ternary plot shown in Fig. 1. The full parameter space is represented by a sequence of vertically stacked ternary plots with increasing p.

For illustration we show the $p = 1.68$ cut in Fig. 2. Here the parameter space is divided into 13 regions, but because of symmetry there are only four types of regions with three critical-curve topologies. In Fig. 3 we present a sequence of ternary plots with increasing p. There are 39 disjoint regions in the 3-D parameter space that can be grouped into 12 types using the symmetry of the problem. The equal-mass triple lens permits nine different critical-curve topologies, sketched at the left side of Fig. 3.

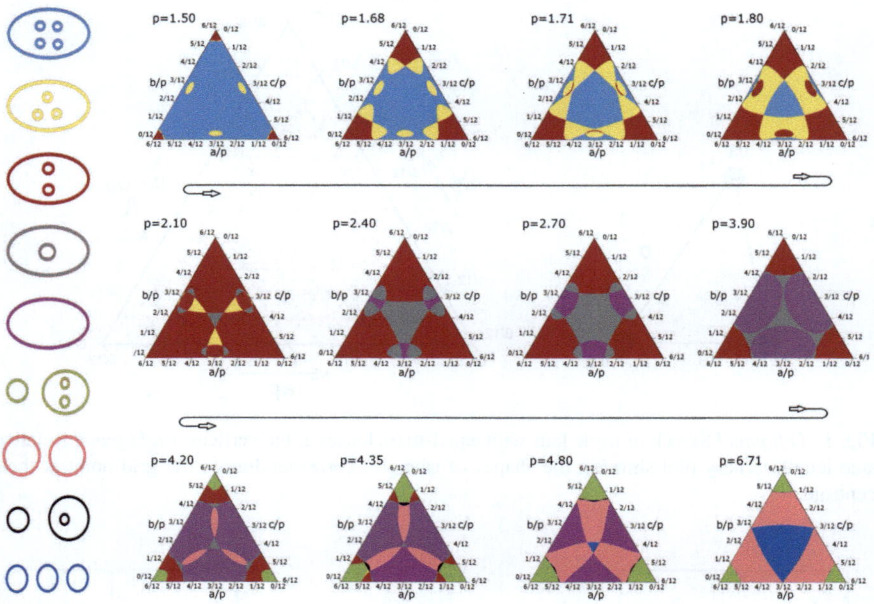

Fig. 3 Constant perimeter cuts and critical-curve topologies: Ternary plots for values of p increasing from *top left* to *bottom right*. Colors correspond to critical-curve topologies sketched in the *left* column

Acknowledgments This research project was supported by Czech Science Foundation grant GACR P209/10/1318 and Charles University in Prague grants GAUK 428011 and SVV 267301.

References

1. Gaudi, B.E.A.: Discovery of a Jupiter/Saturn analog with gravitational microlensing, Science **319**, 927 (2008). doi:10.1126/science.1151947
2. Erdl, H., Schneider, P.: Classification of the multiple deflection two point-mass gravitational lens models and application of catastrophe theory in lensing. Astron. Astrophys. **268**, 453 (1993)
3. Witt, H.: Investigation of high amplification events in light curves of gravitationally lensed quasars. Astron. Astrophys. **236**, 311 (1990)
4. Daněk, K.: Gravitational microlensing by a few-body system. Master's thesis, Charles University, Prague (2010)

Modified Gravity Theories and Dark Matter Models Tested by Galactic Rotation Curves

Marek Dwornik, Zoltán Keresztes and László Árpád Gergely

Abstract Bose–Einstein condensate dark matter model and Randall–Sundrum type 2 brane-world theory are tested with galactic rotation curves. Analytical expressions are derived for the rotational velocities of test particles around the galactic center in both cases. The velocity profiles are fitted to the observed rotation curve data of high surface brightness and low surface brightness galaxies. The brane-world model fits better the rotation curves with asymptotically flat behaviour.

1 Introduction

Luminous matter alone can not explain the observed behaviour of the galactic rotation curves and an invisible, dark matter component is needed [1]. One possibility to explain dark matter is to introduce collisionless dark scalar particles in the form of a Bose–Einstein condensate (BEC) [2, 3]. However, up to now the evidence for dark matter has been only found by its gravitational interaction. It cannot be excluded that general relativity breaks down at scales of galaxies and beyond. Therefore several modified gravity models have been proposed to replace dark matter [4–6]. The Weyl fluid appearing in Randall–Sundrum type 2 (RS2) brane-world models can behave as an effective source for gravity and it is able to replace dark matter in galactic dynamics [7, 8].

We investigate here galactic rotation curves in RS2 brane-world and the BEC model for high surface brightness (HSB) and low surface brightness (LSB) galaxies.

M. Dwornik (✉), Z. Keresztes, · L. Á. Gergely
Departments of Theoretical and Experimental Physics, University of Szeged,
Dóm tér 9., Szeged 6720, Hungary
e-mail: marek@titan.physx.u-szeged.hu

J. Bičák and T. Ledvinka (eds.), *Relativity and Gravitation*,
Springer Proceedings in Physics 157, DOI: 10.1007/978-3-319-06761-2_59,

2 The Baryonic Matter

Distribution of baryonic matter in HSB galaxies is described as the sum of a thin stellar disk and a spherically symmetric bulge component.

We assume that the mass distribution of bulge component with radius r_{bulge} follows the de-projected luminosity distribution with a factor called mass-to-light ratio σ. The surface brightness profile of the spheroidal bulge is described by a generalized Sérsic function [9]: $I_{bulge}(r) = I_{0,bulge} \exp\left[-(r/r_0)^{1/n}\right]$, where $I_{0,bulge}$ and r_0 are the central surface brightness and the characteristic radius of the bulge, respectively, and n is the shape parameter of the magnitude-radius curve. The contribution to the rotational velocity is

$$v_{bulge}^2(r) = \frac{GM_{bulge}(r)}{r}, \tag{1}$$

with gravitational constant G and mass of the bulge $M_{bulge}(r) = \sigma I_{bulge}(r)$ within the radius r projected on the sky.

In a spiral galaxy the radial surface brightness profile of the disk decreases exponentially with the radius [10]: $I_{disk}(r) = I_{0,disk} \exp(-r/h)$, where $I_{0,disk}$ is the disk central surface brightness and h is a characteristic disk length scale. The contribution of the disk to the circular velocity is [10]:

$$v_{disk}^2(x) = \frac{GM_{disk}}{2h} x^2 (I_0 K_0 - I_1 K_1), \tag{2}$$

where $x = r/h$ and M_{disk} is the total mass of the disk. The functions I_m and K_m are the modified Bessel functions of the first and second kind with order m, respectively. The Bessel functions are evaluated at $x/2$.

The best fitting values of $I_{0,bulge}$, n, r_0, r_{bulge}, $I_{0,disk}$ and h are derived from the available photometric data. In case of LSB galaxies the baryonic model only consists of a thin stellar disk component which is the same as for the HSB galaxies.

3 Models for the Dark Matter Component

The mass density distribution of the static gravitationally bounded Bose–Einstein condensate is described by the Lane–Emden equation. An analytical solution for dark matter BEC was obtained in Ref. [3]: $\rho_{BEC}(r) = \rho_{BEC}^{(c)} (\sin(kr)/kr)$, where $k = \sqrt{Gm^3/\hbar^2 a}$ and $\rho_{BEC}^{(c)}$ is the central mass density of the condensate. The mass profile of the galactic halo is $m_{BEC}(r) = 4\pi \int \rho_{BEC}(r) r^2 dr$ giving the following contribution to the rotational velocity [3]:

$$v_{BEC}^2(r) = \frac{4\pi G \rho_{BEC}^{(c)}}{k^2} \left(\frac{\sin kr}{kr} - \cos kr \right). \tag{3}$$

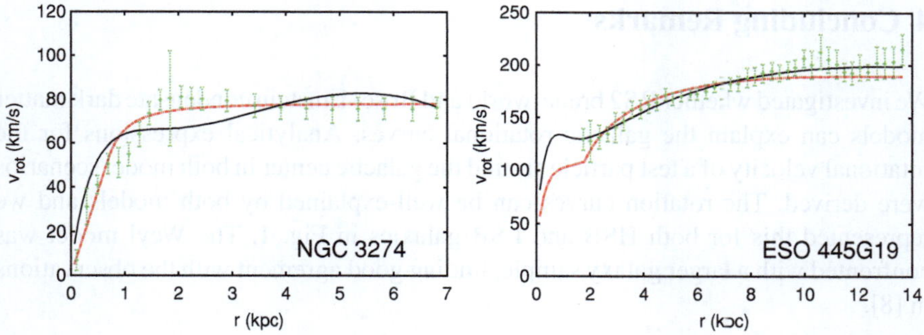

Fig. 1 Best fit curves for the observed velocity profiles of the HSB galaxy ESO445G19 and the LSB galaxy NGC3274. The *solid black* and *dashed red curves* show the BEC and Weyl models, respectively. In case of HSB galaxy, the shape of the curve near the center where baryonic matter dominates is determined by the photometric data

In RS2 brane-world theory the 4-dimensional effective Einstein equation has extra source terms, which arise from the embedding of the 3-brane in the bulk [11]. We assume that the brane embedding is Z_2-symmetric and there is no matter in the 5-dimensional spacetime but there is a cosmological constant. Nevertheless the effect of the brane cosmological constant arising from the brane tension and the 5-dimensional cosmological constant is neglected at the scales of galaxies. Then at low energies there is only one extra source term in the effective Einstein equation arising from the 5-dimensional Weyl curvature, which acts as a fluid (the Weyl fluid). The contribution to the rotational velocity in a Post-Newtonian approximation is derived in [8]:

$$v_{Weyl}^2(r) \approx \frac{G\left(M_0^{tot}\right)}{r} + c^2\beta + c^2 C\left(\frac{r_b}{r}\right)^{1-\alpha}, \quad r > r^*, \tag{4}$$

with constants α, β, C and M_0^{tot} charaterizing the Weyl fluid and velocity of light c. A scaling constant r_b was introduced such that C becomes dimensionless and $r_b = r_{bulge}$ was chosen. The rotational velocity (4) is valid for any $r > r^*$, where r^* represents the radius beyond which the baryonic matter does not extend. We assume that (i) the contribution of the Weyl fluid can be neglected within the bulge radius, and (ii) the observed rotation curves within r_{bulge} can be explained with baryonic matter alone (it is given by the sum of (1) and (2) for HSB and by (2) for LSB galaxies). Outside r_{bulge} we assume the effects of the disk can be handled as perturbation, therefore the rotational curves are generated by the sum of (2) and (4).

The best fit rotational curves of the Weyl and BEC models for the observed velocity profiles of the HSB galaxy ESO445G19 and the LSB galaxy NGC3274 are shown in Fig. 1.

4 Concluding Remarks

We investigated whether RS2 brane-world and Bose–Einstein condensate dark matter models can explain the galactic rotational curves. Analytical expressions for the rotational velocity of a test particle around the galactic center in both model scenarios were derived. The rotation curves can be well-explained by both models and we represented this for both HSB and LSB galaxies in Fig. 1. The Weyl model was confronted with a larger galaxy sample, finding good agrement with the observations in [8].

References

1. Persic, M., Salucci, P., Stel, F.: The universal rotation curve of spiral galaxies—I. The dark matter connection. Mon. Not. R. Astron. Soc. **281**, 27 (1996)
2. Sin, S.: Late-time phase transition and the galactic halo as a Bose liquid. Phys. Rev. D **50**, 3650 (1994). doi:10.1103/PhysRevD.50.3650
3. Böhmer, C., Harko, T.: Can dark matter be a Bose-Einstein condensate? J. Cosmol. Astropart. Phys. **2007**(06), 025 (2007). doi:10.1088/1475-7516/2007/06/025
4. Milgrom, M.: A modification of the Newtonian dynamics as a possible alternative to the hidden mass hypothesis. Astrophys. J. **270**, 365 (1983). doi:10.1086/161130
5. Sanders, R.: Anti-gravity and galaxy rotation curves. Astron. Astrophys. **136**, L21 (1984)
6. Roberts, M.: Galactic metrics. Gen. Rel. Grav. **36**, 2423 (2004). doi:10.1023/B:GERG.0000046830.85831.4a
7. Mak, M., Harko, T.: Can the galactic rotation curves be explained in brane world models? Phys. Rev. D. **70**, 024010 (2004). doi:10.1103/PhysRevD.70.024010
8. Gergely, L., Harko, T., Dwornik, M., Kupi, G., Keresztes, Z.: Galactic rotation curves in brane world models. Mon. Not. R. Astron. Soc. **415**, 3275 (2011). doi:10.1111/j.1365-2966.2011.18941.x
9. Sérsic, J.: Atlas de Galaxias Australes. Observatorio Astronomico, Universidad Nacional de Córdoba, Córdoba (1968)
10. Freeman, K.: On the disks of spiral and so galaxies. Astrophys. J. **160**, 811 (1970). doi:10.1086/150474
11. Shiromizu, T., Maeda, K.I., Sasaki, M.: The Einstein equations on the 3-brane world. Phys. Rev. D. **62**, 024012 (2000). doi:10.1103/PhysRevD.62.024012

Averaging Inside the LRS Family

Petr Kašpar, David Vrba and Otakar Svítek

Abstract Averaging problem in GR and cosmology is of fundamental importance. It is still not clear how to unambiguously average the Einstein equations and the metric tensor. One of the most promising attempts how to deal with averaging in GR are the Buchert equations. However, only scalar part of the Einstein equations is averaged and the system is not closed. Here we will present LRS (locally rotationally symmetric) spacetimes, where one can find preferred spatial direction and the evolution and the constraint equations are described only by scalars. By averaging these scalars we will obtain generalized Buchert equations for LRS spacetimes.

1 Introduction

The averaging problem in general relativity was studied by Ellis [1] and has been investigated many times since then. However, despite some attempts [2], it is not obvious, how to take an average of a tensorial quantity. On the other, hand averaging of scalars according to Buchert [3] is a fully covariant operation, though it has some drawbacks as well. It is performed over some domain on a spatial hypersurface and it depends on the slicing of the spacetime and on the scale.

The key problem in cosmology is that calculating the Einstein tensor from the averaged metric is not the same thing as calculating the Einstein tensor from inhomogeneous metric and taking the average after that:

$$G_{\mu\nu}\left(\langle g_{\mu\nu}\rangle\right) \neq \left\langle G_{\mu\nu}\left(g_{\mu\nu}\right)\right\rangle . \tag{1}$$

This property follows from the nonlinearity of Einstein equations.

P. Kašpar (✉) · D. Vrba · O. Svítek
Institute of Theoretical Physics, Charles University in Prague,
V Holešovičkách 2, 180 00 Prague, Czech Republic
e-mail: petrkaspar@atlas.cz

J. Bičák and T. Ledvinka (eds.), *Relativity and Gravitation*,
Springer Proceedings in Physics 157, DOI: 10.1007/978-3-319-06761-2_60,

2 Buchert Equations

Averaging of scalars was considered by Buchert [3]. The average of a scalar ψ over a domain $\mathscr{D}$ on a spatial hypersurface is defined as

$$\langle \psi \rangle \equiv \frac{1}{V_{\mathscr{D}}} \int_{\mathscr{D}} \psi \sqrt{\det g_{ij}}\, \mathrm{d}^3 x, \tag{2}$$

where $V_{\mathscr{D}}$ is the volume of the domain $\mathscr{D}$. Taking the average of the Raychaudhuri equation and the Hamiltonian for irrotational dust leads to the Buchert equations [3]

$$3\frac{\ddot{a}_{\mathscr{D}}}{a_{\mathscr{D}}} = -4\pi \langle \rho \rangle + \mathscr{Q}, \tag{3}$$

$$3\frac{\dot{a_{\mathscr{D}}}^2}{a_{\mathscr{D}}^2} = 8\pi \langle \rho \rangle - \frac{1}{2} \langle \mathscr{R} \rangle - \frac{1}{2}\mathscr{Q}. \tag{4}$$

Here ρ is the matter density, $\mathscr{R}$ is the Ricci scalar on the spatial hypersurface and $a_{\mathscr{D}}$ is the effective scale factor of the domain $\mathscr{D}$. The quantity $\mathscr{Q}$ in Buchert equations (3)–(4) is called the backreaction and is defined as

$$\mathscr{Q} \equiv \frac{2}{3}\left(\left\langle \theta^2 \right\rangle - \langle \theta \rangle^2\right) - 2\left\langle \sigma^2 \right\rangle. \tag{5}$$

3 LRS Spacetime

Locally rotationally symmetric (LRS) spacetimes are defined by the following characterization [4]: In an open neighborhood of each point p, there is a nondiscrete subgroup of the Lorentz group which leaves the Riemann tensor and its covariant derivatives to the third order invariant. There is, therefore, a preferred direction e^μ (the axis of symmetry) in every point in LRS spacetimes.

We will use the covariant 3+1 splitting of a spacetime with the timelike vector u^μ normalized by the condition $u_\rho u^\rho = -1$ and the projection tensor $h_{\mu\nu} = g_{\mu\nu} + u_\mu u_\nu$. Because of the property of the LRS spacetime, all covariantly defined spacelike vectors orthogonal to u^μ must be proportional to e^μ [5].

$$\dot{u}^\mu = \dot{u} e^\mu, \quad \omega^\mu = \omega e^\mu, \quad h^\sigma{}_\mu \nabla_\sigma \rho = \rho' e_\mu, \quad h^\sigma{}_\mu \nabla_\sigma p = p' e_\mu, \quad h^\sigma{}_\mu \nabla_\sigma \theta = \theta' e_\mu.$$

Dot here denotes the covariant derivative along the flow vector u^μ and the prime denotes covariant derivative along the vector e^μ. With the help of the tensor $e_{\mu\nu} = \frac{1}{3}(3e_\mu e_\nu - h_{\mu\nu})$, we have the relations for the shear tensor and the electric and magnetic parts of the Weyl tensor

$$\sigma_{\mu\nu} = \frac{2}{\sqrt{3}}\sigma e_{\mu\nu}, \quad E_{\mu\nu} = \frac{2}{\sqrt{3}}E e_{\mu\nu}, \quad H_{\mu\nu} = \frac{2}{\sqrt{3}}H e_{\mu\nu}. \tag{6}$$

We will now define the magnitude of the spatial rotation k and the magnitude of the spatial divergence a,

$$k \equiv \left|\eta^{\alpha\beta\gamma\delta}\left(\nabla_\beta e_\gamma\right) u_\delta\right|, \qquad a \equiv h^\alpha{}_\beta\left(\nabla_\alpha e^\beta\right). \tag{7}$$

4 Averaging LRS Spacetime

For simplicity we will restrict to the class II LRS spacetime with the condition $p = 0$ (dust models) which includes LTB spacetimes and their generalizations to spacelike 2-surfaces with negative or zero curvature scalar. Given a preferred spacelike direction, all the equations describing LRS metric are scalar. It means we can perform averaging (which is covariantly defined for scalars). In order to obtain averaged equations we need to derive the commutation relations for the prime (and the time) derivative (with respect to the preferred direction) and averaging,

$$\langle A\rangle'_{\mathscr{D}} = e^\mu \Delta_\mu \left(\frac{1}{V_{\mathscr{D}}}\int_{\mathscr{D}} d^3x\sqrt{\det g_{ij}}\,A\right) = -\langle\xi\rangle_{\mathscr{D}}\langle A\rangle_{\mathscr{D}} + \langle A\xi\rangle_{\mathscr{D}} + \langle A'\rangle_{\mathscr{D}}, \tag{8}$$

where ξ is defined by the relation $\left(\sqrt{\det g_{ij}}\right)' = \sqrt{\det g_{ij}}\,\xi$. Similarly we get the commutation rule for the time derivative. Averaged equations for the dust class II LRS spacetime read as follows:

$$\begin{aligned}
\langle\theta\rangle^{\cdot} &= -\frac{1}{3}\langle\theta\rangle^2 - 4\pi\langle\rho\rangle + \underline{\frac{2}{3}\left(\left\langle\theta^2\right\rangle - \langle\theta\rangle^2\right) - 2\left\langle\sigma^2\right\rangle},\\
\langle\sigma\rangle^{\cdot} &= -\frac{1}{\sqrt{3}}\langle\sigma\rangle^2 - \frac{2}{3}\langle\theta\rangle\langle\sigma\rangle - \langle E\rangle + \underline{\frac{1}{\sqrt{3}}\left(\langle\sigma\rangle^2 - \left\langle\sigma^2\right\rangle\right) + \frac{1}{3}(\langle\theta\sigma\rangle - \langle\theta\rangle\langle\sigma\rangle)},\\
\langle E\rangle^{\cdot} &= -4\pi\langle\rho\rangle\langle\sigma\rangle + \sqrt{3}\langle E\rangle\langle\sigma\rangle - \langle\theta\rangle\langle E\rangle - \underline{4\pi(\langle\rho\sigma\rangle - \langle\rho\rangle\langle\sigma\rangle)},\\
&\quad + \sqrt{3}(\langle E\sigma\rangle - \langle E\rangle\langle\sigma\rangle),\\
\langle\rho\rangle^{\cdot} &= -\langle\rho\rangle\langle\theta\rangle,\\
\langle a\rangle^{\cdot} &= -\frac{1}{3}\langle a\rangle\langle\theta\rangle + \frac{1}{\sqrt{3}}\langle a\rangle\langle\sigma\rangle + \underline{\frac{2}{3}(\langle a\theta\rangle - \langle a\rangle\langle\theta\rangle) + \frac{1}{\sqrt{3}}(\langle a\sigma\rangle - \langle a\rangle\langle\sigma\rangle)},\\
\langle\sigma\rangle' &= \frac{1}{\sqrt{3}}\langle\theta\rangle' - \frac{2}{3}\langle a\rangle\langle\sigma\rangle + \underline{\langle\sigma\xi\rangle - \langle\xi\rangle\langle\sigma\rangle - \frac{1}{\sqrt{3}}(\langle\xi\theta\rangle - \langle\xi\rangle\langle\theta\rangle)}\\
&\quad \underline{-\frac{3}{2}(\langle a\sigma\rangle - \langle a\rangle\langle\sigma\rangle)},
\end{aligned}$$

$$
\begin{aligned}
\langle E\rangle' = &-\frac{2}{3}\langle a\rangle\langle E\rangle + \frac{4\pi}{\sqrt{3}}\langle\rho\rangle' - \underline{\frac{2}{3}\left(\langle aE\rangle - \langle a\rangle\langle E\rangle\right)} \\
&\underline{+\langle\xi E\rangle - \langle\xi\rangle\langle E\rangle - \frac{4\pi}{\sqrt{3}}\left(\langle\xi\rho\rangle - \langle\xi\rangle\langle\rho\rangle\right)}, \\
\langle a\rangle' = &\frac{2}{9}\langle\theta\rangle^2 + \frac{2}{3\sqrt{3}}\langle\theta\rangle\langle\sigma\rangle - \frac{4}{3}\langle\sigma\rangle^2 - \frac{2}{\sqrt{3}}\langle E\rangle - \frac{1}{2}\langle a\rangle^2 - \frac{16\pi}{3}\langle\rho\rangle \\
&\underline{+\langle a\xi\rangle - \langle a\rangle\langle\xi\rangle + \frac{2}{9}\left(\left\langle\theta^2\right\rangle - \langle\theta\rangle^2\right) + \frac{2}{3\sqrt{3}}\left(\langle\theta\sigma\rangle - \langle\theta\rangle\langle\sigma\rangle\right)} \\
&\underline{-\frac{4}{3}\left(\left\langle\sigma^2\right\rangle - \langle\sigma\rangle^2\right) - \frac{1}{2}\left(\left\langle a^2\right\rangle - \langle a\rangle^2\right)}.
\end{aligned}
\tag{9}
$$

The underlined part of the equations denotes the additional terms created by averaging. We can recognize the familiar Buchert equation with the kinematical backreaction term and the mass conservation equation.

5 Conclusion

We have shown how to generalize the Buchert equations for the LRS spacetimes. Averaged Einstein equations consist of evolution equations and constraints that are preserved in time.

Acknowledgments The present work was supported by grants GAUK 398911, GACR 14-37086G and Czech Ministry of Education project SVV-267301.

References

1. Ellis, G.: Relativistic cosmology: its nature, aims and problems, In: Bertotti, B., de Felice, F., Pascolini, A. (eds.) General Relativity and Gravitation, pp. 215–288. Reidel; Kluwer, Dordrecht (1984)
2. Zalaletdinov, R.: Averaging problem in cosmology and macroscopic gravity. Int. J. Mod. Phys. A **23**, 1173 (2008). doi:10.1142/S0217751X08040032
3. Buchert, T.: On average properties of inhomogeneous fluids in general relativity: dust cosmologies. Gen. Relativ. Gravit. **32**, 105 (2000). doi:10.1023/A:1001800617177
4. Ellis, G.: Dynamics of pressure-free matter in general relativity. J. Math. Phys. **8**, 1171 (1967). doi:10.1063/1.1705331
5. van Elst, H., Ellis, G.: The covariant approach to LRS perfect fluid spacetime geometries. Class. Quantum Gravity **13**, 1099 (1996). doi:10.1088/0264-9381/13/5/023

Effect of Magnetic Fields on Equatorial Circular Orbits in Kerr Spacetimes

Ignacio F. Ranea-Sandoval and Héctor Vucetich

Abstract In this work we analyze the effects of an external magnetic field on charged particles on equatorial circular orbits in a Kerr spacetime, both in the black hole and the naked singularity cases. Understanding these phenomena is of great importance because equatorial circular orbits are a key ingredient of simple accretion disc models. We focus on two important magnetic field configurations: (a) a uniform magnetic field aligned with the angular momentum and (b) a dipolar magnetic field. We center our attention on the effect of these external fields on the marginally bound and marginally stable equatorial circular orbits because they give information on observable quantities that could be useful to determine whether the central object is a black hole or a naked singularity. Using a perturbative approach we are able to give analytic results and compare (in the black hole case) with previous results.

1 Introduction

Since Penrose proposed what is now known as the *Cosmic Censorship Conjecture*, spacetimes with naked singularities have been the subject of great debate. In [1] and references therein the authors prove the existence of unstable perturbations for the most relevant nakedly singular spacetimes. These and other works clearly favour black holes over naked singularities to model extremely compact objects.

Following an alternative approach, in [2] the differences between circular geodesics around a Kerr black hole and a Kerr naked singularity are studied.

I. F. Ranea-Sandoval (✉) · H. Vucetich
Grupo de Gravitación, Astrofísica y Cosmología, Facultad de Ciencias Astronómicas y Geofísicas—Universidad Nacional de La Plata, Paseo del Bosque S/N, 1900 La Plata, Argentina
e-mail: iranea@fcaglp.unlp.edu.ar

I. F. Ranea-Sandoval
CONICET, Rivadavia 1917, 1033 Buenos Aires, Argentina

J. Bičák and T. Ledvinka (eds.), *Relativity and Gravitation*,
Springer Proceedings in Physics 157, DOI: 10.1007/978-3-319-06761-2_61,

Accretion processes are usually associated with compact objects and are the main tools astrophysicists possess to study their properties. The presence of magnetic fields in accretion discs produce observable phenomena which are worth examining.

The effects of magnetic fields on accretion discs around a rotating black hole where studied, and the changes in the innermost stable orbit and in the marginally bound orbits are analyzed in [3–5]. As we are only able to observe the effects of the presence of a black hole on particles, changes in these particular radii may give observable quantities that could allow us to distinguish between different theoretical models for compact objects.

In this work we present some of the analytical results of [6] in which the change in the position of the inner edge of an accretion disc in a Kerr spacetime is studied. We generalize previous results by allowing the rotation parameter a to adopt values larger that one. For the sake of space limitation we will not present mathematical details, which are left to the extended work [6].

2 Theoretical Basics

Kerr spacetime represents the exterior gravitational field of a rotating body and is one of the most important exact solutions to Einstein's field equations. Stationary magnetic field configurations in a Kerr background have been studied in detail in [7–9]. The study of the motion of charged particles in the equatorial plane in the presence of fields that preserve the Killing nature of both ∂_ϕ and ∂_t is usually done using variations of the arguments in the pioneering works of Carter [10] for uncharged particles.

3 Results

We use an effective potential and a perturbative approach that allow us to investigate analytically the cases in which the coupling between the external magnetic field and the effective charge of the particles is small. As a result we obtain analytic expressions for the position of the innermost stable and marginally bound radii of equatorial circular orbits.

As explained in [3–5], the coupling between the magnetic field strength and the effective charge of particles is measured by a parameter called λ. Our perturbative approach allows us to study the $\lambda \ll 1$ regime which corresponds to the astrophysically relevant case of a (neutral over large scales) simple fluid disc.

We summarize part of our results in Fig. 1 for the case of a uniform magnetic field, and in Fig. 2 for the dipolar case. These results are in a complete agreement with those obtained for black holes in [3, 5].

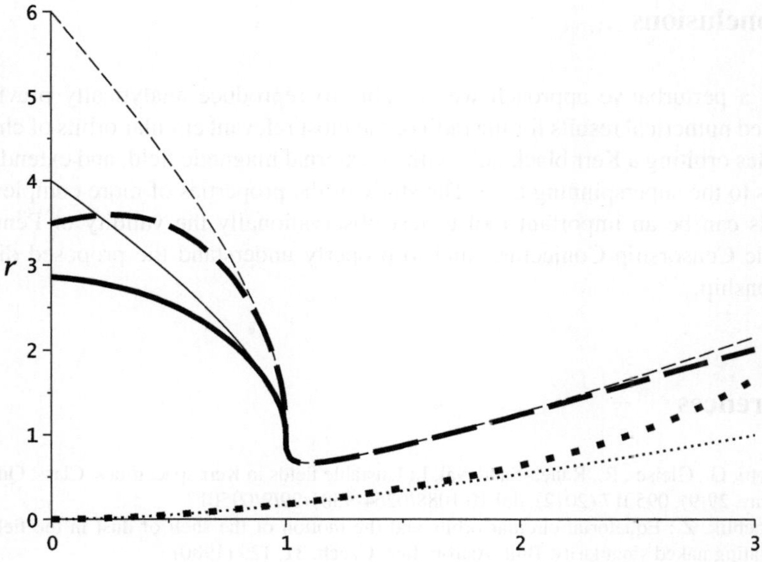

Fig. 1 Radii of the marginally bound orbit (*solid* or *dotted lines*) and of the innermost stable orbit (*dashed lines*) as a function of the rotation parameter a. Equally thick lines correspond to equal values of the parameter λ which measures the coupling between the matter's electric charge and the external magnetic field strength: the thinnest for $\lambda = 0$ and the thickest for $\lambda = 0.1$

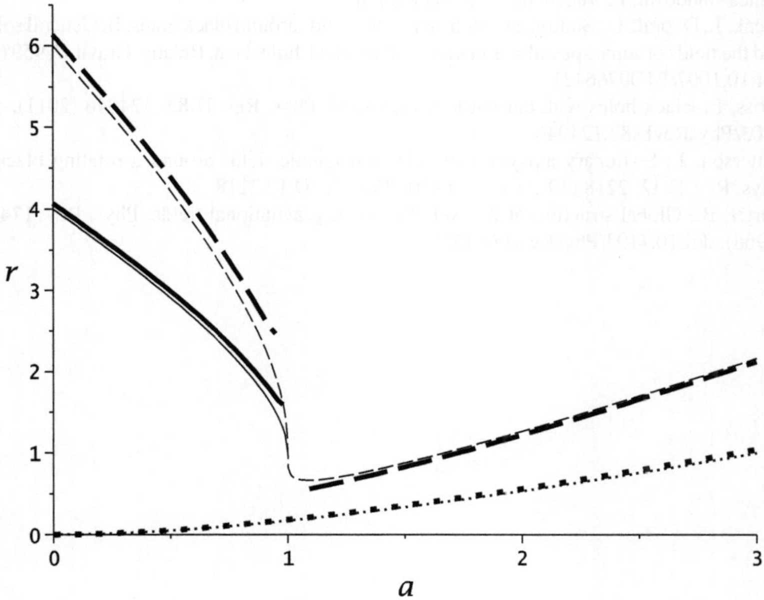

Fig. 2 Radii of the marginally bound orbit and of the innermost stable orbit as a function of the rotation parameter a. The line styles and thickness correspond to the ones presented in Fig. 1

4 Conclusions

Using a perturbative approach we are able to reproduce analytically previously reported numerical results for the radii of the most relevant circular orbits of charged particles orbiting a Kerr black hole with an external magnetic field, and extend these results to the superspinning case. The study of the properties of more complex disc models can be an important tool to test observationally the validity of Penrose's Cosmic Censorship Conjecture, and to properly understand the proposed disc-jet relationship.

References

1. Dotti, G., Gleiser, R., Ranea-Sandoval, I.: Unstable fields in Kerr spacetimes. Class. Quantum Grav. **29**(9), 095017 (2012). doi:10.1088/0264-9381/29/9/095017
2. Stuchlík, Z.: Equatorial circular orbits and the motion of the shell of dust in the field of a rotating naked singularity. Bull. Astron. Inst. Czech. **31**, 129 (1980)
3. Iyer, B., Vishveshwara, C., Wiita, P., Goldstein, J.: Accretion onto a Kerr black hole in the presence of a dipole magnetic field. Pramana **25**, 135 (1985). doi:10.1007/BF02847652
4. Prasanna, A., Vishveshwara, V.: Charged particle motion in an electromagnetic field on Kerr background geometry. Pramana **11**, 359 (1978). doi:10.1007/BF02848160
5. Wiita, P., Vishveshwara, C., Siah, M., Iyer, B.: Magnetic fields and accretion discs around Kerr black holes. J. Phys. A Math. Gen. **16**, 2077 (1983). doi:10.1088/0305-4470/16/9/029
6. Ranea-Sandoval, I., Vucetich, H. In preparation
7. Bičák, J., Dvořák, L.: Stationary electromagnetic fields around black holes. II. General solutions and the fields of some special sources near a Kerr black hole. Gen. Relativ. Gravit. **7**, 959 (1976). doi:10.1007/BF00766421
8. Moss, I.: Black holes with current loops revisited. Phys. Rev. D **83**, 124046 (2011). doi:10.1103/PhysRevD.83.124046
9. Petterson, J.: Stationary axisymmetric electromagnetic fields around a rotating black hole. Phys. Rev. D **12**, 2218 (1975). doi:10.1103/PhysRevD.12.2218
10. Carter, B.: Global structure of the Kerr family of gravitational fields. Phys. Rev. **174**, 1559 (1968). doi:10.1103/PhysRev.174.1559

Exotic (Dark) Eigenspinors of the Charge Conjugation Operator and Cosmological Applications

Roldao da Rocha

Abstract We report about some achievements and developments provided by the ELKO program, in particular the ones recently accomplished [1]. Exotic dark spinor fields have been investigated in the context of inequivalent spin structures on arbitrary curved spacetimes, which induces an additional term in the associated Dirac operator, related to a Čech cohomology class. It implies that the non-trivial topology associated with the spacetime can engender—from the dynamics of ELKO dark spinor fields—drastic constraints on the spacetime metric structure.

1 ELKO (Dark) Spinor Fields

Without loss of generality, an ELKO Ψ can be written as $\Psi(p) = \lambda(\mathbf{p})e^{\pm ip\cdot x}$ where $\lambda(\mathbf{p}) = \binom{i\Theta\phi^*(\mathbf{p})}{\phi(\mathbf{p})}$, and given the rotation generators $\mathfrak{J}$, the Wigner's spin-1/2 time reversal operator Θ satisfies $\Theta\mathfrak{J}\Theta^{-1} = -\mathfrak{J}^*$. Hereon the Weyl representation of γ^μ is used. ELKO spinor fields are eigenspinors of the charge conjugation operator C defined here by its action $C\lambda(\mathbf{p}) = \pm\lambda(\mathbf{p})$, for $C = \begin{pmatrix} \mathbb{O} & i\Theta \\ -i\Theta & \mathbb{O} \end{pmatrix} K$. The operator K $\mathbb{C}$-conjugates 2-component spinor fields appearing on the right. The plus sign stands for self-conjugate spinor fields, $\lambda^S(\mathbf{p})$, while the minus yields anti self-conjugate spinor fields, $\lambda^A(\mathbf{p})$. Explicitly, the complete form of ELKO can be found by solving the equation of helicity $(\sigma\cdot\widehat{\mathbf{p}})\phi^\pm = \pm\phi^\pm$ in the rest frame and subsequently boosting, to recover the result for any $\mathbf{p}$ [2]. Here $\widehat{\mathbf{p}} := \mathbf{p}/\|\mathbf{p}\|$. The boosted four spinor fields result in

R. da Rochao (✉)
Centro de Matemática, Computação e Cognição, Universidade Federal do ABC,
R. Santa Adélia 166, Santo Andre, SP 09210-170, Brazil
e-mail: roldao.rocha@ufabc.edu.br

J. Bičák and T. Ledvinka (eds.), *Relativity and Gravitation*,
Springer Proceedings in Physics 157, DOI: 10.1007/978-3-319-06761-2_62,

$$\lambda^{S/A}_{\{\mp,\pm\}}(\mathbf{p}) = \sqrt{\frac{E+m}{2m}}\left(1\mp\frac{p}{E+m}\right)\lambda^{S/A}_{\{\mp,\pm\}}(\mathbf{0}), \text{ for } \lambda^{S/A}_{\{\mp,\pm\}}(\mathbf{0}) = \begin{pmatrix}\pm i\Theta[\phi^{\pm}(\mathbf{0})]^*\\ \phi^{\pm}(\mathbf{0})\end{pmatrix}. \tag{1}$$

As $\Theta[\phi^{\pm}(\mathbf{0})]^*$ and $\phi^{\pm}(\mathbf{0})$ have opposite helicities, ELKO cannot be an eigenspinor field of the helicity operator.

2 ELKO Dynamics: Exotic Spin Structure

To observe that dark spinor fields are a natural probe of the non-trivial topology one should firstly notice that such exotic spinor fields are parallel transported like standard spinor fields. Meanwhile, an outstanding property distinguishes both kinds of spinor fields: the covariant derivative acting on these exotic spinor fields changes by an additional one-form field that is a manifestation of the non-trivial topology. The exotic structure endows the Dirac operator with an additional term $\frac{1}{2\pi i}\xi^{-1}(x)d\xi(x)$, which is real and closed, but not exact, and defines an integer cohomology class in the Čech sense. Using the relation between Čech and de Rham cohomologies, the integral of $\frac{1}{2\pi i}\xi^{-1}(x)d\xi(x)$ around any closed curve is an integer. The point is that for dark ELKO spinor fields such an exotic term cannot be absorbed by an external gauge field representing an element of $H^1(M,\mathbb{Z}_2)$, inasmuch as ELKO fields cannot carry gauge charges.

In addition to the ELKO spinor fields $\lambda(x)$—that were indeed defined as sections in the bundle $P_{\mathrm{Spin}^e_{1,3}}(M)\times_\rho \mathbb{C}^4$, in Sect. 2—one can get a second type of ELKO $\mathring{\lambda}(x)$, which can be described by sections in the inequivalent spin structure-induced spinor bundle $\mathring{P}_{\mathrm{Spin}^e_{1,3}}(M)\times_\rho \mathbb{C}^4$, with a variation of the covariant derivative, given by

$$i\gamma^\mu\mathring{\nabla}_\mu = i\gamma^\mu\nabla_\mu + \xi^{-1}(x)d\xi(x). \tag{2}$$

The exotic Dirac equation is given then by

$$(i\gamma^\mu\nabla_\mu + (\xi^{-1}(x)\,d\xi(x)) - m\mathbb{I})\psi(x) = 0, \tag{3}$$

ψ denotes a Dirac spinor field. The exotic Dirac spinor fields are annihilated by $\left(i\gamma^\mu\nabla_\mu + (\xi^{-1}(x)\,d\xi(x)) \pm m\mathbb{I}\right)$ where the plus and minus signs stands respectively for particles and antiparticles. Hereon we denote $\xi^{-1}(x)\,d\xi(x)$ by $a(x)$ in order to shorten all formulæ notations.

ELKO spinor field can not be eigenspinors of the exotic Dirac operator $i\gamma^\mu\nabla_\mu + a(x)$. Namely, the mass terms carry opposite signs and consequently ELKO cannot be annihilated by $\left(i\gamma^\mu\nabla_\mu + a(x) \pm m\mathbb{I}\right)$, because the term ε_α^β which implies that $\varepsilon^{\mathrm{S}} = -1$ and $\varepsilon^{\mathrm{A}} = +1$. Furthermore, as comprehensively discussed we can express $\xi(x) = \exp(i\theta(x)) \in \mathrm{U}(1)$. The exotic spin structure term in this way reads

$$\xi^{-1}(x)d\xi(x) = \exp(-i\theta(x))(i\gamma^\mu\nabla_\mu\theta(x))\exp(i\theta(x)) = i\gamma^\mu\partial_\mu\theta(x). \tag{4}$$

The exotic Dirac operator $i\gamma^\mu\nabla_\mu + i\gamma^\mu\partial_\mu\theta - m\mathbb{I}$ annihilates each of the four exotic Dirac spinor fields but does not annihilate ELKO.

There was an extensive discussion about the subtle differences between Majorana and ELKO spinor fields. Both in the Lounesto spinor field classification are type-(5) spinor fields. It would be useful to discuss whether the exotic Dirac operator can be considered as a square root of the Klein–Gordon operator. This feature must remain true for the ELKO and its exotic partner:

$$\left((i\gamma^\mu\nabla_\mu + a(x))\delta_\alpha^\beta \pm m\mathbb{I}\varepsilon_\alpha^\beta\right)\left((i\gamma^\mu\nabla_\mu + a(x))\delta_\alpha^\beta \mp m\mathbb{I}\varepsilon_\alpha^\beta\right) = (g^{\mu\nu}\nabla_\mu\nabla_\nu + m^2)\mathbb{I}\,\delta_\alpha^\beta \tag{5}$$

since the introduction of an exotic spin structure does not modify the Klein–Gordon propagator fulfillment by dark spinor fields.

The corresponding Klein–Gordon equation is given by

$$(\Box + m^2 + g^{\mu\nu}\nabla_\mu\nabla_\nu\theta + \partial^\mu\theta\nabla_\mu + \partial^\mu\theta\partial_\mu\theta)\mathring{\lambda}(x)^{S/A}_{\{\pm,\mp\}} = 0. \tag{6}$$

In order that the Klein–Gordon propagator for the exotic ELKO remains the same it follows that

$$(\Box\theta + \partial^\mu\theta\nabla_\mu + \partial^\mu\theta\partial_\mu\theta)\mathring{\lambda}^{S/A}_{\{\pm,=\}}(x) = 0. \tag{7}$$

The fact that (7) holds for every exotic dark spinor field $\mathring{\lambda}^{S/A}_{\{\pm,\mp\}}(x)$, lets us analyze in particular the solutions of (7) applied to, for instance, $\mathring{\lambda}^{S}_{\{-,+\}}(x)$. Using the expression

$$\nabla_\mu\mathring{\lambda}^{S/A}_{\{\mp,\pm\}} = \partial_\mu\mathring{\lambda}^{S/A}_{\{\mp,\pm\}} - \frac{1}{4}\Gamma_{\mu\rho\sigma}\gamma^\rho\gamma^\sigma\mathring{\lambda}^{S/A}_{\{\mp,\pm\}}, \tag{8}$$

for such case, after some calculation it follows that

$$\begin{aligned}(\Box\theta)\,\mathring{\lambda}^{S/A}_{\{\mp,\pm\}} &+ (\partial_0\theta)\Big[\partial_0\mathring{\lambda}^{S/A}_{\{\mp,\pm\}} - \frac{1}{4}\Big((\Gamma_{000} - \Gamma_{0jj})\mathring{\lambda}^{S/A}_{\{\mp,\pm\}} + i\Gamma_{001}\mathring{\lambda}^{A/S}_{\{\pm,\mp\}}\\ &+\Gamma_{002}\mathring{\lambda}^{S/A}_{\{\pm,\mp\}} \mp \Gamma_{003}\mathring{\lambda}^{S/A}_{\{\mp,\pm\}} \pm i\Gamma_{012}\mathring{\lambda}^{A/S}_{\{\mp,\pm\}} + i\Gamma_{013}\mathring{\lambda}^{A/S}_{\{\pm,\mp\}} \mp \Gamma_{023}\mathring{\lambda}^{S/A}_{\{\pm,\mp\}}\Big)\Big]\\ &-g^{00}(\partial_0\theta)^2\mathring{\lambda}^{S/A}_{\{\mp,\pm\}} = 0,\end{aligned} \tag{9}$$

where $\Gamma_{0jj} = \Gamma_{011} + \Gamma_{022} + \Gamma_{033}$. The equation above couples *again* all the four exotic spinor fields $\mathring{\lambda}^{S/A}_{\{\pm,\mp\}}$, in the case of spacetimes where the associated connections are non zero.

It is possible for cosmological applications to assume that the dark spinor fields depend only on the time variable t via a matter field $\kappa(t)$ compatible with homogeneity and isotropy and act as the only dynamical cosmological variable, in such a way that $\mathring{\lambda}^{S/A}_{\{\pm,\mp\}}(x)$ can be explicitly written as

$$\mathring{\lambda}^{A/S}_{\{-,+\}}(x) = \kappa(t)\,\chi^{A/S}_{\{-,+\}}, \qquad \mathring{\lambda}^{A/S}_{\{+,-\}}(x) = \kappa(t)\,\zeta^{A/S}_{\{+,-\}}, \tag{10}$$

where $\zeta^{S/A}$ and $\chi^{S/A}$ are linearly independent constant spinor fields

$$\chi^S_{\{-,+\}} = \begin{pmatrix}0\\ i\\ 1\\ 0\end{pmatrix}, \quad \chi^A_{\{-,+\}} = \begin{pmatrix}0\\ -i\\ 1\\ 0\end{pmatrix}, \quad \zeta^S_{\{+,-\}} = \begin{pmatrix}1\\ 0\\ 0\\ -i\end{pmatrix}, \quad \zeta^A_{\{+,-\}} = -\begin{pmatrix}1\\ 0\\ 0\\ i\end{pmatrix}. \tag{11}$$

The matter field $\kappa(t)$ satisfies $\frac{\dot{\kappa}}{\kappa} = -\frac{1}{3}\sqrt{\frac{1}{3M_{\rm Pl}^2}\Sigma_{tt}} + \mathscr{O}(\kappa^4)$, where $M_{\rm Pl}^{-2} = 8\pi\,{\rm G}$ is the coupling constant, and total energy-momentum tensor is denoted by Σ_{tt}, involving also the Planck mass, and the Hubble constant. Therefore we can write $\kappa(t) = \exp(\tilde{a}t)$, where $\tilde{a}$ is the constant given in the equation above. Using now (10) and (11) we have $(\partial_0\theta)\Gamma_{012} = 0$, which means that if θ is time dependent, it necessarily means that $\Gamma_{012} = 0$. Otherwise, in the case where θ does not depend on time, it implies that $\partial_0\theta = 0$, and then we obtain the Laplace equation $\nabla^2\theta = 0$. The dark spinor field dynamics can thus be used to probe the topological sector determined by θ. In addition, the above obtained second and third equations imply that

$$\Box\theta + (\partial_0\theta)\left(1 - \frac{1}{2}(\Gamma_{000} - \Gamma_{011} - \Gamma_{022} - \Gamma_{033} - \Gamma_{003})\right) - (\partial_0\theta)^2 = 0, \tag{12}$$

which means that if $\theta = \theta(t)$, so necessarily $4 - (\Gamma_{000} - \Gamma_{011} - \Gamma_{022} - \Gamma_{033}) = \partial_0\theta$.

Acknowledgments R. da Rocha is grateful to Conselho Nacional de Desenvolvimento Científico e Tecnológico (CNPq) grants 472903/2008-0 and 304862/2009-6 for financial support.

References

1. da Rocha, R., Bernardini, A., Hoff da Silva, J.: Exotic dark spinor fields. J. High Energy Phys. **2011**(04), 110 (2011). doi:10.1007/JHEP04(2011)110
2. Ahluwalia-Khalilova, D., Grumiller, D.: Spin-half fermions with mass dimension one: theory, phenomenology, and dark matter. J. Cosmol. Astropart. Phys. **2005**(07), 012 (2005). doi:10.1088/1475-7516/2005/07/012

On Motion of the Magellanic Clouds in the Milky Way Gravitational Field

Zdeněk Stuchlík and Jan Schee

Abstract We demonstrate that the cosmological constant substantially influences motion of both Magellanic Clouds in the gravitational field of Milky Way.

1 Introduction

It is usually assumed that for the motion on the scales given by the distance of neighbouring galaxies the influence of the cosmological constant is negligible. We demonstrate that the role of the cosmic repulsion is crucial even on such relatively small scales, using calculations of trajectories and binding mass for Magellanic Clouds.

2 Magellanic Clouds and Intial Conditions

Large Magellanic Cloud (LMC) and Small Magellanic Cloud (SMC) are dwarf galaxies in the vicinity of Milky Way (MW). We assume both SMC and LMC to be test particles moving (independently) in the MW field, since their mass is substantially smaller than that of the MW and their distance from the visible Galaxy disc is substantially larger than its extension. The Newtonian approach can be used since the GR effects are negligible [1]. Velocity (km/s) and radial vectors (kpc) in galactocentric coordinates are [2] given in Table 1.

Z. Stuchlík (✉) · J. Schee
Faculty of Philosophy and Science, Institute of Physics,
Silesian University in Opava, Bezručovo nám. 13, 746 01 Opava, Czech Republic
e-mail: zdenek.stuchlik@fpf.slu.cz

J. Bičák and T. Ledvinka (eds.), *Relativity and Gravitation*,
Springer Proceedings in Physics 157, DOI: 10.1007/978-3-319-06761-2_63,

Table 1 LMC and SMC coordinates and velocities

	x	y	z
SMC			
x_i	15.3	−36.9	−43.3
v_i	-87 ± 48	-247 ± 42	149 ± 37
LMC			
x_i	−0.8	−41.5	−26.9
v_i	-86 ± 12	-268 ± 11	252 ± 16

3 Milky Way Gravitational Field

The gravitational field of the Milky Way is generated in the standard way by the Galactic disc,

$$U_{disk} = -\frac{\xi G M_{disk}}{\sqrt{x^2 + y^2 + (k + \sqrt{z^2 + b^2})^2}}, \tag{1}$$

the Galactic bulge,

$$U_{bulge} = -\frac{G M_{bulge}}{r + c}, \tag{2}$$

and the CDM halo

$$U_{halo} = v_{halo}^2 \ln(r^2 + d^2), \tag{3}$$

where $\xi = 1, k = 6.5\,\text{kpc}, b = 0.26\,\text{kpc}, c = 0.7\,\text{kpc}$. There is $M_{disc} = 5 \times 10^{10} M_{\odot}$ and $M_{bulge} = 1.5 \times 10^{10} M_{\odot}$. Since $v_{halo} = 114\,\text{km}\,\text{s}^{-1}, d = 12\,\text{kpc}$, the logarithmic halo model implies the halo mass formula

$$M_{halo} = \frac{2 v_{halo}^2 r^3}{G(r^2 + d^2)} \Rightarrow M_{halo}(r = 60\,\text{kpc}) = 3.5 \times 10^{11} M_{\odot}. \tag{4}$$

4 The Role of the Cosmological Constant

Outside the CDM halo we assume that the whole Milky Way halo mass is concentrated in its centre, and its gravitational field is then modelled by the Cosmological Paczynski-Wiita (CPW) potential [3]

$$U_{PN} = -\left(\frac{GM}{r} + \frac{\Lambda c^2}{6} r^2\right)\left(1 - \frac{2GM}{c^2 r} - \frac{\Lambda}{3} r^2\right)^{-1}. \tag{5}$$

Inside the MW halo, the Newtonian cosmological term

$$U_\Lambda = -\frac{\Lambda c^2}{6} r^2, \quad \Lambda = 1.3 \times 10^{-56}\,\mathrm{cm}^{-2} \tag{6}$$

is added to the terms representing gravitational field of the disk, bulge and halo. At the static radius, $r_s = [(3GM)/(c^2\Lambda)]^{1/3}$, gravitational attraction and cosmic repulsion are balanced. It gives boundary of gravitationally bound system detached from cosmic expansion.

5 Influence of $\Lambda > 0$ on the MC Motion

We integrate the equation of motion both to the time when Λ drives the expansion and Milky Way can be considered a fully developed gravitating system, $z \leq 1 \Rightarrow t \simeq 7.5\,\mathrm{Gyr}$. Trajectories are located in the region detached from the cosmological expansion → up to the scale $\simeq Mpc$. Typical trajectories are illustrated in Fig. 1 (for details see [1]).

6 The Binding Mass

Using the effective potential

$$V_{\mathrm{eff}} = \frac{1}{2}\frac{L^2}{r^2} - \frac{GM}{r} - \frac{\Lambda r^2 c^2}{6}, \tag{7}$$

the radius of the unstable circular orbit r_{uc} is given by

$$\frac{\mathrm{d}V_{\mathrm{eff}}}{\mathrm{d}r} = 0 \Rightarrow L^2 = r_{uc}\left(GM - \frac{\Lambda c^2}{3} r_{uc}^2\right). \tag{8}$$

The binding mass M_b is determined by the condition $E(r_0) = E(r_{uc})$:

$$M_b = G^{-1}\left(\frac{1}{r_{uc}} - \frac{1}{r_0}\right)^{-1}\left[\frac{\Lambda c^2}{6}(r_0^2 - r_{uc}^2) + \frac{1}{2}\left(\frac{1}{r_{uc}^2} - \frac{1}{r_C^2}\right) - v_r^2\right], \tag{9}$$

$$L = r_{uc}\left(GM_b - \frac{\Lambda c^2}{3} r_{uc}^2\right). \tag{10}$$

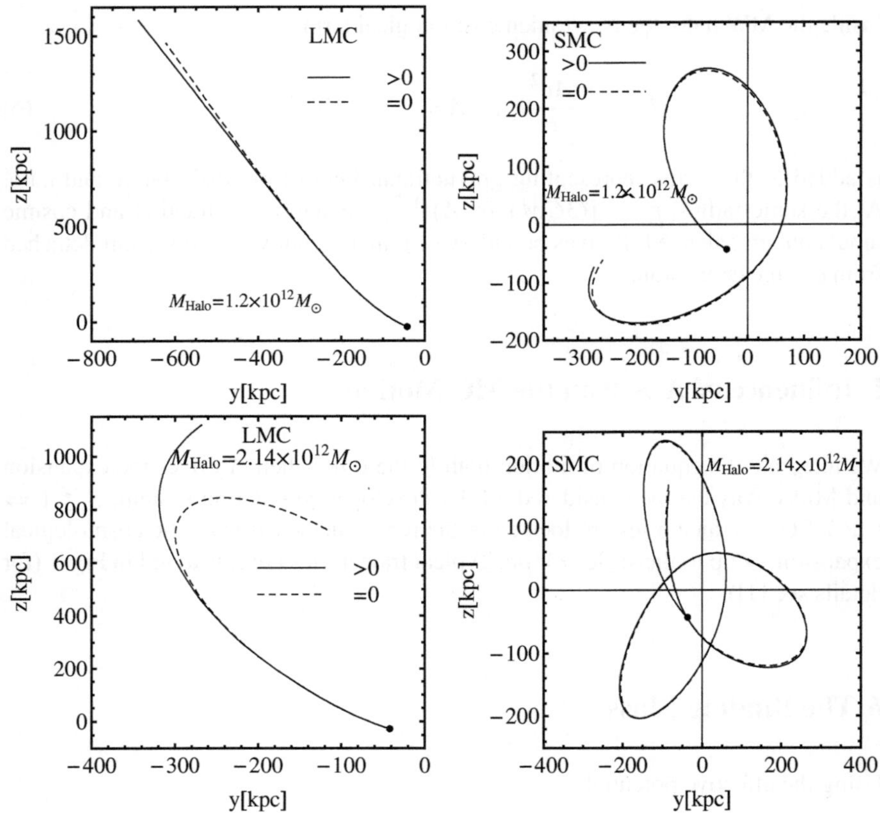

Fig. 1 $Y - Z$ crossections of typical SMC and LMC trajectories

Table 2 The dependence of the binding mass M_b and the radius of the unstable circular orbit r_{uc} on the cosmological constant Λ for LMC and SMC

Λ (cm^{-2})	M_b (10^{11}M$_{\odot}$)	r_{uc} (kpc)
SMC		
1.3×10^{-56}	6.843 (+10.3 %)	887
0	6.200	∞
LMC		
1.3×10^{-56}	8.863 (+7.7 %)	978
0	8.230	∞

The estimates of the binding mass are in Table 2. Calculations including the halo effect imply M_b at least twice as large. For LMC, influence of $\Lambda > 0$ on the value of M_b is about 50 % [4].

7 Conclusion

Comparing the influence of $\Lambda > 0$ on the MCs motion with those of dynamical friction or gravitational effects of M31, we find the influence of $\Lambda > 0$ to be the same or slightly higher [4].

Acknowledgments Supported by the Institutional fund of Silesian University in Opava.

References

1. Stuchlík, Z., Schee, J.: Comparison of general relativistic and pseudo-Newtonian description of Magellanic-clouds motion in the field of Milky Way. Int. J. Mod. Phys. D **21**, 4 (2012). doi:10.1142/S0218271812500319
2. Kallivayalil, N., Besla, G., Sanderson, R., Alcock, C.: Revisiting the role of M31 in the dynamical history of the Magellanic clouds. Astrophys. J. **700**, 924 (2009). doi:10.1088/0004-637X/700/2/924
3. Stuchlík, Z., Kovář, J.: Pseudo-Newtonian gravitational potential for Schwarzschild-De Sitter space-times. Int. J. Mod. Phys. D **17**, 2089 (2008). doi:10.1142/S021827180801373X
4. Stuchlík, Z., Schee, J.: Influence of the cosmological constant on the motion of Magellanic clouds in the gravitational field of Milky Way. J. Cosmol. Astropart. Phys. **2011**(09), 018 (2011). doi:10.1088/1475-7516/2011/09/018

Geodesic Chaos in Perturbed Black-Hole Fields

Petra Suková and Oldřich Semerák

Abstract Dynamics of time-like geodesics in the static and axially symmetric field of a black hole surrounded by a thin disc is studied by two recurrence methods, the recurrence plots (RPs) and the average of directional vectors (ADVs). Their results supplement the information obtained before from Poincaré surfaces of section and from phase-variable evolutions and the corresponding power spectra. The occurrence of chaos due to the presence of ambient matter may be important for evolution and appearance of astrophysical black-hole systems.

Inspired by models of accreting astrophysical black holes, we consider a simple, static and axisymmetric exact configuration of a black hole surrounded by a concentric thin disc or ring. Due to the presence of the additional source, the geodesic dynamics—originally completely integrable in the Schwarzschild field—generally becomes chaotic. Having illustrated this on Poincaré sections and on phase-variable time series and their power spectra [1], we have now turned to two recurrence methods which are based on statistics (i) over recurrences of the orbits to cells of the phase space (the recurrence plots [2, 3], RPs) and (ii) over direction in which the orbits recurrently pass through the cells (the average of directional vectors [4], ADVs). Here just a short glimpse on the results is given obtained for time-like geodesics in the field of a black hole (of mass M) surrounded by the inverted 1st Morgan-Morgan disc with mass $\mathscr{M} = 0.5$M and inner Schwarzschild radius $r = 15$M; see [5–7] for details of our earlier results.

RPs consist in recording the recurrence matrix whose components (zeros or ones) indicate (non-)recurrences of a given orbit to selected cells. This symmetric matrix

P. Suková (✉) · O. Semerák
Faculty of Mathematics and Physics, Institute of Theoretical Physics,
Charles University in Prague, Prague, Czech Republic
e-mail: lvicekps@seznam.cz

O. Semerák
e-mail: oldrich.semerak@mff.cuni.cz

J. Bičák and T. Ledvinka (eds.), *Relativity and Gravitation*,
Springer Proceedings in Physics 157, DOI: 10.1007/978-3-319-06761-2_64,

itself reveals the nature of dynamics, but here we rather show several examples of useful "quantifiers" which can be computed from the recurrence data. The simplest of them is the recurrence rate *RR*, given by the ratio of the recurrence points within all points of the matrix. Another one called *DET* is given by ratio of the points which form a diagonal line longer than a certain minimum $l_{\min}$ within all the recurrence points. The inverse of the longest diagonal $DIV = 1/L_{\max}$ has been shown to provide a rough estimate of the largest Lyapunov exponent. A lower estimate of the sum of positive Lyapunov exponents is given by the correlation (or Rényi's) entropy $\hat{K}_2$, determined by a slope of the cumulative histogram plotted (in log scale) against the diagonal length l (for large l). Statistics over vertical (or horizontal) lines of the matrix brings similar quantifiers, for example, *LAM* is a counter-part of *DET* and *V ENTROPY* is obtained from probability that a chosen vertical line has a prescribed length l. Note that the recurrence matrix and all the quantifiers of course depend on the chosen size of spatial cells and on time step with which the orbits are processed.

We launch geodesic particles with zero radial velocity from different radii between $r = 8$M and $r = 25$M with specific energy $E = 0.955$ and specific angular momentum $L = 4$M. Poincaré section of several hundreds of such orbits coloured according to the value of *DIV* is depicted in Fig. 1 together with the behavior of several RP quantifiers.

The other method, ADVs, is based on monitoring the evolution of tangent to the trajectory in selected boxes of the phase space "reconstructed", for a given data series $x(\tau)$, by taking the delayed replicas $x(\tau), x(\tau - \Delta\tau), x(\tau - 2\Delta\tau), \ldots, x(\tau - d\Delta\tau)$ as its axes, where $\Delta\tau$ is a chosen time shift. The vectors obtained for a large number of transits through the j-th box are summed, the resulting vector is normalised and the norm is averaged over all boxes which were crossed n-times. Finally, the dependence of this average on n and also of the averaged quadratic difference from random-walk case on time lag, $\bar{\Lambda}(\Delta\tau)$, are analysed. For random data the average decreases more quickly than for a deterministic signal; for a regular orbit, it remains 1 theoretically. The ADV method was originally proposed to distinguish between deterministic and random data, but we have found it is also sensitive to different degrees of chaoticity.

The comparison of the above methods is illustrated in Fig. 2. Several trajectories have been followed there for a very long time: the most chaotic trajectories up to $\tau_{\max} = 250000$M while other to $\tau_{\max} = 800000$M (one orbital period represents some 300M of proper time). The particle position is recorded with the step $\Delta\tau = 0.25$M.[1] Each trajectory is plotted in a different colour; regular orbits are green, weakly chaotic orbits range from light blue to orange and yellow and the most chaotic ones are purple. It is seen that the character of orbits is revealed by Poincaré diagram as well as by the values of *DIV* or K_2, and also from the behaviour of $\bar{\Lambda}(\Delta\tau)$. We pay particular attention to comparison of the quantities *DIV* or K_2 (bottom left). Namely, although K_2 is considered a more precise and reliable estimate of orbital divergence, its correct determination is tricky and less suitable for an automatic procedure (see [5]

[1] ADVs require a sufficient number of orbital points in each phase-space box (and the boxes must be small enough, naturally), which leads to quite large data sets. In comparison, for the RPs more sparse data are sufficient (time step 30–50M).

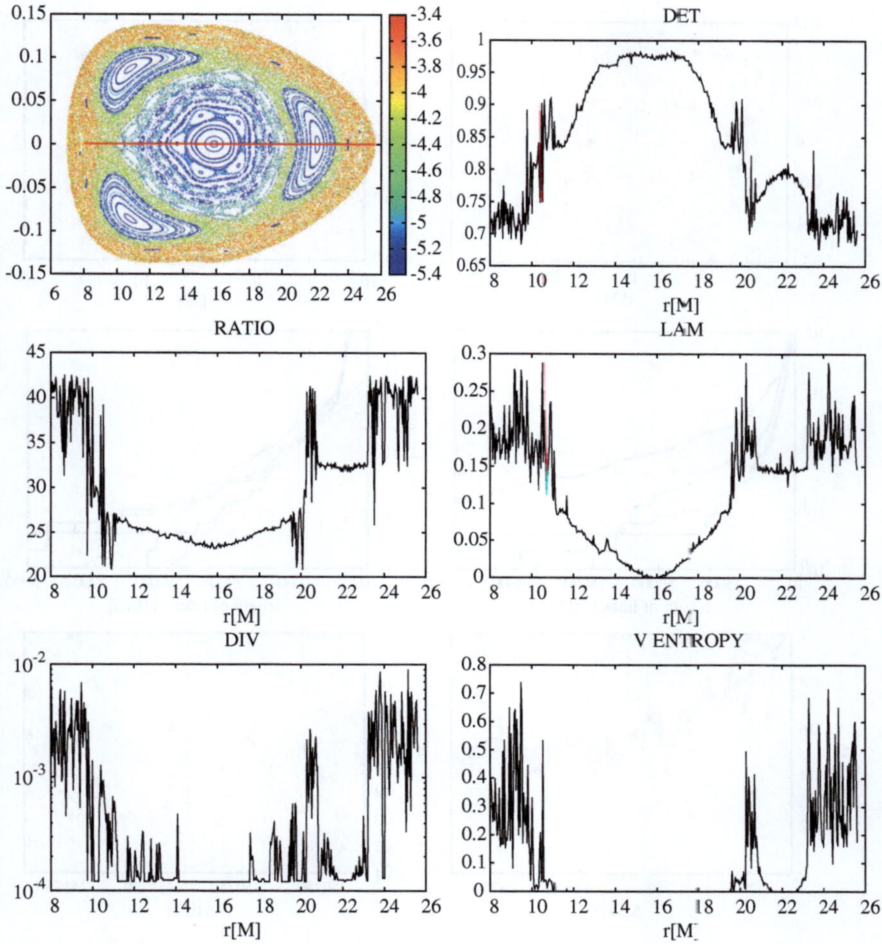

Fig. 1 Poincaré section (*top left*) with each orbit coloured according to the value of *DIV* in log scale. The *red line* indicates starting points. The other plots show several recurrence quantifiers as functions of the initial radius of the orbits

for details). The *DIV* quantifier is obtained much more easily and reliably. The plot shows that the log values of *DIV* and K_2 are roughly proportional and that different types of motion (regular, weakly chaotic and strongly chaotic) yield clearly different values of both quantities.

In the last two plots of Fig. 2, the ADV main result $\bar{\Lambda}$ is shown against the time lag $\Delta\tau$. At bottom left, the time lag ranges from 10 to 2000M (more than 6 orbital periods); $\bar{\Lambda}(\Delta\tau)$ for strongly chaotic trajectories (purple) is rapidly decreasing, while weakly chaotic and regular trajectories look quite similar. At bottom right, $\bar{\Lambda}(\Delta\tau)$ is given for much bigger time lags (250000–252000M); here the difference between weakly chaotic and regular orbits becomes obvious.

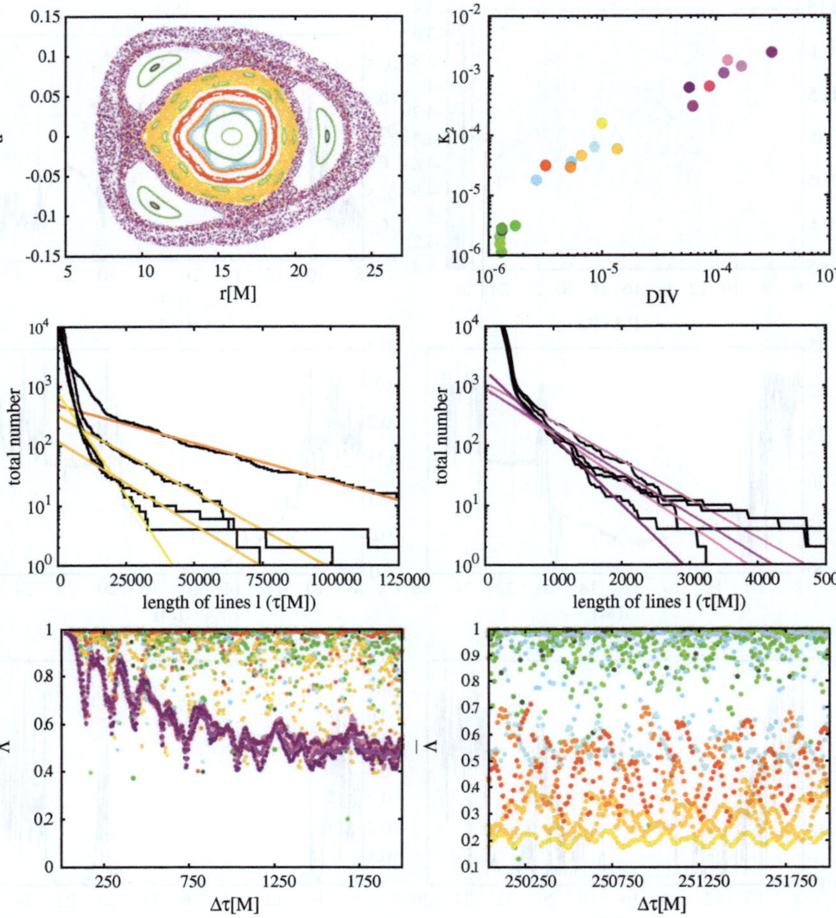

Fig. 2 Poincaré surface of section with each orbit plotted in different colour (*top left*); comparison of K_2 against *DIV* (*top right*); cumulative histograms of diagonal lines in RP for several strongly chaotic (*middle left*) and weakly chaotic (*middle right*) with the linear regression yielding K_2; the ADV parameter $\bar{\Lambda}$ for small (*bottom left*) and big (*bottom right*) time lags

We would like to thank the projects GACR 14-37086G, SVV-267301 and GAUK-428011 (PS), GACR-202/09/0772 and MSM0021620860 (OS).

References

1. Semerák, O., Suková, P.: Free motion around black holes with discs or rings: between integrability and chaos—I. Mon. Not. R. Astron. Soc. **404**, 545 (2010). doi:10.1111/j.1365-2966.2009.16003.x
2. Eckmann, J.P., Oliffson Kamphorst, S., Ruelle, D.: Recurrence plots of dynamical systems. Europhys. Lett. **4**, 973 (1987). doi:10.1209/0295-5075/4/9/004

3. Marwan, N., Romano, M., Thiel, M., Kurths, J.: Recurrence plots for the analysis of complex systems. Phys. Rep. **438**, 237 (2007). doi:10.1016/j.physrep.2006.11.001
4. Kaplan, D., Glass, L.: Direct test for determinism in a time series. Phys. Rev. Lett. **68**(4), 427 (1992). doi:10.1103/PhysRevLett.68.427
5. Semerák, O., Suková, P.: Free motion around black holes with discs or rings: between integrability and chaos—II. Mon. Not. R. Astron. Soc. **425**, 2455 (2012). doi:10.1111/j.1365-2966.2012.21630.x
6. Suková, P.: Chaotic geodesic motion around a black hole and disc. J. Phys. Conf. Ser. **314**, 012087 (2011). doi:10.1088/1742-6596/314/1/012087
7. Suková, P., Semerák, O., Recurrence of geodesics in a black-hole-disc field. In: Beltrán Jiménez, J., Cembranos, J., Dobado, A., López Maroto, A., De la Cruz Dombriz, A. (ed.) Towards New Paradigms: Proceeding of the Spanish Relativity Meeting 2011, AIP Conference Proceedings, vol. 1458, (AIP, Melville, NY, 2012), pp. 523–526. doi:10.1063/1.4734475

Gravitational Waveforms for Black Hole Binaries with Unequal Masses

Márton Tápai, Zoltán Keresztes and László Árpád Gergely

Abstract We derive a post-Newtonian (PN) inspiral-only gravitational waveform for unequal mass, spinning black hole binaries. Towards the end of the inspiral the larger spin dominates over the orbital angular momentum (while the smaller spin is negligible), hence the name Spin-Dominated Waveforms (SDW). Such systems are common sources for future gravitational wave detectors and during the inspiral the largest amplitude waves are emitted exactly in its last part. The SDW waveforms emerge as a double expansion in the PN parameter and the ratio of the orbital angular momentum to the dominant spin.

1 Introduction

Gravitational wave detectors like the Advanced LIGO (aLIGO), or the planned Einstein Telescope (ET), LAGRANGE and eLISA (NGO) space missions will measure gravitational waves from black hole binaries of various total masses m. For astrophysical black hole binaries (with total mass m a few tens of the mass of the sun $M_{\odot}$), the comparable mass and the unequal mass case are both likely. For supermassive black hole binaries (total mass is between 10^6 $M_{\odot}$ and 10^{10} $M_{\odot}$) the typical mass ratio ν is between 0.3 and 0.03 [1, 2].

For unequal masses the mass ratio can stand as a second small parameter. The purpose of this paper is to give an approximation for the gravitational waveforms in the small mass ratio regime.

M. Tápai (✉) · Z. Keresztes · L. Á. Gergely
Departments of Theoretical and Experimental Physics, University of Szeged,
Dóm tér 9, Szeged 6720, Hungary
e-mail: tapai@titan.physx.u-szeged.hu

J. Bičák and T. Ledvinka (eds.), *Relativity and Gravitation*,
Springer Proceedings in Physics 157, DOI: 10.1007/978-3-319-06761-2_65,

2 Spin-Dominated Waveforms

It was shown in Ref. [1], that for rapidly spinning black hole binaries, the smaller spin is of order ν^2 compared to the dominant spin S_1, thus it can be neglected to first order in ν. Furthermore the ratio of the orbital angular momentum L_N and S_1 was also given [1] as

$$\frac{S_1}{L_N} \approx \varepsilon^{1/2}\nu^{-1}\chi_1 , \tag{1}$$

where $\varepsilon = Gm/c^2r \approx v^2/c^2$ (with r the orbital separation and v the orbital velocity of the reduced mass particle $\mu = m_1m_2/m$, G the gravitational constant, c the speed of light) is the post-Newtonian (PN) parameter and $\chi_1 \in [0, 1]$ is the dimensionless spin. For maximally spinning black holes $\chi_1 = 1$.

As the PN parameter increases throughout the inspiral, the relation (1) shows, that S_1 will dominate over L_N at the end of the inspiral (thus the approximated waveforms are called Spin-Dominated Waveforms, SDW). This condition at the technical level is included in the smallness of the parameter $\xi = \varepsilon^{-1/2}\nu$.

PN waveforms were previously calculated to 1.5 PN order [3, 4], and to 2 PN order in Ref. [5]. In order to approximate the waveforms in the small mass ratio regime, we expand the waveforms in both parameters ε and ξ. The waveforms have the following structure [6]:

$$\begin{aligned} h_{\substack{+\\ \times}} = {} & \frac{2G^2m^2\varepsilon^{1/2}\xi}{c^4Dr}\left[h^{0}_{\substack{+\\ \times}} + \beta_1 h^{0\beta}_{\substack{+\\ \times}} + \varepsilon^{1/2}\left(h^{0.5}_{\substack{+\\ \times}} + \beta_1 h^{0.5\beta}_{\substack{+\\ \times}} - 2\xi h^{0}_{\substack{+\\ \times}}\right)\right. \\ & + \varepsilon\left(h^{1}_{\substack{+\\ \times}} - 4\xi h^{0.5}_{\substack{+\\ \times}} + \beta_1 h^{1\beta}_{\substack{+\\ \times}} + h^{1SO}_{\substack{+\\ \times}} + \beta_1 h^{1\beta SO}_{\substack{+\\ \times}}\right) \\ & \left. + \varepsilon^{3/2}\left(h^{1.5}_{\substack{+\\ \times}} + h^{1.5SO}_{\substack{+\\ \times}} + h^{1.5tail}_{\substack{+\\ \times}}\right)\right], \end{aligned} \tag{2}$$

D being the luminosity distance to the source. The terms are of different ε and ξ orders, as indicated in Table 1, and are given in detail in Ref. [6]. The angle β_1 spanned by $\mathbf{J}$ and $\mathbf{S}_1$ is of order ξ too [6].

3 Limits of Validity

We impose the smallness condition $\xi \leq 0.1$. This defines a lower limit of the PN parameter $\varepsilon_1 = Gm/c^2r_1 = 100\nu^2$, implying an upper limit for the mass ratio, $\nu_{\max} = 0.0316 \approx 1{:}32$. The upper limit for ε is defined by the end of the inspiral (chosen here as $\varepsilon_2 = 0.1$ [7]).

From the expression $m = c^3\varepsilon^{3/2}(\pi Gf)^{-1}$ including the gravitational wave frequency f, and the leading order radiative orbital angular frequency evolution [8] an integration leads to the time Δt during which the binary evolves from ε_1 to ε_2:

Table 1 SDW contributions of different ξ and ε orders

	ε^0	$\varepsilon^{1/2}$	ε^1	$\varepsilon^{3/2}$
ξ^0	$h_{+\times}^{0}$	$h_{+\times}^{0.5}$	$h_{+\times}^{1}, h_{+\times}^{1SO}$	$h_{+\times}^{1.5}, h_{+\times}^{1.5SO}, h_{+\times}^{1.5tail}$
ξ^1	$h_{+\times}^{0\beta}$	$h_{+\times}^{0.5\beta}$	$h_{+\times}^{1\beta}, h_{+\times}^{1\beta SO}$	

The SO terms contain the dominant spin

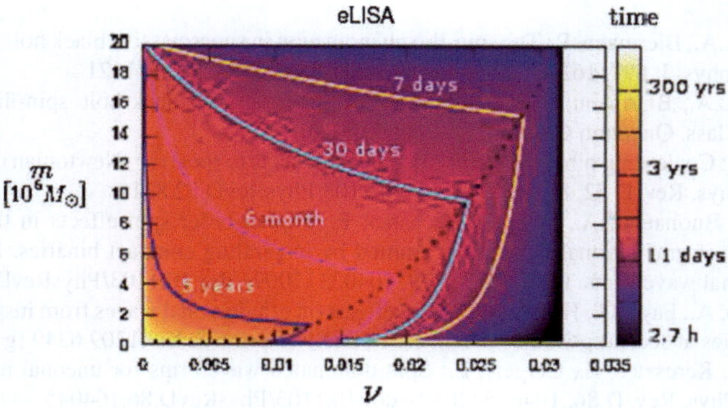

Fig. 1 The time interval Δt until which the SDWs can be detected by eLISA (NGO) as function of the total mass m and mass ratio ν. Δt either begins at the lower bound of the sensitivity range of eLISA ($\varepsilon_{f\,\mathrm{min}}$), or when the SDW approximation begins to hold (ε_1), and ends at the end of the inspiral (chosen here as $\varepsilon_2 = 0.1$). The color code is logarithmic

$$\Delta t = \frac{5Gm}{2^8 c^3}\frac{(1+\nu)^2}{\nu}\left(\varepsilon_1^{-4} - \varepsilon_2^{-4}\right). \tag{3}$$

Δt is shown as function of m and ν on Fig. 1. Even with the SDW approximation holding, the lower sensitivity bound ($f_{\mathrm{min}} = 10^{-4}$ for eLISA [9]) of the instrument may impose a larger value of the PN parameter, as the lower validity bound $\varepsilon_{f_{\mathrm{min}}}$. Hence Δt is calculated from $\max\left(\varepsilon_1, \varepsilon_{f_{\mathrm{min}}}\right)$ to ε_2.

A lower limit for the mass ratio comes from the assumption that the second compact object has at least the mass of a neutron star (1.4 $M_\odot$). The total mass is bounded from above by the lower frequency bound of the detector (for eLISA $m = 2\times 10^7 M_\odot$, hence the minimal mass ratio for the eLISA detector is $\nu_{\mathrm{min}} = 7\times 10^{-8}$).

4 Concluding Remarks

For unequal mass ratios the larger spin dominates over the orbital angular momentum at the end of the inspiral. We have quantified this by the introduction of a second small parameter ξ and computed the respective waveforms as a series expansion in both this and the PN parameter. A comparison between the general waveforms of

Ref. [4] and the SDWs showed that the SDWs are approximately 80 % shorter, due to the smaller parameter space and the second expansion in ξ. We expect the SDWs to be useful tools in gravitational wave detection.

References

1. Gergely, L.A., Biermann, P.: The spin-flip phenomenon in supermassive black hole binary mergers. Astrophys. J. **697**, 1621 (2009). doi:10.1088/0004-637X/697/2/1621
2. Gergely, L.Á., Biermann, P.L., Caramete, L.I.: Supermassive black hole spin-flip during the inspiral. Class. Quantum Gravity **27**, 194009 (2010)
3. Kidder, L.: Coalescing binary systems of compact objects to (post)$^{5/2}$-Newtonian order. V: Spin Effects. Phys. Rev. D **52**, 821 (1995). doi:10.1103/PhysRevD.52.821
4. Arun, K., Buonanno, A., Faye, G., Ochsner, E.: Higher-order spin effects in the amplitude and phase of gravitational waveforms emitted by inspiraling compact binaries: Ready-to-use gravitational waveforms. Phys. Rev. D **79**, 104023 (2009). doi:10.1103/PhysRevD.79.104023
5. Buonanno, A., Faye, G., Hinderer, T.: Spin effects on gravitational waves from inspiraling compact binaries at second post-Newtonian order, ArXiv e-prints arXiv:1209.6349 [gr-qc] (2012)
6. Tápai, M., Keresztes, Z., Gergely, L.: Spin-dominated waveforms for unequal mass compact binaries. Phys. Rev. D **86**, 104045 (2012). doi:10.1103/PhysRevD.86.104045
7. Levin, J., McWilliams, S.T., Contreras, H.: Inspiral of generic black hole binaries: spin, precession, and eccentricity. Class. Quantum Gravity **28**, 175001 (2011)
8. Mikóczi, B., Vasúth, M., Gergely, L.: Self-interaction spin effects in inspiralling compact binaries. Phys. Rev. D **71**, 124043 (2005). doi:10.1103/PhysRevD.71.124043
9. Binétruy, P., Bohé, A., Caprini, C., Dufaux, J.F.: Cosmological backgrounds of gravitational waves and eLISA/NGO: phase transitions, cosmic strings and other sources. J. Cosmol. Astropart. Phys. **2012**(06), 027 (2012). doi:10.1088/1475-7516/2012/06/027

Part III
Quantum Fields and Quantum Gravity

Phenomenology of Quantum Gravity and its Possible Role in Neutrino Anomalies

Mario A. Acero and Yuri Bonder

Abstract New phenomenological models of Quantum Gravity have suggested that a Lorentz-Invariant discrete spacetime structure may become manifest through a nonstandard coupling of matter fields and spacetime curvature. On the other hand, there is strong experimental evidence suggesting that neutrino oscillations cannot be described by simply considering neutrinos as massive particles. In this manuscript we motivate and construct one particular phenomenological model of Quantum Gravity that could account for the so-called neutrino anomalies.

1 Introduction

To construct a theory that reconciles Quantum Mechanics and General Relativity is one of the most challenging problems in Physics. This still unfinished theory is called Quantum Gravity (QG), and we believe that the difficulty in building such theory may be, in part, due to the lack of experimental guidance. Regarding particle physics, the Standard Model of particles (SM) includes three massless neutrinos. The Higgs mechanism [1], through which the masses of all other fermions (as well as bosons) are generated, does not apply to neutrinos because the neutrino fields do not have right handed components [2]. However, the experimental observation of

M. A. Acero (✉)
Departamento de Física, Universidad del Atlántico, Km 7 antigua vía a Puerto Colombia, Barranquilla, Colombia
e-mail: marioacero@mail.uniatlantico.edu.co

M. A. Acero · Y. Bonder
Instituto de Ciencias Nucleares, Universidad Nacional Autónoma de México, Apartado Postal 70-543, 04510 México, D.F., Mexico

Y. Bonder
Physics Department, Indiana University, Bloomington, IN 47405, USA
e-mail: ybonder@indiana.edu

J. Bičák and T. Ledvinka (eds.), *Relativity and Gravitation*, Springer Proceedings in Physics 157, DOI: 10.1007/978-3-319-06761-2_66,

neutrinos changing from one flavor to another, a phenomenon known as *neutrino oscillation*, has motivated people to suggest that neutrinos are actually massive and current research looks for an extension of the SM to include neutrino masses. The simplest extension of the SM (but certainly not the only one, see Refs. [2–4]) is to include right handed components of the neutrino fields, so that they acquire mass through the same mechanism as the other particles.

Nevertheless, to include neutrino masses seems to be insufficient to account for all observations. In this work we argue that the anomalous neutrino oscillations could be regarded as traces of the quantum nature of gravity. More concretely, we propose a modification to the simple extension of the SM described above motivated by a phenomenological model of QG in order to explain all neutrino observations. Before we continue, we warn the reader that we only present the motivation and possible applications of a class of phenomenological models of QG to neutrinos; a deeper study of this issue is needed to test if these models are a feasible explanation of the neutrino anomalies.

2 Neutrino Oscillations

Neutrino oscillations are transition of a neutrino in a definite flavor state into a neutrino with a different flavor. The basic idea is that a neutrino flavor state is a linear combination of states with definite mass. The *oscillation probability*, in the two-neutrinos approximation, is given by [2]

$$P_{\nu_\alpha \to \nu_\beta}(L, E) = \sin^2(2\theta)\sin^2\left(\frac{\Delta m^2 L}{4E}\right), \quad \alpha \neq \beta, \tag{1}$$

where L and E are, respectively, the distance traveled by the neutrino and its energy (both in the laboratory reference frame), and the two fundamental parameters of this process are the *mixing angle* θ and the *masses-squared difference* $\Delta m^2 \equiv m_2^2 - m_1^2$. It is under this effective model that most of the experimental data have been analyzed, given that many experiments are not sensitive to the effects of three-neutrino mixing [2].

Now, from observations of solar, atmospheric, reactor, and accelerator-based neutrino-oscillation experiments, it has been possible to establish firmly the existence of three mixing angles and two separated mass-splitting parameters of order $10^{-5}\,\text{eV}^2$ and $10^{-3}\,\text{eV}^2$ (for an updated combined data analysis, see [5–7]). Yet, there are some results that cannot be explained with these parameters. The Liquid Scintillator Neutrino Detector (LSND) [8, 9] experiment found that their oscillation data point to $\Delta m^2 = O(1\,\text{eV}^2)$, which is much larger than the Δm^2 found by the experiments mentioned before [10–13]. More recently, in an attempt to check this anomalous outcome, the MiniBooNE collaboration [14, 15] found that, with a 99 % confidence level, their analysis leads to a Δm^2 that is consistent with that from

LSND. Note that both experiments, LSND and MiniBooNE, produced the neutrinos in accelerators and have the same L/E (see Eq. (1)). Additional anomalous results have been under study and include the Reactor antineutrino anomaly [16] and the Gallium anomaly [17, 18].

Currently, a great effort is underway to clarify these issues, both from the theoretical and experimental point of view and a number of experiments currently running, and different proposals for the future, are devoted to it [19–25]. On theoretical grounds, perhaps the most popular explanation is the existence of, at least, one additional neutrino which has to have different properties compared to those included in the SM. This (or these) new neutrino is known as *sterile*, given that it does not take part in the weak interactions of the SM [2]. However, there is no further evidence supporting the existence of sterile neutrinos. In this work, we take a different strategy where there is no need to add new particles. In contrast, we propose that gravity, whose fundamental version is still unknown, may couple to the neutrino fields in a non-standard way, producing the anomalous neutrino oscillations. In addition, if gravity is behind neutrino oscillations, it is conceivable that these depend on the gravitational environment, as is suggested by the aforementioned experimental results. In the next section, we briefly present the phenomenological model of QG that gives rise to these couplings.

3 Lorentz Invariant Phenomenology of Quantum Gravity

The phenomenology of QG has been dominated, in the last years, by searching for Lorentz-Invariance (LI) violations. This may be motivated by the fact that a naive discrete spacetime structure naturally selects preferred directions. Besides the significant empirical bounds on LI violations (for the most complete collection of bounds see [26]), Collins [27] have argued that a LI violating discrete spacetime inhabited by quantum fields can be discarded by experiments. Essentially, the radiative corrections would magnify the effects of a LI violating discrete spacetime up to the point where they should have been observed. These arguments motivated a new type of phenomenological models of QG [28, 29] where a LI discrete spacetime structure is sought precisely by using the hypothesis that the symmetry of spacetime building blocks is LI.

It is hard to envision a discrete spacetime structure that respects LI. However, in order to build a phenomenological model, there is no need to have a concrete picture of such a structure. The basic idea is that the presence of a LI discrete spacetime structure may reveal itself when there is a mismatch between the symmetries of spacetime (at a macroscopic scale) and the symmetries of its building blocks. As mentioned above, in these models one assumes that spacetime building blocks are LI, thus, the mismatch with the macroscopic symmetry would occur when the macroscopic spacetime is not LI. This, in turn, happens in curved spacetime regions, suggesting that the effects of a LI discrete spacetime structure could manifest themselves as non-standard couplings of curvature and matter fields.

Studying a coupling of matter and the Ricci tensor (or the curvature scalar) is not interesting phenomenologically because, according to Einstein's equations, these geometrical objects at a given spacetime point are determined by the matter at that same point. Thus, coupling matter with the Ricci tensor can be considered at the phenomenological level a self-coupling. Thus, the Weyl tensor W_{abcd}, which loosely speaking is the part of the Riemann tensor that remains when the Ricci part is subtracted [30], is the object that should be coupled with the matter fields. Moreover, the coupling must vanish in flat spacetime regions where spacetime is actually LI.

In the past, one particular model was extensively studied. It involves fermionic matter fields that couple to the eigenvalues and eigenvectors of two Hermitian operators built out of the Weyl tensor through complicated couplings [29]. This model has been able to produce some bounds in the neutron sector [31] and to motivate an experiment where the effect predicted by the model was sought [32, 33] and bounds on the electron sector were obtained. In the next section, a particular model for neutrinos that may help explain some of the anomalies described above is presented.

4 Neutrino Effective Mass

For simplicity, we only consider Dirac neutrinos with non-vanishing masses and right handed components. The strategy is to generate effective masses that depend on the gravitational environment. Following the motivation discussed above, this effective mass should be obtained through non-minimal couplings of spacetime curvature (Weyl tensor) and the neutrino fields. This coupling should vanish in flat spacetime regions and must respect gauge invariance to have a theory with a well posed initial value formulation (see the discussion on that matter in Ref. [30]).

To define this coupling term we write the Lagrangian density describing Dirac massive neutrinos in a curved background:

$$\begin{aligned}\mathscr{L}_{g+ew} &= ie\bar{\nu}_{L\alpha}e^a_\mu\gamma^\mu D^{(g+ew)}_a\nu_{L\alpha} + ie\bar{\nu}_{R\alpha}e^a_\mu\gamma^\mu D^{(g)}_a\nu_{R\alpha}\\ &\quad - e\Gamma_{\alpha\beta}\left(\bar{\nu}_{L\alpha}\phi\nu_{R\beta} + \bar{\nu}_{R\alpha}\phi\nu_{L\beta}\right), \qquad (2)\end{aligned}$$

where ν_L and ν_R are the left and right neutrinos, ϕ is (one component of) the Higgs field, $\Gamma_{\alpha\beta}$ are the (dimensionless) Yukawa coupling constants, e^a_μ are the tetrads, e is the spacetime natural volume form and $D^{(g+ew)}_a$ is the covariant derivative including gauge interaction and gravity while $D^{(g)}_a$ contains only the gravitational part. The indices α, β label the neutrino flavor. The charged lepton part of the Lagrangian, which should be included to have explicit gauge invariance, is not written since the gauge interaction is not considered in what follows.

To respect gauge invariance, the gravity modification to the mass term must enter into the Lagrangian density (2) though the replacement

$$\Gamma_{\alpha\beta} \to \Gamma_{\alpha\beta} + b_{\alpha\beta} f\left(\frac{W}{M_P^2}\right), \tag{3}$$

where $b_{\alpha\beta}$ are the coupling coefficients, $W(x) \equiv \sqrt{W_{abcd}W^{abcd}}$ and f is a dimensionless real function. The Planck mass, M_P, is introduced in such a way that the argument of f is also dimensionless. The simplest function f that is only suppressed by one power of M_P in the denominator is

$$f\left(\frac{W}{M_P^2}\right) = \frac{\sqrt{W}}{M_P}, \tag{4}$$

which is the function we consider.

Once the gauge symmetry is spontaneously broken, the gravitational part of the Lagrangian density (2) takes the form

$$\begin{aligned}\mathcal{L}_g = & \, i e \bar{\nu}_{L\alpha} e^a_\mu \gamma^\mu D^{(g)}_a \nu_{L\alpha} + i e \bar{\nu}_{R\alpha} e^a_\mu \gamma^\mu D^{(g)}_a \nu_{R\alpha} \\ & - e\left(m_{\alpha\beta} + a_{\alpha\beta}\frac{\sqrt{W}}{M_P}\right)\left(\bar{\nu}_{L\alpha}\nu_{R\beta} + \bar{\nu}_{R\alpha}\nu_{L\beta}\right),\end{aligned} \tag{5}$$

where $m_{\alpha\beta} = \Gamma_{\alpha\beta} < \phi >$ and $a_{\alpha\beta} = b_{\alpha\beta} < \phi >$. Observe that the mass matrix in this case is

$$M_{\alpha\beta}(x) \equiv m_{\alpha\beta} + a_{\alpha\beta}\frac{\sqrt{W(x)}}{M_P}. \tag{6}$$

Typically, $m_{\alpha\beta}$ generates neutrino flavor mixing. In the case we are dealing with, these oscillations would be caused by $M_{\alpha\beta}$ which, in all cases of phenomenological interest can be thought as $m_{\alpha\beta}$ plus a small modulation that depends on the gravitational environment. As neutrinos from different sources (solar, atmospheric, reactor, and accelerator) travel in different gravitational backgrounds, according to the model presented here, they should oscillate slightly differently.

In order to gain some intuition, we consider the effects of this model for neutrinos traveling closely to the Earth's surface, as happens in accelerator and reactor experiments. In this case W can be taken approximately as constant given by $W = \sqrt{48}M/R^3$ where R and M stand for the radius and mass of the Earth, respectively. The numerical value is $\sqrt{W} \approx 10^{-46} M_P$, which would then require $a_{\alpha\beta}$ to be extremely large in order to produce any measurable effect. Thus, at first sight this model seems to be ruled out. However, let us remind the reader that the size of $a_{\alpha\beta}$ is somehow artificial since we put in M_P by hand. Moreover, a different function f could be chosen that could make testable predictions. In any case, a much deeper analysis is required. In particular, one would need to try to fit the free parameters of the model to explain the neutrino anomalies before taking these models seriously. This may be particularly difficult to achieve because, in certain cases, the tidal effects of a wall or a mountain can dominate over the effects of the entire Earth (see

Ref. [34]), thus, a very precise knowledge of the gravitational-source distribution on the neutrinos' path may be needed to correctly model the neutrino oscillations. Also, the effects of matter, which also generate neutrino oscillation [35], must be considered. An intriguing possibility is to try to mimic the well-know MSW [36, 37] effect and search for gravitational environments where resonances could be expected.[1]

To conclude, we want to stress the reasons that motivated us to consider gravity as a possible explanation for the anomalous neutrino behavior. First, we know gravity exists, thus, we do not need to invoke new fields/particles that have not been observed to account for the anomalies. Second, it is conceivable that QG may influence matter in exotic ways and these effects could become manifest at scales below the Planck regime. Third, neutrino experiments are done with particles that have traveled in different gravitational environments, which may account, at least in part, for the different behavior.

Acknowledgments We want to thank D. Sudarsky, J. S. Díaz, R. Lehnert, and V. A. Kostelecký for their comments. MAA acknowledges support by the Mexican RedFAE, CONACyT, during the initial development of this work. YB was partially supported by the research grants CONACyT 101712, PAPIIT-UNAM IN107412 and by the Department of Energy under grant number DE-FG02–91ER40661 and also by the Indiana University Center for Spacetime Symmetries.

References

1. Higgs, P.: Broken symmetries, massless particlees and gauge fields. Phys. Lett. **12**, 132 (1964). doi:10.1016/0031-9163(64)91136-9
2. Giunti, C., Kim, C.: Fundamentals of Neutrino Physics and Astrophysics. Oxford University Press, Oxford (2007)
3. Zuber, K.: Neutrino Physics. Series in High Energy Physics, Cosmology and Gravitation. Institute of Physics, Bristol (2004)
4. Mohapatra, R., Pal, P.: Massive Neutrinos in Physics and Astrophysics. Lecture Notes in Physics, vol. 41, 2nd edn. World Scientific, Singapore (1998)
5. Fogli, G., Lisi, E., Marrone, A., et al.: Global analysis of neutrino masses, mixings, and phases: entering the era of leptonic CP violation searches. Phys. Rev. D **86**, 013012 (2012). doi:10.1103/PhysRevD.86.013012
6. Gonzalez-Garcia, M., Maltoni, M., Salvado, J.: Updated global fit to three neutrino mixing: status of the hints of θ_{13} 0. J. High Energy Phys. **2010**(04), 056 (2010). doi:10.1007/JHEP04(2010)056
7. Schwetz, T., Tórtola, M., Valle, J.: Three-flavour neutrino oscillation update. New J. Phys. **10**, 113011 (2008). doi:10.1088/1367-2630/10/11/113011
8. Athanassopoulos, C., Auerbach, L., Burman, R., et al.: Results on $\nu_\mu \to \nu_e$ neutrino oscillations from the LSND experiment. Phys. Rev. Lett. **81**, 1774 (1998). doi:10.1103/PhysRevLett.81.1774
9. Aguilar, A., Auerbach, L., Burman, R., et al.: Evidence for neutrino oscillations from the observation of $\overline{\nu}_e$ appearance in a $\overline{\nu}_\mu$ beam. Phys. Rev. D. **64**, 112007 (2001). doi:10.1103/PhysRevD.64.112007
10. Eguchi, K., Enomoto, S., Furuno, K., et al.: First results from KamLAND: evidence for reactor antineutrino disappearance. Phys. Rev. Lett. **90**, 021802 (2003). doi:10.1103/PhysRevLett.90.021802

[1] We thank R. Lehnert for this suggestion.

11. Greenwood, Z., Kropp, W., Mandelkern, M., et al.: Results of a two-position reactor neutrino-oscillation experiment. Phys. Rev. D. **53**, 6054 (1996). doi:10.1103/PhysRevD.53.6054
12. Michael, D., Adamson, P., Alexopoulos, T., et al.: Observation of muon neutrino disappearance with the MINOS detectors in the NuMI neutrino beam. Phys. Rev. Lett. **97**, 191801 (2006). doi:10.1103/PhysRevLett.97.191801
13. Wendell, R., Ishihara, C., Abe, K., et al.: Atmospheric neutrino oscillation analysis with sub-leading effects in Super-Kamiokande I, II, and III. Phys. Rev. D **81**, 092004 (2010). doi:10.1103/PhysRevD.81.092004
14. Aguilar-Arevalo, A., Anderson, C., Brice, S., et al.: Search for electron antineutrino appearance at the $\Delta m^2 \sim 1\text{eV}^2$ scale. Phys. Rev. Lett. **103**, 111801 (2009). doi:10.1103/PhysRevLett.103.111801
15. Aguilar-Arevalo, A., Anderson, C., Brice, S., et al.: Event excess in the MiniBooNE search for $\overline{\nu}_\mu \rightarrow \overline{\nu}_e$ oscillations. Phys. Rev. Lett. **105**, 181801 (2010). doi:10.1103/PhysRevLett.105.181801
16. Mention, G., Fechner, M., Lasserre, T., et al.: Reactor antineutrino anomaly. Phys. Rev. D **83**, 073006 (2011). doi:10.1103/PhysRevD.83.073006
17. Giunti, C., Laveder, M.: Short-baseline active-sterile neutrino oscillations? Mod. Phys. Lett. A **22**, 2499 (2007). doi:10.1142/S0217732307025455
18. Acero, M., Giunti, C., Laveder, M.: Limits on ν_e and $\overline{\nu}_e$ disappearance from Gallium and reactor experiments. Phys. Rev. D **78**, 073009 (2008). doi:10.1103/PhysRevD.78.073009
19. Klapdor-Kleingrothaus, H.: Double beta decay and neutrino mass. the Heidelberg-Moscow experiment. Prog. Part. Nucl. Phys. **32**, 261 (1994). doi:10.1016/0146-6410(94)90024-8
20. Lobashev, V.: The search for the neutrino mass by direct method in the tritium beta-decay and perspectives of study it in the project KATRIN. Nucl. Phys. A **719**, 153 (2003). doi:10.1016/S0375-9474(03)00985-0
21. Drexlin, G.: Direct neutrino mass searches. Nucl. Phys. B (Proc. Suppl.) **138**, 282 (2005). doi:10.1016/j.nuclphysbps.2004.11.064
22. Ghoshal, P., Petcov, S.: Neutrino mass hierarchy determination using reactor antineutrinos. J. High Energy Phys. **2011**(03), 058 (2011). doi:10.1007/JHEP03(2011)058
23. Aseev, V., Belesev, A., Berlev, A., et al.: Upper limit on the electron antineutrino mass from the Troitsk experiment. Phys. Rev. D **84**, 112003 (2011). doi:10.1103/PhysRevD.84.112003
24. Abazajian, K., Calabrese, E., Cooray, A., et al.: Cosmological and astrophysical neutrino mass measurements. Astropart. Phys. **35**, 177 (2011). doi:10.1016/j.astropartphys.2011.07.002
25. Blennow, M., Schwetz, T.: Identifying the neutrino mass ordering with INO and NOvA. J. High Energy Phys. **2012**(08), 058 (2012). doi:10.1007/JHEP08(2012)058
26. Kostelecký, V., Russell, N.: Data tables for Lorentz and CPT violation. Rev. Mod. Phys. **83**, 11 (2011). doi:10.1103/RevModPhys.83.11
27. Collins, H.: Gravity's Shadow: The Search for Gravitational Waves. University of Chicago Press, Chicago (2004)
28. Corichi, A., Sudarsky, D.: Towards a new approach to quantum gravity phenomenology. Int. J. Mod. Phys. D **14**, 1685 (2005). doi:10.1142/S0218271805007541
29. Bonder, Y., Sudarsky, D.: Quantum gravity phenomenology without Lorentz invariance violation: a detailed proposal. Class. Quantum Grav. **25**, 105017 (2008). doi:10.1088/0264-9381/25/10/105017
30. Wald, R.: General Relativity. University of Chicago Press, Chicago (1984)
31. Bonder, Y., Sudarsky, D.: Unambiguous quantum gravity phenomenology respecting Lorentz symmetry. Rep. Math. Phys. **64**, 169 (2009). doi:10.1016/S0034-4877(09)90025-8
32. Bonder, Y., Sudarsky, D.: Searching for spacetime granularity: analyzing a concrete experimental setup. In: Morales-Tecotl, H., Urena-Lopez, L., Linares-Romero, R., Garcia-Compean, H. (eds.) Gravitational Physics: Testing Gravity from Submillimeter to Cosmic. AIP Conference Proceedings, vol. 1256, pp. 157–163. American Institute of Physics, Melville (2010). doi:10.1063/1.3473847
33. Terrano, W., Heckel, B., Adelberger, E.: Search for a proposed signature of Lorentz-invariant spacetime granularity. Class. Quantum Grav. **28**, 145011 (2011). doi:10.1088/0264-9381/28/14/145011

34. Landau, S., Teppa Pannia, F., Bonder, Y., Sudarsky, D.: Space-time variation of the electron-to-proton mass ratio in a Weyl model. Astropart. Phys. **35**, 377 (2012). doi:10.1016/j.astropartphys.2011.10.008
35. Kuo, T., Pantaleone, J.: Neutrino oscillations in matter. Rev. Mod. Phys. **61**, 937 (1989). doi:10.1103/RevModPhys.61.937
36. Wolfenstein, L.: Neutrino oscillations in matter. Phys. Rev. D **17**, 2369 (1978). doi:10.1103/PhysRevD.17.2369
37. Mikheyev, S., Smirnov, A.: Resonant amplification of neutrino oscillations in matter and solar-neutrino spectroscopy. Nuovo Cimento C **9**, 17 (1986). doi:10.1007/BF02508049

Loop Quantum Cosmology: Anisotropy and Singularity Resolution

Alejandro Corichi, Asieh Karami and Edison Montoya

Abstract In this contribution we consider the issue of singularity resolution within loop quantum cosmology (LQC) for different homogeneous models. We present results of numerical evolutions of effective equations for both isotropic as well as anisotropic cosmologies, with and without spatial curvature. To address the issue of singularity resolution we examine geometrical and curvature invariants that yield information about the spacetime geometry. We discuss generic behavior found for a variety of initial conditions.

1 Introduction

In general relativity (GR) the singularity theorems of Hawking, Penrose and Geroch tell us that, under reasonable assumptions, singularities are generic. A spacetime is said to be singular if it is not geodesically complete, which may happen when some geometrical curvature invariants diverge. The expectation is that, by quantizing the gravitational degrees of freedom, namely, with a complete theory unifying gravity and the quantum, the singularities shall be resolved. Loop quantization (as in Loop Quantum Gravity) of the homogeneous, isotropic and flat Friedman-Robertson-

A. Corichi (✉) · A. Karami · E. Montoya
Centro de Ciencias Matemáticas, Universidad Nacional Autónoma de México, UNAM-Campus Morelia, Morelia, Mexico
e-mail: corichi@matmor.unam.mx

A. Karami · E. Montoya
Instituto de Física y Matemáticas, Universidad Michoacana de San Nicolás de Hidalgo, Morelia, Michoacán, Mexico
e-mail: karami@ifm.umich.mx

E. Montoya
e-mail: edison@ifm.umich.mx

J. Bičák and T. Ledvinka (eds.), *Relativity and Gravitation*,
Springer Proceedings in Physics 157, DOI: 10.1007/978-3-319-06761-2_67,

Walker (FRW) cosmology coupled to a massless scalar field ϕ, can be exactly solved [1]. For that model it was shown that:

- The matter density operator $\hat{\rho}$ has an absolute upper bound and the expansion θ is also bounded. One can conclude that curvature scalars do not diverge. This is a signal that a singularity is not present.
- All states undergo a bounce and with this, the big bang is replaced by a *big bounce*.
- The GR dynamics is recovered as we go away from the Planck scale, this means that we are recovering the original theory that we want.
- Dynamics of semiclassical states are well captured by an effective theory that retains information about the loop quantum geometry.
- With all these, one can conclude that the singularities are resolved: the geodesics are inextendible, and are well defined on the other side of the would be big bang.

The fact that the *effective theory* provides an accurate description of the dynamics at the Planck scale is strongly used to explore the anisotropic models. The effective theory is obtained from the quantum Hamiltonian operator by taking expectation values on appropriately defined states. The thus obtained effective Hamiltonian then generates the dynamics on a classical phase space. The solutions to the effective theory were shown in [2] to accurately describe the evolution of the expectation value of the observables in the quantum theory when they are considered on semiclassical states. Those results were extended to open, closed and flat FRW models with and without cosmological constant (see [3] for a review). Loop quantum cosmology (LQC) has been extended to the simplest anisotropic cosmological models, namely Bianchi I, II and IX [4–6]. But in none of these cases, the quantum theory has been solved, even numerically. Then, in order to study these models at semiclassical level, one generally assumes that the effective theory reproduces the solutions to the quantum theory when semiclassical states are considered. This is our working hypothesis, which is well justified by the results in the isotropic cases. It would be interesting to know whether the evolution of the semiclassical states reproduces the solutions which we get from the effective theory. From this point of view, the study of the effective theory can be seen as the first step in this direction.

The new issues to consider in the anisotropic models are: is the bounce generic? We now have anisotropy/Weyl curvature, how does it behave near the singularity/bounce? Can we have different kind of bounce, say, dominated by shear σ? Are the geometric scalars absolutely bounded? The goal of this contribution is to answer these questions using the effective theory for Bianchi I which has anisotropies, Bianchi II that has anisotropies and spatial curvature and Bianchi IX which has all the features of Bianchi I, II and is furthermore, spatially compact. Even more, the Bianchi IX model has a non trivial classical limit, in the sense that, Bianchi IX is chaotic in the classical theory and behaves like Bianchi I with Bianchi II transitions as one approaches the singularity.

2 Preliminaries

In this section we briefly review the quantization of some cosmological models which include $k = 0$ and $k = 1$ FRW and Bianchi I, II and IX models by using loop quantum gravity methods. Let us consider the spacetime as $M = \Sigma \times \mathbb{R}$ where Σ is a spatial 3-manifold which can be identified by the symmetry group of the chosen model and is endowed with a fiducial metric ${}^o q_{ab}$ and associated fixed fiducial basis of 1-forms ${}^o\omega^i_a$ and vectors ${}^o e^a_i$. If Σ is non-compact then we fix a fiducial cell, $\mathscr{V}$, adapted to the fiducial triads with finite volume V_o. In GR, the gravitational phase space consists of pairs (A^i_a, E^a_i) on Σ where A^i_a is a SU(2) connection and E^a_i is a densitized triad of weight 1. Since all of the models in which we are interested are homogeneous and, if we restrict ourselves to diagonal metrics, one can fix the gauge in such a way that $A^i_a = \frac{c^i}{V_o^{1/3}} {}^o\omega^i_a$ and $E^a_i = \frac{p_i}{V_o^{2/3}} \sqrt{{}^o q}\, {}^o e^a_i$, where p_i in terms of the scale factors a_i are $|p_i| = V_o^{2/3} a_j a_k$. Note that in isotropic cases, each of phase space variables has only one independent component. The Poisson brackets can be expressed as $\{c^i, p_j\} = 8\pi G\gamma\delta^i_j$ and for isotropic models, the Poisson bracket is $\{c, p\} = 8\pi G\gamma/3$ where γ is Barbero-Immirzi parameter. With this choice of variables and gauge fixing, the Gauss and diffeomorphism constraints are satisfied and the only constraint is the Hamiltonian constraint

$$\mathscr{C}_H = \int_{\mathscr{V}} N \left[-\frac{\varepsilon^{ij}_{\ \ k} E^a_i E^b_j}{16\pi G\gamma^2 \sqrt{|q|}} \left(F^k_{ab} - (1+\gamma^2)\Omega^k_{ab} \right) + \mathscr{H}_{\text{matter}} \right] \mathrm{d}^3x, \qquad (1)$$

where N is the lapse function, $\mathscr{H}_{matter} = \rho V$ and Ω_{ab} is the curvature of spin connection Γ^i_a compatible with the triads.

To construct the quantum kinematics, we have to select a set of elementary observables such that their associated operators are unambiguous. In loop quantum gravity they are the holonomies h_e defined by the connection A^i_a along edges e and the fluxes of the densitized triad E^a_i across surfaces. For our homogeneous models we choose holonomies and p_i. To have the corresponding constraint operator, one needs to express it in terms of the chosen phase space functions h_e and p_i. The first term, $\varepsilon^{ij}_{\ \ k} E^a_i E^b_j / \sqrt{|q|}$, as in loop quantum gravity, can be treated by using Thiemann's strategy [7].

$$\varepsilon_{ijk} \frac{E^{ai} E^{bj}}{\sqrt{|q|}} = \sum_i \frac{1}{2\pi\gamma G\mu} \varepsilon^{abc} \omega^i_c \mathrm{Tr}(h_i^{(\mu)} \{h^{(\mu)-1}, V\} \tau_k), \qquad (2)$$

where $h_i^{(\mu)}$ is the holonomy along the edge parallel to ith vector basis with length μ and V is the volume, which is equal to $\sqrt{|p_1 p_2 p_3|}$. Note that μ is arbitrary. Now to define an operator related to the first term, we can use the right hand side of (2) and replace Poisson brackets with commutation relations. To find an operator related to the curvature F^k_{ab}, for isotropic models and Bianchi I, one can consider a square $\Box_{ij}$

in the i–j plane which is spanned by two of the fiducial triads (for closed isotropic model since triads do not commute, to define this plane we use a triad and a right invariant vector ${}^o\xi_i^a$), each of its sides has length $\bar{\mu}_i$. Therefore, F_{ab}^k is given by

$$F_{ab}^k = 2 \lim_{Ar\Box \to 0} \varepsilon_{ij}{}^k \mathrm{Tr}\left(\frac{h_{\Box_{ij}}^{\bar{\mu}} - \mathbb{I}}{\bar{\mu}_i \bar{\mu}_j} \tau^k\right) {}^o\omega_a^i \, {}^o\omega_b^j. \tag{3}$$

Since in loop quantum gravity, the area operator does not have a zero eigenvalue, one can take the limit of (3) when the area is equal to the smallest eigenvalue of area operator, $\lambda^2 = 4\sqrt{3}\pi\gamma l_p^2$, instead of zero. Then, $V_o^{1/3}\bar{\mu}_i a_i = \lambda$ where a_i is the scale factor. For Bianchi II and IX, we cannot use this method because the resulting operator is not almost periodic, therefore we express connection A_a^i in terms of holonomies and then use the standard definition of curvature F_{ab}^k. The operators corresponding to the connection are given by

$$\hat{c}_i = \frac{\widehat{\sin \bar{\mu}_i c_i}}{\bar{\mu}_i}, \text{ where } \bar{\mu}_i = \lambda \sqrt{\left| \frac{p_i}{p_j p_k} \right|}, \quad i \neq j \neq k. \tag{4}$$

Note that using this quantization method for flat FRW and Bianchi I models, one has the same result as the direct quantization of curvature F_{ab}^k, but for closed FRW it leads to a different quantum theory which is more compatible with the isotropic limit of Bianchi IX. We call the first method of quantization *curvature based quantization* and the second one *connection based*. In Bianchi II and Bianchi IX models the terms related to the curvatures, F_{ab}^k and Ω_{ab}^k, contain some negative powers of p_i which are not well defined operators. To solve this problem we use the same idea as Thiemann's strategy. Write

$$|p_i|^{(\ell-1)/2} = -\frac{\sqrt{|p_i|}}{4\pi G\gamma j(j+1)\tilde{\mu}_i \ell} \tau_i h_i^{(\tilde{\mu}_i)} \{h_i^{(\tilde{\mu}_i)-1}, |p_i|^{\ell/2}\}, \tag{5}$$

where $\tilde{\mu}_i$ is the length of a curve, $\ell \in (0, 1)$ and $j \in \frac{1}{2}\mathbb{N}$ is for the representation. Therefore, for these three different operators we have three different curve lengths $(\mu, \bar{\mu}, \tilde{\mu})$ where μ and $\tilde{\mu}$ can be some arbitrary functions of p_i, so for simplicity we can choose all of them to be equal to $\bar{\mu}$. On the other hand we have another free parameter in the definition of negative powers of p_i, where for simplicity we take $j = 1/2$, and since the largest negative power of p_i which appears in the constraint is $-1/4$ we will take $\ell = 1/2$ to have them directly from (5) and after that to express the other negative powers by them. The eigenvalues for the operator $\widehat{|p_i|^{-1/4}}$ are given by

$$J_i = \frac{h(V)}{V_c} \prod_{j \neq i} p_j^{1/4}, \text{ where } h(V) = \sqrt{V + V_c} - \sqrt{|V - V_c|}, \; V_c = 2\pi\gamma\lambda\ell_p^2.$$

By using these results and choosing some factor ordering, we can construct the total constraint operator. Note that with a different choice of factor ordering we will have different operators but the main results will remain almost the same. By solving the constraint equation $\hat{\mathscr{C}}_H \cdot \Psi = 0$, we have the physical states in physical Hilbert space $\mathscr{H}_{\text{phys}}$. Then we need to identify the physical observables. To test singularity resolution we will study some geometric observables: expansion θ, shear σ^2, curvature scalars and also volume of the universe V and matter density ρ, as relational observables in terms of ϕ, a massless scalar field.

Since working with full quantum theories of the models is difficult and, as shown in [2] for some models, the behavior of the effective or semiclassical equations, which are classical equations with some quantum corrections, are good approximations to the numerical quantum evolutions even near the Planck scale, we will work with the effective equations.

3 Effective Theories

Isotropic Flat and Closed Models

In the FRW model with $k = 0$, the *effective* Hamiltonian is given by

$$\mathscr{H}_{\text{k}=0} = \frac{3}{8\pi G\gamma^2\lambda^2} V^2 \sin(\lambda\beta)^2 - \frac{p_\phi^2}{2} \approx 0, \tag{6}$$

where p_ϕ is the momentum of the field, V is the volume and β its conjugate variable. They are related to the c and p variables by the equations $V = p^{3/2}$, $\beta = c/\sqrt{p}$ and satisfy the Poisson bracket $\{\beta, V\} = 4\pi G\gamma$ and $\{\phi, p_\phi\} = 1$. It was shown [2] that the dynamics of semiclassical states are well captured by the effective Friedman equation $H^2 = \frac{8\pi G}{3}\rho\left(1 - \frac{\rho}{\rho_{\text{crit}}}\right)$, where $H = \dot{V}/3V$ is the Hubble parameter and $\rho = p_\phi^2/2V^2$ is the matter density. The GR dynamics is recovered as we go away from the Planck scale $\rho < \rho_{\text{crit}}/10$, with $\rho_{\text{crit}} = \frac{3}{8\pi G\gamma^2\lambda^2}$.

Now, for the isotropic closed model, as we discussed in previous section, there are two different quantum theories depending on the two different methods of quantization of the curvature F_{ab}^k. The Hamiltonians are,

$$\mathscr{H}^{(1)}_{\text{k}=1} = \frac{3V^2}{8\pi G\gamma^2\lambda^2}[\sin^2(\lambda\beta - D) - \sin^2 D + (1+\gamma^2)D^2] - \frac{p_\phi^2}{2} \approx 0, \tag{7}$$

$$\mathscr{H}^{(2)}_{\text{k}=1} = \frac{3V^2}{8\pi G\gamma^2\lambda^2}[\sin^2 \lambda\beta - 2D\sin\lambda\beta + (1+\gamma^2)D^2] - \frac{p_\phi^2}{2} \approx 0, \tag{8}$$

where $D = \lambda\vartheta/V^{-1/3}$ and $\vartheta = (2\pi^2)^{1/3}$. Since for both effective theories there are some geometric observables which are not absolutely bounded, we go further and use more corrections which come from the inverse triad term in the full theory, to see if the unboundedness of those observables is generic, or whether it improves by adding more corrections. Therefore the Hamiltonian constraints change to

$$\mathcal{H}^{(1)}_{\mathrm{k=1}} = \frac{3A(V)V}{8\pi G\gamma^2\lambda^2}[\sin^2(\lambda\beta - D) - \sin^2 D + (1+\gamma^2)D^2] - \frac{p_\phi^2}{2} \approx 0, \tag{9}$$

$$\mathcal{H}^{(2)}_{\mathrm{k=1}} = \frac{3A(V)V}{8\pi G\gamma^2\lambda^2}[\sin^2 \lambda\beta - 2D\sin\lambda\beta + (1+\gamma^2)D^2] - \frac{p_\phi^2}{2} \approx 0, \tag{10}$$

where $A(V) = \frac{1}{2V_c}(V + V_c - |V - V_c|)$ is a correction term which comes from the operator $\varepsilon_k^{ij} E_i^a E_j^b/\sqrt{|q|}$.

Bianchi I and II

The Hamiltonian for Bianchi I and II can be written in a single expression,

$$\begin{aligned}\mathcal{H}_{\mathrm{BII}} &= \frac{p_1 p_2 p_3}{8\pi G\gamma^2\lambda^2}\left[\sin\bar{\mu}_1 c_1 \sin\bar{\mu}_2 c_2 + \sin\bar{\mu}_2 c_2 \sin\bar{\mu}_3 c_3 + \sin\bar{\mu}_3 c_3 \sin\bar{\mu}_1 c_1\right] \\ &+ \frac{1}{8\pi G\gamma^2}\left[\frac{\alpha(p_2p_3)^{3/2}}{\lambda\sqrt{p_1}}\sin\bar{\mu}_1 c_1 - (1+\gamma^2)\left(\frac{\alpha p_2 p_3}{2p_1}\right)^2\right] - \frac{p_\phi^2}{2} \approx 0,\end{aligned}$$

where the parameter α allows us to distinguish between Bianchi I ($\alpha = 0$) and Bianchi II ($\alpha = 1$). This Hamiltonian together with the Poisson Brackets $\{c^i, p_j\} = 8\pi G\gamma\delta^i_j$ and $\{\phi, p_\phi\} = 1$ gives the effective equations of motion.

Bianchi IX

In the previous Hamiltonians we choose the lapse $N = V$. But now in Bianchi IX, we choose $N = 1$ to include more inverse triad corrections, then the effective Hamiltonian is given by

$$\begin{aligned}\mathcal{H}_{\mathrm{BIX}} = &-\frac{V^4 A(V) h^6(V)}{8\pi G V_c^6\gamma^2\lambda^2}\Big(\sin\bar{\mu}_1 c_1 \sin\bar{\mu}_2 c_2 + \sin\bar{\mu}_1 c_1 \sin\bar{\mu}_3 c_3 \\ &+ \sin\bar{\mu}_2 c_2 \sin\bar{\mu}_3 c_3\Big) + \frac{\vartheta A(V) h^4(V)}{4\pi G V_c^4\gamma^2\lambda}\Big(p_1^2 p_2^2 \sin\bar{\mu}_3 c_3 + p_2^2 p_3^2 \sin\bar{\mu}_1 c_1 \\ &+ p_1^2 p_3^2 \sin\bar{\mu}_2 c_2\Big) - \frac{\vartheta^2(1+\gamma^2)A(V)h^4(V)}{8\pi G V_c^4\gamma^2}\Big(2V\left[p_1^2 + p_2^2 + p_3^2\right] \\ &- \left[(p_1p_2)^4 + (p_1p_3)^4 + (p_2p_3)^4\right]\frac{h^6(V)}{V_c^6}\Big) + \frac{h^6(V)V^2}{2V_c^6} p_\phi^2 \approx 0.\end{aligned}$$

4 Results

Now we will compare the results of the effective theories for the isotropic FRW $k = 0$ and $k = 1$, diagonal Bianchi I, II and IX. All of them with a matter content a massless scalar field satisfying the Klein-Gordon equation. A good starting point to compare the results is to answer the questions that we asked in the introduction, later we will mention other important results:

- Is the bounce generic? Yes. All solutions have a bounce. In other words, singularities are resolved. In the closed FRW and the Bianchi IX model, there are infinite number of bounces and recollapses due to the compactness of the spatial manifold.
- How does anisotropy/Weyl curvature behave near the bounce? These quantities far from the bounce approach their classical values, but when they reach the region near the bounce they behave differently. In Bianchi I, they present only one maximum. In Bianchi II, they exhibit a richer behavior because now they can be zero at the bounce or near to it, and have more than one maximum (for the shear there are up to 4 maxima and for the scalar curvature up to 2 maxima [8]). For Bianchi IX, if we restrict the analysis to one of the infinite number of bounces, it can be shown that anisotropy and curvature behave in the same way as Bianchi I or II. The subject of current research is whether there are new behaviors [9].
- Can we have different kind of bounce, say, dominated by shear σ? Yes, but only in Bianchi II and IX. In Bianchi I the dynamical contribution from matter is always bigger than the one from the shear, even in the solution which reaches the maximal shear at the bounce [8].
- Are geometric scalars θ, σ and ρ absolutely bounded? In the flat isotropic model all the solutions to the effective equations have a maximal density equal to the critical density, and a maximal expansion ($\theta^2_{\text{max}} = 6\pi G\rho_{\text{crit}} = 3/(2\gamma\lambda)$) when $\rho = \rho_{\text{crit}}/2$. For FRW $k = 1$ model, every solution has its maximum density but in general the density is not absolutely bounded. In the effective theory which comes from connection based quantization, expansion can tend to infinity. For the other case, expansion has the same bound as the flat FRW model. However, by adding some more corrections from inverse triad term, one can show that actually in both effective theories the density and the expansion have finite values. For Bianchi I, in all the solutions ρ and θ are upperly bounded by their values in the isotropic case and σ is bounded by $\sigma^2_{\text{max}} = 10.125/(3\gamma^2\lambda^2)$ [10]. For Bianchi II, θ, σ and ρ are also bounded, but for larger values than the ones in Bianchi I, i.e., there are solutions where the matter density is larger than the critical density. With point-like and cigar-like classical singularities [8], the density can achieve the maximal value ($\rho \approx 0.54\rho_{Pl}$) as a consequence of the shear being zero at the bounce and curvature different from zero. For Bianchi IX the behavior is the same as in closed FRW, if the inverse triad corrections are not used, then the geometric scalars are not absolutely bounded. But if the inverse triad corrections are used, then on each solution the geometric scalars are bounded but there is not an absolute bound for all the solutions [9, 10].

- Bianchi I, II and therefore the isotropic case $k = 0$ are limiting cases of Bianchi IX, but they are not contained within Bianchi IX. While the isotropic FRW $k = 1$ is contained within Bianchi IX only if the inverse triad corrections are not included, when they are included then the $k = 1$ universe is a limiting case, like the $k = 0$ universe.
- A set of quantities that are very useful are the Kasner exponents (in classical Bianchi I, the scale factors are $a_i = t^{k_i}$, where k_i are the Kasner exponents), because they can be used to determine which kind of solution is obtained. The Kasner exponents tell us about the Bianchi I transitions (if they exist) and particularly in Bianchi IX, they are used to study the BKL behavior in the vacuum case.

5 Conclusions

One of the main issues that a quantum theory of gravity is expected to address is that of singularity resolution. Loop quantum cosmology has provided a complete description in the case of isotropic cosmological models and singularity resolution has been shown to be generic. A pressing question is whether these results can be generalized to anisotropic models. In this case we lack a complete quantum theory, but one can rely on the existence of an effective description, capturing the main (loop) quantum geometric features. In this contribution we have described the main features of such effective solutions. Singularities seem to be generically resolved as the time evolution of geometrical scalars is well behaved past the would be classical singularity. With the study of these anisotropic models, a question that still arises is whether this behavior is generic for non-homogeneous configurations. That is, are we a step forward toward generic quantum singularity resolution?

References

1. Ashtekar, A., Corichi, A., Singh, P.: Robustness of key features of loop quantum cosmology. Phys. Rev. D **77**, 024046 (2008). doi:10.1103/PhysRevD.77.024046
2. Ashtekar, A., Pawlowski, T., Singh, P.: Quantum nature of the big bang: improved dynamics. Phys. Rev. D **74**, 084003 (2006). doi:10.1103/PhysRevD.74.084003
3. Ashtekar, A., Singh, P.: Loop quantum cosmology: a status report. Class. Quantum Grav. **28**, 213001 (2011). doi:10.1088/0264-9381/28/21/213001
4. Wilson-Ewing, E.: Loop quantum cosmology of Bianchi type IX models. Phys. Rev. D **82**, 043508 (2010). doi:10.1103/PhysRevD.82.043508
5. Ashtekar, A., Wilson-Ewing, E.: Loop quantum cosmology of Bianchi I models. Phys. Rev. D **79**, 083535 (2009). doi:10.1103/PhysRevD.79.083535
6. Ashtekar, A., Wilson-Ewing, E.: Loop quantum cosmology of Bianchi type II models. Phys. Rev. D **80**, 123532 (2009). doi:10.1103/PhysRevD.80.123532
7. Thiemann, T.: Modern Canonical Quantum General Relativity. Cambridge Monographs on Mathematical Physics. Cambridge University Press, Cambridge (2007)

8. Corichi, A., Montoya, E.: Effective dynamics in Bianchi type II loop quantum cosmology. Phys. Rev. D **85**, 104052 (2012). doi:10.1103/PhysRevD.85.104052
9. Corichi, A., Karami, A., Montoya, E.: Loop quantum cosmology: anisotropy and singularity resolution. ArXiv e-prints [arXiv.1210.7248[gr-qc]] (2012)
10. Gupt, B., Singh, P.: Contrasting features of anisotropic loop quantum cosmologies: the role of spatial curvature. Phys. Rev. D **85**, 044011 (2012). doi:10.1103/PhysRevD.85.044011

Tensor Operators in Loop Quantum Gravity

Maïté Dupuis and Florian Girelli

Abstract We discuss how the quantization of the spinorial formalism for Loop Quantum Gravity naturally leads to the notion of tensor operators. These objects encode the natural structure to discuss observables associated to the intertwiner space. They allow in particular to deal with any type of gauge group, classical or quantum. After reviewing the standard case of SU(2), we focus on the specific example of $\mathscr{U}_q(\mathfrak{su}(2))$ and illustrate how dealing with a quantum group leads to the notion of quantum curved geometry.

1 Introduction

The current Loop Quantum Gravity (LQG) theory describes the quantum gravity regime with zero cosmological constant. The kinematical Hilbert space of the theory is spanned by quantum states for spatial geometries, the so-called *spin networks* [1]. Recently, it has been realized that spin network states are the quantization of some classical spinor states [2]. This spinorial formalism is reviewed in the next section, where we recall how it is linked to a discrete version of General Relativity, the *twisted geometries*. Then, focusing on a given vertex, we show how the spinorial structure associated to this vertex can be quantized in terms of tensor operators for the gauge group SU(2). This allows us to embed the $\mathrm{U}(N)$ framework in a new mathematical formalism generalizable to the quantum group case [3].

In the second part, we show how the use of tensor operators in LQG can be generalized to the quantum group $\mathscr{U}_q(\mathfrak{su}(2))$. The use of a quantum group as gauge

---

M. Dupuis
Institute for Theoretical Physics III, University Erlangen-Nuremberg, Erlangen, Germany
e-mail: maite.dupuis@gravity.fau.de

F. Girelli (✉)
Department of Applied Mathematics, University of Waterloo, Waterloo, ON, Canada
e-mail: fgirelli@uwaterloo.ca

J. Bičák and T. Ledvinka (eds.), *Relativity and Gravitation*,
Springer Proceedings in Physics 157, DOI: 10.1007/978-3-319-06761-2_68,

group instead of the Lie group SU(2) is motivated by the idea that this could be a way to introduce a non-vanishing cosmological constant Λ in the LQG framework [4]. From tensor operators, we build observables for $\mathcal{U}_q(\mathfrak{su}(2))$-intertwiner that allow us to identify geometric observables for curved geometries such as the angle and length/area operators.

2 The Spinorial Formalism for LQG

The starting point of LQG is a smeared algebra, the holonomy-flux algebra, associated to graphs embedded into the spatial manifold. The continuum Ashtekar-Barbero variables are replaced by a pair $(g_e, X_e) \in \mathrm{SU}(2) \times \mathfrak{su}(2)$ associated to the edge e of a given graph Γ.

At the quantum level, the kinematical Hilbert space of LQG is spanned by the spin network states, $|\Gamma, \{j_e\}, \{\iota_v\}\rangle$ where $j_e \in \mathbb{N}/2$ is a representation of SU(2) and is associated to each edge e of Γ. ι_v is a SU(2)-intertwiner associated to the vertex v of Γ. The geometrical interpretation of spin network states is provided by the properties of the angle, area and volume operators which are diagonalized by these quantum states. All geometric information is then encoded in the combinatorial aspects of the graphs. Let us now focus on a given graph Γ and consider a truncature of the full continuum theory to a finite Hilbert space $\mathcal{H}_\Gamma = L^2(\mathrm{SU}(2)^E, d^E g)$ with E the number of edges of Γ and dg the Haar measure on SU(2). To understand what the classical degrees of freedom represented by the spin network states in $\mathcal{H}_\Gamma$ are, let us introduce classical *spinor networks*.

2.1 Classical Spinor Networks

We focus on a given edge e of the graph Γ. This oriented edge is decorated by two spinors $|z_e\rangle = \begin{pmatrix} z_e^{(0)} \\ z_e^{(1)} \end{pmatrix} \in \mathbb{C}^2$ and $|\tilde{z}_e\rangle \in \mathbb{C}^2$ respectively associated to the source vertex $s(e)$ of e, and to the target vertex $t(e)$ of e. The phase space is defined by assuming that $|z_e\rangle$ is dual to its conjugate $|\bar{z}_e\rangle$: $\{z_e^{(a)}, \bar{z}_e^{(b)}\} = -i\delta_{ab}$ where $a, b \in \{0, 1\}$. The space $\mathbb{C}^2 \times \mathbb{C}^2$, equipped with its canonical Poisson brackets, allows us to obtain the structure of the phase space of LQG on a given edge e, $\mathrm{SU}(2) \times \mathfrak{su}(2) \simeq T^*\mathrm{SU}(2)$. Indeed, the holonomy-flux algebra can be expressed in terms of the spinor variables[1]:

$$\mathbf{X}_e := \langle z_e|\boldsymbol{\sigma}|z_e\rangle, \qquad g_e := \frac{|z_e\rangle[\tilde{z}_e| - |z_e]\langle\tilde{z}_e|}{\sqrt{\langle z_e|z_e\rangle\langle\tilde{z}_e|\tilde{z}_e\rangle}}, \tag{1}$$

[1] We do not consider here the degenerate configuration $\langle z|z\rangle = 0$ which is equivalent to $|\mathbf{X}| = 0$. See [5] to see how this degenerate case can be treated.

with the additional area matching constraint on the spinors, $\mathscr{M}_e := \langle z_e | z_e \rangle - \langle \tilde{z}_e | \tilde{z}_e \rangle = 0$, where σ^i are the Pauli matrices and $|z_e] := \begin{pmatrix} -\bar{z}_e^{(1)} \\ \bar{z}_e^{(0)} \end{pmatrix}$. Imposing this constraint $\mathscr{M}_e$ ensures that the two 3-vectors $\mathbf{X}_e$ and $\tilde{\mathbf{X}}_e$ have the same norm and that g_e is unitary. The 6-dimensional space $T^*\mathrm{SU}(2)$ and its symplectic structure are recovered by symplectic reduction of $\mathbb{C}^2 \times \mathbb{C}^2$ by the constraint $\mathscr{M}_e$, $T^*\mathrm{SU}(2) \simeq (\mathbb{C}^2 \times \mathbb{C}^2)//\mathscr{M}$.

Let us now go back to the full graph Γ. The spinors are now labelled by a vertex v and one of the edges e which has v for vertex. Thus, Γ is decorated with a set of spinors $|z_{e,v}\rangle$. The components of the corresponding vectors $\mathbf{X}_{e,v}$ can be seen as generating SU(2) transformation on the spinor $|z_{e,v}\rangle$. The classical equivalent of the closure constraint which imposes the global SU(2) invariance at each node of the graph Γ of a spin network state is simply written in terms of the 3-vectors as $\sum_{e \supset v} \mathbf{X}_{e,v} = \mathbf{0}$. This translates into a matricial constraint on the spinor variables, $\mathscr{C}_v \equiv \sum_{e \supset v} \left(|z_{e,v}\rangle\langle z_{e,v}| - \frac{1}{2}\langle z_{e,v}|z_{e,v}\rangle \mathbb{I} \right)$. If E and v denote respectively the number of edges of Γ and the number of vertex of Γ, the symplectic reduction of $(\mathbb{C}^2 \times \mathbb{C}^2)//(\mathscr{M}_e)^E$ by $(\mathscr{C}_v)^V$ gives a symplectic space isomorphic to the gauge invariant phase-space of LQG on a fixed graph. Moreover, Γ decorated by $\{|z_{e,v}\rangle / \mathscr{M}_e = 0 \text{ and } \mathscr{C}_v = 0, \forall e,\, v \subset \Gamma\}$ defines a *spinor network*.

2.2 Twisted Geometries

A nice geometrical interpretation of a spinor network comes from *twisted geometries* [6]. Essentially, a spinor network can be interpreted as a collection of polyhedra glued along their faces, where two shared faces have the same area but not necessary the same shape. The twisted geometry formalism is based on a seminal work by Minkowski which showed how given a vertex and a set of variables defining the twisted geometries, one can reconstruct a unique polyhedron dual to the vertex [7]. These variables and their relationship to the spinor variables are the following: $j_e \equiv \frac{\langle z_e | z_e \rangle}{2}$ the area of the dual surface to the edge e; two unit vectors $N_{e,s(e)}$, $\tilde{N}_{e,t(e)}$ such as $\mathbf{X}_{e,v}(z_{e,v}) = j_e N_{e,v}$; an angle $\xi_e \equiv -2(\arg(z_e^{(1)}) - \arg(\tilde{z}_e^{(1)}))$ the conjugate variable of j_e. $N_{e,s(e)}$, $\tilde{N}_{e,t(e)}$ are the two normals to the dual surface to the edge e as seen from the two vertex frames sharing it. ξ_e is related to the extrinsic curvature between the frames.

The name "twisted geometries" is motivated by this picture where geometries can be discontinuous at the faces connecting the polyhedra. This is related to the fact that the kinematical Hilbert space of LQG has room for torsion. In this sense, twisted geometries are more general than Regge Calculus which is torsion-free. For a graph with 4-valent vertices, Regge Calculus can be recovered by constraining the twisted geometries with some "shape-matching conditions" [8].

2.3 Tensor Operators

The quantization of a spinor $|z\rangle$ and of its dual $|z]$ or more precisely of their conjugate variables give tensor operators of rank $1/2$ for SU(2):

$$\langle z| = (\bar{z}^{(0)}, \bar{z}^{(1)}) \rightarrow T^{1/2} = \begin{pmatrix} a^\dagger \\ b^\dagger \end{pmatrix}, \quad [z| = (-z^{(1)}, z^{(0)}) \rightarrow -\tilde{T}^{1/2} = \begin{pmatrix} -b \\ a \end{pmatrix}, \tag{2}$$

where $\alpha_i = a,\ b$ are harmonic oscillators, $[\alpha_i, \alpha_j^\dagger] = \delta_{ij}$, the other commutators being zero. More generally, a tensor operator of rank $j \in \mathbb{N}/2, \mathbf{t}^j_m$, is an object transforming as a vector $|j, m\rangle$ $(m \in \{-j, \cdots, +j\})$ under the adjoint action, $\triangleright$, of $\mathfrak{su}(2)$. We have $J_z \triangleright \mathbf{t}^j_m = [J_z, \mathbf{t}^j_m] = m\, \mathbf{t}^j_m,\ J_\pm \triangleright \mathbf{t}^j_m = [J_\pm, \mathbf{t}^j_m] = \sqrt{(j \mp m)(j \pm m + 1)}\ \mathbf{t}^j_{m\pm1}$, where $J_z,\ J_\pm$ are the $\mathfrak{su}(2)$-generators.

Note that $J_z,\ J_\pm$ can also be seen as the components of a tensor operator of rank 1 for $\mathfrak{su}(2)$. And just as $|1, l\rangle = \sum_m C^{1/2\,1/2\,1}_{m\ l-m\ l} |1/2, m\rangle \otimes |1/2, m-l\rangle$ $(l \in \{-1, 0, 1\})$ where $C^{1/2\,1/2\,1}_{m\ m-l\ l}$ are Clebsh-Gordan (CG) coefficients, the tensor operator of rank 1 $(-J_+/\sqrt{2}, J_z, J_-/\sqrt{2})$ can be expressed in terms of tensor operators of ranks $1/2$ defined in (2) and we recover the Jordan-Schwinger representation of SU(2). The U(N) formalism developed for $SU(2)$ intertwiners in terms of harmonic oscillators can also be rewritten in terms of tensor operators of rank $1/2$, ${}^{(i)}T^{1/2}$, ${}^{(i)}\tilde{T}^{1/2}$, where i denotes the ith leg of a given vertex v. The observables $E_{ij} = a_i^\dagger a_j + b_i^\dagger b_j$, $F_{ij} = a_i b_j - a_j b_i$ and $F^\dagger_{ij}$ $(i, j \in \{1, \ldots, N\})$ for the space of N valent SU(2)-intertwiners are simply rank 0 tensor operators built from these tensor operators of rank $1/2$ using CG coefficients to combine them into a scalar operator. These operators can be used to generate all observables for the intertwiner [9].

3 Generalization to $\mathcal{U}_q(\mathfrak{su}(2))$ as Gauge Group: Towards Curved Discrete Geometries?

3.1 Tensor Operators

We focus on $\mathcal{U}_q(\mathfrak{su}(2))$ with q real, which representation theory is similar to the one of $\mathfrak{su}(2)$. We refer to the Appendix for the definition and properties regarding $\mathcal{U}_q(\mathfrak{su}(2))$ [10]. As for the $q = 1$ case, a tensor operator is an operator $\mathbf{t}^j_m$ which transforms at the same time under the adjoint action, $\triangleright$, of $\mathcal{U}_q(\mathfrak{su}(2))$ and as the vector $|jm\rangle$,

$$\begin{aligned} K \triangleright \mathbf{t}^j_m &= K \mathbf{t}^j_m K^{-1} = q^m \mathbf{t}^j_m, \\ J_\pm \triangleright \mathbf{t}^j_m &= J_\pm\, \mathbf{t}^j_m\, K^{-1} - q^{\pm\frac{1}{2}} K^{-1}\, \mathbf{t}^j_m\, J_\pm = \sqrt{[j \mp m][j \pm m + 1]}\ \mathbf{t}^j_{m\pm1}, \end{aligned} \tag{3}$$

The decomposition of a tensor operator into a product of tensor operators mentioned in the previous section remains valid. However, the *tensor product* of tensor operators becomes complicated to construct in the $q \neq 1$ case. Indeed, if $\mathbf{t}$ is a tensor operator then ${}^{(1)}\mathbf{t} = \mathbf{t} \otimes \mathbf{1}$ is a tensor operator, but $\mathbf{1} \otimes \mathbf{t}$ is in general *not* a tensor operator (it is however a tensor operator if $q = 1$, i.e. for $\mathfrak{su}(2)$). We need to use the $\mathscr{R}$-matrix to construct an intertwining map from the permutation. We say that the permutation composed with the $\mathscr{R}$-matrix is a q-deformed permutation. Starting from a given $\mathbf{t}$ of rank j, we can build N tensor operators of rank j using consecutive deformed permutations. For all $i \in \{1, \ldots, N\}$,

$$ {}^{(i)}\mathbf{t} = (\mathscr{R}_{ii-1}..\mathscr{R}_{i1}(\mathbf{1} \otimes \ldots \otimes \mathbf{t})\mathscr{R}_{i1}^{-1}..\mathscr{R}_{ii-1}^{-1}) \otimes \mathbf{1} \otimes ..\mathbf{1} \tag{4} $$

is a tensor operator of same rank as $\mathbf{t}$. The fundamental tensor operators are the tensor operators of rank 1/2, the *spinor* operators. Similarly to the $q = 1$ case, a pair of q-harmonic oscillators provides a convenient set of variables to realize these operators [11]. The annihilation and creation operators $\alpha_i = a_i, b_i, \alpha_i^\dagger = a_i^\dagger, b_i^\dagger$, and the number operator N_{α_i} satisfy now the following conditions

$$ [\alpha_i, \alpha_j] = [\alpha_i^\dagger, \alpha_j^\dagger] = 0, \quad [\alpha_i, \alpha_j^\dagger]_{q^{\pm\frac{1}{2}}} = \delta_{ij} q^{\mp N_{\alpha_i}}, \tag{5} $$
$$ q^{N_{\alpha_i}/2}\alpha_i^\dagger = q^{1/2}\alpha_i^\dagger q^{N_{\alpha_i}/2}, \quad q^{N_{\alpha_i}/2}\alpha_i = q^{-1/2}\alpha_i q^{N_{\alpha_i}/2}. $$

We can use this pair of harmonic oscillators to construct the Jordan-Schwinger realization the $\mathscr{U}_q(\mathfrak{su}(2))$ generators [11]

$$ J_z = \frac{1}{2}(N_a - N_b), \quad J_+ = a^\dagger b, \quad J_- = b^\dagger a, \tag{6} $$

which are *not* the components of a tensor operator of rank 1, contrary to the $q = 1$ case. Using (5), we can recover the $\mathscr{U}_q(\mathfrak{su}(2))$ commutation relations provided in (10). We can also use the Fock space of this pair q-harmonic oscillators to generate the representations of $\mathscr{U}_q(\mathfrak{su}(2))$.

Thanks to this realization, we can find the two solutions of (3) for $j = 1/2$, which transform therefore as spinors.

$$ \mathbf{t}^{\frac{1}{2}} = \begin{pmatrix} a^\dagger q^{N_a/4} \\ b^\dagger q^{(2N_a+N_b)/4} \end{pmatrix} \xrightarrow{q\to 1} T^{1/2}, \; \tilde{\mathbf{t}}^{\frac{1}{2}} = \begin{pmatrix} q^{(2N_a+N_b+1)/4} b \\ -q^{(N_a-1)/4} a \end{pmatrix} \xrightarrow{q\to 1} -\tilde{T}^{1/2}. \tag{7} $$

Using the relevant CG coefficients, we construct the operators $\mathbf{t}^1$ which transform as vectors, $\mathbf{t}^1_{\pm 1} = \mathbf{t}^{\frac{1}{2}}_{\pm}\tilde{\mathbf{t}}^{\frac{1}{2}}_{\pm}$, $\mathbf{t}^1_0 = \frac{1}{\sqrt{[2]}}\left(q^{-\frac{1}{4}}\mathbf{t}^{\frac{1}{2}}_{+}\tilde{\mathbf{t}}^{\frac{1}{2}}_{-} + q^{\frac{1}{4}}\mathbf{t}^{\frac{1}{2}}_{-}\tilde{\mathbf{t}}^{\frac{1}{2}}_{+}\right)$. Explicitly, we have,

$$\mathbf{t}^1_{\pm 1} = \mp \frac{q^{-\frac{1}{2}}}{\sqrt{[2]}} q^{\frac{1}{2}(N_a+N_b)} q^{\frac{J_z}{2}} J_\pm, \xrightarrow{q\to 1} \mp \frac{J_\pm}{\sqrt{2}} \tag{8}$$

$$\mathbf{t}^1_0 = \frac{q^{-\frac{1}{2}}}{[2]} q^{\frac{1}{2}(N_a+N_b)} (q^{-1/2} J_+ J_- - q^{1/2} J_- J_+) \xrightarrow{q\to 1} J_z.$$

Any other tensor operator of rank j can be built in a similar way by combining spinor operators and CG coefficients. Then, the construction of tensor operators of rank j from tensor products of a given tensor operator can be done using (4). Note that in general ${}^{(n)}\mathbf{t}^{j_1}$ and ${}^{(m)}\tilde{\mathbf{t}}^{j_2}$ will not commute in the quantum group case, contrary to the $q = 1$ case.

3.2 Observables for the q-Deformed Intertwiner Space

In LQG, the intertwiner $|\iota_{j_1..j_N}\rangle$ is understood as the fundamental chunk of quantum space. We have seen in the previous section that the use of tensor operators allows to construct operators invariant under the adjoint action of $\mathfrak{su}(2)$ and to recover the complete algebra of observables defined in the $\mathrm{U}(N)$ framework. The tensor operator formalism allows us to extend this framework to the quantum group case in a direct manner.

As in the $\mathfrak{su}(2)$ case, each leg k of the intertwiner corresponds to a representation V^{j_k}. We can associate with each leg a tensor operator ${}^{(k)}\mathbf{t}^{\frac{1}{2}}$. Using $\mathcal{U}_q(\mathfrak{su}(2))$ recoupling theory, it is possible to build from these tensor operators a tensor operator of rank 0, i.e. an *observable*. Let us denote $|\iota^q_{j_1..j_N}\rangle$ the intertwiner defined from representations of $\mathcal{U}_q(\mathfrak{su}(2))$. Using spinor operators, as in the $q = 1$ case, there are only three types of observables that can be constructed[2]:

$$\mathcal{E}_{ij} = -\sqrt{[2]} \sum_{m_i} {}_q\mathbf{C}^{\frac{1}{2}\,\frac{1}{2}\,0}_{m_1 m_2 0}\, {}^{(i)}\mathbf{t}^{\frac{1}{2}}_{m_1}\, {}^{(j)}\tilde{\mathbf{t}}^{\frac{1}{2}}_{m_2} \xrightarrow{q\to 1} a_i^\dagger a_j + b_i^\dagger b_j = E_{ij}$$

$$\mathcal{G}^\dagger_{ij} = -\sqrt{[2]} \sum_{m_i} {}_q\mathbf{C}^{\frac{1}{2}\,\frac{1}{2}\,0}_{m_1 m_2 0}\, {}^{(i)}\mathbf{t}^{\frac{1}{2}}_{m_1}\, {}^{(j)}\mathbf{t}^{\frac{1}{2}}_{m_2} \xrightarrow{q\to} a_i^\dagger b_j^\dagger - b_i^\dagger a_j^\dagger = F^\dagger_{ij}$$

$$\mathcal{F}_{ij} = -\sqrt{[2]} \sum_{m_i} {}_q\mathbf{C}^{\frac{1}{2}\,\frac{1}{2}\,0}_{m_1 m_2 0}\, {}^{(i)}\tilde{\mathbf{t}}^{\frac{1}{2}}_{m_1}\, {}^{(j)}\tilde{\mathbf{t}}^{\frac{1}{2}}_{m_2} \xrightarrow{q\to 1} a_i b_j - b_i a_j = F_{ij},$$

where the operators E_{ij}, $F^\dagger_{ij}$, F_{ij} of the $U(N)$ formalism are recovered. Preliminary results indicate that the $\mathcal{E}_{ij}$ can be expressed as functions of the generators of $\mathcal{U}_q(\mathfrak{u}(N))$ [3]. This means that the $\mathcal{U}_q(\mathfrak{su}(2))$ intertwiner can be seen as a repre-

[2] The other ordering choice for ${}^{(i)}\mathbf{t}^{\frac{1}{2}}_{m_1}$ and ${}^{(j)}\tilde{\mathbf{t}}^{\frac{1}{2}}_{m_2}$ is equivalent to our choice modulo a rescaling of the operators.

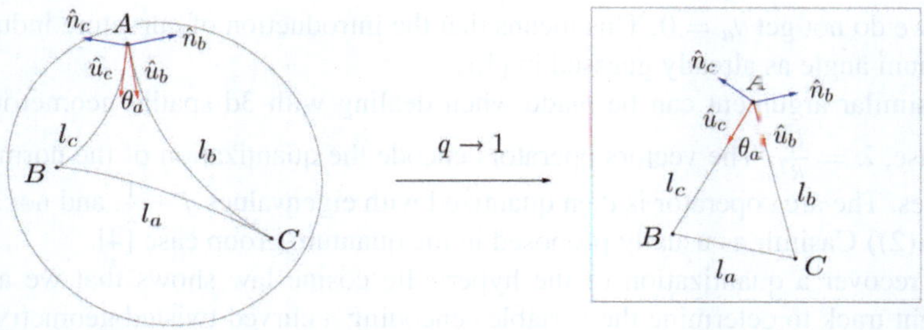

Fig. 1 Hyperbolic and flat triangles represented in the Poincaré disk and the plane respectively

sentation of $\mathscr{U}_q(\mathfrak{u}(N))$, a natural generalization of the classical case, where the E's form a $\mathfrak{u}(N)$ algebra.

Using vector operators we can construct different interesting observables. Considering ${}^{(i)}\mathbf{t}^1$ and ${}^{(j)}\mathbf{t}^1$, we construct the following scalar operator, ${}^{(i)}\mathbf{t}^1 \cdot {}^{(j)}\mathbf{t}^1 \equiv {}_q\mathbf{C}\begin{matrix} 1 & 1 & 0 \\ m_1 & m_2 & 0 \end{matrix} {}^{(i)}\mathbf{t}^1_{m_1} \, {}^{(j)}\mathbf{t}^1_{m_2}$. Using recoupling theory, we calculate the action of this operator on a three legs intertwiner $|\iota^q_{j_b j_c j_a}\rangle$. For simplicity, we assume now a 2d space, i.e. LQG for a 2+1 spacetime. We can then interpret this intertwiner as the quantum state of a triangle. In the limit $q \to 1$, we know that ${}^{(b)}\mathbf{t}^1 \cdot {}^{(c)}\mathbf{t}^1 \propto {}^{(b)}\mathbf{J} \cdot {}^{(c)}\mathbf{J}$ is interpreted as the quantization of the cosine of the angle θ_a between the tangent vectors $\hat{u}_c$, $\hat{u}_b$ (cf Fig. 1).

The action of ${}^{(b)}\mathbf{J} \cdot {}^{(c)}\mathbf{J}$ on $|\iota^{q=1}_{j_b j_c j_a}\rangle$ leads to a quantized version of the flat cosine law [12]. When performing the calculation with $q \neq 1$, we obtain that the action of ${}^{(b)}\mathbf{t}^1 \cdot {}^{(c)}\mathbf{t}^1$, for an appropriate choice of normalization, is diagonal on $|\iota^q_{j_b j_c j_a}\rangle$ with eigenvalue,

$$\frac{\cosh\frac{\lambda}{2}\cosh((j_a+\frac{1}{2})\lambda) - \cosh((j_b+\frac{1}{2})\lambda)\cosh((j_c+\frac{1}{2})\lambda))}{\sinh((j_b+\frac{1}{2})\lambda)\sinh((j_c+\frac{1}{2})\lambda)}, \tag{9}$$

where $\lambda = \frac{l_p}{R}$ with l_p the Planck length, R the cosmological radius, $R = \frac{1}{\sqrt{\Lambda}}$ and Λ the positive cosmological constant. This suggests now that we are dealing with a quantization of the *hyperbolic cosine law*,

$$-\hat{n}_b \cdot \hat{n}_c = \hat{u}_b \cdot \hat{u}_c = \cos\theta_a = \frac{-\cosh\frac{l_a}{R} + \cosh\frac{l_b}{R}\cosh\frac{l_c}{R}}{\sinh\frac{l_b}{R}\sinh\frac{l_c}{R}}.$$

When R goes to infinity, we recover the standard flat cosine law. This strongly suggests that dealing with $\mathscr{U}_q(\mathfrak{su}(2))$ 3-legs intertwiner encodes a quantization of the curved hyperbolic triangle (cf Fig. 1), where the quantized length is $\hat{l}_i = j_i + \frac{1}{2}$. Morevoer, if we consider the case where $j_a = 0$ and $j_b,\ j_c \neq 0$, unlike the classical case, we do not get $\theta_a = 0$. This means that the introduction of curvature induced a minimum angle as already guessed in [13].

A similar argument can be made when dealing with 3d spatial geometries. In this case, $\lambda = \frac{l_p^2}{R^2}$. The vectors operators encode the quantization of the normals to surfaces. The area operator is then quantized with eigenvalues $j + \frac{1}{2}$, and not as the $\mathcal{U}_q(\mathfrak{su}(2))$ Casimir as usually proposed in the quantum group case [4].

To recover a quantization of the hyperbolic cosine law shows that we are on the right track to determine the variables encoding a curved twisted geometry. The natural candidates are the complex variables z_i variables which are the "classical" analogues of the q-harmonic oscillators (5), equipped with a q-calculus. In this case, the relevant differential is given by $D^q f(z) \equiv (f(zq^{\frac{1}{2}}) - f(zq^{-\frac{1}{2}}))/(z(q^{\frac{1}{2}} - q^{-\frac{1}{2}}))$. We leave this for further investigations.

4 Outlook

Twisted geometries and the LQG spinorial reformulation provide an interesting framework to get a better understanding of the classical degrees of freedom carried by a spin network, a quantum state of space. When quantizing this spinorial formalism of LQG, tensor operators arise naturally. These objects are easy to generalize to the quantum group case. Whereas at this stage, we do not know clearly how to introduce the cosmological constant in the LQG framework, we have showed how using the quantum group does encode the presence of a cosmological constant in the geometry dual to an intertwiner. In particular we have obtained a quantization of the hyperbolic cosine law.

Morevoer, identifying these tensor operators as the right building blocks provides new techniques to eventually solve the issue of the quantum groups appearance in the LQG context as well as new directions to explore.

For example, it is now possible using the $\mathcal{E}$'s and $\mathcal{F}$'s observables to generalize the Hamiltonian constraint for 3d gravity proposed in [14] in order to get an invariant Hamiltonian operator under $\mathcal{U}_q(\mathfrak{su}(2))$. This would allow to connect for the first time an LQG Hamiltonian constraint and the $6j$-symbols for $\mathcal{U}_q(\mathfrak{su}(2))$ which appears in the Turaev-Viro model. This would allow to probe how the use of quantum group is a good approximation for encoding the cosmological constant.

Appendix: $\mathcal{U}_q(\mathfrak{su}(2))$

The quasi-triangular Hopf algebra $\mathcal{U}_q(\mathfrak{su}(2))$ is generated by the elements $J_\pm$, $K = q^{\frac{J_z}{2}}$ which satisfy

$$KJ_{\pm}K^{-1} = q^{\pm\frac{1}{2}}J_{\pm}, \quad [J_{+}, J_{-}] = \frac{K^2 - K^{-2}}{q^{1/2} - q^{-1/2}}. \tag{10}$$

The coproduct $\Delta : \mathscr{U}_q(\mathfrak{su}(2)) \to \mathscr{U}_q(\mathfrak{su}(2)) \otimes \mathscr{U}_q(\mathfrak{su}(2))$ and antipode $S : \mathscr{U}_q(\mathfrak{su}(2)) \to \mathscr{U}_q(\mathfrak{su}(2))$ are given by

$$\Delta K = K \otimes K, \; \Delta J_{\pm} = J_{\pm} \otimes K + K^{-1} \otimes J_{\pm}, \; SK = K^{-1}, \; SJ_{\pm} = -q^{\pm 1/2}J_{\pm}.$$

The $\mathscr{R}$-matrix $\mathscr{R} \in \mathscr{U}_q(\mathfrak{su}(2)) \otimes \mathscr{U}_q(\mathfrak{su}(2))$ encodes the quasi-triangular structure, which tells us how much the coproduct is non-commutative. If we note $\psi : \mathscr{U}_q(\mathfrak{su}(2)) \otimes \mathscr{U}_q(\mathfrak{su}(2)) \to \mathscr{U}_q(\mathfrak{su}(2)) \otimes \mathscr{U}_q(\mathfrak{su}(2))$ the permutation, then we have $\psi \circ \Delta X = \mathscr{R}(\Delta X)\mathscr{R}^{-1}, \; \forall X \in \mathscr{U}_q(\mathfrak{su}(2))$. Standard notations are $\mathscr{R}_{12} = \sum R_1 \otimes R_2, \mathscr{R}_{21} = \sum R_2 \otimes R_1, \ldots$ When q is real, the representation theory of $\mathscr{U}_q(\mathfrak{su}(2))$ is essentially the same as that of $\mathfrak{su}(2)$ [10]. A representation V^j is hence generated by the vectors $|j, m\rangle$ with $j \in \mathbb{N}/2$ and $m \in \{-j, .., j\}$. The key-difference is that we use q-numbers $[x] \equiv \frac{q^{x/2} - q^{-x/2}}{q^{1/2} - q^{-1/2}}$.

$$K\,|j, m\rangle = q^m |j, m\rangle, \quad J_{\pm}\,|j, m\rangle = \sqrt{[j \mp m][j \pm m + 1]}\,|j, m \pm 1\rangle.$$

The adjoint action of $\mathscr{U}_q(\mathfrak{su}(2))$ on some operator $\mathscr{O}$ is

$$J_{\pm} \triangleright \mathscr{O} = J_{\pm}\mathscr{O}K^{-1} - q^{\pm 1/2}K^{-1}\mathscr{O}J_{\pm}, \; K \triangleright \mathscr{O} = K\mathscr{O}K^{-1}.$$

References

1. Rovelli, C.: Quantum Gravity. Cambridge Monographs on Mathematical Physics. Cambridge University Press, Cambridge (2004)
2. Dupuis, M., Speziale, S., Tambornino, J.: Spinors and twistors in loop gravity and spin foams. ArXiv e-prints arxiv:1201.2120 [gr-qc] (2012)
3. Dupuis, M., Girelli, F., Livine, E.R.: Deformed spinor networks for loop gravity: Towards hyperbolic twisted geometries, arXiv:1403.7482.
4. Rovelli, C., Smolin, L.: Spin networks and quantum gravity. Phys. Rev. D **52**, 5743 (1995). doi:10.1103/PhysRevD.52.5743
5. Freidel, L., Speziale, S.: From twistors to twisted geometries. Phys. Rev. D **82**, 084041 (2010). doi:10.1103/PhysRevD.82.084041
6. Freidel, L., Speziale, S.: Twisted geometries: a geometric parametrisation of SU(2) phase space. Phys. Rev. D **82**, 084040 (2010). doi:10.1103/PhysRevD.82.084040
7. Bianchi, E., Dona, P., Speziale, S.: Polyhedra in loop quantum gravity. Phys. Rev. D **83**, 044035 (2011). doi:10.1103/PhysRevD.83.044035
8. Dittrich, B., Ryan, J.: Phase space descriptions for simplicial 4d geometries. Class. Quantum Grav. **28**, 065006 (2011). doi:10.1088/0264-9381/28/6/065006
9. Freidel, L., Livine, E.: The fine structure of SU(2) intertwiners from U(N) representations. J. Math. Phys. **51**, 082502 (2010). doi:10.1063/1.3473786
10. Chari, V., Pressley, A.: A Guide to Quantum Groups. Cambridge University Press, Cambridge (1994)

11. Biedenharn, L., Lohe, M.: Quantum group symmetry and q-Tensor algebras. World Scientific, Singapore (1995)
12. Bonzom, V., Freidel, L.: The hamiltonian constraint in 3d riemannian loop quantum gravity. Class. Quantum Grav. **28**, 195006 (2011). doi:10.1088/0264-9381/28/19/195006
13. Bianchi, E., Rovelli, C.: Note on the geometrical interpretation of quantum groups and non-commutative spaces in gravity. Phys. Rev. D **84**, 027502 (2011). doi:10.1103/PhysRevD.84.027502
14. Bonzom, V., Livine, E.: A new Hamiltonian for the topological BF phase with spinor networks. J. Math. Phys. **53**, 072201 (2012). doi:10.1063/1.4731771

Probability Distributions for Quantum Stress Tensors in Two and Four Dimensions

Christopher J. Fewster, L. H. Ford and Thomas A. Roman

Abstract The probability distributions for individual measurements for the smeared energy densities of quantum fields, in the two and four-dimensional Minkowski vacuum are discussed. These distributions share the property that there is a lower bound at a finite negative value, but no upper bound. Thus arbitrarily large positive energy density fluctuations are possible. In two dimensions we are able to give an exact unique analytic form for the distribution. However, in four dimensions, we are not able to give closed form expressions for the probability distribution, but rather use calculations of a finite number of moments to estimate the lower bound, and the asymptotic form of the tail of the distribution. The first 65 moments are used for these purposes. All of our four-dimensional results are subject to the caveat that these distributions are not uniquely determined by the moments. One can apply the asymptotic form of the electromagnetic energy density distribution to estimate the nucleation rates of black holes and of Boltzmann brains.

1 Introduction

There has been extensive work in recent decades on the definition and use of the expectation value of a quantum stress tensor operator. However, the semiclassical theory does not describe the effects of quantum fluctuations of the stress tensor around its expectation value.

C. J. Fewster
Department of Mathematics, University of York, Heslington, York YO10 5DD, UK

L. H. Ford
Department of Physics and Astronomy, Institute of Cosmology, Tufts University, Medford, MA 02155, USA

T. A. Roman (✉)
Department of Mathematical Sciences, Central Connecticut State University, New Britain, CT 06050, USA
e-mail: roman@ccsu.edu

J. Bičák and T. Ledvinka (eds.), *Relativity and Gravitation*,
Springer Proceedings in Physics 157, DOI: 10.1007/978-3-319-06761-2_69,

One way to examine these fluctuations is through the probability distribution for individual measurements of a smeared stress tensor operator. This distribution was given recently for Gaussian averaged stress tensors operators in two-dimensional flat spacetime [1] using analytical methods, and more recently for averaged stress tensors in four-dimensional spacetime from calculations of a finite set of moments. (Throughout our discussion, all quadratic operators are understood to be normal-ordered with respect to the Minkowski vacuum state.)

1.1 Quantum Inequalities

Quantum inequalities are lower bounds on the *expectation values* of the smeared energy density operator in arbitrary quantum states [2–7]. If we sample in time along the worldline of an inertial observer, the quantum inequality takes the form

$$\int_{-\infty}^{\infty} f(t)\,\langle T_{\mu\nu}u^{\mu}u^{\nu}\rangle\,dt \geq -\frac{C}{\tau^{d}}, \tag{1}$$

where $T_{\mu\nu}u^{\mu}u^{\nu}$ is the normal-ordered energy density operator, which is classically non-negative, t is the observer's proper time, and $f(t)$ is a sampling function with characteristic width τ. Here C is a numerical constant, typically small compared to unity, d is the number of spacetime dimensions, and we work in units where $c = \hbar = 1$.

Although quantum field theory allows negative expectation values of the energy density, quantum inequalities place strong constraints on the effects of this negative energy for violating the second law of thermodynamics [2], maintaining traversable wormholes [8] or warpdrive spacetimes [9]. The implication of (1) is that there is an inverse power relation between the magnitude and duration of negative energy density.

For a massless scalar field in two-dimensional spacetime, Flanagan [6] has found a formula for the constant C for a given $f(t)$ which makes (1) an optimal inequality. This formula is

$$C = \frac{1}{6\pi}\int_{-\infty}^{\infty} du \left(\frac{d}{du}\sqrt{g(u)}\right)^{2}, \tag{2}$$

where $f(t) = \tau^{-1}g(u)$ and $u = t/\tau$. In four-dimensional spacetime, Fewster and Eveson [7] have derived an analogous formula for C, but in this case the bound is not necessarily optimal.

2 Shifted Gamma Distributions: 2D Case

In two-dimensional Minkowski spacetime, we determined the probability distribution for individual measurements, in the vacuum state, of the Gaussian sampled energy density

$$\rho = \frac{1}{\sqrt{\pi}\,\tau} \int_{-\infty}^{\infty} T_{tt}\, \mathrm{e}^{-t^2/\tau^2}\, dt. \tag{3}$$

This was achieved by finding a closed form expression for the generating function of the moments $\langle \rho^n \rangle$ of ρ, from which the probability distribution was obtained. The definition of the n'th moment of the distribution of a variable x is given by

$$a_n = \int x^n\, P(x)\, dx. \tag{4}$$

The resulting distribution is conveniently expressed in terms of the dimensionless variable $x = \rho\,\tau^2$ and is a shifted Gamma distribution:

$$P(x) = \vartheta(x + x_0) \frac{\beta^\alpha (x + x_0)^{\alpha - 1}}{\Gamma(\alpha)} \exp(-\beta(x + x_0)), \tag{5}$$

with parameters

$$x_0 = \frac{1}{12\pi}, \qquad \alpha = \frac{1}{12}, \qquad \beta = \pi. \tag{6}$$

Here $x = -x_0$ is the lower bound of the distribution (Fig. 1).

The lower bound, $-x_0$, for the probability distribution for energy density fluctuations in the vacuum is exactly Flanagan's optimum lower bound, (2), on the Gaussian sampled expectation value. As was argued in Ref. [1], this is a general feature, giving a deep connection between quantum inequality bounds and stress tensor probability distributions. The quantum inequality bound is the lowest eigenvalue of the sampled operator, and is hence the lowest possible expectation value and the smallest result which can be found in a measurement. That the probability distribution for vacuum fluctuations actually extends down to this value is more subtle and depends upon special properties of the vacuum state, and is implied by the Reeh-Schlieder theorem.

There is no upper bound on $P(x)$, as arbitrarily large values of the energy density can arise in vacuum fluctuations. Nonetheless, for the massless scalar field, negative values are much more likely; 84 % of the time, a measurement of the Gaussian averaged energy density will produce a negative value. However, the positive values found the remaining 16 % of the time will typically be much larger, and the average first moment of $P(x)$ will be zero.

Furthermore, the probability distribution for the two-dimensional stress tensor is uniquely determined by its moments, as a consequence of the Hamburger moment

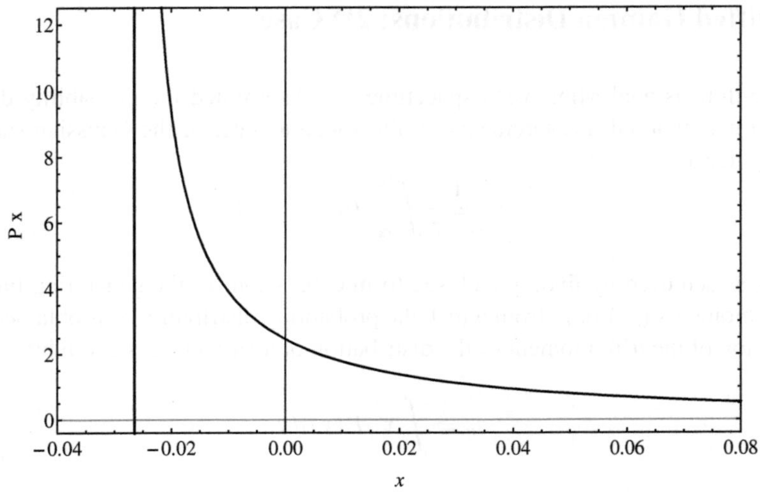

Fig. 1 The graph of $P(x)$ versus x of the probability distribution function for the energy density, ρ, of a massless scalar field sampled in time with a Gaussian of width τ. Here $x = \rho\tau^2$. The distribution has an integrable singularity at the optimal quantum inequality bound $x = -x_0 = -1/12\pi$

theorem [10]. This condition is a sufficient, although not necessary, condition for uniqueness, and is fulfilled by the moments of the shifted Gamma distribution.

3 The 4D Case

In four dimensions, the operators ρ_S, and ρ_{EM} all have dimensions of *length*$^{-4}$. Their probability distributions $P(x)$ are taken to be functions of the dimensionless variable

$$x = (4\pi\,\tau^2)^2\,A, \tag{7}$$

where A is the Lorentzian time average of ρ_S, and ρ_{EM}, where ρ_S and ρ_{EM} are the smeared energy density operators for the massless scalar field, and electromagnetic fields, respectively.

The distributions were calculated numerically from 65 moments [11]. The situation here is less straightforward. In this case, the moments grow too rapidly to satisfy the Hamburger moment criterion. Unfortunately, this means that we cannot be guaranteed of finding a unique probability distribution $P(x)$ from these moments. These probability distributions share some of the main characteristics of their two-dimensional counterparts. They have a lower bound but no upper bound. Our techniques allow us to give approximate lower bounds and the asymptotic forms of the tails of the distributions.

Our estimates for the lower bounds are

$$-x_0(\rho_{EM}) \approx -0.0472, \qquad -x_0(\rho_S) \approx -0.0236. \tag{8}$$

These are also estimates of the optimal quantum inequality bounds for each field. In contrast, the non-optimal bound for ρ_S, given by the method of Fewster and Eveson [7], is $-x_0(FE) = -27/128 \approx -0.21$, which is an order of magnitude larger.

It is of interest to note that the magnitudes of the dimensionless lower bounds, given in (8) are small compared to unity. The fact that the probability distribution has a long positive tail, and must have a unit zeroth moment and a vanishing first moment, implies that the total probability of a negative value to be substantial. The small magnitudes of $x_0(\rho_S)$ and $x_0(\rho_{EM})$ imply strong constraints on the magnitude of negative energy which can arise either as an expectation value in an arbitrary state, or as a fluctuation in the vacuum. They also imply that an individual measurement of the sampled energy density in the vacuum state is very likely to yield a negative value.

One can show that the asymptotic behavior of the tail of the probability distribution is determined by the moments, even if the exact probability distribution is not uniquely determined. Our fitted tail decreases asymptotically as

$$P_{\text{fit}} \sim x^{-2}\, e^{-a x^{1/3}}, \tag{9}$$

where a is a constant. We are also able to show that no distribution with the same moments can have a tail which decreases at a faster rate than ours.

By contrast, the tail of a Boltzmann distribution for thermal fluctuations falls off as

$$P_{\text{Boltzmann}} \sim e^{-\beta x}, \tag{10}$$

where β is a constant. Therefore vacuum fluctuations outweigh thermal fluctuations at high energies (Fig. 2).

3.1 Application: Black Hole Nucleation

The fact that the energy density probability distribution has a long positive tail implies a finite probability for the nucleation of black holes out of the Minkowski vacuum via large, though infrequent positive fluctuations (see Ref. [11]). This probability cannot be too large, of course, or it will conflict with observation. Our estimate of the probability depends only on the asymptotic form of the tail. (One can use similar arguments to estimate the probability of "Boltzmann brains" nucleating out of the vacuum.)

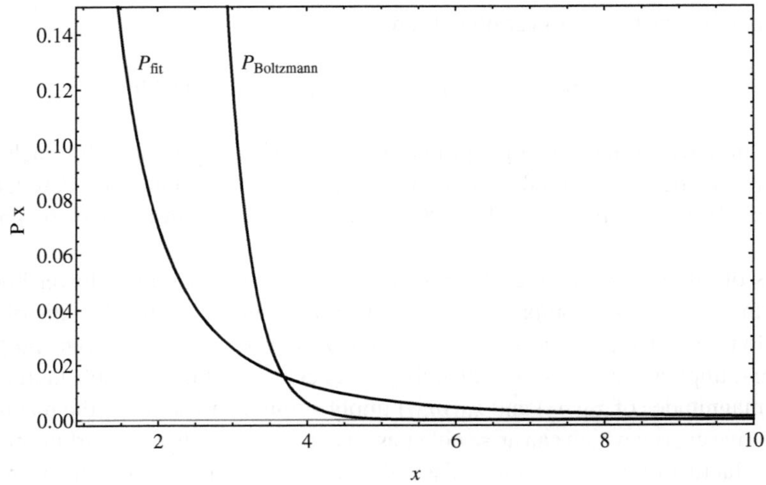

Fig. 2 The figure shows a comparison of the asymptotic form of the tails of both our fitted distribution for vacuum fluctuations and for the thermal fluctuations described by the Boltzmann distribution. At high energies, vacuum fluctuations outweigh thermal fluctuations

4 Summary

We have found that the probability distribution for vacuum fluctuations of the Gaussian-smeared energy density for a massless scalar field in two-dimensional spacetime is *uniquely* defined by a shifted gamma distribution. The distribution has a negative lower bound but no upper bound. It has an integrable singularity (i.e., a "spike") at the lower bound. In addition, we find that there is a deep connection between the lower bound of the distribution and the quantum inequalities. In fact the lower bound of the distribution coincides *exactly* with the *optimal* quantum inequality bound for a Gaussian sampling function, derived earlier by Flanagan.

The lower bound is very small in magnitude, but the probability density is large in the region between zero and the lower bound. As a result, rather surprisingly, the probability of obtaining a negative result in an individual measurement is 84 %! Although the negative fluctuations are very frequent, they are small in magnitude. As a result, one would not expect to see large effects of negative energy (e.g., violations of the second law, wormholes, warpdrives, etc.) nucleating out of the vacuum. However, the distribution has a long positive tail, which guarantees that the frequent but small negative energy density fluctuations are balanced by the much rarer but larger positive energy fluctuations. Therefore, the expectation value of the energy density in the Minkowski vacuum state is zero. It is quite remarkable that the quantum inequalities which are bounds on the *expectation value* of the energy density in an *arbitrary* quantum state, should be so intimately related to the probability distribution of *individual* measurements of the energy density made in the *vacuum* state.

In four dimensions, we find similarities with the two-dimensional case, in that there is a lower bound but no upper bound. We are able to give numerical estimates of the lower bounds, i.e., the optimal bounds, and the asymptotic form of the tails. The lower bounds are negative with small magnitudes. However, our methods do not allow us to determine whether there is a "spike" at the lower bound, as in two dimensions. Nonetheless, the low magnitudes of the lower bounds indicate that a significant fraction of the probability must lie in the negative region. Therefore, as in the two-dimensional case, the probability of obtaining a negative value in an individual measurement is quite high. The long positive tail drops off more slowly than that of a Boltzmann distribution, which implies that vacuum fluctuations dominate over thermal fluctuations at high energies.

Unfortunately, it seems likely that it is not possible to uniquely determine the four-dimensional distributions from the moments alone, as the latter do not obey the Hamburger moment condition. Nonetheless, we are able to glean some information from the moments. For example, we can determine that no distribution with the same moments as ours can have a tail which decreases faster than ours. The asymptotic forms of the long positive tail allow us to estimate the probability of nucleation of (small) black holes and "Boltzmann brains" out of the vacuum.

Clearly further work can be done on this subject. One topic would be to see what additional information can be obtained from our calculated four-dimensional probability distributions, even if they cannot be uniquely determined from their moments. For example, does the "spike" behavior persist in four dimensions as well as in two, and what is its physical significance? Another would be to determine what the optimal quantum inequality bounds actually are. It would also be useful to try various sampling functions. Can the probability distributions and optimal bounds can be obtained by other methods which do not have the limitation of the ambiguities in the moment methods? There is more to do to explore the physical content of stress-tensor fluctuations.

References

1. Fewster, C., Ford, L., Roman, T.: Probability distributions of smeared quantum stress tensors. Phys. Rev. D **81**, 121901 (2010). doi:10.1103/PhysRevD.81.121901
2. Ford, L.: Quantum coherence effects and the second law of thermodynamics. Proc. R. Soc. Lond. Ser. A **364**, 227 (1978). doi:10.1098/rspa.1978.0197
3. Ford, L.: Constraints on negative-energy fluxes. Phys. Rev. D **43**, 3972 (1991). doi:10.1103/PhysRevD.43.3972
4. Ford, L., Roman, T.: Averaged energy conditions and quantum inequalities. Phys. Rev. D **51**, 4277 (1995). doi:10.1103/PhysRevD.51.4277
5. Ford, L., Roman, T.: Restrictions on negative energy density in flat spacetime. Phys. Rev. D **55**, 2082 (1997). doi:10.1103/PhysRevD.55.2082
6. Flanagan, É.: Quantum inequalities in two-dimensional Minkowski spacetime. Phys. Rev. D **56**, 4922 (1997). doi:10.1103/PhysRevD.56.4922
7. Fewster, C., Eveson, S.: Bounds on negative energy densities in flat spacetime. Phys. Rev. D **58**, 084010 (1998). doi:10.1103/PhysRevD.58.084010

8. Ford, L., Roman, T.: Quantum field theory constrains traversable wormhole geometries. Phys. Rev. D **53**, 5496 (1996). doi:10.1103/PhysRevD.53.5496
9. Pfenning, M., Ford, L.: The unphysical nature of 'warp drive'. Class. Quantum Grav. **14**, 1743 (1997). doi:10.1088/0264-9381/14/7/011
10. Simon, B.: The classical moment problem as a self-adjoint finite difference operator. Adv. Math. **137**, 82 (1998). doi:10.1006/aima.1998.1728
11. Fewster, C., Ford, L., Roman, T.: Probability distributions for quantum stress tensors in four dimensions. Phys. Rev. D **85**, 125038 (2012). doi:10.1103/PhysRevD.85.125038

Spontaneous Breaking of Lorentz Symmetry for Canonical Gravity

Steffen Gielen

Abstract In the Ashtekar-Barbero formulation of canonical general relativity based on an SU(2) connection, Lorentz covariance is a subtle issue which has been the focus of some debate. Here we present a Lorentz covariant formulation generalising the notion of a foliation of spacetime to a field of *local observers* which specify a time direction only locally. This field spontaneously breaks the local SO(3,1) symmetry down to a subgroup SO(3); we show that the apparent symmetry breaking to SO(3) is not in conflict with Lorentz covariance. We give a geometric picture of our construction as *Cartan geometrodynamics* and outline further applications of the formalism of local observers, motivating the idea that *observer space*, instead of spacetime, should serve as the fundamental arena for gravitational physics.

1 Introduction

In first order formulations of general relativity one has a notion of local Lorentz invariance, which can be thought of as one way of implementing the equivalence principle.[1]

It is crucial to understand the fate of this gauge symmetry in attempts to quantise gravity, both theoretically and with regard to a possible phenomenology of quantum gravity (including matter). There are strong experimental constraints on many possible types of violation of Lorentz covariance and any proposed theory of quantum gravity must prove itself consistent with such constraints.

In Hamiltonian formulations, in particular the Ashtekar-Barbero connection formulation [1, 2], the issue of Lorentz covariance has been the focus of some debate, since the Ashtekar-Barbero formulation naturally uses the gauge group SU(2) or

[1] Linking my talk at this wonderful conference to Einstein's Prague days.

S. Gielen (✉)
Perimeter Institute for Theoretical Physics, 31 Caroline St. N, Waterloo, ON N2L 2Y5, Canada
e-mail: sgielen@perimeterinstitute.ca

J. Bičák and T. Ledvinka (eds.), *Relativity and Gravitation*,
Springer Proceedings in Physics 157, DOI: 10.1007/978-3-319-06761-2_70,

SO(3),[2] instead of the full Lorentz group. The use of this smaller gauge group is connected to the appearance of second-class constraints in previous attempts to maintain full Lorentz covariance. Here we show how to avoid second class constraints and stay Lorentz covariant by introducing a field of *local observers*. Details are given in the paper [3].

2 Canonical First Order General Relativity

Starting from the Lorentz covariant Palatini-Holst action for vacuum general relativity without cosmological constant

$$S[e,\omega] = \frac{1}{8\pi G}\int \kappa_{abcd}\, e^a \wedge e^b \wedge R^{cd}[\omega]\,, \tag{1}$$

where κ_{abcd} is an SO(3, 1)-invariant bilinear form on $\mathfrak{so}(3,1)$,

$$\kappa_{abcd} = \frac{1}{2}\varepsilon_{abcd} + \frac{1}{2\gamma}(\eta_{ac}\eta_{bd} - \eta_{ad}\eta_{bc})\,, \tag{2}$$

one can perform the usual canonical analysis and find that the 18 momenta π^i_{ab} conjugate to the spatial components of the connection ω_i^{ab} are expressible in terms of only 12 tetrad components e_i^a. This leads to *second class constraints*, which provide an obstacle to quantisation and usually require introducing new variables which are harder to interpret in terms of spacetime geometry.

In Holst's analysis [4] leading to the well-known Ashtekar-Barbero formulation of canonical gravity, one deals with this issue by explicit symmetry breaking to SO(3): Imposing 'time gauge' $e_i^0 = 0$ and defining

$$A^{ab} = \omega^{ab} + \frac{\gamma}{2}\varepsilon^{ab}{}_{cd}\omega^{cd}\,, \tag{3}$$

only the $\mathfrak{so}(3)$ part of A (the *Ashtekar-Barbero connection*) has nonvanishing conjugate momentum, and one avoids second class constraints. However, this comes at the price of losing Lorentz symmetry which is broken explicitly by the gauge choice.

In our formalism we replace time gauge by a condition involving a field of *internal observers* y which specifies a time direction locally, and leads to a spontaneous breaking of symmetry from SO(3, 1) to a subgroup $SO(3)_y$ depending on $y(x)$ at each spacetime point x.

[2] The covering group SU(2) is required if one wants to include spinors. We consider pure gravity; the symmetry groups we discuss arise as the isometry groups of real manifolds or the stabilisers of points in them, and can be taken to be real-valued matrix groups. By expressions such as SO(3, 1), we mean the connected component preserving orientation and time orientation.

3 General Relativity with Local Observers

For a given spacetime manifold with metric g or frame field e, we define a *field of observers* as a unit future-directed timelike vector field u. Using the frame field we can map it to a spacetime scalar $y = e(u)$ valued in the velocity hyperboloid $\mathrm{H}^3 = \mathrm{SO}(3,1)/\mathrm{SO}(3)$. But such a field of *internal observers* can be defined without specifying the metric, and is hence suitable for a framework in which the metric arises dynamically as a solution to the equations of motion.

Our formalism for *generalised canonical gravity* builds on the following variables:

- a field of internal observers y, valued in $\mathrm{H}^3 \subseteq \mathbb{R}^{3,1}$, thought of as giving a local notion of time direction,
- a nowhere-vanishing 1-form $\hat{u}$, thought of as non-dynamical and generalising the normal to a foliation (if $\hat{u} \wedge d\hat{u} = 0$, $\hat{u}$ is of the form $\hat{u} = N\,dt$)—one can always reduce to the case of a foliation by choosing an appropriate $\hat{u}$.
- an $\mathbb{R}^3_y$-valued 'triad' 1-form E, where $\mathbb{R}^3_y$ is the subspace of $\mathbb{R}^{3,1}$ orthogonal to y (this generalises time gauge).

The spacetime coframe field is then simply given by

$$e = E + \hat{u}\, y, \tag{4}$$

analogous to how one reconstructs the spacetime metric in the ADM formulation using lapse and shift. As is usual in first order gravity, we must require e to be nondegenerate. The field of internal observers y defines a field of spacetime observers by $y = e(u)$, and one finds that $E(u) = 0$ so that E is actually *spatial*.

Similarly, we define spatial and temporal parts of the spin connection,

$$\omega = \Omega + \hat{u}\, \Xi, \tag{5}$$

Substituting (4) and (5) into the Palatini-Holst action (1) gives us a generalised Hamiltonian formulation of vacuum general relativity in terms of an action depending on y, E, Ω and Ξ that we give in [3]. Up to this stage everything is Lorentz covariant – we have just changed variables in the action.

The rôle of the field of internal observers y is to give us a local embedding of SO(3) into SO(3, 1). The embedding can be freely changed by applying a Lorentz transformation $y \mapsto y' = \Lambda\, y$; allowing those Lorentz transformations instead of thinking of y as fixed restores Lorentz covariance.

The spatial connection Ω can be projected to its $\mathfrak{so}(3)_y$ part $\boldsymbol{\Omega}$. Then under a local Lorentz transformation

$$\boldsymbol{\Omega} \mapsto \boldsymbol{\Omega}' = \Lambda^{-1}\, \boldsymbol{\Omega}\, \Lambda + \pi_{y'}(\Lambda^{-1}\, d^{\perp} \Lambda)\,, \tag{6}$$

where $\pi_{y'}$ is a projector onto $\mathfrak{so}(3)_{y'}$ and $d^{\perp} = d - \hat{u} \wedge \pounds_u$ is a spatial exterior derivative. Therefore, if one only applies $\mathrm{SO}(3)_y$ transformations which leave y invariant,

$\boldsymbol{\Omega}$ transforms as an $SO(3)_y$ connection, while if one allows for transformations that rotate the local internal observer y to y', the transformed connection $\boldsymbol{\Omega}'$ is in $\mathfrak{so}(3)_{y'}$. This is as it should be.

To understand the dynamical structure of this formalism, we focus on the term in the action that determines the symplectic structure in Hamiltonian general relativity,

$$S = \frac{1}{8\pi G} \int \kappa_{abcd} \hat{u} \wedge E^a \wedge E^b \wedge \pounds_u \Omega^{cd} + \cdots , \tag{7}$$

Since $E \wedge E$ is valued only in $\mathfrak{so}(3)_y$, only half of the components of Ω have nonvanishing conjugate momentum. The number of independent components of E matches the number of conjugate momenta, and no second-class constraints arise—but we did not find it necessary to impose any gauge fixing such as the time gauge employed in Holst's analysis.

One can make the splitting of $\mathfrak{so}(3, 1)$ into a rotational subalgebra $\mathfrak{so}(3)_y$ and a complement $\mathfrak{p}_y$ explicit by choosing local bases J_I^{ab} and B_I^{ab} (depending on y). Then

$$A^I := \boldsymbol{\Omega}^I + \gamma K^I , \tag{8}$$

is conjugate to $(E \wedge E)^I$, where $\boldsymbol{\Omega}$ and K are the $\mathfrak{so}(3)_y$ and $\mathfrak{p}_y$ parts of Ω. Equation (8) is the Ashtekar-Barbero connection, and our formalism is dynamically equivalent to the Ashtekar-Barbero formulation: It has the same phase space variables, subject to the same constraints that define the dynamics. In the form (8) manifest Lorentz covariance is lost; it can be recovered by viewing $\mathfrak{so}(3)_y$ and $\mathfrak{p}_y$ not as fixed (isomorphic) representations of SO(3), but as subspaces of $\mathfrak{so}(3, 1)$ specified by the field y.

4 Cartan Geometrodynamics

Situations of spontaneous symmetry breaking in gravitational theories are geometrically best understood in terms of *Cartan geometry* [5]. A well-known example is the MacDowell-Mansouri formulation [6] of gravity with cosmological constant (we take $\Lambda > 0$ but $\Lambda < 0$ is analogous) in terms of the SO(4, 1) invariant action

$$S_{\text{MM}} = -\frac{3}{32\pi G \Lambda} \int \varepsilon_{abcde} \left(F^{ab} \wedge F^{cd} \right) y^e , \tag{9}$$

where F is the curvature of an SO(4, 1) connection A. The field y takes values in de Sitter spacetime $SO(4, 1)/SO(3, 1) \subseteq \mathbb{R}^{4,1}$; it breaks the symmetry at each point in spacetime to the subgroup $SO(3, 1)_y$ leaving y invariant. Fixing $y = (0, 0, 0, 0, 1)$ in the action breaks the symmetry explicitly.

The Lie algebra $\mathfrak{so}(4, 1)$ splits into a subalgebra $\mathfrak{so}(3, 1)_y$ and a complement $\mathfrak{t}_y$; identifying the $\mathfrak{so}(3, 1)_y$ part of A with the spin connection ω and the $\mathfrak{t}_y$ part with a coframe e,

$$A = \begin{pmatrix} \omega & \sqrt{\frac{\Lambda}{3}}e \\ -\sqrt{\frac{\Lambda}{3}}e & 0 \end{pmatrix}, \tag{10}$$

the action (9) reduces to the Einstein-Hilbert-Palatini action with a cosmological term.

Cartan geometry is about infinitesimally approximating the geometry of a curved manifold by a homogeneous spacetime G/H (in this case de Sitter spacetime) which generalises the tangent space $\mathbb{R}^{p,q}$ used in (pseudo-)Riemannian geometry. The Cartan connection A relates the model spacetimes tangent to different points of the manifold – for a model spacetime of non-zero curvature, A is flat if the manifold is (locally) isomorphic to the model spacetime. This naturally introduces a cosmological constant into gravity, given by the curvature scale of the model spacetime.

Our reformulation of the Ashtekar-Barbero formalism for canonical gravity is best interpreted as describing the geometry of *space* as *Cartan geometrodynamics*: The $\mathfrak{so}(3)_y$ connection $\boldsymbol{\Omega}$ (or, alternatively, the Ashtekar-Barbero connection) and the triad E can be assembled into a Cartan connection

$$\mathbf{A} = \begin{pmatrix} \boldsymbol{\Omega} & \frac{1}{l}E \\ 0 & 0 \end{pmatrix}, \tag{11}$$

taking values in the Lie algebra of the Euclidean group $\mathfrak{iso}(3)$ if we consider a vanishing cosmological constant (l is an (unspecified) length scale put in for dimensional reasons). The appearance of the group ISO(3) is understood as follows: Spacetime is infinitesimally modelled on Minkowski spacetime, with isometry group ISO(3, 1). At a given point in spacetime, picking an observer in the model Minkowski spacetime gives a notion of 'space' in the model spacetime as the maximal totally geodesic hypersurface orthogonal to this observer – in the construction above, we referred to this as the subspace $\mathbb{R}^3_y$ orthogonal to an observer y. This breaks the symmetry to ISO(3), the isometry group of $\mathbb{R}^3_y$. Picking a point in $\mathbb{R}^3_y$ tangent to the spacetime point then breaks the symmetry further to SO(3), giving the splitting (11). For a more detailed discussion of the geometry behind Cartan geometrodynamics we refer to [7].

5 Summary and Outlook

We have given a reformulation of canonical general relativity in first order form which uses local observers that define a local notion of time. These give an embedding of the rotational subgroup SO(3) into the Lorentz group that allows to reconstruct Lorentz covariance from the SO(3) Ashtekar-Barbero formulation of canonical gravity. The geometry behind our constructions is best understood in terms of *Cartan geometrodynamics*. Since this formulation requires only a local choice of time direction not necessarily related to a foliation of spacetime, it links the canonical and covariant formulations of general relativity [8].

It would be important to understand the coupling of matter—which would be necessary to investigate the possibility of physically observable Lorentz violation—and the role of the field of internal observers there. So far they have been treated like lapse and shift, as Lagrange multipliers. Making the observer field dynamical could relate our framework to models with dynamical reference frames, such as Brown-Kuchař dust [9].

Similar constructions could also be useful in approaches to quantum gravity where local Lorentz covariance is not manifest, such as Hořava-Lifshitz gravity, shape dynamics or causal dynamical triangulations.

Taking the idea of local observers one step further, it is natural to consider the space of all possible choices of local observer—*observer space*. In general relativity, this is the direct product of spacetime with the local velocity space H^3 of normalised future-directed timelike vectors, but we consider it as a seven-dimensional manifold in its own right and study its geometry, both in general relativity and in more general settings. This is the viewpoint adopted in the work [7], where we show how the Cartan connection A specified by a frame field e and a spin connection ω as in (10) gives a Cartan geometry on observer space, with model space $SO(4, 1)/SO(3)$, the space of all observers in de Sitter spacetime. Conversely, we investigate integrability conditions that allow the reconstruction of an invariant spacetime starting from an observer space Cartan geometry (i.e. a general Cartan geometry modelled on $SO(4, 1)/SO(3)$); intuitively, such a reconstruction is possible if the connection is flat in the 'velocity' directions of observer space.

Different approaches to quantum gravity and quantum-gravity phenomenology incorporate the idea that spacetime geometry is an observer-dependent (or 'momentum-dependent'), relative concept. From the perspective of observer space, such ideas correspond to observer space Cartan connections that are not flat in velocity directions, so that no invariant spacetime can be reconstructed.

One example is the proposal of relative locality [10] which suggests that 'spacetime' and hence the notion of locality are observer-dependent, but there is an invariant momentum space shared by all observers. In [7] we find that the framework of relative locality corresponds to an observer space connection that is flat in 'spacetime', not 'velocity' directions. For a general observer space geometry, both 'spacetime' and 'velocity space' are only defined relative to an observer.

It will be interesting to see whether other ideas, such as that of an 'effective metric' $\langle g_{\mu\nu}\rangle_k$ (depending on a momentum scale k) that appears in the asymptotic safety scenario for quantum gravity [11], can be discussed in the framework of observer space geometry.

Acknowledgments I would like to thank Derek Wise for collaboration on the papers [3, 7, 8] that discuss the ideas presented in this proceedings contribution in detail and for comments on the manuscript. Research at Perimeter Institute is supported by the Government of Canada through Industry Canada and by the Province of Ontario through the Ministry of Research & Innovation.

References

1. Ashtekar, A.: New variables for classical and quantum gravity. Phys. Rev. Lett. **57**, 2244 (1986). doi:10.1103/PhysRevLett.57.2244
2. Barbero, G.J.: Real Ashtekar variables for Lorentzian signature space-times. Phys. Rev. D **51**, 5507 (1995). doi:10.1103/PhysRevD.51.5507
3. Gielen, S., Wise, D.: Spontaneously broken Lorentz symmetry for Hamiltonian gravity. Phys. Rev. D **85**, 104013 (2012). doi:10.1103/PhysRevD.85.104013
4. Holst, S.: Barbero's Hamiltonian derived from a generalized Hilbert-Palatini action. Phys. Rev. D **53**, 5966 (1996). doi:10.1103/PhysRevD.53.5966
5. Wise, D.: The geometric role of symmetry breaking in gravity. J. Phys. Conf. Ser. **360**, 012017 (2012). doi:10.1088/1742-6596/360/1/012017
6. MacDowell, S., Mansouri, F.: Unified geometric theory of gravity and supergravity. Phys. Rev. Lett. **38**, 739 (1977). doi:10.1103/PhysRevLett.38.739
7. Gielen, S., Wise, D.: Lifting general relativity to observer space, ArXiv e-prints arXiv:1210.0019[gr-qc] (2012)
8. Gielen, S., Wise, D.: Linking covariant and canonical general relativity via local observers. Gen. Relativ. Gravit. **44**, 3103 (2012). doi:10.1007/s10714-012-1443-3
9. Brown, J., Kuchař, K.: Dust as a standard of space and time in canonical quantum gravity. Phys. Rev. D **51**, 5600 (1995). doi:10.1103/PhysRevD.51.5600
10. Amelino-Camelia, G., Freidel, L., Kowalski-Glikman, J., Smolin, L.: Principle of relative locality. Phys. Rev. D **84**(8), 084010 (2011). doi:10.1103/PhysRevD.84.084010
11. Percacci, R.: Asymptotic safety. In: Oriti, D. (ed.) Approaches to Quantum Gravity: Towards a New Understanding of Space, Time and Matter, Chap. 8, pp. 111–128. Cambridge University Press, Cambridge (2009)

The Transfer Matrix in Four-Dimensional Causal Dynamical Triangulations

Andrzej Görlich

Abstract *Causal Dynamical Triangulations* (CDT) is a background independent approach to quantum gravity. In this paper we introduce a phenomenological transfer matrix model, where at each time step a reduced set of quantum states is used. The states are solely characterized by the discretized spatial volume. Using Monte Carlo simulations we determine the effective transfer matrix elements and extract the effective action for the scale factor. In this framework no degrees of freedom are frozen, however, the obtained action agrees with the *minisuperspace* model.

1 Introduction

The model of Causal Dynamical Triangulations (CDT) was proposed some years ago by Ambjørn et al. with the aim of defining a lattice formulation of quantum gravity from first principles [1, 2]. The foundation of this model is the formalism of path-integrals applied to quantize a theory of gravitation. The quantum gravity path integral is regularized by discretizing the spacetime geometry g with piecewise linear manifold $\mathscr{T}$. The building blocks of four dimensional CDT are four-simplices, which properly *glued* along their faces form a simplicial manifold.

An important assumption of CDT is the causality condition. As a consequence of the original Lorentzian signature of spacetime, only causal geometries should contribute to the integral. We will consider globally hyperbolic pseudo-Riemannian manifolds which allow introducing a global proper-time foliation. The leaves of the foliation are spatial three-dimensional Cauchy surfaces called *slices*. Because topology changes of the spatial slices are often associated with causality violation, we forbid the topology of the leaves to alter in time. For simplicity, we chose the spatial slices to have a fixed topology of a three-sphere, and establish periodic boundary

A. Görlich (✉)
Niels Bohr Institute, Blegdamsvej 17, 2100 Copenhagen, Denmark
e-mail: goerlich@nbi.dk

J. Bičák and T. Ledvinka (eds.), *Relativity and Gravitation*,
Springer Proceedings in Physics 157, DOI: 10.1007/978-3-319-06761-2_71,

conditions in the time direction. Therefore, the spacetime topology is $\mathcal{M} = S^1 \times S^3$. Spatial slices of a triangulation are enumerated by a discrete *time coordinate* t. Such time coordinate is assigned to each vertex of the triangulation, bringing on a distinction between space-like links of length a_s and time-like links of length a_t. Because each simplex contains vertices lying in two consecutive spatial slices, there are two kinds of simplices: first of a type {4, 1} with four vertices lying in one spatial slice and one in the neighboring slice, and second of a type {3, 2} with three vertices lying in one spatial slice and two in the adjacent slice. The Wick rotation is performed by the analytic continuation to imaginary lengths of the time-like links $a_t \to ia_t$. The regularized partition function Z is now written as a sum over causal triangulations $\mathcal{T}$,

$$Z = \int \mathcal{D}[g] e^{iS^{EH}[g]} \to \sum_{\mathcal{T}} e^{-S[\mathcal{T}]}. \tag{1}$$

The Einstein-Hilbert action $S^{EH}[g] = \frac{1}{16\pi G} \int \mathrm{d}t \int \mathrm{d}^3x \sqrt{-g}(R - 2\Lambda)$ evaluated on a simplicial manifold $\mathcal{T}$ composed of N_4 simplices, among them N_{41} being of type {4, 1}, and N_0 vertices, gives the discrete Regge action,

$$S[\mathcal{T}] = -K_0\, N_0 + K_4\, N_4 + \Delta\, (N_{41} - 6N_0), \tag{2}$$

where K_0, K_4 and Δ are bare coupling constants, and naively they are functions of G, λ and a_t, a_s.

We applied Monte Carlo techniques, and using the Regge action (2), measured expectation values of observables within the CDT framework. The simplest observable is the scale factor $a(t)$, or more conveniently the three-volume n_t defined as the number of tetrahedra building slice t.

For a certain range of the coupling constants, a typical configuration is bell-shaped, with the average volume profile $\langle n_t \rangle \propto \cos^3(t/B)$. The emerged background geometry behaves like a well defined four-dimensional manifold and is perfectly consistent with an Euclidean de Sitter universe, the classical vacuum solution of a spatially homogeneous and isotropic *minisuperspace* model [3]. In earlier work we have shown [4] that the discretized *minisuperspace* action,

$$S[n_t] = \frac{1}{\Gamma} \sum_t \left(\frac{(n_{t+1} - n_t)^2}{n_{t+1} + n_t} + \mu n_t^{1/3} - \lambda n_t \right), \tag{3}$$

describes well not only the measured $\langle n_t \rangle$ but also the fluctuations

$$\mathbf{C}_{tt'} = \langle (n_t - \langle n_t \rangle)(n_{t'} - \langle n_{t'} \rangle) \rangle. \tag{4}$$

The effective action (3) couples only adjacent slices. Such form suggests that there exists an effective transfer matrix labeled only by the scale factor.

2 The Transfer Matrix

The model of Causal Dynamical Triangulations is completely determined by a transfer matrix $\mathcal{M}$ labeled by three-dimensional triangulations τ. The matrix element $\langle\tau_1|\mathcal{M}|\tau_2\rangle$ denotes the transition amplitude in one time step between states corresponding to triangulations τ_1 and τ_2. It is given by the sum over all four-dimensional triangulations $\mathcal{T}$ of a slab, with boundary triangulations τ_1 and τ_2,

$$\langle\tau_1|\mathcal{M}|\tau_2\rangle = \sum_{\mathcal{T}|_{\tau_1,\tau_2}} e^{-S[\mathcal{T}]}.$$

The transfer matrix $\mathcal{M}$ depends both on the entropy factor, which counts the number of triangulations $\mathcal{T}$ connecting the boundaries in one time step, and the Regge action $S[\mathcal{T}]$. The partition function (1) corresponding to T time steps is then expressed in terms of the matrix $\mathcal{M}$,

$$Z = \sum_{\mathcal{T}} e^{-S[\mathcal{T}]} = \mathrm{Tr}\mathcal{M}^T. \tag{5}$$

The probability of finding a configuration with T spatial slices given by three-dimensional triangulations $\tau_1, \tau_2, \ldots, \tau_T$ is

$$P^{(T)}(\tau_1,\ldots,\tau_T) = \frac{1}{Z}\langle\tau_1|\mathcal{M}|\tau_2\rangle\langle\tau_2|\mathcal{M}|\tau_3\rangle\ldots\langle\tau_T|\mathcal{M}|\tau_1\rangle. \tag{6}$$

We used partition function (5) in Monte Carlo simulations. The measurements performed so far, have been concentrated on the measurement of the three-volume n_t. The probability $P^{(T)}(n_1,\ldots,n_T)$ of finding a configuration with spatial volumes $n_1, n_2, \ldots, n_T$ is given by a proper sum of partial probabilities (6). Let $T_3(n)$ denote the subset of three-dimensional triangulations which are built of exactly n three-simplices. We use the projection operator $\rho(n) \equiv |n\rangle\langle n|$ on the subspace spanned by $T_3(n)$,

$$\rho(n) \equiv |n\rangle\langle n| \equiv \sum_{\tau\in T_3(n)} |\tau\rangle\langle\tau|. \tag{7}$$

to express the probability $P^{(T)}(n_1,\ldots,n_T)$,

$$P^{(T)}(n_1,\ldots,n_T) = \frac{1}{Z}\mathrm{Tr}\,[|n_1\rangle\langle n_1|\mathcal{M}|n_2\rangle\langle n_2|\mathcal{M}|n_3\rangle\,..\,\langle n_T|\mathcal{M}]. \tag{8}$$

In (8) it is misleading to think of the aggregated "state" $|n\rangle$ as a normalized sum of the vectors $|\tau\rangle$, $\tau \in T_3(n)$. Such a vector would again be a single vector located in the space spanned by the $|\tau\rangle$'s. It is more appropriate to interpret the "state" associated with n as arising from a classical uniform *probability distribution* of states $|\tau\rangle$ and in this way to treat $\rho(n)$ as the associated *density operator*.

As mentioned in the Introduction, the form of the effective action (3) obtained from the covariance matrix (4) suggests that there exists an effective transfer matrix $\langle n|M|m\rangle$ whose elements are labeled by the three-volumes and that it is possible to effectively decompose observed distributions $P^{(T)}(n_1, \dots, n_T)$ into a product

$$P^{(T)}(n_1, \dots, n_T) = \frac{1}{Z}\langle n_1|M|n_2\rangle\langle n_2|M|n_3\rangle \cdots \langle n_T|M|n_1\rangle. \tag{9}$$

The effective transfer matrix M depends only on the coupling constants K_0, Δ and K_4 but not on the number of slices T. In analogy to (6), the elements of the effective transfer matrix correspond to transition amplitudes in one time step between states of a given three-volume.

3 Measurements

In the following, we will assume that we can work with an effective transfer matrix $\langle n|M|m\rangle$ and will show that equation (9) provides a very good approximation of measured data [5].

For simplicity, let us define the two-point function,

$$P^{(T)}(n_t, n_{t+\Delta t}) = \frac{1}{Z}\langle n_t|M^{\Delta t}|n_{t+\Delta t}\rangle\langle n_{t+\Delta t}|M^{T-\Delta t}n_t\rangle \tag{10}$$

by summing (9) over all three-volumes except for times t and $t + \Delta t$. The simplest way to measure the matrix elements $\langle n|M|m\rangle$, up to a normalization, is to consider $T = 2$,

$$\langle n|M|m\rangle \propto \sqrt{P^{(2)}(n_1 = n, n_2 = m)}.$$

The effective transfer matrix elements can be measured in various ways. In particular, as follows from (10) for $T = 3, 4$, we have

$$\langle n|M|m\rangle \propto \frac{P^{(3)}(n_1 = n, n_2 = m)}{\sqrt{P^{(4)}(n_1 = n, n_3 = m)}}. \tag{11}$$

We tested, that the elements $\langle n|M|m\rangle$ measured in different ways completely agreed up to numerical noise, supporting validity of equation (9). For technical reasons, most measurements were performed using expression (11).

The coupling constant K_4 in (2) plays a role of a cosmological constant. To correctly perform simulations, we have to approach with K_4 very close to its critical value K_4^{crit}. To efficiently probe the desired range of the three-volume, we added to the Regge action (2) a quadratic term to fix n_t around n_{vol},

$$S \to S + \epsilon \sum_t (n_t - n_{vol})^2.$$

Because it is consistent with the decomposition (9), its effect can be easily canceled. For technical reasons, we measured the transfer matrix M separately for a few overlapping ranges of the three-volume.

4 The Effective Action

The *effective* action obtained from the covariance matrix (4) is directly related to the *effective* transfer matrix M. The *minisuperspace* action (3) suggests that the *effective* transfer matrix given by

$$\langle n|M|m\rangle = \mathcal{N} e^{-\frac{1}{\Gamma}\left[\frac{(n-m)^2}{n+m} + \mu\left(\frac{n+m}{2}\right)^{1/3} - \lambda\frac{n+m}{2}\right]} \tag{12}$$

is a good approximation in the bulk where n_t is large.[1] Further, we will measure the empirical transfer matrix elements $\langle n|M|m\rangle$, extract the parameters Γ, μ and λ, and check that (12) is indeed a good approximation of the data. The measured *effective* transfer matrix M, for range $1200 < n_t < 1600$, is presented in Fig. 1 (left graph). The right graph shows the difference between the measured matrix M and the best fit (12). Indeed, the difference disappears in the numerical noise proving that the approximation (12) is very good.

The measurements presented in this paper were performed for coupling constants $K_0 = 2.2$, $\Delta = 0.6$ and $K_4 = 0.922$.

4.1 The Kinetic Term

To get a better estimation of the parameters associated with the effective action (3) and (12), we first try to fit only the parameters of the kinetic term which is the dominating term from a numerical point of view. We do that by keeping the sum of the entries, i.e. $n + m$, fixed such that the potential term is not changing. In this way we determine Γ with high accuracy. The matrix elements for constant $n + m = c$ show the expected Gaussian dependence on n (see Fig. 2),

$$\langle n|M|m\rangle = \langle n|M|c-n\rangle = \mathcal{N}(c) \exp\left[-\frac{(2n-c)^2}{\Gamma \cdot c}\right], \tag{13}$$

[1] We slightly modified the form of the potential term. Such parametrization is more convenient to extract the parameters of the action.

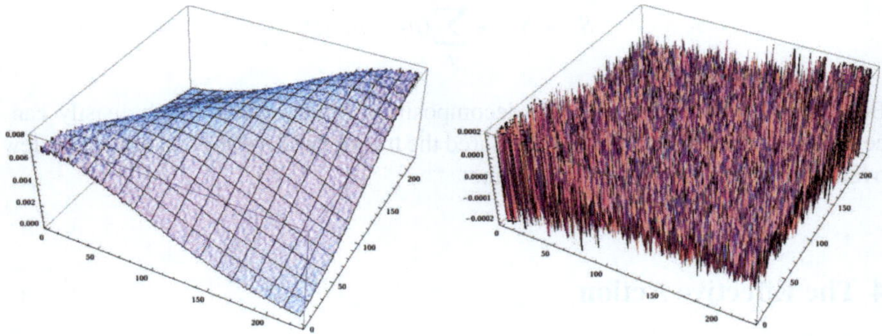

Fig. 1 *Left* The *empirical* transfer matrix for range $1200 < n_t < 1600$. *Right* The difference between the empirical and theoretical matrices disappears in the numerical noise

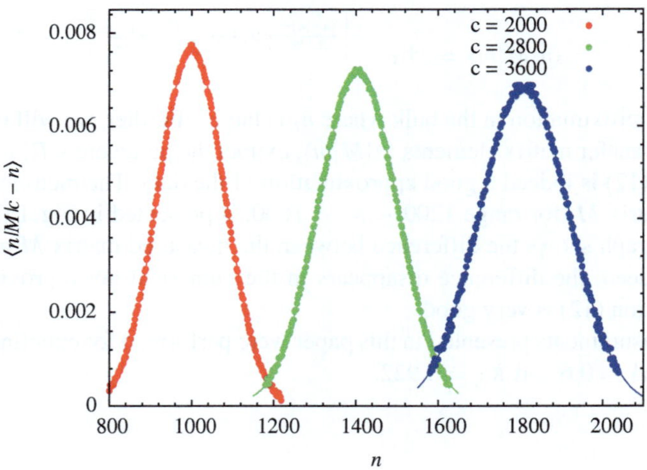

Fig. 2 $\langle n|M|c-n\rangle$ plotted as a function of n for various values of c (dots). Gaussian fits are drawn with a line

where the terms in the effective action which only depend on c are included in the normalization.

We expect the denominator of the kinetic term $k(c)$ to behave like $k(n+m) = \Gamma \cdot (n+m)$. As shown in Fig. 3 this is indeed true and the parameter Γ is constant in the whole range of the three-volumes. The best linear fit gives $\Gamma = 26.07 \pm 0.05$. This result is consistent with the values obtained from the fits for separate ranges of n_t.

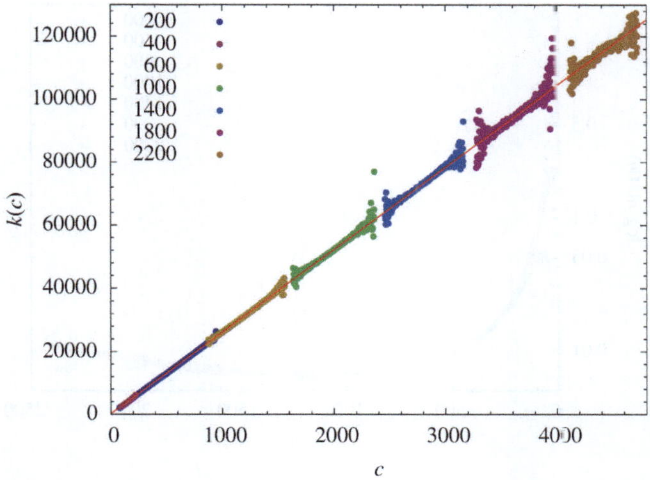

Fig. 3 The coefficient $k(c)$ in the kinetic term as a function of $c = n + m$, (different colors denote different ranges) and a linear fit $k(n + m) = \Gamma \cdot (n + m)$ (*red line*)

4.2 The Potential Term

The potential part of the effective Lagrangian L_{eff} may be extracted from the diagonal elements of the transfer matrix,

$$L_{eff}(n, n) = -\log\langle n|M|n\rangle + \text{const} = \frac{1}{\Gamma}\left(\mu n^{1/3} - \lambda n\right). \tag{14}$$

For technical reasons, we measured the transfer matrix M separately for a few different ranges of the three-volume. Because, the normalization is not uniquely defined, in order to merge the effective Lagrangian, the constant in (14) has to be properly adjusted. The measured merged effective Lagrangian is shown in Fig. 4. The colors denote different ranges for which the transfer matrix was measured. Figure 4 presents also the fit of form (14). In the bulk region, where n_t is large enough, the theoretical expectation (14) fits very well. The measured values are $\mu = 16.5 \pm 0.2$ and $\lambda = 0.049 \pm 0.001$, where we took $\Gamma = 26.07$. Again, this result is consistent with the values obtained from the fits for separate ranges of n_t.

5 Conclusions

The model of Causal Dynamical Triangulations comes with a transfer matrix $\langle \tau_1|\mathcal{M}|\tau_2\rangle$. The measured distributions of the three-volumes n_t, e.g. $P^{(T)}(n_t, n_{t+\Delta t})$, have an exact definition in terms of the *full* transfer matrix $\mathcal{M}$ and the density matrix

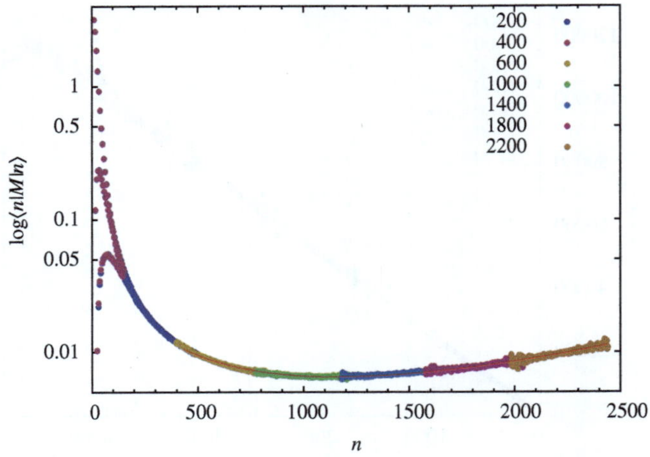

Fig. 4 $\log\langle n|M|n\rangle$ of the scaled transfer matrix (dots, different colors denote different ranges) compared with the fit of the potential term $-L_{eff}$ (*red line*, which stops at $n = 400$)

$|n\rangle\langle n|$. The actual data coming from Monte Carlo simulations seem to allow for a much simpler description in terms of an *effective* transfer matrix M, labeled by *abstract* vectors $|n\rangle$ referring only to the three-volume. The *effective* transfer matrix M allows to directly measure the *effective* action $S[n_t]$. An important advantage of the present method, since the number of slices T is small, is a much faster measurement of the transfer matrix compared to the covariance matrix, which was used previously to extract the effective action. Basically over the whole range of n_t the *effective* transfer matrix elements can be represented as

$$\langle n|M|m\rangle = \mathcal{N} e^{-\frac{1}{T}\left[\frac{(n-m)^2}{n+m} + \mu\left(\frac{n+m}{2}\right)^{1/3} - \lambda\frac{n+m}{2}\right]},$$

with high accuracy. This result is fully consistent with the reduced *minisuperspace* action (3), although in CDT we do not freeze any degrees of freedom.

An issue not addressed in this article, is the problem of small three-volumes. For small n_t we do not observe a Gaussian distribution of the three-volume n_t around the mean value $\langle n_t\rangle$. Because of strong discretization effects, the probability distributions, and consequently the *effective* transfer matrix elements, split into three families [6]. Despite different nature, after the *smoothing* procedure, the effective action for small volumes is basically the same as for large volumes, with a small modification in the potential [5]. It might be interpreted as possible curvature corrections, however, we are not able to measure it accurately in a discretization independent way.

References

1. Ambjørn, J., Jurkiewicz, J., Loll, R.: Dynamically triangulating Lorentzian quantum gravity. Nucl. Phys. B **610**, 347 (2001). doi:10.1016/S0550-3213(01)00297-8
2. Ambjørn, J., Jurkiewicz, J., Loll, R.: Reconstructing the universe. Phys. Rev. D **72**, 064014 (2005). doi:10.1103/PhysRevD.72.064014
3. Görlich, A.: Background geometry in 4D causal dynamical triangulations. Acta Phys. Pol. B **39**, 3343 (2008)
4. Ambjørn, J., Görlich, A., Jurkiewicz, J., Loll, R.: Planckian birth of the quantum de Sitter universe. Phys. Rev. Lett. **100**, 091304 (2008). doi:10.1103/PhysRevLett.100.091304
5. Ambjørn, J., Gizbert-Studnicki, J., Görlich, A., Jurkiewicz, J.: The transfer matrix in fourdimensional CDT. J. High Energy Phys. **2012**(09), 017 (2012). doi:10.1007/JHEP09(2012)017
6. Ambjørn, J., Görlich, A., Jurkiewicz, J., et al.: The semiclassical limit of causal dynamical triangulations. Nucl. Phys. B **849**, 144 (2011). doi:10.1016/j.nuclphysb.2011.03.019

Plane Gravitational Waves and Flat Space in Loop Quantum Gravity

Franz Hinterleitner and Seth Major

Abstract Classically a system of arbitrary plane gravitational waves propagating in the same or opposite directions can be restricted by first-class constraints to unidirectional waves, which travel without dispersion on a flat background. The unidirectionality constraints are formulated as well-defined Loop Quantum Gravity operators, together with criteria for an anomaly-free implantation, which is crucial for the occurrence or non-occurrence of dispersion, and more generally, of local Lorentz invariance violations due to (loop) quantum effects. By a set of further first-class constraints of the same kind we construct a quantum model of a no-wave state, i.e. of empty space.

1 Introduction

The motivation behind this contribution is the search for quantum effects of gravity in the form of dispersion of pure, unidirectional gravitational waves. The existence or non-existence of gravitational wave dispersion, derived for a solvable system from first loop quantum gravity (LQG) principles is an important criterion in the issue of Lorentz invariance at the Planck scale in quantum gravity.

Our approach consists in a symmetry reduction to 1+1 dimensions on the classical level, taken over from [1, 2], vacuum solutions in this model represent plane gravitational waves moving back and forth in one direction. Further reduction to unidirectional waves is achieved by a set of first-class constraints, derived from the Killing equations that describe the special symmetry of space-time with unidi-

F. Hinterleitner (✉)
Department of Theoretical Physics and Astrophysics, Masaryk University, Kotlářská 2, 61137 Brno, Czech Republic
e-mail: franz@physics.muni.cz

S. Major
Department of Physics, Hamilton College, Clinton, NY 13323, USA

J. Bičák and T. Ledvinka (eds.), *Relativity and Gravitation*,
Springer Proceedings in Physics 157, DOI: 10.1007/978-3-319-06761-2_72,

rectional waves [3]. These constraints are to be imposed on the quantum states of one-dimensional, but possibly colliding and interacting waves, and the question is whether or not these constraints can single out waves propagating uniformly at the speed of light.

As a by-product, by imposing one more set of Killing constraints, one can model flat space. A successful quantum solution will show how LQG can predict gravitational fluctuations of Minkowski geometry.

2 Classical Polarized Two-Way Waves in Ashtekar Variables

We are considering plane gravitational waves propagating in the positive and negative z direction, the system is homogeneous in the x and the y directions, all metric components depend only on z and t.

2.1 Variables

The metric is formulated in terms of adapted densitized triad variables with the nonzero components

$$\mathscr{E} = E^z{}_3 \tag{1}$$

along the inhomogeneous direction, orthogonal to the components in the (x, y) plane

$$\begin{aligned} E^x{}_1 &= E^x \cos\eta, \quad E^x{}_2 = E^x \sin\eta, \\ E^y{}_1 &= -E^y \sin\eta, \; E^y{}_2 = E^y \cos\eta. \end{aligned} \tag{2}$$

The mutual orthogonality of these two triad vectors means that we are dealing with polarized waves. In terms of these variables the spatial metric reads

$$\mathrm{d}\sigma^2 = \mathscr{E}\frac{E^y}{E^x}\mathrm{d}x^2 + \mathscr{E}\frac{E^x}{E^y}\mathrm{d}y^2 + \frac{E^x E^y}{\mathscr{E}}\mathrm{d}z^2. \tag{3}$$

The canonically conjugate variables are the connection components $\mathscr{A}$, K_x, K_y, and P with the equal-time Poisson brackets

$$\begin{aligned} \{K_a(z), E^b(z')\} &= \kappa\, \delta_a^b\, \delta(z - z'), \quad a, b = x, y, \\ \{\mathscr{A}(z), \mathscr{E}(z')\} &= \{P(z), \eta(z')\} = \kappa\, \delta(z - z'). \end{aligned} \tag{4}$$

κ is the gravitational constant. The symmetry-reduced model has four phase-space degrees of freedom.

2.2 The Constraints

The standard constraints of canonical general relativity, adapted to the above model, are the Gauß constraint

$$G = \frac{1}{\kappa\gamma}(\mathscr{E}' + P), \tag{5}$$

which generates rotations in the (x, y) plane, the diffeomorphism constraint

$$C = \frac{1}{\kappa}\left[K_x' E^x + K_y' E^y - \mathscr{E}'\mathscr{A} + \frac{\eta'}{\gamma} P\right], \tag{6}$$

and the Hamiltonian constraint

$$H = -\frac{1}{\kappa\sqrt{\mathscr{E}E^x E^y}}\left[E^x K_x E^y K_y + (E^x K_x + E^y K_y)\mathscr{E}\left(\mathscr{A} + \frac{\eta'}{\gamma}\right) - \frac{1}{4}\mathscr{E}'^2 \right.$$
$$\left. - \mathscr{E}\mathscr{E}'' - \frac{1}{4}\mathscr{E}^2\left[\left(\ln\frac{E^y}{E^x}\right)'\right]^2\right] - \frac{\kappa}{4\sqrt{\mathscr{E}E^x E^y}}G^2 - \gamma\left(\sqrt{\frac{\mathscr{E}}{E^x E^y}}\,G\right)'. \tag{7}$$

A prime denotes the derivative with respect to z, γ is the Barbero-Immirzi parameter. H is partially expressed by the Gauß constraint G. These first-class constraints reduce the number of degrees of freedom to one, the correct number for polarized plane waves.

3 Reduction to Unidirectional Waves

Unidirectional waves are characterized by the existence of a null Killing vector field in the direction of propagation. This corresponds to a dependence of the metric functions either on $t - z$ or on $t + z$. To formulate such fields, we add an orthogonal timelike direction and construct a space-time metric with lapse function $N(t, z)$.

On a manifold with this metric we assume a null Killing vector field k^μ with $\nabla_{(\mu}k_{\nu)} = 0$. Two of the Killing equations give rise to nontrivial conditions on the phase space variables, for propagation in the positive z direction they are

$$U_x := E^x K_x - \frac{1}{2}\mathscr{E}' - \frac{1}{2}\mathscr{E}\left(\frac{E^{y\prime}}{E^y} - \frac{E^{x\prime}}{E^x}\right) = 0, \tag{8}$$

$$U_y := E^y K_y - \frac{1}{2}\mathscr{E}' + \frac{1}{2}\mathscr{E}\left(\frac{E^{y\prime}}{E^y} - \frac{E^{x\prime}}{E^x}\right) = 0. \tag{9}$$

In addition to the standard constraints there can be at most one more first-class constraint which, of course, cannot be a gauge generator, because an associated gauge condition would reduce the number of degrees of freedom to zero.

To extract from U_x and U_y a relation that can be added as a set of first-class constraints to the standard constraints, we take the linear combinations

$$U_+ := U_x + U_y \qquad \text{and} \qquad U_- := U_x - U_y, \tag{10}$$

explicitly

$$U_+ = E^x K_x + E^y K_y - \mathcal{E}', \tag{11}$$

$$U_- = E^x K_x - E^y K_y - \mathcal{E}\left(\ln \frac{E^y}{E^x}\right)'. \tag{12}$$

The Poisson brackets of these expressions, smeared out by test functions,

$$U_a[f] := \int \mathrm{d}z\, f(z) U_a(z), \tag{13}$$

are

$$\{U_+[f], U_+[g]\} = \{U_+[f], U_-[g]\} = 0 \tag{14}$$

and

$$\{U_-[f], U_-[g]\} = 2\int \mathrm{d}z (f'g - fg')\,\mathcal{E}. \tag{15}$$

The function U_+ weakly Poisson-commutes also with G, C, and H:

$$\{U_+[f], G[g]\} = 0, \qquad \{U_+[f], C[g]\} = -\frac{1}{\kappa}\, U_+[f'g] \approx 0, \tag{16}$$

$$\{U_+[f], H[g]\} = \frac{1}{\kappa}\, U_+\left[\sqrt{\frac{\mathcal{E}}{E^x E^y}}\, f'g\right] - H[fg] \approx 0. \tag{17}$$

This qualifies $U_+(z)$ as a set of first-class constraints that have to be added to G, C, and H, when we want to restrict counter-current waves to unidirectional ones at the classical level.

Not being a gauge generator, but a restriction of the number of the physical degrees of freedom, the new constraint reduces their number to one half, i.e. to one phase space function. This corresponds to the original formulation [4], which contains two functions, the so-called “wave factor” and the “background factor”, connected by one non-trivial Einstein equation.

4 Preparation for Quantization

After the formulation of unidirectional waves as a classical system with first-class constraints we start the Dirac quantization programme, which distinguishes physical states as those that are annihilated by the constraint operators.

This gives rise to two kinds of problems: The formulation of the constraints as well-defined operators on a suitable Hilbert space of unconstrained states, and the problem of non-trivial structure functions in the constraint algebra. For the standard constraints these problems are solved in general LQG [5], for U_+ they will be dealt with in the following.

Both parts of U_+, $E^x K_x + E^y K_y$ as well as $\mathscr{E}'$ are scalar densities, which can be naturally integrated along z in order to construct an operator. The integral over some interval $\mathscr{I}$ is

$$U_+[\mathscr{I}] = \int_{\mathscr{I}} \mathrm{d}z\,(E^x K_x + E^y K_y) - \mathscr{E}_+ + \mathscr{E}_-, \tag{18}$$

where $\mathscr{E}_\pm$ are the values at the endpoints of $\mathscr{I}$. $\mathscr{E}$ has a meaningful operator equivalent in the adapted LQG framework [2]. In analogy to full LQG the integral can be obtained as the Poisson bracket

$$\int_{\mathscr{I}} \mathrm{d}z (E^x K_x + E^y K_y) = 2\left\{\int_{\mathscr{I}} \mathrm{d}z\, \frac{E^x K_x E^y K_y}{\sqrt{\mathscr{E} E^x E^y}}, \int_{\mathscr{I}} \mathrm{d}z' \sqrt{\mathscr{E} E^x E^y}\right\}. \tag{19}$$

The first expression is part of the kinetic Hamiltonian constraint, denoted by H_1 in the following, the second part is the volume of a slice of space, constructed from a fiducial area in the (x, y) plane and the interval $\mathscr{I}$ in the z direction. Both have an operator interpretation on one-dimensional spin network functions [2].

According to its factor-ordering, the operator formulation of the structure function in (17) raises potentially an anomaly problem. When the factor ordering is chosen analogous to that of the Hamilton constraint operator—connection components to the left of triad components (see [5][1]), then the operator constructed from

$$U_+ \sqrt{\frac{\mathscr{E}}{E^x E^y}} = \frac{(K_x E^x + K_y E^y - \mathscr{E}')\,\mathscr{E}}{\sqrt{\mathscr{E} E^x E^y}} \tag{20}$$

does not obviously annihilate solutions of the gauge constraints and U_+ and its action on them must be examined.

As in the case of U_+, the first step is a consistent operator formulation: The first part of (20) can be written as a Poisson bracket of the second part H_2 of the Hamiltonian constraint (7) with test function 1 and $\mathscr{E}(z)$,

$$-\frac{1}{\kappa}\left(\frac{K_x E^x + K_y E^y}{\sqrt{\mathscr{E} E^x E^y}}\,\mathscr{E}\right) = \{H_2[1], \mathscr{E}(z)\}, \tag{21}$$

[1] A different factor ordering is presented in [6].

so we can write

$$U_+\sqrt{\frac{\mathscr{E}}{E^x E^y}} = \kappa\{\mathscr{E}(z), H_2[1]\} - \frac{\mathscr{E}\mathscr{E}'}{\sqrt{\mathscr{E}E^x E^y}}. \tag{22}$$

Both expressions have operator equivalents, the second term is part of H.

5 Quantum States and the Action of Operators

In LQG a suitable basis of kinematical quantum states is provided by spin network functions, based on three-dimensional graphs. In the present case we have one-dimensional graphs with $U(1)$-holonomies $h_e^{(k)}(\mathscr{A}) = \exp\left(i\frac{k}{2}\int_e \mathscr{A}\right)$, $k \in Z$ associated to its edges. Holonomies along curves in the (x, y) plane are shrunk to "point holonomies" at the vertices v: $h_v^{(\mu)}(X) = \exp\left(i\frac{\mu}{2}X(v)\right)$, $h_v^{(\rho)}(Y) = \exp\left(i\frac{\rho}{2}Y(v)\right)$, and $h_v^{(\lambda)}(\eta) = \exp(i\lambda\eta(v))$. $X = \gamma K_x$, $Y = \gamma K_y$. η is an angular variable, its holonomy has values in $U(1)$, $\mu, \rho \in R$, their holonomies lie in the Bohr compactification of the reals, see [5].

Connection components act in the form of holonomy operators, which add one of the above holonomies to a given state. States, denoted by $|s\rangle$, depend on the graph G and the labels k, μ, ρ, and λ. Triad components and the conjugate variable to η act as flux operators in the following way

$$\hat{\mathscr{E}}(z)\,|s\rangle = \frac{\gamma\ell_{\mathrm{P}}^2}{2}\frac{k_+(z)+k_-(z)}{2}\,|s\rangle, \qquad \int_{\mathscr{I}}\hat{P}\,|s\rangle = \gamma\ell_{\mathrm{P}}^2\sum_v \lambda_v\,|s\rangle$$

$$\int_{\mathscr{I}}\hat{E^x}\,|s\rangle = \frac{\gamma\ell_{\mathrm{P}}^2}{2}\sum_v \mu_v\,|s\rangle, \qquad \int_{\mathscr{I}}\hat{E^y}\,|s\rangle = \frac{\gamma\ell_{\mathrm{P}}^2}{2}\sum_v \nu_v\,|s\rangle. \tag{23}$$

$\mathscr{E}$ is a scalar quantity, the other ones are scalar densities and have to be integrated over an interval $\mathscr{I}$ to give raise to an operator, $k_\pm(z)$ are the representation labels of the edge holonomies left and right to z, ℓ_{P} is the Planck length, the sum is taken over all vertices of G in the interval $\mathscr{I}$.

The Gauß constraint relates the labels k and λ,

$$\lambda_v = -(k_+(v) - k_-(v))/2, \tag{24}$$

so gauge-invariant states are of the form

$$|s\rangle = \prod_e \exp\left[\frac{ik_e}{2}\int_e(\mathscr{A}(z) - \eta'(z))\right]\prod_v\left(\exp\left[\frac{i\mu_v}{2}X(v)\right]\exp\left[\frac{i\rho_v}{2}Y(\rho)\right]\right).$$

In this formula $\int_e \eta' = \eta_+(e) - \eta_-(e)$ was used, where $\eta_\pm$ are the values of η at the endpoints of the edge e.

6 Flat Space

In the case of unidirectional plane waves a null Killing vector field prevents waves in the opposite direction. A second null Killing field in the opposite direction characterizes a no-wave state, namely Minkowski space. The corresponding constraint is

$$\bar{U}_+ = K_x E^x + K_y E^y + \mathcal{E}' = 0. \tag{25}$$

Classically this one additional first-class constraint reduces the number of degrees of freedom to zero, i.e. to one state. From U_+ and $\bar{U}_+$ together follow the constraints

$$\mathcal{E}' = 0 \quad \text{and} \quad K_x E^x + K_y E^y = 0. \tag{26}$$

The operator version of the latter expression is given by (19), the former one is just the derivative of a flux operator.

In the following we consider these constraints separately, which is easier than in the combination U_+. To qualify as a model for flat space, a solution to them must also be a solution to the Hamiltonian constraint and to the Poisson brackets with the Hamiltonian constraint.

$$\{\mathcal{E}'[f], H[g]\} = \int \mathrm{d}z\, f'(z) g(z) \left(\frac{(K_x E^x + K_y E^y)\mathcal{E}}{\kappa \sqrt{\mathcal{E} E^x E^y}} \right)(z), \tag{27}$$

$$\{(K_x E^x + K_y E^y)[f], H[g]\} = H[fg]. \tag{28}$$

From (27) a quantum anomaly may arise. An operator version is already given in (21).

The first one of the constraints (26) is solved by states with the same label k for all edges.

The second one is formulated by replacing the right-hand side of (19) by the commutator of $\hat{H}_1$ with the volume operator. The action of $\hat{H}_1$ on a state $|s\rangle$ of the form (5) with $k_+ = k_- = k$ is given by

$$\hat{H}_1|s\rangle = \frac{\ell_P \gamma^{-\frac{3}{2}}}{2\mu_0\rho_0} \sum_v \sqrt{|\mu_v||\rho_v|} \left(\sqrt{|2k+1|} - \sqrt{|2k-1|} \right) \times \sin(\mu_0 X) \sin(\rho_0 Y)|s\rangle, \tag{29}$$

where μ_0 and ρ_0 are arbitrarily chosen, fixed values. The action of the volume operator is

$$\hat{V}|s\rangle = \frac{\gamma^{\frac{3}{3}}\ell_{\mathrm{Pl}}^3}{2}\sum_v (|\mu_v||\rho_v||k|)^{\frac{1}{2}}\,. \tag{30}$$

Now assume the state function at a certain vertex to be given by the superposition

$$|s\rangle_v = \sum_{m,n} a_{mn}|k, m, n\rangle, \tag{31}$$

where $m = \frac{\mu_v}{\mu_0}$ and $n = \frac{\rho_v}{\rho_0}$. The action of the commutator $[\hat{V}, \hat{H}_1]$ on such a function set equal to zero yields the following difference equation for the coefficients a_{mn},

$$\begin{aligned}
&\sqrt{|m-2||n-2|}\left(\sqrt{|m||n|} - \sqrt{|m-2||n-2|}\right) a_{m-2,n-2} - \\
&\sqrt{|m+2||n-2|}\left(\sqrt{|m||n|} - \sqrt{|m+2||n-2|}\right) a_{m+2,n-2} - \\
&\sqrt{|m-2||n+2|}\left(\sqrt{|m||n|} - \sqrt{|m-2||n+2|}\right) a_{m-2,n+2} + \\
&\sqrt{|m+2||n+2|}\left(\sqrt{|m||n|} - \sqrt{|m+2||n+2|}\right) a_{m+2,n+2} = 0,
\end{aligned} \tag{32}$$

where the k-dependence has dropped out.

The equation resulting from the structure function (27) contains H_2, whose operator version is more ambiguous, compare [1] and [2]. Anyway, after deciding for one version, the action of the structure function operator on solutions of (26) may already lead to an ambiguity, before the Hamiltonian constraint has to be solved. This would indicate gravitational fluctuations of the Minkowski vacuum. Work is ongoing.

References

1. Bojowald, M., Swiderski, R.: Spherically symmetric quantum geometry: Hamiltonian constraint. Class. Quantum Grav. **23**, 2129 (2006). doi:10.1088/0264-9381/23/6/015
2. Banerjee, K., Date, G.: Loop quantization of polarized Gowdy model on T^3: kinematical states and constraint operators. Class. Quantum Grav. **25**, 145004 (2008). doi:10.1088/0264-9381/25/14/145004
3. Hinterleitner, F., Major, S.: Towards loop quantization of plane gravitational waves. Class. Quantum Grav. **29**, 065019 (2012). doi:10.1088/0264-9381/29/6/065019
4. Ehlers, J., Kundt, W.: Exact solutions of gravitational field equations. In: Witten, L. (ed.) Gravitation: An Introduction to Current Research, pp. 33.6, 35.9. Wiley, New York (1962).
5. Thiemann, T.: Modern Canonical Quantum General Relativity. Cambridge Monographs on Mathematical Physics. Cambridge University Press, Cambridge (2007)
6. Rovelli, C.: Quantum Gravity. Cambridge Monographs on Mathematical Physics. Cambridge University Press, Cambridge (2004)

Unruh-DeWitt Detector on the BTZ Black Hole

Lee Hodgkinson and Jorma Louko

Abstract We examine an Unruh-DeWitt particle detector coupled to a scalar field in three-dimensional curved spacetime, within first-order perturbation theory. We first obtain a causal and manifestly regular expression for the instantaneous transition rate in an arbitrary Hadamard state. We then specialise to the Bañados-Teitelboim-Zanelli black hole and to a massless conformally coupled field in the Hartle-Hawking vacuum. A co-rotating detector responds thermally in the expected local Hawking temperature, while a freely-falling detector shows no evidence of thermality in regimes that we are able to probe, not even far from the horizon. The boundary condition at the asymptotically anti-de Sitter infinity has a significant effect on the transition rate.

1 Introduction

Whenever non-inertial observers or curved backgrounds are present in quantum field theory, the notions of vacuum state and particle number become non-unique. For this reason it proves convenient to define particles operationally; that is to say, we couple the field to a simple quantum mechanical system that we think of as our detector and define particles via the field's interaction with the energy levels of this system. Upwards (respectively downwards) transitions can be interpreted as due to absorption (emission) of field quanta, or particles. This is the Unruh-DeWitt model for a particle detector [1, 2].

L. Hodgkinson (✉)
School of Mathematical Sciences, University of Nottingham, Nottingham NG7 2RD, UK
e-mail: lee.hodgkinson@nottingham.ac.uk

J. Louko
Kavli Institute for Theoretical Physics, University of California, Santa Barbara, CA 93106-4030, USA
e-mail: jorma.louko@nottingham.ac.uk

J. Bičák and T. Ledvinka (eds.), *Relativity and Gravitation*,
Springer Proceedings in Physics 157, DOI: 10.1007/978-3-319-06761-2_73,

In this contribution we address a pointlike Unruh-DeWitt detector coupled to a scalar field in three-dimensional spacetime, within first-order perturbation theory. We first find the detector's instantaneous transition rate in an arbitrary Hadamard state. We then specialise to a massless conformally coupled field on the Bañados-Teitelboim-Zanelli (BTZ) black hole, in the Hartle-Hawking vacuum, analysing the thermal response seen by a co-rotating detector and the time evolution of the response of a freely-falling detector. A longer exposition of the results can be found in [3].

2 Transition Rate in (2 + 1) Dimensions

With the Unruh-DeWitt detector, the fundamental quantity of interest is the probability of a transition between the energy eigenstates. In the framework of first order perturbation theory the probability for a transition of energy E is proportional to the response function,

$$\mathscr{F}(E) = 2 \lim_{\varepsilon \to 0_+} \Re \int_{-\infty}^{\infty} \mathrm{d}u\, \chi(u) \int_0^{\infty} \mathrm{d}s\, \chi(u-s)\, \mathrm{e}^{-iEs}\, W_\varepsilon(u, u-s)\,, \quad (1)$$

where χ is a smooth switching function that turns on (off) the detector's interaction with the field and $W_\varepsilon(u, u-s)$ is a one-parameter family of functions that converge to the pull-back of the Wightman distribution to the detector's wordline [4–7]. A related quantity of interest is the transition rate, which can be defined as the derivative of the transition probability with respect to the total detection time. One must take great care when obtaining the transition rate from the response function [8–11]. We will adopt the approach developed in [7, 12, 13] of taking a controlled sharp switching limit.

In three-dimensional spacetime, the Wightman distribution $W(\mathsf{x}, \mathsf{x}')$ of a real scalar field in a Hadamard state can be represented by the $\varepsilon \to 0_+$ limit of a family of functions with the short distance form [14]

$$W_\varepsilon(\mathsf{x}, \mathsf{x}') = \frac{1}{4\pi}\left[\frac{U(\mathsf{x}, \mathsf{x}')}{\sqrt{\sigma_\varepsilon(\mathsf{x}, \mathsf{x}')}} + \frac{H(\mathsf{x}, \mathsf{x}')}{\sqrt{2}}\right], \quad (2)$$

where ε is a positive parameter, $\sigma(\mathsf{x}, \mathsf{x}')$ is the squared geodesic distance between x and x', $\sigma_\varepsilon(\mathsf{x}, \mathsf{x}') := \sigma(\mathsf{x}, \mathsf{x}') + 2i\varepsilon\left[T(\mathsf{x}) - T(\mathsf{x}')\right] + \varepsilon^2$ and T is any globally-defined future-increasing C^∞ function. The branch of the square root is such that the $\varepsilon \to 0_+$ limit of the square root is positive when $\sigma(\mathsf{x}, \mathsf{x}') > 0$ [6, 14]. Here $U(\mathsf{x}, \mathsf{x}')$ and $H(\mathsf{x}, \mathsf{x}')$ are symmetric biscalars which have expansions governed by certain recursion relations [14], and they are regular in the coincidence limit.

Given (2), the detector's instantaneous transition rate can be shown to take the form [3]

$$\dot{\mathscr{F}}_\tau(E) = \frac{1}{4} + 2\int_0^{\tau-\tau_0} ds\,\Re\left[e^{-iEs}\,W_0(\tau,\tau-s)\right], \tag{3}$$

where τ_0 is the proper time at which the detector was switched on, τ is the proper time at which the instantaneous transition rate is read off, and the function W_0 is the pointwise $\varepsilon \to 0_+$ limit of W_ε. We are assuming that any singularities that $W(\mathsf{x},\mathsf{x}')$ may have at $\sigma(\mathsf{x},\mathsf{x}') \neq 0$, not captured by the asymptotic expansion (2), are so mild that taking the pointwise limit is valid. Such singularities will in particular occur in the BTZ spacetime below.

3 Detector in the BTZ Spacetime

We now specialise to a detector in the BTZ black hole spacetime [15–17]. This spacetime can be obtained by periodically identifying AdS_3, and in coordinates adapted to the global isometries the metric takes the form

$$ds^2 = -(N^\perp)^2dt^2 + f^{-2}dr^2 + r^2\left(d\phi + N^\phi dt\right)^2, \tag{4}$$

where $N^\perp = f = \left(-M + \frac{r^2}{\ell^2} + \frac{J^2}{4r^2}\right)^{1/2}$, $N^\phi = -\frac{J}{2r^2}$, ℓ is a positive parameter that sets the AdS_3 curvature scale, ϕ has period 2π, and a non-extremal black hole is obtained when the mass parameter M and the angular momentum parameter J satisfy $|J| < M\ell$. The spacetime has many similarities with the Kerr black hole, but its null infinities are asymptotically AdS, as opposed to asymptotically flat. The conformal diagram of the $J = 0$ case is shown in Fig. 1. The importance of this asymptotic structure for us is that the spacetime is not globally hyperbolic, and to build a sensible quantum field theory one must impose boundary conditions at the infinity. We shall see that the detector response turns out to be highly sensitive to these boundary conditions.

We consider a massless, conformally coupled field. We first introduce on the covering space AdS_3 the three AdS-invariant states whose Wightman functions are given by [17]

$$G_A^{(\zeta)}(\mathsf{x},\mathsf{x}') = \frac{1}{4\pi}\left(\frac{1}{\sqrt{\Delta\mathsf{X}^2(\mathsf{x},\mathsf{x}')}} - \frac{\zeta}{\sqrt{\Delta\mathsf{X}^2(\mathsf{x},\mathsf{x}') + 4\ell^2}}\right), \tag{5}$$

where $\Delta\mathsf{X}^2(\mathsf{x},\mathsf{x}')$ is the squared geodesic distance between x and x' in the flat $\mathbb{R}^{2,2}$ in which AdS_3 can be embedded as a submanifold, the parameter $\zeta \in \{0, 1, -1\}$ specifies whether the boundary condition at infinity is respectively transparent, Dirichlet or Neumann, and we have suppressed the $i\varepsilon$ that controls the short distance form (2). The Wightman function in the induced state on the BTZ spacetime is then given by the image sum [17]

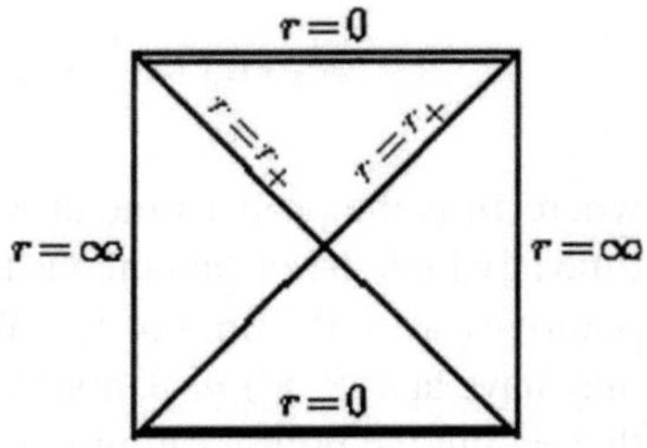

Fig. 1 Conformal diagram for the $J = 0$ BTZ black hole. The Killing horizon of the Killing vector ∂_t is at $r = r_+$, where $r_+ = \sqrt{M}\,\ell$

$$G_{\mathrm{BTZ}}(\mathsf{x}, \mathsf{x}') = \sum_n G_A(\mathsf{x}, \Lambda^n \mathsf{x}') \,, \tag{6}$$

where $\Lambda \mathsf{x}'$ denotes the action on x' of the isometry $(t, r, \phi) \mapsto (t, r, \phi + 2\pi)$, and the notation suppresses the distinction between points on AdS_3 and points on the quotient spacetime. The scalar field is assumed untwisted so that no additional phase factors appear in (6).

The detector's transition rate is obtained by substituting (6) into (3). In Sects. 4 and 5 we discuss the transition rate for selected detector trajectories.

4 Co-Rotating Detector in BTZ

As the first example we consider a detector that is in the exterior region of the BTZ black hole, at constant value of r and co-rotating with the horizon angular velocity Ω_H. In the special case $J = 0$, we have $\Omega_H = 0$ and the detector is static. Unlike in Kerr, these detector trajectories exist at arbitrarily large values of r: this is a consequence of the AdS asymptotics.

As the detector is stationary, we take the switch-on to be in the asymptotic past. The Wightman function turns out to contain singularities between timelike-separated points on the detector's trajectory, but the consequent singularities in the transition rate formula (3) are integrable and the transition rate remains well defined. Further, contour manipulations allow the transition rate to be cast in a manifestly nonsingular form that is amenable to analytic techniques, including asymptotic analyses in a number of asymptotic regimes, as well as to numerical evaluation. We can in particular verify analytically that the transition rate satisfies

$$\dot{\mathscr{F}}(E) = \mathrm{e}^{-E/T_{\mathrm{loc}}} \dot{\mathscr{F}}(-E), \tag{7}$$

where T_{loc} is the co-rotating Hawking temperature at the detector's location [17]. (As the transition rate is stationary, we have dropped the subscript τ.) The transition rate is hence thermal in the local Hawking temperature in the sense of the Kubo-Martin-Schwinger (KMS) property [18, 19], as expected from the general properties of the Hartle-Hawking vacuum [20, 21].

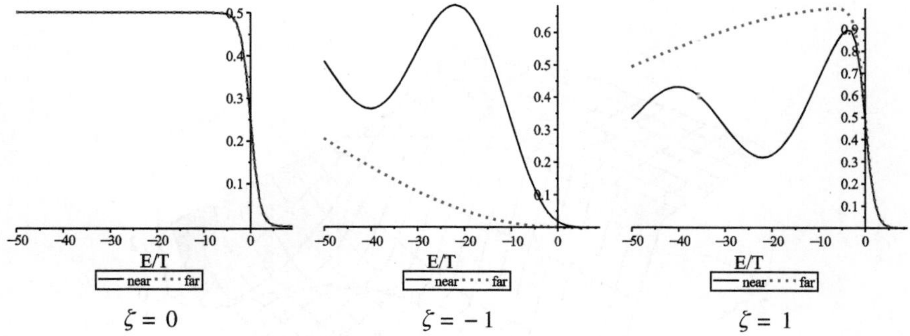

Fig. 2 $\dot{\mathcal{F}}$ for a co-rotating detector, as a function of the detector's energy gap E divided by the local Hawking temperature T, for a large non-spinning hole, with the detector near the hole (*solid*) and far from hole (*dotted*). Note the significant differences between the three boundary conditions

The boundary condition at the infinity is found to have a significant effect on the quantitative properties of the transition rate. The special case of a spinless black hole, with a detector at large and small distances from the hole, is illustrated in Fig. 2.

5 Inertial Detector in BTZ Spacetime

As the second example we consider a detector on a geodesic that falls radially into the spinless black hole. This trajectory is not stationary and the transition rate depends on both the switch-on moment and the switch-off moment. Furthermore the switch-on moment cannot be pushed to the infinite past because the trajectory starts at the white hole singularity at a finite proper time.

We have found no parameter ranges where the transition rate would be thermal in the sense of the KMS property (7). One situation where approximate thermality might have been expected is near the turning point of a trajectory far from the horizon. However, in this case the transition rate just reduces to that of a geodesic detector in AdS_3, which can be verified not to satisfy the KMS property. These observations are compatible with embedding space arguments which suggest that a detector in AdS_3 should respond thermally only when its proper acceleration exceeds $1/\ell$ [22–25].

We were however able to analyse the transition rate by a combination of asymptotic methods and numerical methods. Figure 3 shows a plot of the transition rate when the black hole is large and the switch-on and switch-off moments are not close to the white hole and black hole singularities, with the transparent boundary condition at the infinity.

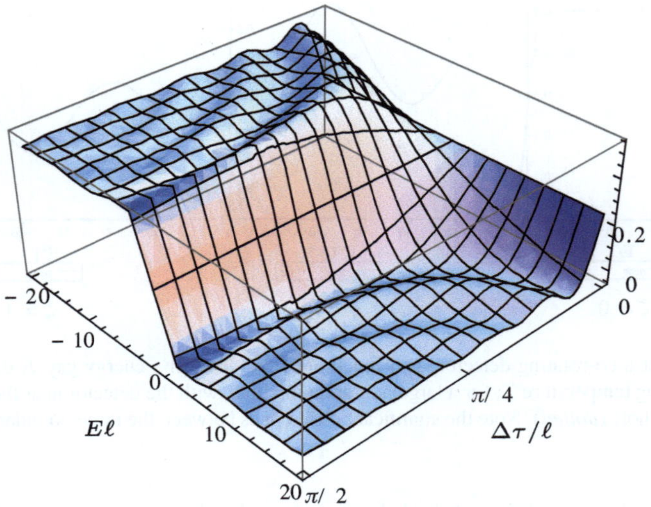

Fig. 3 The transition rate of a detector on a radial geodesic in the spinless BTZ spacetime, assuming that the mass is large and that the switch-on and switch-off moments are not close to the white hole and black hole singularities, with the transparent boundary condition at the infinity. The horizontal axes are the detector's energy gap E and the total detection time $\Delta\tau := \tau - \tau_0$, normalised by the AdS scale ℓ. Note the dominance of the de-excitation rate ($E < 0$) over the excitation rate ($E > 0$) after the transient switch-on effects have died out

6 Concluding Remarks

That the response of a co-rotating detector in the BTZ spacetime is thermal in the co-rotating Hawking temperature was to be expected from the general properties of the Hartle-Hawking vacuum [20, 21]. Our formalism allowed us to analyse this thermal response quantitatively, by a combination of analytic and numerical techniques. We found in particular that the response depends strongly on the choice of the boundary condition at the infinity. We also showed perturbatively that the response loses its thermal character when the detector's angular velocity differs from that of the black hole.

For a detector falling radially into a spinless BTZ hole, we found no parameter space regions where the transition rate would exhibit thermality. The transition rate is again affected by the choice of the boundary condition at the infinity, but this effect appears to be subdominant to those caused by the switching and the motion.

It would be interesting to compare our results for the transition rate in the BTZ spacetime to that in Schwarzschild spacetime. For example, one may expect an inertial detector in Schwarzschild to respond to the Hartle-Hawking vacuum approximately thermally in the asymptotically flat region, owing to the asymptotic flatness of Schwarzschild. We leave these questions subject to future work.

Acknowledgments J. Louko thanks the organisers of the "Bits, Branes, Black Holes" programme for hospitality at the Kavli Institute for Theoretical Physics, University of California at Santa Barbara. This research was supported in part by the National Science Foundation under Grant No. NSF PHY11-25915 and by the Science and Technology Facilities Council.

References

1. DeWitt, B.: Quantum gravity: the new synthesis. In: Hawking, S., Israel, W. (eds.) General Relativity: An Einstein Centenary Survey, pp. 680–745. Cambridge University Press, Cambridge, New York (1979)
2. Unruh, W.: Notes on black-hole evaporation. Phys. Rev. D **14**, 870 (1976). doi:10.1103/PhysRevD.14.870
3. Hodgkinson, L., Louko, J.: Static, stationary and inertial Unruh-DeWitt detectors on the BTZ black hole. Phys. Rev. D **86**, 064031 (2012). doi:10.1103/PhysRevD.86.064031
4. Fewster, C.: A general worldline quantum inequality. Class. Quantum Grav. **17**, 1897 (2000). doi:10.1088/0264-9381/17/9/302
5. Junker, W., Schrohe, E.: Adiabatic vacuum states on general space-time manifolds: definition, construction, and physical properties. Ann. Inst. Henri Poincare A **3**, 1113 (2002). doi:10.1007/s000230200001
6. Kay, B., Wald, R.: Theorems on the uniqueness and thermal properties of stationary, nonsingular, quasifree states on space-times with a bifurcate Killing horizon. Phys. Rep. **207**, 49 (1991). doi:10.1016/0370-1573(91)90015-E
7. Louko, J., Satz, A.: Transition rate of the Unruh-DeWitt detector in curved spacetime. Class. Quantum Grav. **25**, 055012 (2008). doi:10.1088/0264-9381/25/5/055012
8. Langlois, P.: Causal particle detectors and topology. Ann. Phys. **321**, 2027 (2006). doi:10.1016/j.aop.2006.01.013
9. Louko, J., Satz, A.: How often does the Unruh-DeWitt detector click? Regularisation by a spatial profile. Class. Quantum Grav. **23**, 6321 (2006). doi:10.1088/0264-9381/23/22/015
10. Sriramkumar, L., Padmanabhan, T.: Response of finite time particle detectors in noninertial frames and curved space-time. Class. Quantum Grav. **13**, 2061 (1996). doi:10.1088/0264-9381/13/8/005
11. Schlicht, S.: Considerations on the Unruh effect: causality and regularization. Class. Quantum Grav. **21**, 4647 (2004). doi:10.1088/0264-9381/21/19/011
12. Hodgkinson, L., Louko, J.: How often does the Unruh-DeWitt detector click beyond four dimensions? J. Math. Phys. **53**, 082301 (2012). doi:10.1063/1.4739453
13. Satz, A.: Then again, how often does the Unruh-DeWitt detector click if we switch it carefully? Class. Quantum Grav. **24**, 1719 (2007). doi:10.1088/0264-9381/24/7/003
14. Decanini, Y., Folacci, A.: Off-diagonal coefficients of the Dewitt-Schwinger and Hadamard representations of the Feynman propagator. Phys. Rev. D **73**, 044027 (2006). doi:10.1103/PhysRevD.73.044027
15. Bañados, M., Henneaux, M., Teitelboim, C., Zanelli, J.: Geometry of the 2+1 black hole. Phys. Rev. D **48**, 1506 (1993). doi:10.1103/PhysRevD.48.1506
16. Bañados, M., Teitelboim, C., Zanelli, J.: Black hole in three-dimensional spacetime. Phys. Rev. Lett. **69**, 1849 (1992). doi:10.1103/PhysRevLett.69.1849
17. Carlip, S.: The (2+1)-dimensional black hole. Class. Quantum Grav. **12**, 2853 (1995). doi:10.1088/0264-9381/12/12/005
18. Kubo, R.: Statistical-mechanical theory of irreversible processes. I. General theory and simple applications in magnetic and conduction problems. J. Phys. Soc. Jpn. **12**, 570 (1957). doi:10.1143/JPSJ.12.570
19. Martin, P., Schwinger, J.: Theory of many-particle systems. 1. Phys. Rev. **115**, 1342 (1959). doi:10.1103/PhysRev.115.1342

20. Hartle, J., Hawking, S.: Path integral derivation of black hole radiance. Phys. Rev. D **13**, 2188 (1976). doi:10.1103/PhysRevD.13.2188
21. Israel, W.: Thermo-field dynamics of black holes. Phys. Lett. A **57**, 107 (1976). doi:10.1016/0375-9601(76)90178-X
22. Deser, S., Levin, O.: Accelerated detectors and temperature in (anti-) de Sitter spaces. Class. Quantum Grav. **14**, L163 (1997). doi:10.1088/0264-9381/14/9/003
23. Deser, S., Levin, O.: Equivalence of Hawking and Unruh temperatures through flat space embeddings. Class. Quantum Grav. **15**, L85 (1998). doi:10.1088/0264-9381/15/12/002
24. Deser, S., Levin, O.: Mapping Hawking into Unruh thermal properties. Phys. Rev. D **59**, 064004 (1999). doi:10.1103/PhysRevD.59.064004
25. Russo, J., Townsend, P.: Accelerating branes and brane temperature. Class. Quantum Grav. **25**, 175017 (2008). doi:10.1088/0264-9381/25/17/175017

On the Observability of Quantum-Gravitational Effects in the Cosmic Microwave Background

Claus Kiefer and Manuel Krämer

Abstract In any approach to quantum gravity, it is crucial to look for observational effects in order to discriminate between different approaches. Here, we discuss how quantum-gravitational contributions to the anisotropy spectrum of the cosmic microwave background arise in the framework of canonical quantum gravity using the Wheeler–DeWitt equation. From the present non-observation of these contributions, we find a constraint on the Hubble parameter of inflation.

1 Introduction

One of today's most significant tasks in theoretical physics is to find the correct quantum theory of gravity. We have several approaches to such a theory at hand; however, there has not yet been a definite prediction which is testable with today's level of precision by experiment or observation. The reason for this is that quantum effects of gravity should only become sizable in situations where large curvature and very high energies approaching the Planck scale are involved. This effectively makes black hole physics and very early universe cosmology the two main applications for a theory of quantum gravity.

Here, we want to focus on cosmology, and in particular on the Cosmic Microwave Background (CMB), which has opened a new era of precision cosmology ever since its anisotropies have been detected. The power spectrum of these anisotropies has turned out to be a rich source of information about the very early universe and it is therefore a very suitable candidate to look for quantum-gravitational effects.

C. Kiefer (✉) · M. Krämer
Institute for Theoretical Physics, University of Cologne, Zülpicher Str. 77, 50937 Köln, Germany
e-mail: kiefer@thp.uni-koeln.de

M. Krämer
e-mail: mk@thp.uni-koeln.de

J. Bičák and T. Ledvinka (eds.), *Relativity and Gravitation*,
Springer Proceedings in Physics 157, DOI: 10.1007/978-3-319-06761-2_74,

We have chosen to use quantum geometrodynamics as our framework, a direct canonical quantization of gravity. It is unlikely that quantum geometrodynamics turns out to be the ultimate answer to the problem of quantum gravity; however, it should be able to be used at least as an effective theory, since it leads to Einstein's equations in the semiclassical limit, see e.g. [1].

Our aim is to calculate the dominant quantum-gravitational contribution for the primordial power spectrum of cosmological perturbations, which arises from a semi-classical approximation to the Wheeler–DeWitt equation of quantum cosmology.

This conference contribution is based on our papers [2] and [3].

2 The Quantum-Cosmological Model

In order to give a first estimate of how sizable quantum gravity effects for the CMB can be, we choose the simplest model, an inflationary universe with perturbations of only the scalar field ϕ, which plays the role of the inflaton. The background universe is a flat Friedmann–Lemaître universe with a scale factor $a \equiv \exp(\alpha)$. Furthermore, we assume that the slow-roll approximation holds in the form of $\dot{\phi}^2 \ll |\mathscr{V}(\phi)|$, where $\mathscr{V}(\phi)$ is the inflaton potential, which we choose to be $\mathscr{V}(\phi) = \frac{1}{2} m^2\phi^2 \approx$ const. for definiteness.

After setting $\hbar = c = 1$, redefining the Planck mass as $m_{\mathrm{P}} = \sqrt{3\pi/2G} \approx 2.65 \times 10^{19}\,\mathrm{GeV}$ and rescaling the scalar field $\phi \to \phi/\sqrt{2}\pi$, one arrives at the following Wheeler–DeWitt equation in minisuperspace:

$$\mathscr{H}_0\Psi_0(\alpha,\phi) \equiv \frac{\mathrm{e}^{-3\alpha}}{2}\left[\frac{1}{m_{\mathrm{P}}^2}\frac{\partial^2}{\partial\alpha^2} - \frac{\partial^2}{\partial\phi^2} + \mathrm{e}^{6\alpha}m^2\phi^2\right]\Psi_0(\alpha,\phi) = 0\,. \tag{1}$$

Furthermore, we make the assumption that the kinetic term of the ϕ-field can be neglected, as it is small compared to the potential term $\partial^2\Psi_0/\partial\phi^2 \ll \mathrm{e}^{6\alpha}m^2\phi^2\Psi_0$. This allows us to substitute $m\phi$ by $m_{\mathrm{P}}H$, where H denotes the quasi-static Hubble parameter during inflation, and our Wheeler–DeWitt equation for the background becomes

$$\mathscr{H}_0\Psi_0(\alpha) \equiv \frac{\mathrm{e}^{-3\alpha}}{2}\left[\frac{1}{m_{\mathrm{P}}^2}\frac{\partial^2}{\partial\alpha^2} + \mathrm{e}^{6\alpha}m_{\mathrm{P}}^2H^2\right]\Psi_0(\alpha) = 0\,. \tag{2}$$

We include inhomogeneities by adding perturbations to the homogeneous background inflaton field $\phi \to \phi(t) + \delta\phi(\mathbf{x},t)$ and decompose them into Fourier modes, where we assume for simplicity that the space is compact and the spectrum for the wave vector $\mathbf{k}$, $k \equiv |\mathbf{k}|$, discrete: $\delta\phi(\mathbf{x},t) = \sum_k f_k(t)\,\mathrm{e}^{\mathrm{i}\mathbf{k}\cdot\mathbf{x}}$. Note that we use units in which k is a dimensionless quantity. For each of the modes we have a Hamiltonian

$$\mathscr{H}_k = \frac{1}{2}\,\mathrm{e}^{-3\alpha}\left[-\frac{\partial^2}{\partial f_k^2} + \left(k^2\mathrm{e}^{4\alpha} + m^2\mathrm{e}^{6\alpha}\right)f_k^2\right],$$

such that the Wheeler–DeWitt equation that includes the scalar field inhomogeneities reads [4]

$$\left[\mathcal{H}_0 + \sum_{k=1}^{\infty} \mathcal{H}_k\right] \Psi\left(\alpha, \{f_k\}_{k=1}^{\infty}\right) = 0 .$$

Due to the smallness of the fluctuations, one can neglect self-interactions of the respective modes and therefore make a product ansatz for the full wave function including the fluctuation modes: $\Psi\left(\alpha, \{f_k\}_{k=1}^{\infty}\right) = \Psi_0(\alpha) \prod_{k=1}^{\infty} \widetilde{\Psi}_k(\alpha, f_k)$. This ansatz allows us to write out a Wheeler–DeWitt equation for each fluctuation mode $\Psi_k(\alpha, f_k) := \Psi_0(\alpha)\widetilde{\Psi}_k(\alpha, f_k)$, which takes the form:

$$\frac{1}{2}\, \mathrm{e}^{-3\alpha}\left[\frac{1}{m_{\mathrm{P}}^2}\frac{\partial^2}{\partial\alpha^2} + \mathrm{e}^{6\alpha} m_{\mathrm{P}}^2 H^2 - \frac{\partial^2}{\partial f_k^2} + W_k(\alpha) f_k^2\right]\Psi_k(\alpha, f_k) = 0 ,$$

where we have defined $W_k(\alpha) := k^2 \mathrm{e}^{4\alpha} + m^2 \mathrm{e}^{6\alpha}$.

3 The Semiclassical Approximation

We are interested in finding quantum-gravitational correction terms to the standard expressions used to calculate the power spectrum of quantum fluctuations in an inflationary universe. Hence, it suffices to solve the Wheeler–DeWitt equation (3) by performing a Born–Oppenheimer type of approximation, as it was presented for the full Wheeler–DeWitt equation in [5]. The Born–Oppenheimer approximation is widely used in molecular physics, where one can separate the degrees of freedom of the molecules into slow ones (the nuclei) and fast ones (the electrons). In our quantum-cosmological setting, the slow variable is the scale factor, while the fast ones are the fluctuations f_k.

We implement the Born–Oppenheimer approximation by making the ansatz $\Psi_k(\alpha, f_k) = \mathrm{e}^{\mathrm{i}\, S(\alpha, f_k)}$ and expanding S in terms of powers of m_{P}^2: $S(\alpha, f_k) = m_{\mathrm{P}}^2\, S_0 + m_{\mathrm{P}}^0\, S_1 + m_{\mathrm{P}}^{-2}\, S_2 + \ldots$. Inserting this ansatz into equation (3) and comparing terms of equal power of m_{P}, one obtains that at order $\mathcal{O}(m_{\mathrm{P}}^2)$ S_0 obeys the classical Hamilton–Jacobi equation

$$\left[\frac{\partial S_0}{\partial \alpha}\right]^2 - \mathrm{e}^{6\alpha} H^2 = 0 , \tag{3}$$

which describes the classical minisuperspace background on which the quantum fluctuations propagate. At the next order $\mathcal{O}(m_{\mathrm{P}}^0)$, we define $\psi_k^{(0)}(\alpha, f_k) \equiv \gamma(\alpha)\, \mathrm{e}^{\mathrm{i}\, S_1(\alpha, f_k)}$ and impose a condition in order to make $\gamma(\alpha)$ equal to the standard WKB prefactor. At this point, we can introduce a time parameter t that arises from the approximate background defined by the Hamilton–Jacobi equation (3), using the definition

$$\frac{\partial}{\partial t} := -\,\mathrm{e}^{-3\alpha}\frac{\partial S_0}{\partial \alpha}\frac{\partial}{\partial \alpha}\,. \tag{4}$$

This limit can be compared with the limit of geometric optics that arises from wave optics. In that case, light rays result as an approximate concept from the eikonal equation. In our case, an approximate spacetime emerges, and one has an approximate time t at one's disposal.

Consequently, we find that each $\psi_k^{(0)}(\alpha, f_k)$ obeys a Schrödinger equation with respect to t: $\mathrm{i}\,\frac{\partial}{\partial t}\,\psi_k^{(0)} = \mathcal{H}_k\psi_k^{(0)}$.

Hence, the order $\mathcal{O}(m_\mathrm{P}^0)$ corresponds to the limit of quantum theory in an external background. It is at this order where we will obtain the standard results for quantum fluctuations in an inflationary universe.

But before that, we will take the semiclassical approximation one step further, to the order $\mathcal{O}(m_\mathrm{P}^{-2})$, where we use a decomposition of $S_2(\alpha, f_k)$ as follows: $S_2(\alpha, f_k) \equiv \varsigma(\alpha) + \eta(\alpha, f_k)$. After demanding that $\varsigma(\alpha)$ be the standard second-order WKB correction, we find that the wave functions $\psi_k^{(1)}(\alpha, f_k) := \psi_k^{(0)}(\alpha, f_k)\,\mathrm{e}^{\mathrm{i}\,m_\mathrm{P}^{-2}\,\eta(\alpha, f_k)}$ obey a quantum-gravitationally corrected Schrödinger equation of the form

$$\mathrm{i}\,\frac{\partial}{\partial t}\,\psi_k^{(1)} = \mathcal{H}_k\psi_k^{(1)} - \frac{\mathrm{e}^{3\alpha}}{2m_\mathrm{P}^2\psi_k^{(0)}}\left[\frac{\left(\mathcal{H}_k\right)^2}{V}\psi_k^{(0)} + \mathrm{i}\,\frac{\partial}{\partial t}\left(\frac{\mathcal{H}_k}{V}\right)\psi_k^{(0)}\right]\psi_k^{(1)}\,, \tag{5}$$

where $V := \mathrm{e}^{6\alpha}H^2$. The first term in this equation gives the dominant contribution, while the second one corresponds to a small violation of unitarity with respect to the standard inner $\mathcal{L}^2$-product for the modes f_k. Since it is usually negligible with respect to the first term, we will neglect the unitarity-violation term in the following.

4 Calculation of the Power Spectrum

In order to calculate the power spectrum of the scalar field fluctuations, we have to solve the uncorrected Schrödinger equation. We express α in terms of t and use the Gaussian ansatz $\psi_k^{(0)}(t, f_k) = \mathcal{N}_k^{(0)}(t)\,\mathrm{e}^{-\frac{1}{2}\,\Omega_k^{(0)}(t)\,f_k^2}$.

This leads to the following system of equations:

$$\dot{\mathcal{N}}_k^{(0)}(t) = -\frac{\mathrm{i}}{2}\,\mathrm{e}^{-3\alpha}\,\mathcal{N}_k^{(0)}(t)\,\Omega_k^{(0)}(t)\,, \tag{6}$$

$$\dot{\Omega}_k^{(0)}(t) = \mathrm{i}\,\mathrm{e}^{-3\alpha}\Big[-\big(\Omega_k^{(0)}(t)\big)^2 + W_k(t)\Big]. \tag{7}$$

Equation (7) has the following solution

$$\Omega_k^{(0)}(t) = \frac{k^2 a^2}{k^2 + H^2 a^2}\,(k + \mathrm{i}\,Ha) + \mathcal{O}\left(\frac{m^2}{H^2}\right), \tag{8}$$

while (6) together with the normalization of the states yields the solution $|\mathcal{N}_k^{(0)}(t)|^2 = (\Re\mathfrak{e}\,\Omega_k^{(0)}(t)/\pi)^{1/2}$. In order to use equation (7) to calculate the power spectrum, we use the definition of the density contrast in the slow-roll regime given by

$$\delta_k(t) \approx \frac{\delta\rho_k(t)}{\mathcal{V}_0} = \frac{\dot{\phi}(t)\,\dot{\sigma}_k(t)}{\mathcal{V}_0}\,.$$

Here, $\mathcal{V}_0$ represents the scalar-field potential evaluated at the background solution $\phi(t)$, and $\sigma_k(t)$ denotes the classical quantity related to the quantum-mechanical variable $f_k(t)$. We implement this relation by taking the expectation value of f_k with respect to a Gaussian state:

$$\sigma_k^2(t) := \left\langle \psi_k | f_k^2 | \psi_k \right\rangle = \sqrt{\frac{\Re\mathfrak{e}\,\Omega_k}{\pi}} \int_{-\infty}^{\infty} f_k^2\, \mathrm{e}^{-\frac{1}{2}[\Omega_k^*(t)+\Omega_k(t)]f_k^2}\,\mathrm{d}f_k = \frac{1}{2\,\Re\mathfrak{e}\,\Omega_k(t)}\,.$$

The density contrast is then evaluated at the time t_{enter}, when the corresponding mode re-enters the Hubble radius during the radiation-dominated phase. By using a standard relation,

$$\delta_k(t_{\mathrm{enter}}) = \frac{4}{3}\frac{\mathcal{V}_0}{\dot{\phi}^2}\,\delta_k(t_{\mathrm{exit}}) = \frac{4}{3}\frac{\dot{\sigma}_k(t)}{\dot{\phi}(t)}\bigg|_{t=t_{\mathrm{exit}}},$$

we can relate t_{enter} to the time t_{exit}, when the mode exits the Hubble radius during the inflationary phase.

We therefore evaluate $\dot{\sigma}_k^{(0)}(t)$ at the Hubble-scale crossing t_{exit}. Using $\xi(t_{\mathrm{exit}}) = 2\pi$, we arrive at $\left|\dot{\sigma}_k^{(0)}(t)\right|_{t=t_{\mathrm{exit}}} \propto H^2\,k^{-3/2}$. Since the power spectrum is defined as $\Delta^2_{(0)}(k) := 4\pi k^3\,|\delta_k(t_{\mathrm{enter}})|^2 \propto H^4\,\left|\dot{\phi}(t)\right|^{-2}_{t_{\mathrm{exit}}}$, we immediately see that we obtain a scale-invariant power spectrum, which is the standard result for the simplest models of inflation.

5 The Quantum-Gravitationally Corrected Power Spectrum

In order to calculate the quantum-gravitational correction to the power spectrum we determined above, we have to look for an approximate solution to equation (5), where we ignore the unitarity-violating term as mentioned. We assume that we can accommodate the correction by the following modified Gaussian ansatz

$$\psi_k^{(1)}(t, f_k) = \left(\mathcal{N}_k^{(0)}(t) + \frac{1}{m_\mathrm{P}^2}\mathcal{N}_k^{(1)}(t)\right)\exp\left[-\frac{1}{2}\left(\Omega_k^{(0)}(t) + \frac{1}{m_\mathrm{P}^2}\Omega_k^{(1)}(t)\right)f_k^2\right].$$

Inserting this ansatz into equation (5) leads to the following equation for the correction term $\Omega_k^{(1)}(t)$:

$$\dot{\Omega}_k^{(1)}(t) \approx -2\,\mathrm{i}\,\mathrm{e}^{-3\alpha}\,\Omega_k^{(0)}(t)\left(\Omega_k^{(1)}(t) - \frac{3}{4V(t)}\left[\left(\Omega_k^{(0)}(t)\right)^2 - W_k(t)\right]\right). \tag{9}$$

We assume that this correction vanishes for late times, $\Omega_k^{(1)}(t) \to 0$ as $t \to \infty$, and can then solve equation (9) by the method of variation of constants, which reduces the problem to a numerical integration.

The relevant quantum-gravitationally corrected quantity for determining the power spectrum is given by

$$\left|\dot{\sigma}_k^{(1)}(t)\right| = \left|\frac{1}{\sqrt{2}}\frac{\mathrm{d}}{\mathrm{d}t}\left[\left(\mathfrak{Re}\,\Omega_k^{(0)}(\xi) + \frac{1}{m_\mathrm{P}^2}\,\mathfrak{Re}\,\Omega_k^{(1)}(\xi)\right)^{-\frac{1}{2}}\right]\right| \tag{10}$$

and we can incorporate the quantum-gravitational correction into a correction term C_k relating the uncorrected quantity $\dot{\sigma}_k^{(0)}$ to the corrected one $\dot{\sigma}_k^{(1)}$ in the following way $\left|\dot{\sigma}_k^{(1)}\right|_{t_\mathrm{exit}} \simeq |C_k|\left|\dot{\sigma}_k^{(0)}\right|_{t_\mathrm{exit}}$.
The correction term can then be numerically calculated

$$C_k := \left(1 - 43.56\,\frac{1}{k^3}\frac{H^2}{m_\mathrm{P}^2}\right)^{-\frac{3}{2}}\left(1 - 189.18\,\frac{1}{k^3}\frac{H^2}{m_\mathrm{P}^2}\right), \tag{11}$$

which allows us to immediately determine the quantum-gravitationally corrected power spectrum $\Delta^2_{(1)}(k) = \Delta^2_{(0)}(k)\,C_k^2$. Performing a Taylor expansion of C_k with respect to $(H/m_\mathrm{P})^2$ leads to

$$\Delta^2_{(1)}(k) \simeq \Delta^2_{(0)}(k)\left[1 - 123.83\,\frac{1}{k^3}\frac{H^2}{m_\mathrm{P}^2} + \frac{1}{k^6}\,\mathcal{O}\left(\frac{H^4}{m_\mathrm{P}^4}\right)\right]^2. \tag{12}$$

We therefore see that the quantum-gravitational correction explicitly breaks the scale independence of the uncorrected power spectrum and leads to a *suppression of power at large scales* (small k). However, our approximation breaks down when the zero point is approached and one would have to take into account higher orders of $(H/m_\mathrm{P})^2$ to suitably describe this limit.

From equation (12), one also sees that the quantum-gravitational effect only becomes significant if the inflationary Hubble parameter H approaches the Planck scale. From the observational bound of the scalar-to-tensor ratio we can deduce an

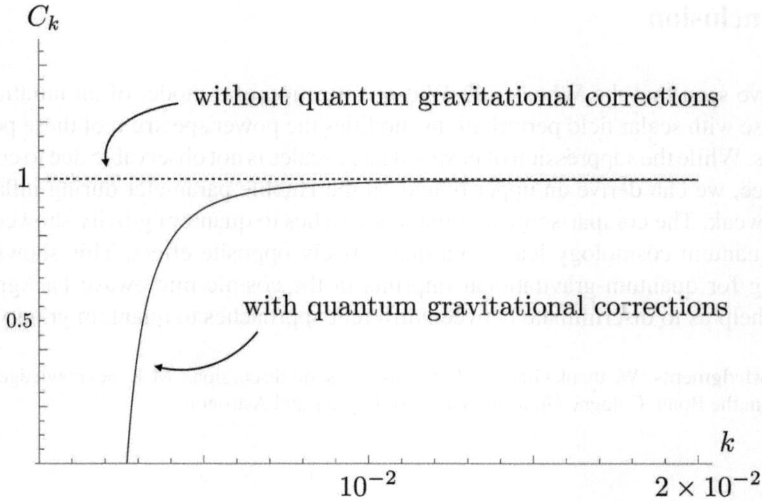

Fig. 1 The function C_k for $H = 10^{14}$ GeV, from [3]

upper bound on H, $H \lesssim 10^{-5} m_{\mathrm{P}} \sim 10^{14}$ GeV. Figure 1 shows the correction term C_k for this value of H. The corrected power spectrum in this limiting case takes the following form:

$$\Delta^2_{(1)}(k) \simeq \Delta^2_{(0)}(k) \left[1 - 1.76 \times 10^{-9} \frac{1}{k^3} + \frac{1}{k^6} \mathcal{O}\left(10^{-15}\right) \right]^2 .$$

We thus see that even in this limiting case the quantum-gravitational effect is extremely small and if one adds that at large scales measurement accuracy is fundamentally limited by cosmic variance, we have to conclude that one will not be able to see this effect even with future, more precise measurements of the CMB anisotropies by satellite missions like PLANCK. A more elaborate discussion of the observable bounds from this calculation can be found in [6].

However, one can use our analysis to derive an upper limit on the Hubble parameter independently of the observational bound based on the tensor-to-scalar ratio. Given the fact that one has not yet unambiguously observed an effect as derived here in the CMB anisotropy spectrum, one can assume for a rough estimate that C_k^2 has to be not less than 0.95 for $k \sim 1$ since one has observed that the power spectrum deviates by less than 5 % from a scale-invariant spectrum [7]. In order to fulfill this condition, one finds that $H \lesssim 1.4 \times 10^{-2}\, m_{\mathrm{P}} \sim 4 \times 10^{17}$ GeV, which is, however, weaker than the bound from the tensor-to-scalar ratio.

Other approaches to quantum gravity also lead to effects in the CMB anisotropy spectrum. While non-commutative geometry and string theory give a similar suppression of power on the largest scales [8–10], loop quantum cosmology predicts an enhancement [11, 12].

6 Conclusion

We have seen that the Wheeler–DeWitt quantization of a model of an inflationary universe with scalar field perturbations modifies the power spectrum of these perturbations. While the suppression of power at large scales is not observable due to cosmic variance, we can derive an upper bound on the Hubble parameter during inflation, albeit weak. The comparison with other approaches to quantum gravity showed that loop quantum cosmology leads to a qualitatively opposite effect. This shows that looking for quantum-gravitational imprints in the cosmic microwave background could help us to discriminate between different approaches to quantum gravity.

Acknowledgments We thank Gianluca Calcagni for useful discussions. M. K. acknowledges support from the Bonn–Cologne Graduate School of Physics and Astronomy.

References

1. Kiefer, C.: Quantum Gravity, International Series of Monographs on Physics, vol. 155, 3rd edn. Oxford University Press, Oxford, New York (2012)
2. Kiefer, C., Krämer, M.: Quantum gravitational contributions to the cosmic microwave background anisotropy spectrum. Phys. Rev. Lett. **108**, 021301 (2012). doi:10.1103/PhysRevLett.108.021301
3. Kiefer, C., Krämer, M.: Can effects of quantum gravity be observed in the cosmic microwave background? Int. J. Mod. Phys. D **21**, 1241001 (2012). doi:10.1142/S0218271812410015
4. Halliwell, J., Hawking, S.: The origin of structure in the Universe. Phys. Rev. D **31**, 1777 (1985). doi:10.1103/PhysRevD.31.1777
5. Kiefer, C., Singh, T.: Quantum gravitational corrections to the functional Schrödinger equation. Phys. Rev. D **44**, 1067 (1991). doi:10.1103/PhysRevD.44.1067
6. Calcagni, G.: Observational effects from quantum cosmology. ArXiv e-prints 1209.0473 [gr-qc] (2012)
7. Komatsu, E., Smith, K., Dunkley, J., et al.: Seven-year Wilkinson microwave anisotropy probe (WMAP) observations: cosmological interpretation. Astrophys. J. Suppl. Ser. **192**, 18 (2011). doi:10.1088/0067-0049/192/2/18
8. Tsujikawa, S., Maartens, R., Brandenberger, R.: Non-commutative inflation and the CMB. Phys. Lett. B **574**, 141 (2003). doi:10.1016/j.physletb.2003.09.022
9. Piao, Y., Tsujikawa, S., Zhang, X.: Inflation in string-inspired cosmology and suppression of CMB low multipoles. Class. Quantum Grav. **21**, 4455 (2004). doi:10.1088/0264-9381/21/18/011
10. Calcagni, G., Tsujikawa, S.: Observational constraints on patch inflation in noncommutative spacetime. Phys. Rev. D **70**, 103514 (2004). doi:10.1103/PhysRevD.70.103514
11. Bojowald, M., Calcagni, G., Tsujikawa, S.: Observational constraints on loop quantum cosmology. Phys. Rev. Lett. **107**, 211302 (2011). doi:10.1103/PhysRevLett.107.211302
12. Bojowald, M., Calcagni, G., Tsujikawa, S.: Observational test of inflation in loop quantum cosmology. J. Cosmol. Astropart. Phys. **2011**(11), 046 (2011). doi:10.1088/1475-7516/2011/11/046

Quantum Singularities in Conformally Static Spacetimes

Deborah A. Konkowski and Thomas M. Helliwell

Abstract After a brief review of the standard definition and analysis of classical singularities in general relativistic spacetimes, and of quantum singularities in static spacetimes with timelike classical singularities, an extension of quantum singularities to conformally static spacetimes is summarized and applied to two test cases. The timelike classical singularities in a Friedmann-Robertson-Walker (FRW) universe with a cosmic string, and in Roberts spacetime, are shown to be quantum mechanically singular when tested by either minimally coupled or conformally coupled scalar waves. In the Roberts case, however, non-minimally coupled scalar waves with a coupling constant $\xi \geq 2$ do not detect the classical singularity.

1 Introduction

We study quantum wave packet propagation in conformally static spacetimes with timelike classical singularities. If the wave propagation turns out to be well defined, the spacetimes are said to be quantum mechanically non-singular.

The order of the paper is as follows: First, classical and quantum singularities are defined with the latter restricted (as usual) to static spacetimes with timelike singularities. Next, the definition of quantum singularity is extended to conformally static spacetimes with a timelike singularity (spacelike singularities, if present, are not tested). In particular, two spacetimes are tested with generally coupled scalar waves: a Friedmann-Robertson-Walker (FRW) spacetime with a cosmic string and the Roberts spacetime. Finally, conclusions are given, together with ideas for further research.

D. A. Konkowski (✉)
Mathematics Department, U.S. Naval Academy, Annapolis, MD 21402, USA
e-mail: dak@usna.edu

T. M. Helliwell
Physics Department, Harvey Mudd College, Claremont, CA 91711, USA
e-mail: helliwell@hmc.edu

J. Bičák and T. Ledvinka (eds.), *Relativity and Gravitation*,
Springer Proceedings in Physics 157, DOI: 10.1007/978-3-319-06761-2_75,

2 Classical Singularities

A spacetime (M, g) is taken to be a paracompact, C^∞, connected, Hausdorff manifold M with a Lorentzian metric g [1]. So what is a classical *singularity*? A spacetime is by definition smooth, so 'singular' points are not part of the spacetime; they must be cut out of the spacetime manifold. This leaves a 'hole', with incomplete curves, a seeming boundary to spacetime. How do we complete spacetime, and how do we define a boundary ∂M to spacetime? There have been a number of attempts, none of them entirely satisfactory. Note that Cauchy completeness works only in Riemannian metrics, not Lorentzian. Boundary definitions have included the a(abstract)-boundary of Scott and Szekeres [2], the b(bundle)-boundary of Schmidt [3], the c(causal)-boundary of Geroch, Kronheimer, and Penrose [4] and the g(geodesic)-boundary of Geroch [5]. In this discussion we will use Geroch's 1968 description of a classical singularity. He states that "a singularity is indicated by incomplete geodesics or incomplete curves of bounded acceleration in a maximal spacetime." This is closest to the definition of classical singularity used in the famous singularity theorems of Hawking and Penrose, which predict that singularities are ubiquitous in exact solutions of Einstein's equations (see, e.g., [6]).

Ellis and Schmidt have classified singular points into three types according to their strength [1]: quasi-regular (mild, topological singularities), non-scalar curvature (diverging tidal forces on curves ending at the singularity; finite tidal forces on some nearby curves) and scalar curvature (diverging scalars—usually one considers only C^0 scalar polynomial invariants). Conical singularities, as in idealized cosmic strings, are a good example of quasiregular singularities. The other two types of singularities are stronger, curvature singularities. Nonscalar curvature singularities include those in whimper cosmologies and certain plane-wave spacetimes, whereas scalar curvature singularities are the best-known, occurring at the centers of black holes or the beginning of big bang cosmologies.

2.1 Quantum Singularities

What happens if instead of classical particle paths (e.g., null and timelike geodesics) one uses quantum mechanical particles (quantum wave packets) to identify singularities? Following pioneering work by Wald [7], Horowitz and Marolf answered this question for static spacetimes with timelike classical singularities. In their 1995 paper they posit that a spacetime is quantum mechanically (QM) *non*singular if the evolution of a test scalar wave packet, representing a quantum particle, is uniquely determined by the initial wave packet, the manifold and the metric, without having to place boundary conditions at the classical singularity. Technically, a static spacetime is QM-singular if the spatial portion of the Klein-Gordon operator is not essentially self-adjoint on $C_o^\infty(\Sigma)$ in the space of square integrable functions $L^2(\Sigma)$, where Σ is a spatial hypersurface.

The term "essentially self adjoint" arises in functional analysis [8]. An operator A is called self-adjoint if (i) $A = A^*$ and (ii) $Dom(A) = Dom(A^*)$, where A^* is the adjoint of A and Dom is short for domain. An operator is *essentially* self-adjoint if (i) is met and (ii) can be met by expanding the domain of the operator A or its adjoint A^* so that it is true.

There are two basic tests for essential self-adjointness [8]. The first uses the von Neumann criterion of deficiency indices [9]; one studies solutions of $A\Psi = \pm i\Psi$, where A is the spatial portion of the Klein-Gordon operator, and finds the number of solutions for each sign that are self-adjoint. The second technique uses the so-called Weyl limit point—limit circle criterion [10], which relates essential self-adjointness of the Hamiltonian operator to the behavior of the "potential" in an effective one-dimensional Schrödinger equation, which in turn determines the behavior of the scalar wave packet. Relevant theorems that simplify the analysis can be found in Reed and Simon [8].

Many authors have used the definition of quantum singularity to study the singularity structure of spacetimes. For a summary, see, for example, the review article by Pitelli and Letelier [11] or the conference proceeding by the authors [12] and the references therein. Also, there is the alternative concept of 'wave regularity' introduced by Ishibashi and Hosoya [13], which is relevant to the discussion. It uses a non-standard Hilbert space, H^1, the first Sobolev space.

3 Conformally Static Spacetimes

A spacetime $g_{\mu\nu}(x^\alpha)$ that is conformally static is related to a static spacetime $\bar{g}_{\mu\nu}(x^a)$ by a conformal transformation $C(\eta)$ of the metric. Here $C(\eta)$ is the conformal factor, where η is the conformal time, related to the time t by $dt = Cd\eta$. Simply put, $g_{\mu\nu}(x^\alpha) = C^2(\eta)\bar{g}_{\mu\nu}(x^a)$. Here Greek letters $\alpha, \beta, \ldots$ label spacetime indices and have the range over 0, 1, 2, 3, and Latin letters $a, b, c, \ldots$ label spatial indicies that range over 1, 2, 3.

The Lagrangian density for a generally coupled scalar field is [14],

$$\mathcal{L} = 1/2(-g)^{1/2}[g^{\mu\nu}\Phi,_\mu \Phi,_\nu - (M^2 + \xi R)\Phi^2], \tag{1}$$

where M is the mass if the scalar particle, R is the scalar curvature, and ξ is the coupling (in particular, $\xi = 0$ for minimal coupling and $\xi = 1/6$ for conformal coupling). Varying the action $S = \int \mathcal{L}\, d^4x$ gives the Klein-Gordon field equation,

$$|g|^{-1/2}\left(|g|^{1/2}g^{\mu\nu}\Phi,_\nu\right),_\mu - \xi R\Phi = M^2\Phi. \tag{2}$$

In the massless case with conformal coupling, the field equation above is conformally invariant under a conformal transformation of the metric and field; in this case the inner product respecting the stress tensor for the field is also conformally invariant.

This led Ishibashi and Hosoya to state [13], in the case of wave regularity, that "the calculation is as simple as that in the static case when singularities in conformally static space-times are probed with conformally coupled scalar fields."

Here we study the quantum particle propagation in spacetimes with massive scalar particles described by the Klein-Gordon equation and the limit point—limit circle criterion of Weyl [8, 10]. In particular, after separating variables we study the radial equation in a one-dimensional Schrödinger form with a 'potential' and determine the number of solutions that are square integrable. If we obtain a unique solution, without placing boundary conditions at the location of the classical singularity, we can say that the solution to the full Klein-Gordon equation is quantum mechanically (QM) nonsingular. The results depend on the spacetime metric parameters and wave equation modes.

After separating variables we take the spatial portion to be an operator equation on a Hilbert space $L^2(\Sigma)$ with inner product (see, e.g., [15]),

$$(\chi, \zeta) = \int d^3x |\bar{g}_3/g_{00}|^{1/2} \chi(x^a)\zeta(x^b), \tag{3}$$

where $\bar{g}_3$ is the determinant of the spatial portion of the static metric, χ and ζ are spatial mode solutions and a, b range over 1, 2, 3. Then we consider the radial portion alone, change variables and write the radial equation in one-dimensional Schrodinger form, $Hu(x) = Eu(x)$, where the operator $H = -d^2/dx^2 + V(x)$ and E is a constant, with the singularity at $x = 0$. The inner product here is simply $\int dx |u(x)|^2$ and the Hilbert space is $L^2(0, \infty)$. At this point one can simply apply the limit point—limit circle criterion as easily as in the static case in order to determine the quantum singularity structure.

3.1 FRW with a Cosmic String

A simple metric modeling a Friedmann-Robertson-Walker cosmology with a cosmic string [16] is given by

$$ds^2 = a^2(t)(-dt^2 + dr^2 + \beta^2 r^2 d\phi^2 + dz^2) \tag{4}$$

where $\beta = 1 - 4\mu$ and μ is the mass per unit length of the cosmic string. This metric is conformally static (actually conformally flat). Classically it has a scalar curvature singularity when $a(t)$ is zero and a quasiregular singularity when $\beta^2 \neq 1$. Here we will consider the timelike quasiregular singularity alone. The Klein-Gordon equation with general coupling can be separated into mode solutions

$$\Phi = T(t)H(r)e^{im\phi}e^{ikz}, \tag{5}$$

where

$$\ddot{T} + 2\left(\frac{\dot{a}}{a}\right)\dot{T} + (M^2a^2 + \xi Ra^2 - q)T = 0, \tag{6}$$

and

$$H'' + \frac{1}{r}H' + (-k^2 - q - \frac{m^2}{\beta^2 r^2})H = 0. \tag{7}$$

The T-equation alone contains M and R. Rewriting the dependent and independent variables as $r = x$ and $H = xu(x)$, we get the correct inner product form and a one-dimensional Schrödinger equation,

$$u'' + (E - V(x))u = 0, \tag{8}$$

where $E = -k^2 - q$ and

$$V(x) = \frac{m^2 - \beta^2/4}{\beta^2 x^2}. \tag{9}$$

Near zero one can show that the potential $V(x)$ is limit point if $m^2/\beta^2 \geq 1$. Therefore any modes with sufficiently large m are limit point, but $m = 0$ is limit circle; thus generically this conformally static space-time is quantum mechanically singular.

3.2 Roberts Spacetime

The Roberts metric [17] is

$$ds^2 = e^{2t}(-dt^2 + dr^2 + G^2(r)d\Omega^2) \tag{10}$$

where $G^2(r) = (1/4)[1 + p - (1 - p)e^{-2r}](e^{2r} - 1)$. The spacetime is conformally static, spherically symmetric, and self-similar (see, e.g., [13]). It has a timelike classical scalar curvature singularity at $r = 0$ for $0 < p < 1$. The Klein-Gordon equation can be solved by separation of variables with mode solutions given by $\Phi = T(t)H(r)Y_{lm}(\theta, \phi)$. The radial operator can be put in one-dimensional Schrödinger form and the limit point—limit circle criterion applied. Details are given in [18]. One finds that the spacetime is quantum mechanically singular if $\xi < 2$ and quantum mechanically non-singular if $\xi \geq 2$. Therefore, the classical timelike singularity remains singular when probed by minimally coupled ($\xi = 0$) waves or by conformally coupled ($\xi = 1/6$) waves.

4 Conclusions

After a brief review of the standard definition and analysis of classical singularities in general relativistic spacetimes, and of quantum singularities in static spacetimes with timelike classical singularities, an extension of quantum singularites to conformally static spacetimes was summarized and applied to two test cases. The timelike classical singularities in a FRW universe with a cosmic string and in Roberts spacetime were shown to be quantum mechanically singular when tested by either minimally coupled or conformally coupled scalar waves. In the Roberts case, however, non-minimally coupled scalar waves with a coupling constant $\xi \geq 2$ did not detect the classical singularity.

Further analysis of the singularity structure of conformally static spacetimes is underway [18]. A class of spherically symmetric conformally static spacetimes is being analyzed; this class includes the spacetimes of HMN [19] and Fonarev [20], as well as the Roberts spacetime.

Acknowledgments One of us (DAK) thanks B. Yaptinchay for useful discussions.

References

1. Ellis, G., Schmidt, B.: Singular space-times. Gen. Relativ. Gravit. **8**, 915 (1977). doi:10.1007/BF00759240
2. Scott, S., Szekeres, P.: The abstract boundary: a new approach to singularities of manifolds. J. Geom. Phys. **13**, 223 (1994). doi:10.1016/0393-0440(94)90032-9
3. Schmidt, B.: A new definition of singular points in general relativity. Gen. Rel. Grav. **1**, 269 (1971). doi:10.1007/BF00759538
4. Geroch, R., Kronheimer, E., Penrose, R.: Ideal points in spacetime. Proc. R. Soc. Lond. Ser. A **327**, 545 (1972). doi:10.1098/rspa.1972.0062
5. Geroch, R.: Local characterization of singularities in general relativity. J. Math. Phys. **9**, 450 (1968). doi:10.1063/1.1664599
6. Hawking, S., Ellis, G.: The Large Scale Structure of Space-Time. Cambridge Monographs on Mathematical Physics. Cambridge University Press, Cambridge (1973)
7. Wald, R.: Dynamics in non-globally hyperbolic, static spacetimes. J. Math. Phys. **21**, 2802 (1980). doi:10.1063/1.524403
8. Reed, M., Simon, B.: Fourier Analysis, Self-Adjointness. Academic Press, New York, London (1972)
9. von Neumann, J.: Allgemeine Eigenwerttheorie Hermitescher Functionaloperatoren. Math. Ann. **102**, 49 (1929)
10. Weyl, H.: Über gewöhnliche Differentialgleichungen mit Singularitäten und die zugehörigen Entwicklungen willkürlicher Funktionen. Math. Ann. **68**, 220 (1910). doi:10.1007/BF01474161
11. Pitelli, J., Letelier, P.: Quantum singularities in static spacetimes. Int. J. Mod. Phys. D **20**, 729 (2011). doi:10.1142/S0218271811019062
12. Konkowski, D., Helliwell, T.: Quantum singularities in static and conformally static spacetimes. Int. J. Mod. Phys. A **26**, 3878 (2011). doi:10.1142/S0217751X11054334
13. Ishibashi, A., Hosoya, A.: Who's afraid of naked singularities? Probing timelike singularities with finite energy waves. Phys. Rev. D **60**, 104028 (1999). doi:10.1103/PhysRevD.60.104028

14. Birrell, N., Davies, P.: Quantum Fields in Curved Space. Cambridge Monographs on Mathematical Physics. Cambridge University Press, Cambridge, New York (1982)
15. Kandrup, H.: Statistical mechanics of the gravitational field in a conformally static setting. J. Math. Phys. **25**, 3286 (1984). doi:10.1063/1.526077
16. Davies, P., Sahni, V.: Quantum gravitational effects near cosmic strings. Class. Quantum Grav. **5**, 1 (1988). doi:10.1088/0264-9381/5/1/009
17. Roberts, M.: Scalar field counterexamples to cosmic censorship hypothesis. Gen. Relativ. Gravit. **21**, 907 (1989). doi:10.1007/BF00769864
18. Helliwell, T., Konkowski, D.: Quantum singularity of spherically symmetric conformally static space-times. Phys. Rev. D **87**, 104041 (2013)
19. Husain, V., Martinez, E., Nuñez, D.: Exact solution for scalar field collapse. Phys. Rev. D **50**, 3783 (1994). doi:10.1103/PhysRevD.50.3783
20. Fonarev, O.: Exact Einstein scalar field solutions for formation of black holes in a cosmological setting. Class. Quantum Grav. **12**, 1739 (1995). doi:10.1088/0264-9381/12/7/016

Granularity in Angle: Observability in Scattering Experiments

Seth A. Major and Jake C. Zappala

Abstract Geometry is quantized in loop quantum gravity. As a step toward building a detailed phenomenology of this discrete geometry a model of an atom of geometry is reviewed. The model, which preserves local Lorentz invariance, exhibits a lever arm that raises the scale at which the granularity in angle becomes apparent. The signature of this effect is a systematic shift of observed angles in processes such as high energy particle scattering experiments. To check assumptions in the model, coherent states of a simple atom of spatial geometry are explored using information intrinsic to the quantum state.

1 Introduction

If space-time or spatial geometry is fundamentally discrete, it will be observationally manifest. The kinematics of Loop Quantum Gravity (LQG) predicts discrete spectra of spatial geometric quantities such as volume, angle, and length. Before the dynamics, and the quantization, is complete we do not know whether the kinematic results extend to the physical state space [1, 2]. However, in the absence of a complete theory and even *because* the complete theory is not yet finished, it is useful to know how the predicted discreteness in spatial geometry could be manifest in observation.

Perhaps the first reaction to granular geometry is that the theory breaks Lorentz invariance. Certainly broken Lorentz symmetry leads to dramatic effects that produce strong constraints (see [3] and [4]), but broken local Lorentz invariance does not necessarily follow from discreteness—the discrete nature of quantum angular momentum does not imply the loss of rotational invariance. Furthermore, the relative

S. A. Major (✉) · J. C. Zappala
Department of Physics, Hamilton College, Clinton, NY 13323, USA
e-mail: smajor@hamilton.edu

J. C. Zappala
Department of Physics, University of Chicago, 5720 South Ellis Avenue, Chicago, IL 60637, USA

J. Bičák and T. Ledvinka (eds.), *Relativity and Gravitation*,
Springer Proceedings in Physics 157, DOI: 10.1007/978-3-319-06761-2_76,

velocity relating inertial reference frames depends on the metric and therefore the measurements of areas of a surface in different frames are not directly comparable [5]. More directly the causal set approach to quantum gravity shows that Lorentz invariance and (space-time) discreteness are compatible.

Without the blunt effects of breaking local Lorentz invariance are there other effects that might reveal the fundamental discreteness of spatial geometry? This contribution reviews a model based on the kinematic predictions of discrete spatial geometry in LQG and demonstrates that the answer is in the affirmative. The model does not break local Lorentz invariance and yet there are effects at a mesoscopic scale above the Planck scale [6]. The 'lever arm' that raises the scale is due to the underlying asymmetry of the angle operator spectrum and the combinatorics of an atom of spatial geometry. At this mesoscopic scale the local geometry differs from flat three dimensional space leading to systematic shift in the distribution of measured angles. To make such a model at this stage of development of LQG and QFT in such a context requires assumptions. These are detailed in Sect. 3.

To begin to test these assumptions we explore coherent states based on the 'hydrogen atom of spatial geometry', a single 4-valent node. Within the limitations of this model we still find a hint of a lever arm, although in this context it is purely due to the uncertainty relations among non-commuting operators.

2 Angle Operator

The angle operator was originally defined in [7] and described further in [6, 8, 9]. In this contribution we focus on the spatial atom. So, briefly, the angle operator is defined on a truncation of the full kinematic Hilbert space, at a single spin network node. The incident links to this node are partitioned into three sets, C_1, C_2, and C_3. Three left-invariant gravitational field operators L_1^i, L_2^i, and L_3^i are defined by these partitions. (The notation follows [10].) In terms of these gravitational field operators for the partitions C_1 and C_2, the quantum angle operator is

$$\hat{\theta}_{(12)} := \arccos \frac{L_1^i L_2^i}{|L_1|\,|L_2|}, \tag{1}$$

in which $|L| = \sqrt{L^2}$. The sum $\sum_{k=1}^{3} L_k^i$ vanishes due to gauge invariance and the exhaustive partitions. The partitioning of links incident to the node selects classes of preferred intertwiner bases. These are labeled by trivalent nodes where each leg of the node, or branch, connects all links in a single partition. This node is the "*intertwiner core*". On this class of bases the spectrum of the angle operator is

$$\hat{\theta}_{(12)}\; |\, j_1\; j_2\; j_3\rangle = \theta_{(12)}\; |\, j_1\; j_2\; j_3\rangle \text{ with} \tag{2}$$

$$\theta_{(12)} = \arccos\left(\frac{j_3(j_3+1) - j_1(j_1+1) - j_2(j_2+1)}{2\,[j_1(j_1+1)\, j_2(j_2+1)]^{1/2}}\right). \tag{3}$$

For more detail, see [6, 9]. With this angular spectrum there are two aspects of flat spatial geometry that are hard to model. The spectrum is sparse for small angles and the frequencies of eigenvalues is far from the $\sin\theta$ distribution of polar angles in flat, 3-dimensional continuum geometry. These are both manifestations of the asymmetry in the angular spectrum. This asymmetry persists even when the spins are very large [8, 11].

It is convenient to visualize the action of the angle operator on polyhedra, with faces dual to the incident links [12]. The areas on the dual surfaces are simply related to the spin j via $\ell_{Pl}\sqrt{j(j+1)}$. The partitioning of C_1, C_2, and C_3 induces a partition of the dual surface into three surface areas S_1, S_2, and S_3. The intertwiner core then represents a decomposition of the polyhedron with "internal faces" determined by j_i.

The notation is as follows. Twice the sum of the representations on the links incident to the node in partition C_k is denoted by the "flux" s_k also denoted **s**. In the dual surface picture this is the flux of spin through the respective surfaces, roughly equal to the face areas. The total flux is the sum of the spins on all the incident edges, denoted s. The quantities $n_k = 2j_k$, the internal areas, uniquely specify the intertwiner core, denoted $|\,\mathbf{n}\rangle$. The fluxes s_k and core labels n_k are distinct and satisfy $n_k \leq s_k$.

3 Combinatorial Phenomenology

The phenomenological model of an atom of 3-geometry is based on the state space described above and three, additional assumptions [6]: (1) The probability measure on the space of intertwiner cores is uniform. (2) All incident links to the node are spin-1/2. (3) The fluxes are large and semi-classical, $1 \ll s_k \ll s_3$, $k = 1, 2$. The last assumption is motivated by the numerical studies of [8, 11] showing that the asymmetry in the angular spectrum shifts the distribution away from the usual $\sin\theta$ distribution of polar angles. To recover the classical distribution it was necessary in these studies to take large fluxes, and, in particular $1 \ll s_j \ll s_3$, $j = 1, 2$. Fluxes **s** that satisfy these relations are called "semi-classical fluxes". This means that we omit terms $O(1/s_i)$, $O(1/n_i)$, and $O(n_i/s_i)$. The assumption (2) is for simplicity, although in statistical studies of large closed surfaces built from oriented areas the spins are about one on average [13].

All physical processes involving angle that we currently observe are on very large scales, many orders of magnitude above the Planck scale. These processes occur in a large effective volume. Since volume scales as the (total flux)$^{3/2} \equiv s^{3/2}$, the scaling defines an effective length $\ell_s = \ell_{Pl} s^{3/2}$ or energy $M_s = M_{Pl}/\sqrt{s}$ scale. The scale is set in the reference frame used in the experimental analysis, such as the CM frame.

With these assumptions, the combinatorics of the model of a spatial atom can be solved analytically [6]. The combinatorics for the number of states may be simply related to a path counting problem with a known solution [11]. For single branch of the basis $|\,\mathbf{n}\rangle$, the probability of an internal spin (or face area) n given a total flux of s is

$$p_s(n) = \frac{n}{s}\exp\left(-\frac{n^2}{2s}\right). \tag{4}$$

This is the Rayleigh distribution for a "distance" n covered in $2s$ steps in an isotropic random walk with unit step size in *two* spatial dimensions. The total probability for the internal state of the atom, $p_{\mathbf{s}}(\mathbf{n})$, is just the product of three of these combinatorial factors. For large, semi-classical spins the normalized probability distribution is simply expressed as

$$p_{\mathbf{s}}(\theta) = \sum_{\mathbf{n}} \delta(\theta - \theta(n))p_{\mathbf{s}}(\mathbf{n}). \tag{5}$$

The partition fluxes $\mathbf{s}$ determine a mixed state, $\rho_{\mathbf{s}} = \sum_{\mathbf{n}} p_{\mathbf{s}}(\mathbf{n})P_{|\mathbf{n}\rangle}$, where $P_{|\mathbf{n}\rangle}$ is the projector on the orthonormal basis of the intertwiner core. The sum is over the admissible integers $\mathbf{n}$ such that $n_i \leq s_i$. The projector is $P_{|\mathbf{n}\rangle} = | \theta_I\rangle\langle\theta_I |$ where $| \theta_I\rangle = \sum_{\mathbf{n}} c_{\theta_I}(\mathbf{n}) \mid \mathbf{n}\rangle$. The probability of finding the angle eigenvalue θ_I in the mixed state $\rho_{\mathbf{s}}$ is

$$\mathrm{Prob}(\theta = \theta_I; \rho_{\mathbf{s}}) = tr\left(\rho_{\mathbf{s}} P_{\theta_I}\right) = \sum_{\mathbf{n}} p_{\mathbf{s}}(\mathbf{n})|\langle n \mid \theta_I\rangle|^2 \equiv p_{\mathbf{s}}(\theta). \tag{6}$$

This procedure can be used to calculate $p_{\mathbf{s}}(\theta)$ for semi-classical fluxes

$$P_{\mathbf{s}}(\theta) := \int d^3n\, p_{\mathbf{s}}(\mathbf{n})|c_\theta(\mathbf{n})|^2\delta\left(\theta - \theta(n)\right). \tag{7}$$

The integration of (7) is straightforward [6]. The key step in the calculation is the identification of the "shape parameter" $\varepsilon := \sqrt{s_1 s_2}/s_3$ that measures the asymmetry in the distribution of angles. As $\varepsilon \to 0$ the continuum distribution of polar angles is recovered.

The resulting angular measure, when expressed in terms of Legendre polynomials, and to $O(\varepsilon^3)$, is [6]

$$\rho_\varepsilon(\theta) \simeq \sin\theta\left(1 - \frac{8}{\pi}P_1(\cos\theta)\varepsilon + \frac{3}{2}P_2(\cos\theta)\varepsilon^2\right). \tag{8}$$

The affect of the modified distribution of polar angles is that the "shape" of space is altered by the atom; the local angular geometry differs from flat 3-space. The total flux s determines the 3-volume of the spatial atom and thus an effective mesoscopic length scale, $\ell_s = \sqrt{s}\ell_P$, greater than the fundamental discreteness scale of ℓ_P. While the shape parameter ε is free of the Planck scale, the effective length scale, determined by the total fluxes s, is tied to the discreteness scale of the theory. For instance, a shape parameter $\varepsilon \sim 10^{-3}$ requires a total flux of at least $s \sim 10^6$, raising the length scale at which this asymmetry would be observed to be nine orders of magnitude above the Planck scale. Any angular measurement involving processes at this scale would be affected by this modified distribution. Thus the angle spectrum and the combinatorics of the intertwiner together provide a lever arm that lifts the fundamental scale of the quantum geometry up to a larger, mesoscopic scale.

4 Example: Scattering

Let's sketch how measurement of angle in an effective atom of geometry for a scattering experiment works. (More detail may be found in [6].) Bhabha scattering is convenient because the e^+e^- scattering process involves "point-like" fundamental particles. The scattering cross section depends on angle and, in a theory that encompasses the quantum state of the geometry, scattering events are measurements of the states that support the geometry. The affects of the discrete geometry will be evident at some energy, modifying the QED vertex. Short distance modifications to QED may be expressed in the Drell parameterization [14], which allows modifications to "switch on" at CM energies corresponding to the short distance structure at the scale ℓ_s. Two kinematic effects were studied in [6], one due to the averaging over angle and the other due to the modified distribution of angle in the state described above. The latter effect is dominant. Assuming that the spatial geometry is homogenous so that each scattering event occurs in the same state described in Sect. 3 then the Bhabha scattering cross section is, using the Drell parameterization [6],

$$\left(\frac{d\sigma}{d\Omega}\right) / \left(\frac{d\sigma}{d\Omega}\right)_{QED} \simeq 1 \mp \left(\frac{3s}{\Lambda_{\pm}^2}\right)\left(\frac{\sin^2\theta}{3+\cos^2\theta}\right)\left(1+\frac{8}{\pi}\cos(\theta)\varepsilon+\cdots\right). \quad (9)$$

A comparison between the model and the data, discussed in [6], shows that the shape correction reduces the observed differential cross section at small angles and increases it at large angles; the shape effect yields a systematic shift in the data.

To check the robustness of this prediction we must both check the framework and the assumptions. In the next section we report on a check of the assumptions of the model and study the most simple atom of geometry, a 4-valent node. We replace the assumption of uniform probability with minimum relative uncertainty and allow higher spin. This allows us to answer the question, Is there evidence of a lever arm in the simplest 4-valent atom of geometry?

5 Coherent States and Angle

To check the robustness of the above model we developed coherent states for the simple 4-valent node, dual to a tetrahedron. Coherent states for semi-classical geometries are based on (some flavor of) $SU(2)$ coherent states, states that are peaked around given directions, $\hat{n}$, normal to the faces. These coherent states are sharply peaked on the scalar products and thus angles, even for moderately large spin, $j > 100$.

With such low spin there is no significant lever arm. However the coherent states require classical information, the directions $\hat{n}$. To avoid adding additional assumptions about these classical directions to our model, we constructed coherent states only from quantum information intrinsic to the atom of geometry.

It has been known for some time that the state of the atom contains enough information to establish dimension of space and the scalar products. The spin geometry theorem of Penrose and Moussouris [15, 16] states that for low relative uncertainties—large spin—the state of the geometric atom yields vectors in Euclidean 3-space. Furthermore, the proof is constructive. By minimizing the relative uncertainties, the distribution of directions $\hat{n}$ is determined. We can construct states that model semi-classical geometry using only information intrinsic to the atom [17].

In this construction the directions in 3-space are well-defined when the relative uncertainties of scalar products are minimized. More precisely, when the state is "δ-classical" $\langle \Delta L_k \cdot L_l \rangle / j_k j_l < \delta$, pairs k, l, then directions associated to the faces may be defined. The angles between normals are determined by the scalar products of these vectors. The spin geometry theorem states that there exists a δ for any approximation of vectors 3-dimensional Euclidean space. We constructed these vectors for a tetrahedron.

We set the maximum spin, typically 20, and took a superposition over all states of the atom of geometry, with amplitudes determined by a complex gaussian. We fixed the parameters of the state (peak value, width, and phase) by minimizing the relative uncertainties. This analysis is straightforward in the basis of one of the angle operators. For the other angles, which do not commute with one another, one must employ recoupling theory. As this results in a lengthy expression to minimize, we used Mathematica to minimize the relative uncertainties.

This process of minimizing δ-constraints for the independent angles yields well-defined directions but low-volume, "squashed" geometries. This is expected since the minimization procedure essentially extremizes the cosine of the angle between outgoing normals times a spin factor, held fixed by our choice of maximum spin. Minimizing the resulting sine yields angles are near 0 and π, producing an elongated shape of the atom ("squashed"), with angles near 0 and π. The minimization produces a distribution of angles that is far from the distribution of classical polar angles.

However simultaneously minimizing the relative uncertainties and maximizing the volume of the atom produces a distribution of angles peaked around the center of the classical distribution, at the cost of increased relative uncertainty in the angle (i.e. larger δ's and a correspondingly worse approximation to the angles of the classical tetrahedron). We see that the "cost" of modeling the classical distribution of angles is an increase in relative uncertainty. As we know from the spin geometry theorem to reduce the uncertainty further we need to increase the total flux, raising the effective scale of the atom of geometry. This "lever arm" is simply a manifestation of the familiar uncertainty relations of non-commuting operators, rather than from the combinatorics of more complex states.

This tetrahedron study shows that (1) It is possible to use quantum information intrinsic to the state to define the coherent states. (Unfortunately the computation grows unwieldily for higher spin and high valence atoms.) (2) There is a "short" lever arm in that the classical distribution of angle is achieved for atoms above the Planck scale. But this lever arm is too short to raise the scale of the effects into experimentally accessible regimes. However, the lever arm in Sect. 3 arises from the combinatorics arising in complex atoms of geometry.

6 Discussion

This contribution reviews a model that explores effects arising from combinatorial structures in the deep spatial quantum geometry of LQG [6]. The model, based on assumptions in Sect. 3, relies on the combinatorics of a discrete model of spatial geometry, a single atom of spatial geometry, the spin network node. This model shows that potentially observable effects of quantum geometry need not be tied to violations of local Lorentz symmetry and that a scale above the fundamental scale of the theory can arise out of the combinatorics of the state.

To test the assumption of the uniform measure we developed coherent states based on information intrinsic to the simplest atom, which is dual to a tetrahedron. Using the spin geometry theorem we developed coherent states of the polyhedron and found evidence for a lever arm. In this case the short lever arm is due to the non-commutivity of operators, rather than the combinatorics of the state. However, this is a very simple atom of geometry that does not have the combinatoric richness of complex higher valence atoms. Work on modeling these more complex structures is ongoing.

Finally, it is important to note that work remains on modeling the vertex modifications in field theory. One possibility is to model the effective metric in, e.g.

$$\mathscr{L}'_\delta(x) = -e \int d^4x \bar{\psi}(x)\gamma_\mu \psi(x) g_\delta^{\mu\nu}(x-z) A_\nu(z), \tag{10}$$

using spin foam techniques. It remains to be seen whether this shape-corrected QED vertex yields the simple shape corrections discussed here.

References

1. Dittrich, B., Thiemann, T.: Are the spectra of geometrical operators in loop quantum gravity really discrete? J. Math. Phys. **50**, 012503 (2009). doi:10.1063/1.3054277
2. Rovelli, C.: Comment on 'Are the spectra...'. ArXiv e-prints arXiv:0708.2481 [gr-qc] (2007)
3. Liberati, S.: Lorentz breaking effective field theory and observational tests. ArXiv e-prints arXiv:1203.4105 [gr-qc] (2012)
4. Girelli, F., S., H.F.M.: Loop quantum gravity phenomenology: linking loops to observational physics. SIGMA **8**, 098 (2012). doi:10.3842/SIGMA.2012.098
5. Rovelli, C., Speziale, S.: Reconcile Planck-scale discreteness and the Lorentz-Fitzgerald contraction. Phys. Rev. D **67**, 064019 (2003). doi:10.1103/PhysRevD.67.064019
6. Major, S.: Shape in an atom of space: exploring quantum geometry phenomenology. Class. Quantum Grav. **27**, 225012 (2010). doi:10.1088/0264-9381/27/22/225012

7. Major, S.: Operators for quantized directions. Class. Quantum Grav. **16**, 3859 (1999). doi:10.1088/0264-9381/16/12/307
8. Major, S., Seifert, M.: Modeling space with an atom of quantum geometry. Class. Quantum Grav. **19**, 2211 (2002). doi:10.1088/0264-9381/19/8/311
9. Major, S.: Quantum geometry phenomenology: angle and semiclassical states. J. Phys.: Conf. Ser. **360**, 012061 (2012). doi:10.1088/1742-6596/360/1/012061
10. Rovelli, C.: Zakopane lectures on loop gravity. ArXiv e-prints arXiv:1102.3660[gr-qc] (2011)
11. Seifert, M.: Angle and volume studies in quantized space. ArXiv e-prints arXiv:gr-qc/0108047 (2001)
12. Bianchi, E., Doná, P., Speziale, S.: Polyhedra in loop quantum gravity. Phys. Rev. D **83**(4), 044035 (2011). doi:10.1103/PhysRevD.83.044035
13. Bianchi, L.: Lezioni sulla teoria dei gruppi continui finiti di trasformazioni. E. Spoerri, Pisa (1918)
14. Drell, S.: Quantum electrodynamics at small distances. Ann. Phys. **4**, 75 (1958). doi:10.1016/0003-4916(58)90038-1
15. Penrose, R.: Angular momentum: an approach to combinatorial space-time. In: Bastin, T. (ed.) Quantum Theory Beyond, pp. 151–180. Cambridge University Press, Cambridge (1971)
16. Moussouris, J.: Quantum models as space-time based on recoupling theory. Ph.D. thesis, Oxford University, Oxford (1983)
17. Zappala, J.: Semiclassical states for an atom of geometry. In preparation (2012)

Signature Change in Loop Quantum Cosmology

Jakub Mielczarek

Abstract The Wick rotation is commonly considered only as a useful computational trick. However, as suggested by Hartle and Hawking already in early eighties, Wick rotation may gain physical meaning at the Planck epoch. While such possibility is conceptually interesting, leading to no-boundary proposal, mechanism behind the signature change remains mysterious. We show that the signature change anticipated by Hartle and Hawking naturally appears in loop quantum cosmology. Theory of cosmological perturbations with the effects of quantum holonomies is discussed. It was shown by Cailleteau et al. [3] that this theory can be uniquely formulated in an anomaly-free manner. The obtained algebra of effective constraints turns out to be modified so that the metric signature is changing from Lorentzian in low curvature regime to Euclidean in high curvature regime. Implications of this phenomenon on propagation of cosmological perturbations are discussed and corrections to inflationary power spectra of scalar and tensor perturbations are derived. Possible relations with other approaches to quantum gravity are outlined. We also propose an intuitive explanation of the observed signature change using analogy with spontaneous symmetry breaking in "wired" metamaterials.

1 Introduction

The metric signature change from Lorentzian to Euclidean is usually performed by the so-called Wick rotation ($t \rightarrow -i\tau$), under which the line element $ds^2 = -dt^2 + d\mathbf{x}^2$ transforms to $ds^2 = d\tau^2 + d\mathbf{x}^2$. The Wick rotation becomes especially important in the path integral formulation of quantum mechanics. It allows to calculate non-perturbative effects by considering *instantons*. Another advantage coming from the

J. Mielczarek (✉)
Theoretical Physics Department, National Centre for Nuclear Research, Hoża 69, 00-681 Warsaw, Poland
e-mail: jakubm@fuw.edu.pl

J. Bičák and T. Ledvinka (eds.), *Relativity and Gravitation*,
Springer Proceedings in Physics 157, DOI: 10.1007/978-3-319-06761-2_77,

Wick rotation is improvement of a convergence property of some path integrals. But these are just useful computational tricks.

However, in 1983 Hartle and Hawking proposed that Wick rotation may gain physical meaning at the Planck epoch [1]. This assumption was crucial for construction of the so-called no-boundary proposal, which was a way to cope with the problem of initial conditions for the Universe. While such possibility is conceptually interesting, mechanism behind the signature change in the very early Universe remains enigmatic.

If such transition from the Lorentzian to Euclidean space has occurred in the early Universe, what could be the origin of this? Can the signature change be due to some quantum gravity effects? So far, there have been no indications supporting such possibility. However, recent results coming from symmetry reduced models of Loop Quantum Gravity (LQG) [2] suggest that indeed the signature change may occur due to the discrete nature of space at the Planck scale [3, 4].

2 Loops

In LQG, granularity of space at the Planck scale is manifested by discrete spectra of geometric operators such as area and volume. The starting point for constructing LQG is the Hamiltonian formulation of General Relativity in language of Ashtekar variables fulfilling $\left\{E^a_j(\mathbf{x}), A^i_b(\mathbf{y})\right\} = 8\pi G\gamma\delta^a_b\delta^i_j\delta^{(3)}(\mathbf{x}-\mathbf{y})$, where γ is a free parameter of the theory called Barbero-Immirzi parameter. In this framework, Hamiltonian of gravity sector can be written as a sum of three constraints: $H_G[N, N^a, N^i] = S[N] + D[N^a] + G[N^i] \approx 0$. Here S is the scalar constraint, D is the diffeomorphism constraint and G is the Gauss constraint. The constraints ($S \rightarrow \mathscr{C}_1, D \rightarrow \mathscr{C}_2, G \rightarrow \mathscr{C}_3$) fulfill the closed algebra $\{\mathscr{C}_I, \mathscr{C}_J\} = f^K{}_{IJ}(A^j_b, E^a_i)\mathscr{C}_K$, where $f^K{}_{IJ}(A^j_b, E^a_i)$ are some structure functions.

Based on the Ashtekar variables, nonlocal variables called holonomies and fluxes are constructed. These new variables are the subject of quantization in LQG. For our purposes, it is sufficient to note that holonomy is defined as parallel transport of $A_a = A^i_a\tau_i$ along some curve e on a spatial hypersurface: $h_e := \mathscr{P}\exp\int_e A_a dx^a$, where $\sigma_j = 2i\tau_j$ are Pauli matrices. The holonomies are elements of $SU(2)$ group.

In LQG, a state of gravity is described by superposition of graphs called *spin networks*. The links of the graphs are labelled by half integers ($j = 1/2, 1, 3/2, \dots$) corresponding to irreducible representations of the $SU(2)$ group. An exemplary spin network is shown in Fig. 1.

Loop Quantum Cosmology (LQC) [5] is a regular lattice model of LQG. In particular, symmetries of isotropy and homogeneity are imposed on the spin network within LQC. In what follows, we will consider an isotropic model with small perturbative inhomogeneities around a flat Friedmann-Robertson-Walker (FRW) background.

In LQC, physical area of the elementary lattice cell $\mathrm{Ar}_\square = \bar{p}\bar{\mu}^2$, where $\bar{p} = a^2$ and a is a scale factor. In general $\bar{\mu} \propto \bar{p}^\beta$, where $-1/2 \leq \beta \leq 0$. For the so-called

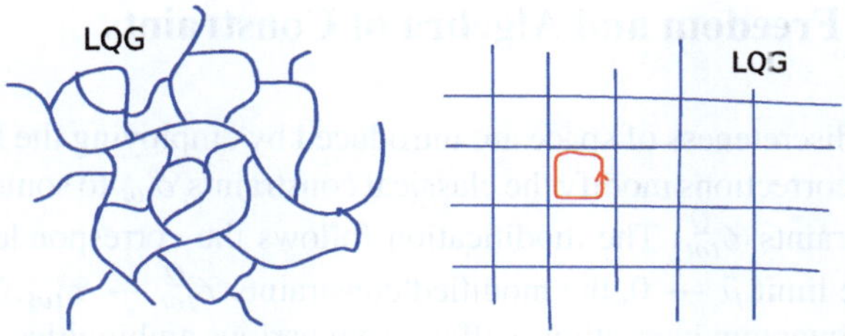

Fig. 1 In LQG, state of geometry is described by the spin network (*left*). In LQC, the spin network takes the form of regular lattice (*right*). The (*red*) loop represents holonomy $h_\Box$ around the elementary cell

$\bar{\mu}$−scheme ("new quantization scheme"): $\bar{\mu} = \sqrt{\frac{\Delta}{\bar{p}}}$, where usually Δ is assumed to be area gap derived from LQG: $\Delta = 2\sqrt{3}\pi\gamma l_{Pl}^2$. The $\bar{\mu}$−scheme, in which physical area of the elementary lattice cell is constant during the cosmological evolution, was shown to be physically favoured.

At the effective level, effects of discreteness can be studied by introducing the so-called holonomy corrections. They are obtained by replacing curvature of Ashtekar connection by holonomy around elementary loop (see Fig. 1). This procedure is called *polymerization*.

3 Cosmological Perturbations

In most cosmological applications, Ashtekar variables can be decomposed into the background (here flat FRW) and perturbation parts: $E_i^a = \bar{E}_i^a + \delta E_i^a$ and $A_a^i = \bar{A}_a^i + \delta A_a^i$, where $\bar{E}_i^a = \bar{p}\delta_i^a$ and $\bar{A}_a^i = \gamma k \delta_a^i$. The perturbations of the Ashtekar variables can be related with the standard metric perturbations: scalar modes (Φ, Ψ, E, B), vector modes (S^a, F_a) and tensor modes (h_{ab}). In total, there are 10 perturbative degrees of freedom. Furthermore, matter degrees of freedom are also subject of perturbative decomposition. In what follows we consider a model with a scalar field, so φ and its canonically conjugated momenta π can be written as: $\varphi = \bar{\varphi} + \delta\varphi$ and $\pi = \bar{\pi} + \delta\pi$. Applying the above decompositions, total constraints $\mathscr{C}_{tot} = \mathscr{C}_G + \mathscr{C}_M$, which take into account contributions from gravity and matter, can be expanded. Our analysis is performed up to the second order in perturbative development: $\mathscr{C}_{tot} = \mathscr{C}^{(0)} + \mathscr{C}^{(1)} + \mathscr{C}^{(2)} + ...$, so the corresponding equations of motion stay linear.

4 Anomaly Freedom and Algebra of Constraints

The effects of discreteness of space are introduced by employing the holonomy corrections. Such corrections modify the classical constraints $\mathscr{C}_{tot}$ to some new effective quantum constraints $\mathscr{C}^{Q}_{tot}$. The modification follows the correspondence principle, such that in the limit $\bar{\mu} \to 0$, the modified constraints $\mathscr{C}^{Q}_{tot} \to \mathscr{C}_{tot}$. The procedure of introducing quantum corrections suffers from various ambiguities. Moreover, the resulting algebra of modified constraints is in general not closed:

$$\{\mathscr{C}^{Q}_{I}, \mathscr{C}^{Q}_{J}\} = g^{K}{}_{IJ}(A^{j}_{b}, E^{a}_{i})\mathscr{C}^{Q}_{K} + \mathscr{A}_{IJ}, \tag{1}$$

where $\mathscr{A}_{IJ}$ are some anomaly terms. Closure of algebra is required by mathematical consistency of the theory. So, the question is: Can we introduce quantum holonomy corrections in the anomaly-free manner (i.e. such that $\mathscr{A}_{IJ} = 0$)? The answer turns out to be "yes". Moreover, there is a unique way of modifying constraints such that the algebra is closed [3]. Additionally, the conditions of anomaly-freedom are fulfilled if and only if $\beta = -1/2$, which corresponds to the $\bar{\mu}$-scheme. Therefore, the only remaining free parameter is the area of elementary lattice cell Δ, which is however expected to be of the order of the Planck area l^2_{Pl}.

The obtained algebra of the effective quantum constraint is [3]:

$$\left\{D_{tot}[N^{a}_{1}], D_{tot}[N^{a}_{2}]\right\} = 0, \tag{2}$$

$$\left\{S_{tot}[N], D_{tot}[N^{a}]\right\} = -S_{tot}[\delta N^{a}\partial_{a}\delta N], \tag{3}$$

$$\{S_{tot}[N_1], S_{tot}[N_2]\} = \Omega D_{tot}\left[\frac{\bar{N}}{\bar{p}}\partial^{a}(\delta N_2 - \delta N_1)\right]. \tag{4}$$

The algebra is closed but deformed with respect to the classical case due to the presence of Ω in (4). Therefore, general covariance is modified. The new factor Ω can be expressed as follows: $\Omega = \cos(2\bar{\mu}\gamma\bar{k}) = 1 - 2\rho/\rho_c \in [-1, 1]$, where ρ is energy density of the scalar matter and the critical energy density $\rho_c := \frac{3}{8\pi G\Delta} \sim \rho_{Pl} := m^4_{Pl}$. What is the interpretation of the above deformation of the algebra of constraints? In order to answer this question let us recall the classical equivalent of the modified bracket (4) for a space with signature σ [2]:

$$\{S_{tot}[N_1], S_{tot}[N_2]\} = \sigma D\left[\frac{\bar{N}}{\bar{p}}\partial^{a}(\delta N_2 - \delta N_1)\right].$$

Here, $\sigma = 1$ corresponds to the Lorentzian signature and $\sigma = -1$ to the Euclidean one. Therefore, we conclude that modification of the effective algebra of constraints (4) means that the space becomes Euclidean for $\rho > \rho_c/2$, while Lorentzian geometry emerges for $\rho < \rho_c/2$. In the regime of high curvatures and high energy densities ($\rho > \rho_c/2$), spacetime becomes 4-dimensional Euclidean space. There is no distinguished time direction in this phase. It is interesting to notice that this model exhibits properties of the Hartle-Hawking no-boundary proposal. However here signature change occurs smoothly with an increase of energy density.

The same effect of signature change was observed also for inhomogeneous spherically symmetric models with holonomy corrections [4]. Therefore, we have a good reason to believe that this phenomenon is a general consequence of quantum polymerization of space. Moreover, we can speculate that the *off-shell* algebra of quantum constraints in LQG should also exhibit such property, i.e. $[\hat{S}, \hat{S}] = i\Omega\hat{D}$.

Among many other comments, the fact that *ultralocal gravity* [6], where $\{S, S\} = 0$, is recovered at the transition point $\rho = \rho_c/2$ is worth stressing.

5 Equations of Motion

The obtained anomaly-free formulation can be now used to derive equations of motion for both background variables and perturbations. The background dynamics is governed by the modified Friedmann equation $H^2 = \frac{8\pi G}{3}\rho(1 - \rho/\rho_c)$, where H is the Hubble factor. Clearly, only $\rho \leq \rho_c$ are physically allowed, which was used to determine the range of Ω.

For the scalar perturbations one can define a gauge-invariant variable v and the corresponding modified Mukhanov equation [3]:

$$\frac{d^2}{d\tau^2}v - \Omega\nabla^2 v - \frac{z''}{z}v = 0, \tag{5}$$

where $z := \sqrt{p}\frac{\dot{\varphi}}{H}$. Here, $\tau = \int dt/a$ is a conformal time. For the considered model with a scalar field, vector modes are pure gauge, and therefore do not contribute [7]. Equation for tensor modes takes the form [8]

$$\frac{d^2}{d\tau^2}h_{ab} + 2\left(aH - \frac{1}{2\Omega}\frac{d\Omega}{d\tau}\right)\frac{d}{d\tau}h_{ab} - \Omega\nabla^2 h_{ab} = 0.$$

The obtained equations are modified by the presence of Ω in front of the Laplace operator. Therefore, transition to the Euclidean domain leads to a change of equation type from hyperbolic to elliptic, as expected. Furthermore, we see that the speed of propagation is varying, since $c_s^2 = \Omega$.

6 Holonomy Corrections to Inflationary Power Spectra

As an application of the obtained equations of motion we will derive holonomy corrections to the inflationary scalar and tensor power spectra. We will focus on the slow-roll inflationary model driven by a single scalar field φ with potential $V(\varphi)$

occurring in the Lorentzian domain. The slow-roll parameters with the holonomy corrections are:

$$\varepsilon := \frac{m_{Pl}^2}{16\pi}\left(\frac{V_{,\varphi}}{V}\right)^2 \frac{1}{(1 - V/\rho_c)} \quad \text{and} \quad \eta := \frac{m_{Pl}^2}{8\pi}\left(\frac{V_{,\varphi\varphi}}{V}\right) \frac{1}{(1 - V/\rho_c)}.$$

Derivation of the scalar and tensor power spectra is based on application of the standard techniques of the quantum field theory in curved spaces. Moreover, normalization is such that in the UV limit the Minkowski vacuum is recovered. Obtained spectra of scalar and tensor (gravitational waves) perturbations are

$$\mathscr{P}_S(k) = A_S \left(\frac{k}{aH}\right)^{n_S - 1} \quad \text{and} \quad \mathscr{P}_T(k) = A_T \left(\frac{k}{aH}\right)^{n_T},$$

where amplitudes and spectral indices are given as follows:

$$A_S = \frac{1}{\pi\varepsilon}\left(\frac{H}{m_{Pl}}\right)^2 \left(1 + 2\frac{V}{\rho_c}\right) \quad \text{and} \quad n_S = 1 + 2\eta - 6\varepsilon\left(1 - \frac{V}{\rho_c}\right),$$

$$A_T = \frac{16}{\pi}\left(\frac{H}{m_{Pl}}\right)^2 \left(1 + 3\frac{V}{\rho_c}\right) \quad \text{and} \quad n_T = -2\varepsilon\left(1 - 3\frac{V}{\rho_c}\right).$$

Furthermore, the consistency relation is

$$r := \frac{A_T}{A_S} \simeq 16\varepsilon\left(1 + \frac{V}{\rho_c}\right). \tag{6}$$

The corrections are introduced by the factors V/ρ_c, which are of the order of 10^{-12} for typical values of parameters. Confrontation of the obtained spectra with the available CMB data will be studied elsewhere [9]. In a more detailed analysis, initial conditions should be established at the transition point $\rho = \rho_c/2$, which can lead to some additional modification of the power spectra. This issue will be investigated in our further research.

7 Towards Understanding the Signature Change

It is tempting to understand the origin of the signature change at a microscopic level. Is this a kind of phase transition occurring at the level of the spin network or, more generally, at the level of some four dimensional spin foam model? One possibility is that some sort of spontaneous symmetry breaking takes place, leading to distinction of the time dimension in low curvature regime.

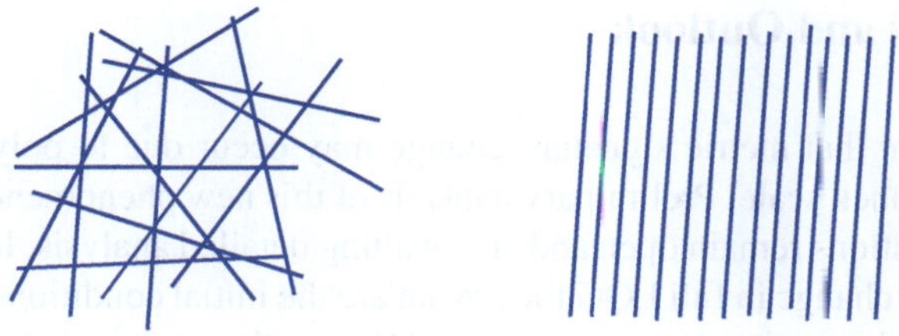

Fig. 2 At high temperatures orientation of nanowires is random (*left*). At low temperatures, some direction is distinguished due to the spontaneous symmetry breaking (*right*). In the distinguished direction, dielectric permittivity is negative, leading to emergence of a time variable

A relevant example of spontaneous symmetry breaking is given by ferromagnets. At high temperatures, ferromagnets are loosing their magnetic properties. The spins (magnetic moments) are randomly orientated, and no direction is distinguished. The system satisfies the $SO(3)$ rotational symmetry, which is also a symmetry of the corresponding Hamiltonian. However, while temperature is decreased, spins start to orientate in some direction and magnetic domains are formed. This process begins when temperature is lowered below the so-called Curie temperature.

One can speculate that an analogous phase transition occurs in case of gravity. Namely, at high energies, the symmetry is, say, $SO(4)$. There is no distinction of any time coordinate. However, while lowering the energy density, which is an analogue of temperature, the symmetry will be broken to $SO(3)$. The energy density $\rho_c/2$ is an analogue of the Curie temperature. It is possible that $SO(3)$ symmetry of the triad rotations in Ashtekar formalism (doubly covered by $SU(2)$ group in LQG) is in fact a residual symmetry of some wider symmetry before spontaneous symmetry breaking.

The phase transition, similar to the one discussed in case of ferromagnets, may occur for the so-called "wired" metamaterials composed of nanowires. It was shown that for such materials, an effective emergence of time variable may occur because of *negative dielectric permittivity* [10]. At high temperatures, dielectric permittivities in all directions are positive. However in a low temperature state, nanowires may align in some direction as spins do. In this distinguished direction, dielectric permittivity becomes negative leading to emergence of "time" direction at the level of equations of motion for electromagnetic field (see Fig. 2). We speculate that the same kind of process occurs in case of gravity, leading to emergence of a time variable. Structure of four dimensional Euclidean space undergoes phase transition, such that some particular direction is picked (presumably domains with different time directions can form). Equations of motion for the fields living on such frame change from elliptic to hyperbolic, which is interpreted as emergence of the time direction. We will explore this interpretation in more details in our further studies.

8 Summary and Outlook

We have shown that metric signature change may occur due to polymerization of space at the Planck scale. Preliminary analysis of this new phenomenon was carried out. Many questions remain open and are awaiting detailed analysis. In particular: Is there signature change in full LQG too? What are the initial conditions at $\rho = \rho_c/2$? What is happening at the microscopic scale? How is the propagation of high energy photons affected? And many, many others.

Summarizing, the paradigm shift seems to be observed in LQC. There is no longer deterministic bouncing phase as was thought for many years. The Big Bounce model in LQC seems to be an artifact of the strong assumption of homogeneity. Due to the Euclidean stage there is no access to the information contained in contacting Universe, which gives answer to the long standing debate on cosmic forgetfulness.

Acknowledgments I would like thank to Aurélien Barrau, Thomas Cailleteau and Julien Grain for excellent collaboration and Włodzimierz Piechocki for his careful reading of the manuscript and useful remarks.

References

1. Hartle, J., Hawking, S.: Wave function of the universe. Phys. Rev. D **28**, 2960 (1983). doi:10.1103/PhysRevD.28.2960
2. Ashtekar, A., Lewandowski, J.: Background independent quantum gravity: a status report. Class. Quantum Grav. **21**, R53 (2004). doi:10.1088/0264-9381/21/15/R01
3. Cailleteau, T., Mielczarek, J., Barrau, A., Grain, J.: Anomaly-free scalar perturbations with holonomy corrections in loop quantum cosmology. Class. Quantum Grav. **29**, 095010 (2012). doi:10.1088/0264-9381/29/9/095010
4. Bojowald, M., Paily, G.: Deformed general relativity and effective actions from loop quantum gravity. Phys. Rev. D **86**, 104018 (2012). doi:10.1103/PhysRevD.86.104018
5. Bojowald, M.: Loop quantum cosmology. Living Rev. Relativ. **11**(4), lrr-2008-4 (2008). http://www.livingreviews.org/lrr-2008-4
6. Isham, C.: Some quantum field theory aspects of the superspace quantization of general relativity. Proc. R. Soc. London, Ser. A **351**, 209 (1976). doi:10.1098/rspa.1976.0138
7. Mielczarek, J., Cailleteau, T., Barrau, A., Grain, J.: Anomaly-free vector perturbations with holonomy corrections in loop quantum cosmology. Class. Quantum Grav. **29**, 085009 (2012). doi:10.1088/0264-9381/29/8/085009
8. Cailleteau, T., Barrau, A., Vidotto, F., Grain, J.: Consistency of holonomy-corrected scalar, vector, and tensor perturbations in loop quantum cosmology. Phys. Rev. D **86**, 087301 (2012). doi:10.1103/PhysRevD.86.087301
9. Mielczarek, J., Kamionka, M., Szydłowski, M.: (2012) (In preparation)
10. Smolyaninov, I., Narimanov, E.: Metric signature transitions in optical metamaterials. Phys. Rev. Lett. **105**, 067402 (2010). doi:10.1103/PhysRevLett.105.067402

Quantum Fields in Gravity

Giovanni Acquaviva

Abstract We give a brief description of some compelling connections between general relativity and thermodynamics through (i) the semi-classical tunnelling method(s) and (ii) the field-theoretical modelling of Unruh-DeWitt detectors. In both approaches it is possible to interpret some quantities in a thermodynamical frame.

1 Introduction

The idea of treating the emission of radiation from black holes as a tunelling process across the horizon traces back to the first path-integral derivation by Hartle and Hawking [1]. As a matter of fact, the null-geodesic method introduced by Kraus, Parikh and Wilczek [2, 3] and the Hamilton-Jacobi method proposed more recently by Padmanabhan and collaborators [4] can be considered as semi-classical versions of the original derivation. On the other hand, the Unruh-DeWitt detector [5, 6] constitutes a field-theoretical approach to the problem, providing a more exact answer to questions regarding the particle content of the field and its thermal features for different observers.

2 The Tunnelling Method(s)

The *null-geodesic* and the *Hamilton-Jacobi* methods mentioned above both rely on the calculation of the classical action S of a particle along a trajectory crossing the horizon. Since such a trajectory is classically forbidden, the action itself developes an imaginary contribution which, in the WKB approximation, allows to calculate the tunnelling probability rate

G. Acquaviva (✉)
Department of Physics, University of Trento, Trento, Italy
e-mail: acquaviva@science.unitn.it

J. Bičák and T. Ledvinka (eds.), *Relativity and Gravitation*,
Springer Proceedings in Physics 157, DOI: 10.1007/978-3-319-06761-2_78,

$$\Gamma_{em} \simeq \exp\left(-2\,\Im S\right), \tag{1}$$

where $\Im$ stands for imaginary part. The use of Kodama-Hayward theoretical results [7, 8], which allow to express observables of interest in terms of invariant quantities, has been a main ingredient.

In [9] this methods has been analysed in detail and the following results have been proven:

- a solid basis for the covariance of the method has been given;
- formal equivalence of the two aforementioned approaches holds at least in stationary cases;
- the method provides an invariant and consistent answer in a variety of situations (higher-dimensional solutions, Taub and Taub-NUT solutions, decay of unstable particles, emission from cosmological horizons and naked singularities).

The calculation can be summarized in the following steps regarding the Hamilton–Jacobi approach:

1. assume that the tunnelling particle's action S satisfies the relativistic Hamilton–Jacobi equation
$$g^{\mu\nu}\partial_\mu S \partial_\nu S + m^2 = 0; \tag{2}$$
2. reconstruct the whole action, starting from the symmetries of the problem; the integration is carried along an oriented, null curve γ with at least one point on the horizon
$$S = \int_\gamma dx^\mu\ \partial_\mu S; \tag{3}$$
3. perform a near-horizon approximation and regularize the divergence in the integral according to Feynman's prescription: the solution of the integral has in general a non-vanishing imaginary part.

The result can be given in the general form

$$\Gamma_{em} = \Gamma_{abs}\ \exp\left(-\frac{2\pi\ \omega_H}{\kappa_H}\right), \tag{4}$$

where ω_H and κ_H are respectively the invariant energy of the tunnelling particle and the invariant surface gravity in Hayward's theory. Through comparison of the transition rate with the Boltzmann factor, we can identify an invariant temperature

$$T_H = \frac{\kappa_H}{2\pi}. \tag{5}$$

3 Unruh-DeWitt Detectors

We consider a conformally flat 4-dimensional metric, a massless scalar field conformally coupled to the metric and a two-level quantum system coupled to the scalar field. The idea is to calculate the probability for the absorption of a scalar quantum and the consequent excitation of the two-level system through the *transition rate*

$$\frac{dF}{d\tau} = \frac{1}{2\pi^2}\int_0^\infty \cos(E\,s)\left(\frac{1}{\sigma^2(\tau,s)}+\frac{1}{s^2}\right)ds - \frac{1}{2\pi^2}\int_{\Delta\tau}^\infty \frac{\cos(E\,s)}{\sigma^2(\tau,s)}, \tag{6}$$

where E is the energy gap of the detector and s is the duration of the detection (see [10] for details on the construction of equation (6)). The second integral is the *finite-time contribution*, generally an oscillating tail exponentially damped. The bulk of the information about the transition rate comes from the geodesic distance between the "switching on" and "switching off" events, evaluated along a fixed trajectory $x(\tau)$

$$\sigma^2(\tau,s) = a(\tau)a(\tau-s)\,[x(\tau)-x(\tau-s)]^2\,, \tag{7}$$

whose inverse is proportional to the positive frequency Wightman function. The $a(t)$ is the conformal factor.

Let's analyze two simple stationary cases: the *Schwarzschild black hole* and the *de Sitter cosmology*. The detector will be placed on a Kodama trajectory, which means that it will sit at a fixed distance from the horizon. Both cases can be treated in the same way, because the function σ^2 can be written in general

$$\sigma^2(s) = -\frac{4V}{\kappa^2}\sinh^2\left(\frac{\kappa}{2\sqrt{V}}\,s\right), \tag{8}$$

where κ is the surface gravity and $\sqrt{V} = \sqrt{-g_{00}}$. A Wightman function which, as in (8), is stationary and periodic in imaginary time is called "thermal" because when Fourier-transformed, it gives a Planckian transition spectrum. In our case, calculating both the stationary and the finite-time contributions,

$$\frac{dF}{d\tau} = \frac{1}{2\pi}\frac{E}{\exp\left(\frac{2\pi\sqrt{V}E}{\kappa}\right)-1} + \frac{E}{2\pi^2}\sum_{n=1}^{\infty}\frac{n\,e^{-n\kappa\Delta\tau/\sqrt{V}}}{n^2+VE^2/\kappa^2}\left(\frac{\kappa}{\sqrt{V}E}\cos(E\Delta\tau)-\sin(E\Delta\tau)\right). \tag{9}$$

4 Conclusions

As regards the tunnelling method, it has been shown that the formalism gives an invariant answer and allows extensions to more general black hole horizons in various dimensions as well as cosmological horizons and naked singularities. Moreover, the extension to dynamical space-times has been carried out: in this framework the radiation seems to originate near the *local trapping horizon*, not the global event horizon.
The Unruh-DeWitt detector constitutes a more exact approach to the Unruh-Hawking effect, relying on a quantum field-theoretical calculation. In stationary cases the response function of the detector is shown to be thermal with temperature given by the surface gravity, just as in the tunnelling approach. The generalization to non-stationary situations gives rise to problems in the analytical resolution and in general, when the background is time-dependent, it is not possible to clearly identify a Boltzmann-like term in the response function and thus the thermal interpretation seems lost.

References

1. Hartle, J., Hawking, S.: Path-integral derivation of black-hole radiance. Phys. Rev. D **13**, 2188 (1976). doi:10.1103/PhysRevD.13.2188
2. Kraus, P., Wilczek, F.: Self-interaction correction to black hole radiance. Nucl. Phys. B **433**, 403 (1995). doi:10.1016/0550-3213(94)00411-7
3. Parikh, M., Wilczek, F.: Hawking radiation as tunneling. Phys. Rev. Lett. **85**, 5042 (2000). doi:10.1103/PhysRevLett.85.5042
4. Srinivasan, K., Padmanabhan, T.: Particle production and complex path analysis. Phys. Rev. D **60**, 24007 (1999). doi:10.1103/PhysRevD.60.024007
5. Unruh, W.: Notes on black-hole evaporation. Phys. Rev. D **14**, 870 (1976). doi:10.1103/PhysRevD.14.870
6. DeWitt, B.: Quantum gravity: The new synthesis. In: Hawking, S., Israel, W. (eds.) General Relativity: An Einstein Centenary Survey, pp. 680–745. Cambridge University Press, Cambridge, New York (1979)
7. Kodama, H.: Conserved energy flux for the spherically symmetric system and the backreaction problem in the black hole evaporation. Prog. Theor. Phys. **63**, 1217 (1980). doi:10.1143/PTP.63.1217
8. Hayward, S.: Unified first law of black-hole dynamics and relativistic thermodynamics. Class. Quant. Grav. **15**, 3147 (1998). doi:10.1088/0264-9381/15/10/017
9. Vanzo, L., Acquaviva, G., Di Criscienzo, R.: Tunnelling methods and Hawking's radiation: achievements and prospects. Class. Quant. Grav. **28**, 183001 (2011). doi:10.1088/0264-9381/28/18/183001
10. Acquaviva, G., Di Criscienzo, R., Tolotti, M., Vanzo, L., Zerbini, S.: Unruh-deWitt detectors in spherically symmetric dynamical space-times. Int. J. Theor. Phys. **51**, 1555 (2012). doi:10.1007/s10773-011-1033-2

Classical and Quantum Scattering in Impulsive Backgrounds

Peter Aichelburg and Herbert Balasin

Abstract We present a simple but rigorous approach to solve the nonlinear distributional scattering problem of particles in impulsive fields. As an illustration we consider a Dirac particle in an impulsive Yang-Mills wave.

1 Introduction

Due to the singular nature of the pulse and the nonlinear character of the problem a generalized framework for singular functions (distributions) has to be used. This will be done in terms of Colombeau's new generalized functions, which form an algebra therefore allowing for nonlinear operations and contain distribution space via an appropriate coarse-graining operation (association).[1] From the physical point of view the Colombeau algebra affords a systematic way to deal with "regularizations" of singular objects, i.e. idealized situations that contain only certain parts of the information (i.e. total charge in case of point-charge distribution) whereas the other regularization-specific information is discarded. In this regard Colombeau objects contain information upon the small-scale (micro-)structure of the physical objects. Upon nonlinear operations part of the micro-structure can be magnified to the macroscopic level. At this level equality is modelled by the so-called association operation which groups together (different) Colombeau-objects that contain the same macro-aspect. However, this coarse-graining operation is in general incom-

[1] We are going to use the so-called simplified algebra [1] which allows a concise treatment and has the advantage of being immediately usable in the manifold context [2].

P. Aichelburg (✉)
Gravitationsphysik Universität Wien, Vienna, Austria
e-mail: aichelp8@univie.ac.at

H. Balasin
Institut für Theoretische Physik, TU-Wien, Vienna, Austria

J. Bičák and T. Ledvinka (eds.), *Relativity and Gravitation*,
Springer Proceedings in Physics 157, DOI: 10.1007/978-3-319-06761-2_79,

patible with nonlinear operations due to the aforementioned magnification effect. The most well-known example of this effect being the product of arbitrary powers of the Heaviside-θ with θ', i.e. $\theta^{n-1}\theta' \approx (1/n)\delta$. Although we have $\theta^n \approx \theta$ and $\theta' \approx \delta$, that is to say although all powers of θ have the same macro-aspect as θ itself and θ' has the same macro-aspect as δ the product $\theta^n\theta'$ nevertheless remembers the microaspect of which power of θ has been multiplied by θ' the resulting object having the same macro-aspect as δ but different pre-factors.

2 The Method

Our approach to solve the nonlinear scattering problems employs precisely this structural interplay between macroscopic and microscopic aspects. Namely, in the first step we construct the solution from the macroscopic form of the scattering-equation, i.e. by gluing free solutions. The gluing of the solution and the singular character of the equations contain the microscopic information. Imposing the nonlinear scattering equation we obtain upon coarse graining, in the second step, magnified micro-aspects which appear in form of undetermined parameters. In order to determine these ambiguities we invoke additional conditions in the form of conservation laws that would be an automatic consequence in the smooth context but do not necessarily follow in the singular context, since they involve nonlinear operations. In this sense we let the equations themselves determine the ambiguities and do not stipulate their values beforehand. Our approach does, due to the use of the Colombeau-algebra, not rely on any particular regularization scheme but rather determines the conditions that physically sensible regularizations have to obey. At the same time keeping most of the simple set-up of the original problem. The scattering by impulsive fields initiated by 't Hooft [3] using Penrose's [4] cut-and-paste method has been studied previously in a number of papers e.g. the classical aspects of test particles (without spin) in gravitational pp-waves first by Balasin [5], and then by Kunzinger and Steinbauer [6], making use of the full Colombeau machinery. The scattering of Dirac particles was considered e.g. by Sanchez and deVega [7], however with the drawback of an arbitrary choice of the arising ambiguities.

3 Impulsive Yang-Mills Scattering

In the following we will consider scattering of quantum particles in an impulsive pp-wave Yang-Mills field. The latter is treated in analogy to the gravitational case

$$A^i_a = f^i p_a, \quad (p\partial)f^i = 0, \; p^2 = 0, \; \partial_a p^b = 0.$$

Borrowing terminology from the gravitational case the Lie-algebra-valued function f^i will be referred to as the wave-profile. For the quantum particle we will consider the corresponding Dirac equation with respect to a pp-wave Yang-Mills field, where we take the spinor field to transform with respect to the canonical representation of the Lie-group G on its Lie-algebra

$$(\gamma^a D_a - m)\psi = 0, \qquad \{\gamma_a, \gamma_b\} = 2\eta_{ab}, \qquad D_a = \partial_a - [A_a, .].$$

Taking the pp-wave character of the Yang-Mills field into account the Dirac equation becomes

$$(\gamma^a \partial_a - m)\psi - [f, \not{p}\psi] = 0.$$

The impulsive nature of the wave reflects itself in the appearance of a delta function in the wave profile concentrated on the null hyperplane with generator p^a, i.e. $f^i = \delta(px)\tilde{f}^i$, where $\tilde{f}^i$ is the reduced profile that depends only on the spacelike directions along the null plane. Within the Colombeau-algebra we may glue the Dirac spinor ψ from two solutions of the free Dirac equation "above" and "below" the pulse-plane

$$\psi = \theta_+\psi_+ + \theta_-\psi_- \qquad \begin{matrix} \theta_+ = \theta \\ \theta_- = 1-\theta \end{matrix}.$$

Together with $\theta\delta \approx A\delta$ the weak equation $(\gamma^a\partial_a - m)\psi - [f, \not{p}\psi] \approx 0$ therefore entails

$$\not{p}\psi_+ - \not{p}\psi_- = A[\tilde{f}, \not{p}\psi_+] + (1-A)[\tilde{f}, \not{p}\psi_-],$$

or equivalently

$$(id - A[\tilde{f}, .])\not{p}\psi_+ = (id + (1-A)[\tilde{f}, .])\not{p}\psi_-. \tag{1}$$

The "ambiguity" (macroaspect of the product between θ and δ) A will be determined by a physical condition. The smooth Dirac equation implies the existence of a conserved current J^a

$$J^a = [\bar{\psi}, \gamma^a\psi] \qquad \bar{\psi} := \psi^\dagger B \quad -B\gamma^a B^{-1} = \gamma^{a\dagger} \Leftrightarrow \bar{\gamma^a} = -\gamma^a,$$

via

$$\begin{aligned} D_a[\bar{\psi},\gamma^a\psi] &= [\overline{D_a\psi}, \gamma^a\psi] + [\bar{\psi}, \gamma^a D_a\psi] \\ &= -[\overline{\gamma^a D_a\psi}, \psi] + [\bar{\psi}, \gamma^a D_a\psi] = -m[\bar{\psi}, \psi] + m[\bar{\psi}, \psi] = 0, \end{aligned}$$

where in the first equality we took the reality of D_a with respect to YM-inner-product and in the second the imaginary character γ^a (anti-hermiticity with respect to the Dirac adjoint) into account. In the impulsive pp-wave context conservation involving non-linear operations no longer follows from the weak equation. Therefore requiring the weak conservation equation

$$D_a[\bar{\psi}, \gamma^a \psi] = \partial_a[\bar{\psi}, \gamma^a \psi] + [f, [\bar{\psi}, \not{p}\psi]]$$
$$= \partial_a[\bar{\psi}, \gamma^a \psi] + \delta[\tilde{f}, [\bar{\psi}, \not{p}\psi]] \approx 0$$

yields, together with $\theta^2 \delta \approx B\delta$,

$$[\bar{\psi}_+(0), \not{p}\psi_+(0)] - [\bar{\psi}_-(0), \not{p}\psi_-(0)] + B[\tilde{f}, [\bar{\psi}_+(0), \not{p}\psi_+(0)]] + (A - B)\times$$
$$([\tilde{f}, [\bar{\psi}_+(0), \not{p}\psi_-(0)]] + [\tilde{f}, [\bar{\psi}_-(0), \not{p}\psi_+(0)]]) + (1 - 2A + B)[\tilde{f}, [\bar{\psi}_-(0), \not{p}\psi_-(0)]] = 0. \quad (2)$$

By taking the inner product of (2) with $\tilde{f}$ we find

$$tr(\tilde{f}[\bar{\psi}_+(0), \not{p}\psi_+(0)]) - tr(\tilde{f}[\bar{\psi}_-(0), \not{p}\psi_-(0)]) = 0,$$

which yields together with (1)

$$tr(\bar{\psi}_-(0)(id + (1 - A)[\tilde{f}, .])^\dagger (id - A[\tilde{f}, .])^{-1\dagger} (id - A[\tilde{f}, .])^{-1}\times$$
$$(id + (1 - A)[\tilde{f}, .])[\tilde{f}, \not{p}\psi_-(0)]) = tr(\bar{\psi}_-(0)[\tilde{f}, \not{p}\psi_-(0)])$$

fixing $A = 1/2$. This result, however, implies together with (2), $B = 1/4$, thereby fixing both ambiguities. Therefore the junction condition reads in its final form

$$\not{p}\psi_+(0) = (id - \frac{1}{2}[\tilde{f}, .])^{-1}(id + \frac{1}{2}[\tilde{f}, .])\not{p}\psi_-(0),$$

where $U = (id - 1/2[\tilde{f}, .])^{-1}(id + 1/2[\tilde{f}, .])$ is unitary with respect to the inner product, being the Cayley-transform of the anti-hermitean operator $1/2[\tilde{f}, .]$ and no further restriction on $\psi_\pm$.

References

1. Colombeau, J.: Multiplication of Distributions: A Tool in Mathematics, Numerical Engineering and Theoretical Physics. Multiplication of Lecture Notes in Mathematics, vol. 1532. Springer, Berlin (1992)
2. Balasin, H.: Distributional aspects of general relativity: the example of the energy-momentum tensor of the extended Kerr-geometry. In: Grosser, M., Hörmann, G., Kunzinger, M., Oberguggenberger, M. (eds.) Nonlinear Theory of Generalized Functions, Research Notes in Mathematics, pp. 231–239. Chapman & Hall/CRC, Boca Raton (1998)
3. Hooft, G.'t: Graviton dominance in ultra-high-energy scattering. Phys. Lett. B **198**, 61 (1987). doi:10.1016/0370-2693(87)90159-6
4. Penrose, R.: The geometry of impulsive gravitational waves. In: O'Raifeartaigh, L. (ed.) General Relativity: Papers in Honour of J.L. Synge, pp. 101–115. Clarendon Press, Oxford (1972)
5. Balasin, H.: Geodesics for impulsive gravitational waves and the multiplication of distributions. Class. Quantum Grav. **14**, 455 (1997). doi:10.1088/0264-9381/14/2/018

6. Kunzinger, M., Steinbauer, R.: A rigorous solution concept for geodesic and geodesic deviation equations in impulsive gravitational waves. J. Math. Phys. **40**, 1479 (1999). doi:10.1063/1.532816
7. Sanchez, N., de Vega, H.: Particle scattering at the Planck scale and the Aichelburg-Sexl geometry. Nucl. Phys. **317B**, 731 (1989)

Effective Vacuum Bianchi IX in Loop Quantum Cosmology

Alejandro Corichi, Asieh Karami and Edison Montoya

Abstract In classical general relativity, the chaotic behavior of the Bianchi IX model can be useful to describe the generic local evolution near a singularity. However, one expects that quantum effects can modify it. In this contribution we show that the modifications which come from Loop Quantum Cosmology imply a non-chaotic effective behavior.

1 Introduction

Bianchi models are spatially homogeneous models such that the symmetry group $\mathscr{S}$ acts simply and transitively on a space manifold $\Sigma \cong \mathscr{S}$. The symmetry group for Bianchi IX model is generated by three spatial rotations on a 3-sphere. We identify this group with SU(2) to define fiducial frames and co-frames. The fiducial cell is a 3-sphere with radius a_o (=2) and its volume is $V_o = 2\pi^2 a_o^3$. We define $\ell_o = V_o^{1/3}$ and $\vartheta = \ell_o/a_o = (2\pi^2)^{1/3}$. In terms of the phase space variables used in loop quantum gravity (LQG) [1–3], a connection A_a^i and a densitized triad E_i^a, the classical constraint is

A. Corichi (✉) A. Karami · E. Montoya
Centro de Ciencias Matemáticas, Universidad Nacional Autónoma de México, UNAM-Campus Morelia, Morelia, Mexico
e-mail: corichi@matmor.unam.mx

A. Karami · E. Montoya
Instituto de Física y Matemáticas, Universidad Michoacana de San Nicolás de Hidalgo, Morelia, MICH, Mexico
e-mail: karami@ifm.umich.mx

E. Montoya
e-mail: edison@ifm.umich.mx

J. Bičák and T. Ledvinka (eds.), *Relativity and Gravitation*,
Springer Proceedings in Physics 157, DOI: 10.1007/978-3-319-06761-2_80,

$$\mathscr{C}_H = \int_{\mathscr{V}} N \left[-\frac{E_i^a E_j^b}{16\pi G \gamma^2 \sqrt{|q|}} \epsilon^{ij}_{\ k} \left(F_{ab}^k - (1+\gamma^2) \Omega_{ab}^k \right) + \mathscr{H}_{\text{matter}} \right] \mathrm{d}^3 x \,, \quad (1)$$

where N is the lapse function, $\mathscr{H}_{\text{matter}} = \rho V$ and Ω_{ab} is the curvature of the spin connection Γ_a^i which is compatible with the triads. In what follows, we take $N = 1$ and since we work in vacuum, ρ is equal to zero. If we restrict ourselves to diagonal Bianchi IX model, we can parametrize A_a^i and E_i^a as $A_a^i = c^i {}^o\omega_a^i / \ell_o$ and $E_i^a = p_i \sqrt{{}^o q}\, {}^o e_i^a / \ell_o^2$, the p_i's in terms of scale factors a_i are $|p_i| = \ell_o^2 a_j a_k$, and the volume is $V = \sqrt{|p_1 p_2 p_3|}$. The nonzero Poisson brackets are given by $\{c_i, p_j\} = 8\pi G \gamma \delta_{ij}$, where γ is the Barbero-Immirzi parameter.

To quantize the Hamiltonian constraint in (1), we find an operator corresponding to F_{ab} and we express the connection A_a^i in terms of holonomies [4]. The operators associated to the connection are then given by $\hat{c}_i = \widehat{\sin \bar{\mu}_i c_i / \bar{\mu}_i}$, where $\bar{\mu}_i = \lambda \sqrt{p_i / p_j p_k}$, $i \neq j \neq k$ and $\lambda^2 = 4\sqrt{3}\pi \gamma l_p^2$ is the smallest eigenvalue of area in LQG.

For the term which contains the inverse of the metric determinant, and for those which contain the negative powers of p_i's, we use the Thiemann strategy [3]. The idea is to find some classical equivalent expression for them in terms of holonomies and positive powers of p's and then quantize this expression. For instance, to quantize a negative power of p_i we know that, classically, there is the following identity

$$|p_i|^{(\ell-1)/2} = -\frac{\sqrt{|p_i|}}{4\pi G \gamma \mu_i \ell j(j+1)} \tau_i h_i^{(\mu_i)} \{ h_i^{(\mu_i)-1}, |p_i|^{\ell/2} \} \,, \quad (2)$$

where μ is the length of a curve which is used for calculating the holonomy, ℓ is a number between 0 and 1 and $j \in \frac{1}{2}\mathbb{N}$ labels the representation. For simplicity we take $j = 1/2$ and choose $\mu_i = \bar{\mu}_i \ell_o$ because they appear in the terms corresponding to curvature. Since the largest negative power of p's which appears in the constraint is $-1/4$ we will take $\ell = 1/2$ to obtain it directly from the above equation. After that, we express the other negative powers by it. The eigenvalues for the operator $\widehat{|p_i|^{-1/4}}$ are

$$J_i = \frac{h(V)}{V_c} \prod_{j \neq i} p_j^{1/4} \,, \text{ where } h(V) = \sqrt{V + V_c} - \sqrt{|V - V_c|}, \; V_c = 2\pi \gamma \lambda \ell_p^2 .$$

The correction term which comes from the $\epsilon_k^{ij} E_i^a E_j^b / \sqrt{|q|}$ is $A(V) = (V + V_c - |V - V_c|)/2V_c$. Hence, with these definitions one obtains the corresponding constraint operator.

In this work, we are interested in studying the classical effective Hamiltonian which has some modifications from the quantum theory to gain qualitative insights into the leading order quantum effects. Since the Hamiltonian is invariant under parity, we restrict ourselves to positive p_i's. The effective Hamiltonian which is derived from the quantum theory is

$$\mathscr{H}_{\text{eff}} = -\frac{V^4 A(V) h^6(V)}{8\pi G V_c^6 \gamma^2 \lambda^2}\Bigg(\sin\bar{\mu}_1 c_1 \sin\bar{\mu}_2 c_2 + \sin\bar{\mu}_1 c_1 \sin\bar{\mu}_3 c_3$$
$$+ \sin\bar{\mu}_2 c_2 \sin\bar{\mu}_3 c_3\Bigg) + \frac{\vartheta A(V) h^4(V)}{4\pi G V_c^4 \gamma^2 \lambda}\Bigg(p_1^2 p_2^2 \sin\bar{\mu}_3 c_3 + p_2^2 p_3^2 \sin\bar{\mu}_1 c_1$$
$$+ p_1^2 p_3^2 \sin\bar{\mu}_2 c_2\Bigg) - \frac{\vartheta^2 (1+\gamma^2) A(V) h^4(V)}{8\pi G V_c^4 \gamma^2}\Bigg(2V\Big[p_1^2 + p_2^2 + p_3^2\Big]$$
$$- \Big[(p_1 p_2)^4 + (p_1 p_3)^4 + (p_2 p_3)^4\Big]\frac{h^6(V)}{V_c^6}\Bigg).$$

2 The Effective Potential

It is helpful to use the potential term of the constraint to study the solutions. The classical potential which comes from the spin connection's curvature in the classical constraint, in terms of Misner variables is [5]

$$W = \frac{1}{2}e^{-4\Omega}\left(e^{-4\beta_+} - 4e^{-\beta_+}\cosh\sqrt{3}\beta_- + 2e^{-2\beta_+}[\cosh 2\sqrt{3}\beta_- - 1]\right), \quad (3)$$

where $\Omega = -\frac{1}{3}\log V$ and the anisotropies $\beta_\pm$ are defined via $a_1 = e^{-\Omega+(\beta_+ + \sqrt{3}\beta_-)/2}$, $a_2 = e^{-\Omega+(\beta_+ - \sqrt{3}\beta_-)/2}$ and $a_3 = e^{-\Omega-\beta_+}$. Since the Ω dependence factorizes, one can obtain an anisotropy potential $\mathscr{V}(\beta_+, \beta_-)$ which exhibits exponential walls for large anisotropies. The universe can be seen as a particle moving in such a potential (W) that presents reflections at the walls. An infinite number of these reflections implies that the system behaves chaotically. When the volume becomes small, the quantum effects become important and one should work with the full quantum theory, but one can use the effective equations to have a qualitative view of what happens near the classical singularity. From the effective Hamiltonian, the modified potential can be derived as a function of p_i:

$$W = -\frac{V^2 A(V) h^4(V)}{V_c^4}\left(p_1^2 + p_2^2 + p_3^2 - \Big[(p_1 p_2)^4 + (p_1 p_3)^4 + (p_2 p_3)^4\Big]\frac{h^6(V)}{2V V_c^6}\right).$$

For a simple case, when $\beta_- = 0$ and $\beta_+ \to -\infty$, the classical potential is $W(\beta_+, \Omega) \sim \frac{1}{2}e^{-4\Omega-4\beta_+}$. If we rewrite the modified potential in terms of Misner variables we can see that in this limit, the modified potential behaves as $\frac{1}{2V_c^9}e^{-52\Omega-4\beta_+}$ where the β_+-dependency of both classical and modified potential are the same, so we have an infinite wall for the modified potential too, (see Fig. 1, left). Note however that, for small volumes, the modified potential can be negative at some points.

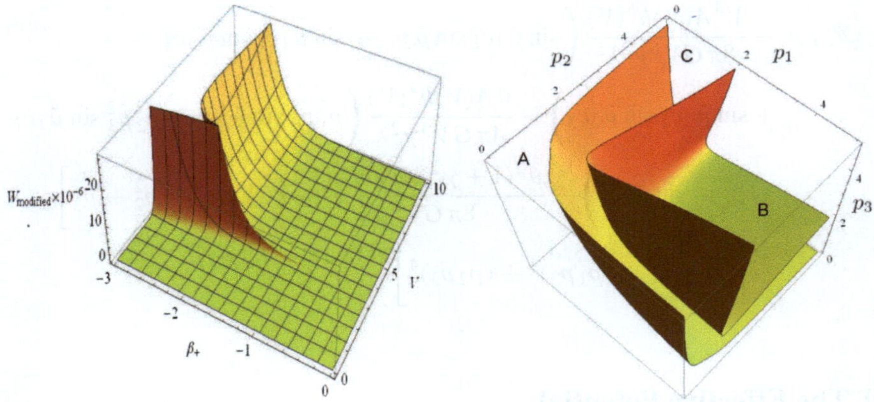

Fig. 1 *Left* modified potential when $\beta_- = 0$. *Right* zero surfaces of the maximum allowed density; density in regions A and B can be non-negative but in region C it is always negative; both in Planck units

3 Density

In the general case, as can be seen in Fig. 1 (right), the maximum allowed density (which arises from the modified Hamiltonian by choosing the sine functions equal to -1), has two distinct disconnected regions with positive values. Therefore, if we impose the weak energy condition and start the evolution within one region, the universe cannot reach the other region. To study the vacuum Bianchi IX, we start from large volumes which lie in region B of Fig. 1 (right) and, as we go to smaller volumes, we cannot reach zero volume because 'crossing' to region A is not allowed. Therefore, there is a smallest reachable volume in region B and, since very large anisotropies are not allowed near this smallest volume, and the modified potential is not too large there, then we have, at most, finite oscillations before reaching the bounce. On the other hand, in the internal region A, the anisotropies are very large when some of the p_i are very small, and then the volume of the universe cannot be large enough to start the evolution from there.

4 Conclusions

We have studied the behavior of a modified potential for the Bianchi IX model when quantum effects become important. We showed that the potential wall does not disappear and we have potential chaotic behavior near the classical singularity. However, if the weak energy condition holds and if we start from large volumes and evolve the equations into small volumes, there will be a lower bound for volumewithin region B, and one does not reach region A (connected to zero volume). Since there are no large anisotropies near the smallest allowed volume, the solutions will *not* exhibit chaotic behavior.

References

1. Ashtekar, A., Lewandowski, J.: Background independent quantum gravity: a status report. Class. Quantum Grav. **21**, R53 (2004). doi:10.1088/0264-9381/21/15/R01
2. Rovelli, C.: Quantum Gravity. Cambridge Monographs on Mathematical Physics. Cambridge University Press, New York (2004)
3. Thiemann, T.: Modern Canonical Quantum General Relativity. Cambridge Monographs on Mathematical Physics. Cambridge University Press, New York (2007)
4. Ashtekar, A., Wilson-Ewing, E.: Loop quantum cosmology of Bianchi type II models. Phys. Rev. D **80**, 123532 (2009). doi:10.1103/PhysRevD.80.123532
5. Misner, C.: Mixmaster universe. Phys. Rev. Lett. **22**, 1071 (1969). doi 10.1103/PhysRevLett. 22.1071

Coupling Dimers to CDT to Obtain Higher Order Multicritical Behavior

Lisa Glaser

Abstract This contribution reviews some recent results on dimers coupled to CDT. A bijective mapping between dimers and tree-like graphs allows for a simple way to introduce dimers to CDT. This can be generalized further to obtain different multicritical points.

1 Introduction

Causal Dynamical Triangulations (CDT) is a proposed theory of quantum gravity. In CDT the path integral for gravity is regularized through simplices as in dynamical triangulation. CDT introduces a preferred time slicing to provide for a well-defined Wick rotation. This preferred time slicing leads to a better behaved continuum theory [1] (see [2] for a review).

For matrix models it is well-known that random lattices can be coupled to matter, like dimers or the Ising model, to find quantum gravity coupled to conformal field theories [3, 4]. It is then an interesting prospect to try and couple matter to the random lattices of CDT.

In this article we review the results obtained in [5] and present a simple extension of the model which allows for higher order multicritical points.[1]

[1] Quite similar results have been obtained simultaneously in [6, 7].

L. Glaser (✉)
The Niels Bohr Institute, Copenhagen University, Blegdamsvej 17,
2100 Copenhagen Ø, Denmark
e-mail: glaser@nbi.dk

J. Bičák and T. Ledvinka (eds.), *Relativity and Gravitation*,
Springer Proceedings in Physics 157, DOI: 10.1007/978-3-319-06761-2_81,

2 Coupling CDT to Dimers

Durhuus et al. [8] proved that there is a bijective mapping between rooted tree graphs and CDT (see Fig. 1). This bijection makes it possible to determine the critical exponents of CDT using recursive equations as in [9]. It also makes it possible to consider the easier problem of coupling dimers to a rooted tree graph instead of directly placing them on the CDT. The simple rule of placing any number of hard dimers on the tree will lead to a partition function which allows for new multicritical behavior [5].

Hard dimers are a type of matter with local interactions. One can imagine a dimer like a fixed rod that can be added to the lattice on any link whose neighboring links are not taken up with dimers. This is illustrated in Fig. 1 on the right.

The partition function for CDT with dimers reads

$$Z(\mu, \xi) = \sum_{BP} e^{-\mu} \sum_{\mathrm{HD(BP)}} \xi^{|\mathrm{HD(BP)}|}, \tag{1}$$

where BP is the set of all tree-like graphs, HD(BP) the set of all dimer configurations on that graph and HD(BP) the number of dimers in a given configuration. This partition function can be solved using recursive equations which arise for the tree like graphs and are discussed in detail in [9]. The recursion depicted in Fig. 2 leads to the equations

$$Z = e^{-\mu}\left(\frac{1}{1-Z} + W\frac{1}{(1-Z)^2}\right), \qquad W = e^{-\mu}\xi\left(\frac{1}{1-Z}\right), \tag{2}$$

where Z is the partition function for a tree with a normal link at the root and W is the partition function for a tree rooted in a dimer. At a n-multicritical point the first $n-1$ derivatives of the coupling μ by the partition function Z are zero

$$\left.\frac{\partial \mu}{\partial Z}\right|_{Z_c} = \cdots = \left.\frac{\partial^{n-1} \mu}{\partial Z^{n-1}}\right|_{Z_c} = 0. \tag{3}$$

We can then solve equations (2) to find the third multicritical point at $Z_c = \frac{5}{8}$, $\xi_c = -\frac{1}{12}$ and $e^{\mu_c} = \frac{32}{9}$. The critical exponents at this point are $\gamma = \frac{1}{3}$, $d_H = \frac{3}{2}$ and $\sigma = \frac{1}{2}$. In pure CDT, not coupled to dimers, one finds $\gamma = \frac{1}{2}$, $d_H = 2$ and σ is not defined, so it is clear that CDT coupled to dimers lies in a different universality class than pure CDT. Therefore it represents an interacting system of matter and gravity. However the negative weight ξ does make the physical interpretation of the results less clear [10]. It is easy to generalize this model to higher order multicritical points. To do so one introduces different types of dimers, denoted as type a with weight ξ and type b with weight ζ. The rule is then that a vertex with an incoming dimer can not spawn any type of dimer, while a vertex with an incoming empty link can spawn at most one dimer of each color. This is illustrated in Fig. 2. This model leads to the partition function

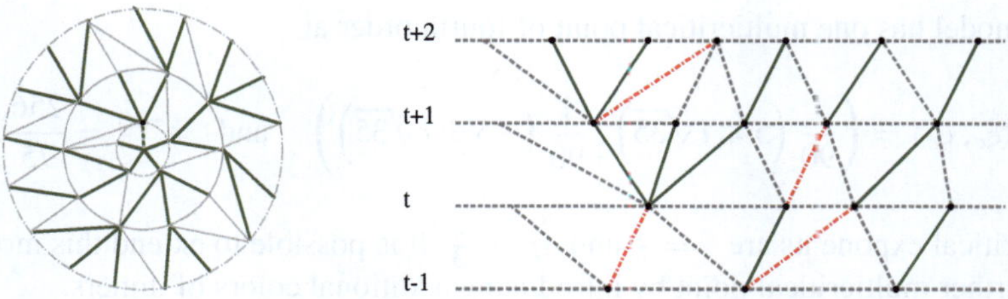

Fig. 1 On the *left* is an embedding of a simple CDT. The *green marked lines* on it are those that are also part of the graph. These suffice to characterize the entire CDT. If we only had the *green lines* we could get back to the full CDT by just reintroducing the space-like links and the leftmost link at every vertex. The figure on the *right* are three time slices of a CDT. The tree graph is marked *green*. The *red* markings indicate dimers, which can be placed on the tree so as to be non-touching

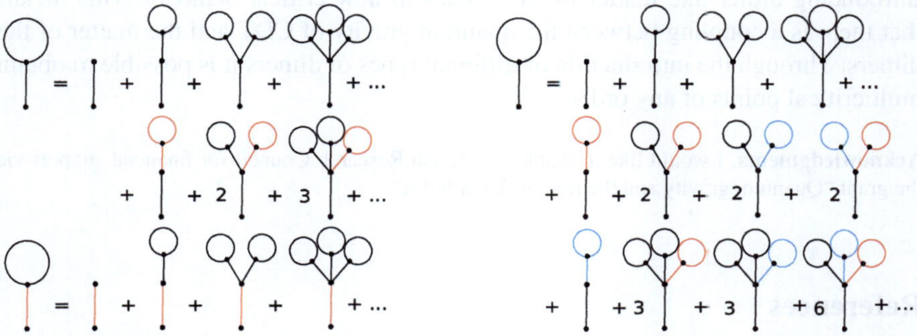

Fig. 2 These figures are pictorial representations of equations (2), (5). Dimers are depicted as *red* (*blue*) links. The *upper left figure* is equivalent to the first equation in (2) while the *lower left* is the second equation there. The *right figure* corresponds to equation (5)

$$Z(\mu, \xi, \zeta) = \sum_{\text{BP}} e^{-\mu} \sum_{\text{HD(BP(a,b))}} \xi^{|a|} \zeta^{|b|}, \tag{4}$$

where HD(BP(a, b)) denotes the set of configurations of hard dimers of type a and b and $|a|$ ($|b|$) is the number of dimers of type a (b) in the configuration. It can again be solved using recursive equations for the tree graphs

$$Z = e^{-\mu}\left(\frac{1}{1-Z} + \frac{1}{(1-Z)^2}(W+V) + \frac{WV}{(1-Z)^3}\right), \tag{5}$$

where W denotes the partition function starting in a dimer of type a and V for dimers of type b. For W and V we obtain equations like (2).

This model has one multicritical point of fourth order at

$$(\xi_c, \zeta_c) = \left(\frac{1}{90}\left(5 \mp i\sqrt{35}\right), \frac{1}{90}\left(-5 \pm i\sqrt{35}\right)\right) \quad \text{and} \quad e^{-\mu_c} = \frac{256}{75}.$$

The critical exponents are $\gamma = \frac{1}{4}$ and $d_H = \frac{4}{3}$. It is possible to extend this model to any further multicritical point by introducing additional colors of dimers.

3 Summary

Introducing dimer-like matter to CDT leads to new critical behavior. This means that there is a coupling between the quantum gravity of CDT and the matter of the dimers. Through the introduction of different types of dimers it is possible to obtain multicritical points of any order.

Acknowledgments I would like to thank the Danish Research Council for financial support via the grant "Quantum gravity and the role of Black holes".

References

1. Ambjørn, J., Loll, R.: Non-perturbative Lorentzian quantum gravity, causality and topology change. Nucl. Phys. B **536**, 407 (1998). doi:10.1016/S0550-3213(98)00692-0
2. Ambjørn, J., Görlich, A., Jurkiewicz, J., Loll, R.: Nonperturbative quantum gravity. Phys. Rep. **519**, 127 (2012). doi:10.1016/j.physrep.2012.03.007
3. Kazakov, V.: The appearance of matter fields from quantum fluctuations of 2d-gravity. Mod. Phys. Lett. A **04**, 2125 (1989). doi:10.1142/S0217732389002392
4. Staudacher, M.: The Yang-Lee edge singularity on a dynamical planar random surface. Nucl. Phys. B **336**, 349 (1990). doi:10.1016/0550-3213(90)90432-D
5. Ambjørn, J., Glaser, L., Görlich, A., Sato, Y.: New multicritical matrix models and multicritical 2d CDT. Phys. Lett. B **712(1–2)**, 109 (2012). doi:10.1016/j.physletb.2012.04.047
6. Atkin, M.R., Zohren, S.: An analytical analysis of CDT coupled to dimer-like matter. Phys. Lett. B **712**, 445 (2012). doi:10.1016/j.physletb.2012.05.017
7. Atkin, M., Zohren, S.: On the quantum geometry of multi-critical CDT. J. High Energy Phys. **2012**(11), 037 (2012). doi:10.1007/JHEP11(2012)037
8. Durhuus, B., Jonsson, T., Wheater, J.: On the spectral dimension of causal triangulations. J. Stat. Phys. **139**(5), 859 (2009). doi:10.1007/s10955-010-9968-x
9. Ambjørn, J., Durhuus, B., Jonsson, T.: Quantum Geometry: A Statistical Field Theory Approach. Cambridge Monographs on Mathematical Physics. Cambridge University Press, Cambridge (1997)
10. Glaser, L.: Coupling dimers to CDT. In: Rosquist, K., Jantzen, R., Ruffini, R. (eds.) Thirteenth Marcel Grossman Meeting on Recent Developments in Theoretical and Experimental General Relativity, Astrophysics, and Relativistic Field Theories. World Scientific, Singapore (2013)

A Sheet of Graphene: Quantum Field in a Discrete Curved Space

Nikodem Szpak

Abstract The dynamics of electrons in a sheet of graphene can be described as a quantum field living in a discrete space—the graphene honeycomb lattice. As this space can be curved in various ways, the system offers a fascinating tool for studying and simulating the impacts of non-trivial geometries on quantum fields living in it. Local and global deformations as well as defects of the lattice can be mapped, via a discrete differential geometry, onto curvature and torsion in the continuous analog model. This allows for physical simulation and observation of quantum evolution and scattering in curved geometry and interaction with torsion. Time-dependent lattice perturbations, such as sound waves, can be interpreted as dynamical geometry and mimic gravitational waves. The immanent quantum character of the lattice structure—composed of carbon atoms—can be used for proposing a physical simulator of quantum geometry. We discuss the main ideas constituting these analogies, the latter being the topic of our ongoing project.

1 Introduction

Graphene, probably the most intensively studied material in the last decade, consists of a regular two-dimensional lattice of carbon atoms. Scientists try to produce, measure and model graphene in order to better understand its many original features. But graphene can also be seen as a laboratory offering the unique possibility of quantum simulations of relativistic phenomena which are known from particle physics but are still not accessible in experiments. Its crucial feature is the energy dispersion relation for electronic excitations which is similar to that of the relativistic Dirac equation. Thus, the excitations of the ground state of graphene behave like relativistic elementary particles and offer new insights into some ultra-relativistic processes. One

N. Szpak (✉)
Faculty of Physics, University Duisburg-Essen, Essen, Germany
e-mail: nikodem.szpak@uni-due.de

J. Bičák and T. Ledvinka (eds.), *Relativity and Gravitation*, Springer Proceedings in Physics 157, DOI: 10.1007/978-3-319-06761-2_82,

of the new directions of graphene research, made possible by the newest scanning techniques, is related to the investigation of its non-flat (curvature) and non-perfect (defects) crystalline structures and their influence on its macroscopic properties [1].

The dynamics of quasi-particles hopping on the crystal lattice can be effectively described as a class of discrete Hubbard Hamiltonians [2], which have been extensively studied in the literature in the past half of the century [3]. Universality of the Hubbard class allows for matching together physical systems of very different origin but having the same mathematical description and designing new simulators in which one system mimics another. Deep connections exist between the lattice Hamiltonians and discretized fundamental relativistic field theories [4–7]. In the current project, we are working to extend this analogy to curved and dynamically perturbed lattices by involving the quickly developing language of discrete differential geometry [8]. Similarly, as is the case in numerical methods based on finite difference schemes, it should allow for the approximation of the dynamics of continuous quantum fields by their evolution in discrete spaces. These relations offer a new exciting possibility of designing lattice quantum simulators for continuous quantum fields evolving in curved geometries. We plan to conduct a systematic investigation of various types of lattice perturbations and to assign them to geometric objects such as metric, connection, curvature or torsion. The advantage of this approach should be two-fold: firstly, it should provide a new elegant language for the effective description of the curved and defective structures in graphene, which should help to better understand their properties since the continuous analog field equations are mathematically much more treatable than the corresponding discrete systems. Secondly, the language should allow to interpret these types of defects as quanta of curvature or torsion and give rise to a new, very interesting type of simulators of quantum fields living in curved spaces or even of quantum geometry.

2 Discrete Differential Geometry from Hubbard Models

The basic Hubbard Hamiltonian has the form

$$\hat{H} = \sum_{\langle n,m\rangle} T_{nm}\, \hat{\psi}_n^\dagger\, \hat{\psi}_m + \sum_n V_n\, \hat{\psi}_n^\dagger\, \hat{\psi}_n + \text{h.c.}, \tag{1}$$

where $\hat{\psi}_n^\dagger$ and $\hat{\psi}_n$ stand for creation and annihilation of a (quasi)particle at site n. Usually, the first sum is carried over all neighboring pairs of sites denoted as $\langle n, m\rangle$. The hopping parameters T_{nm} represent probabilities for a (quasi)particle to jump between the sites n and m in a given time and V_n represents a local potential at site n.

The simplest example of a relation between the Hubbard model and a field dynamics is the scalar field $\phi_n = \phi(x_n)$ in one dimension discretized on a non-uniform lattice $x_n = (n+\varepsilon_n)\ell$ where ε_n parameterize small displacements of the regular grid. The Laplace-Beltrami operator Δ_g, defined in a Riemannian geometry with metric g_{ij}, has the continuous representation

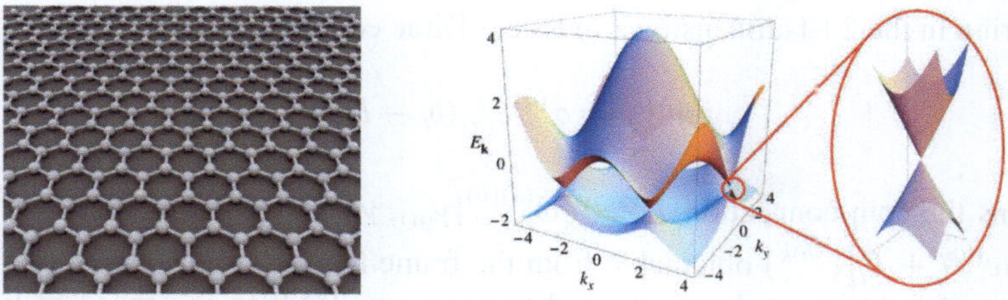

Fig. 1 The hexagonal lattice of graphene (*left*) and the dispersion relation for the Hubbard model with one Dirac cone magnified (*right*)

$$\Delta_g = \frac{1}{\sqrt{|g|}} \partial_i \left(\sqrt{|g|} g^{ij} \partial_j \phi \right), \tag{2}$$

where $|g| = |\det g_{ij}|$. Going from the continuum to the lattice, as is also done in numerical calculations, one needs a discrete scheme for calculating derivatives, e.g. $\partial_x \phi_n = (\phi_{n+1} - \phi_n)/a_n$ where $a_n = x_{n+1} - x_n = (1 + \varepsilon_{n+1} - \varepsilon_n)\ell$ is the distance between two neighboring lattice sites. Then, a discretized Laplacian on the lattice reads

$$\bar{\Delta}\phi_n = \frac{1}{a_n \ell} \bar{\partial}_x \left(\frac{1}{a_n \ell} \bar{\partial}_x \phi_n \right). \tag{3}$$

The dynamics of a quantum particle satisfying the nonrelativistic Schrödinger equation $H\phi = -\Delta\phi + V\phi$ can be thus approximated on the lattice by a Hubbard Hamiltonian (1) with the hopping parameters $T_{n,n+1} \equiv J_n = [-1 + 2(\varepsilon_{n+1} - \varepsilon_n)]/\ell^2$ and $V_n = [2 - 2(\varepsilon_{n+1} - \varepsilon_{n-1})]/\ell^2$. Since the latter describes crystalline systems in solid state physics it allows for the construction of a class of analog models in which lattice systems with given J_n and V_n can mimic quantum physics in curved space with the effective metric $(g_{11})_n = (a_n)^2 = (1 + \varepsilon_{n+1} - \varepsilon_n)^2 \ell^2$.

In higher dimensions the situation is more complex—in addition to parameters representing distances between the neighbors the angles between the lattice links also enter the formula for the discretized Laplace-Beltrami operator. The language of non-orthonormal frames, in which the metric $g_{(a)(b)} = \mathbf{e}_{(a)} \cdot \mathbf{e}_{(b)}$ is related to the co-frame $\mathbf{e}_{(a)}$ spanned by the links attached to each lattice vortex, is then more suitable. In this approach, perturbations of the metric correspond to the strain tensor $\delta g_{(a)(b)} = \varepsilon_{ab} = \frac{1}{2}(\partial_a u_b + \partial_b u_a)$ where u_a is the deformation field describing perturbations of the lattice. The same strain tensor appears in models describing electrons in deformed crystals [9], whenever the dispersion relation is quadratic at low energies, i.e. $E(\mathbf{k}) \sim \delta_{ab} k^a k^b$. It allows for the construction of analog models in which the lattice strain ε_{ab} gives rise to curved geometry with an effective metric $g_{ab} = \delta_{ab} + \varepsilon_{ab}$.

In contrast, in hexagonal lattices like graphene, the dispersion relation is linear at low energies (Fig. 1), i.e. $E(\mathbf{k}) \sim \pm|k|$, and the effective discrete field behaves like a Dirac spinor and the metric does not couple directly to it. As is known from the continuous Dirac equation defined on a curved space, the covariant derivative appearing in the 2+1-dimensional massless Dirac equation

$$i\partial_t \Psi = \bar{c}\,\sigma^{(a)} e^{\;i}_{(a)}(\partial_i - \Gamma_i)\Psi$$

contains the spin connection $\Gamma_i = \frac{1}{8}\omega_i^{(a)(b)}[\gamma_{(a)}, \gamma_{(b)}]$ with coefficients $\omega_i^{(a)(b)} = e^{(a)}_{\;j}(\partial_i e^{(b)j} + \Gamma^j_{ik} e^{(b)k})$ obtainable from the frame field $e^{\;i}_{(a)}$ only. $\bar{c}$ is the effective velocity of light in graphene, about 300 times smaller than the speed of light in vacuum c.

There also exists a mechanism generating mass in the Dirac field based on breaking the hexagonal symmetry. The so called Kekule perturbations [10] modify the hopping parameters in such a way that the energy spectrum splits into two gapped bands and the Dirac points disappear leaving relativistic dispersion relation for massive particles $|E(\mathbf{k})| \approx \pm\sqrt{m^2 + \mathbf{k}^2}$ at low energies. The mass of the excitations can be freely tuned by the strength of the Kekule perturbation.

This language enables for the description of several types of plastic deformations of graphene like strain, ripples or folding by means of external curvature from embedding in the third dimension and has been applied to study the electronic properties of curved graphene (for a review see Vozmediano et al. [11] and the references therein).

3 Curvature and Torsion from Lattice Defects

All lattice deformations discussed so far have had a character of small perturbations, not changing its global structure. However, it is virtually impossible to have real crystals with macroscopic curvature without structural defects. Therefore, in our current project, we focus on lattice irregularities that have an impact on the long range lattice structure. This requires an extension of the current approach by the description of structural defects like dislocations and disclinations. These effects can introduce effective curvature and torsion fields, giving rise to Riemann-Cartan geometries in which the simulated quantum fields can evolve.

Geometrization of the defects has been initiated by Kleinert [12] and applied to graphene by Vozmediano et al. [11]. Extending the geometric language based on non-orthonormal frames allows for a direct interpretation of defects in terms of (a discrete) differential geometry: disclinations appear to be quanta of curvature while dislocations appear to be quanta of torsion [12, 13].

4 Open Problems

The elegant geometric language has one drawback, presenting an obstacle in applications to physical systems. The framework proposed so far assumes that the frame field $e^{\;i}_{(a)}$, defined directly by the lengths and angles between the lattice links, gives rise to a natural connection $\Gamma^{(a)}_{(b)(c)} = e^{(a)}_{i}\partial_{(b)}e^{i}_{(c)}$ which defines parallelism on the lattice (cf. [14]). These connection coefficients tell the particles moving in the curved geometry which lines are straight and which are bent. While in the discretized differential geometry that approach seems to be quite natural, when applied to physical crystals it appears inaccurate. Electrons in crystals satisfy a fundamental *ab initio* quantum description and the notion of parallelism (straight lines) must derive from its properties, not from the geometry of the lattice links only. In other words, it is an idealizing assumption that the effective continuous geometric fields (frame, metric, curvature, torsion) are identical to those of the underlying lattice geometry. However, in the quantum simulators these are the Hamiltonians that have to match, not the geometries. Therefore, we see an urgent need to improve the current state-of-the-art by developing methods for deriving the real connection coefficients from the underlying quantum physical models.

5 Applications

The presented language opens the way for the modeling and simulation of very interesting effects in which quantum fields interact with curvature and torsion either distributed across the lattice or in the form of localized singularities. Once it has been fully developed, specific examples of quantum fields and their coupling to curvature can be considered. The simplest examples which can be studied are graphene sheets with point or line defects. Already a single dislocation defect (quantum of a torsion in the analog model) is theoretically very interesting. In some sense, it corresponds to a vortex on which the scattering of quantum fields poses a very complex problem with many surprises, e.g. recently discovered bound states [15, 16]. The vortex background is also related to rotating black-hole spacetimes (e.g. the Kerr black hole) for which many conjectures about the field behavior (energy conservation, boundedness) remain open.

Line defects in two dimensions correspond to grain boundaries and occur commonly in graphene. They can be simulated in the analog models with extended torsion density (cf. [17]) and bear some similarities to cosmic strings.

Further, the so called "5-7" defects introduce significant elastic distortions in the lattice around [18], bending the nearby "straight lines". Especially interesting are the defects giving rise to the global angle defects, such as the "5" or "7"-rings (Fig. 2), which correspond to a curvature quanta in the analog model. They bend the "geodesic" lines around the curvature centers in a way corresponding to gravity or anti-gravity (positive or negative mass, respectively).

Similar ideas to those described here, have been also proposed for optical lattices—another discrete system with very promising features as an analog system for quantum fields. Analogs of cosmological spacetimes [19] or torsion [20] have been discussed but have not been realized yet.

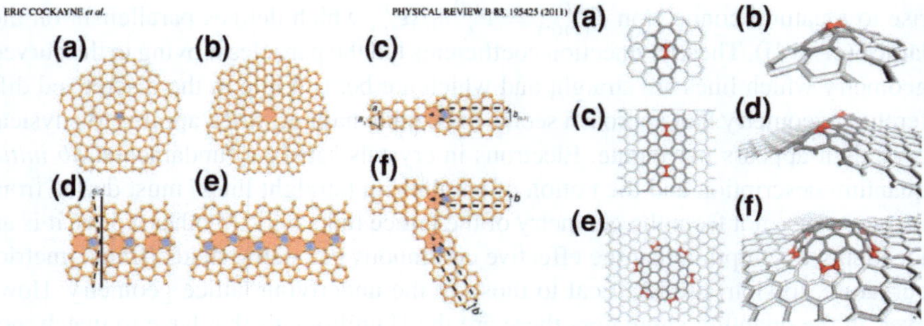

Fig. 2 Flat (*left*) and bumpy (*right*) graphene defects

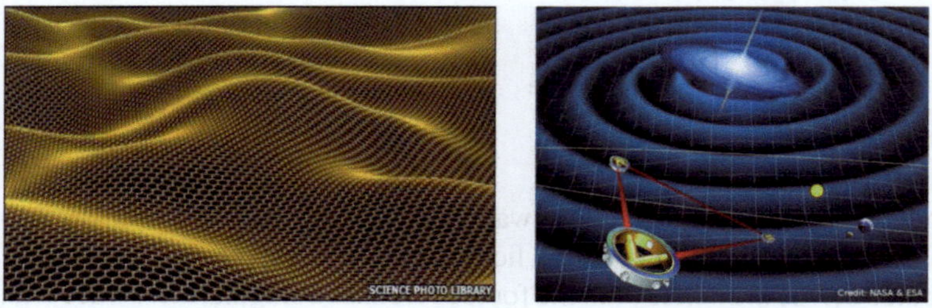

Fig. 3 Graphene ripples (*left*) compared to gravitational waves modifying space geometry (*right*)

6 Outlook

Graphene also allows quantum simulators with dynamical geometries mimicking the evolution of (quantum) fields in the presence of passing gravitational waves to be considered. Sound waves in graphene, causing tiny changes in the carbon atom positions, can represent time-dependent perturbations of the lattice geometry and mimic the metric waves. This idea seems especially attractive since it aims to provide a simulator with truly time-dependent geometry as we know it only from General Relativity (Fig. 3).

Another very attractive idea appearing to be within reach is the simulation of quantum geometry. Considering slightly displaced atoms in a crystal as a base for the effective geometry in which the valence electrons live and treating the atoms themselves as quantum objects should lead to some form of quantum geometry (cf. spin net and spin foam models [21, 22])—a very hotly discussed topic in Quantum Gravity. The phenomenon of defect dynamics, in which defects slowly propagate on the lattice, change type or disappear, adds another perspective for studying dynamical quantum geometries.

7 Conclusions

Graphene is a very flexible material with rich geometric possibilities. It provides an ideal model for implementing both types of curvature appearing in the lattice models: (a) the extrinsic curvature present due to folding and embedding of the two-dimensional graphene sheets in three dimensions and (b) the intrinsic curvature and torsion coming into play due to structural defects. Moreover, graphene allows for the design of closed surfaces with nontrivial topologies, as is the case for fullerenes, nanotubes and similar structures.

Configurations in which the effective Hamiltonians, in the continuous limit, correspond to quantum fields interacting with curvature or torsion open up the possibility of designing a new type of quantum simulators with some hope of proposing the first physical simulator of quantum geometry for quantum fields. It needs to be mentioned that the problem of the evolution of quantum fields in the presence of curvature and torsion still poses some fundamental questions of a theoretical nature and as such has never been observed in any laboratory experiment. Any access to such systems would be highly desirable.

The increasing interest in the technological applications of graphene requires the development of an efficient and easy-to-use geometric language (toolbox) for the classification and modeling of various types of graphene geometries as well as the prediction, description and study of new phenomena appearing in defective graphene structures. Understanding the implications of defects on its physical properties and purposive usage of the defects can be another big step in the engineering of new materials.

References

1. Sun, G., Jia, J., Xue, Q., Li, L.: Atomic-scale imaging and manipulation of ridges on epitaxial graphene on 6H-SiC(0001). Nanotechnology **20**, 355701 (2009). doi:10.1088/0957-4484/20/35/355701
2. Hubbard, J.: Electron correlations in narrow energy bands. Proc. R. Soc. Lond. Ser. A **276**(1365), 238 (1963). doi:10.1098/rspa.1963.0204
3. Montorsi, A. (ed.): The Hubbard Model. World Scientific, Singapore (1992)
4. Giuliani, A., Mastropietro, V., Porta, M.: Lattice quantum electrodynamics for graphene. Ann. Phys. (N.Y.) **327**, 461 (2012). doi:10.1016/j.aop.2011.10.007
5. Katsnelson, M., Novoselov, K.: Graphene: new bridge between condensed matter physics and quantum electrodynamics. Solid State Commun. **143**, 3 (2007). doi:10.1016/j.ssc.2007.02.043
6. Szpak, N., Schützhold, R.: Quantum simulator for the Schwinger effect with atoms in bichromatic optical lattices. Phys. Rev. A **84**, 050101(R) (2011). doi:10.1103/PhysRevA.84.050101
7. Szpak, N., Schützhold, R.: Optical lattice quantum simulator for quantum electrodynamics in strong external fields: spontaneous pair creation and the Sauter-Schwinger effect. New J. Phys. **14**, 035001 (2012). doi:10.1088/1367-2630/14/3/035001
8. Bobenko, A., Schröder, P., Sullivan, J., Ziegler, G. (eds.): Discrete Differential Geometry, Oberwolfach Seminars, vol. 38. Birkhäuser, Basel (2008)
9. Bir, G., Pikus, G.: Symmetry and Strain-Induced Effects in Semiconductors. Wiley, New York (1974)

10. Chamon, C.: Solitons in carbon nanotubes. Phys. Rev. B **62**, 2806 (2000). doi:10.1103/PhysRevB.62.2806
11. Vozmediano, M., Katsnelson, M., Guinea, F.: Gauge fields in graphene. Phys. Rep. **496**, 109 (2010). doi:10.1016/j.physrep.2010.07.003
12. Kleinert, H.: Gauge Fields in Condensed Matter. Differential Geometry of Defects and Gravity with Torsion, vol. 4. World Scientific, Singapore (1989)
13. de Juan, F., Cortijo, A., Vozmediano, M.: Dislocations and torsion in graphene and related systems. Nucl. Phys. B **828**, 625 (2010). doi:10.1016/j.nuclphysb.2009.11.012
14. Mesaros, A., Sadri, D., Zaanen, J.: Parallel transport of electrons in graphene parallels gravity. Phys. Rev. B **82**, 073405 (2010). doi:10.1103/PhysRevB.82.073405
15. Marecki, P.: Propagation of sound on line vortices in superfluids: role of ergoregions. J. Phys. A: Math. Theor. **45**, 295501 (2012). doi:10.1088/1751-8113/45/29/295501
16. Marecki, P., Schützhold, R.: Whispering gallery like modes along pinned vortices. ArXiv e-prints [arXiv:1110.5928[cond-mat.other]] (2011)
17. Rodrigues, J., Peres, N., Lopes dos Santos, J.: Scattering by linear defects in graphene: a continuum approach. Phys. Rev. B **86**, 214206 (2012). doi:10.1103/PhysRevB.86.214206
18. Warner, J., Young, N., Kirkland, A., Briggs, G.: Resolving strain in carbon nanotubes at the atomic level. Nat. Mater. **10**, 958 (2011). doi:10.1038/nmat3125
19. Boada, O., Celi, A., Latorre, J., Lewenstein, M.: Dirac equation for cold atoms in artificial curved spacetimes. New J. Phys. **13**, 035002 (2011). doi:10.1088/1367-2630/13/3/035002
20. Ablowitz, M., Ilan, B., Schonbrun, E., Piestun, R.: Solitons in two-dimensional lattices possessing defects, dislocations, and quasicrystal structures. Phys. Rev. E **74**, 035601(R) (2006). doi:10.1103/PhysRevE.74.035601
21. Bahr, B., Dittrich, B., Ryan, J.: Spin foam models with finite groups. ArXiv e-prints [arXiv:1103.6264[gr-qc]] (2011)
22. Oriti, D.: The microscopic dynamics of quantum space as a group field theory. ArXiv e-prints [arXiv:1110.5606[hep-th]] (2011)

MIX
Papier aus verantwortungsvollen Quellen
Paper from responsible sources
FSC® C105338

If you have any concerns about our products,
you can contact us on
ProductSafety@springernature.com

In case Publisher is established outside the EU,
the EU authorized representative is:
Springer Nature Customer Service Center GmbH
Europaplatz 3, 69115 Heidelberg, Germany

Printed by Libri Plureos GmbH
in Hamburg, Germany